动力机械工作过程及其测试技术研究

Investigation of Power Machinery: Its Working Processes and Measurement Techniques

李德桃教授论文选集

Selected Papers of Professor Li Detao

李德桃等　著

江苏大学出版社

图书在版编目（CIP）数据

动力机械工作过程及其测试技术研究：李德桃教授论
文选集 / 李德桃等著. —镇江：江苏大学出版社，2008.11
ISBN 978-7-81130-029-1

Ⅰ.动… Ⅱ.李… Ⅲ.动力机械—文集 Ⅳ.TK05-53

中国版本图书馆 CIP 数据核字（2008）第 169332 号

动力机械工作过程及其测试技术研究：李德桃教授论文选集

著　　者／李德桃等
责任编辑／汪再非　　徐云峰
出版发行／江苏大学出版社
地　　址／江苏省镇江市梦溪园巷 30 号（邮编：212003）
电　　话／0511-84446464
排　　版／镇江文苑制版印刷有限责任公司
印　　刷／丹阳市教育印刷厂
经　　销／江苏省新华书店
开　　本／787 mm×1 092 mm　1/16
印　　张／38.75
字　　数／1 100 千字
版　　次／2008 年 11 月第 1 版　2008 年 11 月第 1 次印刷
书　　号／ISBN 978-7-81130-029-1
定　　价／120.00 元

本书如有印装错误请与本社发行部联系调换

李德桃　教授

Professor　Li Detao

序 一

　　李德桃教授是我国著名的动力机械专家,他从 20 世纪 50 年代中期开始动力机械的研究和教学生涯。作为老一辈内燃机权威专家戴桂蕊教授的助手,李德桃教授参与了当时的重大项目内燃机水泵的研究和我国第一个排灌机械专业的筹建。20 世纪 60 年代,他主持涡流室式柴油机的设计和开发,研制成功的 185 型柴油机性能达到国际同类机型的水平。70 年代初,他瞄准涡流室式柴油机燃烧过程及燃烧系统这一研究方向,经过多年的理论探索和实验验证,发明了低油耗、低污染、低爆压的柴油机涡流燃烧室,提高了国产柴油机的转速和功率并降低了油耗,使当时的涡流室式柴油机有了国际竞争力;国家资助他出版的 3 部专著堪称该领域的奠基之作。80 年代初他就关注和开展了内燃机排放有害成分的研究,并在 90 年代初与史绍熙院士(现已故)一起为建立我国内燃机排放法规和解决汽车污染问题提出重要建议;90 年代中期和无锡油泵油嘴研究所开展柴油机电控共轨喷油系统的产学研合作,在数值模拟和实验研究方面处于国内先进水平;90 年代末在国内率先开展了微动力机电系统的研究,时至今日,该领域已成为动力机械的一个研究热点。

　　据我所知,李教授一直在非常艰苦的工作条件下开展科学研究,但他 50 年如一日地将全部精力投入到科研和教学中,并获得了卓越的成绩;退休后仍为科研团队提供力所能及的帮助,令人敬佩。

　　这本论文选集是从他及其科研团队发表的 200 余篇论文中选录的,凝结着李教授一生的心血。我推荐它出版,不仅能使有关的科技人员学习和借鉴他的学术思想和研究成果,而且能使读者领略他的钻研和创新精神。

　　值李德桃教授论文选集出版之际,作为他的老朋友,谨表衷心的祝贺。我相信,本书的出版对我国动力机械和工程热物理的研究与发展将起到积极的促进作用。

中国工程院院士,浙江大学教授、博士生导师
国务院学位委员会动力工程与工程热物理评议组召集人

2008 年 3 月 6 日

序　二

江苏大学出版社今年推出第一本精装版学术专集——《动力机械工作过程及其测试技术研究：李德桃教授论文选集》，这是一件很有意义的工作。

作为科教战线上的一位老兵，看到我们下一代学者获得丰硕的劳动成果，看到改革开放带来的欣欣向荣的繁荣景象，内心感到由衷的喜悦和欣慰。

学校建立之初，我就与德桃相识，他是我的老朋友、已故原镇江农业机械学院副院长戴桂蕊教授的得意门生。他爱国敬业，严谨治学，善于创新，甘为人梯。几十年来，他的工作和生活条件都很差，但他数十年如一日将全部精力投入到科研与教学中，取得了包括国家发明奖在内的10多项国家和省部级奖项，发表了200余篇学术论文，培养了约50名硕士生、博士生、高级访问学者。他是一位实事求是、努力做学问的人，是一位勇攀科学高峰的学术带头人，是一位德才兼备的科技专家，是为内燃机与工程热物理事业作出重要贡献的著名专家。

我希望，也深信，随着改革开放事业的进一步发展，一定会一代一代地出现更多的优秀科技人才和优秀科技成果。我校出版园地亦必更加满园春色。

中国农业机械学科奠基人之一
我国首批博士生导师、原镇江农业机械学院副院长

高良润

2008 年 3 月 26 日

序　三

  早在 2004 年，当国内外久负盛名的内燃机专家李德桃教授迎来 70 喜寿之际，我和亦曾受到李教授培养和指导多年、现在国立新加坡大学机械系任教的杨文明博士酝酿过编辑出版一本汇李教授几十年如一日教书、育人、研究于一体的文集。光阴似箭，提笔之际，不觉又过了 3 个多春秋。

  李教授出生于湖南省茶陵县一个贫苦的农家。4 岁丧父，童年和少年是由祖母和母亲含辛茹苦抚养大的。小学毕业后，以优异成绩考取了中学。由于家庭经济拮据，曾屡次几近辍学。为了照顾祖母和母亲，1952 年他就近考取了湖南大学。院系调整后，又以优异的成绩毕业于吉林工业大学并留校任教，为早年留英归来的我国著名内燃机专家戴桂蕊教授所赏识，亲点他作为科研助手，并让他协助戴教授筹办了我国第一个排灌机械专业。1963 年整个专业南迁至镇江农业机械学院，即现今的江苏大学。幼年生活的艰辛，练就了他坚毅、勤劳、多思、敬业的性格。良师戴教授的言传身教，培养了他严谨和奋发的精神。

  客观地说，在中国像李教授这一代的科学家是极不幸的。在他们最年富力强、成果迭出的年代，遭遇了史无前例的文化大革命。十年动乱，人妖颠倒，摧残了多少优秀的科技人才。然而，即便在这样恶劣的环境中，李教授仍然坚持钻研业务。20 世纪 60 年代他与上海内燃机研究所等单位成功研制了 185 型柴油机。70 年代在常州柴油机厂主持涡流燃烧室的研究，首次把国产小型柴油机的转速提高到 3 000 r/min，油耗达到国际水平。1978 年，春暖花开，当人们迎来中国科学的又一个春天的时候，他已获得许多既有广泛应用价值又达到国内领先水平的成果，一跃成为当时国内内燃机行业中出类拔萃的年轻专家之一。1979 年经国家选拔，派往罗马尼亚深造，并于 1982 年获得罗马尼亚蒂米什瓦拉工业大学工学博士。他的人品和论文水平，受到导师贝林单教授和评审委员会的高度评价。学校还特地向中国驻罗使馆对他进行了表扬。

  从罗马尼亚回国后，李教授毫不懈怠，在当时的江苏工学院筹建了工程热物理研究室并主持工作。从那以后的 20 多年里,李教授在极其艰难的条件下,把他全部的智慧、经验和精力倾注在研究室的成长和发展中。年复一年,科研开发成果累累,教书育人桃李芬芳,从而奠定了他在国内外内燃机和工程热物理界的学术地位。李教授的突出成果可以归纳为以下三大研究方向:

(1) 研究涡流室式柴油机燃烧过程和燃烧系统。李教授主持并完成了 5 个国家自然科学基金项目。首次提出了以热力学和经验模型组成的复合计算模型,发展和完善了计算方法和程序,完成了涡流室式柴油机从零维、准维到多维模型的模拟计算。在实验研究中,精确地测量出涡流燃烧室的示功图和压差图,提出了 M 过程在涡流燃烧室中合理应用的条件、方法和限度,为改善其燃烧过程提供了科学依据。这一系列全面深入的研究工作,为该领域奠定了理论、计算和分析的基础。

(2) 研究柴油机冷起动过程。李教授建立了柴油机冷起动过程四个阶段的模型和热力参数,揭示了起动孔的作用机理,为改善冷起动过程,增加冷起动的可靠性奠定了理论基础,并成功地将涡流室式柴油机最低起动温度降低了 5～10 ℃,由此减轻了操作人员的劳动强度,打开了若干制造企业的产品销路。该研究被列为国家自然科学基金委的重要成果之一。

(3) 发展内燃机测温、测速、测压和测流量技术。早在 20 世纪 70 年代,李教授就与上海工业自动化仪表研究所合作,主持了漩涡流量计的研究,该成果获江苏省科学大会奖。在 20 世纪 80—90 年代,主持研究非稳态压力和温度的数据记录和处理系统,并设计开发了柴油机涡流室内激光测速的光学系统,利用高速纹影技术拍摄涡流室内的喷油和燃烧过程。首次将激光莫尔偏折技术应用于涡流室内气流温度场的测试。

要想在工程热物理和内燃机工程学术领域中脱颖而出,需要有丰富扎实的数理功底和坚韧不拔的钻研精神。而要成为该领域中一流的代表人物,还需要有独具一格的战略眼光和魅力四射的组织才能。作为一代学术精英,李教授无疑在这两方面是出类拔萃的。

李教授对科研发展趋势和方向的把握,有着极强的敏感度和前瞻力。早在 20 世纪 80 年代初,当人们正热衷于降低内燃机油耗,改善经济性能的时候,李教授就开始了柴油机排气有害成分的研究,并致力于呼吁内燃机排放法规的建立。他取得了降低涡流室式柴油机 NO_x 排放 20% 的研究成果,首创“低油耗、低公害、低爆压柴油机涡流燃烧室”。这比后来英国提出的所谓“优化涡流燃烧室”早了 6 年。1994 年他参与美国德州大学奥斯汀分校负责制定天然气排放法规的研究,并公布了与法规有关的若干新的测试结果,这在当时美国州立排放法规中是极具前瞻性的。此后不久,他又与史绍熙院士联名发表文章,呼吁我国重视汽车排放问题,并提出了许多解决此问题的合理化建议。在把地球环境保护列为世纪重大课题的今天,我们无不感叹李教授深邃的战略眼光。

进入 20 世纪 90 年代中期,微机电系统的研究和开发正逐渐成为欧美等许多发达国家的热点课题。1999 年李教授应邀到新加坡国立大学作学术交流。在一系列的学术讨论中,他对我们科研小组刚刚展开的微流动和传热的课题提出了许多建设性的建议,并鼎力支持和倡导对微动力系统的探索。翌年,他在国家自然科学基金主办的"燃烧学科发展学术讨论会"上提出了开展微型燃烧器的可行性研究的设想,并在《世界科技研究与发展》期刊上撰文介绍微动力机电系统和微发动机的研究进展。2002 年,他的"微型发动机燃烧过程和燃烧室的基础研究"课题得到国家自然科学基金资助。由于当时国家自然科学基金是首次资助这种高度创新但风险巨大的项目,只承诺了一年的经费。然而,数年后,当李教授迈入古稀之年,把科研的接力棒交给一线的年轻人的时候,他欣喜地看到微动力的研究和开发已在国内许多大学和研究所如火如荼地展开。

综观李教授的科研历程,他从国内生产需要出发,在全面系统地总结国内外有关研究成果的基础上逐步深入,从理论到实践,由创新到提高,相辅相成,相得益彰,并不断把研究成果转化为生产力,直至达到了国际先进水平。

作为动力机械及工程热物理学科的带头人,李教授治学严谨,为人师表,以他独特的魅力,吸引了许多优秀的学生和学者。20 多年来,他的科研团队,人才济济,成果斐然。他的学生有的获得了"史绍熙科技教育基金奖",江苏省优秀博士论文;有的成了"拔尖人才培养工程"培养对象;有的则已成为国际知名的中青年专家。他倡导的"跨学科、跨单位、跨地区、跨国家"的研究团队,更是汇四海英才,集八方智慧。多年来,这个研究团队,在交换信息、共享资源和互相合作等方面独树一帜,颇具特色。

在几十年的教学和科研岁月中,李教授共出版了 4 部专著,发表论文 200 多篇,培养了近 50 名硕士生、博士生和高访学者。他先后应邀赴美、日、欧和东南亚等许多国家和地区的知名大学和研究所进行讲学和学术交流,成果蜚声海外。李教授赢得了许多重要的奖项,其中有国家发明奖,多项省、部级科技进步奖,全国机械科学大会奖、省科学大会奖。他还被评为江苏省优秀博士生导师、江苏大学杰出研究生导师。在学术界,他在国家自然科学基金委、中国汽车工程学会等机构兼任职务,同时担任过许多主要的专业刊物如《内燃机学报》、《内燃机工程》和《燃烧科学与技术》的编委。他作为第六、七届全国人民代表大会代表,为改善地方的生活环境和教育提出了许多积极的建议。

熟悉李教授的人都知道他生活简朴,淡泊名利。他曾经多次婉谢了担任江苏省镇江市副市长或市政协副主席等职,他认为自己的专长是教学和科研,而教学和科研需要全身心的投入。他常以"岂为图报方言善,非盼功名才读书"自勉,而对学术界的一些钻营投机,急功近利,甚至于整人害人的风气则深恶痛绝。

不管是当年动乱岁月那艰苦奋斗的日日夜夜,还是被授予享受政府特殊津贴的专家的那一刻,让李教授宽慰的是他那同甘共苦的幸福家庭。师母相夫教子,堪称贤妻良母。李教授和师母有两个女儿和三个孙女。大女儿一家近在镇江,周末假日经常回家团聚。小女儿一家远居美国德州达拉斯,但时常邀请退休之后的李教授夫妇越洋小住。2007年初夏,小女儿一家四口偕李教授夫妇游览加州。当我和李教授的学生兼好友、旅居美国多年的钱冀平先生前去洛杉矶机场向他们道别的时候,两个可爱的孙女正依偎在李教授的身旁,刹那间,我的脑海里浮现起李教授作的"七十岁有感"一诗中意味深长的诗句"科技献身情未了,忠良笃信效先贤"。李教授一生为科技献身,即便退休后身在海外,每次我们和李教授促膝详谈或电话联系,他还总是离不开科研和他的"四跨"团队。对于这样一位笃信忠良的良师,我坚信上帝一定是慷慨地施与了他的圣明。

是啊!我们衷心地为李教授祝福,祝愿他身体康健,阖家欢乐。

加州州立工业大学终身教授　薛玄

2007 年 12 月于洛杉矶

Contents 目录

1

Ⅱ. 柴油机冷起动的基础研究和改善措施

Ⅲ. 柴油机共轨喷油系统的研究

Ⅳ. 微型动力机电系统的研究

V. 其他相关研究

I. 涡流室式柴油机的燃烧过程和燃烧系统

The Combustion Process and Combustion System of Swirl Chamber Diesel Engine

涡流燃烧室高速适应性的研究[①]

李德桃

[摘要]　本文是对涡流燃烧室的结构设计进行大量的、系统全面的试验研究的一个总结。探讨了涡流燃烧室的改进方向,报道了所获得的新型涡流燃烧室的结构设计,提出了涡流燃烧室的一些新概念设计。

众所周知,涡流室式燃烧室,与直接喷射式相比,它的制造工艺简单,对喷油系统要求不高,排出的有害气体 NO_x 较少,工作柔和;与预燃室式燃烧室相比,它有较高的平均有效压力和较低的油耗。因此,它在工农业生产和交通运输中获得广泛的应用。我国自行研制出了一些性能较好的涡流室式燃烧室[1]。

但是,当前我国广泛使用的涡流燃烧室,是在小型农用动力($n=1\,500\sim2\,000$ r/min)上试验得出的[2]。为了适应柴油机向高速强化、多种用途和提高经济性的方向发展,要求对涡流燃烧室进一步加以研究和改进。特别对于柴油机转速提高到 $3\,000$ r/min 以上时,涡流燃烧室的适应性如何,在我国是一项必须探索的迫切问题。

1　改进方向和试验方法

涡流燃烧室应该沿着什么方向改进呢?我们认为,首先必须认真总结和分析国内外涡流燃烧室的实践经验,特别有必要先分析一下油膜式混合气形成和燃烧过程(M过程)应用于这种燃烧室所取得的成绩和存在的问题,然后结合涡流燃烧室的特点,探讨它适应高速化的有利条件和克服有关困难的基本途径。在此基础上,提出改进方向和具体措施。

大家知道,M过程应用于涡流燃烧室的可能性和现实性,不同研究者都曾根据各自的高速摄影照片作了明确的说明[3—5]。性能良好的常柴Ⅰ号和常柴Ⅱ号,也可以看成M过程应用于这种燃烧室的具体例证。实践业已表明,M过程应用于涡流燃烧室的结果,使它获得了良好的混合气形成和燃烧,从而使涡流室式柴油机具有较好的扭矩特性,降低了敲击声,提高了经济性。但是,也带来一些明显的缺陷:一个是冷起动性能差;另一个是在燃烧后期缺乏有效的涡流,致使完全燃烧延迟,不利于热效率的提高。常柴Ⅰ号和常柴Ⅱ号涡流燃烧室在克服上

[①]　李德桃教授于20世纪60—70年代(时处"文革"中),在常州柴油机厂主持涡流燃烧室的研究,仅试验就耗时近3年,人工划线加工了不同结构设计的缸头30多个,活塞20多个,镶块数以千计。加工量之大,工作之艰辛,生活之艰苦,实属罕见。项目结束后,写成的技术总结报告《常柴涡流燃烧室的发展研究》,载于1974年第1期的《常柴技术简报》。当时全国80多家涡流室式柴油机厂都向龙头企业常柴厂索取这份简报,并用于本企业产品的改进。此后,当时国内唯一的内燃机杂志,邀李教授在上述报告的基础上,写成此文,研究结果全部奉献给社会。

——编者注

述缺点方面取得了一定效果，但是，面临高速强化等要求，人们自然会提出这样的问题：应用 M 过程的涡流燃烧室能否适应高速化？实现高速化的同时能否获得较好的经济性？对此我们试作如下的分析。

(1) 按照茅瑞尔(Meurer J S)关于混合气形成力学的最新观点[6]，M 过程的主要特性是对蒸发和混合起决定性作用的混合气形成手段——油粒和空气的相对速度，在燃烧过程中始终都保持较高的值。这就给燃料的蒸发和混合气形成创造了极有利的条件。拿这个观点来解释涡流室内的混合气形成和燃烧过程，我们还可以进一步地推断：在相同的油粒速度下，同任何其他类型的燃烧室相比，具有强烈空气涡流的涡流室有着更高的油粒-空气相对速度。这是涡流室应用 M 过程的一个特点，也是它易于实现高速化的一个优点。

关于涡流的特性，现在还没有一个一致的说法。但是，不论是"半自由涡"也好，是自由涡同其他形式的涡相结合也好，各种形状的涡流室，都较其他类型的燃烧室有着更高的涡流强度这一点，是没有疑问的。彗星 V 号涡流室具有平底，虽然涡流有所衰减，仍能适应高速化(5 000 r/min)就是明证[7]*。

顺便指出，至今人们还沿用 30 年代提出的所谓涡流比(空气涡流转数与发动机转数之比)来反映涡流强度对工作过程的影响。我们认为，随着 M 过程应用于涡流燃烧室，随着人们对混合气形成和燃烧过程的理解加深，有必要探讨新的特性参数来综合反映涡流强度和油注运动对工作过程的实际影响。

(2) 如何有效地利用主燃室的空气，将涡流室喷出的火焰再次燃烧，早就是人们注意的一个问题。彗星Ⅲ号的双涡型凹坑和常柴Ⅱ号的铲击型凹坑，都是在这个方面探索的结果[1,8]。长尾不二夫和奥尔科克(Alcock J F)等人曾用高速摄影来显示双涡型凹坑的燃烧情况，说明它对性能有重大影响[3,9]。随着发动机向高速强化发展，可以预计到主燃室的形状对性能的影响会增加。因为如前所述，M 过程应用于涡流室的缺陷之一，即燃烧后期由于涡流室内缺乏有效的涡流，致使完全燃烧延迟；这就意味着加强主燃室的空间混合，促进喷出火焰的迅速燃烧，对弥补这个缺陷有着重要的意义。从反应动力学的观点看，主燃室中的燃烧速度在很大程度上取决于火焰与氧气通过相互扩散而混合的速度。这个速度，显然同主燃室的结构形式关系甚大。

关于主燃室的结构形式，铲击型主燃室给我们的启发是：双涡型主燃室的框框是可以打破的，采用铲击型凹坑，使火焰由两侧向整个活塞顶面铺围过来，可获得同双涡型一样良好(或更好)的性能。然而这种主燃室在单缸机上的试验情况表明，虽然发动机转速也能达到 3 000 r/min，但油耗和排温都比较高(参看文献[1]图 9)。正如本文的下一部分所指出，这是由于在该主燃室中，气流能量损失较大和空间混合速度不够高所致。如果我们从这两方面着眼进一步改进这种主燃室，则对提高涡流燃烧室的高速适应性和经济性是很有利的。实践证明，设计合理的主燃室形状，确实是缩短燃烧过程总时间，获得高的循环效率，弥补 M 过程应用于涡流燃烧室所造成的缺陷的有效途径之一；也是涡流燃烧室适应高速化时可以而且必须利用的另一有利因素。

(3) 关于涡流燃烧室的混合气形成和燃烧过程的方式，大体说来，在涡流室中是复合式；在主燃室中是空间式。然而这两种方式以什么比例配合才能确保实现高速化时获得高的经济性，则值得研究。这里涉及到两个问题：一个是涡流室中油膜式和空间式两部分燃料量的比

---

* 文献[7]还报道，M 型燃烧室用在较大的发动机上，其转速也已达 3 000 r/min。

例;一个是涡流室和主燃室两部分空气量的比例。关于头一个问题,我们认为,为了使涡流室的混合气形成和燃烧过程主要按油膜式进行,除了油注顺气流喷射外(这方面已作了许多研究[3,10]),应该适当增加油注偏移 e(图1)。因为很明显,随着 e 的增加,涂散在室壁上的燃料量就增加,被气流带走的燃料量就减少;而且 e 的增加使分布在室壁上的燃油受更强烈的制动而增加油粒-空气的相对速度。关于第二个问题,我们认为,为了使更多的燃油能在涡流室中按油膜式进行混合和燃烧,除增加油注偏移 e 外,还必须让更多的空气进入涡流室。因为这样可使更多的空气同更大面积的油膜接触;而且从反应动力学的观点看,增加涡流室的过量空气系数,可使燃料的裂化速度减慢[6],从而保证涡流室的燃烧质量。长尾不二夫主张尽可能加大容积比 δ_K(涡流室容积同压缩室容积之比)[11],其原因可能就在此。我国自行设计研制的一些性能较好的涡流燃烧室,其容积比也有逐渐增加的趋势*(图2)。

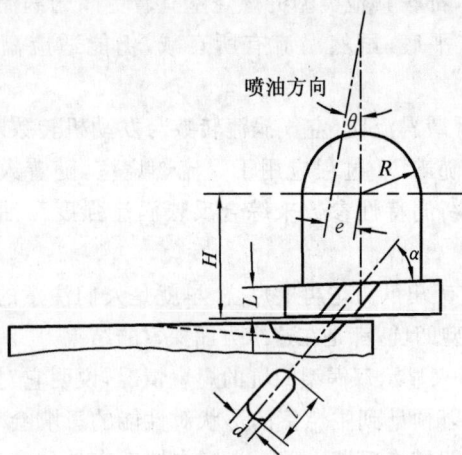

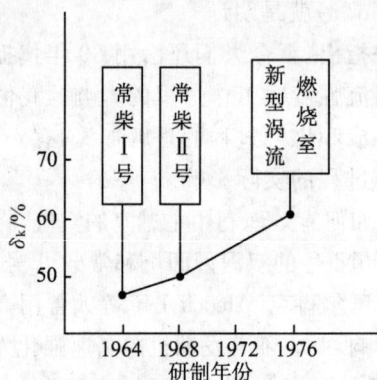

图 1　涡流燃烧室示意图　　　　图 2　我国自行研制的涡流燃烧室的容积比

但是任何事物都是充满矛盾的。油注偏移 e 的增加,会使本来起动性较差的涡流室式柴油机起动更困难。这个矛盾,通过采用起动孔、品陶(Pintaux)型喷嘴和电热塞等方法可以解决。容积比增加会带来流动损失和热损失增加等不良影响。然而由于 δ_K 的适当增加能较大地改善燃烧质量,其结果,经济性是随之提高的。

(4) 一个性能良好的涡流燃烧室,连接通道截面面积与燃烧室容积、通道面积比(通道截面面积与活塞面积之比)与容积比,必定配合较好。例如,采用大的容积比时,面积比也要相应地较大。这样配合的目的是,一方面可使压缩空气经通道流入涡流室时产生完善燃烧所需要的速度,另一方面可减少流动损失和热损失。这种良好配合对于适应高速化的重要性是不言而喻的。

用上述观点分析对比常柴 I 号和常柴 II 号就会发现:虽然后者比前者的容积比大,而且采用后者的发动机转速比采用前者的发动机转速高,然而它们的通道是一样的。这显然于常柴 II 号的性能不利。常柴 II 号用在单缸试验机上的试验情况表明,仅使通道长度 l 从 14.5 mm 增至 18 mm(面积比从 1.2% 增至 1.53%),在 3 000 r/min 时最低油耗就可降低 9.52 g/(kW·h)[即 7

---

*　对球柱形涡流室而言,容积比 δ_K 主要随球半径 R 而定。以常柴 I 号为例,R 从 15 mm 增至 16 mm,δ_K 则从 50% 增至 56%。

克/(马力·小时)]。这个具体数据生动地说明了上述配合对性能的重大影响。

此外,实践还表明:根据不同缸径和转速,选择最佳通道倾角 α 和位置,也有利于改善涡流燃烧室的性能。

从以上分析可知,应用 M 过程的涡流燃烧室,不仅能适应高速化,而且可以在适应高速化的同时争取获得较好的经济性。根据以上分析,我们选择涡流燃烧室的改进方向是:让更多的燃油量和空气量在涡流室内按 M 过程组织混合气形成和燃烧;与此同时,加强主燃室中的空间混合以实现完全燃烧。

为了朝这个方向努力,需要采取的具体措施是:

a. 适当增加油注偏移和尽可能增加容积比;

b. 设计合理的主燃室形状;

c. 选择最有利的连接通道的形状、尺寸和位置。

这些措施的实现,不可能用解析方法而只能用系统试验的方法来探求。因此,我们的研究方法是:在单缸试验机上进行系统试验,即系统地观察和记录燃烧室各部分的形状、尺寸和结构参数改变时,其性能指标如功率、排温、油耗等的变化。为了弄清它们各自改变时对发动机性能的影响,我们在每次试验中仅改变一个因素。最后进行组合试验,据此以确定各部分的合理形状与最佳尺寸。

我们的试验除针对高速工况(3 000 r/min)外,考虑到柴油机的多种用途和便于对比分析,也记录了 2 500 r/min 和 2 000 r/min 时的数据。

为了进行系统试验,我们设计制造了一种"三化"程度较高而又能满足试验要求的单缸试验机(图 3)。采用改进后的涡流燃烧室,该机在 3 000 r/min 时,最大功率可达 14.71 kW[即 20 马力],最高平均有效压力可达 0.81 MPa [即 8.3 公斤/厘米²]。试验机的基本参数如下:

卧式四冲程水冷柴油机;缸径,95 mm;行程,115 mm;燃烧方法,涡流室式;燃油喷射系统,Ⅰ 号泵,4S1 喷油器;配气定时,

图 3　单缸试验机外形

进气门开　　　　上死点前 15°;

进气门关　　　　下死点后 61°;

排气门开　　　　下死点前 58°;

排气门关　　　　上死点后 18°。

从外形看来,该试验机与东风12型195柴油机几乎没有区别,而实际上它却有自己的特点:

(1) 缸头和机体是按试验要求专门设计制造的。缸头厚为 95 mm,以便能承受较高的气体压力和试验不同形状与尺寸的涡流室。加大了进排气通道面积。进气门直径 44 mm,排气门直径 36 mm。为防止进排气门之间开裂,鼻梁处铣有一小圆弧凹坑(图 4)。

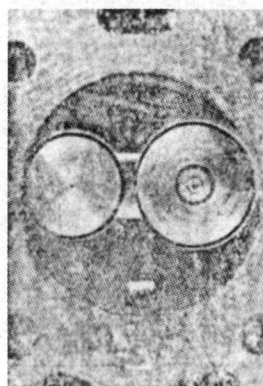

图 4　气缸头底面

(2) 单缸机的运动件,如活塞、连杆和进排气门等与 495Q 型高速车用柴油机通用;曲轴、

平衡轴、齿轮等与东风12型195柴油机通用;配气凸轮形状与495Q型柴油机的相同。由此可见,该单缸机的易损零件更换性好,这样大大节省了试验费用和时间。飞轮是用295型柴油机飞轮改装的,其直径为380 mm。

(3)采用强制冷却装置。冷却水经电动循环水泵、水箱、散热器进入机体,由水箱上盖流出。水箱上盖处装有温度计。

该机在3 000 r/min,平均有效压力0.69~0.81 MPa[即7~8.3公斤/厘米²]下累计运行近1 000小时,证明它能满足试验的要求。

2 涡流室形状和尺寸试验

涡流室的形状不但与涡流强度有关,而且对缸头的工艺性影响较大。具有代表性的涡流室形状,归纳起来,不外乎球形、圆柱形、球柱形等几种。常柴Ⅱ号所采用的球柱形涡流室,不仅工艺性好,而且与球形相比,球柱形的平底可使火焰集中在底部,因而对喷油方向不太敏感,其起动性也较好。根据流体力学的理由,我们曾考虑把涡流室做成图5的A型,但因其工艺性差,而且与B型的性能很接近,因此,我们仍然保留了常柴Ⅱ号的涡流室形状,而集中力量寻求其最佳尺寸。

涡流室的各尺寸参数都处于一定的关系中。譬如,从保证获得必要的涡流强度的角度来考虑,在一定条件下,随着容积比的变化,通道面积和位置等尺寸参数也应相应的变化。这一点可以从表征涡流强度的涡流比Ω公式中定性地看出[12]

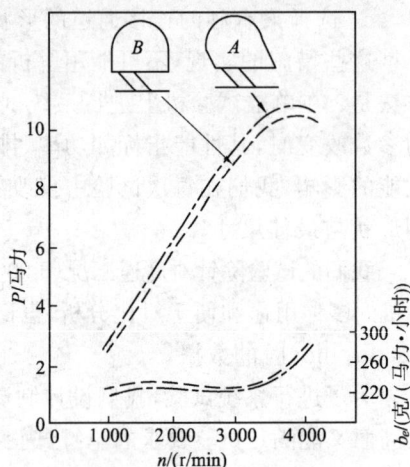

图5 两种不同形状涡流室的性能对比

$$\Omega = \frac{R_1}{R_2^2} \cdot \frac{\delta_K V_h}{(\varepsilon-1)4\mu f_K}\left[\frac{1}{\varepsilon-1}+f(\varphi)\right]\int_{\varphi_0}^{\varphi_x}\frac{[\psi(\varphi)]^2}{\left[\frac{1}{\varepsilon-1}+f(\varphi)\right]^3}d\varphi$$

式中R_1为通道中心线至涡流室球中心的垂直距离;R_2为涡流室容积的惯性半径;δ_K为容积比;V_h为气缸工作容积;f_K为通道截面面积;μ为流量系数;ε为压缩比;φ为曲轴转角。

此式尽管没有考虑发动机速度工况的影响,不完全反映过程进行的条件,但它却表明了各尺寸参数间的相互关系。譬如,随着容积比δ_K的增加,通道面积f_K应相应地增加,才能使涡流比保持在所要求的范围内。下面我们就通过系统试验来考查涡流室尺寸对性能的影响。

2.1 球半径 R

为了增加容积比,我们在单缸试验机上做了$R=15,16,17,17.5$ mm(相应的$\delta_K=50\%$,57%,65%,69%)的对比试验。图6示出了前三个尺寸在三个速度下的试验结果。试验的条件是:通道面积比为1.39%,圆柱部分高度$H=17$ mm,$\varepsilon=20$。图6表明:在各种转速下,随着δ_K的增加,不仅最低油耗下降,而且最大马力增加。这就证明了我们在上面所作的分析的正确性。不过,根据我们的经验,对于95系列柴油机,$R>17.5$ mm,即$\delta_K>70\%$要受到主燃室的余隙高度的限制。

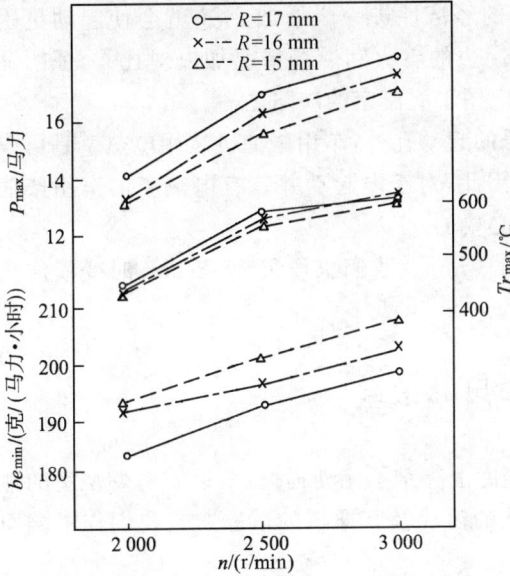

图 6　球半径对性能的影响

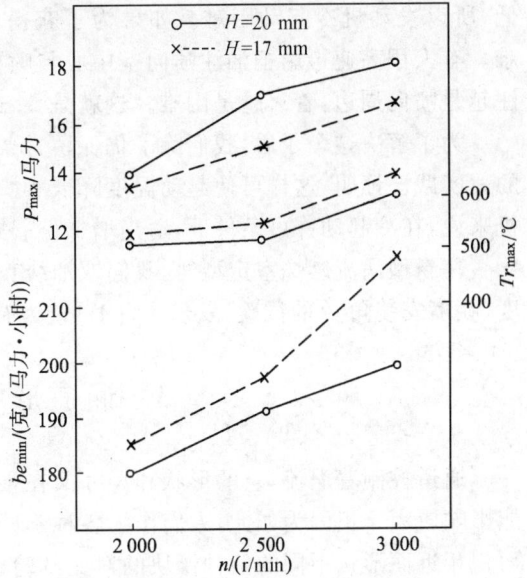

图 7　圆柱高度对性能的影响

2.2　圆柱高度 H

为了查明 H 对性能的影响,我们作了 $H=17$, 18.5,20 mm(相应的 $\delta_K=57\%$,58.5\%,66.5\%)的对比试验,试验条件是 $R=16$ mm($R>16$ mm,则增加 H 的试验受结构的限制),$\varepsilon=20$,$f_K/F=1.18\%$,主燃室为直沟型。图 7 示出了两种 H 的试验结果。该图表明,随着 H 的增加,性能有较大的改善,高速时更甚。分析其原因有:

（1）H 增加,δ_K 增加了;

（2）H 增加,油膜蒸发表面积增加,油膜燃烧的燃料量增加;

（3）H 增加,油注前端的飞驰油粒因碰撞燃烧室底面反跳入空气中的可能性减小。

这些因素都促使性能提高。当然,H 增加会使流动损失和热损失增加,但是在这里,它们不是起支配作用的因素。

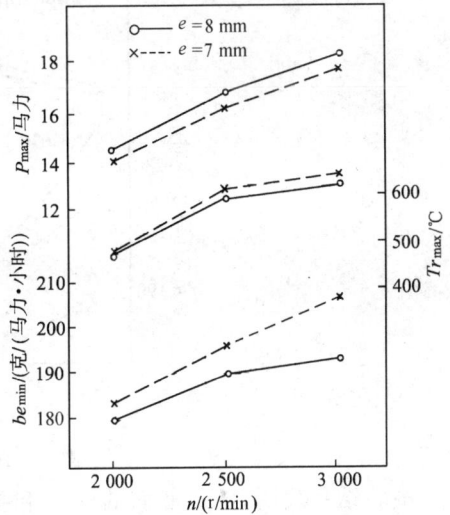

图 8　油注偏移对性能的影响

2.3　油注偏移 e

从文献[3]中图 21 的高速摄影照片 4 可以清楚看出,油注偏移的燃烧图形是:火焰从壁面向中央流动,空气从中央被扔向壁面。这就给壁面处的迅速燃烧造成极有利的条件,获得了 M 过程的主要效果。常柴Ⅱ号的油注偏移为 5 mm。当时由于考虑发动机手摇起动困难而没有进一步增加 e。我们作了 $e=7,8$ mm 的对比试验(图 8)。试验的条件是 $R=17$ mm,喷油器与涡流室中心线的夹角 $\theta=20°$。试验结果表明,在各种转速下,增加 e 都能较大的提高发动机的经济性和动力性,而且越是高速,作用越大。很易理解,这是混合气形成和燃烧过程更加 M 过程化的必然结果。

但是,在有强烈涡流的涡流室内,由于气流与室壁之间存在较强的热传导而使压缩温度降低,所以发动机冷起动困难。如果为了获得较好的经济性进一步增加 e,这就会使起动更困难。有人用高速摄影把油注喷向室中心与喷向室周边的着火和燃烧情况加以对比[9],证明油注越是喷向周边,着火越是困难。这就是 M 过程应用于涡流室的缺陷之一。

为了解决这个矛盾,我们除了仍保留常柴Ⅱ号的起动孔外,车用柴油机装用电热塞是可取的。实践已证明,这样可使起动温度降低 40 ℃。因此,对于电起动并装有电热塞的车用柴油机来说,在冷起动许可条件下,适当增加偏心距是必要的。

在新型涡流燃烧室定型时,我们取油线偏移为 8 mm。我们也作了 $e=8$ mm 时变喷注角度(喷嘴安装角)θ 的试验,发现 θ 为 10°较为有利。

3　通道形状和尺寸试验

通道控制着涡流,它的形状和尺寸又在很大程度上决定了流动损失,所以它对燃烧室的性能影响较大。正因为如此,人们作了各种不同形状的镶块及其通道的试验[8]。我们作了截面面积相近而形状不同的两种镶块的对比试验(图 9)。

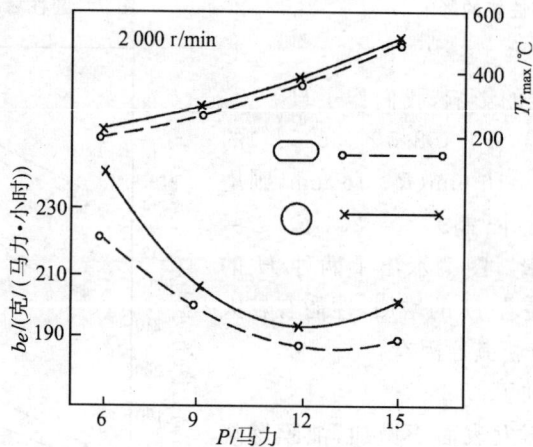

图 9　通道形状对性能的影响

试验结果表明,圆形截面的油耗和排温都较圆矩形截面差。此外,根据工厂制造圆矩形通道的镶块的经验,其工艺性也较好。因此,我们沿用了圆矩形通道截面的镶块。

3.1　通道截面长宽比

首先,我们使截面的宽度一定($d=6.5$ mm),作了各种长度 $l=14.5,16.5,18,19.5,21$ mm(相应的 $f_K/F=1.2\%,1.39\%,1.53\%,1.66\%,1.8\%$)的对比试验(图 10),发现:当 l 从 14.5 mm 增至 18 mm 时,油耗随 l 的增加而下降,而排温则变化甚微;当 l 从 18 mm 继续增至 21 mm 时,油耗几乎保持不变,但排温增加较多。从前述涡流比的公式来分析,在其他参数不变时,f_K 增加,则涡流比下降;不过,在所论情况下,根据近似计算涡流比偏高,上死点的 Ω_C 值达 40~45,所以 f_K 增加些,仍能达到所要求的涡流强度,然而这时流动损失却减少了,结果是经济性提高。当然,当 f_K 增加超过某个限度时,由于涡流强度太弱而使油耗增加,由于节流效应过小而使排温显著增加。

随后,我们使截面的长度一定($l=18$ mm),做了两种宽度 $d=6.0,7.0$ mm 的对比试验(图 11)。三个速度工况下的试验数据表明,当 l 较长时,d 增加不仅使油耗和排温都增加,而且使最大马力下降。这个试验同上一个试验一起表明,用增加 l 来增加 f_K 较用增加 d 来增加 f_K 有利。

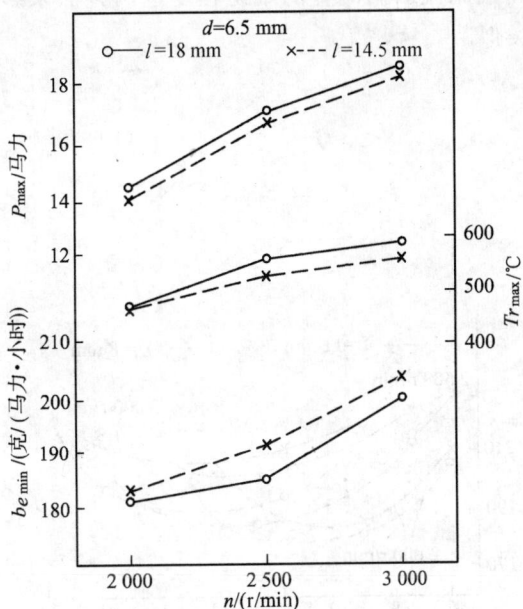

图 10　通道截面长宽比对性能的影响(一)

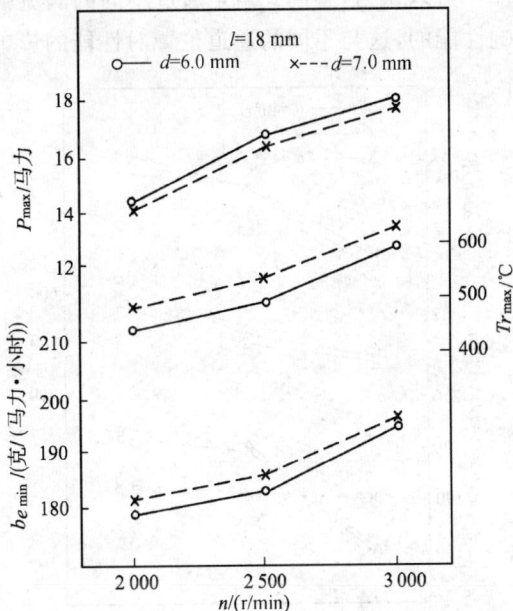

图 11　通道截面长宽比对性能的影响(二)

3.2　通道位置

国外有人认为[8],通道位置以其上口锐角对准涡流室中心线为最好。常柴Ⅱ号通道的上口锐角偏左约 4.3 mm(相当于产品图纸上的通道下口中心向左偏离涡流室中心 8.1 mm)。我们针对这两种情况做了对比试验(图 12)。试验结果表明:在三个速度工况下,后一种情况都比前一种情况好。从前述涡流比的公式来看,由于前者的 R_1 大于后者的 R_1,这样,在其他参数都不变时,前者的涡流比应大于后者的涡流比。这就是说,锐角左移后,涡流强度有些下降,这一点也为试验所证实[8]。但是,通道上口朝涡流室中心位移,比起它偏于一边的情况来,气流流出涡流室的阻力小些,能量损失少些;路程短些,燃烧持续时间就短些。所以在保持一定的涡流强度时,通道向中心位移,性能反而有所提高。

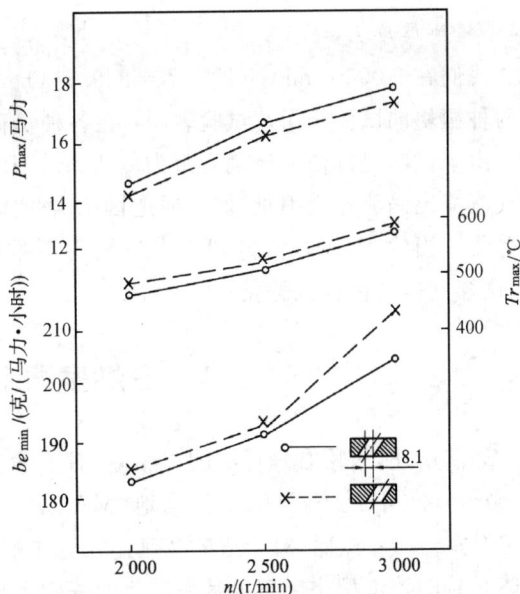

图 12　通道位置对性能的影响

3.3 通道倾角 α

常柴Ⅱ号的通道倾角参考彗星Ⅴ号定为40°。我们做了 $\alpha=35°,40°,45°$ 三个倾角的对比试验(图13)。从试验结果可以看出,在 2 000 r/min 时,35°和45°较好;在 2 500~3 000 r/min 时,以45°为好。

可以认为,不同 α 角是通过不同的涡流强度和气流流出涡流室的难易程度这两方面来影响性能的,这与不同的通道位置对性能的影响相类似。

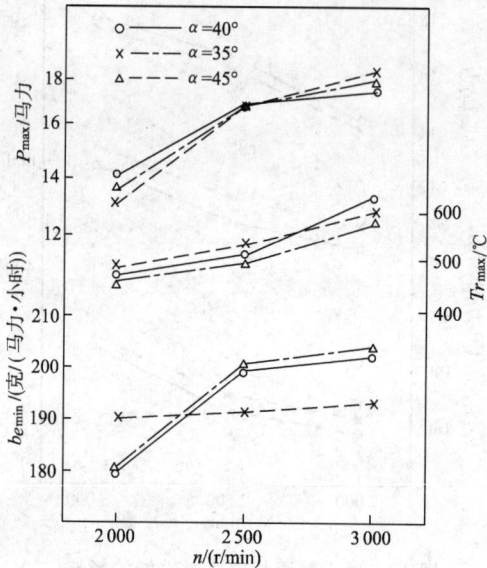

图 13　通道倾角对性能的影响

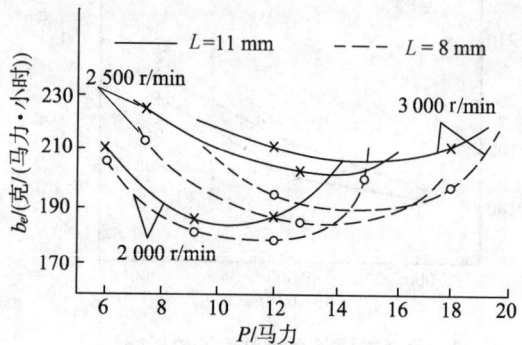

图 14　镶块厚度对性能的影响

3.4 镶块厚度 L

我们在 3 000 r/min 下做了 $L=8,9.5,11$ mm 三种厚度的镶块的对比试验。图14表示前后两种镶块的试验结果。试验表明:在各种负荷下,都以 $L=8$ mm 时的性能最好。

由于试验时都使通道偏心距为 8.1 mm,因此随着 L 的增加,通道上口向左偏离中心,这样气流流出涡流室就困难,加之通道因 L 增加而变长,流动损失增加。这两个因素都使得性能恶化。当然,若使 L 减薄到进出口锐角不重叠,就可能发生因实际的 R_1 太小而使涡流太弱,这样同样会使性能恶化。

4　主燃室表状和尺寸试验

如前所述,采用合理形状的主燃室,对于涡流燃烧室适应高速强化有着重要意义。李卡图(Ricordo)公司、纳齐(HATH)、马地(МАДИ)、大发公司等虽然通过试验得到一些性能较好的主燃室[13-15];然而,对于适应高速化究竟主燃室应具怎样的形状才比较合理,则仍需研究。常柴Ⅱ号的铲击型主燃室也是在试验的基础上得到的。但是,铲击型凹坑同双涡流凹坑一样,一般都是在低速时性能较好[1,11]。我们认为,为了适应高速化,主燃室仅形成局部涡流是不够的,应力求使火焰尽快向整个主燃室扩展;同时也应尽量避免气流的能量损失。我们就是本着这种想法来寻求适应高速化的主燃室的形状和尺寸的。

通过对铲击型、双涡型、直沟型等主燃室的对比分析,我们拟对铲击型作以下改进:

(1) 为了减少气流从涡流室喷出时的损失,活塞顶可增加一段倾角和形状都与通道相适应的导向气道,以便因势利导地把气流引入凹坑(图 15b 和图 17)。

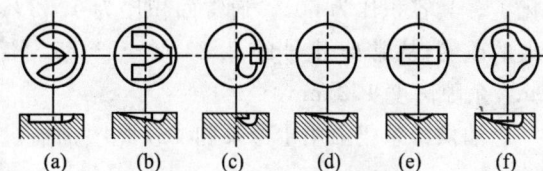

图 15　不同形状的活塞顶凹坑

(2) 为了使从涡流室喷出的火焰迅速向整个主燃室传播,我们把铲击型的同一深度的凹坑改为渐浅凹坑。这样两股气流就成"楔形"并高速向前扩展。

(3) 为了减少气流在活塞顶部由于流经渐缩渐扩气道而引起的能量损失(这在铲击型凹坑上是存在的),我们把气道设计成近似等宽度的。

(4) 为了减少两股气流在凹坑尾部起翘后相互碰撞而引起能量损失的可能性,把分流后的气道改成近似直沟型。

根据以上观点,我们设计了如图 15b 所示的活塞顶凹坑,并把它定名为"双楔型"主燃室。它与铲击型的对比试验结果示于图 16。该图表明,几乎在全部试验转速和负荷范围内,b 型都较 a 型为佳。特别是在高转速、高负荷下,b 型的排温和油耗都有较大幅度的下降。这表明以上所作的改进分析基本上是正确的。

随后,我们又作了四种形状的主燃室(图 15c~f)的试验:中间带浅槽的双涡型(c),直沟型(d),圆弧型(e)和让气门型(f)。试验结果表明,其性能都比 b 型差[1]。

为了考查活塞顶凹坑的尺寸对性能的影响,我们作了两个形状相似而尺寸不同的 b 型主燃室的对比试验。其中一个的凹坑容积为 7.6 cm³,另一个为 5 cm³。试验所得结果是:在三个速度工况和各种负荷下,两者的性能都

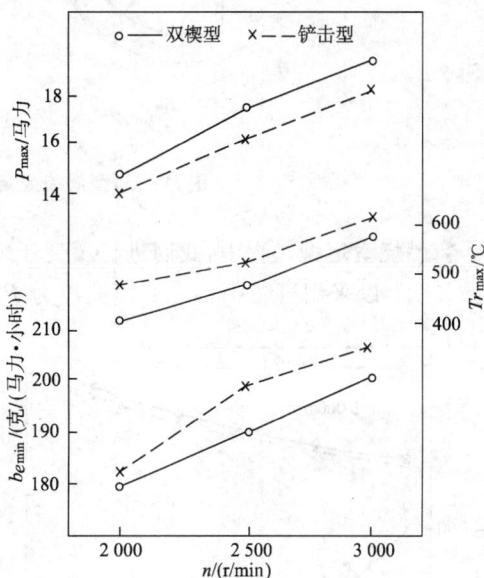

图 16　双楔型和铲击型的性能对比

很接近,这说明凹坑容积对性能的影响不大。由此可见,适当减少活塞顶凹坑容积以增加容积比,不仅是必要的,而且是可能的。

综合对主燃室的试验研究,我们得到这样一个概念:我们对主燃室所谋求达到的目标,是提高其中的空气利用率,以实现完全燃烧;而达到这个目标的主要途径是设计合理的主燃室形状。

5　新型燃烧室的试验结果

我们把改变燃烧室的形状和尺寸的试验研究结果,进行综合分析和组合试验,最后得出如图 17 所示的涡流燃烧室的结构形式和具体尺寸。

这种燃烧室的涡流室形状仍然同常柴Ⅱ号一样是球柱形,但它的基本尺寸参数几乎全部

变化了。其容积比为 60％,通常面积比为 1.53％。为了适应高速化,通道面积的长宽比作了较大的改变,从常柴Ⅱ号的 2.23 增至 2.77。活塞顶凹坑采用图 15b(双楔型),其容积为 5 cm³,余隙高度 1.2 mm。

燃油供给系仍沿用Ⅰ号泵和 4S1 型喷油器。

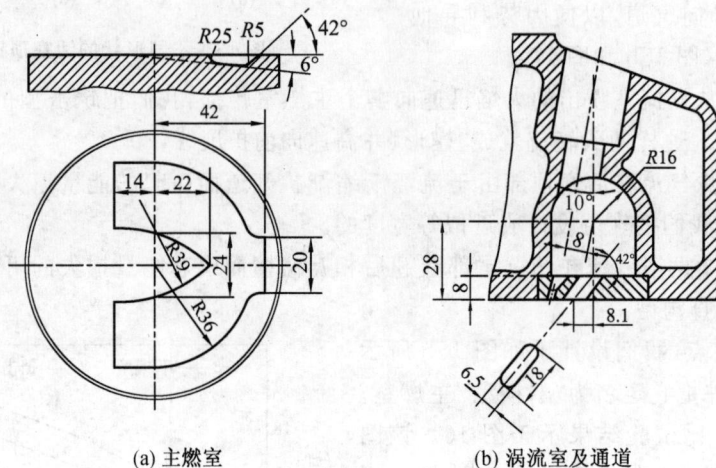

(a) 主燃室　　　　　(b) 涡流室及通道

图 17　新型涡流燃烧室的结构型式和具体尺寸

新燃烧室定型过程中,我们进行了变压缩比试验(图 18)和变喷油压力试验(图 19)。试验结果表明:以采用压缩比 ε＝20,喷油压力 P_ϕ＝13.72 MPa[即 140 公斤/厘米²]为宜。

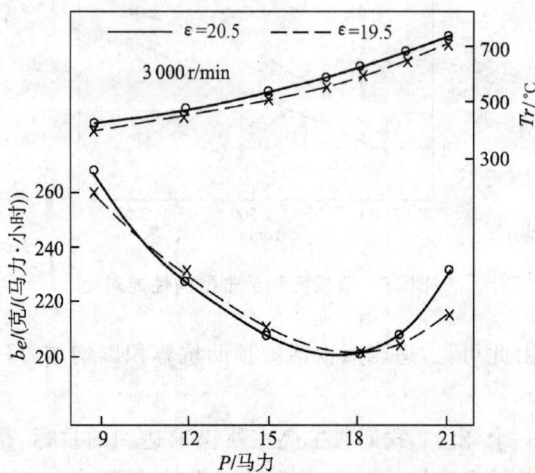

图 18　压缩比对性能的影响　　　　图 19　喷油压力对性能的影响

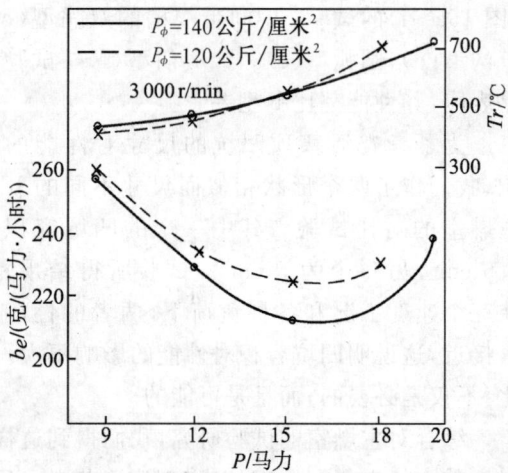

图 20 为燃烧室定型后,在单缸试验机上所测得的万有特性。

现把该特性曲线上的若干特性点的测量数据摘录如下:

在 3 000 r/min,13.24 kW[即 18 马力]时,油耗 252.96 g/(kW·h)[即 186 克/(马力·小时)],排温 530 ℃;

在 3 000 r/min,14.71 kW[即 20 马力]时,油耗 265.20 g/(kW·h)[即 195 克/(马力·小时)],排温 580 ℃;

最高平均有效压力在转速 2 000 r/min 时达 0.81 MPa[即 8.3 公斤/(厘米)²],油耗 258.40 g/(kW·h)[即 190 克/(马力·小时)],排温 500 ℃;

最低油耗为 239.36 g/(kW·h)[即 176 克/(马力·小时)]。

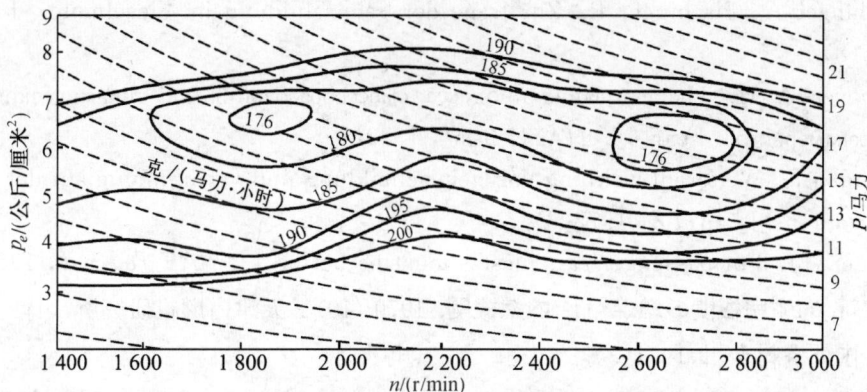

图 20　采用新型涡流燃烧室的单缸试验机的万有特性

定型后的涡流燃烧室应用于产品试验样机(495Q 型汽车柴油机)上,在 100 小时强化试验中,以 15 分钟功率连续运行。现将每两小时记录一次的性能指标抄录如下表。

功　率 P	转　速 n	油　耗 be	排　温 Tr
/kW[马力]	/(r/min)	/(g/(kW·h))[克/(马力·小时)]	/℃
50.01[68]	3 000	252.96~257.04[186~189]	450~460

这些数据表明,495Q 型柴油机的功率、油耗、排温等指标,与单缸试验机的有关指标相吻合,并且很稳定。

6　结　语

(1)通过上述试验研究表明,M 过程可以在涡流燃烧室中实现,而且有许多更有利的条件。我们关于 M 过程涡流燃烧室适应高速化的分析与改进方向,看来基本上是正确的。

(2)试验结果表明,采用较大的油注偏移和尽可能大的容积比,采用双楔型主燃室,采用长宽比较大的圆矩形通道并配以最佳位置和倾角尺寸,能在实现高速化的同时获得高的经济性。

(3)通过系统试验研制成的新型涡流燃烧室,用在 95 单缸机和 495Q 型车用柴油机上,功率可提高 15%~20%,油耗可降低约 10%,最大升功率可达 17.65 kW/L[即 24 马力/升],最低油耗降至 239.36 g/(kW·h)[即 176 克/(马力·小时)]。

(4)采用新型涡流燃烧室后,有关燃烧噪音、排烟和热负荷等问题,还有待今后作进一步的研究。

参 考 文 献

[1] 李德桃.常柴涡流燃烧室的发展研究[J].常柴技术简报,1974(1).

［２］常州柴油机厂．常柴Ⅰ号燃烧室［G］．参加柴油机燃烧室经验交流会议资料，1962．

［３］Nagao F，Kakimato H．Swirl and combustion in divided combustion chamber type diesel engines［J］．*SAE Trans*，1962，70．

［４］Pischinger A，Pischinger R．Zur frage der gemischildung im dieselmotor［J］．*MTZ*，1965(8)．

［５］Миронов А П．Исследование продесса смесеобразовония на динамической модели вихрекамерного дизедя［J］．*НАТИ Вып*，1959(19)．

［６］Meurer J S．Weiterentwicklung von gemischbildung und verbrennung auf der basis des mverfahrens［J］．*MTZ*，1972(8)．

［７］Walder C J．For passenger cars diesels must be better［J］．*SAE Journal*，1965(11)．

［８］住江新．ちず室機関の燃焼［J］．内燃機関，1970（10）．天津内燃机研究所，译．小型内燃机：国外资料专辑，1971(3)．

［９］Alcock A F，Scott W M．Some more light on diesel combution［J］．*IME Proc Auto Div*，1962-63(5)．

［10］Pischinger A，Pischinger F．Bombversuche über die diesel-verbrenung unter motorischen bedingungen［J］．*MTZ*，1959(1)．

［11］长尾不二夫．压燃式发动机的燃烧［J］．内燃机快报，1965(19/20)

［12］Дьяченко Н Х，и Др．*Теорня Двигатедей Внугреннего Сгорания*［M］．Машиностроение，1965．

［13］Модчанов К К．Исследование возможности повышения топливной экономичности тракторного дизеля Д54［J］．*НАТИ Вып*，1959(19)．

［14］Ховах М З．Исследование смесеобразования в двигателях с разделенными вихревым камерами сгорания［C］∥*Сгорание и Смесеобразование в Дизелях*，АН СССР，1960．

［15］自動車用デイーゼル機関の出力向上の一方法［J］．内燃機関，1971(5)．

（本文原载于《内燃机》1975 年第 11 期）

Investigation on high speed adaptability
of swirl combustion chamber

Abstract：The experimental investigations on the swirl combustion chamber design are summarized．The paper describes the future trends of the swirl chamber's improved design，presents a new type of swirl chamber construction and introduces some new design concept for the swirl chamber．

关于 M 过程在涡流燃烧室上的应用[①]

李德桃

1 任务的提出

涡流室式柴油机是我国当前主要的农用动力。出现这种情况决非偶然。涡流室式柴油机的工作过程易于组织,对燃油品质和喷雾质量不敏感,在较大的速度范围内具有较高的空气利用率,供油系统故障少,排出的有害气体少,高速性好,可靠性高,因而它便于县社工厂制造和维修,便于农村综合利用,便于使用不同品质的燃油。这些都是促使它在我国迅速发展和广泛使用的重要因素。

当然,涡流燃烧室式柴油机也存在一些众所周知的缺点,如燃油消耗率较直接喷射式高,起动性能较差等。生产的发展,向我们提出了克服这些缺点和进一步改进这种燃烧室的任务。

本文试图从 M 过程合理应用于涡流燃烧室的角度,探索改进和提高这种燃烧室性能的途径。

2 M 过程应用于涡流燃烧室的必要性和现实性

M 过程应用于涡流燃烧室的必要性,可以从 M 过程已成为柴油机工作过程的一种合理模式来说明。众所周知,M 过程的一些最有意义的特征是:

(1) 混合气形成过程。燃料大都涂于室壁,分层汽化并与气流混合。燃烧开始后,已燃气体沿螺旋轨道向燃烧室中心移动;与此同时,新鲜空气向室壁移动。这样,分层汽化的燃料不断与新鲜空气混合燃烧。这种混合气形成的宏观图形已为许多高速摄影照片所证实[1,2]。毕兴格(Pischinger A)等人对此进行过理论研究[3]。实践和理论都证明,这种混合气形成的方式对柴油机工作较有利,因为分层汽化与"热力混合"易于实现较完全的燃烧。

(2) 燃烧过程。初始反应速度很低,一旦着火,燃烧便加速进行。M 过程和普通过程两种放热规律的比较(图 1)清楚地显示出 M 过程的这一特征[4],说明按 M 过程工作时初始压力升

---

① 该文是作者对涡流燃烧室进行了 10 年的实验和理论研究之后,对涡流燃烧室内的空气运动、喷雾和燃烧过程提出的认识,它反映了当时国内外该领域的研究水平,此文在 1978 年召开的全国内燃机燃烧过程讨论会上宣读。此后,作者及其科研团队继续进行了近 20 年的试验和理论研究,发表了一系列的论文,建立了具有独创性的理论体系,参见科学出版社出版的专著《涡流室式柴油机的燃烧过程和燃烧系统》《柴油机冷起动的基础研究和改善措施》。

——编者注

高率和燃烧噪音都较低。

（3）热物理过程。油膜相对汽化速度与空气相对涡流速度大致成正比关系[5]（图2），这就说明了在油膜汽化过程中，不断形成新混合区，而空燃比则近似恒定。

在滞燃期内，喷注与室壁之间形成的一层蒸汽，使壁面和燃油之间的传热系数减小[6]，导致初期蒸发速度缓慢。着火后，火焰辐射占重要地位，使蒸发加速进行[4]。

M方式的热物理过程的上述特征，使我们有可能利用蒸发过程来控制放热过程。

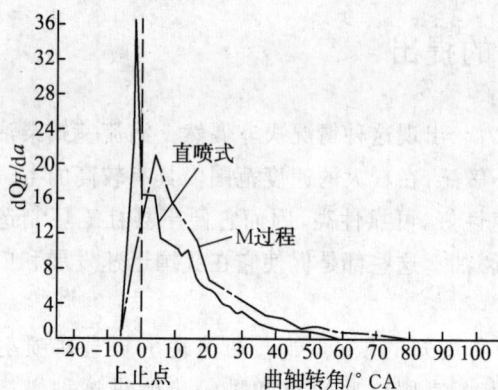

图 1　两种放热规律的比较

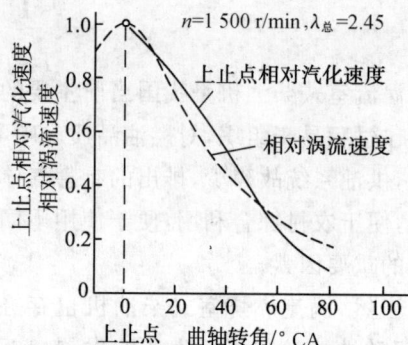

图 2　油膜相对汽化速度和空气相对涡流速度随曲轴转角变化的情况

显而易见，以上三个特征是相互关联，相互依存，相互影响的。正是这些最有意义的特征，构成了 M 过程这一柴油机混合气形成和燃烧过程的合理模式。这种模式，不仅在提高空气利用率和降低燃烧噪音等方面已经取得了专业界公认的实际效果，而且人们从混合气形成力学、燃烧反应动力学和热物理学等方面为它找到了越来越多的理论根据[6-8]。正因为如此，最初在 MAN-M 型燃烧室上实现的 M 过程，必然超出个别的具体的燃烧室结构型式的范围，在其他型式的燃烧室上获得不同程度的普遍应用[3,4,10-12]。

对于将 M 过程应用于涡流燃烧室的可能性和现实性，不少研究者都曾根据各自的高速照片和理论分析作了明确的说明[2,3,9]。我们近年研制成功的高速涡流燃烧室（图3）也在某种程度上应用了 M 过程[10]。采用 4S1 型喷油器，喷油角 θ 和喷注偏移 e 如图 3 所示时，油膜表面积约占整个涡流室表面积的一半。我们曾在单缸试验机上进行仅仅改变喷注偏移 e 这个结构参数（e 分别为 5,6,7,8 mm）的性能对比试验。图 4 示出了 $e=7,8$ mm 的对比试验的结果。试验时，采用 4S1 型单孔轴针式喷油器，喷油提前角都调整至最佳值。涡流室的主要几何参数是：$R=17$ mm，$\theta=20°$。试验结果表明，在各种试验转速下，增加 e 可不同程度地提高发动机的经济性和动力性（我们发现低速、低负荷时情况与此相反）。

很明显，这是增加涡流室中油膜燃烧的燃料量的结果。当然，应用 M 过程的涡流燃烧室与 MAN-M 型燃烧室的工作情况存在着区别，但是 M 过程的一些最有意义的特征在前一种燃烧室上也可以显示出来[2,3,9]。

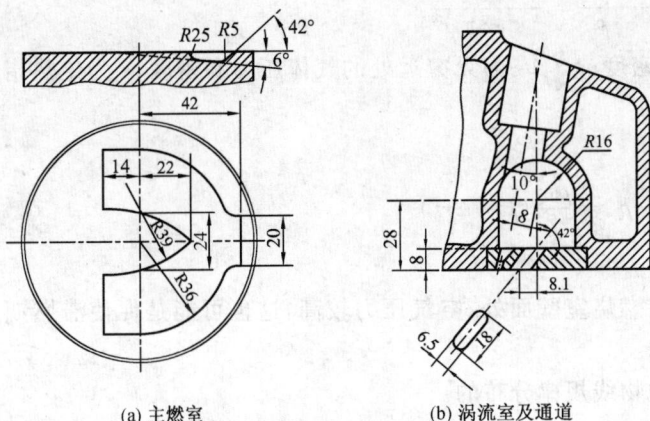

(a) 主燃室　　　　　(b) 涡流室及通道

图 3　新型涡流燃烧室的结构形式和具体尺寸

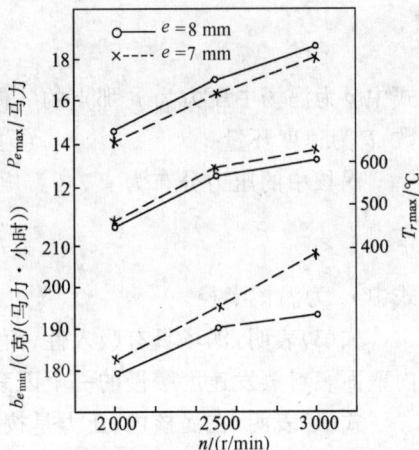

图 4　喷注偏移对性能的影响

3　M 过程在涡流燃烧室上的合理应用

如前所述,M 过程作为柴油机混合气形成和燃烧过程的合理模式,无疑会促使人们在涡流燃烧室上也应用它。当然,应用时也应考虑涡流燃烧室的特点。也就是说,必须将 M 过程的一般原理同涡流燃烧室的具体条件合理地结合起来。

在涡流燃烧室上如何恰当地、合理地应用 M 过程呢? 我们认为,就 M 过程本身来说,诚如茅瑞尔(Meurer J)所指出,把燃料涂布于燃烧室壁上与沿涂油表面的空气运动居首位。这两种方法不能分开来应用,因为它们有机地联在一起[4]。涡流燃烧室应用 M 过程时,在一定程度上仍应遵循这一原则。在分析这一具体应用之前,有必要先对涡流室内气流的速度场、压力场和温度场作些说明。

关于涡流的特性,现在还没有一个一致的说法。但是,随着测试技术的发展,其中包括激光多普勒效应测速仪的应用,这个问题的解决已为时不远了。根据目前已有的资料[13,14],在上止点附近,空气涡流是接近势涡的(图 5),即涡场中的速度分布为[15]

$$v = \frac{C_1}{r} \tag{1}$$

式中 C_1 为积分常量。

由此可知,涡场中某一点的速度同该点至涡核中心的距离成反比。

涡核部分的速度分布为

$$v = r\omega \tag{2}$$

式中 ω 为涡核绕中心轴转动的角速度。

式(1)、式(2)和图 5 表明,在靠近涡核处,涡流的速度是比较高的。

相应地,涡场中的压力分布为

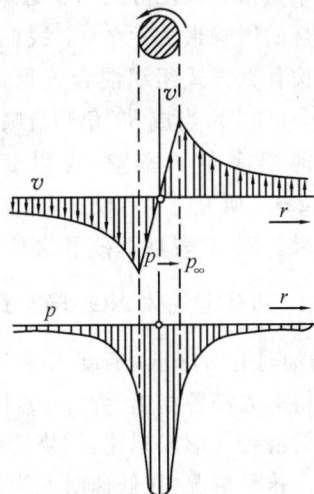

图 5　涡核内部和外部的速度分布和压力分布

$$\frac{p}{\rho} - \frac{p_\infty}{\rho_\infty} = \frac{\kappa-1}{\kappa} \frac{\Gamma}{8\pi^2 r^2} \qquad (3)$$

式中 ρ 为涡场中压力为 p 那点的气体密度；p_∞，ρ_∞ 为无穷远处的气体压力和密度；κ 为绝热指数；Γ 为速度环量。

涡核中的压力分布为

$$p - p_\infty = \frac{\rho\omega^2}{2}(r^2 - 2r_0^2) \qquad (4)$$

式中 r_0 为涡核半径。

式(3)表明，涡场具有吸入性，而在燃烧室壁面处，空气压力较高，这也可能是促使滞燃期内壁面燃料蒸发速度缓慢的一个因素。

式(4)表明，在涡核内，压力是按抛物线规律分布的。

涡流室的温度分布如图 6 所示[16]。Ⅰ，Ⅲ 两种连接通道的布置形式，在涡流室截面上都有一圈高温带，而室中心及靠近室壁部分的温度较低。Ⅲ 的高温带（410 ℃）温度稍低，范围较窄；Ⅰ 的高温带（420 ℃）温度稍高，范围也较宽。

现在，我们可以针对涡流室的一些具体情况来讨论如何把 M 过程合理地应用于涡流燃烧室。我们认为，必须通过以下三个方面把 M 过程的一般原理同涡流燃烧室的具体条件结合起来。

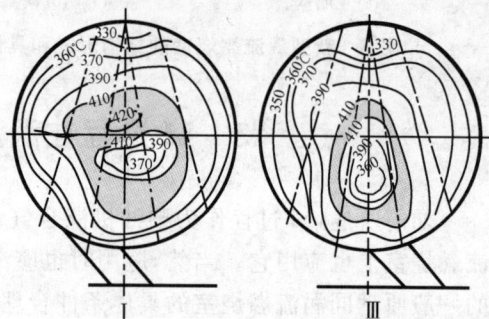

图 6　涡流室内的平均温度分布

3.1　适当的喷油方向

毕兴格和长尾等人早就通过各自的试验研究发现，喷油方向对燃烧的影响最大[2,17,18]。我们认为，不同的喷油方向实质上反映了按什么方式组织混合气形成和燃烧过程。根据长尾等人的研究[2]（图7），当成80°角顺流喷油时，喷注成锐角喷到室壁上，如一层薄膜涂布于室壁，可以获得 M 过程的效果。根据郎格（Lange）的研究[19]，当成75°角顺流喷油时（图8），也可获得同样的效果。图 8 表明，采用这样的喷油方向，最大压力升高率 $\left(\dfrac{\mathrm{d}p}{\mathrm{d}\varphi}\right)_{\max}$ 和有害气体 NO_x 降至最小值，但是平均有效压力 p_e 比

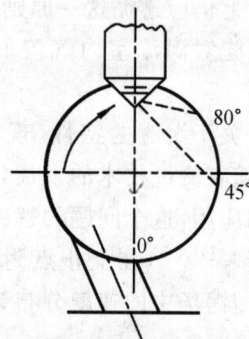

图 7　喷油方向和涡流方向

45°喷油时有所降低，烟度 R 和有害气体 HC 值却增加了。长尾和郎格等人都曾指出[2,19,20]，燃料大都涂布于壁面后，由于燃烧后期缺乏有效的涡流，致使壁面处的混合气形成减速，燃烧持续期增加，过后燃烧量增加，影响发动机的动力性和经济性的提高。这种情况，在低速时尤为显著。

采用这样的喷油方向，还有一个缺点，就是冷起动性能差。虽然采用一些辅助装置（如电热塞），可以使起动性能稍有改善，但由于在滞燃期内壁面蒸发量太少，使起动性能得不到明显改善。

鉴于上述情况，我们认为，在涡流室内采用容积-油膜式过程（或复合式过程）是合理的，即把一部分的燃料喷于室壁，按 M 方式组织混合气形成和燃烧；一部分燃料离散于空间，按

"空间式"组织混合气形成和燃烧。图 8 所示的约 45°喷油就是这种情况。此时增加了喷注自由混合表面积,燃料通过涡流速度较大的区域和高温带(参看图 5 和图 6),热交换和质量交换的条件都较好。轻油滴在这里迅速蒸发,并与空气混合。这样,可以在一定程度上弥补一部分燃料按 M 过程组织混合气形成和燃烧所带来的一些缺点。

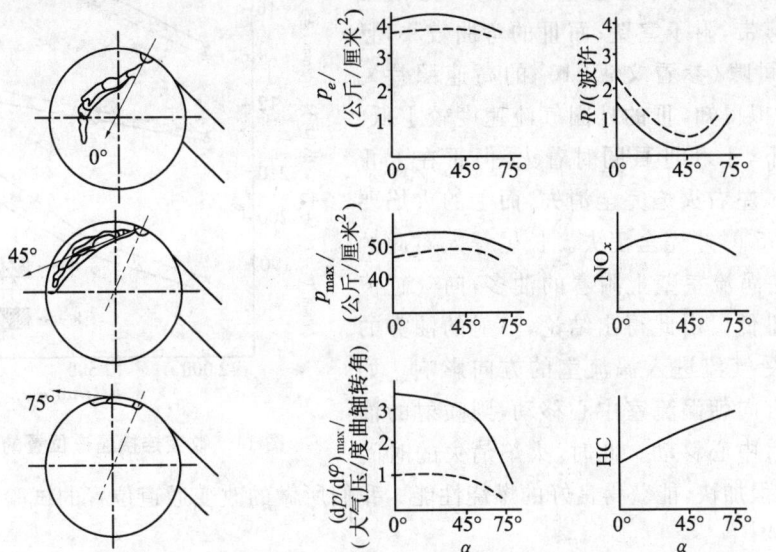

(a) 不同喷油角度的喷注扩展情况　(b) 改变涡流室的喷油方向和壁温对性能的影响

——$T_\omega = 150$ ℃　----$T_\omega = 350$ ℃

图 8　喷油方向对性能的影响

对于涡流燃烧室来说,适当的喷油方向是由喷油角度 θ 和喷注偏移 e 这两个尺寸来确定的(图 9),因为自由喷注长度、喷注落点和喷注经历的温度场都取决于这两个尺寸。为了在涡流燃烧室上合理地应用 M 过程,国外一般采用较大的喷油角度和较小的喷注偏移(图 10)[21]。我们所研制成功的新型涡流燃烧室,由于具有起动孔而采用较小的喷油角度和较大的喷注偏

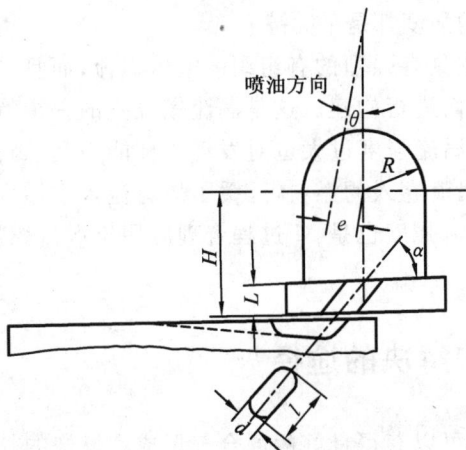

图 9　涡流燃烧室的主要尺寸参数

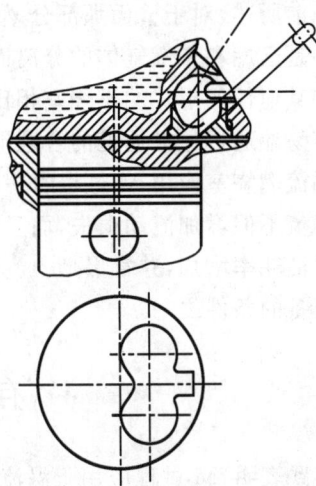

图 10　VW 型柴油机的涡流燃烧室

移。实践表明,在这两种情况下,都能获得较好的性能。

3.2 适当的气流进入涡流室的方向

长尾等人通过高速摄影观察连接通道不同位置(如图6)时的混合气形成和燃烧状况[16],发现通道形式Ⅰ因气流靠近室壁,喷油后,形成强烈弯曲的油雾带,附于室壁;而Ⅲ的弯曲较小,且与室壁略有间隙(参看文献[16]的高速照片)。此外,由水模拟得知,Ⅲ的外圈气流速度较Ⅰ低。在这种情况下,Ⅰ和Ⅲ虽同时着火,但Ⅲ在上止点后$20°\sim25°$左右火焰完全消失,而Ⅰ的火焰要在上止点后$35°$时才完全消失。长尾等人对此的解释是:Ⅰ在涡流室壁上附着的油多,向气缸流出的速度较Ⅲ低。由此得出结论,喷射的油雾的路线和形状受气流进入涡流室的方向影响。如果气流进入方向朝涡流室中心移动,则喷射的油雾也向涡流室中心移动。此时,火焰消失的时间

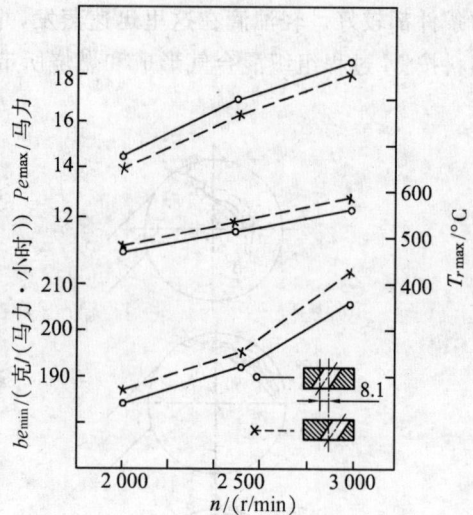

图 11 改变连接通道位置的对比试验

缩短,燃烧过程加快,能获得良好的燃烧性能。我们所做的改变通道位置的试验也证实了这个结论(图 11)[12]。

上述情况表明,要实现将 M 过程合理应用于涡流燃烧室,把适当数量的燃料涂布在室壁上,除了喷油方向要适当外,还必须匹配以适当的气流进入方向。我们认为,使喷注横穿部分涡流室容积,让较轻的油滴在较高的涡流速度区和较高的温度带蒸发并同空气混合,而较重的油滴涂布在室壁上,这样,既能保证迅速着火和具有较短的持续燃烧时间,又能获得较高的空气利用率和较低的压力上升率。

根据我们的经验,适当的气流流入涡流室的方向,是由正确地选择通道的位置和倾角来得到的[12]。

3.3 适当的气流强度

如前所述,对于壁面那部分燃料来说,空气运动是至关重要的。因为无论对燃料与空气的混合还是已燃和未燃气体的分离而言,空气涡流的强度都是个关键。

同其他任何形式的燃烧室相比较,涡流燃烧室具有强烈的有组织的空气涡流;而且,它可以较容易地通过选择通道的合理形式与最佳尺寸而进行构建。这是涡流燃烧室的一个特点,也是涡流燃烧室应用 M 过程的一个优点。当然,涡流强度过大也对发动机性能不利,因为过强的涡流不但增加流动损失,而且往往还增加被涡流卷入的燃料,延缓了燃料流入主燃室,结果燃料消耗率增加,并冒黑烟[2]。因此,适度的气流强度也是 M 过程合理应用于涡流燃烧室必须具备的条件之一。

4 存在的问题和解决的途径

实践表明,M 过程应用于涡流燃烧室的结果,可以获得良好的混合气形成和燃烧的效果,从而使涡流燃烧室式柴油机具有较好的扭矩特性,降低了燃烧噪音,提高了经济性。但是,这

也会带来一些缺陷,一个是恶化了起动性能,另一个是延长了燃烧持续时间,后者在低速低负荷时更显著。虽然通过上述的合理应用的各种措施,这些缺陷得到一定程度的克服,但是,我们仍针对这些问题,进一步采用以下的措施来解决[12]。

4.1 采用起动孔以改善起动性能

M 过程在涡流燃烧室上的应用,使本来起动性能较差的涡流室式柴油机起动更困难。对这个问题,国外是采用品陶(Pintaux)型喷油嘴和电热塞的方法来解决[22]。我国常州柴油机厂的工人和技术人员创造的起动孔(图 3),为改善涡流室式柴油机的起动性能作出了自己的贡献。此起动孔是配合喷注较集中的 4S1 型喷油器使用的。起动时,发动机转速低,涡流强度弱,因而喷注弯曲小。这样,较多的燃油经过起动孔直接喷入温度较高的主燃室,改善着火条件。起动后,转速增加,涡流增强,喷注在涡流的作用下靠壁弯曲。此时,经过起动孔直接喷入主燃室的燃油较少,起动孔对性能的影响减弱。多次试验表明,采用起动孔后,发动机能在 0 ℃时顺利起动。

4.2 设计合理的主燃室形状以实现完全燃烧

如前所述,M 过程应用于涡流室的缺陷之一,即燃烧后期由于涡流室内缺乏有效的涡流,致使完全燃烧延迟。这就意味着加强主燃室的空间混合,促进喷出火焰的迅速燃烧,对弥补这个缺陷有着重大意义。从反应动力学的观点看,主燃室中的燃烧速度在很大程度上取决于火焰与氧气通过相互扩散而混合的速度。这个速度,显然同主燃室的结构形式关系甚大。

关于主燃室的结构形式,铲击型主燃室给我们的启发是:双涡型主燃室的框框是可以打破的。采用铲击型凹坑,使火焰由两侧向整个活塞顶面铺围过来,可获得同双涡型一样良好的性能。然而这种主燃室在单缸机上的试验情况表明,虽然发动机也能达到 3 000 r/min,但燃料消耗率和排温都比较高[23]。铲击型凹坑和双涡型凹坑一样,一般都是在低速时性能较好[18,23]。我们认为,为了缩短燃烧持续时间,主燃室仅形成局部涡流是不够的,应力求使火焰尽快向整个主燃室扩展;同时也应当尽量避免气流的能量损失。

通过对铲击型、双涡型、直沟型等主燃室的对比分析,我们对铲击型作了以下的改进:

(1) 为了减少气流从涡流室喷出时的损失,活塞顶可增加一段倾角和形状都与通道相适应的导向气道,以便因势利导地把气流引入凹坑(图 3 和图 12);

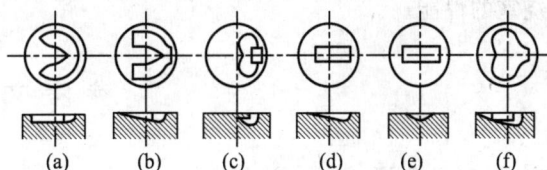

图 12　不同形状的活塞顶凹坑

(2) 为了使从涡流室喷出的火焰迅速向整个主燃室传播,把铲击型的同一深度的凹坑改为渐浅凹坑,这样,两股气流就成"楔形"高速向前扩展;

(3) 为了减少气流在活塞顶部由于流经渐缩渐扩气道而引起的能量损失(这在铲击型凹坑上是存在的),把气道设计成近似等宽度的;

(4) 为了减少两股气流在凹坑尾部起翘后相互碰撞而引起能量损失的可能性,把分流后的气道改成近似直沟型。

根据以上观点,我们设计了如图 12 所示的活塞顶凹坑,并把它定名为"双楔型"主燃室。它与铲击型的对比试验结果示于图 13。

该图表明,几乎在全部试验转速和负荷范围内 b 型都较 a 型为佳。特别是在高转速、高负荷下,b 型的排温和燃料消耗率都有较大幅度的下降。这表明以上所作的改进分析基本上是正确的。

随后，我们又做了四种形状的主燃室（图 12c～f）的试验。它们是中间带浅槽的双涡型（c）、直沟型（d）、圆弧型（e）和让气门型（f）。试验结果表明，其性能都比 b 型差[23]。

为了考查活塞顶凹坑的尺寸对于性能的影响，我们做了两个形状相似而尺寸不同的 b 型主燃室的对比试验。其中一个凹坑容积为 7.6 cm³；另一个为 5 cm³。试验所得的结果是在三个速度工况和各种负荷下，两者的性能都很接近，这说明凹坑容积对性能的影响不大。由此可见，适当减少活塞顶凹坑容积以增加容积比，不仅是必要的，而且也是可行的。

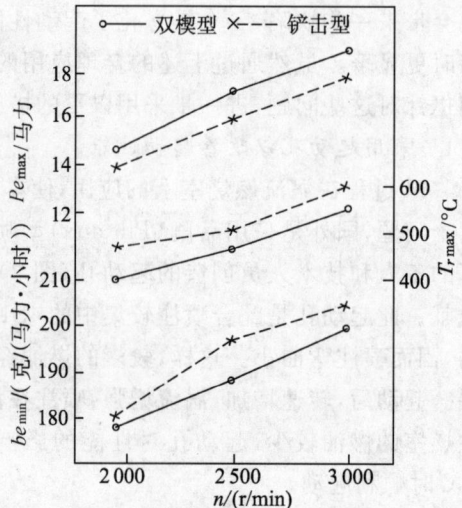

图 13　双楔型和铲击型的性能对比

综合对主燃室的试验研究，我们得到这样一个思路：我们对主燃室所谋求达到的目标，是提高空气利用率，以实现完全的燃烧；而达到这个目标的主要途径是设计形状合理的主燃室。

我们关于涡流燃烧室，特别是其主燃室的改进试验的指导思想和试验结果，在九江动力机厂等单位的试验研究中也得到了证实[24]。

5　结　束　语

通过分析和实践，我们确认了 M 过程应用于涡流燃烧室的必要性、可能性和现实性；提出了合理应用的方向和实现合理应用的具体措施；探讨了应用时存在的问题和解决的途径。我们获得了一种性能较好的涡流燃烧室。我们相信，通过进一步的研究，可以更全面地提高这种燃烧室的性能。

参 考 文 献

[1] Alcock A F, Scott W M. Some more light on diesel combustion[J]. *IME Proc Auto Div*, 1962, 63(5).

[2] Nagao F, Kakimoto H. Swirl and combustion in divided combustion chamber type diesel engines[J]. *SAE Trans*, 1962, 70.

[3] Pischinger A, Pischinger R. Zur frage der gemischildung im dieselmotor[J]. *MTZ*, 1965(8).

[4] Meurer J S. Der wandel in der vorstellung vom ablauf der gemischbildung und verbrennung im dieselmotor[J]. *MTZ*, 1966(4).

[5] Urlaub A, Müller E. M 过程柴油机排气质量的试验和理论研究[C]//第十届国际内燃机会议论文选（柴油机部分）. 北京：机械工业出版社，1976.

[6] Flatz W. Das verdumpfen von dieselkraftstoffen an der wand[J]. *MTZ*, 1965, 26(1).

[7] Meurer J S. Weiterentwicklung von gemischbildung und verbrennung auf der basis des M-verfahrens[J]. *MTZ*，1972(8).

[8] Соколик А С. M 过程的动力学解释[C]∥内燃机译文集.北京：国防工业出版社,1965.

[9] Миронов А П. Исследование продесса смесеобразования на динамической модели вихрекамерного дизеля[J]. *НАТИ Вып*，1959(19).

[10] 史绍熙,等. 柴油机复合式燃烧系统的研究[J].农业机械学报,1965,8(5).

[11] 魏恭隆.柴油机新产品研制中的若干问题[J].江苏柴油机,1976.

[12] 李德桃.涡流燃烧室高速适应性的研究[J].内燃机,1975(11).

[13] Бондаренко Г П. О стесеобразовании и горении в вихревой камере двигателя с воспламенением от сжатня[J]. *Автомобильная Промышленмость*，1961(9).

[14] 天津内燃机研究所译.国外资料专辑[J].小型内燃机,1971(3).

[15] Фабрикант Н Я. *Аэродинамика*[M]. Наука,1964.

[16] Nagao F，Kegami M，Shinzoto T，Bamba T. Air motion and combustion in a swirl type diesel engine[J]. *Bulletin of JSME*，1967：833.

[17] Pischinger A，Pischinger F. Bombversuche über die diesel-verbrenung unter motorischen bedingungen[J]. *MTZ*，1959(1).

[18] 长尾不二夫.压燃式发动机的燃烧[J].内燃机快报,1965(19/20)

[19] Lange K. Diesel motorische verbrennung unter besonderer berücksichtigung der wandauftragung des brennstoffes[J]. *MTZ*，1974(2).

[20] Lange K. Untersuchung der verbrennung im motor mit optischen methoden[J]. *MTZ*，1973(1).

[21] Hofbouer P. Der dieselmotor für das kampaktauto VW Goff[J]. *MTZ*，1977(6).

[22] Ricardo. *The High Speed Internal Combustion Engine* [M]. Landon：Blackie & Son，1953.

[23] 常州柴油机厂技术科,镇江农机学院内燃机教研组.常州柴油机厂涡流燃烧室的发展研究[J]. 常柴技术简报,1974(1).

[24] 九江动力机厂技术科.95 系列柴油机燃烧室的改进及高速性能试验[R].95 系列技术交流会资料,1977.

（本文曾在 1978 年《全国内燃机燃烧过程》学术讨论会上宣读,后收入沈恒荣、刘急、高政冠主编的《内燃机燃烧性能研究》,机械工业出版社,1982 年）

Study of air movement in a separate swirl chamber by means of a bidimensional dynamic liquid model[①]

Li Detao , Vasile Berindean

Abstract

The paper deals with the movement of the air and the structure of the swirl in a separate bell-type swirl chamber, in which, up to the present, the movement of the swirl and its structure have not been studied on bidimensional dynamic models.

The main part of the experimental equipment consists of a bidimensional dynamic model made of transparent plastic material(stiplex).

By using, for the approximate similitude of the Reynolds criterion, different particles in order to visualize the currents and by having them photographed, the influence of the shaft radius angle and of the revolution on the movement and structure of the swirl has been established.

The form and the structure of the swirl depends on the shaft radius angle and is practically independent of the revolution.

1　Generalities

This research was meant to establish the air movement, the way the swirl is made up and structured in the separate bell-type swirl chamber. From references published worldwide it results that the method using the bidimensional similarity with water is very adequate in the study of the way the swirl is structured and made up.

The importance of the studies on models lies in the fact that the process develops

---

① 利用精确的水模型试验,发现吊钟型涡流室内存在两个副涡,是本文的贡献,也与后来的模拟计算结果一致。

——编者注

24

on the model just like in reality. Thus, the studies obtained on a model may be taken up again in the actual process, if their similitude criteria are equal two by two.

Paper [1] presents the research into air movement for separate spherical and cylindrical swirl chambers. The paper presents the criteria of similitude used, the bidimensional dynamic model and the method of research. The experimental conditions and the model were dimensioned by taking into account the condition of equality between the Froude criterion established on a model and the Mach criterion determined for compressed air before the end of the compression stroke in the engine cylinder.

The movements in unitary combustion chambers by means of tridimensional dynamic single stroke models are presented on models in the paper [2].

Thus we can assert that the swirl movement and its structure have not been studied so far on bidimensional dynamic models for separate bell-type swirl chambers. Therefore, the present work is such an attempt. We have also conceived and accomplished a bidimensional dynamic model of a cylinder with a separate bell-type swirl chamber.

2   Establishing criteria of similitude[3-8]

Quantities specific to the operation of internal combustion engines, like the duration of one rotation of the crankshaft τ_r, stroke S and average speed of piston c_m were introduced in order to establish criteria of similitude.

Density, pressure difference Δp and viscosity υ are correlated to the initial state, thus changing into nondimensional quantities.

The fluid is considered incompressible, that is, the density variation at high flow speeds is neglected. Further, at the beginning of the flowing process, the fluid is considered to be at rest(the process of gas exchange is neglected and no change of heat by conductivity is produced).

On account of the above-mentioned simplifying hypotheses we may write the following criteria:

– *Strouhal's Number*

$$\left(\frac{S}{c\tau_r}\right)_M = \left(\frac{S}{c\tau}\right)_m \tag{1}$$

– *Froude's Number*

$$\left(\frac{c_m^2}{SF}\right)_M = \left(\frac{c_m^2}{SF}\right)_m \tag{2}$$

– *Euler's Number*

$$\left(\frac{P_0}{\rho_0 c_m^2}\right)_M = \left(\frac{P_0}{\rho_0 c_m^2}\right)_m \tag{3}$$

– *Reynolds Number*

$$\left(\frac{c_m S}{\upsilon_0}\right)_M = \left(\frac{c_m S}{\upsilon_0}\right)_m \tag{4}$$

where: M—model

m—engine

From references it is known that Froude's Number is little used in researches because the influence of the gravity force, as an accelerating force, is negligible due to the low density of gases. If viscosity forces and gravity forces are not important, and provided the inertia force produced by the movement of the piston is neglected, Euler's Number equals the value of 1. Thus, the inertia force is produced only due to the pressure force. Reynolds criterion[9] is used with these simplifications of approximate similitude, in the case of the bidimensional model used.

3 The experimental research method

3.1 The experimental equipment

The main part of the experimental equipment is the bidimensional dynamic model made of transparent plastic material(stiplex), Fig. 1 and Fig. 2.

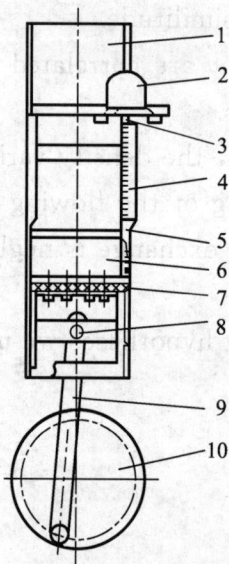

1—cylinder head   2—swirl chamber   3—connecting hole
4—graduated ruler   5—cylinder liner   6—piston   7—rubber packing   8—piston-pin   9—connecting-rod   10—disk

Fig. 1 Dynamic model

Fig. 2 The image of the dynamic model

The geometric scale compared with engine 4. 95 Q is 2 ∶ 1 and the height of the model wall $h=100$ mm. The separate bell-type swirl chamber is provided only with the connecting hole, because the diameter of the auxiliary starting hole is of only $\Phi=2$ mm. Thus, its section represents 3. 89 percent of the main hole section. Therefore its influence on the air movement is negligible.

The sealing of the piston is assured by a rubber plate mounted transversely at a distance of 2. 5 mm from the piston head wall.

The position of the piston with respect to T. D. C. while taking the photograph is established using a ruler in mm attached to the model cylinder.

The model is fed by a continuous current electric engine EP—$2 \sim 24$ V, variable speed, depending on voltage($10 \sim 24$ V).

The photos were taken with an Exacta-Varex type camera(made in the German Democratic Republic).

3. 2 Visualization of streams[10,11]

The research into spatial and plane flow processes for high Reynolds Numbers must be carried out on wholly immersed bodies. Water channels are quite adequate in rendering the image of spatial processes by introducing solid particles, paint or gas bubbles in the water.

In the present case, in order to visualize the streams, one can use:

—mustard seeds in salt water, in order to increase water density;

—potassium permanganate powder;

—leaves of black tea;

—hot yoghurt;

—aluminium powder;

—blue ink.

The best results were obtained with mustard seeds in salt water and with potassium permanganate powder.

The density of the substances used was somewhat lower than water density, so that the velocities field was not influenced by the particle mass.

Direct observation of streams by visualisation was completed when their photos were taken.

4 Experimental results

The following results were obtained by taking the photo of the model combustion engine:

Fig. 3 renders the image when visualizing the water movements with tea. The photo was taken after several strokes of the piston, and was exposed for 0.001 s. In the image one can distinguish particles of tea, their majority being grouped in the central part next to the connecting hole. This proves that the movement is considerably influenced by the hole direction and by the shape of the opposite wall. Some of the particles are stopped by the plane wall of the chamber in the chamber corners.

Visualization of water movements was further carried out with mustard seeds immersed in salt water in order to keep the mustard seeds floating in the water.

The engine rotation speeds, corresponding to the model, were deduced from the equality between Reynolds Number for the model and for the engine:

$$n_M = n_M \frac{S_m}{S_m} \frac{v_{\text{water}}}{v_{\text{air}}} \qquad (5)$$

Figs 4 ~ 12 render the images when visualizing the water movements with mustard seeds; exposure time 1/10 s in order to reveal the stream lines. The photo of the chamber was taken towards the end of the piston stroke when the particles were introduced in order to distinguish the particles participating in the movements of the model swirl chamber. Each figure specifies the conditions under which the water movements took place; if the photo is taken without taking out particles from the swirl chamber, the legend of the figure reads "repeated".

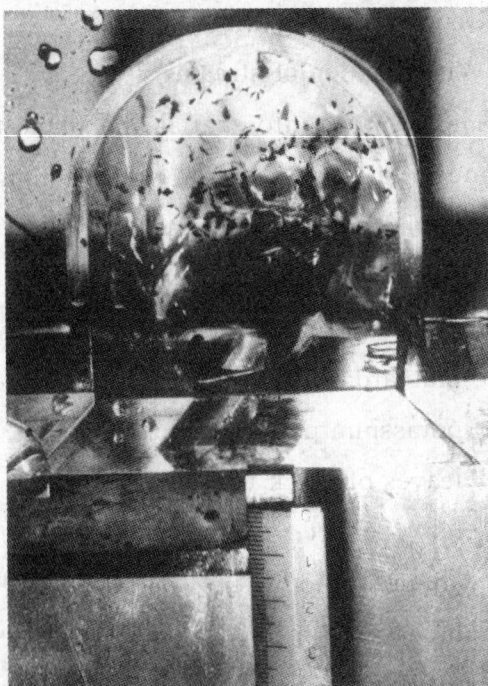

Fig. 3 The water movement in the swirl chamber for: $n_M = 10$ r/min; $n_m = 1\ 496$ r/min; the angle of crankshaft radius $\alpha = 10°$ CA before T. D. C.

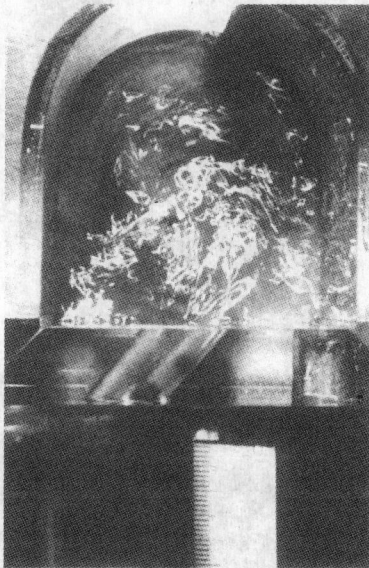

Fig. 4   The water movement in the swirl chamber
for: $n_M = 10$ r/min; $n_m = 1\ 544$ r/min;
$\alpha = -3°$ CA before T. D. C.

Fig. 5   The water movement in the swirl
chamber for: $n_M = 10$ r/min; $n_m = 1\ 546$ r/min; $\alpha = -2°$ CA before
T. D. C.

Fig. 6   The water movement in the swirl chamber
for: $n_M = 10$ r/min; $n_m = 1\ 550$ r/min;
$\alpha = 0°$ CA at T. D. C.

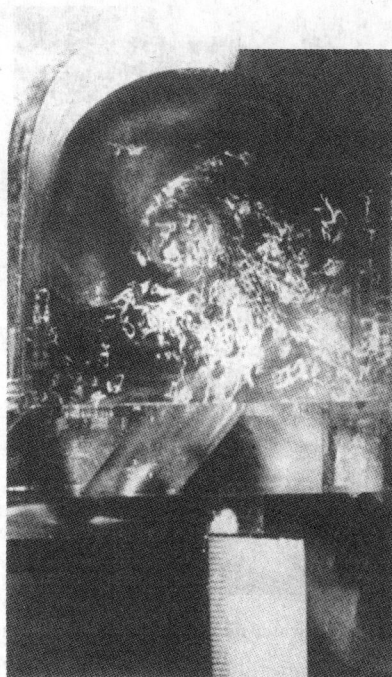

Fig. 7   The water movement in the swirl chamber
for: $n_M = 10$ r/min; $n_m = 1\ 550$ r/min;
$\alpha = 0°$ CA at T. D. C.

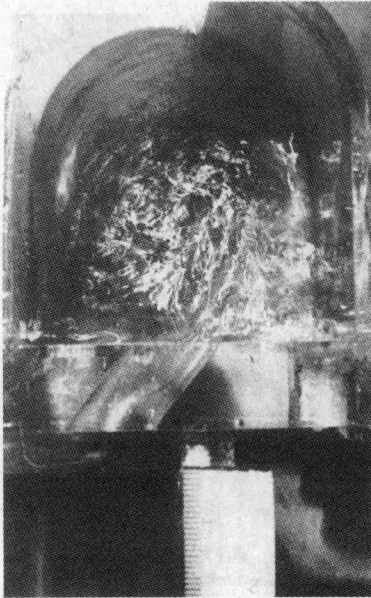

Fig. 8　The water movement in the swirl chamber for: $n_M = 14$ r/min; $n_m = 2\ 167$ r/min; $\alpha = -1°$ CA before T. D. C.

Fig. 9　The water movement in the swirl chamber for: $n_M = 15$ r/min; $n_m = 2\ 235$ r/min; $\alpha = -17°$ CA before T. D. C.

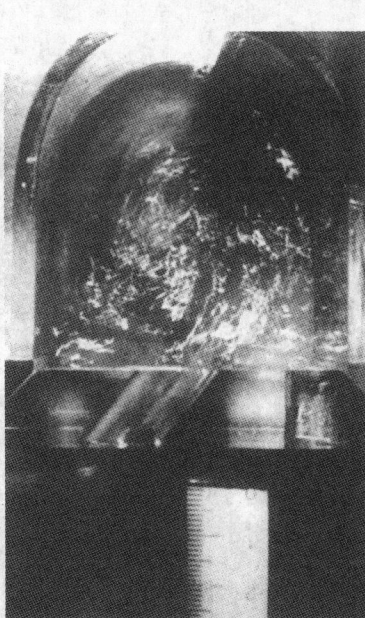

Fig. 10　The water movement in the swirl chamber for: $n_M = 15$ r/min; $n_m = 2\ 310$ r/min; $\alpha = -10°$ CA before T. D. C.

Fig. 11　The water movement in the swirl chamber for: $n_M = 15$ r/min; $n_m = 2\ 312$ r/min; $\alpha = -7°$ CA before T. D. C.

From the above-mentioned figures we may draw the conclusion that the flow direction at inlet is determined only by the direction of the connecting hole

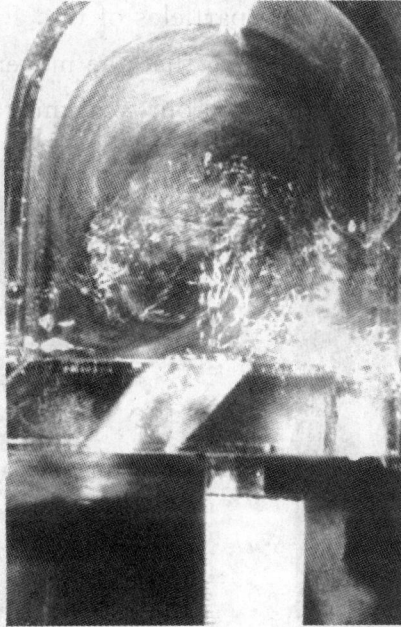

Fig. 12 **The water movement in the swirl chamber for: $n_M = 20$ r/min; $n_m = 3\,084$ r/min; $\alpha = 0°$ CA at T. D. C.**

irrespective of the shape and dimensions of the chamber. When the fluid is in, the flow direction is determined by the shape and dimensions of the separate swirl chamber. In the model bell-type swirl chamber, the following movements appear:

—towards the side wall on the left(in the figure) of the connecting hole;

—along the side wall on the left of the fi-gure;

—an important number of particles move swirl-like;

—the movements are stopped at the plane wall on the right(in the figure) of the connec-ting hole.

A black spot may be detected on the left, in all images, due to photo-graphic conditions.

From the images presented above we can establish the following influences for:

(1) The angle of crankshaft radius. By comparing the swirl shape, for speeds $n_m = 2\,167 \sim 2\,312$ r/min(the average value of rotation speeds is $n_m = 2\,167$, compared with the maximum speed variations which are 6. 57 percent) for $\alpha \in (-17°, -1°, 0°)$ it results that the movement shapes do not vary considerably.

(2) Rotation speed. For $\alpha = 0°$ CA(T. D. C.) and $n \in (1\,544, 1\,544, 3\,084)$ it is concluded that the movements have a clearer image for low speeds than for high speeds, but on the whole, the shape of the movements does not vary with respect to the rotation speed.

The influence of the particles used for visualization was studied with particles

31

of potassium permanganate and with particles of tea，Fig. 13 and Fig. 14.

From Figs 13 and 14 we can notice that the movements develop just like in the case when the visualization is carried out with mustard seeds.

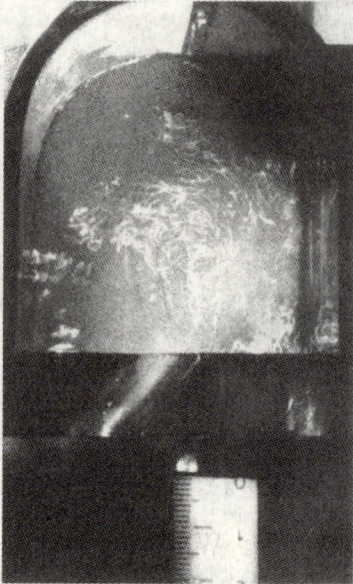

Fig. 13　The water movement in the swirl chamber for: $n_M = 14$ r/min; $n_m = 2\ 162$ r/min; $\alpha = 3°$ CA before T. D. C.

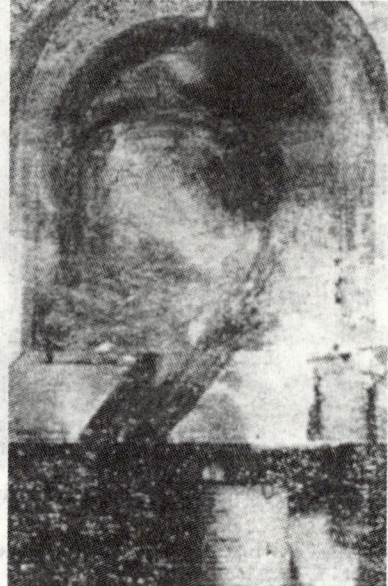

Fig. 14　Visualization with tea for: $n_M = 13$ r/min; $n_m = 2\ 012$ r/min; $\alpha = 0°$ CA at T. D. C.

5　Conclusions

Taking into account the research into air movements in the separate bell-type swirl chamber by bidimensional modelling with water, the following conclusions can be drawn:

(1) There exists a swirl movement in the bell-type swirl chamber, in which the greatest part of the air mass participates. The direction of the movement is determined by the position of the connecting hole. The swirl movement is slowed down by the plane wall of the chamber.

(2) The shapes of the movements in the swirl chamber resemble those established by plasma visualization[12].

(3) The shapes of the movements do not depend essentially on the rotation speed.

(4) A secondary movement is detected in the corners of the chamber.

References

[1] Nagao F, Ikagami M. Air motion and combustion in swirl chamber type Diesel engine [J]. *Bulletin of JSME*(in English), Japan, 1967,10:41.

[2] Knoht B. Primenenie modelirovania dlia isledovanie vozduşnih potokov i dalinoboisti fakela topliva v otkritih sgorania cetirelaktnih dvigatelei[J]. *CIMAC*, 1978, Xi, M.

[3] Böhme L. Untersuchung der Strömungsverhältnisse im Brennraum eines schnel-laufenden Zweitakl-Oltomotors[J]. *Kraftzeugtechnik*, 1980,7.

[4] Vasilescu ai A. Analiza dimensională şi teoria similitudinii[J]. *Ed Academiei*, Bucureşti, 1969.

[5] Kruglov M G. Termodinamika i gazodinamika dvuhtatnih dvigatelei vnutrennego sgorania[J]. *Maşghiz*, Moskva,1963.

[6] Konakov P M. Teoria podopia i primenenie teplotehnike M—1[J]. *Gosenergoizdat*,1959.

[7] Lustgarten G. *Untersuchung der Gemischbildung und Verbrennung im Dieselmotor unter Anwendung der Modell-theorie*[M]. Diss Nr 5166, ETH,Zürich,1979.

[8] Apostolescu N, taraza D. *Bazele Cercetării Experimentale a Masinilor Termice*[M]. Ed didactică şi pedagogică, Bucuresti,1979.

[9] Willis D A, a. o.. Mapping of airflow patterns in engines with induction swirl[J]. *SAE*, 660093.

[10] Berindean V. Studiul şi cercetarea vizualizării mişcării apei de răcire in blocul şi chiulasa motorului D—103[C]//*Comunicare*, *Institutul Politehnic "Traian Vuia"*, Timişoara, aprilie, 1971.

[11] ** Vizualizarea miscării fluidului[J]. *Bulletin JSF Japan*, 1968,23:6.

[12] Nakajima K, Kajiya S. An experimental investigation of the air swirl motion and combustion in the swirl chamber of Diesel engines[C]// *XII Congress International Barcelona*, FISITA, 1968.

(From: *Rev. Roum. Sci. Techn.*——*Mec. Appl.* Tome 29, No 2, Bucarest, 1984)

利用二维液流动模型研究吊钟型涡流室内的空气运动

[摘要] 作者利用有机玻璃制成的二维液流动模型首次研究了吊钟型涡流室内的空气运动。试验研究结果表明,在吊钟型涡流室内,绝大部分空气处于涡流运动之中,在涡流室两个拐角处存在副涡。涡流的方向与连接通道的位置关系极大。涡流受室壁制动而减弱。涡流的形式与转速无关。

A review of the paper "Study of air movement in a separate swirl chamber by means of a bidimensional liquid model", by Li Detao and Vasile Berindean.

This paper deals with a fundamental study of the processes ocurring in the swirl chamber of the S195 Type Diesel engine manufactured in China, A water model of a single cylinder and swirl chamber was constructed using perspex slides for visualisation purposes. The rotational speed of the resulting model was obtained via Reynolds Number Modelling.

The resulting flow visulisation photographs clearly show the type of fluid motion occurring in the swirl chamber over a wide range of crank angle. In general a large stable vortex structure is obtained over much of the sections of the chamber, together with smaller vortices in the corners. Clear conclustions are drawn as to the positioning of the connecting hole between the main part of the engines cobustion chamber and the swirl chamber.

It is clear that the information produced by this experimental programme has laid the foundations for professor Li's recent work on predicting the rates of heat release in the S195 Type Diesel. The two papers together make an excellent contribution to the field, providing a very useful design tool for engines working on this and similar engines. This is the type of work which brings great credit to the Authors and their Institution and should be further encouraged being a nice blend of experimental and theoretical work. In particular professor Li should be encouraged to extend his water modelling work to cover a wider range of crank angles.

Prof. *Nick Syred*

N. Syred
Dept Mech Eng & Energy Studies,
University College, UK

频谱分析在研究柴油机燃烧过程时的应用

李德桃,姜　哲,郭晨海,薛　宏,孙　颖

[摘要]　在内燃机燃烧过程的研究中,通常把燃烧过程在时域上展开,即制取示功图。本文从另一个侧面即在频域上对燃烧过程进行研究。通过初步实践,可以使我们从频域这个新的角度对燃烧过程有所了解,对寻求改进燃烧系统有着积极的指导意义。

在内燃机燃烧过程的研究中,为了得到燃烧信息(指示功、最高爆发压力、压力上升率、放热率等),通常把燃烧过程在时域上展开,即制取示功图。然而另外一些信息,如压力波动的强度和频率成分等,则在示功图上难以精确地观测出来。对于燃烧室存在连接通道的涡流室柴油机,其压力波动的精确观测,对于了解燃烧过程的进行情况和分析燃烧室形状的合理性,都有较重要的意义。因此在研究涡流燃烧室时将燃烧过程不仅在时域上展开,而且应在频域上展开,即对气缸压力进行频谱分析。通过对 10 余种涡流燃烧室方案所制取的示功图、气缸压力频谱图、噪声频谱图和发动机性能曲线的综合分析,找到了改进这种燃烧室的一些有价值的规律,同时也证明把频谱分析用于研究燃烧过程的尝试是有效的。

1　理论依据和试验装置

为了将气缸压力 $p(t)$ 在频域上展开,可以将 $p(t)$ 进行傅氏变换,以求得气缸压力的能谱 $G_p(f)$(即功率谱)。这里能谱 $G_p(f)$ 表示气缸压力在频率域上的能量分布。

进行气缸压力频谱分析时,采用丹麦 B&K 公司制造的 2131 型频谱分析仪。这时气缸压力在频域上的能量分布用分贝(dB)表示,它是气缸压力能谱的对数表示形式。

仿照声学上声压级的定义,定义气缸压力级为

$$L_p = 10 \lg \frac{(p^2)}{p_0^2} \tag{1}$$

式中 (p^2) 为气缸压力 $p(t)$ 的均方值;p_0 为参考压力,$p_0 = 2 \times 10^{-5}$ Pa;L_p 为气缸压力级。

设 $l_p(f)$ 为气缸压力级 L_p 在频率 f 上的压力级分量,那么

$$L_p = 10 \lg \left(\int_0^\infty 10^{l_p/10} \mathrm{d}f \right) \tag{2}$$

于是有

$$\frac{(p^2)}{p_0^2} = \int_0^\infty 10^{l_p/10} \mathrm{d}f \tag{3}$$

根据巴什凡(Parseval)原理

$$(p^2) = \int_0^\infty G_p(f) \mathrm{d}f \tag{4}$$

式中 $G_p(f)$ 为 $p(t)$ 的半边能谱。

因而有

$$10^{l_p/10} = G_p(f)/p_0^2 \tag{5}$$

两边取对数得

$$l_p = 10\lg \frac{G_p(f)}{p_0^2} \tag{6}$$

由此可见,气缸压力级在频域上的分量 $l_p(f)$ 表示了气缸压力的平均能量在频域上的分布,不同之处只不过是采用了对数的表现形式而已。因此,通过对频域上气缸压力级的分析,可以了解气缸压力的能量分布情况,从而为燃烧过程的研究提供一种手段。

B&K 公司 2131 型频谱分析仪的频率坐标是用频带(1/3 倍频程)表示的,于是对应于式 (6)有

$$l_{pi} = 10\lg \frac{G_{pi}(f)}{p_0^2} \tag{7}$$

式中 $l_{pi}(f)$ 分别和 $G_{pi}(f)$ 表示第 i 频带上的压力级和能谱(即气缸压力的能量)。

测量气缸压力级频谱与制取气缸压力示功图同时进行,其试验装置如图 1 所示。

图 1 测量装置示意图

测量时压力传感器从缸头伸入到主燃烧室,并与缸头平齐。接收主燃烧室的压力信号,经电荷放大器放大,分别送入频谱分析仪及示波器,从而获得气缸压力的频谱图及示功图。测量中所使用的部分仪器如图 2 所示。

图 2 测量使用的部分仪器

由于 B&K2131 型频谱分析仪的动态范围为 50 dB,而所测量的气缸压力信号的动态范围达 70 dB 以上,因此为了测量上的方便,在进行频谱分析时采用 A 计权。

为了与声学上的声压级一致,参考压力 p_0 的选取与声压级的基准压力一样,即取 $p_0 = 2 \times 10^{-5}$ Pa。

2 气缸压力频谱的标定和测量结果的修正

由于气缸压力级的参考压力 p_0 的选取与声压级的基准压力一样,即 $p_0 = 2 \times 10^{-5}$ Pa,因此为了使测量结果满足式(1),有必要对测量系统进行标定或修正。

在我们的试验过程中,采用了上海内燃机研究所制造的 SYC 型压力传感器,扬州无线电二厂制造的 FDH-2 型电荷放大器以及丹麦 B&K 公司的 2131 型数字频谱分析仪,由此构成进行频谱分析的测量系统(参见图1)。

根据电荷放大器的归一化要求,将放大器的"传感器灵敏度"档位调至与所采用的压力传感器灵敏度相同的数值,那么电荷放大器的输出电压为

$$u = AKX \tag{8}$$

式中 u 为电荷放大器的输出电压,mV;A 为电荷放大器的输出增益档数,mV/unit;K 为对应于传感器灵敏度的放大倍率;X 为被测量的物理量,unit。

电荷放大器的输出电压作为频谱分析仪的输入电压,此输入电压与频谱分析仪的内部参考电压 u_0 比较,因此频谱分析仪的测量结果为

$$L_p = 20\lg \frac{u}{u_0} \tag{9}$$

式中 L_p 为气缸压力级。

为了使式(9)与气缸压力级的定义式(1)一致,应该使频谱分析仪的内部参考电压 u_0 等于由 p_0(2×10^{-5} Pa)作用于压力传感器时电荷放大器所输出的电压量。因此利用式(3)有

$$u_0 = AKp_0 \tag{10}$$

由此式可以求出频谱分析仪内部的参考电压 u_0,调整频谱分析仪的内部参考电压,并使之等于式(10)所计算出的数值。这样频谱分析仪的测量结果就是以 p_0(2×10^{-5} Pa)为参考压力的气缸压力级 L_p。

事实上,对于 B&K2131 型频谱分析仪,其内部参考电压通常为 1 μV。虽然该仪器内部参考电压可以适当调整,但幅度有限,很难满足式(10)的要求。这时可对测量结果进行修正而达到与式(1)一致的要求。

将式(9)改写成

$$L_p = 20\lg \left[\frac{输入电压 u}{频谱仪实际参考电压} \cdot \frac{频谱仪实际参考电压}{参考电压 u_0} \right]$$

$$= 20\lg \frac{输入电压 u}{频谱仪实际参考电压} + 20\lg \frac{频谱仪实际参考电压}{参考电压 u_0} = L_p' + \beta \tag{11}$$

式中 L_p' 为频谱分析仪以某一实际参考电压为基准而测量出的压力级;β 为以参考电压 u_0 为基准的修正值。

$$L_p' = 20\lg \frac{输入电压 u}{频谱仪实际参考电压} \tag{12}$$

$$\beta = 20\lg \frac{频谱仪实际参考电压}{参考电压 u_0} \tag{13}$$

2131 型频谱分析仪内部参考电压通常为 1 μV。那么此时式(13)中频谱仪实际参考电压也等于 1 μV,这样对频谱分析的测量结果的修正值为

$$\beta = 20\lg \frac{1 \times 10^{-3}}{u_0} = 20\lg \frac{1 \times 10^{-3}}{AKp_0} \tag{14}$$

因此利用式(11)和式(14)就可以得到满足式(1)定义的气缸压力级 L_p。本文的有关试验就是采用这种修正方法。

3 气缸压力频谱的分析和应用

利用图 1 所示的测量装置,所测得的 S195 型柴油机主燃烧室的气缸压力级频谱如图 3 所示。此时发动机工况为标定工况,即 2 000 r/min、12 马力(8.83 kW),图中频率坐标为 1/3 倍频程中心频率(以下均同)。

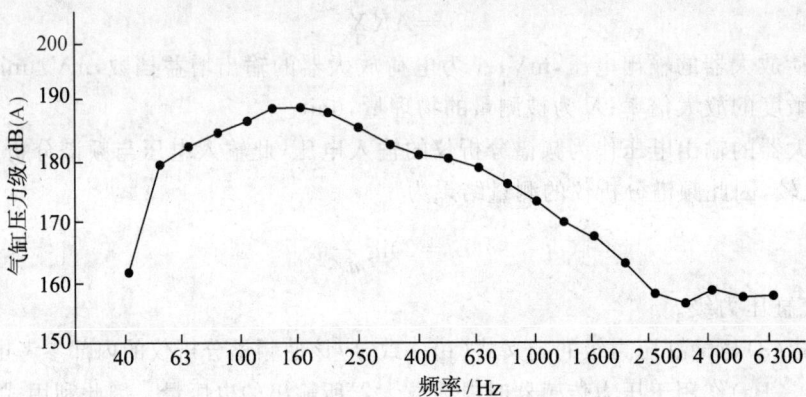

图 3 标定工况时,主燃烧室气缸压力级频谱,其中 dB(A)指 A 计权压力级

分析此图时可将频谱曲线分成三部分:低于 250 Hz 以下的低频部分,曲线有较高的峰值;在 250~2 000 Hz 之间的中频部分,频谱曲线近似于直线,随频率的增加而迅速下降;在高于 2 000 Hz 以上的高频部分曲线略有起伏,并较为平坦。

频谱曲线中的三个部分与燃烧性能具有紧密的联系[2]。在气缸压力频谱图中曲线的低频部分峰值由最高爆发压力 p_z 决定,气缸的最高爆发压力 p_z 越高,频谱曲线的低频峰值就越高。频谱曲线的中频部分受气缸压力上升率 dp/da 控制,dp/da 越大图中直线部分就越平坦;反之 dp/da 越小,直线部分就越陡。频谱曲线的高频部分(高于 2 000 Hz)由气缸压力的高频波动产生,主要与 d^2p/da^2 有关。关于燃烧压力波动的机理,目前众说不一[3]。不过有一点是肯定的:由于燃烧紊流脉动速度的增加,势必会导致压力脉动的增强。因此采用能导致紊流增强的结构措施,是产生高频波动的原因之一。显然,增强紊流脉动,会使燃烧更加充分、完全,而有益于提高发动机的动力性和经济性。

所谓频率实际上就是信号的变化率。如果气缸压力信号 $p(t)$ 为一谐波,那么在频域上该信号的能量都集中在谐波的频率上。如果压力信号为一脉冲波,那么它将具有丰富的频率成分,其能量在整个频域上均匀分布。对于实际的气缸压力信号,含有着火基频的低频谐波成分,也具有由于压力上升率 dp/da 而产生的脉动成分。由最高爆发压力确定的谐波成分表现在频谱图中的低频部分。对于由 dp/da 产生的脉动成分,当 dp/da 越大时,该成分越接近脉冲,所含频率成分越丰富,表现在频谱图中直线部分就比较平坦,反之亦然。

综上所说,可将气缸压力频谱曲线分成三部分,对此进行分析研究,可从一个新的角度对燃烧过程有所了解。

前已说明气缸压力级频谱图表示了气缸压力的平均能量在频率域上的分布。如果将频谱图各频率成分上的能量合成（对数合成）起来就表示了气缸压力在一个循环中的平均能量。为了提高发动机的动力性及经济性，从能量的观点看，应使频谱图上各频率成分的能量均有所提高。下面针对频谱图上所划分的三个部分进行讨论。

如果使低频部分的能量提高，就必须提高发动机的最高爆发压力 p_z。然而这对于发动机的寿命和柔顺性等都带来不良的影响。在不影响发动机的动力性前提之下，应尽可能降低最高爆发压力，这已成为一台优质发动机指标之一，因此提高低频部分的能量受到了限制。

不改变低频部分的能量，使频谱图上中频区域直线部分趋于平坦，可提高中频部分的能量，这就必须增大压力上升率 $\mathrm{d}p/\mathrm{d}a$。这导致燃烧噪声的增加，目前一致认为 $\mathrm{d}p/\mathrm{d}a$ 不宜增大。因此想以增大 $\mathrm{d}p/\mathrm{d}a$ 来提高中频部分的能量也是不可取的。

提高高频部分的能量对改善发动机的经济性是有益的。根据 S195 柴油机涡流燃烧室的特点，采用一些结构上的措施可以提高主燃烧室的紊流强度和标度，增加频谱图上高频部分的能量。这时由于燃烧室紊流强度和标度增加，燃烧速度加快，相应地燃料燃烧更加充分，从而使发动机的动力性和经济性得到改善。

一般地说，提高气缸压力频谱会影响发动机的燃烧噪声。对燃烧噪声的研究表明它主要受气缸压力上升率 $\mathrm{d}p/\mathrm{d}a$ 及最高爆发压力 p_z 的影响，即燃烧噪声主要受频谱图中的低频和中频部分的能量的影响，而提高频谱图中的高频部分的能量对燃烧噪声的影响较小。我们的试验进一步表明，S195 柴油机的主要噪声为机械噪声，而燃烧噪声在发动机的整机噪声中所占的比例很小。即使气缸压力级略有变化，对发动机的总噪声几乎没有什么影响。

试验时改变 S195 柴油机的供油提前角，使气缸压力级在频谱图上有明显变化，测量对应的噪声情况，其结果示于图 4。

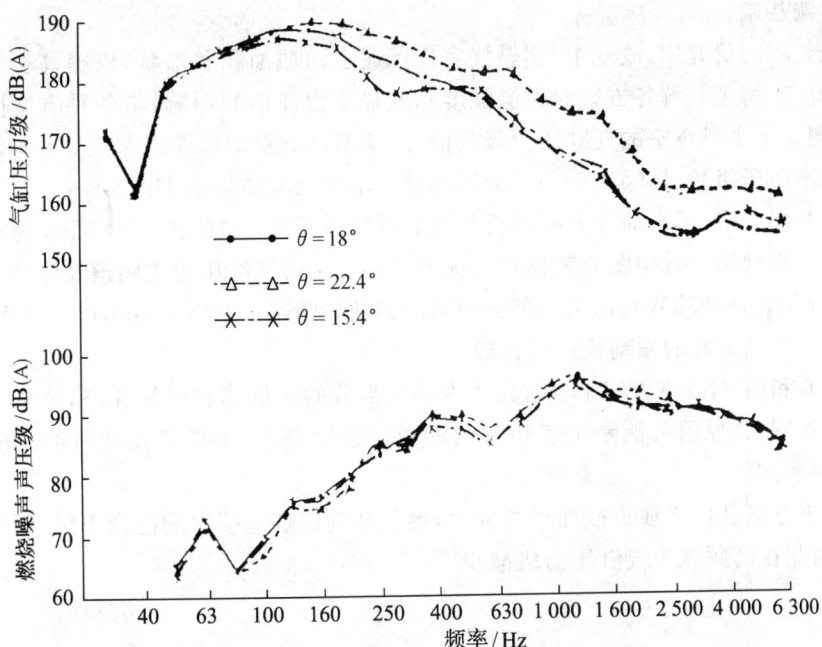

图 4　改变供油提前角时气缸压力级与噪声的关系（2 000 r/min，8.83 kW）

测量时,考虑到试验是在普通内燃机实验室进行,房间四周均无吸声措施、具有一定的反射。因此噪声采用近场测量,目的是作相对比较。测量时传声器距发动机前端面 130 mm,与缸头平齐。仪器为 B&K2131 频谱分析仪以及该公司制造的传声器和前置放大器。频谱图的气缸压力取自主燃烧室。发动机的工况为标定工况(2 000 r/min,8.83 kW)。由图 4 可见,当静态供油提前角从正常 $\theta=18°$ 改变到 $\theta=22.4°$ 时,气缸压力级在中频和高频部分都有较大差异,而对应的噪声声压级(A)却几乎没有什么差别。另一方面改变发动机的转速,测量的噪声情况如图 5 所示。此时随着转速的改变,噪声却有明显的差异。

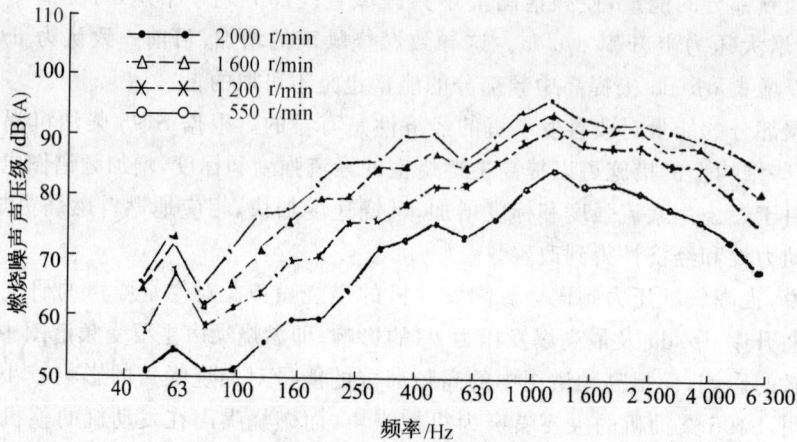

图 5　噪声与转速的关系

由此可见该机的主要噪声为机械噪声,因此即使将频谱图的高频部分略加提高,对发动机的总噪声影响甚微。

在燃烧过程的研究中,改变主、副燃烧室连接通道的截面积和形状,以提高主燃烧室的紊流强度和标度。为了了解各方案对紊流强度和燃烧压力波动的影响,将各种方案同产品方案进行比较,测量了主燃烧室的气缸压力级频谱、示功图以及发动机性能参数等。同时还测量了各方案的噪声声压级频谱,噪声仍采用近场测量。部分试验结果示于图 6~8。

在图 6 和图 7 中,下半部分为发动机前端的近场声压级(A)频谱;上半部分为不同改进方案的气缸压力级频谱。其中图 6 为标定工况点时各方案的气缸压力级频谱及噪声频谱的比较情况(2 000 r/min,8.83kW);图 7 为部分负荷时的比较情况(2 000 r/min,6.62 kW)。图 8 为改进方案 I 与产品方案的发动机性能比较。

对比图 6 和 7,改进方案 I 气缸压力在高频部分的能量都有明显提高,这就意味着燃烧的压力波动加剧,其原因可能是主燃烧室内的紊流强度提高,结果使发动机的经济性有所改善,如图 8 所示。

对于改进方案 II 由于通道截面增加太大,燃气从副燃烧室喷出的能量下降,削弱了主燃烧室的紊流,因此在高频域中气缸压力级减少了。

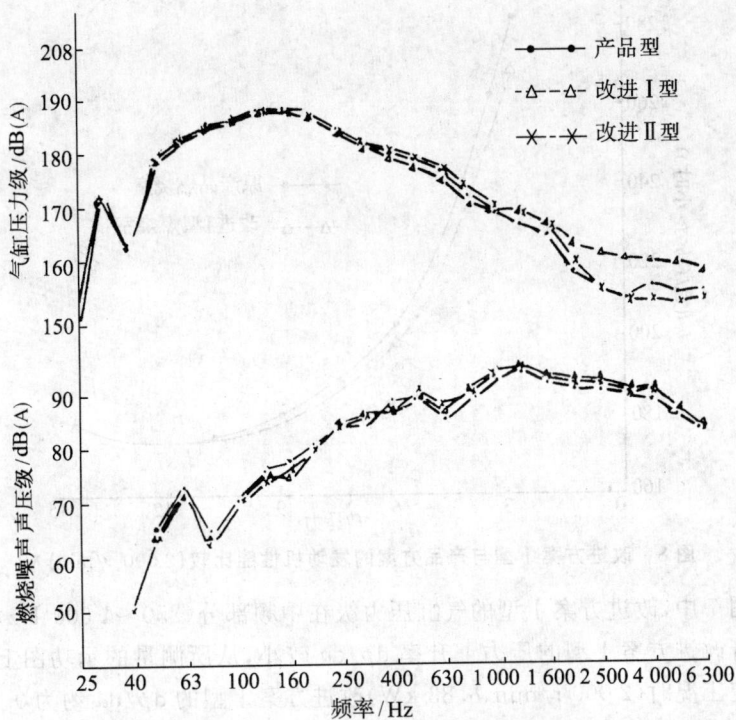

图 6 标定工况点时气缸压力级频谱及噪声频谱的比较

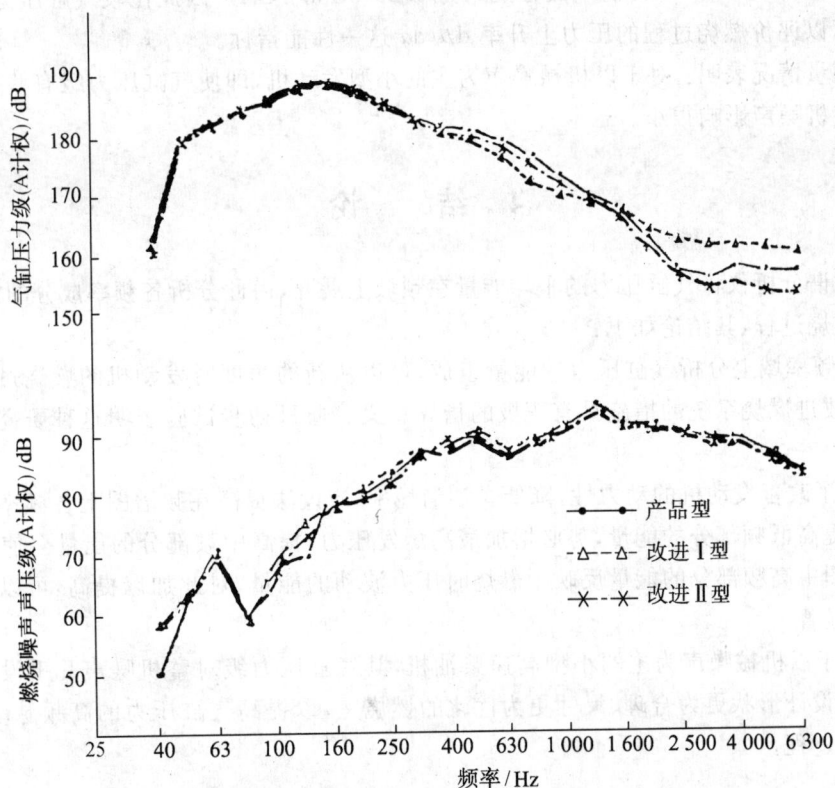

图 7 部分负荷时气缸压力级频谱及噪声频谱的比较(2 000 r/min,6.62 kW)

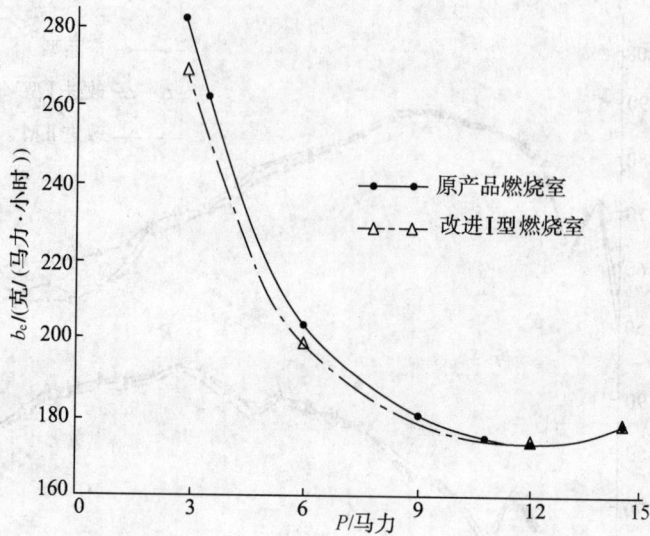

图8 改进方案Ⅰ型与产品方案的发动机性能比较(2 000 r/min)

在图6和图7中,改进方案Ⅰ型的气缸压力级在中频部分(250~1 600 Hz)比产品方案要低。这就意味着改进方案Ⅰ型的压力上升率 dp/da 较小,从所测量的示功图上也得到了同样的结论。在标定工况时(2 000 r/min、8.83 kW)改进方案Ⅰ型的 dp/da 约为 0.20 MPa/°CA,而产品方案约为 0.23 MPa/°CA。在部分负荷(2 000 r/min、6.62 kW)时改进方案Ⅰ型的 dp/da 约为 0.21 MPa/°CA,而产品方案约为 0.25 MPa/°CA。因此比较气缸压力频谱图的中频部分可以评价燃烧过程的压力上升率 dp/da 这一性能指标。

噪声测量情况表明:对于以机械噪声为主的小型发动机,即使气缸压力级有些变化,但对发动机的整机噪声影响很小。

4 结 论

利用频谱分析仪将气缸压力的平均能量在频域上展开,由此分析各频率成分的能量,研究发动机的燃烧过程,其结论如下:

(1)在频率域上分析气缸压力的能量组成,可以从新的角度对发动机的燃烧过程有所了解,对寻求改进燃烧系统的措施具有积极的指导意义。通过初步试验表明这种研究方法是有效的。

(2)为了改善发动机的动力性,降低燃油消耗率,应设法提高在频谱图上各频率成分的能量。但是,提高低频部分的能量,势必增加最高爆发压力;提高中频部分的能量会使 dp/da 增大;而频谱图中高频部分的能量反映了燃烧时压力波动的能量,对此加以提高,可以改善燃烧过程,效果显著。

(3)对于以机械噪声为主的小型高速柴油机,其气缸压力级对整机噪声几乎没有什么影响。因此可设计形状更为合理、尺寸更为优化的燃烧室,以提高气缸压力的高频脉动,改善柴油机的燃烧过程。

参 考 文 献

［1］李德桃,姜哲,等. 提高涡流室式柴油机部分负荷时的经济性的研究(成果鉴定会资料)
    ［G］.镇江:江苏工学院,1984.

［2］池上詢. 圧縮着火機関の燃焼騒音の発生機構［J］.内燃機関,1975,14：162.

［3］Gupta H C，Bracco V. The origin of pressure oscillations in divided chamber engines
    ［J］. *Combustion and Flame*,1982(48)：33-49.

(本文原载于《江苏工学院学报》1985 年第 3 期)

Applicaion of spectrum analysis to study of
diesel combustion process

Abstract：On the study of combustion process of internal combustion engines，the combustion process is usually studied in the time region. This paper does this from another side，in the frequency region. On our elementary practice，we can further understand combustion process from the new view-point of frequency region and this is of active guiding significance in search of improving combustion system.

Studies on improving the economy of swirl chamber diesel at partial load and reducing the emission

Li Detao , Guo Chenhai , Xue Hong

1　Introduction

Based on some previous studies in swirl chamber, we made a further approaching on the mechanism of combustion and found out some ways to improve the combustion system of Diesel. By a synthetical study on the pressure-crank-angle diagram, pressure-spectrum diagram, noise spectrum diagram and character curves, we have gained the results of reducing the specific fuel consumption(SFC) $2 \sim 16$ g/(PS・h) at partial load and 20% NO_x at rating power[1].

2　Study methods and test apparatus

We took Model S195 Diesel as a test engine(Fig. 1).

The main engine is

Cylinder bore	95 mm
Piston replacement	115 mm
Compression ratio	20

Rating horsepower/Rating speed 12 PS/2 000 r/min.

Within a dozen of improving design in swirl chamber. We worked out the pressure-crank-angle diagram, pressure-spectrum diagram, noise spectrum diagram and character curves. By a synthetical study, we chose two rational design so that the SFC at the rating power can be reduced $2 \sim 16$ g/(PS・h). At different loads corresponding to several speed, NO_x is reduced $4\% \sim 20\%$.

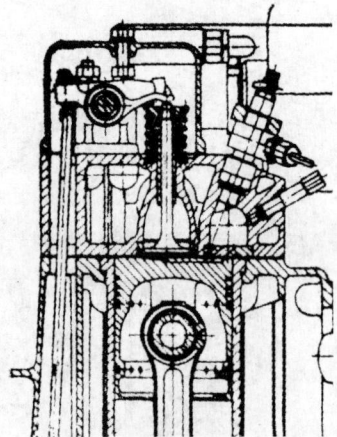

Fig. 1　Model S195 diesel test engine

To spread cylinder pressure $p(t)$ out, apply a Fourier trans-formatim to $p(t)$ to get the energy spectrum of cylinder pressure $Q_p(f)$ (or power spectrum).

The energy spectrum represents the cylinder pressure distribution on the frequency region.

For doing this, we used Model 2131 spectrum analyzer made by B&K Company Ltd.. Here the energy distribution is represented by decibel(dB). It is the logarithem expression of energy spectrum of cylinder pressure[2].

The frequency coordinate of B&K Model 2131 frequency spectrum analyzer is represented by frequency band(1/3 frequency multiplication). Based on theoretical considerations, we have

$$L_{pi} = 10\lg \frac{Q_i(f)}{p_0^2}$$

Here L_{pi} and $Q_i(f)$ represent the cylinder pressure level and energy spectrum respectively.

While measuring cylinder pressure spectrum, we made the pressure-crankangle diagram. The testing apparatus is showed in Fig. 2.

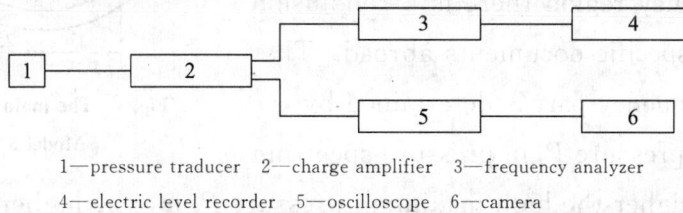

1—pressure traducer   2—charge amplifier   3—frequency analyzer
4—electric level recorder   5—oscilloscope   6—camera

Fig. 2   Testing apparatus sketch map

The pressure sensor is put into the main combustion chamber from cylinder head, on the same level as the bottom of cylinder head, to receive pressure signal. The signal is amplified by charge amplifier and send to frequency analyzer and oscilloscope respectively. The cylinder pressure spectrum and pressure-crankangle diagram are gained.

The dynamic scope of B&K Model 2131 frequency analyzer is 50 dB, while the one of cylinder pressure signal is more than 70 dB, for the convenience of measuring. We did the frequency analysis by means of(A) dB.

To go with sound pressure level(SPL) in acoustics, the reference pressure p_0 is selected the same as the datum pressure of sound pressure level, that is $p_0 = 2 \times 10^{-5}$ Pa.

3　The analyzer of cylinder pressure

The cylinder pressure spectrum in the main chamber(Fig. 3) of Model S195 Diesel is showed in Fig. 4. The engine was in rated working condition, that is 2 000 r/min, 12 PS. In Fig. 4, the frequency coordinate is 1/3 centre frequency in frequency multiplication band(Following figures are the same).

Fig. 4 shows, under the situation we discussed at the lower frequency part under 250 Hz, the curve reached a higher peak value, but between $250 \sim 2\,000$ Hz, the curve is almost near a straight, which declines as the frequency increases. At the higher frequency part over 2 000 Hz, the curve slightly goes up and down. As to the relationship of the cylinder pressure on both time and frequency region there is a conolusion made by some specific documents abroad. That is the lower frequency part is determined by the high maximum pressure P_z in pressure spectrum

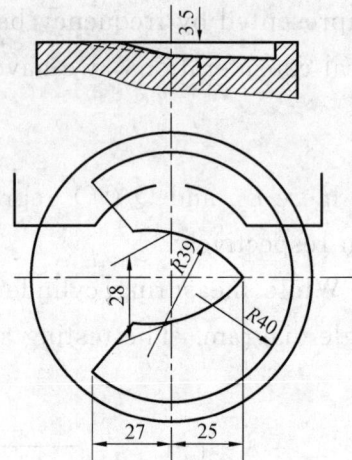

Fig. 3　The main chamber of Model S195

diagram. The higher the high maximum pressure P_z is, the higher the lower frequency peak value will be. The middle frequency part or the straight part is determined by the rates of pressure rise $dp/d\alpha$. The faster$(dp/d\alpha)$ rises, the flatter the slope of straight part will be. On the contrary, the slower$(dp/d\alpha)$ rises, the steeper the slope will be. The higher frequency part(over 2 000 Hz)is caused by frequency wave motion of cylinder pressure, mainly related to $d^2p/d\alpha^2$. About the mechanism of combustion pressure wave motion, the explanation is quite different, but one of them which has been made sure is the increasing of pulsation velocity of combustion turbulence definitly will lead to the increasing of pressure wave motion. Therefore, a method to cause the high frequency pulsation will surely increase turbulence. Generally speaking, increasing turbulent pulsation is benifit to the form of air-fuel mixture in combustion chamber, furthermore will be benifit to both dynamic and economic properties.

Frequency actually is the changing rate of the signal. If pressure signal $p(t)$ is a harmonic, the energy of signal in frequency regime will be concentrated on the harmonic.

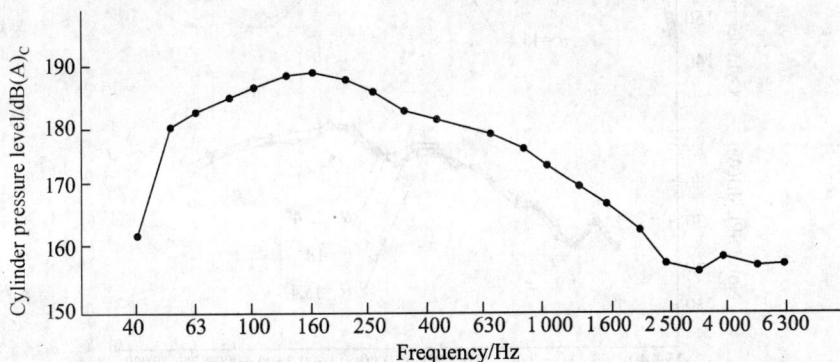

Fig. 4   The cylinder pressure spectrum

It will contain a variety of frequency and its energy will be evenly distributed in overall frequency region if the pressure signal is a pulse. As to the actual cylinder pressure curve, it contains lower frequency harmonic which took firing frequency as its basic one, and also the pulsation caused by $dp/d\alpha$. The harmonic caused by high maximum pressure is just expressed in the lower part of pressure spectrum diagram. About the pulsation component caused by $dp/d\alpha$ the bigger $dp/d\alpha$ is, the nearer to the pulse it is, the more it contains frequency, the flatter the straight part, will be and vice versa. The combustion pressure fluctuation expressed by $d^2 p/d\alpha^2$ belongs to higher frequency part.

Through the experiments we found out there is some relationship between the energy from the higher frequency part and economic property of an engine.

Generally speaking, the raising of the energy of both middle and higher frequency part will effect the combustion noise. The main noise in Model S195 Diesel is a mechanical one, the combustion noise holds few in overall noise.

Fig. 5 shows the relationship between the cylinder pressure level and noise SPL by adjusting the timing. Fig. 6 shows the relationship between noise SPL and speed. From Fig. 5 and Fig. 6 we can see clearly that the main noise is mechanical one. Therefore, although we raised the high frequency part a little, there was almost no effection on overall noise.

47

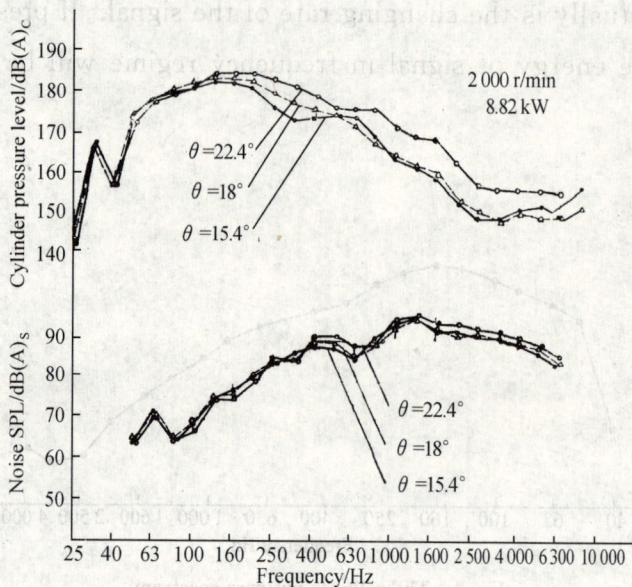

Fig. 5   Relationship between the cylinder pressure level and noise SPL by adjusting the timing

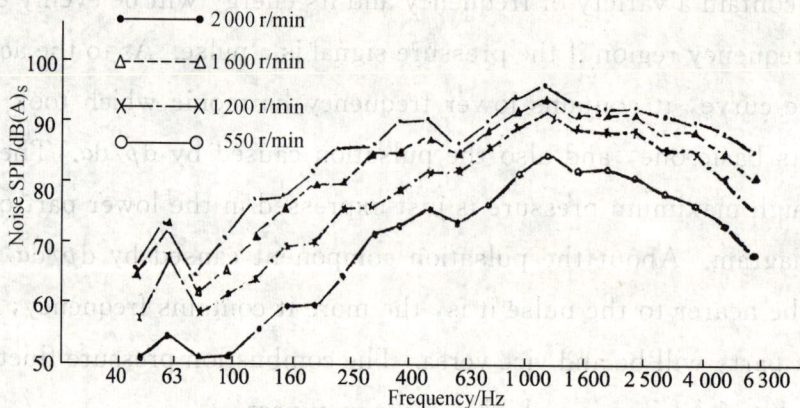

Fig. 6   Relationship between noise SPL and speed

4   The testing result and discussion on different combustion chamber design

To let vortex cure current a swirl chamber be regular and symmetry and increase the intensity and scale of turbulence, in main chamber, we optimized the shape and cross section area of the passage(Fig. 7) thus, design Ⅰ and Ⅱ are obtained.

Fig. 7   The shape and cross section area of the passage

Fig. 8 and 9 made a comparison of cylinder pressure level and sound pressure level between rated and partial load respectively.

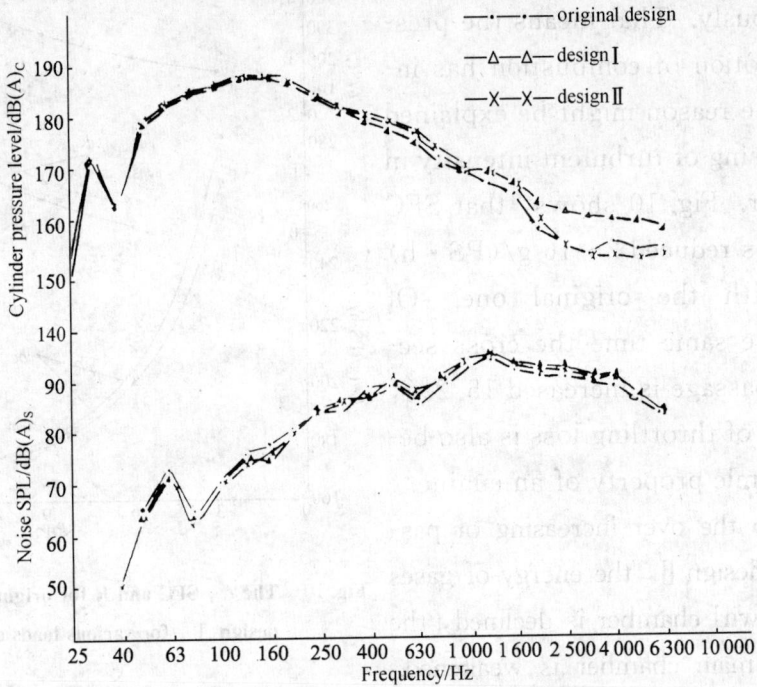

Fig. 8　Cylinder pressure level and sound pressure level for various design

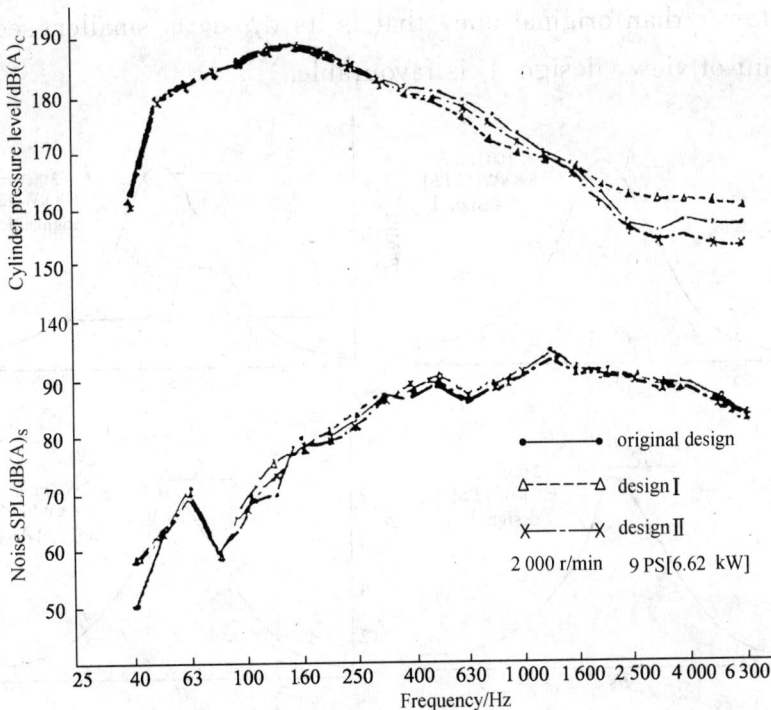

Fig. 9　Cylinder pressure level and sound pressure level for 2 000 r/min and 9 PS

The result shows the energy of higher frequency part in design Ⅰ increased obviously. That means the pressure wave motion of combustion has intensified. The reason might be explained by the increasing of turbulent intensity in main-chamber. Fig. 10 shows, that SFC of design Ⅰ is reduced 2～16 g/(PS • h) compared with the original one. Of course, at the same time the cross section area of passage is increased 15. 2%, the reduction of throttling loss is also benifit to economic property of an engine.

Owing to the over increasing of passage area, in design Ⅱ, the energy of gases ejected from swirl chamber is declined, the turbulence in main chamber is weakened.

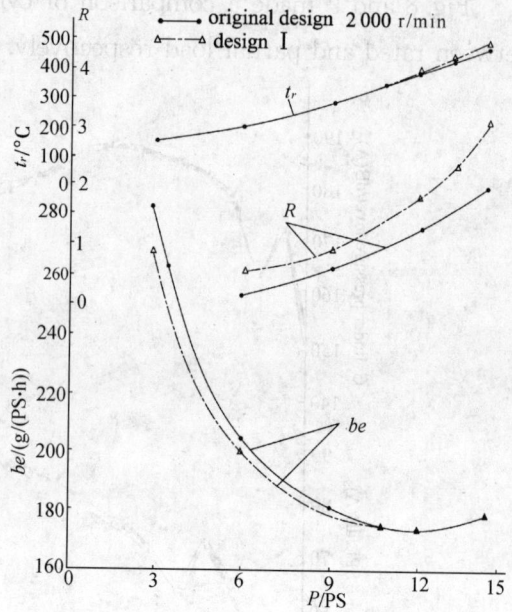

Fig. 10　The t_r, SFC and R for original design and design Ⅰ, for various loads at 2 000 r/min

Therefore the cylinder pressure level in high frequency part is reduced.

Fig. 8 and Fig. 9 also show the pressure level in the middle frequency part in design Ⅰ is lower than original one, that is its $dp/d\alpha$ is smaller(see Fig. 11). From this point of view, design Ⅰ is favourable.

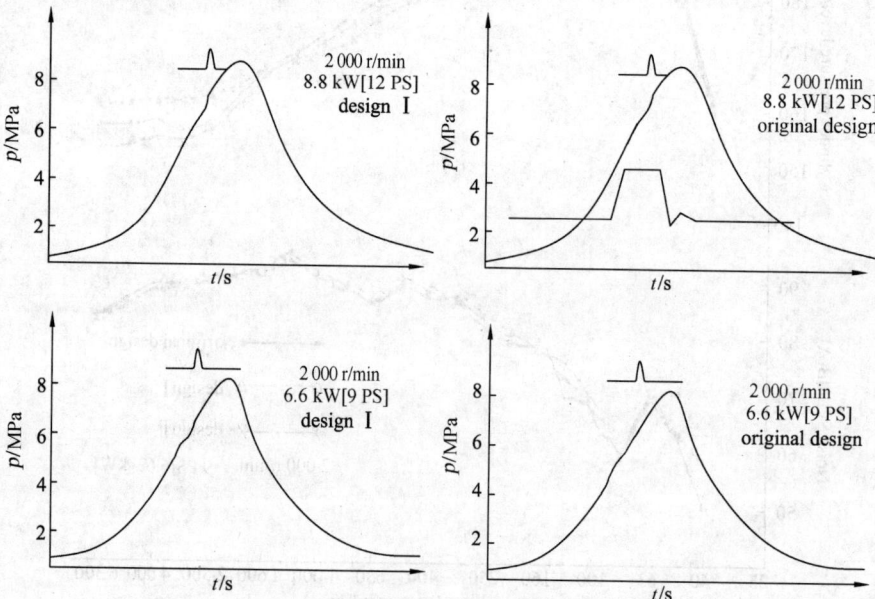

Fig. 11　The p-t diagrams at 9 PS and 2 000 r/min

As to the overall noise, the difference between design I,II and original one is little.

Fig. 12 shows the comparison between cylinder pressure level and noise at 4. 8 PS, 1 600 r/min. The energy at higher frequency part in design I is still much bigger than others. The vortex intensiby in a swirl chamber Diesel is proportional to its speed. The cuiginal design, with reduction of the speed, the vortex intensity in swirl chamber will certainly decrease. However, to design I, the amplitude of reduction is much smaller. So the adaptability of swirl chamber to speed with design I turned out to be wider, the property of swirl chamber is improved.

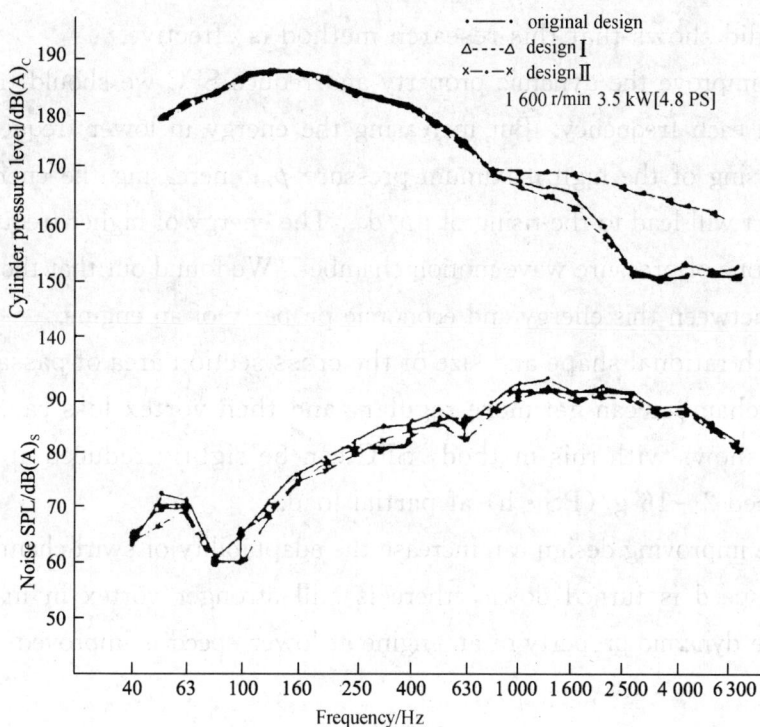

Fig. 12 The comparison between cylinder pressure level and noise at 4. 8 **PS and** 1 600 **r/min**

By the emission comparison between different passages, we have established that at full load, the weight of specific displacement of NO_x in design I and II is 21% lower than original one within two kinds of speed ranges. The results indicate that properly changing the shape of passage and sightly enlarging the cross section area can effectively turn NO_x emission down. The explanation might be given in the following two aspects.

1. By enlarging the cross section area of passage, by reducing the sting time of gases in swirl chamber, the product of NO_x can be reduced. This is according with testing result made by Makoto Ikegami.

2. The shape of passage we selected can widen the vortex amplitude in swirl chamber, so that the well-organized vortex can be concentrated in the plane parallel to the passage and avoid the high temperature area in swirl chamber which mainly cause NO_x.

5 Conclusions

(1) To analyze the energy component of cylinder pressure in frequency region and understand the combustion processing of an engine from a new point of view play an important role in improving the combustion system. The primary practice we did shows that this research method is effective.

(2) To improve the dynamic property and reduce SFC we should try to increase the energy in each frequency. But increasing the energy in lower frequency part will lead to the rising of the high maximum pressure p_z; increasing the energy in middle frequency part will lead to the rising of $dp/d\alpha$. The energy of higher frequency part represents the one of pressure wave motion chamber. We found out that there is a certain relationship between this energy and economic property of an engine.

(3) With rational shape and size of the cross section area of passage, the vortex in swirl chamber can get more regular, and then vortex loss can be reduced. The results shows with this method, SFC can be sightly reduced at rated power and to reduced $2\sim16$ g/(PS·h) at partial load.

(4) The improving design can increase the adaptability of swirl chamber to speed. Even if the speed is turned down, there is still stronger vortex in main chamber. Therefore the dynamic property of an engine at lower speed is improved.

References

[1] Li D T. Application of spectrum analysis to study of diesel combustion process[J]. *Journal of Jiangsu institute of technology*, 1985,3(3):43-52.

[2] Gupta H C, Bracco V. The origin of pressure oscillations in divided chamber engines [J]. *Combustion and Flame*, 1982,48:33-49.

(From: *Buletinul Stiinţific şi Tehnical Institutului Politehnic "Traian Vuia"*, Timişoara, Tom 30(44), Mecanica, 1985. 此文曾于 1984 年 10 月在天津召开的国际燃烧大会上宣读)

Studii asupra imbunatatirii economicitatii camerei separate de virtej a motorului diesel la sarcini partiale si a reducerii

［**Rezumat**］　In partea 1-a a lucrării se prezintă metodica，aparatura de cercetare şi analiza presiunii din cilindru．Experimentele au fost efectuate cu ajutorul unni stand echipat cu un motor monocilindru S 195，cu o putere de 12 CP la 2 000 rot/min，iar pentru analiza spectrului presiunii din cilindru s-a utilizat un analizor model 2131，B & K Comp．Ltd．Din diagrama înregistrată a variaţiei nivelului presiunii din cilindru în funcţie de frecvenţă，se deduc măsurile necesare pentru îmbunătăţirea procesului de ardere şi de reducere a nivelului de zgomot al motorului．In conti-nuare，se analizează relaţialintre nivelul presiunii şi zgomot，pentru diferite valori ale avansului la începerea injecţiei şi diferite turaţii，din care se deduce că，in pincipal，natura zgomotului este mecanică．

In partea a 2-a a lucrării se discută şi se interpretează rezultatele încercărilor，obţinute pentru diferite concepţii de camere de ardere，privitoare la corelaţia dintre nivelul presiunii şi al zgomotului şi consumul specific de combustibil，emisiile de CO şi NO_x，pentru diferite sarcini．Lucrarea se încheie cu stabilirea conclu-ziilor cercetărilor，din care rezultă îmbunătăţirea performanţelor motorului cercetat．

改善涡流室式柴油机部分负荷时的经济性和降低有害排放物的研究

[摘要]　涡流室式柴油机是我国当前生产量最大,制造厂最多,使用面最广的一类内燃机。因此设法降低其燃油消耗率和减少其排出的有害气体,对节约能源和减少环境污染具有重大的现实意义。通过多年的研究和改进,一些单缸机在标定工况下的油耗已达到国际先进水平。但在保证标定工况时的先进指标的前提下,如何降低部分负荷时的油耗,扩大使用经济性,目前尚未很好解决。至于减少排污,则更未引起人们重视。为此,我们在对涡流燃烧室作了 10 余年的基础上,对燃烧机理进行了进一步探讨。通过对发动机示功图、气缸压力频谱图、噪声频谱图和性能曲线图的综合分析,找到了一些改善燃烧系统的途径,获得了降低部分负荷时的油耗 2～16 g/(PS·h),和减少了标定工况的有害气体 NO_x 达 20% 的效果。

涡流室式柴油机放热率理论模型的探讨

李德桃,杨本洛,张伟夫

[摘要] 本文导出并分析了计算涡流燃烧室放热率的完整方程组。除考虑温度对混合气比热的影响外,还考虑了由于浓度差异产生的扩散流动对混合气比热的影响,从而使理论模型更精确。

涡流室式柴油机放热率的计算,迄今尚有一些问题未能解决。Woschni G、池上询[1,2]等人采用过于简化的理论模型进行计算,显然存在较大的误差。我们按 Trada ko 最近提供的计算方法和程序进行过计算,所得结果与他的结果仍相差较大[3],估计可能是由于 Trada ko 的方法也采用了一些与实际情况不符的假定所致。为了使涡流燃烧室放热率的计算精确化,本文对放热率计算的理论模型作了进一步的分析,导出了计算该型燃烧室放热率的完整方程组。同时,在理论模型中还考虑到扩散流的影响,从而使理论模型更精确。

1 基本方程的导出

从热力学观点来看,整个涡流燃烧室系统可以视作如图 1 所示的热力学模型。它由主、副燃烧室以及通道组成,图中下标 1,2 分别对应于主、副室。在不考虑实际的燃烧过程时,将燃烧放热量 Q_B 和工质与壁面交换的热量 Q_T 的差值作为加热量 Q,并认为主、副室中驰豫时间充分小,主、副室中混合气处于平衡状态时,可以从热力学的基本定律出发,分别建立相应的基本方程。

1.1 能量方程

首先考虑副燃烧室,在热力学开口体系不计势能情况下:

$$dQ_2 = dU_2 - \left(h^* + \frac{c_f^2}{2}\right) \cdot dm_{2f} + dW_n$$

式中 dQ_2 表示副燃烧室在某一瞬时 τ 的加热量的增量,按上述约定,则

$$dQ_2 = dQ_{B_2} - dQ_{T_2}$$

dU_2 表示副燃烧室内工质的内能增量,一般不计燃料燃烧前本身的吸热量,故此内能增量为副室内空气和燃气混合气内能增量,即 $dU_2 = d(m_2 c_{v2} T_2)$,其中,m_2、c_{v2}、T_2 分别为副燃烧室中混合气(不计未燃烧的燃料量)的质量、定容比热及温度。

$\left(h^* + \frac{c_f^2}{2}\right)$ 是对应于流经通道的副室混合气质量增量 dm_{2f} 的比焓和比动能之和。考虑到通道的表面积和两个燃烧室相比如此之小,可以忽略流经通道的

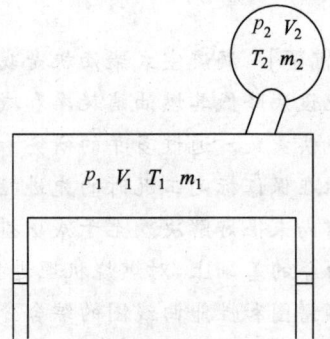

图 1 涡流燃烧室系统热力学模型示意图

燃烧放热量以及对壁面的散热影响。再若忽略通道内混合气质量变化的话,则 $\left(h^* + \dfrac{c_f^2}{2}\right)$ 显然就等于通道中实际来流处滞止焓值。也就是说,当 $p_2 > p_1$,工质从副室中流出,即 $\mathrm{d}m_{2f} < 0$ 时

$$h^* + \frac{c_f^2}{2} = h_2 = c_{p2} \cdot T_2$$

反之,当 $p_2 < p_1$ 时,工质流入副室,即 $\mathrm{d}m_{2f} > 0$ 时

$$h^* + \frac{c_f^2}{2} = h_1 = c_{p1} \cdot T_1$$

式中 c_p 为定压比热。

最后考虑到副室里没有轴功输出,即

$$\mathrm{d}W_n = 0$$

从而得到副燃烧室的能量方程

$$c_{pi} \cdot T_i \cdot \mathrm{d}m_{2f} + \mathrm{d}Q_2 = \mathrm{d}(m_2 \cdot c_{v2} \cdot T_2) \tag{1}$$

式中:$p_1 > p_2$ 时,$i = 1$;$p_1 < p_2$ 时,$i = 2$。

主燃烧室情况有所不同,它除了和外界有质量交换 $\mathrm{d}m_{1f}$ 以外,随着活塞的移动,边界也在变化。从热力学的观点来看,它既不是一个单纯的封闭体系,也不是单纯的开口体系,故不可直接应用这两种体系所对应的能量公式,而应从能量平衡的角度推导主燃烧室的能量方程。

图 2 分别表示了主燃烧室在 τ 时刻和 $\tau + \mathrm{d}\tau$ 时刻的状态。

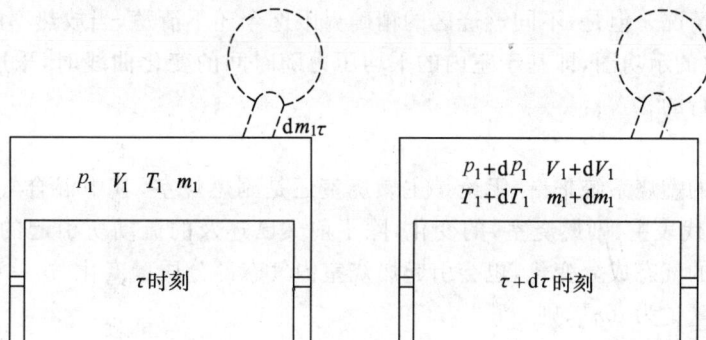

图 2　主燃烧室的热力学状态示意图

如果完全不计活塞与缸壁之间的泄漏,则 $\mathrm{d}\tau$ 间隔内能量变化 $\mathrm{d}U_1$ 为

$$\mathrm{d}U_1 = (m_1 + \mathrm{d}m_1) \cdot (c_{v1} + \mathrm{d}c_{v1}) \cdot (T_1 + \mathrm{d}T_1) - (m_1 \cdot c_{v1} \cdot T_1)$$

略去高阶无穷小量得

$$\mathrm{d}U_1 = \mathrm{d}(m_1 \cdot c_{v1} \cdot T_1)$$

相同的时间间隔 $\mathrm{d}\tau$ 中,体系对外做功量为

$$\mathrm{d}W_1 = p_1 \cdot \mathrm{d}V_1{}^*$$

由系统的质量增量 $\mathrm{d}m_{1f}$ 所携带的能量为 $\left(h^* + \dfrac{c_f^2}{2}\right)\mathrm{d}m_{1f}$,按能量守恒定律,系统能量的变化应为系统得到的能量和付出的能量之差,因此可以得到

---

* 此功量不是轴功量 $\mathrm{d}W_n$,此间隔中体系对外作的轴功 $\mathrm{d}W_n = (p_1 - p_0) \cdot \mathrm{d}V_1$,其中 p_0 为大气压。

$$dQ_1 + (h^* + \frac{c_i^2}{2})dm_{1f} = d(m_1 \cdot c_{v1} \cdot T_1) + p_1 \cdot dV_1$$

这和对副室的分析基本相同，对应于流经通道的质量增量 dm_{1f} 的比焓与比动能之和为：

当 $p_1 > p_2$ 时，工质从主燃烧室流出，即 $dm_{1f} < 0$ 时，$h^* + \frac{c_i^2}{2} = h_1 = c_{p1} \cdot T_1$；

当 $p_1 < p_2$ 时，工质流入主燃烧室，即 $dm_{1f} > 0$ 时，$h^* + \frac{c_i^2}{2} = h_2 = c_{p2} \cdot T_2$。

故主燃烧室的能量方程可以表示为

$$c_{p1} \cdot T_i \cdot dm_{1f} + dQ_1 = d(m_1 \cdot c_{v1} \cdot T_1) + p_1 \cdot dV_1 \tag{2}$$

式中 $p_1 > p_2$ 时，$i = 1$；$p_1 < p_2$ 时，$i = 2$。

1.2 状态方程

从热力学考虑，一个气态的热力系统只要远离液态，通常就可以视作理想气体，适用理想气体的状态方程，即

$$p_1 V_1 = m_1 \cdot Rg_1 \cdot T_1 \tag{3}$$

$$p_2 V_2 = m_2 \cdot Rg_2 \cdot T_2 \tag{4}$$

由于燃气和空气的分子量相比差别很小[1]，气体常数 Rg_1，Rg_2 在燃烧室空间取同一常数。

诚然，在燃烧室中进行燃烧的情况下，各处的压力、温度和化学浓度都不同，并随时间变化，而且它们沿火焰面是不连续的，因此在均匀气体状态下成立的方程(3)和(4)并不适用于内燃机燃烧室中的情况。但是，不同燃烧区的精确界限迄今还不清楚，当放热率的计算是基于从主、副燃烧室制取的示功图，即基于室内的平均压力随时间的变化曲线时，采用上述两方程进行近似计算是可行的。

1.3 质量方程

对图 1 所示的燃烧系统来讲，无论是主燃烧室还是副燃烧室，其中混合气质量 m_i（此处 i 取值 1 和 2，分别代表主、副燃烧室）的变化，除了前文已述及的流动所引起的增量 dm_{if} 以外，因燃烧前后工质的气态成分变化，也会引起燃烧室内气态部分质量变化，这一部分由于燃烧所产生的混合气增量记为 dm_{ig}，则

$$dm_i = dm_{ig} + dm_{if} \qquad (i = 1, 2) \tag{A}$$

先考虑第一部分质量增量 dm_{1g}。燃烧产生的混合气体增量显然和燃料的消耗增量相等，即

$$dm_{ig} = \frac{1}{H_u}dQ_{iB}$$

式中 H_u 为燃料的低热值，dQ_{1B} 为燃烧放热量。

许多研究结果表明，气缸内工质向室壁的散热量占燃料总放热量 Q_B 的 $18\% \sim 22\%$[4]，上、下限差别不大。按前述假设 $dQ_{iB} = dQ_i + dQ_{iT}$，将散失热量占总发热量的份额视为常数以后，dm_{ig} 可以直接表示成加热量增量 dQ_i 确定的线性函数，即

$$dm_{ig} = K_u \cdot dQ_i \tag{B}$$

例如，设 $dQ_{iT} = 20\% \cdot dQ$ 时，$K_u = 1.25/H_u$。

确定质量增量的另一部分 dm_{if}（流经通道的质量增量）要困难得多。图 3 表示了连接通道中流动的情况。通道两端两个燃烧室内的压力随时间发生很大的变化，而这两个燃烧室压力的差异正是通道中产生混合气流动的主要因素。高温燃气和空气混合气流经通道时，除了向壁面散热以及燃烧等实际影响，难以确定的边界条件也使流场更加复杂，因此对于这样一个变

化非常迅速并带有化学反应的非稳态混合流动过程,即使作了一些近似的假定以后,希望比较准确地求解这个流场,依然十分困难。而且,在将两燃烧室当作平衡态的假设下,其解亦十分不可靠 *。

对于这部分流动质量,各种资料常用的计算公式,是采用定常、绝热、等熵和定比热等假设下得到的理想气体流量公式:

在 $p_1 > p_2$ 时,

$$W = \frac{\mathrm{d}m_{2f}}{\mathrm{d}\tau} = \mu \cdot A \cdot \rho_2 \sqrt{2 \frac{\kappa}{\kappa-1} p_1 \frac{V_1}{m_1} \left[1 - \left(\frac{p_2}{p_1}\right)^{\frac{\kappa-1}{\kappa}}\right]}$$

在 $p_1 < p_2$ 时,

$$W = \frac{\mathrm{d}m_{1f}}{\mathrm{d}\tau} = \mu \cdot A \cdot \rho_1 \sqrt{2 \frac{\kappa}{\kappa-1} p_2 \frac{V_2}{m_2} \left[1 - \left(\frac{p_1}{p_2}\right)^{\frac{\kappa-1}{\kappa}}\right]}$$

在 $p_1 > p_2$ 的情况下,如果压力比 p_2/p_1 不小于临界压力比 $(p_2/p_1)_c$,即

$$\frac{p_2}{p_1} \geq \left(\frac{p_2}{p_1}\right)_c = \left(\frac{2}{\kappa+1}\right)^{\frac{\kappa}{\kappa+1}} 时,$$

$$W = \frac{\mathrm{d}m_{2f}}{\mathrm{d}\tau} = \mu A \sqrt{2 \frac{\kappa}{\kappa-1} \left(\frac{2}{\kappa+1}\right)^{\frac{2}{\kappa-1}}} \sqrt{\frac{p_1 m_1}{v_1}} \tag{C}$$

同样,在 $p_1 < p_2$ 的情况下,如果压力比 p_1/p_2 不小于临界压力比 $(p_1/p_2)_c$,即

$$\frac{p_1}{p_2} \geq \left(\frac{p_1}{p_2}\right)_c = \left(\frac{2}{\kappa+1}\right)^{\frac{\kappa}{\kappa+1}} 时,$$

$$W = \frac{\mathrm{d}m_{1f}}{\mathrm{d}\tau} = \mu A \sqrt{2 \frac{\kappa}{\kappa-1} \left(\frac{2}{\kappa+1}\right)^{\frac{2}{\kappa-1}}} \sqrt{\frac{p_2 m_2}{v_2}} \tag{C'}$$

图 3　连接通道流动示意图

式中 A 为通道截面积;κ 为不计化学反应时的绝热指数,$\kappa = c_p/c_v$;μ 为流量系数,文献[6]已作了介绍。

在燃烧过程中,混合气的温度 T 和各种组分的浓度 c_i 都在变化。显然作为 T 和 c_i 函数的 κ,在空间不同点处其值不同,同一空间点上在不同的时刻 κ 值也不一样,因此在求解上述能量方程时,应考虑 κ 随温度 T 变化,反复迭代,以提高流量的计算精度。但仔细考察此方程的应用条件和实际工况存在的较大差异,会发现在使用如此粗糙的公式基础上,着眼于 $\kappa = c_p/c_v$ 在不同温度下的一些不甚大的差异,其意义不大,更何况在存在着化学反应的实际流动中[5],

$$\kappa = \frac{c_p/c_v}{\left(\frac{\partial \ln \gamma}{\partial \ln \rho}\right)}$$

---

* 假定管流是一维、绝热、没有化学反应的流动,可得一微分方程组[5]:

$$\begin{cases} \rho t + \omega \rho x + \rho \omega = 0 \\ p \omega t + \rho \omega \omega_x + p_x = 0 \\ p_t + \omega p_x + \kappa rt(\rho t + u \rho x) = 0 \\ p = \rho rt \end{cases}$$

这是一个封闭的一阶偏微分方程组。假定入流处为滞止状态,并以前面方程(1)~(4)确定的主副燃烧室内工质的状态参数作为边界条件,理论上可以数值求解。但实际上入流口处混合的初始速度和通道出口处的混合气流速不是一个大数,故在尚未精确确定燃烧室内流场的情况下,这种"精确的计算"亦是没有意义的。

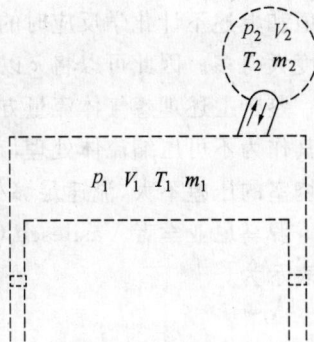

其值和上述不计化学反应时的 $\kappa=c_p/c_v$ 的差别本身就比 $\kappa=c_p/c_v$ 在不同的温度范围内的变化幅度大得多。因此可以将 κ 以某一合适的平均值代入方程,求解流量。

鉴于上述理想气体流量方程物理模型本身的不精确性,甚至可以忽略混合气的可压缩性,将其作为不可压缩流体处理,直接使用伯努利方程求解流量,而不失其可靠性。事实上,主、副燃烧室的压差不大,流速足够小,对可压缩性的忽略从流体力学的观点来分析是完全可以的。

罗马尼亚学者 Vasilescu C A 按照不可压缩流流量公式进行试验得到公式中的流量系数 μ,表示为[7]

$$W=\frac{\mathrm{d}m_f}{\mathrm{d}\tau}=\mu A \sqrt{\frac{2(p_o-p_c)}{\rho}} \tag{D}$$

式中 ρ 为平均密度,$\rho=\dfrac{\rho_1+\rho_2}{2}$;$p_o$,$p_c$ 为对应较大和较小一侧的燃烧室内的压力。

只要试验可靠,笔者认为式(D)比式(C)要可靠而简便得多。这样将式(D)和式(B)代入式(A)可得质量方程

$$\mathrm{d}m_i=K_u\mathrm{d}Q_i\pm\mu A \sqrt{\frac{2|p_1-p_2|}{\rho}} \tag{5}$$

式中 $i=1,2$,分别对应于主、副燃烧室;$p_1>p_2$ 时,$i=1$ 对应"—"号,$i=2$ 对应"十"号;$p_1<p_2$ 时,$i=1$ 对应"十"号,$i=2$ 对应"—"号。

至此,得到式(1)~(4)和(5)(包含 $i=1,2$)6 个方程,其中包含 6 个未知数:m_i,Q_i,T_i $(i=1,2)$。当 p_i 由示功图给出以后,就可以求得放热率。

2 混合气成分变化的影响与计算

前面在质量方程的推导中已经指出,既然理想气体流量公式(C)本身和实际物理模型的偏差很大,因而不计公式中 $\kappa=c_p/c_v$ 的变化对流量的影响(即 κ 以常数处理)是可以的,甚至可以不考虑其压缩性,而直接引用以伯努利方程为基础得到的方程(D)。但是,在论及主、副燃烧室中的内能变化时,比热 c_v 和 c_p 的变化必须给予考虑,它们的变化直接影响所要计算的加热量大小。温度变化对混合气比热的影响可以从直接使用热力学中的不同经验公式 $c_p=f(T)$ 中得到。不少文献都已论及并进行了计算,但对混合气中空气和已燃气体所占比例对比热 c_p,c_v 的影响,尚未进行深入研究和探讨。

根据资料[8],空气和燃气的比热相差达 0.027 kJ/(kg·K),与温度对比热的影响相比显然不是一个小量,故应对混合气成分对比热的影响进行分析和计算。

用下标 a 和 b 分别表示混合气中的空气和燃气,和质量方程推导中的讨论一样,任一燃烧室中混合气成分的变化也由两部分引起。第一是燃烧室内的燃烧使混合气中空气成分减少,燃气成分增多;第二是伴随着两燃烧室压差作用,混合气整体流动的同时,由于两燃烧室的浓度的差异而产生扩散流。

现以两燃烧室中的任一个为研究对象,讨论它内部的空气和燃气的变化。显然

$$\mathrm{d}m_{ai}=\mathrm{d}m_{aig}+\mathrm{d}m_{aif}$$
$$\mathrm{d}m_{bi}=\mathrm{d}m_{big}+\mathrm{d}m_{bif}$$

式中下标 i 表示所论对象,g 和 f 分别对应于燃烧生成项和流动项。设完全燃烧时,燃烧 1 kg

的燃料需 L kg 的空气,写成质量方程为

$$1\ \mathrm{kg(fuel)}+L\ \mathrm{kg(air)}=(L+1)\ \mathrm{kg(burnt\ gases)}$$

设空气增量 $\mathrm{d}m_{aig}=-K_a\mathrm{d}Q_i$,同建立基本质量方程的假设一样,若认为燃烧放热量近似为 $\mathrm{d}Q_{ib}=K_u\mathrm{d}Q_i$ 时,则显然系数 K_a 为

$$K_a=K_uL/H_u$$

相仿,燃气增量记为

$$\mathrm{d}m_{big}=K_b\cdot\mathrm{d}Q_i$$

式中系数 $K_b=K_u(L+1)/H_u$。

另一方面,可以建立混合气各组分的扩散微分方程[9]:

$$\frac{\mathrm{d}m_{aif}}{\mathrm{d}\tau}=\frac{\rho_a}{\rho}\frac{\mathrm{d}m_{if}}{\mathrm{d}\tau}-DA\rho\nabla(\rho_a/\rho)$$

$$\frac{\mathrm{d}m_{bif}}{\mathrm{d}\tau}=\frac{\rho_b}{\rho}\frac{\mathrm{d}m_{if}}{\mathrm{d}\tau}-DA\rho\nabla(\rho_b/\rho)$$

上两式中等号左边分别对应于空气和燃气的质量流量;右边的 $\dfrac{\mathrm{d}m_{if}}{\mathrm{d}\tau}$ 则是前面公式(C)或 (D)所表示混合气总体质量流;ρ_a,ρ_b 和 ρ 分别表示空气、燃气及混合气的当地密度;A 为通道截面积;D 为菲克扩散系数,对于理想气体取作常数。

理论上应该相应地建立一组基本的动量、能量等方程和上两式联立求解。但同样基于前述分析流量方程时的理由,并考虑到通道足够小,可假设浓度在通道内沿质量流动方向作线性分布,即认为浓度梯度在整个通道中为一常数,则

$$\nabla(\rho_a/\rho)=\frac{(\rho_a/\rho)_i-(\rho_a/\rho)_j}{l}\quad(i\neq j;i,j\subset1,2)$$

$$\nabla(\rho_b/\rho)=\frac{(\rho_b/\rho)_i-(\rho_b/\rho)_j}{l}\quad(i\neq j;i,j\subset1,2)$$

式中 l 为通道长度。

因此可得

$$\mathrm{d}m_{ai}=-K_a\mathrm{d}Q_i+\left(\frac{\rho_a}{\rho}\right)\mathrm{d}m_{if}-DA_\rho\frac{(\rho_a/\rho)_i-(\rho_a/\rho)_j}{l}\mathrm{d}\tau$$

$$\mathrm{d}m_{bi}=-K_b\mathrm{d}Q_i+\left(\frac{\rho_b}{\rho}\right)\mathrm{d}m_{if}-DA_\rho\frac{(\rho_b/\rho)_i-(\rho_b/\rho)_j}{l}\mathrm{d}\tau$$

τ 时刻燃烧室中混合气的成分为

$$m_{ai}=m_{aio}+\int_0^\tau\mathrm{d}m_{ai}$$

$$m_{bi}=m_{bio}+\int_0^\tau\mathrm{d}m_{bi}$$

式中 m_{aio} 和 m_{bio} 分别为空气和燃气在对应燃烧室中任定初始时刻的质量。τ 时刻对应的比热为

$$c_{pi}=\frac{m_{ai}}{m_i}\cdot c_{pa}(T)+\frac{m_{bi}}{m_i}\cdot c_{pb}(T)$$

$$c_{vi}=\frac{m_{ai}}{m_i}\cdot c_{va}(T)+\frac{m_{bi}}{m_i}\cdot c_{vb}(T)$$

将这两式代入基本方程组中迭代可求解放热率。

鉴于混合气成分对比热的影响,计算时不仅要考虑主燃室容积等初始参数的影响,还要考虑燃烧前主、副燃烧室中空气和燃气的分布,以使其尽量符合实际工况。

3 结 束 语

要使涡流燃烧室放热率的计算方法精确化,必须首先使理论模型精确化。本文导出并分析了涡流燃烧室放热率的完整方程。除考虑温度对混合气比热的影响外,还考虑了由于浓度差异产生的扩散流动对混合气比热的影响,从而在理论模型精确化方面推进了一步。利用这种理论模型求解一台具体的涡流式柴油机的放热率,将是我们下一步要做的工作。

参 考 文 献

［ 1 ］ Woschui G. 内燃机内部热力过程研究［G］. 武汉:华中工学院,译. 1981.

［ 2 ］ 池上詢, 川合悦蔵, 藤村裕二. デイーゼル機関にぉけろ室素酸化物の生成［J］. 日本機械学会論文集,1972,39(327).

［ 3 ］ Li Detao. Contributii la studiul adaptarii camerelor de ardere ale motoarelor cu aprindere prin comprimare cu camere separate de virtej la turatii inalte［D］. Teza de doctorat,Timisoara,1982.

［ 4 ］ Взров Ъ А. *Тракторные Дизели Справочник*［M］. Машиностроенйе,1981.

［ 5 ］ Joe D Hoffman. *Gas Dynamics*［M］. John Wiley & Son,1977.

［ 6 ］ 刘炽棠. 预燃室柴油机燃烧放热规律计算［J］. 上海交通大学学报,1978(1).

［ 7 ］ Vasilescu C A. 压燃式发动机预燃室流通孔流量系数的研究［J］. 林文进,译. 内燃机译丛,1965(3).

［ 8 ］ Герасимова С Г. *Теплотехнический Справочник*［M］. Госэнергоизлат,1957.

［ 9 ］ James R Welty. *Fundamentals of Momentum Heat & Mass Transfer*［M］. John Wiley & Son ,1976.

(本文曾在全国内燃机燃烧放热规律讨论会上宣读,原载于《江苏工学院学报》1986 年第 3 期)

Approach to theoretical model of heat release rates in swirl chamber diesel

Abstract:A complet system of equations is derived and analyzed for calculating heat release rates in swirl chamber. Besides the influence of temperature on specific heat of mixture, considered is also that of diffusion flow due to concentration discrepancies, thus making the theoretical model more exact.

A study on accurate calculating method of rates of heat release in swirl chamber diesel engines

Li Detao

Abstract

In this paper, the research progresses in calculating method of rates of heat release (ROHR) of swirl chamber (indirect injection, IDI) diesel engines over the past fifteen years have been reviewed systematically. It has been presented that calculations based upon a combination of thermodynamic model and experimental model are of great importance. The experimental model for calculating passage discharge coefficient has been developed. The p-φ diagram of the S195 Type Diesel engine made in China has been strictly measured and recorded, and ROHR has been calculated by using our program. The major factors affecting the calculating precision of ROHR of IDI diesel engine have been analysed. The research results show that ROHR of IDI diesel derived from this calculation method is obviously more reasonable.

1 Introduction

Up to now, the precision for calculating ROHR of IDI diesel is not satisfactory. Lyn W T had studied ROHR of IDI diesel[1] without separate consideration of main and swirl chamber. Nakajima K[2] and Jankov R[3] had presented the model for calculating ROHR of divided chamber engine, but the model they described was too simple and they didn't consider the influence of the heat transfer and the flow through connecting passage. Anisits F[4] had set up a comparatively detailed model for calculating ROHR of the main and swirl chambers, however, he made some assumption that didn't accord with the actual situation, such as taking the discharge coefficient of conneting passage for the same constant in different flow directions. Mansourt S H and Kort R T did the same as Anisits F[5,6]. Watson N arrived at the conclusion, through a steady flow test, that the discharge coefficient of connecting passage is a function of flow directions, the pressure ratio, Piston Proximity[7]. Recently, we found from the flow tests that the discharge coefficient has something to do with the factors above, besides, the shape, the position and the structure size of connecting passage have a greater influence upon the discharge coefficient[8]. The existance of the start hole (an obvious feature of the products in China) has an obvious influence on the discharge coefficient in two flow directions.

2　A combination model

Our research shows that only in this way, i. e., combining the thermodynamic model with the experimental model that is based on a lot of experiment datum, can the calculation method become a more effective and accurate tool to study the cycle performance of IDI diesel engine.

From the view of thermodynamic, the whole swirl combustion system can be shown by Fig. 1. The corresponding equation can be set up individually[9].

(1) The equation of conservation of energy

For the swirl chamber, we have:

$$c_{pi}T_i\mathrm{d}m_f+\mathrm{d}Q_2=\mathrm{d}(m_2c_{v2}T_2)$$

$$\begin{cases} \text{when } p_1 > p_2, \ i=1 \\ \text{when } p_1 < p_2, \ i=2 \end{cases} \tag{1}$$

For the main chamber, we have:

$$c_{pi}T_i\mathrm{d}m_f+\mathrm{d}Q_1=\mathrm{d}(m_1c_{v1}T_1)+p_1\mathrm{d}V_1$$

$$\begin{cases} \text{when } p_1 > p_2, \ i=1 \\ \text{when } p_1 < p_2, \ i=2 \end{cases} \tag{2}$$

(2) State equation

$$p_1V_1=m_1R_1T_1 \tag{3}$$

$$p_2V_2=m_2R_2T_2 \tag{4}$$

(3) The equation of conservation of mass

$$\mathrm{d}m_1=\mathrm{d}m_{1g}\pm\mathrm{d}m_f \begin{cases} \text{when } p_1 > p_2, \ \text{take negative} \\ \text{when } p_1 < p_2, \ \text{take positive} \end{cases} \tag{5}$$

$$\mathrm{d}m_2=\mathrm{d}m_{2g}\pm\mathrm{d}m_f \begin{cases} \text{when } p_1 > p_2, \ \text{take positive} \\ \text{when } p_1 < p_2, \ \text{take negative} \end{cases} \tag{6}$$

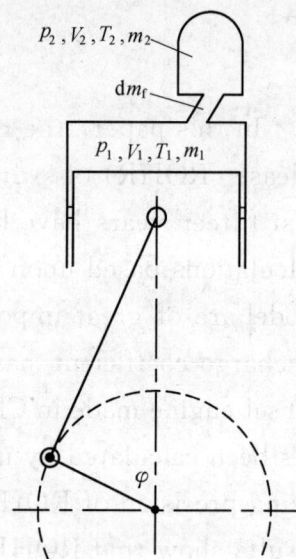

Fig. 1　Diesel thermodynamic simulation system

here, $dm_{ig} = dQ_{bi}/H_u \quad (i=1,2)$ \hfill (7)

Supposing the flow through connecting passage is one-dimensional, quasi-steady and heat insulation flow, there are:

when $p_1 > p_2$, $\dfrac{dm_f}{d\varphi} = \dfrac{\mu_{1\cdot 2f}}{6n}\dfrac{p_1}{\sqrt{R_1 T_1}}\sqrt{\dfrac{2\kappa_1}{\kappa_1-1}\left[\left(\dfrac{p_2}{p_1}\right)^{2/\kappa_1} - \left(\dfrac{p_2}{p_1}\right)^{(\kappa_1+1)/\kappa_1}\right]}$ \hfill (8)

when $p_1 < p_2$, $\dfrac{dm_f}{d\varphi} = \dfrac{\mu_{2\cdot 1f}}{6n}\dfrac{p_2}{\sqrt{R_2 T_2}}\sqrt{\dfrac{2\kappa_2}{\kappa_2-1}\left[\left(\dfrac{p_1}{p_2}\right)^{2/\kappa_2} - \left(\dfrac{p_1}{p_2}\right)^{(\kappa_2+1)/\kappa_2}\right]}$ \hfill (9)

Put the equation (7), (8), (9), into (5), (6), we can obtain:

$$dm_1 = dQ_{b1}/H_u \pm \dfrac{\mu_{1\cdot 2f}}{6n}\dfrac{p_1}{\sqrt{R_1 T_1}}\sqrt{\dfrac{2\kappa_1}{\kappa_1-1}\left[\left(\dfrac{p_2}{p_1}\right)^{2/\kappa_1} - \left(\dfrac{p_2}{p_1}\right)^{(\kappa_1+1)/\kappa_1}\right]}\,d\varphi \quad (10)$$

$\begin{cases} \text{when } p_1 > p_2, \text{ take negative} \\ \text{when } p_1 < p_2, \text{ take positive} \end{cases}$

$$dm_2 = dQ_{b2}/H_u \pm \dfrac{\mu_{2\cdot 1f}}{6n}\dfrac{p_2}{\sqrt{R_2 T_2}}\sqrt{\dfrac{2\kappa_2}{\kappa_2-1}\left[\left(\dfrac{p_1}{p_2}\right)^{2/\kappa_2} - \left(\dfrac{p_1}{p_2}\right)^{(\kappa_2+1)/\kappa_2}\right]}\,d\varphi \quad (11)$$

$\begin{cases} \text{when } p_1 > p_2, \text{ take positive} \\ \text{when } p_1 < p_2, \text{ take negative} \end{cases}$

There are 6 unknown variables, m_i, Q_i, $T_i (i=1,2)$ in the 6 equations. After p-φ values are gained from the indicator pressure diagram, the 6 equations above will be solved and ROHR can be gained.

The discussions on the experimental calculation model of the discharge coefficient and the heat transfer coefficient are presented below.

2. 1　Discharge coefficient

Flux coefficient influences the calculation precision of direct injection type diesel engine. Different fluid directions and the configurable designs of jlutting channel influence flux coefficient significantly(Tab. 1).

The results of the typical experiment is shown in Fig. 2.

We have dealt with the data with the regression technique and obtained experimental formula for calculating discharge coefficient:

For the flow from main chamber to swirl chamber:

$\mu_{1\cdot 2} = Re/(8\times 10^4 + 0.94Re)$ \hfill (12)

For the flow from swirl chamber to main chamber:

$\mu_{2\cdot 1} = Re/(3\times 10^4 + 1.1Re)$ \hfill (13)

Here, $2\times 10^4 < Re < 1\times 10^6$.

Tab. 1　Flux coefficient in different fluid directions and jlutting models

jlutting model $\diagdown$ μ	fluid direction	
	main→swirl	swirl→main
$L18$	0.68~0.70	0.74~0.76
$L16$	0.71~0.73	0.75~0.77
$L13$	0.70~0.73	0.75~0.78
$L17$ ($\theta=30°$)	0.66~0.68	0.73~0.75
$R16$	0.66~0.67	0.73~0.74
$R18$	0.63~0.65	0.69~0.70

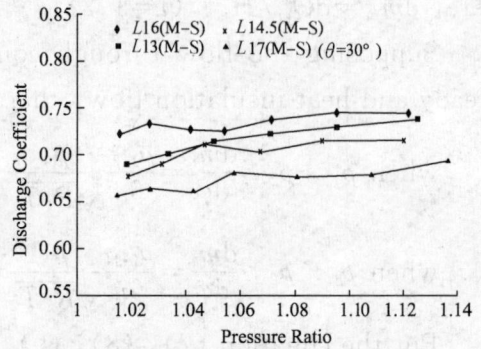

Fig. 2　Relation of discharge coefficient and pressure ratio

2.2　The heat transfer coefficient

If the heat transfer in the way of convection and radiation from two chambers is considered, the quantity of heat which is passed to the cooling water can be calculated as the following general formula:

$$dQ_w/d\tau = \alpha F(T - T_w) + \varepsilon F\sigma(T^4 - T_w^4) \tag{14}$$

Convective heat transfer coefficient is assumed to be related to Reynolds Number: $Nu = \alpha Re^b$ (15)

From (7): $Nu = \alpha Re^{0.7}$ (15a)

To the swirl chamber　$\alpha = 0.023$

To the main chamber　$\alpha = 0.012$

Hence, $\alpha = 0.04(1+\alpha) \cdot (pC_m)^{0.7} T^{-0.2} d_e^{-0.3}$ (16)

3　The measure and record of indicator pressure diagram and the calculation of ROHR

The measure and record system is shown in Fig. 3.

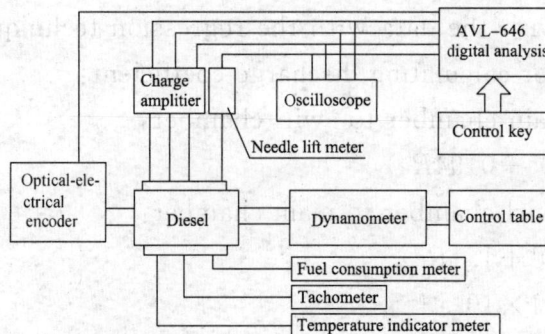

Fig. 3　Installation of measuring and recording indicate pressure diagram

We only used "DIG" program of that system to measure and record the pressure data of two chambers. ROHR were calculated and the diagram was drawn on the micro-computer by our program.

Fig. 4 shows the indicator pressure diagram of S195 Type Diesel. Fig. 5,6,7 shows the pressure difference and pressure hoist rate and pressure ratio between main and swirl chamber.

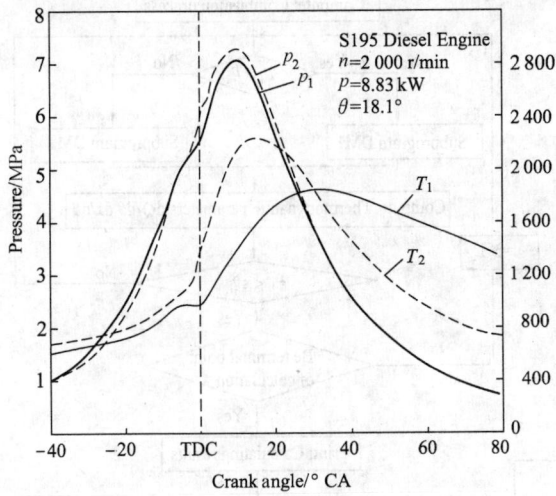

Fig. 4   Indicator diagrams and temperature

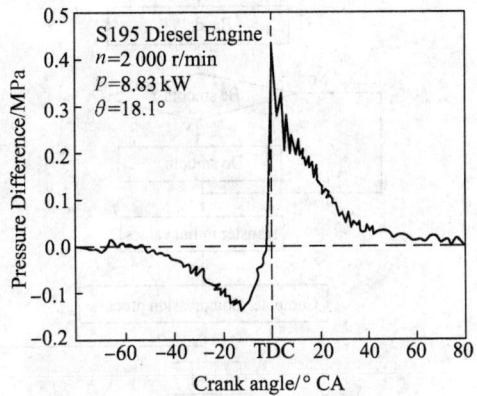

Fig. 5   Pressure difference between main and swirl chamber

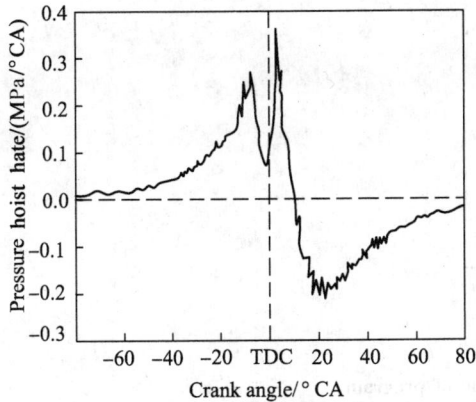

Fig. 6   Pressure hoist rate between main and swirl chamber

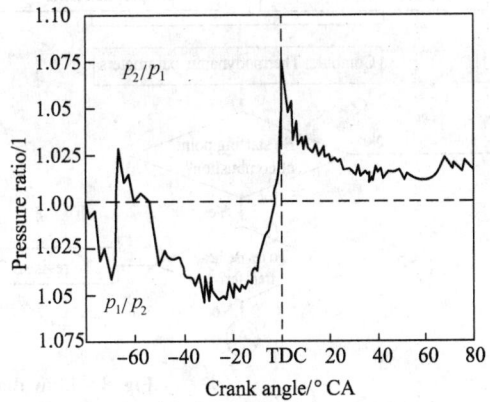

Fig. 7   Pressure ratio between main and swirl chamber

The combination model and measured pressure, ROHR of IDI diesel can be calculated. We have edited the program of calculation by ourselves and the flow diagram has been shown in Fig. 8. The main importing parameters are shown in Tab. 2.

Tab. 2　The main imported parameters

Bore /mm	Stroke /mm	Power /kW	Speed of revolution /(r/min)	Compression ratio	Volume ratio /%	Injection pressure /MPa
95	115	8.83	2 000	19	47	12.5

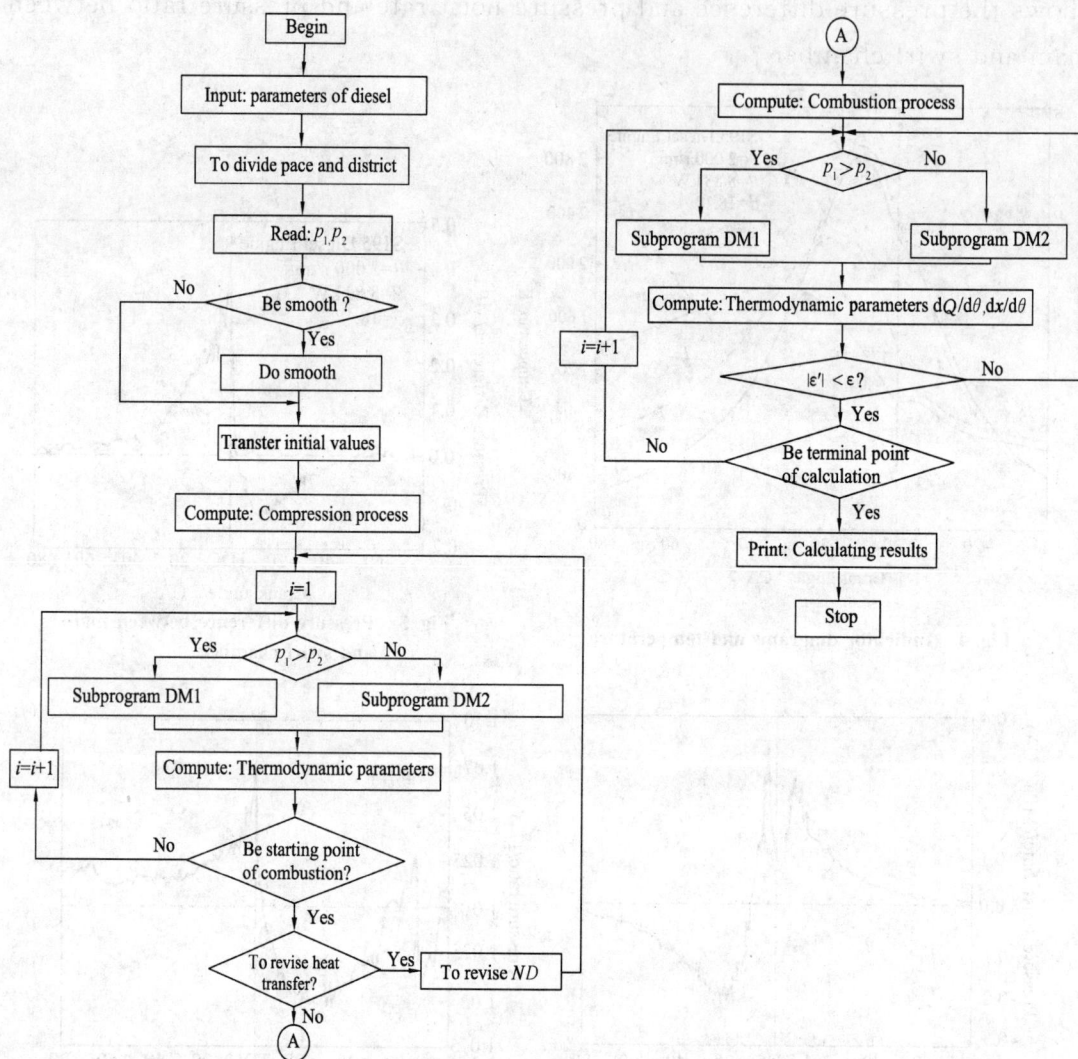

Begin

Input: parameters of diesel

To divide pace and district

Read: p_1, p_2

Be smooth ? — No / Yes

Do smooth

Transter initial values

Compute: Compression process

$i=1$

$p_1 > p_2$ — Yes / No

Subprogram DM1　　Subprogram DM2

$i=i+1$

Compute: Thermodynamic parameters

Be starting point of combustion? — No / Yes

To revise heat transfer? — Yes → To revise ND / No

A

A

Compute: Combustion process

$p_1 > p_2$ — Yes / No

Subprogram DM1　　Subprogram DM2

Compute: Thermodynamic parameters $dQ/d\theta, dx/d\theta$

$i=i+1$

$|\varepsilon'| < \varepsilon$? — No / Yes

Be terminal point of calculation — No / Yes

Print: Calculating results

Stop

Fig. 8　Flow diagram of program

4　Analysis of factors influencing calculation precision of ROHR

4.1　The pace and interval for calculation

For calculating ROHR, we have adopted the method of changing pace. Near TDC, the calculation pace is 0.5° CA to guarantee the necessary calculating precision; and in the intervals ($-37 \sim -17°$ CA and $+23 \sim +43°$ CA), it is regarded as a transition interval where the pace is 1° CA; in the part of being far away from TDC, the calculation pace is 2° CA.

For direct injection diesel, the starting point for calculating ROHR may be chosen at any point before ignition. But to IDI diesel, the calculation should begin at the begining of the compression. Thus the temperature, mass and component of mixture in the main and swirl chamber at ignition point can be obtained exactly.

4. 2   The influence of the position error of TDC

The calculation results shows that when the position of TDC moves 1° CA forward, the peak value of ROHR increases by 15. 10%(Fig. 9), and when the position of TDC is late for one degree, the peak value of ROHR will reduces by 14. 80%.

4. 3   The influence of the zero point error of pressure

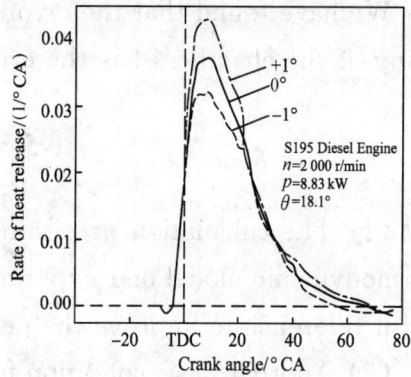

Fig. 9   The effect of position error of TDC on ROHR

Fig. 10 illustrates the influence of pressure charged +0. 1 MPa in indicator pressure diagram on ROHR. Calculated Error is shown in Table 3.

Tab. 3   Calculated error

error/% \ item \ Δp	$\Delta p = +0.1$ MPa			$\Delta p = -0.1$ MPa		
	overall error	main chamber	swirl chamber	overall error	main chamber	swirl chamber
cumulate rates of heat release	9. 67	15. 63	3. 48	−6. 13	−10. 27	−2. 75
rates of heat release	1. 63	3. 64	1. 29	−1. 59	−3. 61	−1. 24

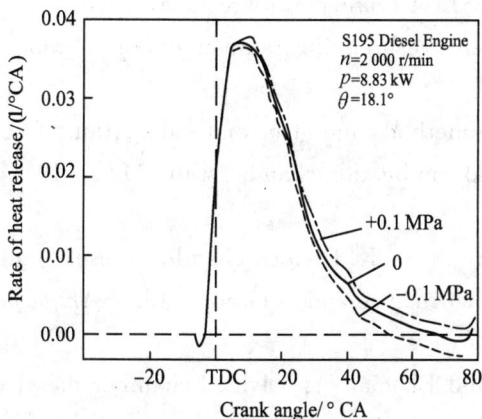

Fig. 10   The effect of zero point error of pressure on ROHR

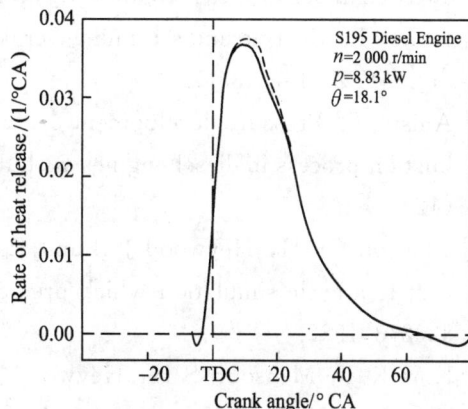

Fig. 11   The effect of wall temperature on ROHR

4. 4　The influence of the wall temperature of the chambers

Fig. 11 has shown the calculation result for different wall temperatures. The result indicates that the value deviation of overall ROHR is 2. 33%. Thus it can be drawn that the influence may be ignored.

4. 5　The influence of the pressure asynchronous of main and swirl chamber

We have found that the error of ROHR can double when the pressure phase of any of the chambers has the error of $\pm 1°$ CA.

5　Conclusions

(1) The calculation precision of ROHR can be improved by combining the thermodynamic model and experimental model, and the results can be used as a useful reference to improve the performance of IDI diesel engine.

(2) Another basic condition for calculating ROHR precisely is measuring and recording indicatior pressure diagram and the pressure difference diagram.

(3) It has a main function to the improvement of calculation precision to reduce the position error of TDC, to remove the zero point error of pressure, to make the pressure change process of two chambers in synchronism and to elect the fit pace and interval for the calculation.

References

[1] Lyn W T, Samaga B S and Bowden C M. Rate of heat release in highspeed indirect-injection diesel engines[J]. *Proc I Mech E*, 1969/70,184(3J).

[2] Nakajima K. An experimental research on the air swirl motion and combustion in the swirl chamber of diesel engines[C]∥*XII FISITA Congress*, Barcelona, 1968.

[3] Jankov R. Energie und strömungsverältnisse in verbrenningsräumen von diesel-motoren [J]. *MTZ*, 1972,33(7).

[4] Anisits F. Pressure development evaluation methods and electronic calculation of combustion process in diesel engines with divided combustion chambers[J]. *MTZ*, 1971,32 (12).

[5] Mansouri S H, Heywood J B and Radhakrishnan K. Divided-chamber diesel engine, part I: a cycle simulation which predicts performance and emissions[J]. *SAE paper*, 820273,1982.

[6] Kort R T, Mansouri S H, Heywood J B and Ekchian A. Divided-chamber diesel engine, part II: experimental validation of a predictive cycle simulation and heat release analysis[J]. *SAE paper*, 820274,1982.

[7] Watson N and Kamel M. Thermodynamic efficiency evaluation of an indirect injetion diesel engine[J]. *SAE paper*, 790039, 1979.

[8] Li Detao. A study of discharge coefficient of connecting passage on swirl chamber diesel engines[C]// *Proce of Power Eng*. Beijing: [s. n.], 1985.

[9] Li Detao and Yang Benluo. A discussion of theoretical model of heat release rates on swirl chamber diesel engines[C]. *Symposium on combustion of CSCI*, 1984.

MAIN NOTATION

c_v Specific heat at constant volume

d_e Equal quantity diameter of combustion chamber

f Section area of connecting passage

m_f Gas mass through passage

m_g Fuel mass in combustion chamber

n Engine speed

Q Heat release to gas

Q_b Heat release of fuel combustion

V Volume

α Heat transfer coefficient

ε Enissivity

μ Discharge coefficient

σ Stephen-Boltzmann Constant

φ Crank angle

(From: *Ith International Symposium on Power*, 1985. 本文参见《内燃机学报》1986 年第 4 期)

涡流室式柴油机放热率的计算方法和计算程序的精确化研究

[摘要] 本文系统而简要地评述了十多年来涡流室式柴油机放热率计算方法的研究进展。提出以热力学模型和经验模型组成的复合模型作为计算基础具有重要的实际意义。建立了计算通道流量系数的经验模型,精确地测录了国产 S195 型柴油机的 p-φ 图,并按自编程序进行了放热率计算。根据作者的实践,对影响涡流室式柴油机放热率的因素进行了分析。研究结果表明,用这种方法得出的涡流室式柴油机的放热率具有明显的合理性。

涡流室式柴油机在不同条件下的
放热特性和性能的对比分析

李德桃,林德嵩,沈俊贤,蔡忆昔,单春贤,苏仕清

[摘要]　作者应用复合模型计算 S195 型柴油机在不同条件下的放热特性,并对计算结果和性能试验结果进行综合分析和评估,从而为这种生产量大、使用面广的柴油机性能的改善提供科学依据。

众所周知,国产优质涡流室式柴油机的燃料经济性接近或达到了世界先进水平。然而从放热特性的角度分析和评估这种柴油机的各种性能的研究却很少。这主要是因为:从建立柴油机放热特性的精确计算模型来看,对涡流室式柴油机的研究要比直喷式困难得多;精确测量主、副燃烧室的压力差很困难。

Woschni[1]、徐锡洪[2]、Nakajima K[3]、Jankov R[4]、Terada K[5]等虽然提出了分隔式柴油机放热特性的计算模型,但这些模型都过于理想,几乎都没有考虑连接通道的流量系数对主、副室质量分配的影响。根据作者最近的研究[6],这种影响对精确计算放热特性来说是完全不能忽视的。Anisits F 虽然建立了一个比较详细的模型[7],然而他还是作了一些与实际情况相差较大的假定,如在两个不同流动方向上连接通道流量系数取为同一常值。前不久 Mansourt S H 和 Kort R T 等人发表的文章仍然是这样处理的[8,9]。

最近作者对吊钟型涡流燃烧室的连接通道流量系数做了大量的试验研究,在此基础上提出了精确计算涡流室式柴油机放热特性的复合模型[6]。本文应用这种模型计算 S195 型柴油机在不同条件下的放热特性,并把计算结果和性能试验结果进行综合分析和评估,从而提出进一步提高这种柴油机性能的若干合理途径。

1　试验装置和试验方法

为了精确计算涡流室式柴油机的放热特性,笔者用 AVL-646 数字分析仪反复测录了一台技术状态良好的 S195 型柴油机的主、副燃烧室的示功图、喷油嘴针阀升程及高压油管嘴端压力波。整个测试系统的示意图如图 1 所示。选用的两个石英压力传感器的线性误差小于 $\pm 0.3\%$FSO,压力传感器的安装见图 2。传感器测取的信号经 AVL 放大器后输出,在 Tektronix5113示波器监视下进行数据采集。该系统的采样频率为 700 kHz,幅值分辨率为 10 bit,最大角度分辨率为 0.1° CA。为了提高精度,采用 0.5,1.0,2.0° CA 的变步长测量表。由于该系统没有计算涡流室式柴油机的放热特性的程序,因此,只需用该分析仪的"DIG"程序测取两室的压力数据,而放热率、累积放热率、燃烧室温度等,则按自编程序在微机上进行计算和绘图。

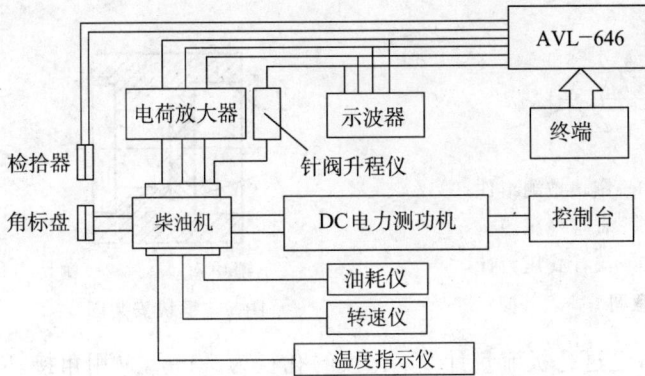

图 1 示功图测录系统示意图

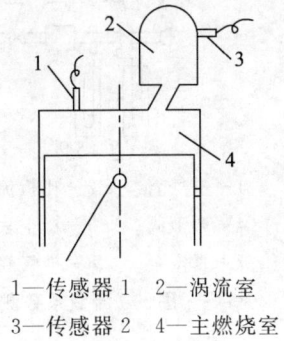

1—传感器 1 2—涡流室
3—传感器 2 4—主燃烧室

图 2 压力传感器安装图

图 3 所示的示功图是经过 64 个循环的平均化处理后得到的,它清楚地显示了主、副燃烧室的压力变化过程。为了精确地测录示功图,传感器的安装做到尽可能消除通道效应和避免热冲击的影响。测录前后对测试系统进行了标定[10]。

除此之外,还要精确测定连接通道的流量系数和柴油机的换热系数。对于吊钟型涡流室式柴油机的传热问题,笔者还在进行研究。因此,在计算放热特性时,引用比较接近所论具体机型的放热系数公式[6],而流量系数则是根据大量模拟试验测定的。由于

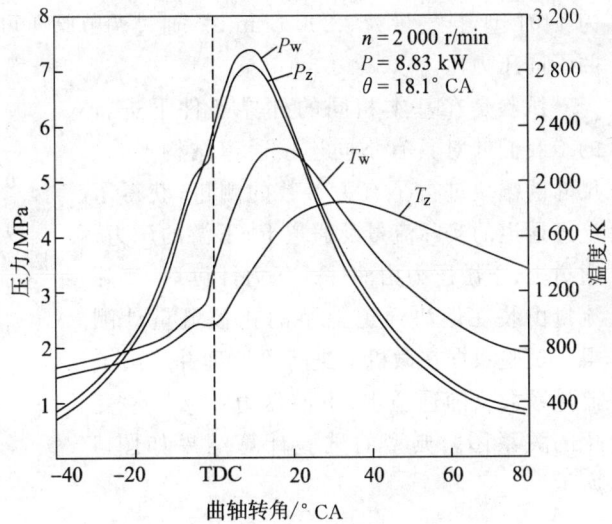

图 3 主、副燃烧室的压力和温度

柴油机工质流经镶块连接通道的流动属于非定常流动,因而很难测量其状态参数和流动参数。为测试方便,采用了稳定流动近似模拟实际流动。

模拟试验中,在保持 Reynolds 准则与实际情况相同的条件下,根据所测得的主、副燃烧室之间的压力比,在稳流试验台上,把镶块通道上、下游的压力比控制在同样的变化范围内。在测定流量系数的模拟试验中,须保持试验情况与实际情况的 Re 数相同,即 $\left(\dfrac{WL}{\nu}\right)_{模型}=\left(\dfrac{WL}{\nu}\right)_{柴油机}$。在所论情况下,由于在稳流试验台上采用了柴油机上的镶块,故特征尺寸 $L_{模型}=L_{柴油机}$。在压力比相同的条件下,如忽略绝热指数和气体性质的影响,可近似地认为通道流速 $W_{模型}=W_{柴油机}$。如果进一步忽略粘性的变化,则可近似认为 $Re_{模型}=Re_{柴油机}$,这样就可近似实现稳流试验台内与柴油机内流体流经通道时的动力学相似。

模拟试验是在一台标准的稳流试验台上进行的。图 4 是试验装置的示意图。

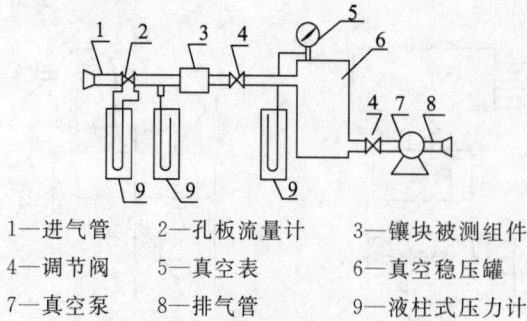

1—进气管　　2—孔板流量计　　3—镶块被测组件
4—调节阀　　5—真空表　　　　6—真空稳压罐
7—真空泵　　8—排气管　　　　9—液柱式压力计

图 4　流量系数测量装置示意图

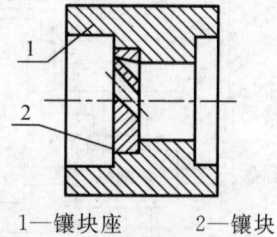

1—镶块座　　2—镶块

图 5　镶块安装图

空气自进气管 1 进入试验装置,经过孔板流量计 2。孔板的孔径为 7 mm,采用角接环室取压。流量的大小可由液柱式压力计 9 读出。根据试验的要求,不同结构设计的镶块作为被测组件可方便地更换(见图 5)。真空稳压罐 6 直径为 400 mm,起稳定背压的作用。真空泵 7 为 1401 型,排气量为 3 200 L/min。通过调节阀 4 可方便地调节背压的变化,以获得试验所需的不同压力比。

试验是在基本相同的外界条件下进行的。我们针对具有不同连接通道结构形状、尺寸的镶块进行了有关参数的测量,获得了流经通道的实际流量及通道上、下游的压力。通道上、下游压力由液柱式压力计读取,实际流量由装在镶块通道上游的孔板流量计测得。试验数据在微机上进行了处理并绘出了流量系数 μ 随通道上、下游压力比 p_1/p_2 变化的关系图。典型的试验计算结果如图 6 所示。

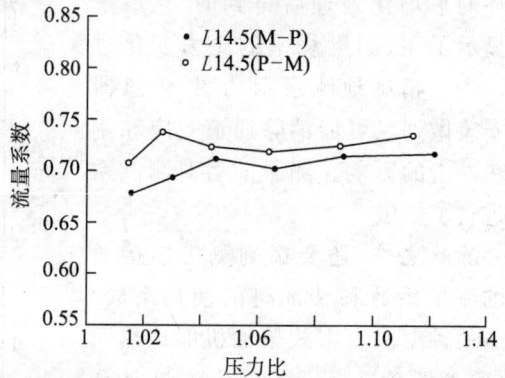

图 6　流量系数与压力比的关系

2　不同条件下的放热特性和性能的对比分析

2.1　在不同负荷时

S195 型柴油机在标定转速(2 000 r/min)下的 50%负荷和全负荷特性的示功图如图 7 所示。从试验结果得到:

(1)主、副燃烧室的最高爆发压力都随负荷的增加明显上升;全负荷与 50%负荷相比,最高爆发压力上升约 0.4 MPa。

(2)不同负荷下的燃烧始点在上止点前 3～4.5°CA;随着负荷增加,燃烧始点略向前移。

(3)最高爆发压力点随负荷的增加往后移动,与 50%负荷相比,全负荷后移约 2°CA。

(4)低负荷时针阀关闭不迅速,但没有发现二次喷射现象。

基于上述示功图,我们进行了放热特性的计算,计算结果表明:

(1)当负荷变化时,累积放热率在主、副燃烧室中的分配发生有规律的变化(见图 8)。随着负荷的减小,主室的累积放热率 χ 不断减小,涡流室的累积放热率 χ' 不断增加。由此可见,改善涡流室内混合气形成和燃烧过程,对于提高低负荷的性能有重要意义;改善主燃室内的混合气形成和燃烧过程,对于提高大负荷性能指标有重要意义。主、副室的最大累积放热率见表1。

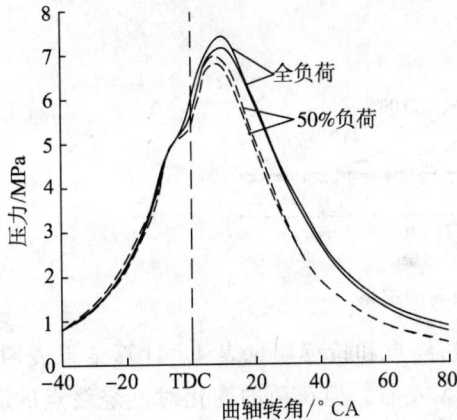

图 7　50%负荷和全负荷的示功图

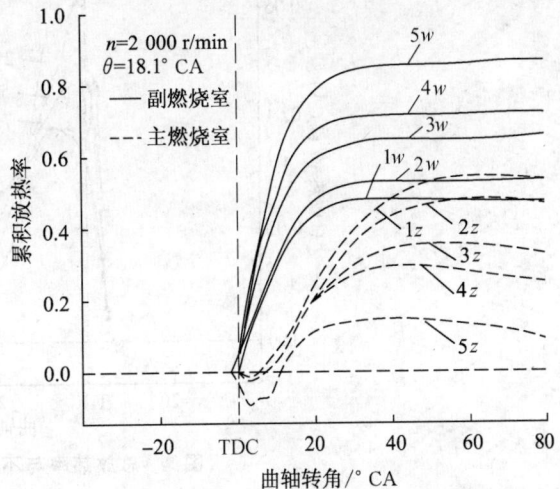

1—110%负荷　2—100%负荷　3—75%负荷
4—50%负荷　5—25%负荷

图 8　不同负荷时主、副燃烧室的累积放热率

表 1　主、副燃烧室的最大累积放热率

项目	负　荷				
	110%	100%	75%	50%	25%
χ_{max}	0.537 2	0.475 8	0.356 1	0.290 5	0.148 7
χ'_{max}	0.472 7	0.526 3	0.654 9	0.719 2	0.857 2

(2) 主、副燃烧室内的最高燃烧温度随负荷的减小而逐渐减小。最高温度点的位置随负荷的减小向前移动,如表2所示。

表 2　主、副燃烧室的最高温度及其相应位置　　　　　　　　　　　K

项目	负　荷				
	110%	100%	75%	50%	25%
T_{max}	1 993(32° CA)	1 855(32° CA)	1 629(26° CA)	1 477(20.5° CA)	1 287(17.5° CA)
T'_{max}	2 236(17° CA)	2 173(17° CA)	2 167(14° CA)	1 940(13° CA)	1 754(13° CA)

(3) 随着负荷的减小,主、副室的最大放热率及最大总放热率均不断增加(见表3),总放热率随负荷的变化关系如图9所示。

表 3　燃烧室放热率与负荷的变化关系　　　　　　　　　　1/° CA

项目	负　荷				
	110%	100%	75%	50%	25%
$(d\chi/d\varphi)_{max}$	0.017 5	0.017 6	0.017 6	0.017 7	0.021 9
$(d\chi'/d\varphi)_{max}$	0.029 1	0.029 1	0.045 5	0.050 8	0.066 1
$\left(\dfrac{d\chi}{d\varphi}+\dfrac{d\chi'}{d\varphi}\right)_{max}$	0.036 7	0.036 8	0.047 8	0.059 4	0.067 3

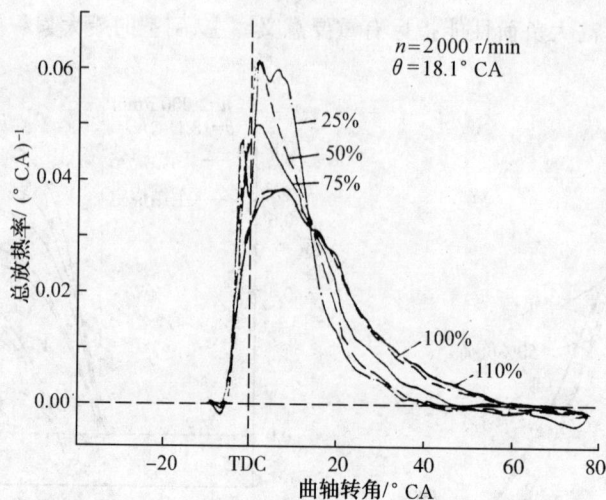

图 9　总放热率与不同负荷的关系

（4）计算所得标定转速下的负荷特性的燃烧始点、终点和持续期见表 4。计算结果表明，负荷不同对燃烧始点的影响不大，一般只相差 $1.5°$ CA 左右。但负荷的变化对燃烧终点的影响比较大。随着负荷的减小，主燃烧室的燃烧终点逐渐前移，燃烧持续期缩短；涡流室的燃烧终点逐渐后移，燃烧持续期有所延长。当然总的燃烧持续期还是缩短的，如图 10 所示。

表 4　标定转速下的负荷特性的燃烧始点、终点和持续期

$°$ CA

负　荷	主燃烧室			涡流室		
	始　点	终　点	持续期	始　点	终　点	持续期
110%	−2.0	64	66.0	−4.5	32	36.5
100%	−1.5	61	62.5	−4.0	35	39.0
75%	−1.5	49	50.5	−4.0	37	41.0
50%	−0.5	43	43.5	−3.5	38	41.5
25%	0	40	40.0	−3.0	43	46.0

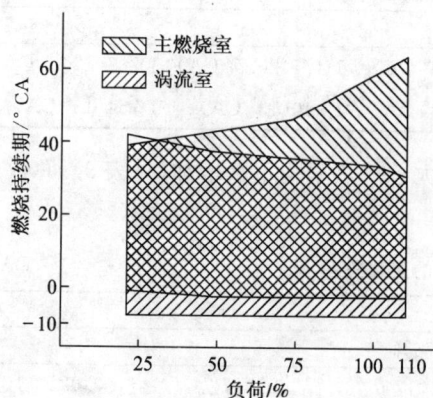

图 10　燃烧持续期与负荷的关系

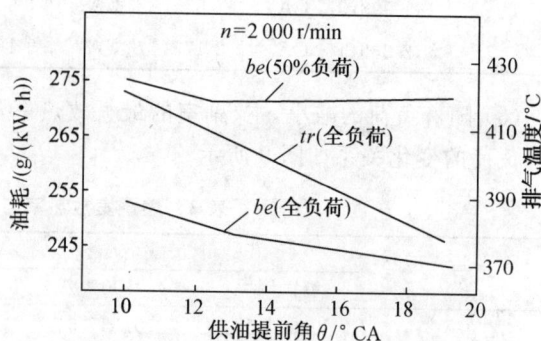

图 11　供油提前角对油耗、排温的影响

2.2　在不同供油提前角时

S195 型柴油机的供油提前角 θ 对油耗和排温的影响如图 11 所示。该图表明，在 $\theta=15\sim$

19°CA 范围内,供油提前角的变化对 50% 负荷时的油耗的影响甚微。对标定工况来说,θ 每减小 1°CA,油耗约升高 1.0 g/(kW·h),排温约升高 5 ℃。当 $\theta < 13$°CA 时,油耗上升得很快,约为 2.5~4.0 g/(kW·h)。不同供油提前角时所测取的示功图数据可看出:在 $\theta = 16 \sim 18$°CA 范围内,最高爆发压力 p_z 变化缓慢,θ 每增加 1°CA,p_z 仅增加 0.1 MPa;但当 $\theta > 18$°CA 时,p_z 即急剧上升,每增加 1°CA,p_z 约增加 1 MPa;当 13°CA $< \theta <$ 16°CA 时,p_z 值约以 0.4~0.5 MPa/°CA 随 θ 减小而下降;当 $\theta = 13.4$°CA 时,燃烧始点出现在上止点后 2°CA,气缸压力曲线呈明显的双峰,油耗上升加快,这表明供油已偏晚。

表 5 列出了计算所得 $\theta = 18.1$°CA 和 $\theta = 19.2$°CA 时的最大放热率数据。

表 5 计算所得特定 θ 下的最大放热率及其相应 °CA 位置

项 目	θ	
	18.1° CA	19.2° CA
$(d\chi/d\varphi)_{max}$	0.017 6(18° CA)	0.023 7(16.5° CA)
$(d\chi'/d\varphi)_{max}$	0.029 1(2° CA)	0.044 0(−2.0° CA)
$\left(\dfrac{d\chi}{d\varphi}+\dfrac{d\chi'}{d\varphi}\right)_{max}$	0.036 8(9° CA)	0.048 2(−1.5° CA)

计算结果表明,当供油提前角从 18.1°CA 增加至 19.2°CA 时,主、副室最大放热率和总的最大放热率均有明显增大。这就是当 $\theta > 18$°CA 时,p_z 急剧上升的原因。

由此看来,既要保证高的经济性,又要降低最高爆发压力,则 θ 取约 18°CA 是适宜的。

2.3 不同供油速率时

我们将高压油泵的柱塞直径由 Φ8.5 mm 改为 Φ8.0 mm,以改变供油速率。多次的性能试验表明,采用前一种柱塞时的标定工况油耗通常要比采用后一种柱塞时低 2.5~4.0 g/(kW·h)。两种情况的放热率计算表明,供油速率较小的,最大总放热率低 10.5%,而且燃烧持续期延长了约 6°CA(图 12),这就是油耗较高的原因。但在燃烧初期,两种柱塞的放热率差别不大,因此适当增加供油速率,不会对初期的压力升高率有太大的影响。

2.4 不同喷油压力时

我们将 S195 型柴油机的喷油压力从 12.5 MPa 提高到 14.0 MPa,放热率的计算结果表

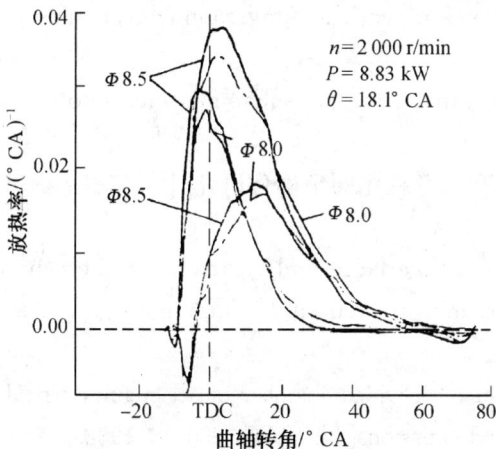

图 12 不同供油速率时的放热率 图 13 不同喷油压力时的放热率

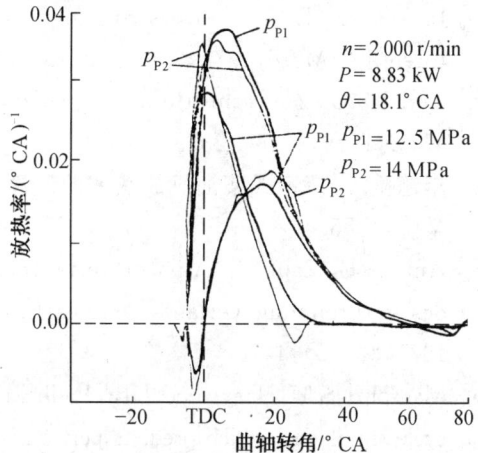

明，主室的最大放热率增大 10.2%，涡流室的最大放热率增大 21.3%（见图 13），但总的放热率变化幅度不大，而且这两种喷油压力下的放热率变化规律相近。

测取两种不同喷油压力下的示功图的试验表明，在喷油压力 $p_P = 12 \sim 14$ MPa 的范围内，改变喷油压力对气缸压力、油管压力、针阀升程等没有明显的影响。但当喷油压力低于 12 MPa 时，油耗有所升高，最高爆发压力有所下降；当 $p_P > 14$ MPa 时，油耗和最高爆发压力均有所增加。因此，从性能的角度来看，喷油压力取 12 $\sim$ 14 MPa 为宜。

总之，试验和计算结果表明，喷油压力在 12 $\sim$ 14 MPa 范围内的性能和放热特性是相近的。

3 结 语

根据以上分析，可以得出如下几点主要结论：

(1) 随负荷的减小，主室的累积放热率减小，燃烧持续期缩短；涡流室的累积放热率增加，燃烧持续期延长。不同负荷对燃烧始点影响不大。

(2) 为了既要保证高的经济性，又要降低最高爆发压力，涡流室式柴油机必须取适当的供油提前角。S195 型涡流室式柴油机 θ 约为 18° CA。

(3) 采用直径为 $\Phi 8.5$ mm 的柱塞能降低油耗，而对初期的压力升高率影响很小。因此，采用 $\Phi 8.5$ mm 的柱塞对改进性能有利。

(4) 喷油压力在 12 $\sim$ 14 MPa 范围内对性能和放热特性影响不大。

参 考 文 献

[1] Woschni G. Die berechnung der wandverluste und der thermischen belastung der bauteile von dieselmotoren[J]. *MTZ*,1970,31(12):491.

[2] 徐錫洪. 副室付きデイーゼル機関の熱力学特性[J]. 内燃機関,1965(5).

[3] Nakajima K. An experimental research on the air swirl motion and combustion in the swirl chamber of diesel engines[C]// *XII FISITA Congress*，Barcelona，1968.

[4] Jankov R. Energie-und strömungsverhältnisse in verbrennungsräumen von dieselmotoren[J]. *MTZ*,1972,33(7)：281.

[5] Terada K. Zur ermittlung der heizverläufe in einem nebenkammer-dieselmotor[J]. *MTZ*,1979,40(5):237.

[6] 李德桃. 涡流室式柴油机放热率的计算方法和计算程序的精确化研究[R]. 镇江：江苏工学院,1985.

[7] Anisits F，Zapf H. Auswertverfahren der druckverläufe und elektronische berechnung des verbrennungsverlaufs in dieselmotoren mit uterteilten brennräumen[J]. *MTZ*,1971,32(12):447.

[8] Mansourt S H，Heywood J B，Radhakrishnan K. Divided-chamber diesel engine，part I：a cycle simulation which predicts performance and emissions[J]. *SAE*，820273,1982.

[9] Kort R T，Mansourt S H，Heywood J B，Ekchian A. Divided-chamber diesel engine，

part II: experimental validation of a predictive cycle simulation and heat release analysis [J]. *SAE*, 820274, 1982.

[10] 高先声,林德嵩,陈彦生. 柴油机示功图的测定及其对放热规律计算精度的影响[C]// 中国内燃机学会节能、净化、燃烧专业委员会专题讨论会,1982.

（本文原载于《农业机械学报》1987 年第 1 期）

Comparative analysis of properties of heat release and performance of swirl chamber diesel engine

Abstract: This paper describes the calculation of properties of heat release of S195 type diesel engine under different conditions by the combination model. Calculation results and performance experimental results are analysed and evaluated synthetically. Therefore, a scientific basis for improving the performance of this diesel engine are provided.

涡流室式柴油机在冷起动条件下
非稳态燃烧过程的研究

李德桃,何晓阳

[摘要] 本文报道了利用高速摄影和连续测录瞬态燃烧示功图的方法,对涡流室式柴油机在冷起动条件下初次着火后的非稳态燃烧过程进行研究的结果。阐述了三种典型的非稳态燃烧型式与燃烧压力的关系;讨论了非稳态燃烧时的着火特性、空气涡流运动特性和柴油机的变速特性;建立了非稳态燃烧过程的物理模型。在此基础上,阐明了涡流室式柴油机冷起动困难的基本原因,为改善这类柴油机的冷起动性能指出了方向和途径。

涡流室式柴油机冷起动过程的深入研究对于进一步改善其起动性能是十分必要的。

一些研究者对冷起动着火过程做过一些研究[1-4]。但是,着火并不一定自持[5]。柴油机起动时自持与否和初次着火后的非稳态燃烧过程的进展有密切关系。小林昭夫等曾利用高速摄影对直喷式柴油机在冷起动条件下的燃烧进行观测,弄清了获得较好的起动性能所必备的条件[6]。然而,据作者所知,迄今对涡流室式柴油机初次着火后的非稳态燃烧过程研究甚少。本文通过对吊钟型涡流室内初次着火后的非稳态燃烧进行高速照相和测录非稳态燃烧示功图,论述了该种柴油机的非稳态燃烧特性,分析了涡流室式柴油机的冷起动机理,建立了非稳态燃烧过程的物理模型,为提高这类柴油机的冷起动性能寻求合理的途径。

1 试验装置与试验方法

表 1 列出了试验发动机的技术规格和试验条件。试验机的压缩比受到控制,使燃烧室内的空气温度在压缩终了时约为 650 K,以模拟发动机在 -5 ℃的环境温度条件下起动。由于条件的限制,机油、冷却水均未作冷却处理。

为了尽可能地拍摄实际发动机的燃烧照片,我们在原产品气缸盖上直接开窗观察涡流室内的燃烧情况。观察用气缸盖如图 1 所示。图 2 示出了观察用涡流室形状和观察范围。此范围约为涡流室最大平面的 81.87%。

本试验使用 E-10 型 16 mm 旋转棱镜式高速摄影机,拍摄频率为 300～10 000 f/s。摄影的光学系统和照明布置见图 3。试验装置简图见图 4。主、副燃烧室内都装有压力传感器,用16 线光电示波器连续记录燃烧压力,同时检测喷油器的针阀升程,测定动态喷油提前角。把磁电式转速传感器传来的脉冲信号,即周期性电压转换成频率而测出发动机的瞬时转速。试验机是在室温下用电动机拖动的,并利用摩擦副控制拖动转速。

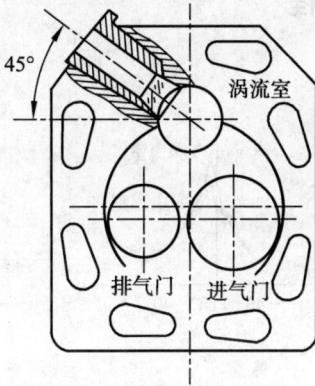

图 1　观察用气缸盖简图

表 1　试验发动机的技术规格和试验条件

项　目	规　格
发动机型号	S195
发动机型式	水冷、四冲程、涡流室
缸径×行程/(mm×mm)	95×115
工作容积/L	0.815
涡流室容积/cm³	25.758
压缩比	19.16(可变)
标定功率/kW	8.8
标定转速/(r/min)	2 000
喷油泵型号	0 号齿条泵
柱塞直径/mm	8
喷油嘴型式	4S1 轴针式
喷油器开启压力/MPa	12.75
静态喷油提前角/° CA	BTDC 17.5
润滑油	14 号机油
燃油	0 号柴油
环境温度/K	278

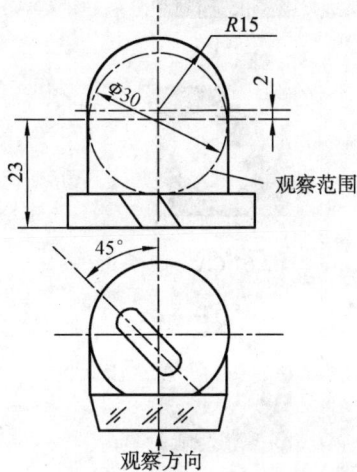

图 2　观察用涡流室形状和观察范围

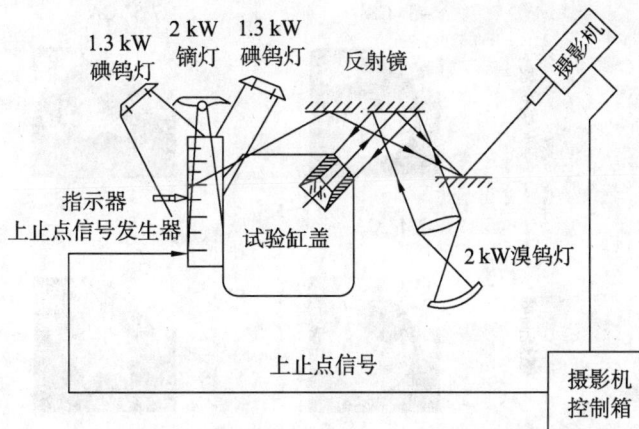

图 3　摄影的光学系统和照明布置简图

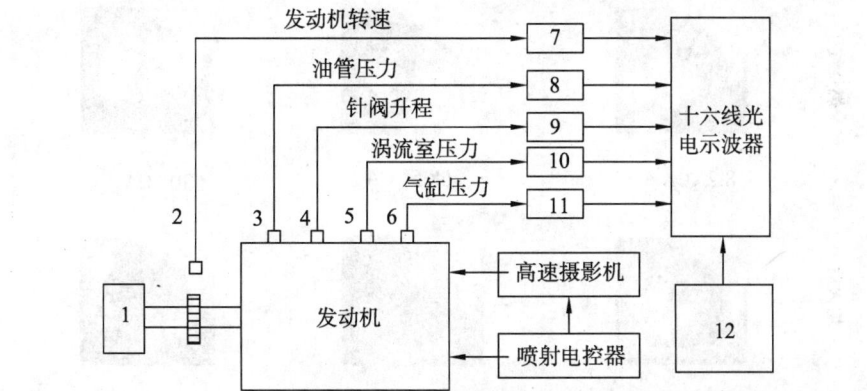

1—电动机　2—磁电式转速传感器　3—油管压力传感器　4—针阀升程传感器
5,6—SYC-1000 型压力传感器　7—FDH-4 型电荷放大器　8—FDH-2 型电荷放大器
9—针阀升程仪　10,11—FDH-7 型电荷放大器　12—XD7 低频信号发生器

图 4　试验装置简图

2 典型的非稳态燃烧过程

2.1 初次着火时的燃烧过程——H 型燃烧过程

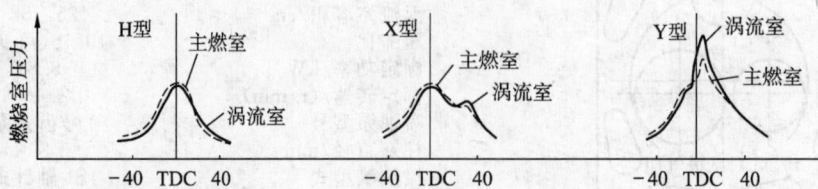

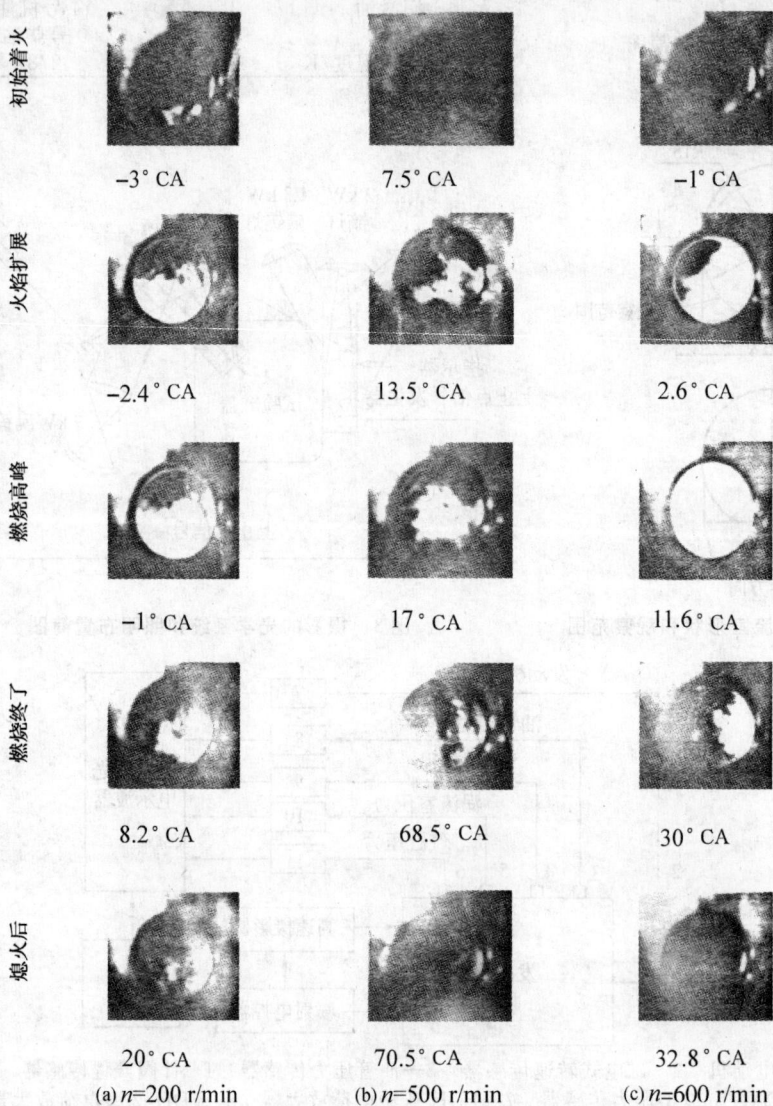

图 5 典型的非稳态燃烧型式

由图 5a 可见:

（1）以毫秒计的着火延迟较长，约为正常运转情况下的 6 倍左右。涡流室内在上止点前 $3°CA$ 开始着火，着火仅在靠近连接通道和起动孔附近的燃烧室底部发生，并出现多个着火核。这同新村惠一对慧星 V 号燃烧室的观察结果相吻合[7]。在这种燃烧过程中，主燃室可能比涡流室内先着火，因为主燃室的燃烧起点比涡流室早 $1.5°CA$。

（2）由于此时的燃烧室壁温较低，燃烧条件不好，仅有部分燃油参与燃烧。从照片上看，火焰始终覆盖不了涡流室上部，燃烧仅在涡流室的部分空间内进行。火焰在上止点就开始衰减，并在上止点后 $8.2°CA$ 就较早地熄灭。

（3）主燃室的燃烧峰值压力大于涡流室峰值压力，发生在连接通道处的二次窒息现象，延缓主燃室的二次燃烧，导致缸内压力升高较少。此时的摩擦力矩又较大，燃烧产生的爆发力矩不足以克服摩擦力矩，发动机转速不会有明显的升高（约等于拖动转速）。

2.2　不稳定燃烧过程——X 型燃烧过程

由图 5b 可以看出以下几点：

（1）以曲轴转角计的着火延迟期长，主、副室都在上止点后 $7.5°CA$ 才开始着火。由于涡流室内的着火条件随柴油机转速升高而发生变化，着火核出现在涡流室中、下部。

（2）由于燃烧在活塞下行时才开始，气缸内的传热损失急剧增加，燃烧速率降低。照片上，涡流室内的火焰呈纤维状；主副室间的压差极小，使涡流室内的高温燃气的喷出速率较低，引起主燃室的燃烧变差。后燃严重，火焰一直持续到上止点后 $68.5°CA$ 才熄灭。

（3）燃烧压力曲线呈马鞍型。这时的爆发力矩显然较小，不能克服摩擦力矩。这样，加速力矩趋近零甚至为负值，其外部特征是发动机滞速甚至转速降低。

2.3　旺盛燃烧过程——Y 型燃烧过程

Y 型燃烧过程照片和燃烧压力曲线示于图 5c 中。它表示：

（1）着火向上止点靠近。以毫秒计着火延迟缩短，为正常燃烧时的 1/2 左右。

（2）涡流室内常出现多个着火核，并分布在靠近室壁的高温环带（图 6[8]）区域内。

（3）着火后，火焰迅速扩展并覆盖全部涡流室。这种燃烧照片分布在 TDC—11°CA 之间且火焰亮度大。说明涡流室内燃烧旺盛，能产生较高的火焰喷出速率，使主燃室内的燃烧完善。从燃烧曲线上可以看到，最大燃烧峰值压力是最大压缩压力的 1.5 倍以上，表明 Y 型燃烧过程能产生较高的爆发力矩；另外，旺盛燃烧能产生较高的温度，使柴油机的水温和油温升高，从而降低了柴油机的摩擦力矩。上述两者的结合，可使转速有较大升高。

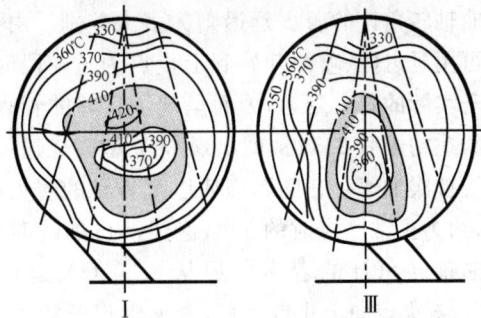

图 6　涡流室内的平均温度分布

（4）燃烧持续期适中。在上止点后 $30°CA$，涡流室内的火焰熄灭，后燃较少。所喷射的燃油大部分参与燃烧，熄火后已几乎观察不到白烟。

3　非稳态燃烧过程中的燃烧特性

图 7a 和 7b 分别示出了吊钟型涡流室内初次着火后的非稳态燃烧过程的燃烧照片和燃烧

特性。现详细分析如下：

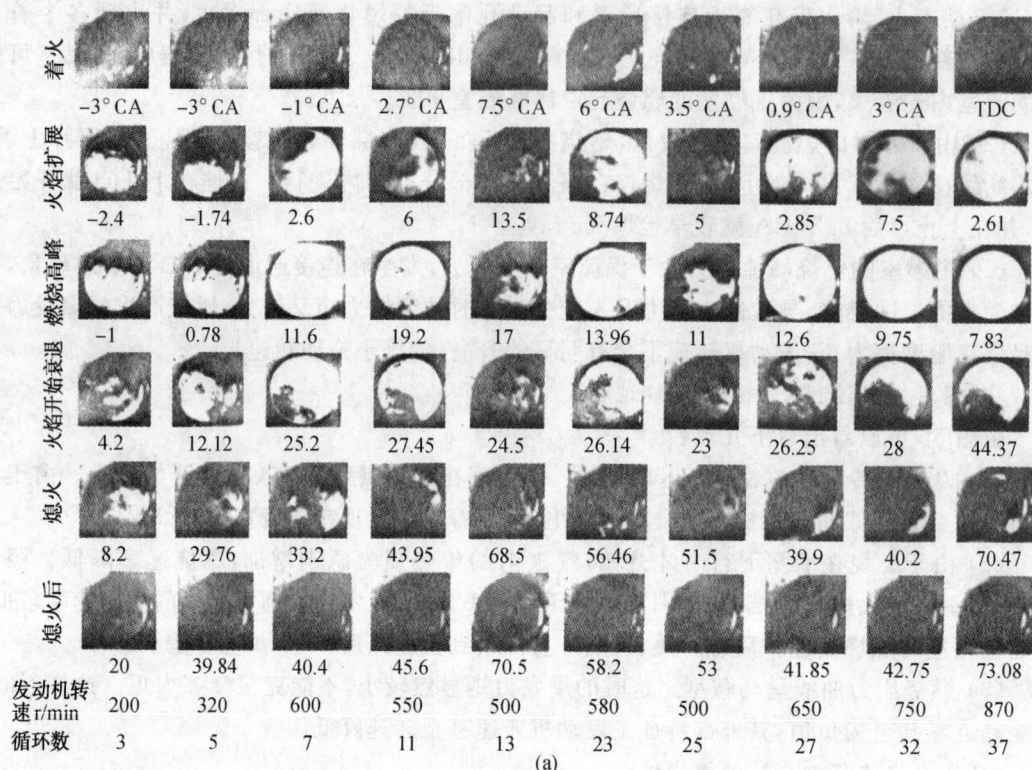

(a)

3.1　着火特性

如图7b所示，在起动初期，柴油机转速较低，以毫秒计的着火延迟比柴油机正常运转时长得多，但以曲轴转角计的着火延迟则无多大差别。由于此时的可用延迟（从喷油开始到上止点为止一段时间[9]）大于实际的着火延迟，因而呈现了良好的着火特性。大约在初爆后2.35 s，就出现了Y型燃烧过程，使柴油机首次加速。旺盛燃烧使以毫秒计的着火延迟缩短，约为正常燃烧时的1/4，但柴油机的瞬时升速却以曲轴转角计的着火延迟从8.5° CA延长到18° CA。着火点向上止点后移，着火延迟开始大于可用延迟。因此，自第11个循环后，出现了在上止点后7.5° CA才开始着火的X型燃烧过程，看不到燃烧压力上升；衰弱的燃烧使燃烧室壁温不会有明显升高，着火条件得不到改善，着火延迟再度变长。着火时刻的落后，使着火延迟与可用延迟之间的差距增大，着火特性恶化，从而连续出现了在上止点后4～7° CA才开始着火的不稳定燃烧循环，发动机滞速甚至转速下降。此时，以毫秒计的着火延迟在±3～

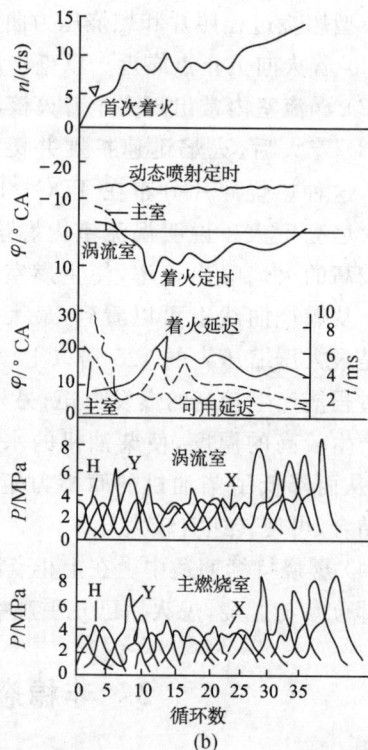

(b)

图7　非稳态燃烧过程的照片和燃烧特性

±1 ms,以曲轴转角计的着火延迟在±6°CA的范围内随转速的波动而上下变动,但其平均值大致保持一定(约4.2 ms)。随着燃烧循环的累积,燃烧室内的温度会有所提高,化学延迟大大缩短,使着火延迟自第25个循环开始持续下降到2.01 ms,渐渐趋近于可用延迟,出现连续的Y型燃烧过程,使发动机再次加速,再次呈现了良好的着火特性。

从燃烧照片上可以看到,起动初期循环,涡流室内的着火核都是在质交换和热交换条件较好的、靠近主燃室连接通道和起动孔附近的涡流室底部。着火位置逐渐顺气流向连接通道上方的涡流室高温环带处移动。这可能是涡流室内的温度分布发生变化,从而最宜着火的地点也发生了变化的缘故。

从上述分析可知,涡流室式柴油机冷起动时的着火特性对初次着火后非稳态燃烧过程的燃烧影响很大。作者认为,良好的燃烧将由较好的着火特性所提供。涡流室式柴油机之所以冷起动困难,原因之一是起动初期燃烧循环的着火特性较差。良好的着火特性是,在初次着火后的所有燃烧循环,都应在上止点前开始着火,即应使燃料喷雾的实际着火延迟始终小于或等于可用的着火延迟。这可以通过燃油的喷射特性与着火特性的优化匹配,改善起动过程初期的着火条件,缩短该阶段的实际着火延迟来实现。

3.2 涡流运动特性

燃烧室内的空气涡流与混合气形成质量和燃烧效率是密切相关的。所以,深入研究非稳态燃烧时的空气涡流运动有着重要的意义。由于起动时的火焰传播速度比正常运转时弱得多,因而我们把非稳态燃烧过程中的火焰旋转速度近似地看作空气的运动速度,这样就可以从高速照片中测得着火后的涡流特性瞬时值,以帮助定性地分析涡流强度和涡流形态。

3.2.1 涡流强度

从表2列出的涡流比数据,我们得到以下几点认识:

表2 涡流特性数据

循环数	发动机转速 $n/(r/min)$	曲轴转角 ATDC/(°CA)	测点在燃烧室直径上的位置	N/n (粗略值)
5	320	16	0.5r	24.7
			0.8r	19.46
			1.0r	7.23
7	600	16	0.5r	41.40
			0.8r	26.84
			1.0r	12.33
11	550	15	0.5r	33.72
			0.8r	22.39
			1.0r	10.19
13	500	17	0.5r	31.08
			0.8r	20.20
			1.0r	9.34

循环数	发动机转速 $n/(r/min)$	曲轴转角 ATDC/($°$CA)	测点在燃烧室直径上的位置	N/n（粗略值）
25	500	15	$0.5r$	31.48
			$0.8r$	19.51
			$1.0r$	8.96
27	650	16	$0.5r$	43.80
			$0.8r$	27.92
			$1.0r$	13.14
35	750	17	$0.5r$	47.63
			$0.8r$	30.10
			$1.0r$	14.13
37	870	15	$0.5r$	52.44
			$0.8r$	37.10
			$1.0r$	16.14

注：N/n ——空气旋转角速度/发动机转速（涡流比）；

r ——燃烧室中心至壁面的距离，mm。

（1）涡流比随柴油机转速增加而增加，其变化范围约在 7～55 之间。

（2）在非稳态燃烧过程中，涡流比的变化规律与柴油机转速的变化相当一致（见图 8）。

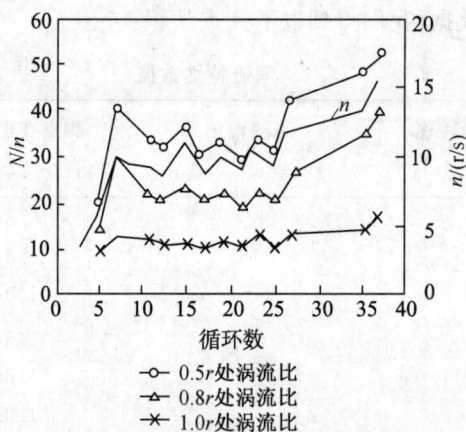

图 8　非稳态燃烧过程中的涡流比变化

（3）涡流比大小影响火焰的扩展形态。从燃烧照片中可以看到，初次着火后的若干个循环的弱涡流火焰多呈块状；此时的火焰径向扩散速度约为 35～45 m/s，周向扩散速度仅为 5～15 m/s。这时涡流室内的混合和燃烧均不好，对起动不利。随着涡流转速的提高，燃烧室内的空气利用率变好，顺利地进行了热混合，此时的火焰呈旋涡状；火焰的径向扩散速度约减少到 20～25 m/s，而火焰的周向扩散速度则增加到 25～60 m/s，此时的燃烧效率高，对起动有利。

3.2.2　涡流形态

从非稳态燃烧过程中选择了两组特征速度下的燃烧照片进行测量，测得的数据如表 3 所示。可知：

表 3 不同特征速度下涡流特性数据

特征速度 （r/min）	曲轴转角/° CA	测点在燃烧室 直径上的位置	N/n	涡流旋转速度 V_t/(m/s)
320	BTDC 4	0.5r	32.44	8.15
		0.8r	26.14	10.51
		1.0r	22.51	11.43
320	TDC	0.5r	27.45	6.89
		0.8r	23.78	9.56
		1.0r	20.00	7.52
320	ATDC 16	0.5r	24.70	6.21
		0.8r	19.46	7.83
		1.0r	7.23	8.63
650	ATDC 0.9	0.5r	52.39	26.75
		0.8r	34.87	28.48
		1.0r	14.97	15.28
650	ATDC 16	0.5r	48.80	22.36
		0.8r	29.92	22.81
		1.0r	13.14	13.42

注：N/n——空气旋转角速度/发动机转速；

　　r——燃烧室中心至壁面的距离,mm。

（1）涡流旋转速度大致与测量点的半径成反比,亦即线速度大致不变(见图 9)。

（2）涡流形态随曲轴转角而变化(见图 10)。

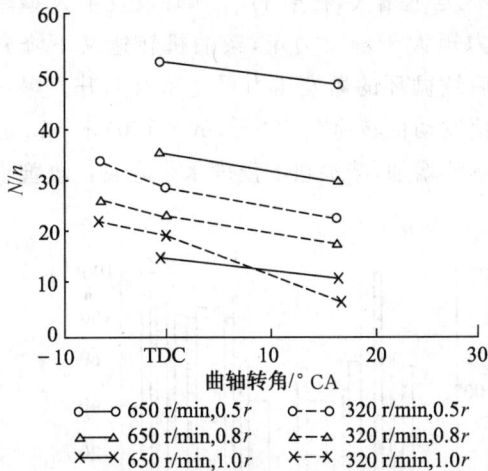

图 9 吊钟型涡流室内的涡流比(非稳态)

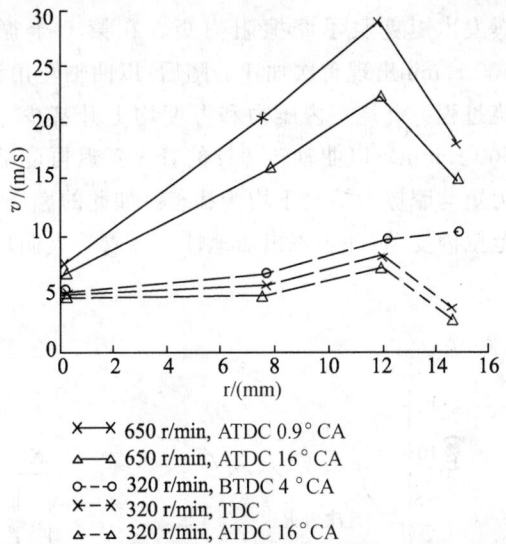

图 10 吊钟型涡流室中的涡流运动结构

（3）涡流形态与柴油机转速无关。

（4）涡流形态在整个非稳态燃烧过程中基本保持不变(图 11)。

3.3 变速特性

曲轴转速的变化,是柴油机起动时一个重要的外部特征。它既受燃烧的制约,又反作用于燃烧。

试验柴油机在非稳态燃烧过程中的速度变化示于图12。现根据其特征划分成几个阶段:初始段、第1次加速段、滞速段和第2次加速段。

柴油机自持过程中诸力矩的关系式为

$$加速力矩＝爆发力矩－摩擦阻力矩$$

Ⅰ	转动曲轴
Ⅱ	初始段
Ⅲ	第1次加速段
Ⅳ	滞速段
Ⅴ	第2次加速段

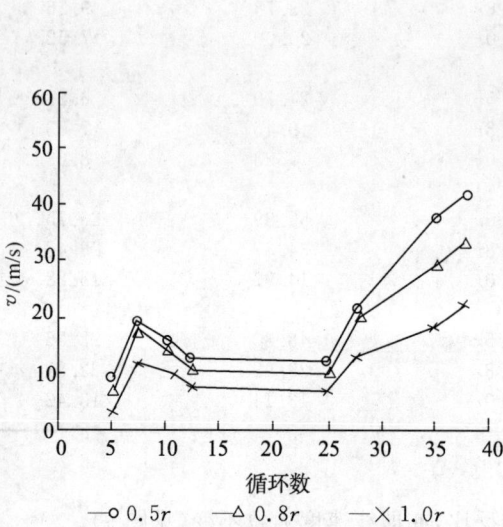

—○0.5r   —△0.8r   —×1.0r

图11 非稳态过程中的涡流运动形态

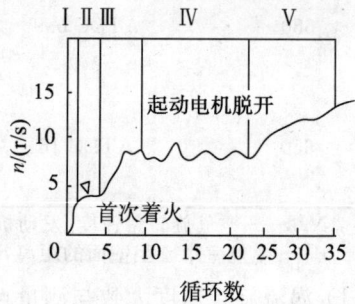

图12 非稳态燃烧过程中的速度变化

如图13所示,在初始段发生着火循环时,着火是在上止点前,立即出现了 Y 型燃烧过程。爆发力矩克服了摩擦阻力矩。在第七个循环就使柴油机转速从 200 r/min 迅速增加到 600 r/min出现首次加速。随后,以曲轴转角计的着火延迟增大,在第 11 个循环就发生 X 型燃烧过程。燃烧室内压力和温度均上升甚少。摩擦力矩大于爆发力矩,柴油机转速又下降到 500 r/min。以曲轴转角计的着火延迟再度缩短。后续循环的爆发压力再度稍有上升。爆发力矩与摩擦力矩处于均衡状态。如此的燃烧变动,使发动机转速在 $500\pm(60\sim100)$ r/min 的范围内变动,即表现出滞速性。随着着火循环数的继续增加,柴油机的温度水平升高,改善了

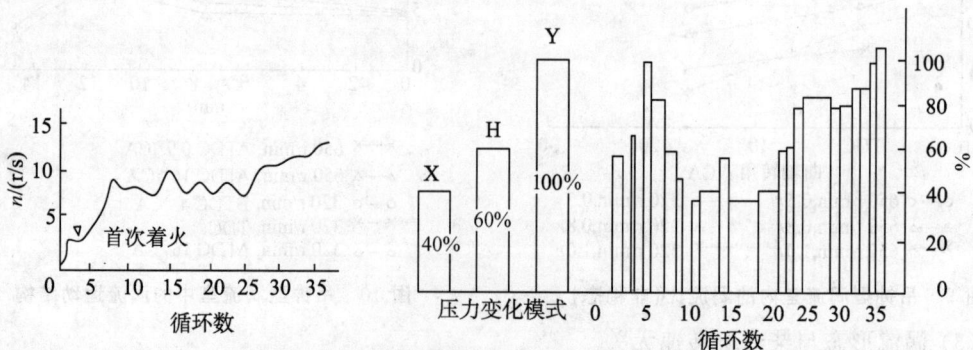

图13 非稳态燃烧过程中的速度变化及其压力变化模式

着火条件,使着火延迟缩短,着火始点向上止点处移动,发生连续的 Y 型燃烧过程,爆发力矩增加,大大超过了摩擦阻力矩,柴油机又开始加速。

由此看来,非稳态燃烧时升速是由于爆发力矩不断超过摩擦阻力矩;滞速的发生,则是两者达到均衡状态的缘故;柴油机转速变化呈现波动性特征,是由于燃烧产生的爆发力矩和柴油机的摩擦力矩之差的变化而产生的。作者将非稳态燃烧过程中涡流室式柴油机与文献[6]所介绍的直喷式柴油机的速度变化(图 14)比较得知,在直喷式柴油机中,即使较差的起动过程,也仅在第 20 个循环才发生滞速现象。而涡流室式柴油机在冷起动过程中,一般都在第 7~10 个循环就会发生滞速现象。由此可以推论,涡流室式柴油机之所以冷起动困难,主要是由于滞速发生在柴油机温度水平很低的较早循环,此时燃烧室内的着火条件更差,燃烧效率低,从而使滞速阶段的时间增长,导致起动时间延长。所以,改善滞速阶段的着火条件,强化滞速阶段的燃烧,缩短滞速时间,是我们改善涡流室式柴油机冷起动性能的重要途径。

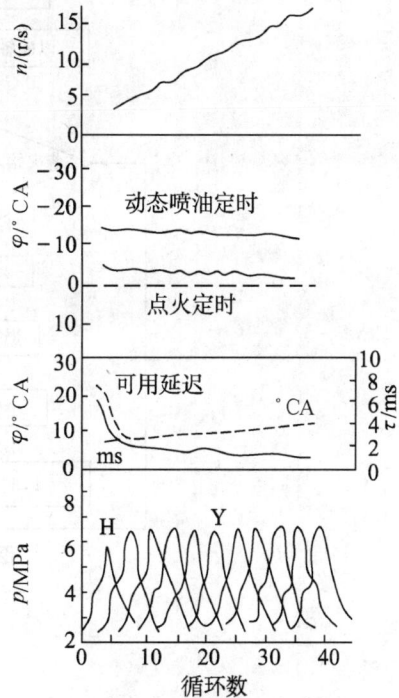

综合上述的分析可知,理想的非稳态燃烧特性是:过程中所有循环的实际着火延迟都应短于或等于可用着火延迟;初次着火后要连续发生 Y 型燃烧,柴油机转速要持续上升,如图 15 所示。这正是提高涡流室式柴油机冷起动性能的努力方向。

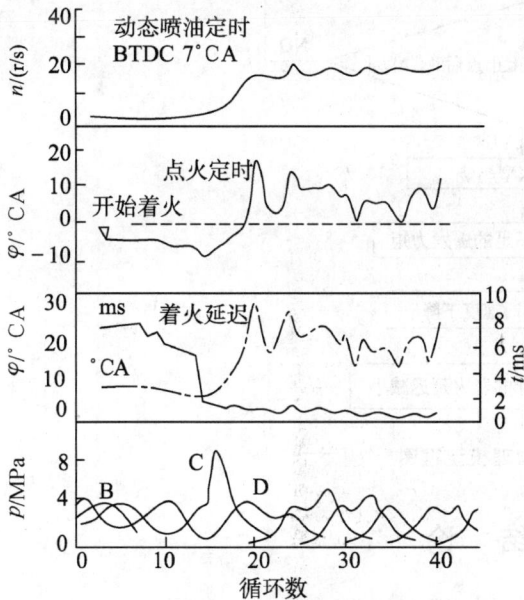

图 14　较差的直喷式柴油机的起动性能　　　图 15　理想的非稳态燃烧特性

4　非稳态燃烧过程的物理模型

根据上节的分析,我们可以建立涡流室式柴油机初次着火后的非稳态燃烧过程的物理模型,如下面的图 16 所示:

```
                    ┌──────────────┐
                    │   起动开始    │
                    └──────┬───────┘
                           │
         ┌────────►┌──────────────┐
         │         │ 以转速 n₀ 拖动 │
         │         └──────┬───────┘
         │                │
    NO   │            ◇─────────◇
    └────┴───────────  发生H型燃烧  ◄──────────┐
                       ◇─────────◇            │
                           │ YES              │
                           │                  │
                    ◇──────────────◇          │
                    以毫秒计的着火延迟小于3 ms ──NO─┘
                    ◇──────────────◇
                           │ YES
                    ┌──────────────┐
              ┌────►│  发生Y型燃烧   │
              │     └──────┬───────┘
              │     ┌──────────────┐
              │     │ 燃烧产生足够的爆发力矩 │
              │     └──────┬───────┘
              │     ┌──────────────┐
              │     │ 发动机瞬时速度升高 │
              │     └──────┬───────┘
              │     ┌──────────────┐
              │     │ 以曲轴转角计的着火延迟增加 │
              │     └──────┬───────┘
              │            │
              │     ◇──────────────◇
              │     着火始点发生在上止点后3°CA以上 ──NO─┘
              │     ◇──────────────◇
              │            │ YES
              │     ┌──────────────┐
              │     │  发生X型燃烧   │
              │     └──────┬───────┘
              │     ┌──────────────┐
              │     │ 燃烧产生不足的爆发力矩 │
              │     └──────┬───────┘
              │     ┌──────────────┐
              │     │ 发动机瞬时速度下降 │
              │     └──────┬───────┘
              │     ┌──────────────┐
              │     │ 以曲轴转角计的着火延迟减小 │
              │     └──────┬───────┘
              └────────────┘
```

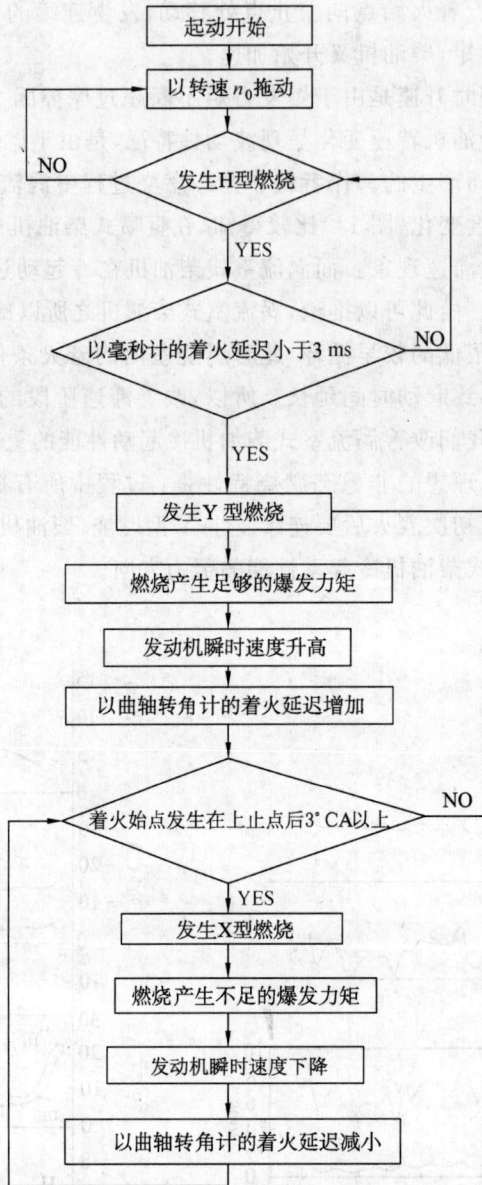

图 16　物理模型框图

5　结　论

（1）在非稳态燃烧过程中，产生较高爆发力矩的良性循环和产生低爆发力矩甚至熄火的不良循环往往是交替进行的。这是由于着火条件随转速变化而引起的。

（2）旺盛燃烧必须有良好的着火特性。良好的着火特性是，所有循环的实际着火延迟都小于或等于可用着火延迟。

（3）非稳态燃烧过程中涡流室内的空气涡流形态，与柴油机正常运转情况下基本一致。

（4）连续的旺盛燃烧是发动机持续升速而达到自持的重要保证。

（5）非稳态燃烧过程中柴油机的速度变化具有波动性特征，这是由于燃烧产生的爆发力矩与柴油机的摩擦力矩之差的变化而产生的。非稳态燃烧过程中存在着初始、加速和滞速时期。初始时期和滞速时期的延长将导致起动时间的增加。

（6）涡流室式柴油机冷起动困难的基本原因是，在初次着火后初期循环的着火特性较差；滞速发生在起动时着火特性较差的早期循环，导致了滞速时间的增加。强化初期循环的燃烧，缩短滞速时间，是改善涡流室式柴油机冷起动性能的合理途径。

对常州柴油机厂为试验加工提供方便，对单春贤、朱广圣、康志新、李捷辉和许金顺同志在试验中的大力协助，在此一并鸣谢。

参 考 文 献

[1] Austen A E W, Lyn W T. Some investigations on cold-starting phenomena in diesel engines [J]. *Proc Inst Mech Engrs (Auto, Div)*, 1959–60, 172(5).

[2] BiddulphT W, Lyn W T. Unaided starting of diesel engines[J]. *Proc Inst Mech Engrs*, 1966–67, 181(1).

[3] 深沢正一. デイーゼル機関の始動性について[C]//第 1 報:始動としリアションの関係. 日本機械学会論文集, 1961, 27(180): 1361–1369.

[4] 深沢正一. デイーゼル機関の始動性について[C]//第 2 報:始動時の着火に影響する諸因子について. 日本機械学会論文集,, 1979, 33(9): 1369–1376.

[5] Komiyama K, et al. Investigations on cold-starting mechanism in diesel engine [C]. 13*th* *CIMAC Congress*, Vienna, 1979: 1369–1376.

[6] 小林昭夫, ほか. 高速度写真による自動車用デイーゼル機関の始動時燃焼解析[J]. 自動車技術, 1979, 33(9): 789–796.

[7] 新村惠一. 渦流室式デイーゼル機関のアイドル燃焼観測[J]. 内燃機関, 1981, 20(4).

[8] Nagao F, et al. Air motion and combustion in a swirl type diesel engine[J]. *Bulletin of JSME*, 1967, 10(4): 833.

[9] Lyn W T. 柴油机燃烧研究(二)[J]. 国外内燃机, 1980(3): 8–9.

（本文原载于《内燃机学报》1987 年第 3 期）

Investigation on unstable state combustion in swirl chamber diesel engine under cold-starting condition

Abstract：This paper presents the results of investigation on unstable state combustion in a swirl chamber diesel engine under cold-starting condition by high-speed photography and continuously recording the transient indicator diagram. The relation between three types of unstable state combustion and combustion pressure is discussed. The characteristics of ignition, swirl motion and variable engine speed are analysed. The physical model of unstable state combustion is introduced.

This article also gives reasons for the difficulty in starting swirl chamber diesel engines in cold-starting condition, points out the way to improve the cold startability.

Determination of heat transfer coefficient in the swirl-chamber diesel cylinder

Li Detao, *Zhu Xiaoguang*, *Shan Chunxian*

Abstract

Measurement and analysis of the temperature field of the cylinder liner were conducted for a single-cylinder, water-cooled swirl-chamber diesel engine. A new equation is obtained to calculate the heat transfer coefficient inside the cylinder liner. A new method of welding thermocouples is described. The effects of ceramic coating on the outer surface of the cylinder liner on the temperature field and fuel consumption are discussed.

1 Introduction

The heat transfer process of IC engines has been investigated for over fifty years. Only during the last twenty years, an equation was suggested for calculating the heat transfer coefficient in the swirl chamber diesel engine. Knight B E[1], Hassan H[2], Kamel M[3], Quo Qiyi[4] measured the transient temperatures on the wall of the swirl chamber and cylinder head, and obtained the equations for calculating the local average transient heat transfer coefficients in the main and swirl chambers. But there is a big temperature difference between the cylinder head and the inner surface of cylinder liner. Even for the same cylinder liner, there are differences among areas. It would be ideal to calculate the heat transfer coefficient in the main chamber using local transient temperature of the cylinder head. However, it is impossible to measure the transient temperature with the thermocouple on the inner surface due to the piston movement. It is also very difficult to accurately measure the steady state temperature distribution for the cylinders with and without ceramic coating.

To solve the problems described above, a new method of welding the thermocouples to the cylinder was developed to ensure the precision of repeated measurements. The steady state temperature field of cylinder liner is evenly divided. The transient temperature field of the cylinder liner is analyzed. A new equation is developed to calculate the heat transfer coefficient in the main chamber. Applying the equation to a single-cylinder, water-cooled swirl chamber diesel engine shows good results.

2　Thermocouple welding and temperature measurement

2.1　Thermocouple welding

In order to weld the 0.35 mm copper-constantan thermocouples to the cylinder, a special Stored-Energy Welder S1250 was developed. It has the advantages of high welding quality, lower power consumption, easy operation and no space restriction, etc. The circuit diagram of the welder is shown in Fig. 1.

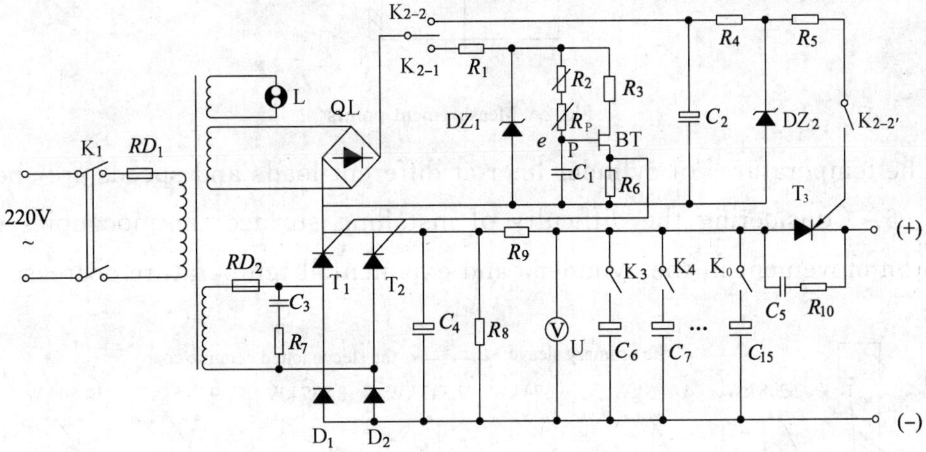

Fig. 1　Circuit diagram of stored energy welder S1250

2.2　Temperature measurement

The temperature field is closely related to engine type, cylinder bore, engine speed, and cooling method, etc. In order to understand the heat transfer in diesel engine S195 which is widely used in China, its cylinder liner temperatures were measured. Another S195 cylinder liner with the zirconium oxide coating at the outer surface was also tested. The temperature fields and fuel consumptions of two cylinders were compared to see the effect of ceramic coating.

Since the temperatures along the cylinder periphery are distributed with the maximum difference not more than 12 ℃, the cylinder liner temperature field is simplified as an axis symmetrical one.

Eighteen thermocouples were installed on the cylinder liner in both radial and axial directions, as shown in Fig. 2. Ten of the thermocouples were placed at the depth of 1.5 mm from the inner surface; others were evenly spaced on the outer surface.

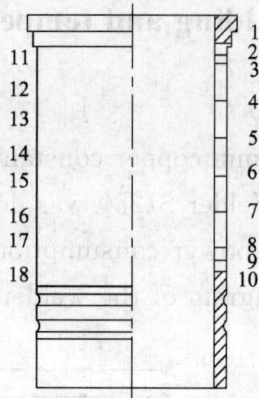

Fig. 2　Measurement points

The temperatures of cylinder liner at different loads and speeds are shown in Fig. 3 – 6. Considering the difficulty of installing surface thermocouples caused by piston movement in the cylinder, and exponential temperature decrease in the

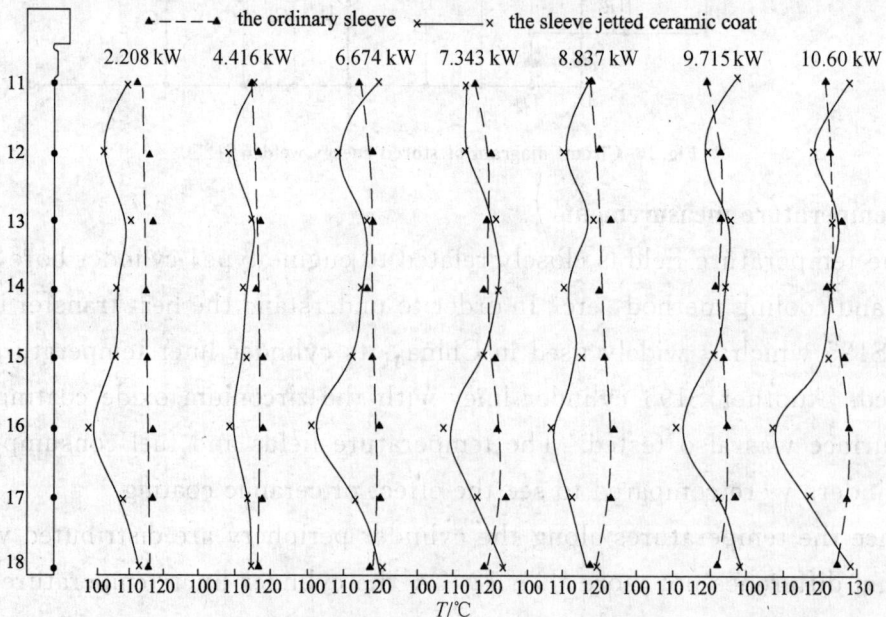

Fig. 3　Temperature distribution of out surface with different loads

wall (the fluctuation value drops to zero at 0.8~1.0 mm from the inner surface),
high welding-quality thermocouples were used to measure the steady state temperatures of the cylinder liner.

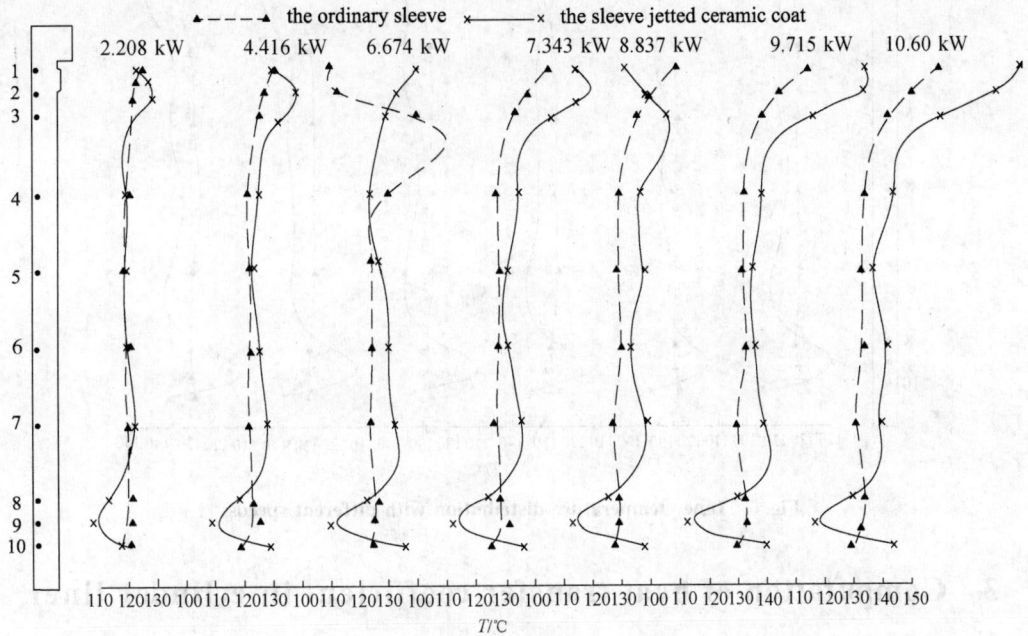

Fig. 4　Inner temperature distribution with different loads

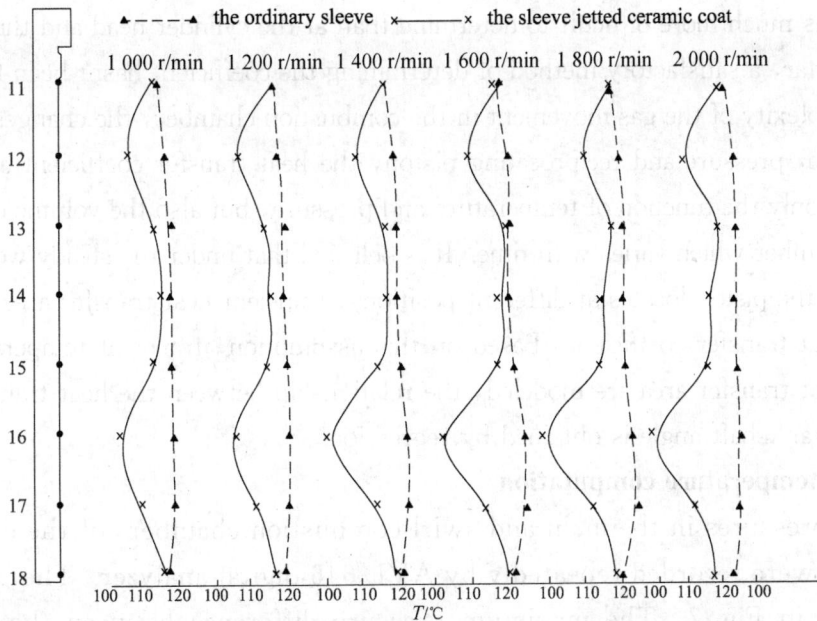

Fig. 5　Temperature distribution of out surface with different speeds

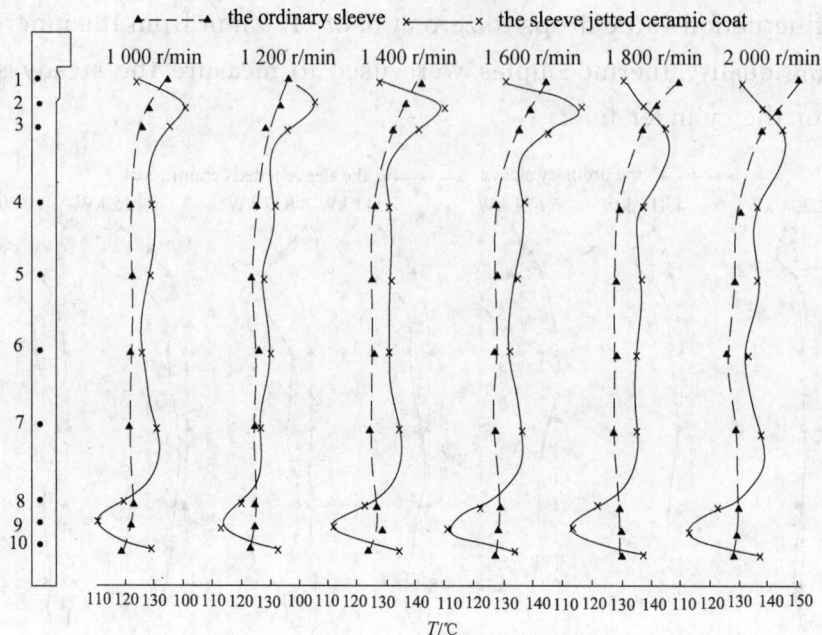

Fig. 6　Inner temperature distribution with different speeds

3　Computation of heat transfer coefficient in cylinder liner

3.1　Analysis of boundary condition

Because of the periodic covering of piston on the inside wall of cylinder, heat transfer coefficient is much more difficult to determine than at the cylinder head and the top of the piston. So far, a satisfactory method of determining the coefficient hasn't been found. Due to the complexity of the gas movement in the combustion chamber, the changes of the gas temperature, pressure and reciprocating piston, the heat transfer coefficient at the inner wall is not only the function of temperature and pressure, but also the volume of the combustion chamber which varies with time. It is believed that under the steady working conditions, as the piston locates at different positions, transient heat transfer area has a local uniform heat transfer coefficient. Based on this assumption, transient temperature fields on local heat transfer area are modeled, the relationship between the heat transfer coefficient and crankshaft angle is obtained by regression.

3.2　Gas temperature computation

The pressures in the main and swirl combustion chambers of the experimental engine were recorded repeatedly by AVL-646 digital analyzer. The p-φ curves are shown in Fig. 7. The maximum pressure difference between the main and swirl combustion chamber is 5×10^{-1} MPa as shown.

To understand the effect of the connecting channel between the main and swirl chambers on the mass distribution of gas mixture in two chambers, the flow coefficient was also measured. The test was done on a steady flow experiment bench. The flow through the connecting channel was simulated. Regressing the test data, the following empirical formula for the flow coefficient is derived:

$$M_{1,2} = \frac{Re}{8 \times 10^4 + 0.94 \cdot Re}$$

$$M_{2,1} = \frac{Re}{3 \times 10^4 + 1.1 \cdot Re}$$

here, $2 \times 10^4 < Re < 1 \times 10^6$.

The transient temperature of combustion gases is calculated using $M_{1,2}$, $M_{2,1}$, gas state equation, and the gas mixture changes in compression, combustion and expansion processes. The relationship between the temperature and crankshaft angle is shown in Fig. 7.

3.3 Finite element analysis of cylinder liner

The temperature field of the chosen area varying with crankshaft angle is modeled by finite element method for the cylinder liners with and without ceramic coating. Fig. 8 shows a finite element grid obtained by superposition of local grids. The computation flow chart is shown in Fig. 9. The combustion temperature is given in Fig. 7, the first boundary condition is used on the outer surface of cylinder liner, and the temperature fields of cylinder liner with and without ceramic coating, under the standard working conditions, are shown in Fig. 10 and 11.

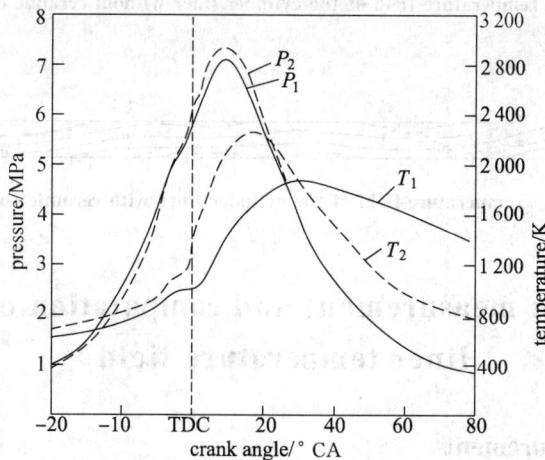

Fig. 7　The *P-φ* & *T-φ* diagram in two chambers of the S195 diesel

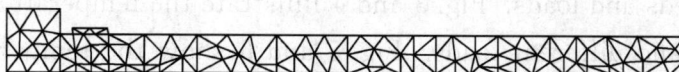

Fig. 8　Finite element grid of cylinder liner

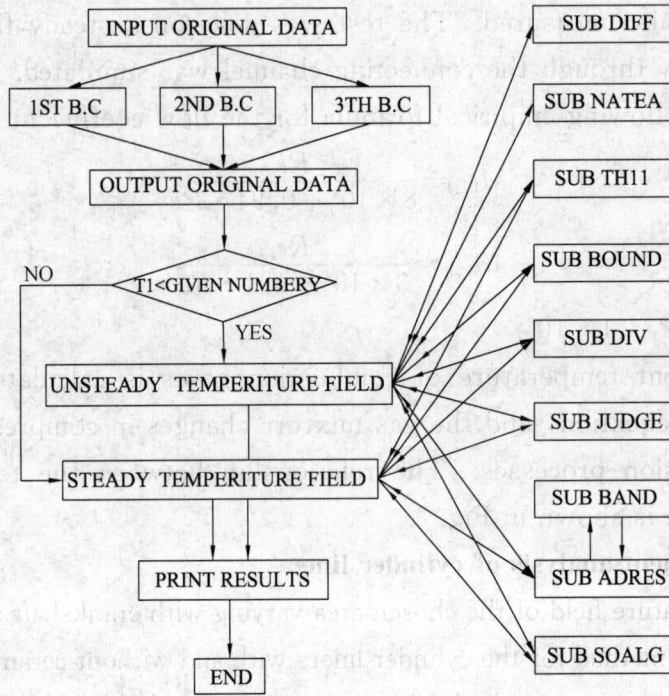

Fig. 9　Computation diagram of finite element method

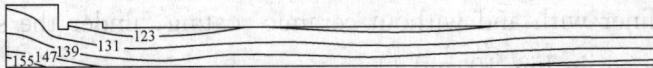

Fig. 10　Temperature field of the cylinder liner without ceramic coating

Fig. 11　Temperature field of the cylinder liner with ceramic coating

4　Results from measurement and computation of the cylinder liner temperature field

4.1　Results of measurement

Cylinder liner temperatures with and without ceramic coating were measured at different speeds and loads. Fig. 3 and 4 illustrate the temperature distributions at 2 000 r/min and different loads. Fig. 5 and 6 show the temperature distributions at different speeds.

As shown in Fig. 3 and 4, the cylinder liner temperature increases with load. For the cylinder liner without ceramic coating, the temperature increases linearly with loads. The highest temperature is 161.5 ℃ at point 1 in Fig. 4. The temperature change rate is also the biggest at the point. The decreasing rate of the temperature inside the cylinder wall reduces from the cylinder top to the bottom. The temperature distribution at the outer surface is almost uniform because it is determined primarily by cooling style and cylinder structure. The higher temperature at the top is due to the contact with burning gas for a long time without direct water cooling. The other parts of the cylinder liner are of even wall thickness with the outer surface directly contacting with the cooling water in evaporation state, the temperature distribution tends to be homogeneous.

For the cylinder liner with ceramic coating, the outer surface has more thermal resistance, thus the temperature at the inner surface is higher than the one without coating. It is also found that the temperature distribution is distorted by the change of the heat flux path. The mechanism of this phenomenon needs to be further studied.

Fig. 5 and 6 show that the cylinder temperature also increases with the speed, but its changing rate is less than that with the load. Under the different speeds and loads, the temperature distributions are similar in shape.

4.2　Analysis of steady temperature field of cylinder liner

Based on temperature measurements, the heat transfer coefficient in cylinder liner is determined with optimization. The temperature field is calculated by finite element method. The maximum difference between modeling and testing is less than 10%. Fig. 10 and 11 illustrate the modeling results of the temperature fields for both cylinder liners at rated conditions, respectively.

It is shown in Fig. 10 and 11 that the temperature inside the cylinder liner without ceramic coating decreases gradually from top to bottom, while the temperature inside the cylinder liner with the ceramic coating fluctuates. Different coating thickness and different ceramic materials result in different temperature distributions. Further work is still being done.

The temperature fields in cylinder liners with and without ceramic coating are calculated with different speeds and loads. The results are similar to those in Fig. 10 and 11. The results confirmed that the temperature increases with the increase of engine speed and load.

4.3 Determination of heat transfer coefficient in cylinder

At rated conditions, the relationship between heat transfer coefficient and crankshaft angle is shown in Fig. 12. The temperature fluctuation at node 4 of the inner surface is depicted in Fig. 13.

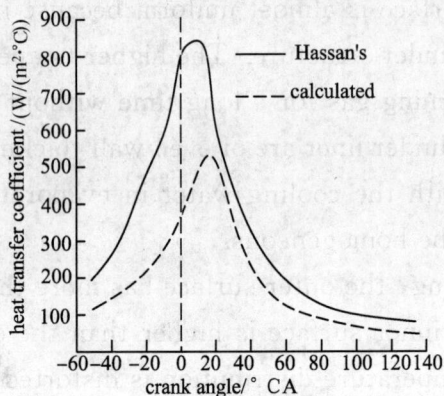

Fig. 12 **Relationship between heat transfer coefficient and crankshaft angle**

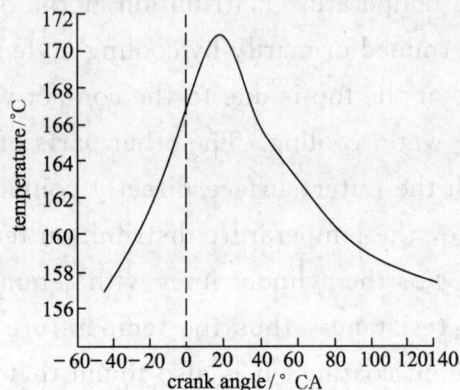

Fig. 13 **Temperature fluctuation at inner surface of the cylinder liner**

From Fig. 12, one can see that the calculated results differ from that of Hassan's. Fig. 13 shows that the calculated temperature fluctuation at the inner surface of the cylinder liner, about 10 to 20 ℃, agrees with the results measured by the film thermocouple[4].

The calculated results are regressed as the following equation:

$$Nu = 0.073\ 5\ Re^{0.65}$$

Using this equation, the heat transfer coefficient of the main chamber is determined for the water-cooled swirl-chamber diesel engine.

References

[1] Knight B E. The problem of predicting heat transfer in diesel engines[J]. *Proc I M E*, 1964, 179(3).

[2] Hassan H. Unsteady heat transfer in a motored I C engine cylineder[J]. *Proc I M E*, 1970, 185.

[3] Kamel M, Watson J. Heat transfer in the indirect injection diesel engine[J]. *SAE Paper*, 790826, 1979.

[4] Quo Qiyi. Experiemental investigation of instantaneous heat transfer coefficient in the indirect injection diesel engine[J]. *Transaction of CSICE*，(324).

[5] Li Detao. An accurate study on calculation method and problem of heat release rate in swirl chamber diesel engines[J]. *Transactions of CSICE*，1986，4.

（From：*Procceding of the International centre for Heat and Mass Transifer*，Yougosl. New York：Hemisphere，1987)

涡流室式柴油机缸套传热系数的确定

[摘要]　通过对气缸套温度场的测量、分析和计算,建立了一个适用于单缸、水冷涡流室式柴油机传热系数的计算公式,给出了电容储能式热电偶焊接机的设计原理图,该焊接机可确保热电偶的焊接质量,并对原气缸套和外表面喷涂陶瓷层的气缸套温度场进行了测量和分析,探讨了气缸套温度的变化对燃油消耗率的影响。

ZS4S1 喷油嘴在无旋流场中喷注贯穿的研究

李德桃，朱亚娜

[摘要]　本文介绍了作者利用高速摄影在常温、有背压、无旋流条件下 ZS4S1 喷油嘴喷注贯穿距离的研究结果。在对广安博之公式进行修正的基础上，提出了此型喷油嘴喷注贯穿距离的计算公式。通过计算表明，按该公式预测的喷注贯穿距离与试验结果相吻合。

符　号

A——常数

B——常数

C——速度系数

K——常数

L——喷注开始分裂时的贯穿距离，m

L'——喷注开始分裂时贯穿距离试定值，m

n——油泵转速，r/min

N_1——喷注初始段的试验样本数

N_2——喷注第二阶段试验样本数

u_0——燃油喷射初速，m/s

d_0——轴针直径，m

G——循环供油量，mL/cyc

α——喷注长度修正系数

p_a——介质压力，Pa

Δp——喷孔前后压差，Pa

p_j——针阀开启压力，Pa

ρ_f——燃油密度，kg/m³

ρ_a——介质密度，kg/m³

t——从喷油开始计算的时间，s

t_b——从喷油开始到喷注分裂所经历的时间，s

t_b'——从喷油开始到喷注分裂所经历时间试定值，s

S——喷注贯穿距离，m

迄今，研究孔式喷嘴喷注贯穿特性者较普遍[1-3]，而研究轴针式喷嘴喷注贯穿特性者甚少。这可能是由于过去认为分开式燃烧室之轴针式喷嘴的喷注贯穿特性与柴油机性能关系不大。但随着柴油机燃烧研究的不断深入，人们已逐渐认识到研究轴针式喷油嘴的喷注贯穿特性对建立间喷式柴油机燃烧模型和研究其燃烧过程的重要性。例如，日本广安博之提出的预测间喷式柴油机放热和排放的数学模型就是以喷嘴喷注贯穿距离作为计算浓度分布、空燃比、放热率的原始输入数据[4]。然而作者所得的高速摄影照片表明，广安模型与 ZS4S1 喷油嘴的喷注贯穿试验差别较大，这可能是广安模型的研究对象仍为孔式喷嘴所致[3]。ZS4S1 喷油嘴应用广泛，研究该种喷油嘴的喷注贯穿特性具有十分重要的意义。

1 计算模型

根据 ZS4S1 喷油嘴喷注贯穿过程的特点,作者认为,基于广安方法推导 ZS4S1 喷油嘴的喷注贯穿模型比较适合;但是,广安的假设仍需商榷。例如,当喷注分裂成油滴后,是否能像广安所表述的喷注贯穿距离 S 正比于 $\sqrt{t}$[3],即能否将喷注看成"气对气"的喷射。从射流原理知:只有在喷注充分发展阶段,射流束与周围介质完全混为一体,射流横截面上每一点的密度都等于介质密度时,才能推导出 S 正比于 $\sqrt{t}$[5]。对 ZS4S1 喷嘴喷注过程而言,喷注分裂为油滴后,其索特平均直径较大;而且由于它的喷雾锥角很小,喷注卷入空气较少,燃油所占比重较大,因此很难将其喷注贯穿过程看成"气对气"的紊流沉浸射流。若设 S 正比于 t 的 K 次方(指数 K 通过试验求得)或许能得到令人满意的结果。

基于上述考虑,作者在建立 ZS4S1 喷油嘴喷注贯穿模型时假设:

(1) 密度为 ρ_f 的液体喷注在密度为 ρ_a 的气体介质中运动,$\rho_a \ll \rho_f$,且液体喷注和气体介质之间的相对速度很大。

(2) 喷注表面的扰动幅度因气体压力扰动而增加,并且随着喷注扰动幅度的增加,喷注倾向于不稳定,最终分裂成油滴。

(3) 在喷注尚未分裂的贯穿距离 L 内,喷注速度等于喷射初速。从而可得喷注未分裂前的喷注贯穿速度

$$u_0 = c \sqrt{\frac{2\Delta p}{\rho_f}} \tag{1}$$

此段的贯穿距离

$$S = c \sqrt{\frac{2\Delta p}{\rho_f}} t \tag{2}$$

根据 Levich 喷注长度理论[6],分裂长度

$$L \approx u_0 t_b \approx \alpha \sqrt{\frac{\rho_f}{\rho_a}} \cdot d_0 \tag{3}$$

从喷射开始到喷注开始分裂所经历的时间可由式(1)和式(3)得

$$t_b = \frac{\alpha \rho_f d_0}{\sqrt{2c^2 \rho_a \Delta p}} \tag{4}$$

在喷注的第二阶段,设 $S = Bt^k$.

可由 $S = L$ 和 $t = t_b$ 得到 B 的表达式:

$$B = 2^{0.5k} \alpha^{1-k} d_0^{1-k} c^k \rho_f^{0.5-k} \rho_a^{0.5(k-1)} \Delta p^{0.5k} \tag{5}$$

所以,在喷注分裂为油滴后,其喷注贯穿距离计算公式为

$$S = 2^{0.5k} \alpha^{1-k} d_0^{1-k} c^k \rho_f^{0.5-k} \rho_a^{0.5(k-1)} \Delta p^{0.5k} t^k \tag{6}$$

公式(2),(4)和(6)就是作者提出的预测 ZS4S1 喷油嘴喷注贯穿距离的计算模型,其中常数 c,α,k 由试验确定。而如何确定这些常数正是使用该计算公式的关键。

按照广安的方法[3],首先根据 0.3 mm 直径单孔喷嘴的稳定喷注及间隙喷注的试验结果,求与时间成正比增加的喷注距离与时间,然后在此基础上求得 α 和 c 的值。在广安模型中,$k = 0.5$ 是从自由射流理论导出的。由此可知,速度系数 c 和液体喷注长度修正系数 α 完全由

分裂长度确定,整个计算模型的准确度也完全依赖于分裂长度的精度。实际上,在目前的测试条件下要从试验中直接获得分裂长度和分裂时间还是比较困难的。广安等人将与时间成正比例增加的喷注距离定为分裂长度,虽然理论上可以这样处理,但由于试验的随机误差,使得与时间成正比增加的喷注距离难以精确确定。特别是对 ZS4S1 喷油嘴的喷注贯穿过程来说,喷注分裂成油滴后,单颗油粒的粒度较大,相应地动量也较大,其贯穿速度衰减较慢,而与时间成正比增加的贯穿终点不明显,因此要确定它更加困难。作者是这样处理这个问题的:

令常数

$$A = c\sqrt{\frac{2\Delta p}{\rho_f}} \tag{7}$$

则式(2)变为

$$S = At \tag{8}$$

常数 A 可采用线性回归的方法求得

$$A = \frac{\sum\limits_{i=1}^{N_1} S_i t_i}{\sum\limits_{i=1}^{N_1} t_i^2} \tag{9}$$

当 $t > t_b$ 时,

$$S = Bt^k$$

常数 B, k 可用幂回归的方法从试验中求得

$$k = \frac{N_2 \sum\limits_{i=1}^{N_2} \ln S_i \ln t_i - \sum\limits_{i=1}^{N_2} \ln S_i \sum\limits_{i=1}^{N_2} \ln t_i}{N_2 \sum\limits_{i=1}^{N_2} (\ln t_i)^2 - \left(\sum\limits_{i=1}^{N_2} \ln t_i\right)^2} \tag{10}$$

$$B = \exp\left[\left(\sum\limits_{i=1}^{N_2} \ln S_i - k\sum\limits_{i=1}^{N_2} \ln t_i\right)/N_2\right] \tag{11}$$

首先根据试验结果粗略地试定喷注分裂长度 L' 以及相应的分裂时间 t_b'。然后以 t_b' 为分界点将喷注贯穿过程分为两个阶段。将 t 小于 t_b' 的试验数据按式(9)进行回归计算,可求出常数 A;再将 t 大于 t_b' 的试验数据按式(10)和式(11)进行计算,可求出 k 及 B;由式(5)求解 α;然后回过来将 α 值代入式(3)可求得 L。作者将按式(3)计算得到的 L 值定义为计算值 L。将计算值 L 与试定值 L' 相比较,如果 L 等于 L' 则说明试定 L' 是适合的;否则不适合,即两条回归曲线的交点并不在试定点 (t_b', L'),这就必须以计算值 (t_b, L) 作为新的试定值,将喷注分为两个阶段,重复上述修正步骤,直到 L 与 L' 相等为止,此 L 值方可认为喷注分裂长度,从而可相应求出 c, α, k。

按照上述方法,作者经过反复计算,最后得到 $c=0.507, \alpha=6.60, k=0.842$。

这样,就导出了预测 ZS4S1 喷油嘴喷注贯穿距离的计算模型。

当 $0 < t < t_b$ 时,

$$S = 0.507\sqrt{\frac{2\Delta p}{\rho_f}}\, t \tag{2'}$$

当 $t > t_b$ 时,

$$S = 1.018\rho_f^{-0.342}\rho_a^{-0.079}d_0^{0.158}\Delta p^{0.421}t^{0.842} \tag{6'}$$

$$t_b = 9.205 \frac{\rho_f d_0}{\sqrt{\rho_a \Delta p}} \tag{4'}$$

图 1 到图 4 是计算值和实测值的对比。由图可知,作者提出的计算公式在广泛的试验条件下能与试验结果良好吻合。这说明此计算公式能有效地预测 ZS4S1 喷油嘴在无旋流条件下的喷注贯穿距离。

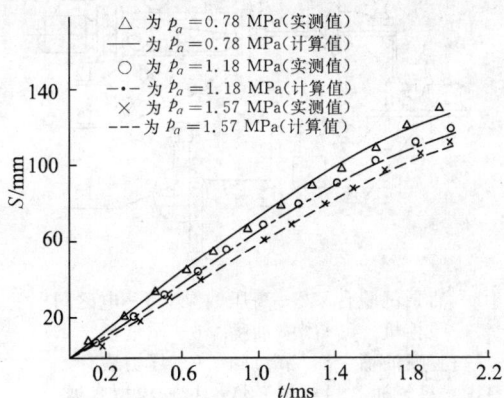

图 1　背压对喷注贯穿距离的影响

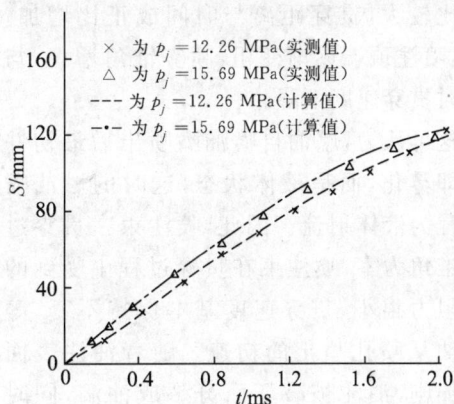

图 2　针阀开启压力对贯穿距离的影响

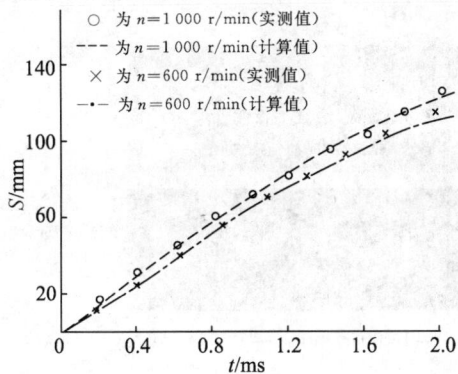

图 3　转速对贯穿距离的影响

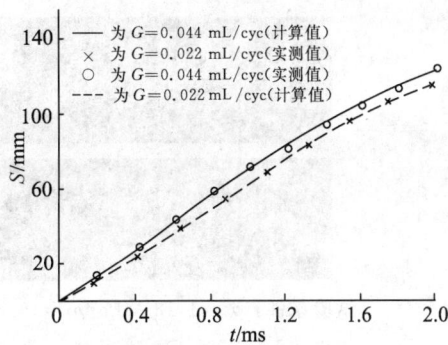

图 4　循环供油量对贯穿距离的影响

2　试验装置和方法

本试验在模拟装置上进行,其布置如图 5 所示。喷射背压可在 $0 \sim 1.96$ MPa 范围内调节,容器内气体处于常温、无旋流条件下。ZS4S1 喷油嘴安装在容器左端的中央,仍由柱塞泵完成供油,并利用专门喷射机构使之作一次性喷油,以保证摄影清晰。同时通过同步机构使一次喷射系统与高速摄影机同步。当进行一次喷射时,高速摄影机将完成喷注全过程的拍摄工作。拍摄使用日本 NAC 公司生产的 E-10 型旋转棱镜式高速摄影机,试验时的拍摄频率为 $4\,500 \sim 6\,000$ p/s。

3 试验结果及其分析

作者使用日本 NAC 公司生产的 PH-160B 胶片运动分析仪对喷注贯穿发展过程进行了分析。

从所摄照片（图 6）看到：燃油自 ZS4S1 喷油嘴喷出，初始喷雾锥角为零，而且贯穿速度比较大，贯穿距离与时间成正比增加。然后，喷注顶端逐渐变粗，喷雾锥角增大，与此同时贯穿速度下降。

这可认为，燃油自喷油嘴喷出后最初并未立即雾化，而呈液体状态，这时的燃油喷射过程为液体射流。因此，喷注束在贯穿过程中锥角为零，喷注束在贯穿过程中受到的介质阻力很小，贯穿速度基本保持不变，等于燃油从喷孔喷出的初速。随着油注表面波动加剧，油注被破坏而分裂成油滴，同时空气卷入喷注束顶端，因此喷注束逐渐变粗，受到的介质阻力增加，贯穿速度随之降低。

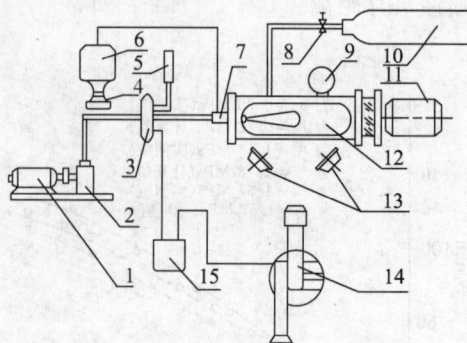

1—油泵试验台 2—高压油泵 3—电磁阀
4—照相机 5—副喷油嘴 6—示波器
7—主喷油嘴 8—节流阀 9—压力表
10—氮气瓶 11—聚光灯 12—模拟容器
13—碘钨灯 14—高速摄影机 15—同步控制箱

图 5 试验装置示意图

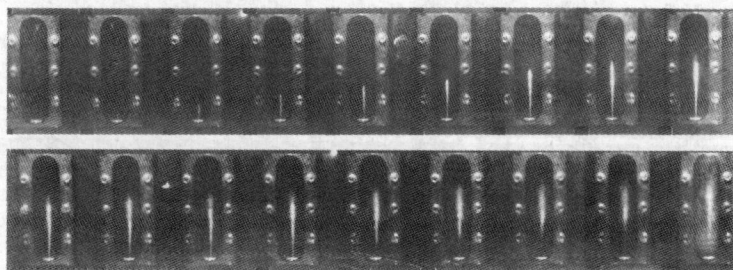

试验条件：$p_a=1.18$ MPa，$p_j=15.69$ MPa，$n=1\,000$ r/min，$G=0.044$ mL/cyc
拍摄效率：4 500 p/s

图 6 喷注贯穿过程的高速摄影照片

图 7 为 ZS4S1 喷油嘴和孔式喷油嘴的喷注贯穿距离的比较。从该图可知：ZS4S1 喷油嘴的喷注贯穿距离大于孔式喷油嘴的贯穿距离。

与孔式喷油嘴相比，ZS4S1 喷油嘴的针阀开启压力小，针阀开启面积大，燃油喷射压力相应较小，雾化质量差。再加上它的喷雾锥角很小，已分裂的小油滴均集中在喷射轴线附近的小区域内，油滴间的内聚力会使这些小油滴重新聚合成大油滴。

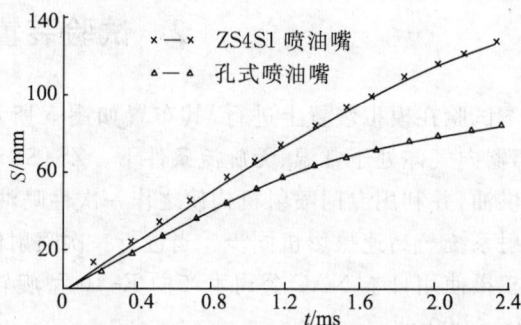

图 7 ZS4S1 喷油嘴和孔式喷油嘴喷注贯穿距离的比较

这样,喷注分裂成油滴后,油滴的平均索特直径较大,单颗油粒的动量和惯性也相应较大。因此,ZS4S1 喷油嘴喷注贯穿速度远大于孔式喷油嘴的喷注贯穿速度。

4 结 论

(1) 研究轴针式喷油嘴的喷注贯穿特性对建立间喷式柴油机的燃烧模型和研究燃烧过程具有重要意义。

(2) 作者以广安公式为基础并根据试验结果,提出了预测 ZS4S1 喷油嘴喷注贯穿距离的计算公式。

(3) 通过计算和试验结果相比较,表明作者提出的计算公式在广泛的试验条件下能与试验结果良好吻合。

参 考 文 献

[1] Wakuri Eral Y. Study the penetration of fuel spray in a diesel engine[J]. *Bulletin of JSME*, 1960,3(9).

[2] Dent J C. A basis for the comparison of various experimental methods for studying spray penetration[J]. *SAE Trans*, 710571,1971.

[3] 广安博之. 柴油机的油注贯穿距离和喷雾锥角[J]. 车用发动机,1982(3).

[4] 广安博之. 预测非直喷式柴油机的放热率和废弃排放的数学模型[J]. 国外内燃机,1985(5).

[5] 刘峥. 直喷式柴油机混合气形成过程中喷注物理及数学模型的研究概况[M]. 清华大学内燃机教研室,1980.

(本文原载于《内燃机工程》1988 年第 2 期)

An investigation on penetration of injector type ZS4S1 without turbulent air flow

Abstract: In this paper the investigative results of the spray penetration of injector type ZS4S1 at ambient temperature ,with back pressure and without turbulent air flow ,by using the high speed photograph are given. Based on revising Hiroyasu equation a formula to calculate this type injector fule spray penetration has been proposed. It has been shown that results calculated by this way is coincide with experimental results.

吊钟型涡流室内喷油和燃烧过程的研究

李德桃,朱广圣

[摘要] 介绍了利用高速摄影技术对 S195 型柴油机吊钟型涡流室内燃料不着火状态下的喷油过程和工作状态下的燃烧过程进行的详细研究。对比了实际涡流室内的喷油过程和无旋流压力模拟容器中的喷油过程;结合测录的示功图和计算的放热规律对燃烧过程进行了分析;探讨了转速对喷油和燃烧过程的影响。通过分析,揭示了吊钟型涡流室内喷油和燃烧过程的基本特征。

有些研究者利用高速摄影技术对涡流室中的喷雾和燃烧过程进行过研究[1]。然而,不同形式的涡流室内的混合气形成和燃烧过程的特征不尽相同[2]。据作者所知,利用高速摄影技术对吊钟型涡流室内混合气形成和燃烧过程进行研究者还很少。此外,在建立燃烧模型时,目前往往只考虑刚体涡流对静态模拟容器中油束的影响,来模拟实际燃烧室内的喷油过程。但在实际涡流室内,喷油期间的涡流更倾向于势涡流结构,且涡流室内存在着高温高压环境、油束同燃烧室壁碰撞和油束扩散等实际情况。因此,在帮助认识混合气形成特征的同时,研究实际涡流室内的喷油过程,对于今后精确建立实际燃烧室内的喷油模型也是十分必要的。

1　试验装置简介

为了揭示该型涡流室内喷雾和燃烧过程的基本特征,作者用我国产量最大、使用最广的吊钟型涡流室式柴油机,改装成高速摄影试验机,同采集的相应工况下的燃烧过程数据相配合,观察和分析了涡流室内喷油和燃烧过程。试验装置和涡流室观察范围如图 1 所示。

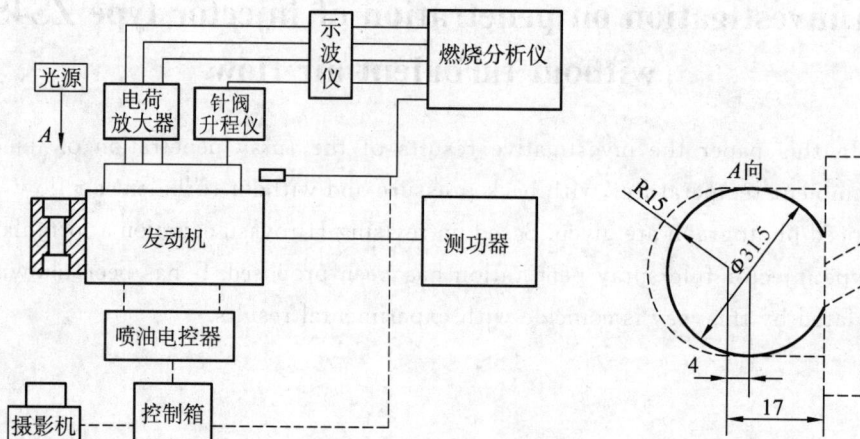

图 1　试验装置和涡流室的观察范围示意图

2 试验结果及其分析

2.1 吊钟型涡流室内的喷油过程

拍摄喷油过程时,发动机压缩比降为 12.8,在燃料处于不着火状态下拍摄喷油全过程。图 2 表示转速为 1 230 r/min、循环供油量为 23 mm³ 工况下涡流室内的实际喷油过程。在照片上标注有相应的曲轴转角。

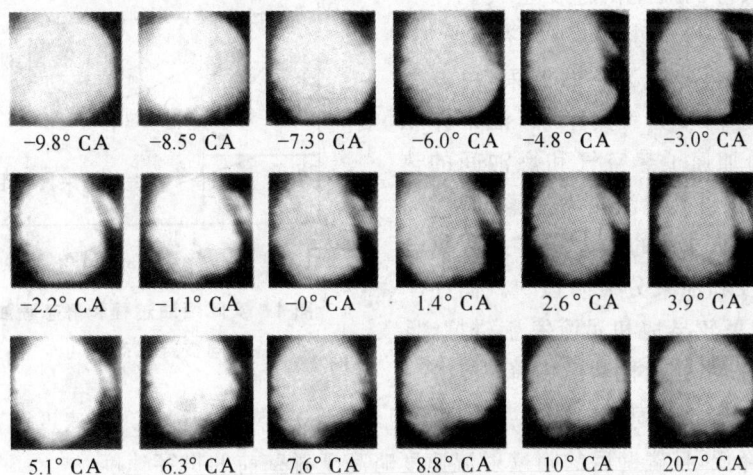

−9.8° CA	−8.5° CA	−7.3° CA	−6.0° CA	−4.8° CA	−3.0° CA
−2.2° CA	−1.1° CA	−0° CA	1.4° CA	2.6° CA	3.9° CA
5.1° CA	6.3° CA	7.6° CA	8.8° CA	10° CA	20.7° CA

图 2　吊钟型涡流室内的喷油过程

由照片可见,上止点前 9.8° CA 开始喷油。在喷油开始阶段,喷油嘴节流较强,燃油喷注动量小,油束顶部靠近涡流室边缘,边缘动量较高的空气使燃油一出现即被强烈吹偏。照片上表现为油束和室壁间无间隙。

随着喷油持续进行,油束因动量增加而偏离壁面,内侧油粒顺涡流向室壁扩散开来,同时可以看到,由于油束顶部贯穿速度降低,随后喷出的燃油赶上同一喷油过程先喷出的燃油,油束顶部明显加厚(见上止点前 6° CA 照片)。

油束在上止点前 4.8° CA 开始碰撞涡流室圆柱壁面。碰撞时燃油空间反射不明显,少许飞溅出的油滴立即加入空气涡流运动;绝大部分燃油碰撞后首先沿柱面向下移动,经拐角转弯后顺涡流在镶块表面扩展开来。在扩展过程中不断蒸发而消失,其中部分燃油沿右侧壁面继续顺涡流向顶部上行。在此阶段,油束动量较高,空气涡流对其影响较小,燃油主要集中在涡流室底部。至上止点后 2.6° CA 的喷油后期,油束的动量明显减小,结构松散,受空气涡流作用沿涡流室壁弯曲变化。由照片可见,这时油束几乎不和壁面直接碰撞。上止点后 6.3° CA 喷油结束后,这部分燃油沿燃烧室壁缓慢移动,并集中在涡流室底部,最终从底部消失。

图 3 是与试验机同类型的供油系统在常温无旋流压力模拟容器中的喷油过程(以下简称静态喷油过程)。对应转速为 1 200 r/min,循环供油量是 22 mm³。由照片可见,燃油以很小的锥角自喷油嘴喷出,贯穿速度较大,整个喷油持续期内油束外形变化较小,呈锥体对称结构,仅在喷油后期,喷雾锥角才略有增加。

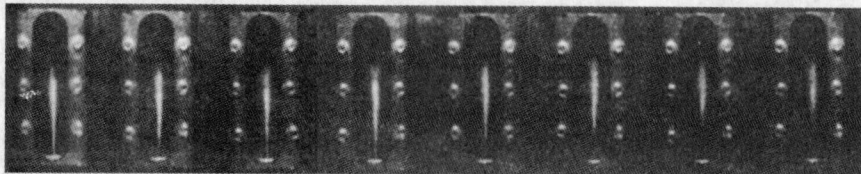

图 3　静态喷油过程

对比静态喷油过程,可对实际涡流室内的喷油过程有以下认识:

(1) 由于实际涡流室内存在空气涡流和较高的温度和压力,油束中空气含量增加。对比图 2 和图 3 两种情况下的油束,可以看出在实际涡流室内喷雾锥角增加,油束体积明显加大,从而加速了热空气和燃油间的热交换和质交换,促进了燃油的快速蒸发。

(2) 静态喷油过程的油束贯穿过程可分为等速贯穿阶段和油束分裂阶段[3]。而在实际涡流室内,油束较早地和涡流室壁碰撞,且阻力系数加大,油束前缘移动速度降低。图 4 表示这两种情况下的不同贯穿状态。由图可见,实际涡流室内油束贯穿速度和碰壁后的附壁射流速度逐渐降低,且碰壁后的附壁射流速度略高于碰壁前的贯穿速度。

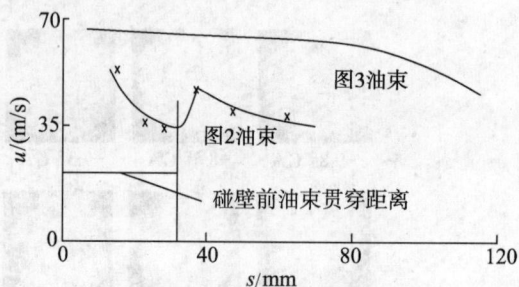

图 4　实际喷油过程和静态喷油过程的不同贯穿状态

(3) 空气涡流对油束的影响使得油束结构不对称,外层油粒顺涡流扩散。在喷油开始时,油束被空气涡流强烈吹偏;在喷油末期,涡流又使油束沿燃烧室壁弯曲;而在喷油中期,油束受空气涡流影响较小。

通过试验还研究了转速对喷油过程的影响。图 5 表示循环喷油量相同,而转速不同情况下的油束轨迹形式。在转速较低时,由于空气涡流减弱,油束受空气涡流影响相对较小。如图所示,喷油早期,油束和室壁间有明显间隙;喷油中期油束略被吹偏,且在接近涡流室拐角处和室壁碰撞,部分燃油滞留在拐角处,燃油有明显向涡流室空间反射的倾向。图 6 示出了这种燃油碰撞情况。

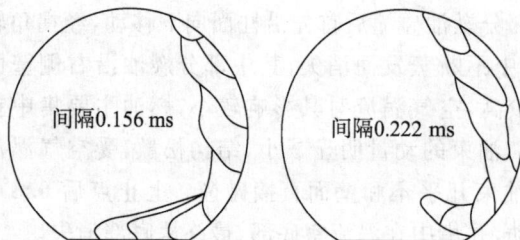

图 5　涡流室内的油束轨迹形式

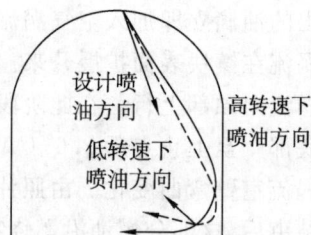

图 6　油束碰壁反射示意

2.2　吊钟型涡流室内的燃烧过程

图 7 是试验机转速为 1 350 r/min、循环供油量为 77 mm^3 工况下吊钟型涡流室内的喷油和燃烧过程照片(着火点前照片底部白色为照明光)。图 8 为相应工况下的压力、温度和放热率曲线。现将燃烧过程分为四个阶段进行分析:

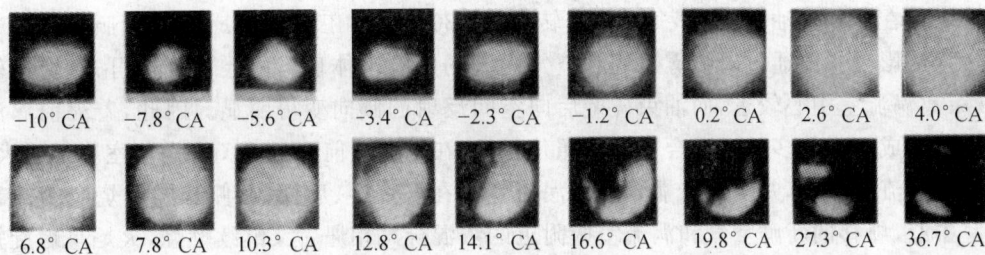

| −10° CA | −7.8° CA | −5.6° CA | −3.4° CA | −2.3° CA | −1.2° CA | 0.2° CA | 2.6° CA | 4.0° CA |

| 6.8° CA | 7.8° CA | 10.3° CA | 12.8° CA | 14.1° CA | 16.6° CA | 19.8° CA | 27.3° CA | 36.7° CA |

图 7　吊钟型涡流室内燃烧过程(1 350 r/min)

（1）着火延迟阶段：燃油于上止点前 10° CA 喷出。在喷油的早期，油束亦受空气涡流的影响而弯曲，这可能是由于涡流室内的温度和压力较高，油束密度降低，空气涡流对油束影响相对增加所致。在实际燃烧工况下，油束仍是以液注形式和室壁碰撞；碰撞后燃油发生快速分散和蒸发，部分照明光被蒸发前的分散油雾遮挡；大部分燃油顺涡流在镶块表面移动，并很快蒸发消失。至上止点前 2.3° CA，涡流室底部通道口附近先着火。该着火位置在各工况和

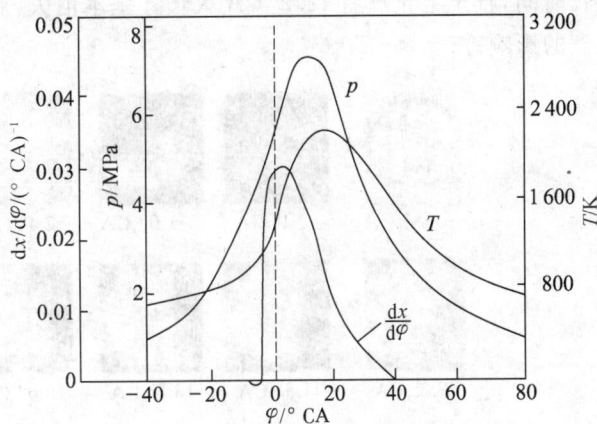

图 8　涡流室内压力、温度和放热率曲线

各循环中基本不变。依据放热曲线可确定着火点是在上止点前 4° CA。

（2）速燃阶段：着火后火焰首先在涡流室底部传播，并迅速向涡流室顶部发展，至上止点后 2.6° CA 充满整个涡流室。放热速率在上止点后 2° CA 达到峰值，压力、温度也迅速上升，上止点后 8° CA 上升至最高压力。喷油在此阶段后期结束，较多的燃油能在这个阶段形成可燃混合气并完全燃烧。从燃烧照片上可以看出，火焰持续充满整个涡流室，并伴随着较高的发光度。从放热曲线也可以看出，在此阶段燃烧放热率保持一段高峰值，而不像直喷式柴油机燃烧那样仅出现一个放热尖峰。

（3）缓燃阶段：由照片可见，在涡流室左侧首先显露燃烧产物，火焰和燃烧产物一起旋转。此时火焰仍充满大部分涡流室，但活塞已加速下行，燃烧室内燃料分子和氧分子浓度降低，燃烧速率相对较低，放热率下降，压力下降，而温度仍在升高，至上止点后 16° CA 达到最高温度。

（4）后燃阶段：这时火焰已明显变暗，燃烧室内工质的缓慢旋转同时伴随火焰面积的迅速减小，至上止点后 36.7° CA 火焰基本从涡流室内消失，此时燃烧室内工质仍维持一定的旋转速度。涡流室内放热也在上止点后 40° CA 结束。

一些研究者在研究球型涡流室中的燃烧过程时曾指出，球型涡流室中浓混合气呈环状分布，火焰迅速旋转及伴随的强烈热混合作用是该型涡流室内具有良好燃烧过程的重要条件[4]。从我们对吊钟型涡流室内混合气形成特点的分析可知，该型涡流室底部形成较多的混合气，着火后火焰由底部向顶部扩展，热力混合作用不强，但由于涡流室内浓混合气聚集在通道口附近，浓混合气较早地进入主燃室，因此，吊钟型涡流室同样也具有良好的燃烧过程。

图 9 是在循环供油量与图 7 相同时,转速为 830 r/min 工况下的燃烧过程照片。转速对燃烧过程的影响主要通过对喷雾过程和空气涡流的影响来体现。由上节分析可知,转速较低时,油束离涡流室中心较近,而油束碰壁后向空间反射的倾向变得明显。因此,着火前涡流室中间部位形成相对较多的浓混合气。由照片可见,在上止点前 2.1° CA 自涡流室底部着火后,火焰卷向涡流室中心,并迅速传播到整个涡流室。在燃烧中、后期,左侧外围燃烧产物被卷向燃烧室中心,外移的新鲜空气和涡流室壁附近的浓混合气相遇后又继续燃烧,火焰呈螺旋辐射状。上止点后 37.6° CA 工质涡流运动停止,但这时燃烧室中部浓混合气仍在燃烧,并持续了较长时间,直至上止点后 42.2° CA 火焰才基本消失,后燃期延长。显然,这给发动机性能带来不利的影响。

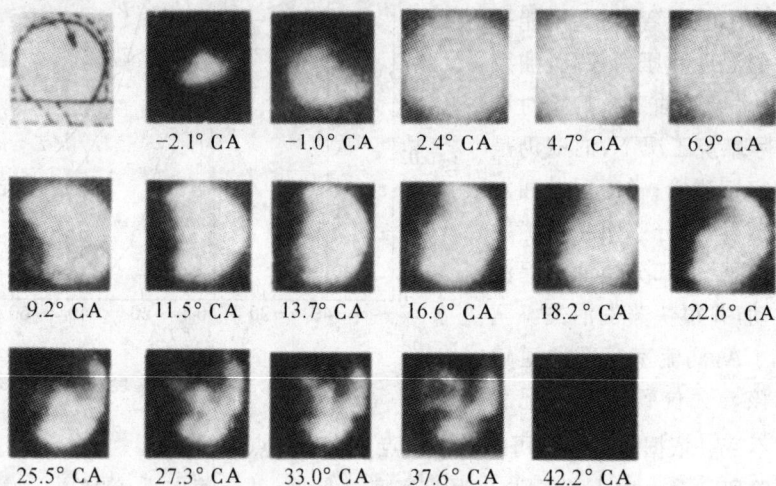

图 9　吊钟型涡流室内燃烧过程(830 r/min)

3　结　论

(1) 在喷油初期和末期,油束受空气涡流影响被强烈吹偏并顺应燃烧室壁弯曲;在喷油中期,空气涡流对油束影响不大。与喷入模型装置静态空气中的油束相比,涡流室内油束中空气含量较多,喷雾锥角加大,贯穿速度较低,并较早地和室壁碰撞,碰撞后部分燃油发生快速分散和蒸发,大部分燃油顺涡流沿燃烧室壁扩展。

(2) 转速较低时油束离涡流室中心较近,碰壁后,部分燃油向空间反射,致使在涡流室中心附近形成一些浓混合气。此外,转速较低时燃烧后期涡流趋于消失,延长了这部分浓混合气在燃烧室中心区的燃烧时间,因此对发动机性能造成不利的影响。

(3) 吊钟型涡流室与球型和 Comet V 型涡流室的混合气形成和燃烧过程不同。在吊钟型涡流室中,燃烧室底部形成较多的浓混合气,燃烧初期火焰没有明显的旋转运动,涡流室内热力混合效应较弱。由于浓混合气较早地进入主燃室,同样可获得良好的燃烧性能。

(4) 和直喷式柴油机燃烧不同,涡流室内喷油持续期较短,大部分循环供油量在速燃期内燃烧完毕,燃烧室内保持持续的高峰放热,随后放热率迅速降低,火焰面积迅速减少,燃烧较早结束,放热曲线接近直角三角形。

参 考 文 献

［1］李德桃. 涡流燃烧室内混合气形成和燃烧过程研究进展[J]. 拖拉机，1984(5).

［2］Meintjes K，Alkidas A C. An experimental and computational investigation of the flow in diesel prochambers[J]. *SAE Trans*，820275：1158-1164.

［3］李德桃，朱亚娜. 无旋流场中 ZS4S1 油嘴喷雾贯穿度计算模型的研究[J]. 内燃机工程，1988(2)：1-6.

［4］Miwa K，Ikemagi M. Combustion and pollutant formation in an indirect injection diesel engine[J]. *SAE Trans*，800026：118-119.

(本文原载于《内燃机学报》1989 年第 1 期)

Investigation into the fuel injection and combustion process in bell type swirl chamber

Abstract：This paper presents the results of investigation on fuel injection under motoring condition and combustion process under working condition in S195 type diesel engine bell type swirl chamber by way of high speed photography. The fuel injections in actual swirl chamber was compared with which in model apparatus with still air. The combustion process was analysed by means of the measured pressure diagram and calculated heat release rate and the effects of engine speed on fuel injection and combustion were discussed.

用激光全息术研究轴针式喷油嘴的
早期喷雾特性[*]

李德桃,康志新,钱冀平

[摘要] 作者利用同轴全息照相术研究了 ZS4S1 轴针式喷油嘴的早期喷雾机理和油滴空间分布。在常温、常压、无涡流条件下,拍摄了不同喷油压力下油雾场的全息照片。通过再现全息片并对已微粒化油滴的直径大小和数目进行判读后发现,在早期喷雾中,油注的头部区域就发生破碎;中心液注则一直持续到外围完全微粒化。本研究对加深喷雾过程的理解和建立轴针式喷油嘴精确的喷雾模型具有一定的价值。

1 试验装置和试验方法

拍摄系统:整个拍摄系统布置于暗室中,图 1 所示的是拍摄装置示意图。

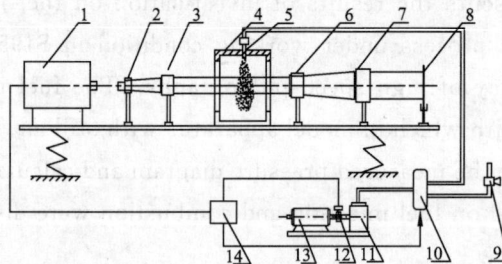

1—DJ-1 型激光器 2—扩束光学镜头 3—准直光学镜头 4—配有石英玻璃窗的定容燃烧室
5—试验用 ZS4S1 型喷油嘴 6,7—富氏扩束放大装置 8—全息干板 9—辅助喷油嘴
10—电磁阀 11—油泵试验台 12—喷油始点传感器 13—调速电机 14—同步控制系统

图 1 拍摄装置示意图(同轴全息光路)

DJ-1 型激光器的主要技术指标如下:

工作物质:红宝石;激光波长:6 943 Å;脉冲方式:双脉冲;能量指标:每个光脉冲所携带的能量为 100 mJ;相干长度:大于 1 m;脉冲宽度:小于 50 ns;工作方式:反射全息,透射全息,同轴全息,离轴全息均可;触发方式:手动触发或同步触发。

油泵转速由调速电机控制。同步系统包括单次喷射控制器以及控制喷油和拍摄同步的 Z8080 单板机延时系统,此外还包括与此有关的元器件。Z8080 单板机采用 PIO 接口技术达到延时并输出信号的目的。

再现系统:本文使用同轴全息光路拍摄,所以也使用同轴型再现光路再现,图 2 为其示意图。

---

[*] 试验在南京航空学院(现南京航空航天大学)进行,并得到王家骅教授的关心和帮助。

再现系统在再现过程中把照片上油滴的尺寸放大。考虑到拍摄时的放大率,最后油滴直径在屏幕上的放大倍数为 190。

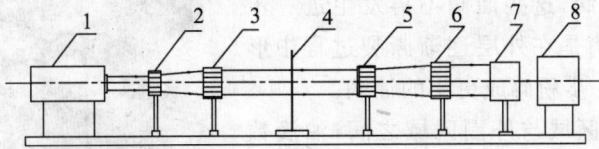

1—He-Ne 激光器,激光波长为 6 328 Å　2—扩束镜　3—准直镜　4—放置全息片的三维拖板
5,6—放大光学装置　7—摄像机系统　8—电视屏幕

图 2　再现系统示意图(同轴光路)

试验方法:作者为了获得早期喷雾的全息片,选择了喷油开始后 0.4,0.6 和 0.7 ms 的喷雾场进行拍摄。

试验系在油泵转速控制为 1 000 r/min,采用 0 号柴油,并在室温、大气压力和无涡流及改变喷油压力的条件下进行。

进行拍摄时,先启动 DJ-1 型激光器,调整好光路。然后启动调速电机,使油泵达到测试要求的转速。一切准备就绪后,按下同步系统的按钮。此时同步系统控制喷油嘴单次喷射,并使激光的触发符合所需拍摄的时刻。脉冲激光使全息干板曝光,完成一次全部的拍摄过程。

2　试验结果及其分析

将拍摄的全息片在再现系统上再现,从纵向和横向进行扫描。从屏幕上放大的再现像可以看出:

(1)喷油开始后不久,外围油注刚被撕裂时,油注的头部已被明显地撕裂。空气卷入头部并出现了许多小油滴。这就证实了早期喷雾中油注头部的局部区域存在首先撕裂和粉碎的现象(如图 3 所示)。

(2)在油注的轴线方向上,油滴的空间分布不是严格的轴对称,某侧的油滴密度较大;撕裂和分散过程也不是严格的轴对称,油注在分散过程中,外层被逐渐撕裂,形成外围的雾状油滴区;中心部分则是一连续的液注区域。文献[1]中指出这一液注存在一个空心的内区。笔者认为,空心内区的存在只限于一个较小的区域,即喷孔附近,而在液注的大部分区域内,则是实心的。

(3)中心液注一直存在到外围已基本微粒化的时期。早期喷雾场就像是一根吸附了无数大小不同的细微铁粒的长棒形磁铁一样。

(4)中心液注区域的边缘是片状或膜状结构(如图 4 所示)。片状结构中有许多空气卷入后形成的空洞,如蜂窝状;这些片状结构中夹杂着一些油滴,这是撕裂和扭曲的结果。

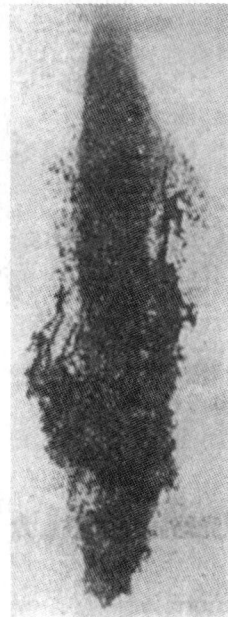

喷油压力:14 MPa;喷油时间:喷油开始后 0.4 ms;油泵转速:1 000 r/min

**图 3　油注头部发生撕裂和
粉碎的现象**

113

（5）外围的油滴总是呈球形或椭球形，而且直径大小分布不同。作者认为，这些油滴，有的是喷射开始即形成的单个油滴，这些油滴不再发生进一步的破碎，还有一些油滴是在外层逐渐撕裂过程中形成的，这些油滴能进一步粉碎成更小的油滴。

（6）在中心液注区域与外围区域之间，油滴数最多，密度较大，但分布则呈集群状态，即某一区域油滴较多，其他区域则较少。有时存在若干个这样的油滴数目密集区，这是外围与中心液注的片状燃油发生撕裂的结果。

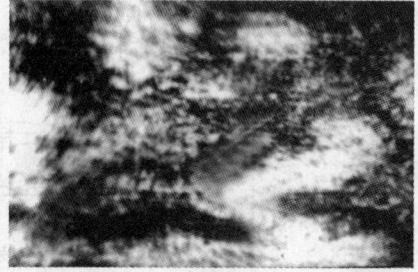

图4 片状油雾示意图

（7）轴针式喷油嘴喷出的燃油确实需要一段滞后时间才能完全微粒化。

（8）喷油压力对喷雾形成有影响。喷油压力越高则喷油的扩展范围越大。但喷油压力过高则中心液注变粗，雾化反而变差。

作者依据所拍摄的全息片，作出了图5所描述的喷雾场形成和发展机理的示意图。

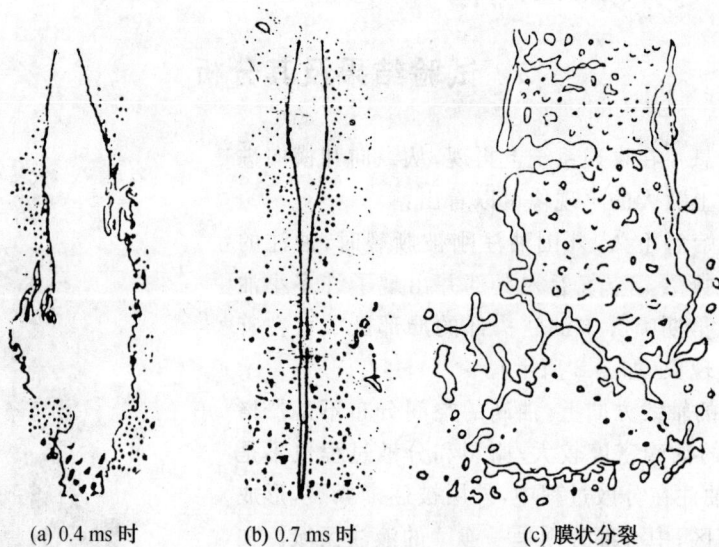

(a) 0.4 ms 时　　(b) 0.7 ms 时　　(c) 膜状分裂

图5 喷雾场的形成和发展机理示意图

将全息片再现后判读油滴，计算油注长度上某一直径范围内的油滴数目 N，测量其直径 d，从而可确定油滴的空间分布。图6是判读的几何位置示意图和单元划分法，划分方法与文献[2]提出的划分法相同。

图6a 中，取 $\Delta h = 6$ mm，$l = 60$ mm，O 点代表油注初始位置。图6b 中的单元呈正方形，每边长为 2 mm，即三维拖板每次在 2 mm 的范围得到扫描。在 R 方向的长度范围，随着油注在此区域的扩展范围而定。

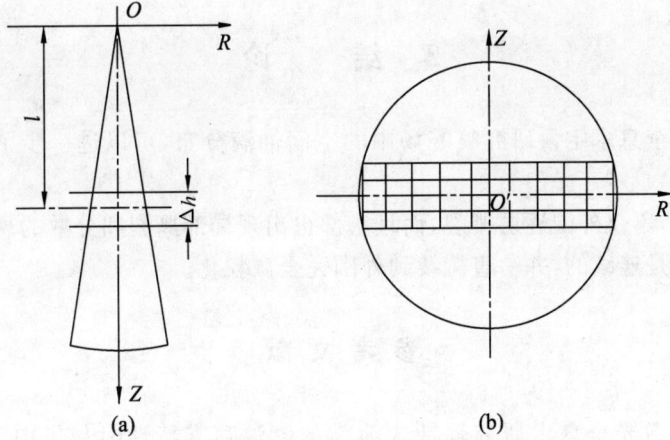

图 6　判读的几何位置(a)和单元划分法示意图(b)

本文取喷油压力为 18 MPa 情况下的全息片进行判读,图 7 和图 8 是用图形表示的处理结果。图中 $\sum N$ 为总油粒数;n 为某直径范围的油粒数占总油粒数的百分比。

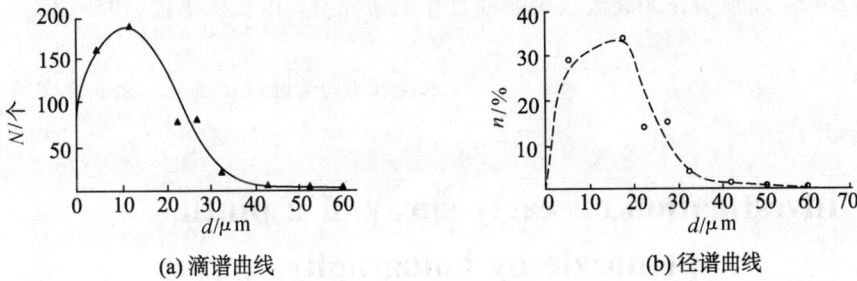

(a) 滴谱曲线　　　　　(b) 径谱曲线

图 7　喷油压力为 18 MPa,$l=60$ mm 时的 N-d 滴谱曲线与径谱曲线

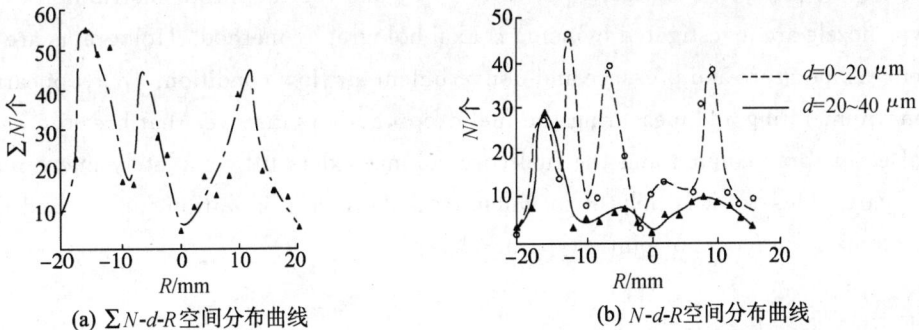

(a) $\sum N$-d-R 空间分布曲线　　　(b) N-d-R 空间分布曲线

图 8　喷油压力为 18 MPa,$l=60$ mm 时 $\sum N$-d-R 与 N-d-R 空间分布曲线

从图 8 中可看出如下几点:

(1) 在喷雾场内已微粒化的油注中,某种直径范围内的油滴数最多,其他直径范围内的油滴数均少于此直径范围内的油滴数。

(2) 在所判读的已微粒化的空间区域,沿 R 轴即油注的横向区域,油滴数是起伏变化的,油滴数曲线有时有若干个峰值。

(3) 在 R 轴方向,不同直径范围的油滴在同一区域出现的机会和数目各不相同,有些区域甚至不存在某种直径范围的油滴,所以喷雾场的微粒化将随时间和空间的不同而有所不同。

3 结 论

（1）应用激光全息术定量研究喷雾场中的空间油滴分布，可以进一步了解油注的发展过程和形成机理。

（2）早期油注，不仅外围逐层撕裂，而且头部也出现局部撕裂和分散的现象。

（3）中心液注是连续的，并一直持续到外围完全微粒化。

参 考 文 献

［1］张煜盛.应用激光全息术研究轴针式喷嘴的燃油喷雾特性［J］.华中工学院学报,1987（3）.

［2］何万详,等.应用激光全息术对柴油机喷雾场分布规律的测量与研究［J］.内燃机学报,1985,3（1）.

［3］李德桃,等.ZS4S1 喷油嘴在无旋流场中喷注贯穿的研究［J］.内燃机工程,1988（2）.

（本文原载于《内燃机工程》1989 年第 3 期）

Investigation on early spray of a pintle type nozzle by holography

Abstract：In this paper, the characters of early spray and the fuel drops distribuion of ZS4S1 pintle type nozzle are investigated by using coaxal holography method. Holograms are gained at ambient temperature and pressure and nonturbulent air flow condition. By reappearing the holograms, numerating and measuring the fuel drops, authors discover that breakup occurs at the tip of early spray column and the fuel core column exists till completely micro-nizing of the outer part. This work is helpful to understand the fuel atomization further and develop an exact atomization model of pintle type nozzle.

涡流室式柴油机镶块材料的研究

钱冀平,李德桃,张吉庆,蒋春华

[摘要] 利用具有国际先进水平的热疲劳试验机,对我国当前大量使用和可能推广的涡流室式柴油机镶块材料进行了模拟试验研究。又对这些材料的镶块,在农村使用条件下进行了试验和考核,为国产农用柴油机寻求成本低、可靠性高的镶块材料,提供科学依据。

涡流室式柴油机镶块承受的热负荷居柴油机各受热零件之首,而且还承受着腐蚀性高速燃气流的冲刷。近年来,国内外对镶块材料(包括陶瓷材料)进行了大量的研究[1~3]。作者针对国内当前大量使用的镶块材料 45 钢和可能推广使用的 HT20-40 铸铁,进行了较深入的试验研究和分析,为面广量大的农用柴油机寻求出成本低、可靠性好的镶块材料,提供科学的参考依据。

1 试验设备和条件

本试验采用先进的 EHF-ES-TF 型热疲劳试验机。

1.1 试验条件

(1) 采用光滑圆柱试样;

(2) 热循环温度范围 100~750 ℃;

(3) 加热、冷却方式采用高频感应装置加热和压缩空气冷却;

(4) 热循环波形为梯形波(如图 1),750 ℃保温 60 s,循环周期 200 s(频率 $f = 0.005$ Hz)。

1.2 失效确定方法

按试样表面出现肉眼可见裂纹的热循环次数作为该材料热疲劳抗力失效的标准。

2 试验结果及分析

2.1 热疲劳试验及分析

热疲劳试验的结果列于表 1,45 钢 2 号试样和 HT20-40 铸铁 2 号试样表面照片见图 2。

由表 1 及图 2 可见,45 钢试样在 205 次循环后就出现明显龟裂现象,而 HT20-40 铸铁

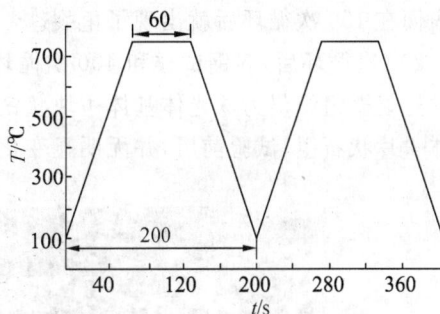

图 1 热循环波形

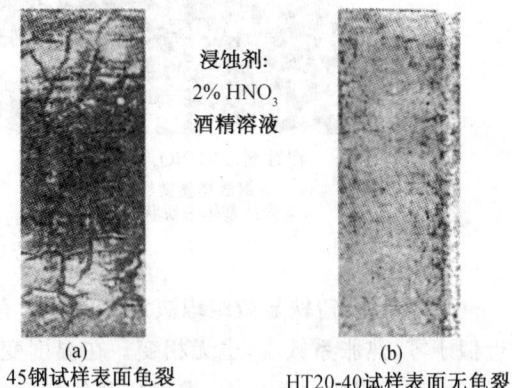

浸蚀剂:
2% HNO₃
酒精溶液

(a) 45钢试样表面龟裂

(b) HT20-40试样表面无龟裂

图 2 热疲劳试验后试样表面(12×)

试样在 401 次循环时,仅形成表面氧化膜和发纹。

表 1　热疲劳试验结果

材料	试样号	失效循环数	试样状况
45 钢	1	210	试样热变形、严重龟裂、氧化剥落
	2	205	试样热变形、明显龟裂
	3	150	试样热变形、明显龟裂
HT20-40	1	296	表面氧化剥落、试样上部断裂
	2	401	表面形成氧化膜、多条发纹
	3	400	表面形成氧化膜、多条发纹

HT20-40 铸铁 1 号试样,在 296 次循环时发生断裂,断口严重烧损。断裂部位位于试样加热段(平行段)与冷却端的过渡区,估计是由铸铁材料内部缺陷(铸造缺陷,材质不均匀等)造成的。而对于铸铁 2 号、3 号试样,虽经 400 次循环,均未出现类似现象。

45 钢试样在热疲劳试验中,当高频加热温度超过 A_{r1} 时,表层呈现奥氏体化,产生体积收缩,加之材料处在高温,强度较低,晶界被氧化变脆,这种收缩易产生晶界开裂;在激冷时,有可能产生淬火或其他转变,冷却过程中表层先是热应力造成收缩,然后,在低温中,组织转变又造成表层膨胀,在亚表面产生拉应力,这种组织应力循环,也可能造成亚表面累积损伤,并形成裂纹。由于这种相变产生的组织应力和试样表里温度不均产生的热应力作用叠加的结果[4],试样表面在 150 次循环后就出现了龟裂。

200 次循环后 45 钢试样和 400 次循环后 HT20-40 铸铁试样的金相组织如图 3 所示。45 钢试样显微组织仍为珠光体基体＋块状铁素体,HT20-40 铸铁试样的显微组织也仍为铁素体基体＋片状石墨,试验前后,并无明显变化。

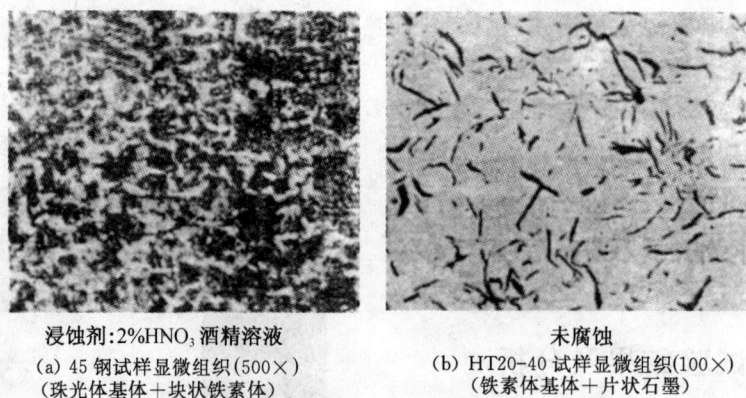

浸蚀剂:2%HNO₃ 酒精溶液　　　　　　　　　　未腐蚀
(a) 45 钢试样显微组织(500×)　　　(b) HT20-40 试样显微组织(100×)
(珠光体基体＋块状铁素体)　　　　　(铁素体基体＋片状石墨)

图 3　热疲劳试验后试样金相组织

HT20-40 铸铁显微组织决定了试样具有较好的热疲劳强度,这是因为石墨的强度和硬度近似于零,热胀系数小,也无相变。在温度变化时,表层的基体金属发生热胀冷缩,石墨区可以变形缓和或吸收这些应力,推迟疲劳龟裂的发生。此外,石墨周围是铁素体或铁素体＋少量珠光体,这些组织的线胀系数小,导热系数大,有利于减小热应力梯度,避免热应力集中。

从试验和上述分析可知,HT20-40 铸铁热疲劳抗力要明显优于 45 钢。

2.2 实用试验

将上述两种材料的涡流室镶块,装于东风 12 型手扶施拉机上,经过约 3 000 h 的田间作业考核试验,至今未发现任何异常现象。在同一作业条件下,每小时燃油消耗量相差不超过 5～20 g。因此,可认为,两种涡流室镶块材料的发动机具有同样良好的技术状态。

3 结 论

(1)热疲劳试验和装机实用考核相结合,是优选涡流室式柴油机镶块材料的基本手段。通过它,可根据柴油机强化程度和使用条件,选择合适的镶块材料。

(2)HT20-40 铸铁的热疲劳强度优于 45 钢,而且价格低,工艺性好,可作为农用柴油机的镶块材料推广使用。

(3)对强化程度高的柴油机,镶块材料有更高、更多方面的要求。在选择镶块材料时,应同时考虑线膨胀系数、抗氧化性、热疲劳强度、导热系数和组织应力等诸因素。

参 考 文 献

[1] 龙祖高,等.涡流室镶块快速试验方法[J].拖拉机,1989(2).

[2] 李惠民.X195 柴油机涡流室镶块奇特失效金相分析[J].拖拉机,1986(6).

[3] Sumio Kamiya, Mikio Murachi, et al. Silicon nitride swirl lower-chamber for high power turbocharged diesel engines[J]. *SAE*, 850523,1985.

[4] 赵建康.铸铁合金及其熔炼[M].北京:机械工业出版社,1985.

(本文原载于《内燃机工程》1990 年第 4 期)

Research on insert materials of swirl chamber deisel engine

Abstract:Modelling experimental research on swirl chamber insert materials which are widely used and possibly spread in domestic has been carried out on an advanced thermal fatique testing machine. The inserts made of these materials were tested and examined under rural application conditions. Then a scientific basis for searching a low cost and high reliability insert material used in agricultural diesel engine is provided.

The experimental and computational investigation of the flow in the diesel swirl chamber

Wang Qian, Luo Tiqian, Li Detao, Yang Weijia

Abstract

In this paper, the air flow field in the swirl chamber of a motored diesel engine was measured by a Laser Doppler Anemometry(LDA). The tested diesel engine was S195 diesel engine with a bell type swirl chamber. This study concentrated on the flow fields during the compression stroke. Measurements were made at engine speed of 900 r/min. The experiments data was acquired and processed by a quick data acquisition and processing system. Based on the measurements and modified k-ε turbulence model, a flow numerical model for the bell type swirl chamber of the tested engine was established. To compare the measured data with the calculated results showed that they were well agreed each other.

1 Introduction

The air-motion in the swirl chamber of a diesel engine is one of key factors to control air-oil mixing. Although the study on it has been carried out for several years, the measurement porcceding was limited to mechanical methods and water-simulation[1,2] without correspondent calculated results. In recent years, new methods and computer technology have been empolyed in studies, but the measurement were conducted in fixed volume model and the results did not denote the real flow. In this paper, the air flow field in the bell type swirl chamber of the S195 diesel engine was measured by LDA. Based on the measurement results and modified k-ε turbulence model, a numerical model was tested and verified.

2 Experimental apparatus and methods

The test engine is a reformed model S195 single cylinder bell type swirl chamber diesel engine, with bore 95 mm, stroke 100 mm, displacement 815 cm³, swirl chamber volume 25.7 cm³ and compression ratio 19. Two quartz windows are installed on each side of the bell type swirl chamber as the laser beam accesses unchanging volume ratio and compression ratio of the original engine(Fig. 1). The measured area is about 80.2 percent of the section of the swirl chamber. The engine is motored via gear box.

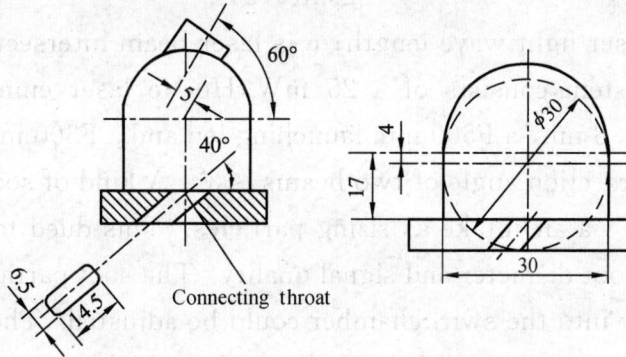

Fig. 1   The size of swirl chamber and the measured area

The experimental arrangement consisting of three parts: (1) mechanic system driving the test engine; (2) LDA; (3) data acquisition and procession system. These are shown in Fig. 2.

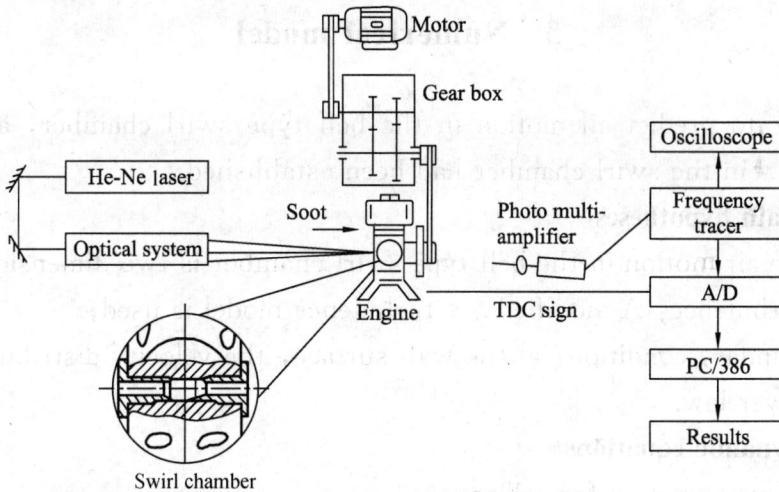

Fig. 2   The experimental arrangement

The LDA is a two demensional differential system. A laser beam is passed through a diffraction fringe rotated in a stable speed and diffracted to 0, ±1, ±2,⋯order diffraction beams. The ±1 order beams go through polalizers separately and become linear polalized laser beams. Then the 0 and ±1 order diffraction beams are focused in the measurment point and interfere each other to form two groups of interference fringes vertical each other with preset frequency

121

shift. This system could measure the two velocity components and judge the flow direction. When a particle suspended in air passes the fringes, it reflects laser lights with Doppler frequencies F_d. Then F_d is measured and the velocity component V_f in vertical direction of the fringes could calculated by the formula:

$$V_f = \frac{\lambda F_d}{2\sin(\alpha/2)}$$

Where λ is laser light wave length; α is laser beam intersection angle.

The LDA system consists of a 25 mW He-Ne laser emmiting light with wavelength of 632.8 nm, a F500 mm launching len and a F300 mm receiving Fourier len. The intersection angle of two beams is 3°. A kind of soot is sprayed into the swirl chamber via air intake as sizing particles. This dued to considering the relation between soot diameter and signal quality. The soot particle size about 1.2 μm and its density into the swirl chamber could be adjusted. The velocity of soot could approximately represents air velocity and could be actually calculated by computer:

$$V_f = 1.342 \times 10^{-3} \times (127.589 + 0.078\,778V)$$

Where V is mean A/D converting simulating voltage.

3　Numerical model

In order to predict air-motion in the bell type swirl chamber, a numerical model of flow in the swirl chamber had been established.

3.1　The main hypotheses

(a) The air-motion in the bell type swirl chamber is two dimensional;

(b) Turbulence: A modified k-ε turbulence model is used;

(c) Boundary condition: at the wall surface, the velocity distribution agrees with the power law.

3.2　Aerodynamic equations

Continuous equation for cell m is:

$$\frac{\delta\rho_m}{\delta t} + \nabla(\rho_m u) = \nabla\left[\rho D\nabla\left(\frac{\rho_m}{\rho}\right)\right] \tag{1}$$

Where ρ_m is mass density of m; ρ is total mass density; u is air velocity; D is the diffusivity of m.

The momentum equation for the fluid mixture is:

$$\frac{\partial(\rho u)}{\partial t} + \nabla(\rho uu) = +\frac{1}{\alpha^2}\nabla p - A_0\nabla(2/3\rho k) + \nabla\sigma + \rho g \tag{2}$$

Where ρ is air pressure; α is undimensional value; $A_0 = 0$ if the flow is laminal and $A_0 = 1$ if the flow is turbulent; σ is the viscous stress tensor given by:

$$\sigma = \mu [\nabla u + (\nabla u)^{\mathrm{T}}] + \lambda \nabla u Q \tag{3}$$

Where Q is the unit dyadic.

The internal energy equation is:

$$\frac{\partial(\rho I)}{\partial t} + \nabla(\rho u I) = -\rho \nabla u + (1 - A_0)\sigma \nabla u - \nabla J + A_0 \rho \varepsilon \tag{4}$$

Where I is the specific internal energy of the fluid (chemical energy exclusive); J is the heat flux vector given by:

$$J = -K \nabla T - \rho D \sum_m h_m \nabla(\rho_m / \rho) \tag{5}$$

Where T is the absolute temperature and h_m is the specific enthalpy of cell m.

The state relations for ideal gas are:

$$p = R_0 T \sum_m (\rho_m / W_m) \tag{6}$$

$$I(T) = \sum_m (\rho_m / \rho) I_m(T) \tag{7}$$

$$C_p(T) = \sum_m (\rho_m / \rho) C_{pm}(T) \tag{8}$$

$$h_m(T) = I_m(T) + R_0 T / W_m \tag{9}$$

Where $I_m(T)$ is the specific internal energy for cell m; $C_{pm}(T)$ is the specific heat at constant pressure for cell m; W_m is the molecular weight of cell m; R_0 is the universal gas constant.

The transport coefficients are:

$$\mu = (1.0 - A_0)\rho \nu_0 + \mu_{\mathrm{air}} + A_0 C_\mu k^2 / \varepsilon \rho \tag{10}$$

$$\lambda = A_3 \mu \tag{11}$$

$$K = \frac{\mu C_p}{Pr} \tag{12}$$

$$D = \frac{\mu}{\rho Sc} \tag{13}$$

Where ν_0 is a constant decided by uniform back-ground turbulent diffusivity; $\mu_{\mathrm{air}} = A_1 T^{3/2}(T + A_2)$, where A_1 and A_2 are constant, $A_3 = 2/3$ if the flow is turbulent, Pr is the Prandtl number and Sc is the Schmidt number.

This numerical model could be used to calculate the flow in various type swirl chamber. In this paper (Fig. 3), only the flow in the bell type swirl chamber during the

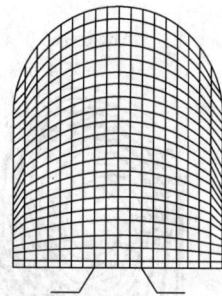

Fig. 3　The meshed grid of the swirl chamber

compression stroke was calculated. The bell type swirl chamber was divided into
22×19 cells.

4 Calculated results

Fig. 4 shows the calculated results of the flow in whole bell type swirl chamber with 40° inclination angle of connecting throat.

During the compression stroke period, the flow velocity in the swirl chamber goes up with the crank angle increasing. The velocity increases rapidly after about BTDC 130° CA. The velocity in the connecting throat achieves maximum one at BTDC 23° CA, and at TDC it drops down to zero. In this working period the flow in the bell type swirl chamber center is more powerful than that in the other area. There is a flow detention area in the up-stream side near the wall due to the friction between the air and the wall. In addition the position of vortex core in the swirl chamber moves towards the left-below gradually with crank angle rising from BTDC 100° CA to TDC and it is under the center of the swirl chamber at last. As shown in Fig. 4, there are two subvortex at the corner of the bell type swirl chamber. This agrees with the results of water-simulation[2].

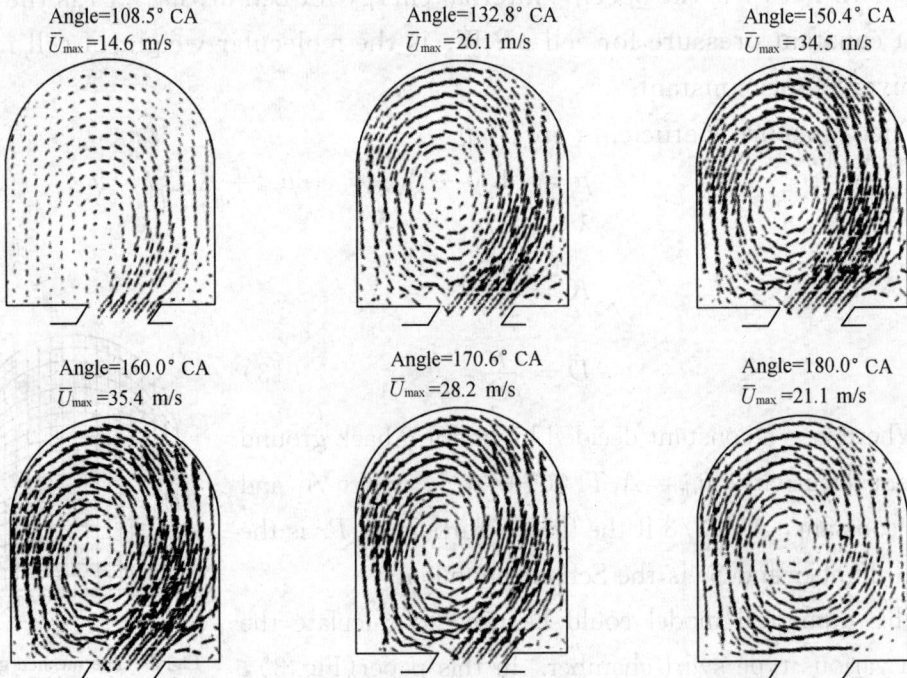

Angle=108.5° CA
$\bar{U}_{max}$=14.6 m/s

Angle=132.8° CA
$\bar{U}_{max}$=26.1 m/s

Angle=150.4° CA
$\bar{U}_{max}$=34.5 m/s

Angle=160.0° CA
$\bar{U}_{max}$=35.4 m/s

Angle=170.6° CA
$\bar{U}_{max}$=28.2 m/s

Angle=180.0° CA
$\bar{U}_{max}$=21.1 m/s

Fig. 4 The calculated results

5 Analysis of results

Fig. 5 shows that the velocity in point 1 and point 2 locates at $r/r_0 = 0.7$ (where r_0 is radius of the swirl chamber) of the bell type swirl chamber change with the crank angle. There the actual line denotes the measured results and the dotted line denotes the calculated results. Two curves have the same shape and nearly same value. The maximum velocity in point 1 occurs at BTDC 23° CA(the calculated value $\overline{U}_{max}$ is 38.3 m/s and the measured value $\overline{U}'_{max}$ is 35.4 m/s). In point 2 the maximum velocity occurs at BTDC 10° CA(the calculated value $\overline{U}_{2max}$ is 21.2 m/s and the measured value $\overline{U}'_{2max}$ is 18.6 m/s).

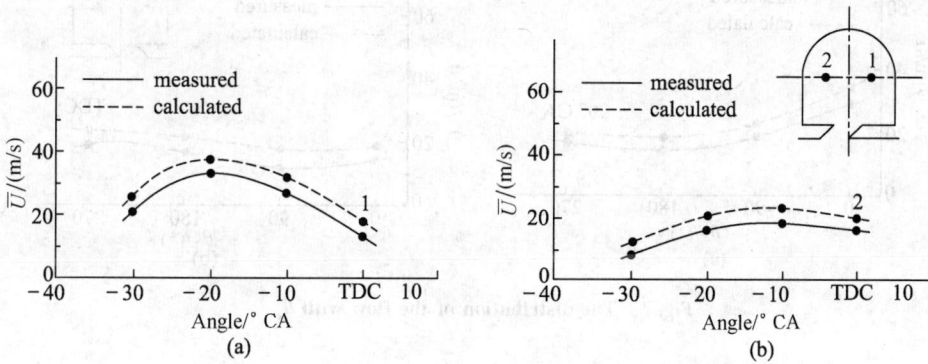

Fig. 5 **The velocity of flow changes with crank angle**

Fig. 6 shows the measured and calculated velocity $\overline{U}$ change along the r/r_0 at BTDC 20° CA and at TDC. The maximum velocity occurs between $r/r_0 = 0.7$ and 0.9. When the crank angle increases from BTDC 20° CA to TDC, the maximum volecity position moves towards the center of the chamber. Comparing the measured results with calculated ones in Fig. 6a and Fig. 6b, the two results show fairly good agreement.

Fig. 7 shows the distribution of velocity along $r/r_0 = 0.7$ circular in the bell type swirl chamber at TDC and at BTDC 20° CA. The angle θ_S nominates the position. The mean velocity achieves maximum when $\theta_S = 0$ at BTDC 20° CA. The measured $\overline{U}_{max}$ is 33.5 m/s and the calculated $\overline{U}_{max}$ is 29.8 m/s. When $\theta_S = 0$ the concerned point locates near the connecting throat and is directly influenced by the flow through the throat from the main chamber, especially at BTDC 20° CA. When θ_S goes up to 90°, $\overline{U}$ achieves its low limit and then keeps constant. At TDC $\overline{U}$ keeps a steady value because the air velocity in the connecting throat is nearly zero, shown in Fig. 7b clearly.

125

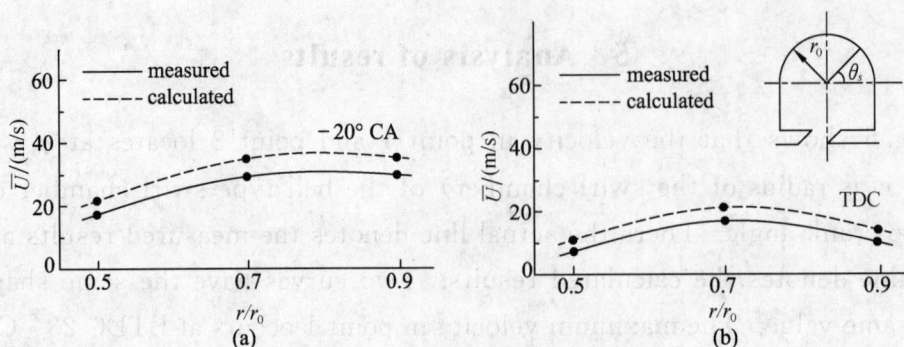

Fig. 6 The velocity changes with r/r_0

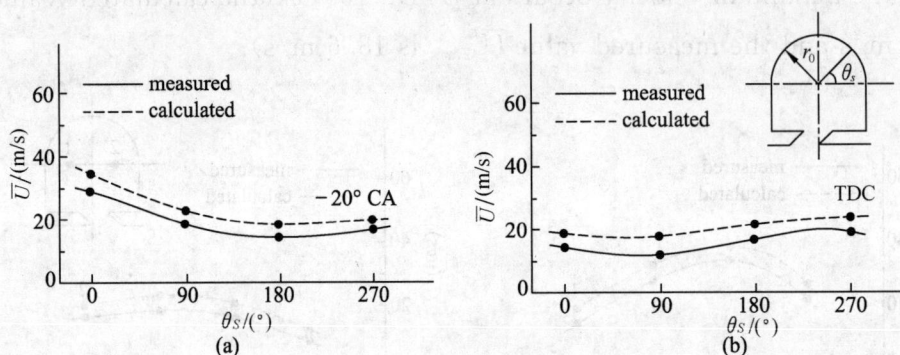

Fig. 7 The distribution of the flow with θ_s

6 Conclusions

(1) In compression process, the tangential velocities in the bell type swirl chamber vary with the crank angle. The velocity in the upstream side goes up with rising of the crank angle. At BTDC 23° CA the velocity gets maximum value. In the downstream side velocities nearly keep constant after BTDC 20° CA and is lower than in upstream side. There are two subvertexes in the corners of the chamber.

(2) In compression process, there are strong vortex motion. In the half compression period, it is a quasi-solid vortex, then changes to quasisolid vertex in the end of the compression period. The vertex core is located near the chamber center at the begining and in the middle of the compression stroke, then moves down left and at TDC it reaches the down-center of the chamber.

7. Nomenclature

λ——laser light wave length, mm

α——laser beam intersection angle, deg

F_d——Laser Doppler frequency, Hz

V_f——the velocity of soot, m/s

V——mean A/D converting simulating voltage, V

ρ_m——mass density of m, kg/m^3

ρ——total mass density, kg/m^3

$\overline{U}$——air velocity, m/s

D——the diffusivity, m^2/s

p——pressure, N/m

σ——the viscous stress tensor

Q——the unit dyadic

I——the specific internal energy, J

J——the heat flux vector

T——the absolute temperature, K

h_m——the specific enthalpy of cell m, kJ/(kg · K)

$I_m(T)$——the specific internal energy for cell m, kJ/(kg · K)

$C_{pm}(T)$——the specific heat at constant pressure for cell m, kJ/(kg · K)

W_m——the molecular weight of cell m

R_0——the universal gas constant

ν_0——constant

μ_{air}——diffusivity of the molecular, m^2/s

A_0, A_1, A_2, A_3——constant of the model

Pr——the Prandtl number

Sc——the Schmidt number

r——radius coodinate

r_0——radius, mm

$\overline{U}_{max}$——the maximum flow velocity, m/s

θ_S——angle, deg

∇——Hamilton operator

Subscrip

air——air flow

Σ——total

m——cell m

References

[1] Ricardo H P. *The High Internal Combustion Engine* [M]. London: Blakie And

Sons，1953.

[2] Li Detao，Bemdean V. Study on air movement in separate swirl chamber by means of a bidimensional dynamic liquid mode[J]. *Rev Rom Sci Tech Mec Appl*，1984.

[3] Gen'ichi KOMATSU，Makoto IKEGAMI. Numerical simulation of air motion in the swirl chamber of diesel engines[C]. *Proceeding of the 5th Internal Combustion Engine Symp of Japan*，86—0693A：1480.

[4] Noboru Miyamoto, et al. Air flow analysis in swirl chamber with hot-wire anemometry [J]. *Proceeding of the JSME Annual meeting*，1990(45)：9.

[5] Wu Shanmou, et al. The investigation of swirl and turbulence in diesel swirl chamber [J]. *Transaction of CSICE*，1990,8(3)：27.

(From：*International Symposinm COMODIA*，Japan，1994)

柴油机涡流室内流场的实验和计算研究

[摘要]　本文报导了采用激光技术对柴油机涡流室内流场进行实测的结果。用计算机及高速数据采集系统完成了试验数据的采集和处理。作者还模拟实验发动机建立了吊钟型涡流室内流场的计算模型，并进行了模拟计算。通过试验结果和计算结果的对比分析，研究了涡流室内流场及其随曲轴转角的变化规律。试验结果与计算结果有较好的一致性。

运用韦柏函数分析发动机燃烧过程时
若干问题的探讨

熊　锐,李德桃,朱亚娜,黄跃欣

[摘要]　探讨了韦柏函数的内涵及其在实际应用中的若干问题;应用双韦柏函数对涡流室式柴油机的燃烧过程进行分析,加深了对分隔室式发动机燃烧过程的认识。

从发动机实测示功图分析放热规律是目前研究燃烧过程最行之有效的方法之一。描述放热过程最常用的韦柏函数正是基于化学反应动力学基本方程和大量实测示功图而推导出的,它与示功图的紧密联系,较好地逼近实际放热率曲线,其参数又具有明确的物理含义,便于对燃烧过程深入分析,因此进一步加强对韦柏函数在实际应用中的研究很有意义。

自韦柏函数问世以来[1],不断有应用该函数分析发动机燃烧过程的报道,但据我们所知,对该函数在实际应用中的发展还缺乏系统的分析。本文简单介绍韦柏函数模型,对韦柏函数的内涵及其在实际应用中的若干问题进行了探讨,还应用双韦柏函数分析涡流室式柴油机的燃烧过程,并据此提出了几个提高发动机性能的措施。

1　韦柏函数及其内涵

1.1　单韦柏燃烧放热模型

韦柏基于有关链式反应速度的原理,推导出在燃烧期间烧掉的燃料百分数 χ 与从着火时刻算起的无量纲曲轴转角 φ/φ_z 的关系为

$$\chi = 1 - \exp[c(\varphi/\varphi_z)^{m+1}] \tag{1}$$

式中 φ_z 为以曲轴转角表示的燃烧延续期;m 为燃烧品质指数;系数 c 由在反应实际结束时烧掉的燃料百分数 χ_z 确定,$c = \ln(1-\chi_z)$,韦柏假定 $\chi = 99.9\%$,算得系数 c 为 -6.908。

由公式(1)得到韦柏放热速率函数表达式:

$$\frac{dQ}{d\varphi} = 6.908 \frac{Q_0}{\varphi_z}(m+1)\left(\frac{\varphi}{\varphi_z}\right)^m \exp\left[-6.908\left(\frac{\varphi}{\varphi_z}\right)^{m+1}\right] \tag{2}$$

式中 Q_0 为每循环燃料燃烧放热量。

决定韦柏燃烧放热函数的参数是燃烧品质指数 m 和燃烧持续期 φ_z。用韦柏函数拟合实际放热率时,需要由实际燃烧放热规律来确定韦柏函数中的 m 和 φ_z 值,两参数的准确性直接影响着韦柏函数的拟合精度,文献[1-4]中详细地介绍了参数 m 和 φ_z 不同的确定方法。

单韦柏函数是针对汽油机近似均匀混合气的情况下,综合大量试验数据以反应动力学推导出的半经验公式,应用在中低速柴油机上计算结果与试验数据比较吻合,但却不能反映高速柴油机实际放热规律的双峰特点。

1.2 韦柏函数内涵

在推导韦柏函数时,韦柏假定在燃烧反应结束($\varphi=\varphi_z$)时烧掉的燃料百分数为 $\chi_z=$ 99.9%,由此推算出公式(1)中的系数 $c=\ln(1-\chi_z)=-6.908$。作者认为这一系数可进一步探讨。其一,影响发动机燃烧过程完善程度的因素很多,涉及发动机的结构设计、各系统的优化匹配、运行工况等,不同机型在不同工况下的燃烧程度有所不同,不能绝对地认为在所有情况下燃烧持续期内烧掉的燃料份额皆为 99.9%。当某台发动机在某种工况下 $\chi=99.8\%$ 时,韦柏函数的系数则应为 $c=-6.215$,即 χ_z 的假定只要有 0.1% 的误差,将导致系数 c 出现 11.3% 的偏差。其二,由韦柏函数拟合实际放热率时,韦柏函数中最佳燃烧延续期为 $\varphi_z=(50\pm 10)°\text{CA}^{[1]}$,然而在这么短的燃烧延续期内烧掉的燃料百分数不一定能达到 99.9%。有研究结果表明[5]:如果取 $\chi_z=95\%$,则能与实际放热特性更好地吻合,在这种情况下系数 $c=-2.996$。因此在应用韦柏函数时,需要根据具体的机型和工况对韦柏函数的系数作适当的修正。

韦柏函数中参数的确定是建立在由实测示功图得到的放热规律基础上的,韦柏函数模型能否真实地反映发动机中实际燃烧放热过程,首先取决于实测示功图的测试精度,同时也涉及如何确定燃烧期间缸内工质向气缸壁的传热损失。韦柏[1]假定在整个燃烧期间导向气缸壁的热损失所占燃料燃烧热的百分比不变,其值由试验确定,其他学者多通过计算传热系数来求解传热损失,文献[6]考察了 Sitkei, Jyen V, Eichelberg 和 Woschni 四个传热系数公式对同一试验机放热规律计算结果的影响,研究表明由不同传热系数计算得的放热规律相差极小,这点也是应用韦柏函数时应该考虑到的。运用现代测试技术获得发动机的示功图,使所记录的压力、上止点位置都具有很高的精确度,对韦柏函数的拟合精度至关重要。

2 韦柏函数应用中的几个问题

从发动机燃烧过程来看,放热过程实际上是预混合燃烧和扩散燃烧放热的综合叠加[7]。发动机(特别是高速机)的这一双相燃烧特征使人们不可能用单韦柏函数准确地描述整个燃烧过程的各个部分。因此有必要改进韦柏函数,从总放热率曲线中分出预混合和扩散燃烧放热,从而能更好地分析、评价发动机的燃烧过程。

2.1 双韦柏函数燃烧放热模型

采用双韦柏函数作为高速机燃烧放热率的模型(见图1),其形式如下:

$$\frac{\text{d}Q}{\text{d}\varphi}=6.908\frac{Q_p}{\varphi_p}(m_p+1)\left(\frac{\varphi}{\varphi_p}\right)^{m_p}\times$$

$$\exp\left[-6.908\left(\frac{\varphi}{\varphi_p}\right)^{m_p+1}\right]+6.908\frac{Q_d}{\varphi_d}\times$$

$$(m_d+1)\left(\frac{\varphi}{\varphi_d}\right)^{m_d}\exp\left[-6.908\left(\frac{\varphi}{\varphi_d}\right)^{m_d+1}\right] \qquad (3)$$

式中下标 p,d 分别表示预混合燃烧和扩散燃烧部分,Q_p,Q_d 为燃烧放热值;m_p,m_d 为燃烧品质指数;φ_p,φ_d 为燃烧持续期。若两个韦柏函数始点相同而燃烧持续期不同,

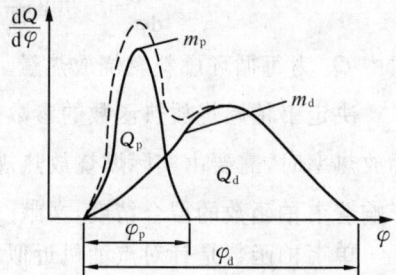

图 1 双韦柏函数模拟放热规律图

则 φ_{d} 即为燃烧过程的总持续期。

用双韦柏函数拟合放热率曲线的问题,就是要根据实际放热率曲线确定双韦柏函数中的参数,使双韦柏函数准确地表达被拟合放热率曲线。取 $\mathrm{d}Q/\mathrm{d}\varphi$ 与实际放热率曲线的误差平方和函数作为双韦柏函数曲线与被拟合曲线逼近程度的度量,由于被拟合曲线由离散的数值构成,双韦柏函数的参数用最小二乘法由下式确定:

$$F(Q_{\mathrm{p}},Q_{\mathrm{d}},m_{\mathrm{p}},m_{\mathrm{d}},\varphi_{\mathrm{p}},\varphi_{\mathrm{d}}) = \min \sum_{i=1}^{n} \left[f_{i(\varphi)} - \mathrm{d}Q/\mathrm{d}\varphi \right]^2 \qquad (4)$$

式中 n 为拟合点总数, $f_{i(\varphi)}$ 为实际放热速率数据,考虑到一定的发动机实际工况时其预混合与扩散燃烧放热之总和等于总放热量,即 $Q_{\mathrm{p}} + Q_{\mathrm{d}} = \mathrm{const}$,因此,式(4)中需确定的独立变量有5个。

双韦柏函数能较精确地描述发动机放热率曲线的形态,对其参数的讨论有助于加深对燃烧过程双相燃烧特征的认识,但用最小二乘法确定双韦柏函数的参数时计算工作繁重,对试验数据准确性要求很高[1]。故有文献[8,9]采用简化形式的韦柏函数来拟合高速机的实际放热率。

2.2 其他函数加韦柏函数燃烧放热模型

文献[10]用等腰三角形近似描述预混合燃烧,扩散燃烧则用单韦伯函数来近似模拟(图2)。

$$\frac{\mathrm{d}Q_{\mathrm{p}}}{\mathrm{d}\varphi} = \begin{cases} 4Q_{\mathrm{p}} \dfrac{\varphi}{\varphi_{\mathrm{p}}^2}, & 0 \leqslant \varphi \leqslant \dfrac{\varphi_{\mathrm{p}}}{2} \\ 4Q_{\mathrm{p}} \dfrac{(\varphi_{\mathrm{p}}-\varphi)}{\varphi_{\mathrm{p}}^2}, & \dfrac{\varphi_{\mathrm{p}}}{2} \leqslant \varphi \leqslant \varphi_{\mathrm{p}} \end{cases} \qquad (5)$$

$$\frac{\mathrm{d}Q_{\mathrm{d}}}{\mathrm{d}\varphi} = 6.908 \frac{Q_{\mathrm{d}}}{\varphi_{\mathrm{d}}} (m_{\mathrm{d}}+1) \left(\frac{\varphi}{\varphi_{\mathrm{d}}}\right)^{m_{\mathrm{d}}} \exp\left[-6.908\left(\frac{\varphi}{\varphi_{\mathrm{d}}}\right)^{m_{\mathrm{d}}+1}\right], \quad 0 \leqslant \varphi \leqslant \varphi_{\mathrm{d}} \qquad (6)$$

$$\frac{\mathrm{d}Q}{\mathrm{d}\varphi} = \frac{\mathrm{d}Q_{\mathrm{p}}}{\mathrm{d}\varphi} + \frac{\mathrm{d}Q_{\mathrm{d}}}{\mathrm{d}\varphi} \qquad (7)$$

用等腰三角形加韦柏函数作为高速机模拟放热规律的基函数,能较准确地反映高速机放热规律的双峰特性,其函数形式较双韦伯函数简练,函数中包含的待定参数也较双韦柏函数的少。模拟计算表明,等腰三角形加韦柏函数模型比较适用于高速直喷式柴油机的实际状况[8],但在拟合分隔室式柴油机实际放热率工线时有较大偏差。作者认为,还需更深入地探讨等腰三角形加韦伯函数模型对不同机型的普适性。

文献[9]采用一个韦柏函数和两个Guass函数的组合来模拟发动机实际放热规律,拟合结果表明所采用的模型能较准确地预示发动机性能的变化规律。

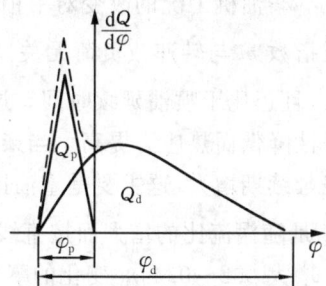

图2 等腰三角形加韦柏函数模拟放热规律图

2.3 对韦柏函数拟合扩散燃烧放热的探讨

柴油机中喷雾的扩散燃烧很大程度上取决于喷雾特性和燃油的蒸发、混合过程。受混合速率的限制,在扩散燃烧初期实际燃烧放热率要比韦柏函数放热率值低;而在燃烧后期,由于初期所喷燃料的未燃部分同后期喷入的燃料一同燃烧,使实际放热率值比韦伯函数放热率值高,这造成在严格的意义上只适用于均质预混合气体燃烧过程的韦柏函数在拟合扩散燃烧时存在不足。可用修正双韦伯函数中扩散燃烧相系数的方法予以改进[7]：

$$\frac{\mathrm{d}Q_\mathrm{d}}{\mathrm{d}\varphi}=\frac{k}{1-\mathrm{e}^{-k}}\frac{Q_\mathrm{d}}{\varphi_\mathrm{d}}(m_\mathrm{d}+1)\left(\frac{\varphi}{\varphi_\mathrm{d}}\right)^{m_\mathrm{d}}\exp\left[-k\left(\frac{\varphi}{\varphi_\mathrm{d}}\right)^{m_\mathrm{d}+1}\right] \tag{8}$$

式中 k 为反映扩散燃烧与预混合燃烧差异的参数，k 越大，表示喷雾特性、混合过程等因素对燃烧造成的放热率曲线与预混燃烧放热率的差异越小，当 k 为 6.908 时，由于 $\mathrm{e}^{-k}\ll 1$，则式(8)即为韦柏函数。

为了减少韦柏函数拟合扩散燃烧放热时的偏差，也可对双韦柏函数中预混和扩散两个燃烧相的系数和形式同时作出修正[10]：

$$\frac{\mathrm{d}Q}{\mathrm{d}\varphi}=m_\mathrm{p}A_\mathrm{p}\frac{Q_\mathrm{p}}{\varphi_\mathrm{cp}}\left(\frac{\varphi}{\varphi_\mathrm{cp}}\right)^{m_\mathrm{p}}\exp\left[-\frac{m_\mathrm{p}}{m_\mathrm{p}+1}\left(\frac{\varphi}{\varphi_\mathrm{cp}}\right)^{m_\mathrm{p}+1}\right]+$$

$$m_\mathrm{d}A_\mathrm{d}\frac{Q_\mathrm{d}}{\varphi_\mathrm{cd}}\left(\frac{\varphi}{\varphi_\mathrm{cd}}\right)^{m_\mathrm{d}}\exp\left[-\frac{m_\mathrm{d}}{m_\mathrm{d}+1}\left(\frac{\varphi}{\varphi_\mathrm{cd}}\right)^{m_\mathrm{d}+1}\right] \tag{9}$$

式中 A_p、A_d 为决定放热特性的系数，$A_\mathrm{p}+A_\mathrm{d}\approx 1$；$\varphi_\mathrm{cp}$、$\varphi_\mathrm{cd}$ 分别为预混和扩散两个燃烧期内放热率达到最大值时的曲轴转角。对某柴油机放热率拟合表明，计算与实测值具有较好的一致性。

2.4 特殊工况下韦柏函数应用中的问题

在模拟发动机放热率时，不同机型所采用的韦柏函数形式应该有所不同。单韦柏函数在模拟汽油机和中低速柴油机放热率曲线时已能满足工作过程计算的要求，在拟合压缩机式柴油机放热率时也能达到较高的精度[1]，这是因为压缩机式柴油机是用压缩空气把燃料送入燃烧室中，不仅保证了燃油雾化得细小，同时也产生了特殊形式的混合气形成和燃烧过程。但在模拟具有双峰特征的高速柴油机放热规律时，则应采用双韦柏函数模型或变燃烧品质指数的单韦柏函数模型。对 285F 压缩式双燃料发动机放热规律的研究表明[9]，实际放热率曲线除了可用双韦柏函数或韦柏函数与其他函数的组合模型来模拟之外，还可用单韦柏函数较好地拟合，但其燃烧品质指数 m 应随燃烧进程的变化而变化，燃烧初期 m 很小，主燃期内 $m\leqslant 0.5$，后燃期后 m 开始迅速增大。

柴油机工况的改变对韦柏函数中参数的变化产生影响[1]。在直接喷射的情况下，燃烧品质指数 m 与转速或负荷无关。以曲轴转角表示的燃烧延续期 φ_z 随转速或负荷的提高而增大，且正比于喷油延续时间。过大的 φ_z 值同数值小的 m 相配合会引起燃烧延迟，使柴油机经济性降低而热应力提高。当柴油机用重油为燃料时，与轻柴油相比，其燃烧特征指数不变而燃烧延续期增大，这主要是重油比重大、蒸发慢的缘故。对于有涡流作用的柴油机，燃烧延续期 φ_z 是随涡流比的增大而按直线规律减少的。柴油机增压度提高时 φ_z 减小，而燃烧品质指数 m 增大 0.2～0.4，m 变化的事实可以解释为什么随着增压度的提高，柴油机工作比较柔和。

柴油机工作环境的改变也使韦伯函数的参数发生变化[8]。与平原工况相比，柴油机在高原工况工作时压缩过程压力下降，滞燃期增加，滞燃期内形成的可燃混合气明显增多，致使燃烧过程预混燃烧部分占有很大比例，因而用韦柏函数拟合的预混燃烧部分显得更为重要。预混燃烧量增加使燃烧开始后压力、温度上升很快，这一方面使高原工况整个燃烧持续期缩短，另一方面扩散燃烧重心前移，扩散燃烧品质指数 m_d 减小。

运用韦柏函数分析分隔室式发动机燃烧过程的工作，目前多局限于对发动机总燃烧放热率进行分析[11]。事实上，主燃室与副燃室的燃烧过程既互相联系又有所区别，两室的燃烧特性及其对发动机性能的影响有着不同的规律，这些不同规律在仅对总放热速率进行分析时得不到反映。因此,应该分别研究主、副燃烧室燃烧特性及总放热规律的变化对发动机性能的影响,这方面尚有大量工作要做。

3 用韦柏函数分析涡流室式柴油机燃烧过程

运用双韦柏函数对 S195 涡流室式柴油机燃烧速率进行拟合计算,得到双韦柏函数燃烧速率曲线的特征参数随负荷的变化(如图 3 所示)。

由图 3a 知,在副燃室中,φ_p,m_p,m_d 基本保持不变。预混合燃烧放热量 Q_p 随负荷变化在 9 马力附近出现最大值后,随着负荷的增加,其在副燃室总放热量 Q_2 中所占的比例则缓慢下降。φ_d 随负荷的增加略有上升,但在上止点后 45° CA 前较早结束。

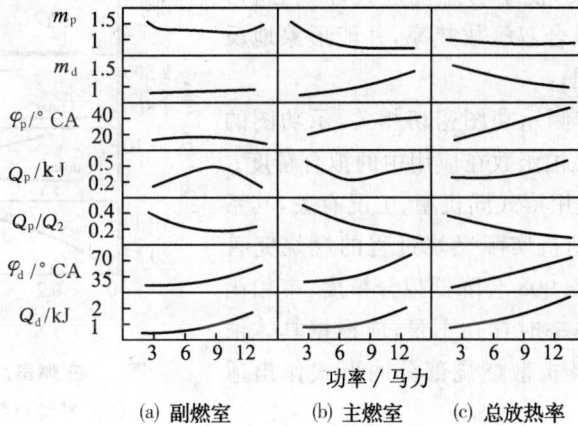

图 3 双韦柏函数参数随负荷的变化特性

如图 3b 所示,在主燃室中,m_p 基本上不随负荷变化,Q_p 在 6 马力附近出现最低值,随着负荷的增加,Q_p 在主燃室总放热量中所占的比例逐渐减少。在小负荷时,主燃室的燃烧较早结束,主、副室的燃烧持续期较接近;在大负荷时,主燃室 Q_d,φ_d 上升幅度较大,后燃延长。而主燃室预混合燃烧持续期较副燃室的长,在 30~45° CA 范围内随负荷的增加而延长。

通过分析涡流室式柴油机主、副燃室燃烧速率曲线特性参数对柴油机性能的影响发现:

在副燃室燃烧特性参数不变时,增加主燃室的预混合燃烧放热量 Q_1,可较多地提高指示热效率,但同时主燃室的最大压力 p_{1max}、最高温度 T_{1max} 和最大压力升高率 $dp_1/d\varphi$ 也随之增加。当降低主燃室预混合燃烧特征指数 m_{1p} 以使预混合燃烧重心提前,或是在预混合燃烧持续期 φ_{1p} 不大的情况下缩短 φ_{1p},虽使指示热效率略有提高,但造成燃烧粗暴,主燃室的 p_{1max} 和 $dp_1/d\varphi$ 急剧上升。而缩短扩散燃烧持续期 Q_{1d} 或降低扩散燃烧特征指数 m_{1d},可在 p_{1max},T_{1max} 和 $dp_1/d\varphi$ 略有增加的情况下,有效地提高指示热效率。

在主燃室燃烧特性参数不变时,增加副燃室预混燃烧量 Q_{2p} 或降低副燃室扩散燃烧特征指数以使副室扩散燃烧重心提前,虽然指示热效率略有提高,但使副室的最大压力 p_{2max}、最大压力升高率 $dp_2/d\varphi$ 及主室的 $dp_1/d\varphi$ 增加较大。在副燃室燃烧持续期 φ_{2d} 小于 60° CA 时,适当推迟副燃室燃烧结束时间,可以在指示热效率降低甚微而主、副燃室温度变化不大的情况下,大大降低主、副燃室的最大压力和最大压力升高率。

主、副燃室的放热量分配对发动机的性能有很大影响。S195 柴油机主燃室放热量占总放热量的比例 Q_1/Q 对发动机性能的影响如图 4 所示。随着主燃室放热比例的增加,指示热效率 η_i 增加较大,主燃室 p_{1max},T_{1max} 和 $dp_1/d\varphi$ 稍有增加,而副燃室内的 p_{2max},T_{2max} 和 $dp_2/d\varphi$ 迅速降低。因此,增加涡流室式柴油机主燃室的放热量,可以在不降低发动机其他性能指标的前提下有效地提高发动机的指示效率。

4 结 论

(1) 韦柏函数是一种可靠的代用燃烧模型,其参数具有明确的物理意义,便于对燃烧过程深入分析,双韦柏函数能较好地拟合双峰放热率,并能形象地反映预混和扩散燃烧过程;

(2) 韦柏函数与实测示功图密切相关,示功图的测试精度直接影响着韦柏函数在应用中的拟合精度;

(3) 韦柏函数的应用形式同机型、工况有关,其系数值的大小取决于发动机实际燃烧过程的燃烧完善程度,笼统采用系数 −6.908 会降低拟合精度,韦柏函数在拟合扩散燃烧放热率时存在不足,应根据具体情况对韦柏函数的系数及扩散燃烧部分的形式作出适当的修正;

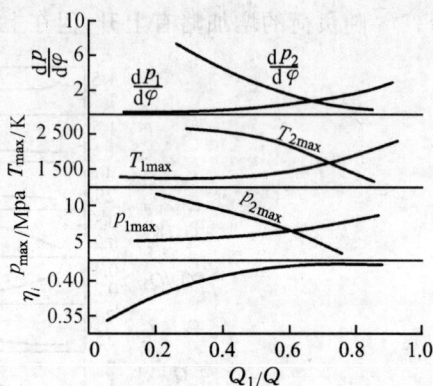

图 4 主燃室放热比例 Q_1/Q 对发动机性能的影响

(4) 运用韦柏函数分析分隔室式发动机燃烧过程时,应分别研究主、副燃室燃烧特性及总放热规律对发动机性能的影响,控制主、副燃室燃烧延续时间小于 $60°\mathrm{CA}$ 及增加主燃室的放热量,有助于提高发动机性能,改进经济性。

参 考 文 献

［1］Вибе Н Н. *Новое о рабочем цикле дыигателей（скорость сгорашия рабоций цикл двигателя）*［M］. Москва,1962.

［2］Wotson N, et al. A combustion correlation for diesel engine simulation［J］. *SAE Paper*,800029,1980.

［3］Miyomoto N, et al. Description and analysis of diesel engine rate of combustion and performance using Wiebe's Functions［J］. *SAE Paper*,850107,1985.

［4］Stas M,Wajand J A. Bestimmung der vibe-parameter fur den zweiphasigen brennverlauf in direkteinspritz-dieselmoteren［J］. *MTZ*,1988,49(7/8).

［5］奥林 A C,等.活塞式及复合式发动机原理［M］.北京:机械工业出版社,1987.

［6］严新娟,等.涡流室式柴油机工作过程计算中几个问题的探讨［C］// 全国高校工程热物理第四届学术会议论文集,1992.

［7］纪丽伟,等.单峰放热率曲线的双韦柏函数拟合探讨［J］.内燃机工程,1993(1).

［8］李照东,等.高速直喷式柴油机高原工况模拟放热规律的研究[J].内燃机工程,1989(3).

［9］费少梅.压燃式双燃料发动机燃烧模型的研究[D].杭州:浙江大学,1993.

［10］刘佳才,等.内燃机振动声学特性的预测[J].内燃机学报,1993(3).

［11］朱广圣,李德桃.运用韦柏函数分析分隔室式发动机的燃烧过程[C]∥中国内燃机学会燃烧专业委员会第五届年会论文,1989.

（本文原载于《兵工学报：坦克装甲车与发动机分册》1994 年第 2 期）

Study on several problems of analysing engine combustion processes using Wiebe's Functions

Abstract：Intension and several problems in the actual use of Wiebe's Functions are studied. The combustion processes of a swirl chamber diesel engine are analyzed by using two Wiebe's Functions，getting a deeper understanding in combustion processes of separated chamber diesel engines.

用 LDA 研究柴油机涡流室内空气运动规律

黄跃欣，王　谦，吴志新，李德桃

[摘要]　利用二维氩离子激光多普勒测速仪(LDA)实测柴油机涡流室内空气运动，揭示柴油机倒拖工况下涡流室内空气运动随曲轴转角的变化规律以及涡流室内流场的空间分布特点。对深入理解和认识柴油机涡流室内空气运动规律及进一步建立涡流室式柴油机燃烧模型具有重要意义。

柴油机燃烧室和气缸中的空气运动对混合气的形成及燃烧过程都有很大影响。涡流室中的空气运动对燃烧过程的影响尤为显著[1]。测量涡流室中的空气运动不仅对深入了解涡流室中空气运动有益，也是建立精确的涡流室式柴油机燃烧模型的基础。激光多普勒测速仪是一种区别于热线风速仪，既能测量速度又能测量湍流的仪器，它不受流场内温度、压力的影响，对流场也无干扰，而且还有很好的方向辨别能力和较高的时空分辨率[2]。作者使用二维氩离子激光多普勒测速仪(LDA)对倒拖工况下柴油机吊钟型涡流室内空气运动进行了测量，并利用测量结果分析其平均速度和湍流强度，为涡流室式柴油机工作过程的数值模拟提供可靠的数据，也为实际柴油机涡流室的设计提供依据。

1　实验装置与实验方法

实验采用 DANTEC 2020 型三光束双色二维氩离子测光器、55N11-12 型频移器、57N10BSA 频谱分析仪、IBM-AT 微型计算机、TK-360 型轴编码器等仪器设备来测量气缸盖上开有石英玻璃窗的单缸柴油机涡流室内的空气运动。实验时在进入燃烧室的空气中加入一定浓度的二氧化锆微粒，作为示踪粒子。检测系统框图见图 1。

实验所用柴油机主要参数有

冲程数：4　　压缩比：19

缸径/行程：95 mm/110 mm

涡流室型式：吊钟型

工作容积：850 cm³

涡流室容积：25.7 cm³

容积比：47%

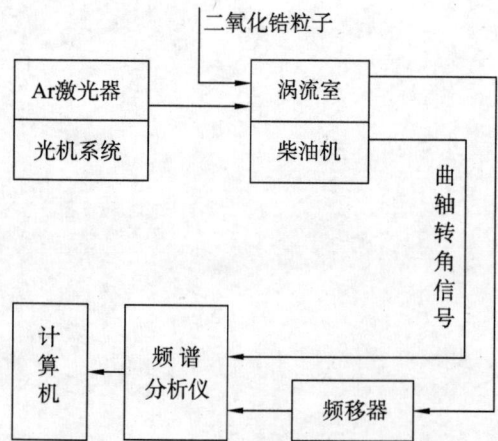

图 1　检测系统框图

测量点位于涡流室中心截面上,测点布置如图 2 所示。激光透过石英玻璃窗测量涡流室中心截面上各测点位置 x, y 方向的速度。数据采集由微机控制,瞬时速度值与曲轴转角信号同时送入计算机中,从而得到每曲轴转角对应的空气速度数据。

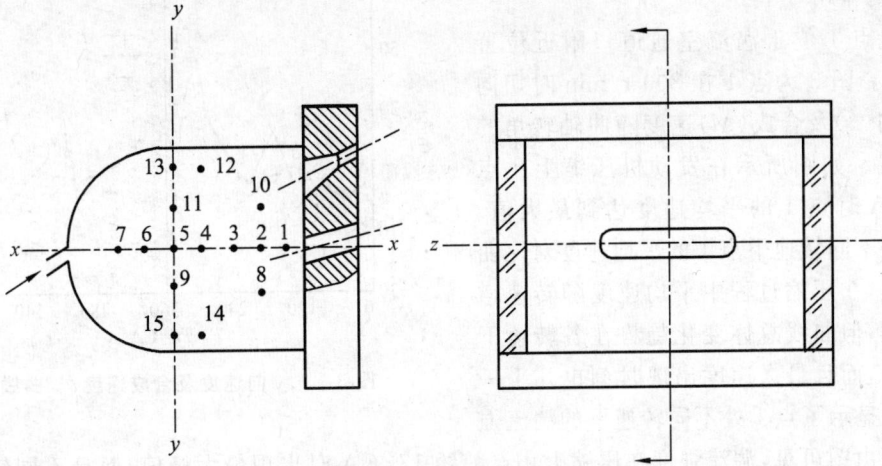

图 2　测量点布置简图

2　数据处理

根据 LDA 所测数据特点,以总体实验数据为依据,采用相平均对实验数据进行处理[3]。有关参量定义如下:

① 平均速度

$$\bar{u}(\varphi) = \frac{1}{N_\varphi} \sum_{i=1}^{N_\varphi} u_i(\varphi) \tag{1}$$

式中 $u_i(\varphi)$ 是瞬时速度;N_φ 是同一曲轴转角下所测得的瞬时速度的次数。

② 湍流强度

$$\tilde{u}(\varphi) = \left\{ \frac{1}{N_t} \sum_{i=1}^{N_t} \left[u_i(\varphi) - \bar{u}(\varphi) \right]^2 \right\}^{\frac{1}{2}} \tag{2}$$

式中 N_t 是所测得的总瞬时速度的次数。

③ 相对湍流强度

$$\varepsilon = \frac{\tilde{u}(\varphi)}{\bar{u}(\varphi)} \times 100\% \tag{3}$$

④ 合成速度与湍流强度

$$\bar{u}_m = \sqrt{\bar{u}_x^2 + \bar{u}_y^2} \tag{4}$$

$$\tilde{u} = \frac{1}{2} \sqrt{\tilde{u}_x^2 + \tilde{u}_y^2} \tag{5}$$

式中 $\bar{u}_x, \bar{u}_y, \tilde{u}_x, \tilde{u}_y$ 分别表示 x 与 y 方向上单向平均速度和湍流强度。

3 实验结果及其分析

3.1 通道口附近

测量点 1 位于涡流室通道口附近位置（见图 2）。图 3 为点 1 在 800 r/min 时切向 (y)、径向 (x) 及合成 (M) 速度随曲轴转角的变化曲线。如图所示在发动机压缩上止点前 26°CA 时点 1 的平均速度达到最大值。图 4 给出不同转速下点 1 的平均速度对比曲线。显然，在压缩过程中平均速度随转速增加而增大，但曲线总体变化趋势在各转速下是相同的，而且最大速度出现时刻也几乎一致。图 5 显示了点 1 处不同转速下的湍流强度曲线。由图可见，湍流强度在压缩上止点前约 17°CA 时出现最大峰值，而且不同转速时曲线总体变化趋势也是相同的。实验数据显示在发动机转速 800 r/min 情况下在压缩上止点前 26°CA 时点 1 的瞬时速度达到峰值 60.29 m/s。而此时点 1 的湍流强度为 8.15 m/s，相对湍流强度为 13.5%。点 1 处最大湍流强度值为 10.01 m/s，发生在 800 r/min、压缩上止点前 17°CA，此时平均速度为 48.8 m/s，相对湍流强度为 20.5%。

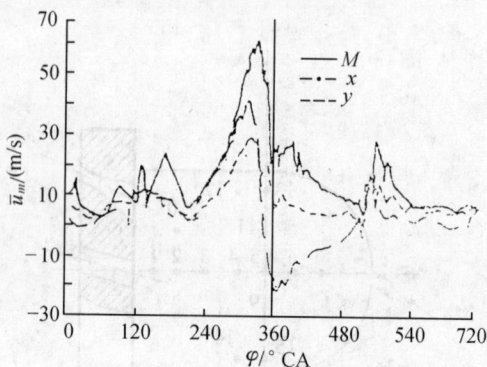

图 3 x, y 向速度及合成速度 $\bar{u}_m$ 曲线

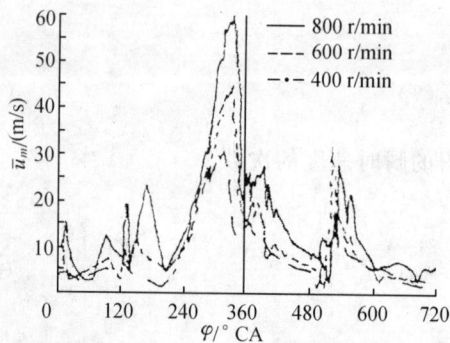

图 4 不同转速时点1的合成速度 $\bar{u}_m$ 曲线

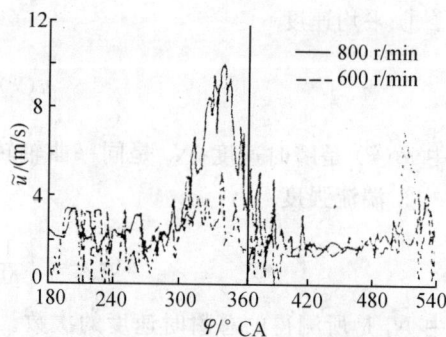

图 5 不同转速时点1的湍流强度曲线

3.2 喷油嘴一侧

测量点 11 位于喷油方向一侧（见图 2）。图 6 和图 7 分别显示点 11 在 800 r/min 时的平均速度及湍流强度曲线。此处最大湍流强度为 11.11 m/s，相应曲轴转角为压缩上止点前 22°CA，而此时平均速度为 19.75 m/s，相对湍流强度为 56.3%，速度的脉动程度已相当高。而在喷油时刻，点 11 的平均速度为 19.25 m/s，湍流强度为 6.02 m/s，相对湍流强度为 31.3%，脉动程度也较强，这对涡流室内燃油与空气的混合十分有利。

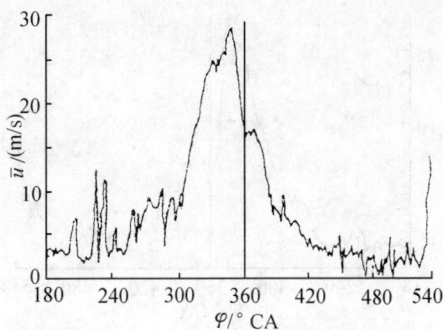

图 6   800 r/min时点11的平均速度$\bar{u}$曲线

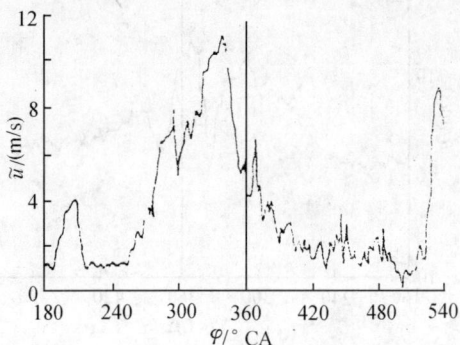

图 7   800 r/min时点11的湍流强度曲线

3.3   涡流室中心附近

测量点 3,4,5 位于涡流室中心附近(见图 2)。图 8 给出 800 r/min 时点 3,4,5 的平均速度对比曲线。可见,点 5 的平均速度在压缩上止点前 40°CA 以前为最小,而当曲轴转角增大至上止点时点 4 的平均速度变为最小,并且在上止点前 22°CA 时其平均速度几乎为 0,其他转速下也有类似结果。这就证实了随曲轴转角向压缩上止点移动,涡流中心由涡流室的几何中心向涡流室下方移动。800 r/min 时湍流强度曲线见图 9,可以看出,在压缩上止点前 40°CA 以前点 3 的湍流强度最大,而当曲轴转角继续增大至上止点时点 4 处湍流强度变为最大。其他转速下实验结果也表明,涡流室中心附近各点湍流强度的相对关系随曲轴转角的变化而变化。图 10 和图 11 分别是 800 r/min 时点 4 的平均速度与湍流强度曲线。在压缩上止点前 22°CA 时,平均速度达到最小,为 0.85 m/s,但此时湍流强度为 13.72 m/s,脉动程度相当高。

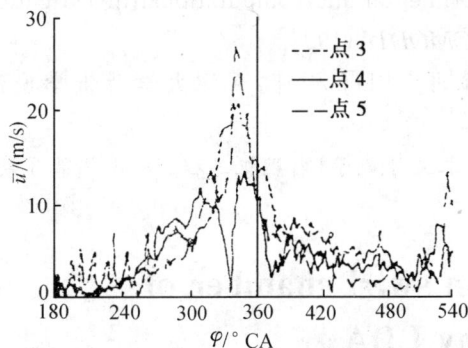

图 8   800 r/min时点3,4,5的平均速度曲线

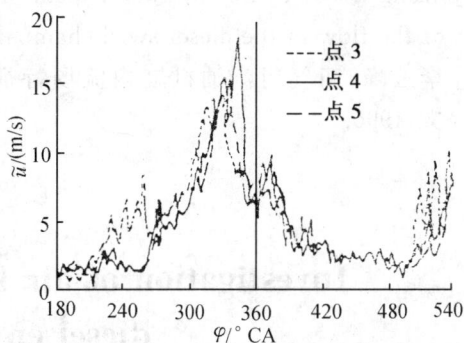

图 9   800 r/min时点3,4,5的湍流强度曲线

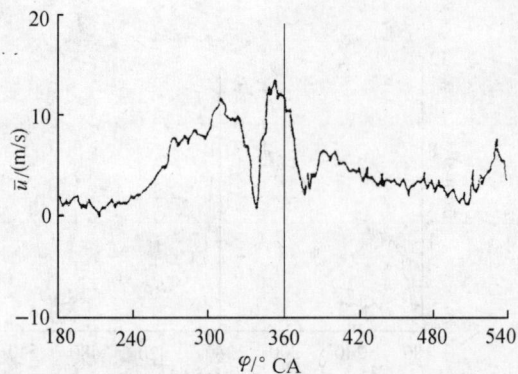

图 10　800 r/min时点4的平均速度曲线

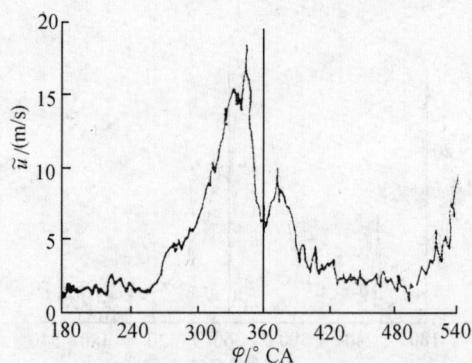

图 11　800 r/min时点4的湍流强度曲线

4　结　　论

（1）本文采用 LDA 测试涡流室内空气运动所得的实验结果比其他测量方法所得结果准确、可靠，采用的数据处理方法也是合理的。

（2）涡流室内空气运动呈非均匀的湍流场，其速度在涡流室内的空间分布有较大的差异，而且速度最大值出现的时刻也不一致。

（3）在涡流室内喷油方向一侧区域，具有较大的相对湍流强度，这有利于油和气的混合。

（4）随活塞向压缩上止点运动，涡流运动中心逐渐从涡流室几何中心向下方移动；上止点时，涡流中心在涡流室的下方。

参 考 文 献

［1］李德桃.柴油机涡流燃烧室的研究与设计［M］.北京:机械工业出版社,1986.

［2］Wang Qian，Li Detao，Luo Tiqian. The experimental and computational investigation of the flow in the diesel swirl chamber［J］. *COMODIA*,1994.

［3］李玉峰.四气门汽油机缸内流场特征的实验研究［D］.天津:天津大学博士学位论文,1995.

（本文原载于《内燃机工程》1996 年第 3 期）

Investigation of air flow in swirl chamber of diesel engine by LDA

Abstract：The air flow in a diesel engine's swirl chamber was measured by 2D-Ar-ion LDA. The variation of the air flow in the swirl chamber with crank angle and the spacial distribution of the flow field in the chamber under motored condition are revealed. The results obtained are contributive to understand the air flow characteristics in the swirl chamber and to establish the combustion model of swirl chamber diesel engines.

涡流室式柴油机非稳态燃烧
过程的不稳定性分析

严新娟,李德桃

[摘要]　运用 S195 型涡流室式柴油机冷起动过程的实测数据,分析了非稳态燃烧过程的不稳定性;针对 3 种不同的燃烧过程,进行了放热规律计算和分析.

众所周知,涡流室式柴油机非稳态燃烧过程是很复杂的,其复杂性可从燃烧过程的不稳定性及放热规律的多样性这两方面看出,下面就从这两方面具体阐述。

1　试验及结果

试验机是 S195 型涡流室式柴油机,起动试验是在冷起动实验室内进行的,起动过程中压力、转速等数据的连续变化历程由磁带记录仪采录。冷起动过程是典型的非稳定燃烧过程,其燃烧过程可分为三类[1]：① 初次着火时的 H 型燃烧过程;② 不稳定的 X 型燃烧过程;③ 旺盛的 Y 型燃烧过程。表 1 示出了一次成功起动过程加速期连续 18 个循环的试验结果。

表 1　冷起动过程加速期的试验结果

循环序数	主室最高压力/MPa 及对应角度/° CA	副室最高压力/MPa 及对应角度/° CA	燃烧过程	转速/(r/min)
8	8.345/13	7.832/14	近 Y	796
9	6.205/22	5.992/21	近 X	828
10	8.885/9	8.660/10	Y	908
11	8.728/11	8.380/11	Y	960
12	8.094/16	7.775/15	X-Y	1 012
13	8.116/14	7.739/14	Y	1 056
14	8.991/14	8.743/13	Y	1 102
15	8.412/15	8.329/14	近 Y	1 138
16	9.237/9	8.990/11	Y	1 178
17	9.549/11	9.354/10	Y	1 218
18	7.287/19	7.096/18	X	1 252
19	9.899/10	9.693/9	Y	1 290
20	6.810/21	6.592/20	X	1 320
21	9.705/7	9.304/9	Y	1 356
23	9.792/8	9.266/9	Y	1 422
24	8.112/13	8.013/14	X-Y	1 448
25	9.632/8	9.403/9	Y	1 486

2 燃烧过程的不稳定性分析

非稳态燃烧过程的不稳定性可以从两个方面来阐述。

2.1 压力的不稳定性

从表 1 中可以看出,随着转速的升高,主、副室的最高压力波动地上升,同一循环的主室最高压力都略大于副室最高压力。主、副室最高压力对应的曲轴转角有时相同,有时主室的大,有时副室的大;主室最高压力对应的曲轴转角在 7~22° CA ATDC 之间,副室最高压力对应的曲轴转角则在 9~21° CA ATDC 之间。

这些都反映了气缸内燃烧过程压力的不稳定性。

2.2 燃烧过程的不稳定性

参看表 1,该起动过程加速段的燃烧过程有 Y 型、X 型及 X-Y 型 3 种,中间型——X-Y 型严格讲要归到 Y 型中,图 1 是 3 个实测的不同类型燃烧压力图。

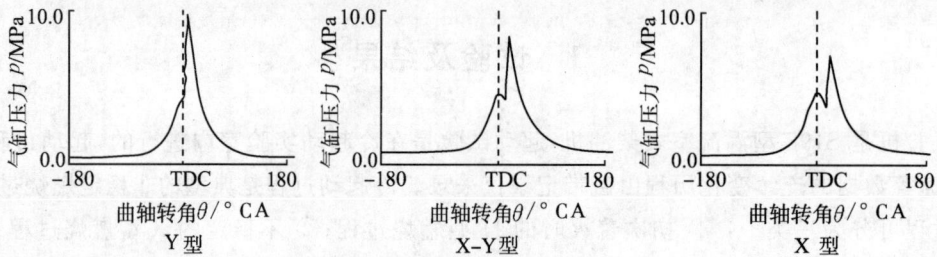

图 1 不同类型燃烧的压力图

从图 1 可以看出,Y 型示功图是细窄的单峰型,压力峰值大;X 型示功图是平缓的双峰型,压力峰值小;X-Y 型示功图介于二者之间。

该非稳态过程各循环燃烧过程有很大的差异,Y 型、X 型和 X-Y 型 3 种燃烧过程交替出现,其中 Y 型居多。大约每隔 2 个 Y 型燃烧,就出现 1 个非 Y 型燃烧,这与文献[2]所述的起动初期的燃烧特点——二次着火循环之间有一两个循环或更多循环不着火有相似之处,反映了不稳定性也是有一定规律的。

在第 25 个循环后,燃烧过程都近乎 Y 型,表明非稳态起动过程已过渡到稳定运转阶段。

3 实测压力的放热分析

3.1 计算模型

放热规律的计算采用参考文献[3]中提出的模型。模型的计算简图如图 2 所示。

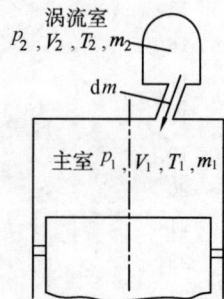

图 2 放热分析计算简图

燃烧放出的总热量变化率 $\dot{Q}_T$ 由下式计算:

$$\dot{Q}_T = \frac{1}{(\gamma-1)}(V_1\dot{p}_1 + V_2\dot{p}_2) + \frac{\gamma}{\gamma-1}p_1\dot{V}_1$$

式中 $\dot{p}_1$、$\dot{p}_2$ 为主、副室压力的变化率;$\dot{V}_1$ 为主室容积的变化率;γ 为比热比;$\dot{Q}_T$ 为燃烧热 $\dot{Q}_{HR}$、传热率 $\dot{Q}_{HT}$ 及燃油蒸发加热率 $\dot{Q}_V$ 之总

和,即

$$\dot{Q}_T = \dot{Q}_{HR} - \dot{Q}_{HT} - \dot{Q}_V$$

3.2 放热分析

运用上述模型,对各类燃烧过程进行了计算分析。

3.2.1 Y 型燃烧过程放热分析

本文就第 10、第 19 和第 23 循环这 3 个典型的 Y 型燃烧过程进行了放热规律分析,结果如图 3 所示。

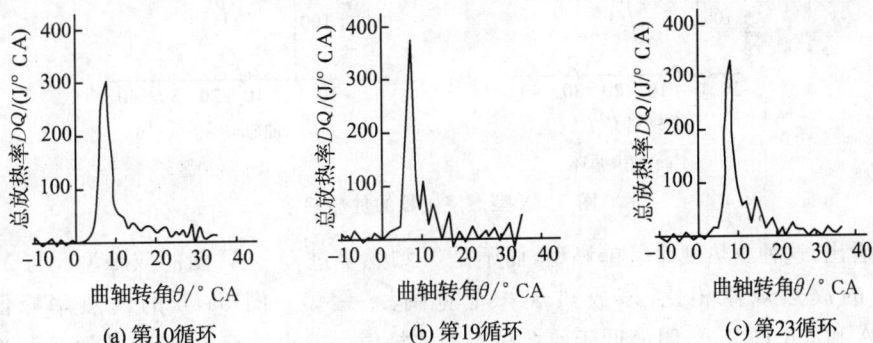

(a) 第10循环　　　　　　(b) 第19循环　　　　　　(c) 第23循环

图 3　Y 型燃烧过程的放热规律

由图 3 可以看出,Y 型燃烧过程的放热规律是细窄型的,与压力图形相似。图 3a,b,c 中,放热率峰值分别为 300 J/°CA、395 J/°CA 和 350 J/°CA;着火始点在 −5～0°CA 之间,这样的着火始点可保证燃烧在上止点附近迅速进行;燃烧持续期在 30°CA 左右。比较表 1 和图 3,可以看出放热率峰值早于压力峰值出现,且压力峰值越大,放热率峰值也越大。与稳定工况下的放热规律[4]比较,非稳态 Y 型燃烧过程的放热率大得多,燃烧持续期也短得多。

3.2.2 X-Y 型燃烧过程的放热分析

图 4 示出了第 8、第 12 及第 24 循环 3 个 X-Y 型燃烧过程的放热规律。

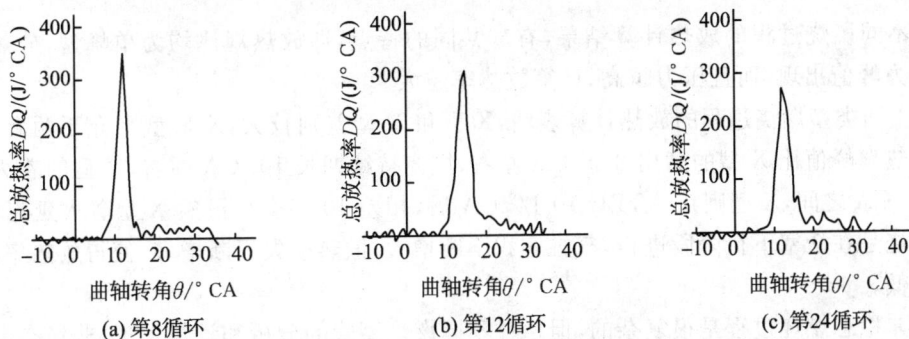

(a) 第8循环　　　　　　(b) 第12循环　　　　　　(c) 第24循环

图 4　X-Y 型燃烧过程的放热规律

可以看出,图 4 的各放热规律也是细高型的,但其放热率峰值较之 Y 型燃烧的放热率峰值要低些,图 4a,b,c 的放热率峰值分别为 364 J/°CA、318 J/°CA 及 285 J/°CA,着火始点在 5～8°CA 之间;燃烧持续期为 23～26°CA。由于着火较迟,燃烧主要发生在活塞下行阶段,故放热率峰值减小。比较表 1 和图 4 也可看出放热率峰值早于压力峰值出现。比较图 4b,c,虽然图 4c 的放热率峰值低于图 4b,但着火始点早于图 4b,相当于放热规律左移,故燃烧性能好,对应的最高压力也就大些。

3.2.3 X型燃烧过程的放热分析

图5所示的放热规律分别由第9和第20循环——两个典型的X型燃烧过程计算而得。

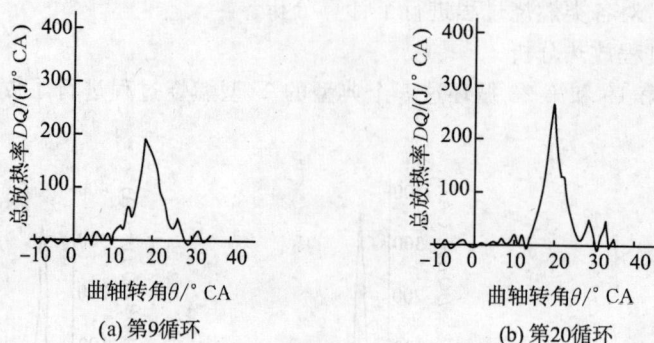

(a) 第9循环　　　(b) 第20循环

图5　X型燃烧过程的放热规律

可以看出,这种放热规律呈单峰型,而不是X型燃烧过程 $p\text{-}V$ 图的双峰型。与Y型、X-Y燃烧过程的放热规律相比,该放热规律呈宽阔、平缓型。图5a,b的放热率峰值分别为200 J/°CA和286 J/°CA,明显低于前2种类型的峰值。着火始点也更迟,推至ATDC 10°CA后,燃烧持续期不足20°CA。由于这类燃烧过程的着火太迟,燃烧在气缸容积不断增加的情况下进行,燃烧必须很快才能使气缸压力稍有增加或保持不变,因而这类燃烧过程的最高压力都较低。

4　结　论

(1)该非稳态燃烧包含3种不同的燃烧的过程,三者交替出现,呈一定的规律。相同燃烧过程的不同循环,其压力峰值和对应的曲轴转角也有很大的变化。这就表明燃烧过程是非稳态的。

(2)不同燃烧过程的放热计算结果,有着共同的特点,即放热规律均为单峰型,放热率峰值早于压力峰值出现,而且压力越高,计算放热率越大。

(3)不同类型燃烧过程的放热计算表明,X型和Y型差别较大,X-Y型则介于两者之间。Y型的放热率峰值比X型的高出200°CA左右,燃烧持续期长10°CA左右,Y型的着火始点在-5~0°CA之间,X型则在ATDC 10 J/°CA后,相差10多度。虽然X型着火延迟较长,但由于燃烧是在活塞下行阶段进行,气缸容积不断增大,散热损失不断增多,使得放热率减小,压力也较低。

虽然非稳态燃烧过程是很复杂的,但通过不同燃烧型式的分析和计算发现,非稳态燃烧过程仍有一定的规律性。

参 考 文 献

[1]李德桃,何晓阳.涡流室式柴油机在冷起动条件下非稳态燃烧过程的研究[J].内燃机学报,1987(3):25-31.

[2] Henein Naeim A，et al. Diesel engine cold starting：combustion instability[J]. *SAE Paper*，1992：23-38.

[3] Kort Raja T，et al. Divided-chamber diesel engine，part II：experimental validation of a predictive cycle-simulation and heat release analysis[J]. *SAE Paper*，1982：1133-1147.

[4] 严新娟，李德桃，等. 涡流室式柴油机工作过程计算中几个问题的探讨[C]. 全国高等学校工程热物理第四届学术会议论文集. 杭州：浙江大学出版社，1992：180-183.

（本文原载于《江苏理工大学学报》1996 年第 3 期）

Analysis on instability of unsteady-state combustion in a swirl-chamber diesel engine

Abstract：On the basis of the measured data for the cold start of swirl-chamber diesel engine, Type S195, the instability during the unsteady-state combustion is analysed. Furthermore, the heat release laws are calculated and analysed for three types of unsteady state combustion.

涡流室式柴油机非稳态燃烧过程
理论模型和放热分析

朱亚娜，李德桃，熊　锐

[摘要]　建立了一个热力学模型，用来研究非稳态情况下间接喷射式、4冲程、单缸、非增压柴油机的瞬态特征。该模型包括详细的子模型，能模拟发动机的整个循环。最后，在测得的示功图上分析了反拖和起动情况下，间接喷射式发动机的热释放特性。

柴油机非稳态燃烧过程是指柴油机在变工况或过渡工况时的不稳定燃烧过程。由于柴油机在非稳态过程中将从一个稳定的状态变化到另一个稳定状态，其间发动机的转速、燃油喷射率和负荷在每一循环中均不相同，而且循环结束时的条件也不等于循环开始时的条件，所以不能用稳态模型研究非稳态过程。基于上述考虑，本文采用了在"充满和排空"概念基础上提出的准稳态模型来研究柴油机非稳态过程。该模型最初是为模拟涡轮增压柴油机在变速或变载的情况下的性能特征而提出的[1]，近年来也被用来计算直喷式柴油机在冷起动情况下的瞬态特性[2,3]。但据作者所知，目前还尚未有人利用该模型研究涡流室式柴油机的瞬态特性。本文试图在这一方面作一探讨，利用这种方法建立适合于涡流室式柴油机非稳态燃烧过程的热力学模型，为进一步研究其瞬态特性打下基础。

1　基本方程及其求解

1.1　方程的提出

本模型是建立在以下假设之上的：

（1）气缸内工质的状态是均匀的，亦即不考虑气缸内各点的压力、温度和浓度的差异。

（2）工质为理想气体，满足理想气体状态方程式。

（3）考虑到可能存在的燃烧不完全性和燃油的高温裂解，认为气体常数 R 为工质压力 p、温度 T 和过量空气系数 Φ 的函数；工质比内能 u 为 p, T 和 Φ 的函数。即

$$R = R(p, T, \Phi) \qquad u = u(p, T, \Phi)$$

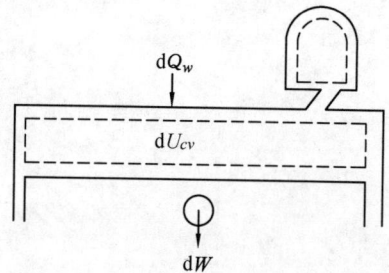

图1　发动机热力系统

根据以上假设可以得到如下热力学方程式（见图1）：

状态方程
$$pV = mRT \tag{1}$$

能量方程
$$dU = -p\,dV + \sum_i dQ_i + \sum_j h_j\,dm_j \tag{2}$$

质量方程
$$dm = \sum_j dm_j \tag{3}$$

又
$$dU = m\,du + u\,dm \tag{4}$$

$$\frac{\mathrm{d}u}{\mathrm{d}\theta} = \left(\frac{\partial u}{\partial T}\right)\frac{\mathrm{d}T}{\mathrm{d}\theta} + \left(\frac{\partial u}{\partial p}\right)\frac{\mathrm{d}p}{\mathrm{d}\theta} + \left(\frac{\partial u}{\partial \Phi}\right)\frac{\mathrm{d}\Phi}{\mathrm{d}\theta} \tag{5}$$

$$\frac{\mathrm{d}R}{\mathrm{d}\theta} = \left(\frac{\partial R}{\partial T}\right)\frac{\mathrm{d}T}{\mathrm{d}\theta} + \left(\frac{\partial R}{\partial p}\right)\frac{\mathrm{d}p}{\mathrm{d}\theta} + \left(\frac{\partial R}{\partial \Phi}\right)\frac{\mathrm{d}\Phi}{\mathrm{d}\theta} \tag{6}$$

联立以上方程求解,可得

$$\frac{\mathrm{d}T}{\mathrm{d}\theta} = \frac{A - \dfrac{p}{D}\dfrac{\partial u}{\partial p}\left(\dfrac{1}{m}\dfrac{\mathrm{d}m}{\mathrm{d}\theta} - \dfrac{1}{V}\dfrac{\mathrm{d}V}{\mathrm{d}\theta} + \dfrac{1}{R}\dfrac{\partial R}{\partial \Phi}\dfrac{\mathrm{d}\Phi}{\mathrm{d}\theta}\right) - \dfrac{\partial u}{\partial \Phi}\mathrm{d}\Phi}{\dfrac{\partial u}{\partial T} + \dfrac{pC}{TD}\dfrac{\partial u}{\partial p}} \tag{7}$$

$$A = \frac{1}{m}\left(\sum_j h_j \frac{\mathrm{d}m_j}{\mathrm{d}\theta} + \sum_i \frac{\mathrm{d}Q_i}{\mathrm{d}\theta}\right) - \frac{u}{m}\frac{\mathrm{d}m}{\mathrm{d}\theta} - \frac{p}{m}\frac{\mathrm{d}V}{\mathrm{d}\theta} \tag{8}$$

$$C = 1 + \frac{T}{R}\frac{\partial R}{\partial T} \tag{9}$$

$$D = 1 - \frac{p}{R}\frac{\partial R}{\partial p} \tag{10}$$

式中 p,V,T 分别为控制体内的压力、体积和温度;R 为气体常数;m,u,h 分别为控制体内工质的质量、比内能和比焓;Φ 为过量空气系数。

方程(7)描述了系统内部温度随曲轴转角变化的规律,它是描述发动机非稳态燃烧过程最基本的微分方程式。考虑到柴油机具体工作条件,可以将公式(7)简化。由于柴油机的压缩比高,过量空气系数较大,因此可不考虑燃油的高温分解,亦即内能 u 和气体常数 R 均与压力 p 无关。

$$u = u(T,\Phi) \qquad R = R(\Phi)$$

于是式(7)可化简为

$$\frac{\mathrm{d}T}{\mathrm{d}\theta} = \frac{\left(\sum_j h_j \dfrac{\mathrm{d}m_j}{\mathrm{d}\theta} + \sum_i \dfrac{\mathrm{d}Q_i}{\mathrm{d}\theta}\right) - u\dfrac{\mathrm{d}m}{\mathrm{d}\theta} - p\mathrm{d}V - m\dfrac{\partial u}{\partial \Phi}\dfrac{\mathrm{d}\Phi}{\mathrm{d}\theta}}{m\dfrac{\partial u}{\partial T}} \tag{11}$$

1.2 气缸容积

可以认为涡流室的容积 V_k 为常量,即

$$V_k = 常量 \tag{12}$$

$$\mathrm{d}V_k = 0 \tag{13}$$

主室容积 V_m 的变化可以依据发动机的几何形状和转速来计算:

$$V_m = \frac{V_h}{2}\left[\frac{2}{\varepsilon - 1} + 1 - \cos\theta + \frac{1}{\lambda_s}(1 - \sqrt{1 - \lambda_s^2 \sin^2\theta})\right] \tag{14}$$

$$V_h = \frac{\pi D^2 r}{2} \tag{15}$$

式中 V_k,V_m 分别为涡流室容积和主室容积;D 为气缸直径;r 为曲轴半径;λ_s 为曲径比;V_h 为活塞排量。

1.3 工质质量及成分计算

将柴油机的燃烧室分成两个相互关联的热力系统,即主室和涡流室,并规定输入系统的能量和质量为正值,离开系统的能量和质量为负值;当从一个系统过渡到另一个系统时必须考虑符号的变换,主室参数加下标 m,涡流室参数加下标 k。

设 m_{bk} 为涡流室已燃燃料的质量,m_{lk} 是涡流室中空气的质量,m_k 为涡流室中工质质量,L_0 为 1 kg 燃料完全燃烧所需的理论空气量。m_{bm} 和 m_{lm} 分别是主燃烧室中已燃燃料的质量和空气量,则涡流室内的质量平衡方程为

$$\frac{\mathrm{d}m_k}{\mathrm{d}\theta} = \frac{\mathrm{d}m_u}{\mathrm{d}\theta} + \frac{1}{Hu}\frac{\mathrm{d}Q_{bk}}{\mathrm{d}\theta} \tag{16}$$

$$\frac{\mathrm{d}m_{bk}}{\mathrm{d}\theta} = \frac{1}{Hu}\frac{\mathrm{d}Q_{bk}}{\mathrm{d}\theta} + \frac{m_{bi}}{m_i}\frac{\mathrm{d}m_u}{\mathrm{d}\theta} \tag{17}$$

$$\frac{\mathrm{d}m_{lk}}{\mathrm{d}\theta} = \frac{\mathrm{d}m_u}{\mathrm{d}\theta}\left(1 - \frac{m_{bi}}{m_i}\right) \tag{18}$$

主室内的质量平衡方程为

$$\frac{\mathrm{d}m_m}{\mathrm{d}\theta} = \frac{\mathrm{d}m_u}{\mathrm{d}\theta} + \frac{1}{Hu}\frac{\mathrm{d}Q_{bm}}{\mathrm{d}\theta} + \frac{\mathrm{d}m_t}{\mathrm{d}\theta} \tag{19}$$

$$\frac{\mathrm{d}m_{bm}}{\mathrm{d}\theta} = \frac{1}{Hu}\frac{\mathrm{d}Q_{bm}}{\mathrm{d}\theta} + \frac{m_{bi}}{m_i}\frac{\mathrm{d}m_u}{\mathrm{d}\theta} + \frac{m_{bm}}{m_m}\frac{\mathrm{d}m_t}{\mathrm{d}\theta} \tag{20}$$

$$\frac{\mathrm{d}m_{lm}}{\mathrm{d}\theta} = \frac{\mathrm{d}m_m}{\mathrm{d}\theta} - \frac{\mathrm{d}m_{bm}}{\mathrm{d}\theta} \tag{21}$$

$$\varPhi_i = \frac{m_{li}}{L_0 m_{bi}} \tag{22}$$

式中 Hu 为燃料的低热值;$\dfrac{\mathrm{d}Q_{bk}}{\mathrm{d}\theta}$ 为涡流室的燃烧放热率;$\dfrac{\mathrm{d}Q_{bm}}{\mathrm{d}\theta}$ 为主燃烧室的燃烧放热率;$\dfrac{\mathrm{d}m_u}{\mathrm{d}\theta}$ 为通过镶块通道主室和涡流室交换的工质质量;$\dfrac{\mathrm{d}m_t}{\mathrm{d}\theta}$ 为通过活塞环端口的漏气量。

由于在不同的阶段通过镶块通道的工质流向是不同的,所以下标 i 在不同的阶段(压缩、燃烧、膨胀)所代表的意义也不同。当工质从主室流向涡流室时,$i = m$;当工质从涡流室流向主室时,$i = k$。

1.4　工质热物理性质的计算

如果不考虑工质的高温分解,工质的内能、焓和比热之间存在以下关系:

$$u = h - RT \tag{23}$$

$$c_v = \left(\frac{\partial u}{\partial T}\right)_v \tag{24}$$

$$c_p = \left(\frac{\partial h}{\partial T}\right)_p \tag{25}$$

内能 u 可按下列公式计算[4]:

$$u = 0.144\,55\Bigg[-\left(0.097\,5 + \frac{0.048\,5}{\varPhi^{0.75}}\right)(T-273)^3\times10^{-6} + \left(7.768 + \frac{3.36}{\varPhi^{0.8}}\right)\times$$

$$(T-273)^2\times10^{-4} + \left(489.6 + \frac{46.4}{\varPhi^{0.93}}\right)(T-273)\times10^{-2} + 1\,356.8\Bigg] \tag{26}$$

1.5　工质对缸壁的热交换

气体与壁面间交换的热量可以由下式计算:

$$\frac{\mathrm{d}Q_w}{\mathrm{d}\theta} = \frac{h_c A_t}{\omega}(T_w - T_g) \tag{27}$$

式中 h_c 为气体对壁面的热交换系数;A_t 为参与热交换的气缸壁面面积;T_g 为工质温度;T_w 为

气缸壁面温度；ω 为发动机角速度。

经过计算比较，我们选用下列公式计算热交换系数[5]。

主燃烧室：
$$(Nu)_m = 0.048\,2(Re)_m^{0.792\,4} \tag{28}$$

涡流室：
$$(Nu)_k = 0.035(Re)_k^{0.771\,2} \tag{29}$$

1.6 通过活塞环的漏气量

对于柴油机某些非稳态过程，比如冷起动起程，由于活塞平均速度和壁温都较低，导致漏气损失较大，所以必须考虑通过活塞环的漏气损失。

在如图 2 所示的漏气模型中，假设活塞环的端口间隙是唯一的泄漏途径，油环不密封气体。气体的流动假设为理想气体的一元准稳态绝热流动。对于一个 3 道气环的发动机，考虑到不同的压力比及气体倒流情况，得到如下公式：

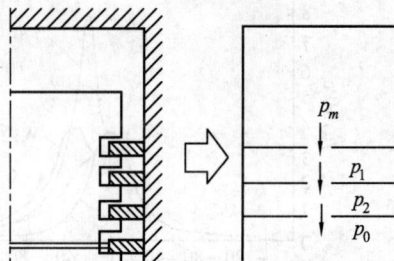

图 2　活塞漏气模型

$$\frac{\mathrm{d}m_t}{\mathrm{d}\theta} = \frac{C_d A_e}{\omega}\sqrt{\frac{2k}{k-1}\frac{p_m}{v}\left[\left(\frac{p_1}{p_m}\right)^{\frac{2}{k}} - \left(\frac{p_1}{p_m}\right)^{\frac{k+1}{k}}\right]} \tag{30}$$

式中 C_d 为方形小孔的流量系数（大约为 0.65）。依次研究每一对相邻的容积，可求得通过活塞环的漏气质量流量。

1.7 主室和涡流室交换的工质质量

为了计算通过镶块通道主室和涡流室之间交换的工质质量，采用一维、非稳态、可压缩、绝热流动的计算公式：

$$\frac{\mathrm{d}m_u}{\mathrm{d}\theta} = \frac{\mu A}{\omega}\sqrt{\frac{2k}{k-1}\frac{p_i}{v_1}\left[\left(\frac{p_j}{p_i}\right)^{\frac{2}{k}} - \left(\frac{p_j}{p_i}\right)^{\frac{k+1}{k}}\right]} \tag{31}$$

式中 p_i 为上游系统的压力；p_j 为下游系统的压力；A 为镶块通道截面积；v_1 为工质比容；μ 为镶块通道的流量系数，它可从文献[6]选用。

2　放热率计算示例

为了探讨柴油机非稳态过程的放热特性，我们在实测示功图的基础上以前述热力学模型计算了 S195 柴油机的放热率。根据文献[7]可知冷起动过程中示功图的形状大致可分为 3 类，如图 3 所示。

（1）V 型：它的形状类似于不着火示功图，燃烧压力没有明显升高。通过计算发现其循环指示功较小，大约只有标定工况的 10%～50%，甚至有时循环指示功为负值。该种形式的示功图基本上出现在起动初期转速较低的情况下，而且随着环境温度的降低出现的可能性越大。

（2）W 型：整个压力变化过程出现两个峰值，最高爆发压力和压力升高率均不大。通过计算发现其循环指示功大约为标定工况的 80%～120%。该种形式的示功图大多发生在冷起动过程的中期。

（3）Y 型：相似于正常燃烧的示功图，最高爆发压力和压力升高率均较大。将之与稳态过程标定工况的示功图相比较后发现（图 4）：两者的最大爆发压力接近，有时起动过程的最大爆发压力甚至更高些;压力升高率则比标定工况大些,这主要是由于起动过程燃烧室内工质温度低,燃烧滞燃期长,着火较迟的缘故而引起的。Y 型示功图的最高爆发压力超过稳态运行标定工况时的最高爆发压力,这一点应该在发动机设计时加以考虑。Y 型示功图在整个起动过程中都有出现,但以后期最多。

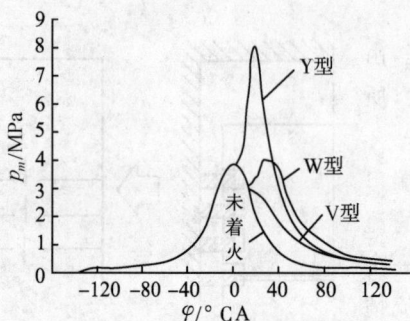

图 3　三种典型示功图

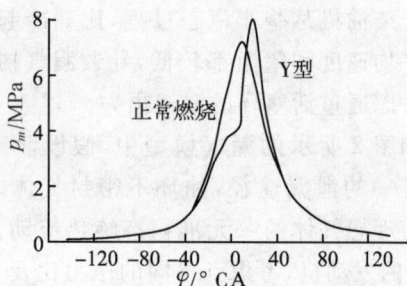

图 4　Y 型和正常燃烧过程示功图比较

从示功图知,在起动过程中气缸内燃烧情况波动较大,尽管每一循环的喷油量差不多,但实际参加燃烧的燃油量却相差较大,亦即每循环参与燃烧的燃油量并不等于循环供油量,因此给燃烧放热过程终点的确定带来了困难。笔者采用双对数曲线法确定了始燃点和终燃点[8],即以 $\lg p$-$\lg V$ 图中膨胀线上由直线转变为曲线的过渡点为始燃点,以曲线转变为直线的过渡点为终燃点。将得到的终燃点带入以上述热力学模型为基础而编写的计算程

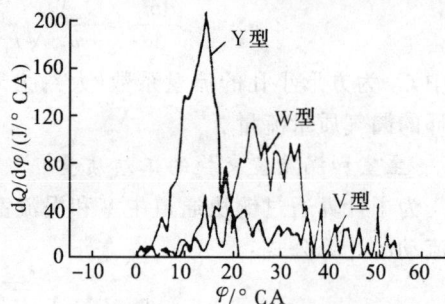

图 5　三种示功图的放热率

序进行计算,可得到 3 种不同形式示功图的放热率(图 5 为主室的放热率,涡流室的放热率与之类似),参与燃烧的循环油量以及燃烧始点。将用两个不同方法确定的燃烧始点相比较得知,二者趋于一致。通过对放热率的计算发现:3 种形式的放热过程有一共同的特点即燃烧均在上止点后才发生,Y 型在上止点后 $5°$CA 内开始着火,它的初期放热率较大,放热结束很早,整个放热持续期大约只有 $16°$CA,在上止点后 $20°$CA 左右燃烧就基本上结束,参与燃烧的燃油量大约为稳定运动时标定工况的 $110\%\sim130\%$,这可能是由于冷起动过程中燃烧室的壁温较低,使得上一个循环喷在燃烧室壁的燃油直到下一循环才完全燃烧的缘故;W 型在上止点后 $10°$CA 左右着火,燃烧放热期大约 $25°$CA,由于放热期较 Y 型长,所以虽然最大放热率不大,但参与燃烧的燃油量也不少,大约为标定工况的 $80\%\sim120\%$;V 型燃烧最不稳定,放热率波动较大,参与燃烧的燃油量也波动较大,大约为标定工况的 $0\%\sim50\%$ 不等,放热过程也拖得最长。

总之,涡流室式柴油机的非稳态过程是一个非常复杂的过程,作者提出的模型以及以此计算出的放热特性还只是其一个方面,要真实地预测瞬态特性需做进一步的工作。

3　结　论

(1) 文章首次建立了涡流室式柴油机非稳态过程的热力学模型。

（2）利用所提出的热力学模型，计算和分析了 S195 柴油机冷起动过程的放热特性。

（3）在冷起动过程中，虽然每循环的供油量差不多，但实际参与燃烧的燃油量却相差很大，亦即循环供油量并不等于循环燃烧油量。

（4）在冷起动过程中，最大爆发压力有时甚至比稳定运行时的标定工况的最大爆发压力还要大，这是值得发动机设计者注意的。

参 考 文 献

[1] Neil Waston, Maged Marzouk. A non-linear digital simulation of turbocharged diesel engines under transient conditions[J]. *SAE Trans* 770123，1977.

[2] Gardner Timothy P, Henein Naein A. Diesel starting—a mathematical model[J]. *SAE Trans*，880426,1988.

[3] Gardner Timothy P, Henein Naein A. Compression ratio optimization in a direct injection diesel engine—a mathematical model[J]. *SAE Trans*，880427,1988.

[4] 林杰伦. 内燃机工作过程数值计算[M]. 西安：西安交通大学出版社，1986.

[5] 郭七一，肖永宁，潘克煜. 涡流室式柴油机瞬时传热的实验研究[J]. 内燃机学报，1988,6（2）.

[6] 朱广圣，计微斌，李德桃. 涡流室式柴油机连接通道流量系数的计算研究[J]. 江苏工学院学报，1992,13(3).

[7] 范永忠，贾大锄，李德桃. 带有不同形式起动孔的涡流室式柴油机冷起动全过程的研究[J]. 内燃机工程，1994,15(1).

[8] 何学良，李疏松. 内燃机燃烧学[M]. 北京：机械工业出版社，1990:285-287.

（本文原载于《内燃机学报》1996 年第 4 期）

A theoretical model of the unsteady-state combustion process of the swirl chamber diesel engine and the analysis of its heat release

Abstract：A thermodynamic model is developed to study the transient behavior of a four-stroke，single cylinder naturally-aspirated ，IDI diesel engine under unsteady-state conditions. The model including detailed sub-models can simulate the full thermodynamic cycle of the engine. At last ，the characteristics of heat release of the IDI engine during cranking and starting is analyzed based on some existing indicator diagrams.

An investigation on measuring temperature distribution in the swirl chamber of a diesel engine during compression stroke by laser-moire deflectometry[①]

Li Detao, *Xiong Rui*, *Tian Dongbo*, *Wu Zhixin*

Abstract

The ignition and combustion of fuel-air mixture are affected by temperature distribution in the swirl chamber of a diesel engine during compression stroke. The authors have applied the laser-moire deflection technique for the first time to measure and study the temperature distribution of diesel engine swirl chamber, and acquired a lot of distinct moire stripe patterns. The axisymmetric numerical model to describe the temperature field of swirl chamber is established, and the relative numerical calculation method is also developed. The temperature distribution and its variation during compression stroke in the swirl chamber of a real diesel engine have been calculated in this paper.

1　Introduction

The temperature distribution of the engine chamber is usually measured by contact measurement method[1]. However, due to the influence of sensor's hot-inertia and the interference with flow field by temperature-measuring probe, the measured results of these methods are not satisfying. When temperature distribution of flow field is measured by a non-contact method, it is usually supposed that light beam can pass through the measured flow field in straight line, and the deflection effect of light in temperature field is neglected. So the imagery lens should be used to correct the deflection of ray

---

①　作者首次采用激光莫尔偏折法测量 IDI 柴油机燃烧室内的温度场。此法从概念的提出至本文的发表间隔仅 20 年。可以说这是迄今 IDI 柴油机燃烧室内温度场的最精确的测量和分析的结果。

——**编者注**

when actual survey is carried out[2]. Some researchers have tried moire deflectometry for transient measurement of temperature field since 1980's[3-6]. Owing to its many advantages such as simplicity in apparatus, convenient for regulation, the moderate demand on coherency of light source and good interference-free property, the authors tried for the first time to apply this method in measuring the temperature distribution of diesel engine swirl chamber. However, the authors were faced with the difficulties: (1) the measuring object is a fluid with high temperature, high pressure and high speed. (2) unlike the exhaust jet field of a rocket engine, the gas is sealed in engine chamber, and there is no boundary field with uniform parameters which can be used as the test reference field. (3) it is difficult for the observers to gain the multidirection moire deflection patterns about the chamber. Although there have been such difficulties to manifest the temperature field quantitatively in diesel engine chamber by moire deflectometry, the authors have achieved relatively satisfying results.

2　Experimental apparatus and methods

The test engine is a reformed model S195 single cylinder bell type swirl chamber diesel engine. Two quartz windows are installed on each side of the bell type swirl chamber as the laser beam access while not changing volume ratio and compression ratio of the original engine. The effective light-passing diameter of quartz window is $\Phi 30$ mm, the measured area is about 80. 2% of the section of the swirl chamber(Fig. 1).

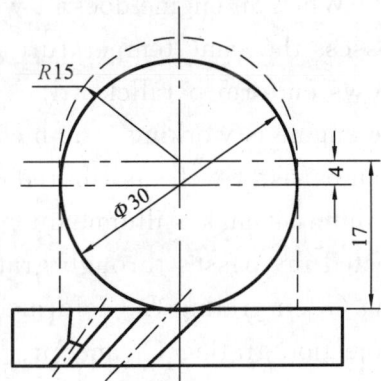

Fig. 1　The size of swirl chamber

To avoid Newton interference ring when parallel beam passes quartz windows, the surface crudeness of two quartz window glasses is minor than one tenth of measuring light wavelength. The included angle between two planes of a quartz glass is made into $10' \sim 20'$ to dispel the effect of exerting on moire stripe by glass backround stripe. Fig. 2 shows the laser-moire deflection test system.

3　Temperature measuring principle of moire deflectometry

According to Fermat principle, the change of medium refracting power will cause ray to produce deflection effect while the ray passes through a flow field.

This deflection effect is exactly used by moire deflectometry to measure the refracting power of medium and then to calculate the other parameters of the flow field.

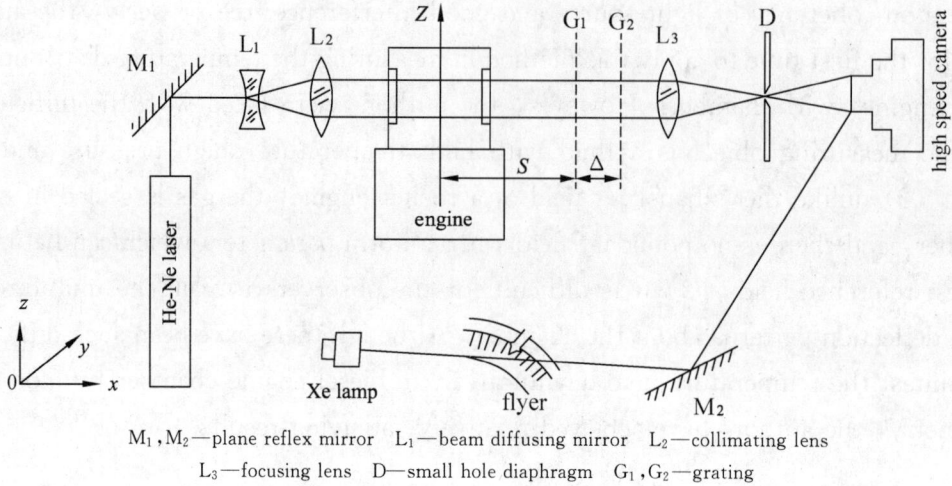

M₁,M₂—plane reflex mirror   L₁—beam diffusing mirror   L₂—collimating lens
L₃—focusing lens   D—small hole diaphragm   G₁,G₂—grating

Fig. 2   The laser-moire deflection test system

When an engine doesn't work, the mediums of various area of swirl chamber possess the same temperature and density, the projection of gratings G₁ ang G₂ shows uniform parallel stripes on the film of high speed camera(Fig. 3). When the engine is working, the medium refracting power of different area in the swirl chamber isn't well-distributed owing to the changes of temperature and density of medium, it makes illuminant parallel ray exert corresponding deflection. The deflected ray passes through grating G₁ and poroduces the projective image of grating G₁ on grating G₂(deformed grating stripes). Then real grating G₂ is piled on projection grating G₁ and produces moire stripes, mirroring in moire deflection drawing formed on high speed camera, which is shown in Fig. 4, the original parallel stripes occur drift and engender distortion.

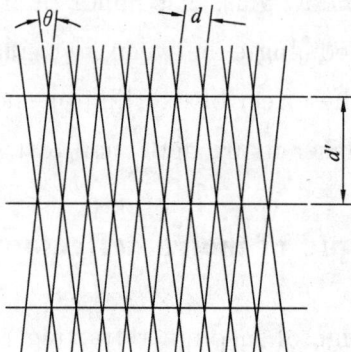

Fig. 3   Parallel moire stripes

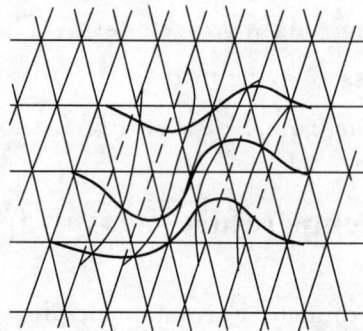

Fig. 4   The drift of moire stripes

The coordinate system established in the test installation drawing shown in Fig. 2 is as follows, the swirl chamber axis direction is adopted as z axle, the ray propagation direction is chosen as x axle and y axle is defined by right hand regulation. The relevant deflection angle ray can be known by reading the displacement of every point on every moire stripe from the moire deflection pictures which have been taken, and the deflection angle is related to refracting power[5].

$$\varphi = \frac{1}{n_0}\int_0^{x_0}\left(\frac{\partial n}{\partial y}\right)\mathrm{d}x \tag{1}$$

where the subscript 0 refers to swirl chamber boundary layer; φ is ray deflection angle; n is refracting power of medium.

The deflection may also be showed as

$$\varphi = \frac{h \cdot \theta}{\Delta} \tag{2}$$

where θ is deflection angle of gratings; Δ is the distance between two gratings; h is the displacement of stripe.

The bell type swirl chamber of S195 diesel engine is the axisymmetric structure. So its flow field is treated approximately as axisymmetric flow field and $\sqrt{x^2+y^2}=r$ is taken into account, then

$$\varphi(y,z) = 2y\int_y^{r_0}\frac{\partial n(r,z)}{\partial r}\cdot\frac{1}{\sqrt{r^2-y^2}}\mathrm{d}r \tag{3}$$

After Abel transformation is applied to equation (2), the following equation can be got

$$n - n_0 = -\frac{1}{2\pi}\int_y^{r_0}\frac{\varphi(r,z)}{\sqrt{r^2-y^2}}\mathrm{d}r \tag{4}$$

Considering equation (2), then

$$n(y,z) = n_0 - \frac{\theta}{2\pi\Delta}\int_y^{r_0}\frac{h(r,z)}{\sqrt{r^2-y^2}}\mathrm{d}r \tag{5}$$

Provided the refracting power n_0 of boundary reference layer and the displacement quantity $h(r,z)$ of moire stripe have been known, the refracting power distribution of the flow can be calculated by equation (5).

When the wavelength of test light source is 632.8 mm, the relation between medium refracting power and temperature may be described accurately if the medium of the swirl chamber is taken as perfect gas[7].

$$n-1 = \frac{0.292\,015\times10^{-3}}{1+0.368\,184\times10^{-2}t} \tag{6}$$

where t is centigrade temperature.

155

4 Reproduction of temperature field and analysis of the results

Fig. 5 shows the moire stripe photos at four representative crank angles during pure-compression condition. The displacement of moire stripe is read by two-moire-stripe-drawing-method. The negative of moire stripe photo which is taken first when the engine doesn't work is overlaped on the negative of moire deflection picture which is to be read. Then the displacement quantities of every point in deflection stripe can be determined by reading equipment from the overlap drawing. After the displacement data are approached by three-sample-strip function, the distribution of refracting power of swirl chamber and the temperature field of the swirl chamber can be calculated by equation (5) and (6).

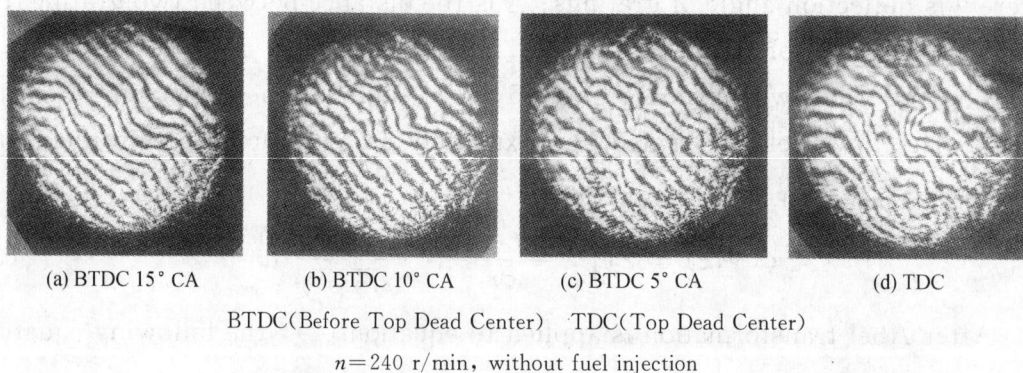

(a) BTDC 15° CA   (b) BTDC 10° CA   (c) BTDC 5° CA   (d) TDC

BTDC(Before Top Dead Center) TDC(Top Dead Center)

$n=240$ r/min, without fuel injection

Fig. 5 Moire deflection stripe photos of the swirl chamber during compression stroke

Since the medium in the swirl chamber of a diesel engine is sealed, the atomosphere can't be treated as the boundary reference field of quantitative calculation like measuring the exhaust jet field of a rocket engine[7,8]. The authors discovered that there exists a quite thin air boundary layer possessing even density close to the surface of swirl chamber. So the authors take this air boundary layer as the reference boundary layer. When the measurement is made using moire deflectometry, the transient wall surface temperature of the swirl chamber is measured by transient wall surface thermocouple at the same time. This wall surface temperature is used as the temperature of boundary reference layer, and the quantitative reproduction of transient temperature distribution of swirl chamber is realized. Fig. 6 shows the compression temperature distribution of the swirl chamber which is calculated from Fig. 5 quantitatively.

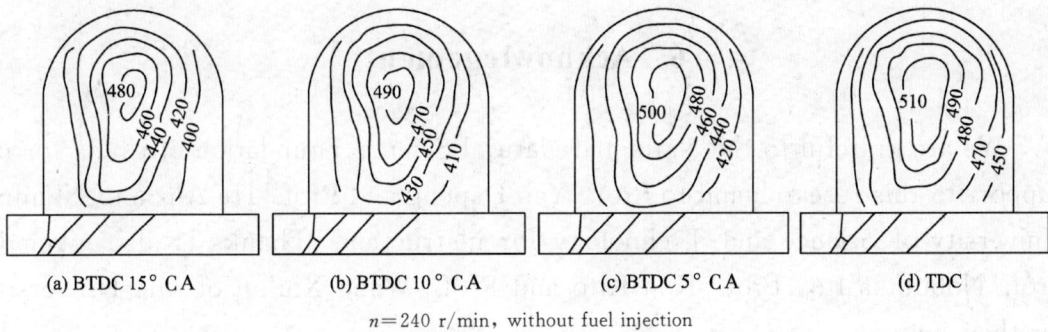

(a) BTDC 15° CA   (b) BTDC 10° CA   (c) BTDC 5° CA   (d) TDC

$n=240$ r/min, without fuel injection

Fig. 6   The temperature distribution of the swirl chamber during compression stroke(℃)

It can be seen from Fig. 6 that the gas temperature at the center of swirl chamber is higher than those at the wall surface and bottom of swirl chamber during compression stroke. With the progress of compression, the total temperature of the swirl chamber rises and the high temperature area at the center of the swirl chamber gradually moves down. The gas temperature at two sides of swirl chamber exist discrepancy during compression stroke, the discrepancy is considerable and the gas temperature of the side which fuel injection is toward to is obviously higher than that of the other side during the initial stage of compression stroke, but the difference in temperature between the two sides reduces during the later stage of compression stroke.

5   Conclusions

(1) Laser-moire deflectometry technique have many advantages such as high precision, fast response, large volume of information and non-disturbance on flow field. It is an advanced measurement method.

(2) The transient wall surface temperature of a swirl chamber is put forward for the first time to be used as the boundary refernce layer of flow field, and the difficulty of determining the boundary reference conditions has been overcomed when the temperature field of swirl chamber is measured by moire deflectometry.

(3) During the compression process, the gas temperature at the side which fuel injection is toward to is higher than that of the other side during the initial compression stroke. With the progress of compression, the difference between two sides reduces gradually. But the high temperature area is located at the center of swirl chamber from beginning to end during compression stroke.

6　Acknowlegement

We are grateful to the Nationnal Natural Science Foundation of China for its support to this research and to Prof. Yan Dapeng and Prof. He Anzhi of Nanjing University of Science and Technology for instruction. Thanks is also given to Prof. Nobutaka Ito，Prof. Koji Kito and Prof. Wang Xiulun of Mie University for their help.

References

[1] Li Detao. *Study and Design on the Swirl Chamber of a Diesel Engine*[M]. Beijing：Press of Machinery Industry，1986：127.

[2] Vest C M. Interferometry of strong refracting axisymmetric phase object[J]. *Appl Opt*，1975，14(7)：1601-1607.

[3] Kafrif O. Morie deflecto-metry a ray deflection approach to optical testing[J]. *Optics Letters*，1980，5(12)：444-464.

[4] Keren E，et al. Measurements of temperature distribution of flame by moire deflecto-metry[J]. *Appl Opt*，1981，20(12)：189-204.

[5] Stricker J. Analysis of 3-D phase object by moire deflectometry[J]. *Appl Opt*，1984，23：3657-3659.

[6] Bar Ziv E，Sgulim S，Kafri O，Keren E. Temperature mapping in flames by moire deflectometry[J]. *Appl Opt*，1983，22(5)：543-549.

[7] Lian W Y，et al. The structure and internal properties of underexpanded exhaust jets [C]∥ *International Symposium on Refined Flow Modeling and Turbulence Measurements*，Iowa，USA，1985(9)：53-57.

[8] Trolinger J D. Holography for aerodynamics[J]. *Astronautics and Aeronautics*，1982，10(8)：56-61.

(From：*The First Asia-Pacific Conference on Combustion*，Japan，1997)

柴油机燃烧过程涡流室内瞬态温度分布的研究

[摘要]　采用激光莫尔偏折技术，结合高速摄影方法，对 S195 柴油机燃烧过程涡流室内的瞬态温度分布进行了测试研究，定量描述并详细分析了燃烧过程涡流室内的温度分布规律及其变化历程。

Quasi-dimensional coherent flamelet model in an IDI diesel engine

Xia Xinglan, *Li Detao*, *Xue Hong*, *Dong Gang*, *Yang Wenming*

Abstract

A simplified turbulent combustion model based on the coherent flamelet model is developed in this study and is used in S195 type IDI diesel engine. The combustion physics involved in each phase of the IDI diesel engine is described and modeled. The combustion events in the swirl chamber is broken into five phases: low temperature ignition kinetics, transition to high temperature chemical kinetics, high temperature premixed burn kinetics, transition to mixing controlled diffusion burn and flamelet diffusion burn phase; while the combustion in the main chamber is only considered as flamelet diffusion combustion. The Shell ignition model, global Arrhenius equation and coherent flamelet model are used to model low temperature ignition, high temperature premixed burn and diffusion burn respectively. The transition to the high temperature premixed burn is accomplished by using a criterion based on temperature and the transition to diffusion burn is based on a critical Damkohler number. The combustion model is used in quasi-dimensional model. The magnitude and location of the pressure and heat release rate predicted by model are coincided with those of experiment. The effects of the stretch parameter and the destruction parameter in model on the pressure are also studied in the paper and the results are used to pick the optimum settings for the model.

1 Introduction

Coherent flamelet combustion model, which is recently developed and is used in turbulent, is a new research method. In turbulent field, when

Damkohler number is very large, chemical reaction duration is very short compared to scale of turbulent time and chemical reaction zone is very thin compared to scale of turbulent length, the reaction zone thus forms a flame sheet. The local behavior of the flame sheet is laminar and turbulent fluctuation induces a strain on the flame in its own plane and produces many wrinkles other than effect on the local structure of the flame. The turbulent reaction zone is considered as an ensemble of flamelet element, which is stretched by the turbulent flow field but maintains the laminar flamelet structure locally. The concepts of the flame area density and turbulent combustion velocity are used to describe combustion process. In transport equation of the flame area density, the source terms, which accounted for the production and destruction of flame area due to strain induced by turbulence, is determined by physical mechanisms of flame production and flame destruction.

In the case of spark ignition engines, the flame structure is now quite well understood and the coherent flamelet model has been used to calculate combustion in such engines successfully[1,2,12]. The cylinder pressure and the heat release rate predicated by the model are coincided with the results of experiments. In addition, the coherent flamelet model has been applied to direct injection diesel engines by Dillies et al. [3]. The cylinder pressure and heat release rate are calculated with some successes by three dimensional Kiva II code, but the premixed combustion in the diesel engine is not described satisfactorily. The examples in the study of Dillies are limited to low compression ratio, and the peak pressure predicted is lower than that of experiment. However, in indirect injection engines the problems are more complex and there is great difference in the physical phenomena involved in the main chamber and swirl chamber: the formation and evolution of spray droplets, mixing of fuel and air, auto-ignition of the mixture, propagation of the flame as well as influence between the main chamber and swirl chamber by means of passenger. Therefore, the coherent flamelet model is not used in the indirect injection engines.

The main chamber, swirl chamber and link passage are now treated differently in indirect injection engine by means of the coherent flame model. The analysis of quasi-dimensional coherent flamelet model is described in this paper.

2　The coherent flamelet model

2.1　The chemical reaction rate

The chemical reaction takes place in a group of sheet that are imbedded in a turbulent flow field. The sheets are very thin and are called flamelets(Fig. 1). In the premixed combustion, the unburnt mixture and the burnt gas are devided by the flame sheet. In the diffusion combustion, fuel and oxygen are devided by the flame sheet. The chemical reaction rate in the sheet is speeded up by the strain of the flamelet. The chemical reaction rate per unit flame area can be calculated by the analysis of flame sheet and the overall chemical reaction rate W_i($i=0$, f referring to oxygen and fuel respectively) can then be found by integrating the local reaction rate per unit flame area over all of the flame area and is given by:

$$W_i = \rho_{i\infty} U_{Di} \Sigma \tag{1}$$

Where $\rho_{i\infty}$ is the reactant species density far from the flame front, Σ is the local flame area density, U_{Di} is the volume rate of consumption of reactant per unit flame area. Since U_{Di} has unit of velocity, it is often referred to as a diffusion combustion velocity[8].

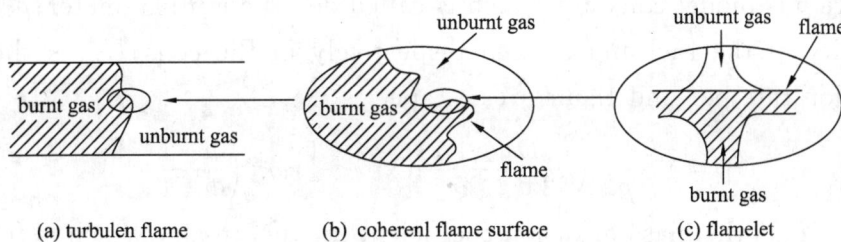

| (a) turbulen flame | (b) coherenl flame surface | (c) flamelet |

Fig. 1　General organization of the flamelet model

2.2　Flame area density equation

The strain rate imposed by turbulent flow has the effect of increasing the available flame area as well as the flame area density. The flame generation process is balanced by several destruction mechanisms such as flame shortening with mutual annihilation of adjacent flamelet by consumption of one of the reactants, wall quenching, and an excessively large strain rate. The flame area density balance equation can be written as:

$$\frac{d\Sigma}{dt} = \frac{d\Sigma}{dt}\bigg|_{pro} + \frac{d\Sigma}{dt}\bigg|_{des} \tag{2}$$

It has been proposed that the behavior of an increment per unit of flame area can be represented by:

161

$$\left.\frac{d\Sigma}{dt}\right|_{pro} = \alpha\varepsilon_s\Sigma \tag{3}$$

where α is a model constant which is called stretch parameter, ε_s is the mean strain rate induced on the flame surface by the fluid.

The strain rate term ε_s can be found by several methods. The turbulent strain rate is usually dominant, and is most often the factor used in determining the strain rate. If the strain term is based on the large scale strain rate predicted by the k-ε model, the form of the strain rate will be[3]:

$$\varepsilon_s = \frac{\varepsilon}{k} \tag{4}$$

Where k is turbulent kinetic energy, ε is the dissipation rate of turbulent kinetic energy.

The flame area destruction term $\left.\dfrac{d\Sigma}{dt}\right|_{des}$ of Eq. (2) can be due to many factors. The flame shortening with mutual annihilation of adjacent flamelet by consumption of reactants can be represented by $\left.\dfrac{d\Sigma}{dt}\right|_{comb}$ [4]:

$$\left.\frac{d\Sigma}{dt}\right|_{comb} = -\beta\left[\frac{\overline{\rho_{f\infty}}U_{Df}}{\overline{\rho_{f\infty}}} + \frac{\overline{\rho_{o\infty}}U_{Do}}{\overline{\rho_{o\infty}}}\right]\Sigma^2 \tag{5}$$

Where β is model constant which is called destruction parameter; $\overline{\rho_{f\infty}}$, $\overline{\rho_{o\infty}}$ are bulk density of the fuel and oxygen respectively in the cells; U_{Di} is the volume consumption rate per unit flame area of species i.

$$U_{Di} = Y_{i\infty}\frac{\overline{\rho}}{\rho_{i\infty}}\sqrt{\frac{\varepsilon_s D}{2\pi}}\frac{(\Phi+1)}{\Phi}\exp\left\{-\left[rerf\left(\frac{\Phi-1}{\Phi+1}\right)\right]^2\right\} \tag{6}$$

Where D is the mass diffusion coefficient for fuel in air, which is function of temperature; Φ is the fuel equivalence ratio; $rerf$ is the inverse of the error function; $\overline{\rho}$ is the density of the mixture.

It is possible that if the strain is too high, the flame may be quenched due to overstretching, and the flame area density may be decreased. This overstretching is usually defined as occurring when the strain rate ε_s exceeds some critical value. It has been observed in experiments that this critical strain rate $\varepsilon_{s,crit}$ is on the order of $10^4 \sim 10^5$ s^{-1}[2]. The destruction term due to overstretching can be represented by:

$$\left.\frac{d\Sigma}{dt}\right|_{over} = -2\alpha(\varepsilon_s - \varepsilon_{s,crit})h(\varepsilon_s - \varepsilon_{s,crit}) \tag{7}$$

Where h is the heaviside function.

Wall quenching is a phenomenon often occurring near the wall, which is found not to be important in the place of the combustion chamber. This term is neglected in this study.

Considering the phenomena above, the balance equation for flame area density may be written as:

$$\frac{d\Sigma}{dt} = \alpha \varepsilon_s \Sigma - \beta \left(\frac{\rho_{f\infty} U_{Df}}{\rho_{f\infty}} + \frac{\rho_{o\infty} U_{Do}}{\rho_{o\infty}} \right) \Sigma^2 - 2\alpha(\varepsilon_s - \varepsilon_{s,crit}) h(\varepsilon_s - \varepsilon_{s,crit}) \tag{8}$$

2.3 The calculation of turbulent flow

In equation (4), the turbulent kinetic energy and its dissipation rate must be calculated. In this study, a simplified method, which is based on experimental results, is used.

$$k = \frac{1}{2}(u_1^2 + u_2^2 + u_3^2) \tag{9}$$

$$\varepsilon = C_\mu \frac{k^{1.5}}{l} \tag{10}$$

Where C_μ is the empirical coefficient, in this study, $C_\mu = 0.09$; u_1, u_2, u_3 are the components of fluctuation velocity in turbulent flow, which is determined by experiment[10]; l is length scale of turbulence, $l = 0.05B$; B is equivalent diameter in the cell[9].

3 Quasi-dimensional coherent flamelet model in an IDI diesel engine

The combustion events in distinct place and distinct stage of a swirl type IDI diesel engine are different. It is necessary to describe these events with different models. In this study the combustion in the swirl chamber is broken into five phases: 1) ignition delay, which is also called low temperature ignition kinetics, 2) transition to high temperature chemical kinetics, 3) high temperature chemical kinetics controlled premixed combustion kinetics, 4) transition to mixing controlled diffusion burn, and 5) flamelet diffusion burn phase. However, in the main combustion chamber, the combustion is only considered as flamelet diffusion burn, while the combustion in the passenger is neglected.

3.1 Low temperature ignition kinetics

The Shell ignition model which was proposed by Halstead et al.[5] was a multi-step Arrhenius kinetic model of low temperature chemical reaction leading to ignition. The model employs five generic species, each of which represents an entire family of radicals presented during the ignition delay, and eight reaction equations which based on chain reaction are obtained. In ignition combustion of the diesel engine, many successful uses had achieved by this model[6]. The Shell model is unaltered for use here.

3.2 The transition to high temperature kinetics

When the low temperature chemical reactions reach a point where the fuel which has already mixed with the air begins to react quickly, the local gas temperature increases rapidly and the chemical kinetics become very fast. The critical temperature of high temperature premixed chemical kinetics is defined as $T_{crit} = 920$ K in this study. Thus in transition both low temperature ignition kinetics and high temperature premixed chemical kinetics model are used, then the reaction rate and consequently the total heat release rate is the sum of results predicted by two models.

3.3 High temperature premixed combustion

The high temperature premixed combustion occurs only on the oxygen side of the diffusion flame, and the mixture which is formed during ignition delay is burnt. During the post-ignition premixed burn, the chemical reactions are dominated by high temperature kinetics. The high temperature premixed burn model is a single one step global Arrhenius reaction which has been turned for high temperature kinetics:

$$W_f = A \bar{\rho}^2 \bar{Y_f}^m \bar{Y_o}^n \exp(-E/T) \tag{11}$$

Where $\bar{Y_f}$ is mean mass fraction of fuel vapor in calculation cell, $\bar{Y_o}$ is mean mass fraction of oxygen in calculation cell, A is frequency factor, E is activation temperature.

The frequency factor and the activation temperature is assumed to be $A = 5 \times 10^{10}$ m^3/(kg·s), and $E = 1.2 \times 10^4$ K respectively in this study. The exponent m and n are taken to be $m=1$ and $n=5$ respectively so as to maximize W_f at the stoichiometric fuel concentration ($\bar{Y_f} = 0.22$ in the fuel vapor-oxygen mixture)[9].

3.4 Transition to the diffusion burn

The transition to the diffusion burn occurs when the chemical reactions become very fast compared to the rate of mixing, and the fuel no longer has time to mix with the oxygen before combustion. This transition is accomplished by using a critical Domkoler number. The Da is defined as:

$$Da = \frac{A\exp(-E/T)}{\varepsilon/k} \tag{12}$$

The critical Da_{crit} used in this study is 50. Once $Da > Da_{crit}$ the diffusion burn begins. The diffusion flame has strictly limited fuel on one side and oxygen on the other. However, due to ignition delay in the IDI diesel engine, fuel vapor mixes with the oxygen before the beginning of the diffusion burn. Thus in transition there may be premixed fuel which is burning on the oxygen side of the diffusion flame. New fuel which vaporizes from the liquid fuel droplets is trapped on the fuel side of the diffusion flame. It will react very quickly when it reaches the flame surface and can not diffuse to the other side, and the premixed fuel is eventually exhausted some time after the diffusion burn is initiated.

3.5 The flamelet diffusion burn

The reaction rate in the flamelet diffusion burn is obtained from equation (1) and (8). After the premixed burn turns to the flamelet diffusion burn, the flame area must be initialized. Since the droplets are the sources of fuel in the diffusion burn, it is reasonable to use the area of the droplets as an initial flame area.

3.6 The gas injection diffusion burn in the main combustion chamber

After combustion in swirl chamber begins, a fraction of fuel enters the main chamber through passenger. Then, this fuel mixes with the air in the main chamber to form new mixture and burns in the main chamber. The combustion in the main chamber is considered as the flamelet diffusion burn and the reaction rate is also obtained from Eq. (1).

The combustion model presented above is used in quasi-dimensional combustion model of swirl type IDI diesel engine. The details of the quasi-dimensional combustion model is described in reference[11].

4 Results and comparison with experiments

The engine modeled in this study is a swirl type IDI diesel engine with the following specification and operation conditions.

single cylinder, four stroke

bore/stroke: 95/115 mm

compression ratio: 19.6 : 1

volume ratio of the swirl chamber: 0.49

injector operation pressure: 12.75 MPa

injection timing/duration: 12° CA BTDC/18° CA

The cylinder pressure and swirl chamber pressure were measured in the experiment, and heat release rate curves were derived from the experimental pressure through the use of a zero-dimensional simulation code[7]. The cylinder and swirl chamber averaged results were compared to those predicted by the model.

The pressure comparisons are shown in Fig. 2. The heat release rate comparisons are shown in Fig. 3. The measured heat release rate is derived from the measured pressure cure using a simulation program, and is not actually measured directly.

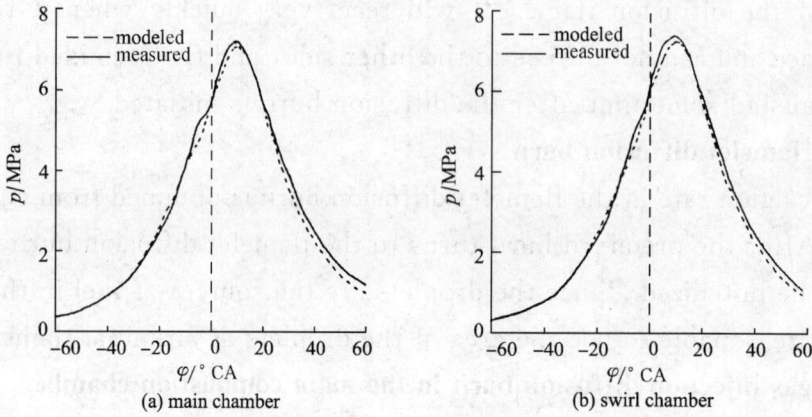

(a) main chamber

(b) swirl chamber

Fig. 2 Modeled and measured pressure traces

The results predicted by the model are well matched with those of the measurement. The magnitude of the premixed burn heat release rate and the diffusion burn heat release rate can be distinguished in Fig. 4.

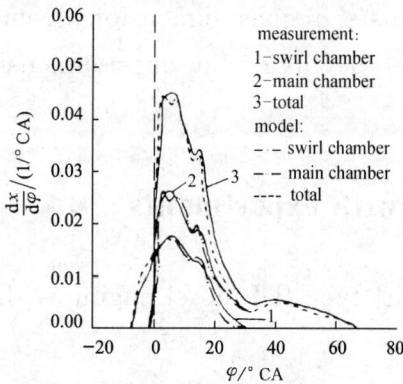

Fig. 3 Modeled and measured heat release rate traces

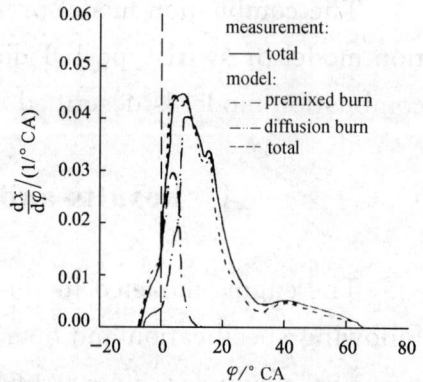

Fig. 4 Modeled heat release rate traces

The stretch parameter α was varied from its standard value of 20. The results of this sensitivity study are shown in the pressure trace of Fig. 5. Increasing α leads to higher pressures in both swirl chamber and the main chamber. Doubling α increases the peak pressure by approximately 10% while the locations of the peak pressure have almost no change.

The destruction parameter β was also varied from its standard value of 5, and the resulting pressure traces are shown in Fig. 6. The magnitudes in the peak pressure change are approximately 10% for factor of two changes in β while the locations of the peak pressure have also almost no change. This is similar to the effects of the stretch parameter α.

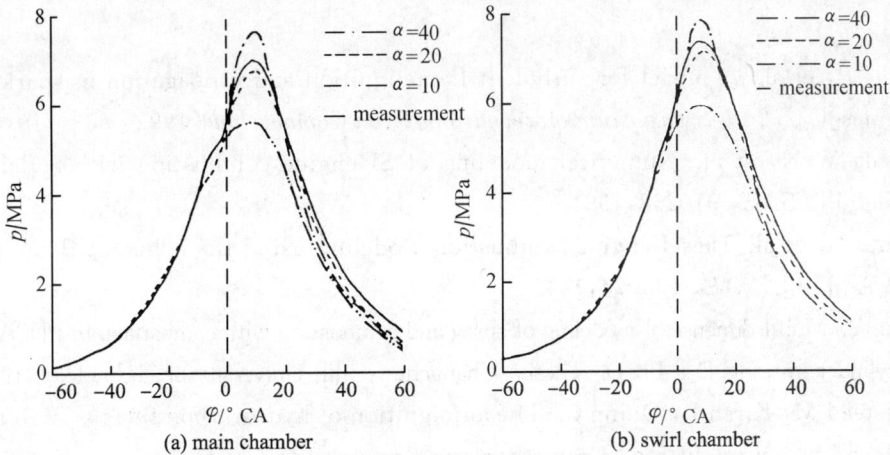

Fig. 5   Effect of the stretch parameter α on the pressure traces

(a) main chamber   (b) swirl chamber

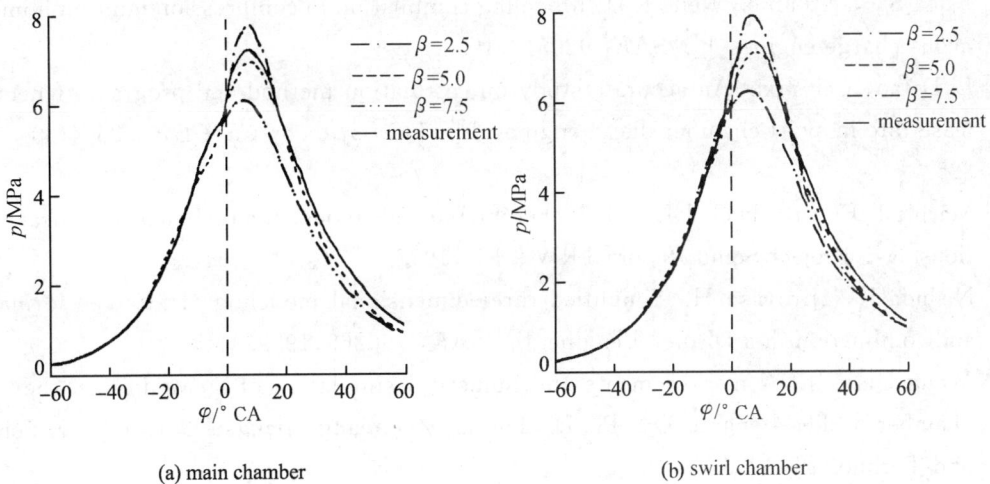

(a) main chamber   (b) swirl chamber

Fig. 6   Effect of the destruction parameter β on the pressure traces

5   Summary and conclusions

（1）The coherent flamelet model is a kind of new combustion model, and the flame structure can be better described by it. The pressure and the heat release rate in the swirl chamber and the main chamber are coincided with those of the measurement.

（2）In the swirl type IDI diesel engine the premixed burn can be qualitatively distinguished from the diffusion burn by the coherent flame model.

（3）Increasing stretch parameter α and decreasing destruction parameter β have similar effects on the pressure traces in both the swirl chamber and the main chamber.

References

[1] Bodie P, et al. A model for turbulent flame ignition and propagation in spark ignition engines[C]// *24th Symposium (International) on Combustion*, 1992: 503-510.

[2] Cheng W K, et al. Numerical modeling of SI engine combustion with a flame sheet model[J]. *SAE*, 910268,1991.

[3] Dillies B, et al. Diesel engine combustion modeling using the coherent flame model in KIVA- II [J]. *SAE*, 930074,1993.

[4] Dong Y. Multil-dimensionl modeling of spray and combustion with comparison to PDPA data and in-cylinder imaging[D]. Ph. D. Thesis, Changchun: Jilin University of Technology,1996.

[5] Halstead M, Kirsh L, Quinn C. The autoignition of hydrocarbon fuels at high temperature and pressures-fitting of a mathematical model[J]. *Combustion and Flame*, 1997, 30: 45-60.

[6] Kong S C, Ayoab N, Reitz R D. Modeling combustion in compression ignition homogeneous charge engines[J]. *SAE*, 920512,1992.

[7] Li Detao, Cai Yixi. An accurate study on calculation method and program of heat release rate in swirl chamber diesel engines[J]. *Transation of CSICE*, 1986,4(4): 313-325.

[8] Marble F E, Broadwell J E. The coherent flamelet model for turbulent chemical reactions[R]. Project Squid Report TRW-9-PU,1997.

[9] Nishida K, Hiroyasu H. Simplified three-dimensional modeling of mixture formation and combustion in a DI diesel engine[J]. *SAE*, 890269,1989.

[10] Wang Qian. LDA measurements and numerical simulation of air motion in the swirl chamber of diesel engine[D]. Ph. D. Thesis, Zhenjiang: Jiangsu University of Science and Technology,1996.

[11] Wu Zhixin. Investigation on quasi-dimensional combustion model of swirl chamber diesel engine[D]. Ph. D. Thesis, Zhenjiang: Jiangsu University of Science and Technology, 1997.

[12] Zhao X W, Matthews R D, et al. Three dimensional numerical simulation of flame propagation in spark ignition engines[J]. *SAE*, 932713,1993.

(From: *Procceding of TPTSPE*, Japan, 1997)

非直喷柴油机准维相关微元火焰模型

[摘要]　本文提出一种基于相关火焰模型的简化湍流模型,并应用于 S195 型非直喷式涡流室式柴油机。描述了该类柴油机燃烧的每个阶段。将涡流室内的燃烧过程分为五个时期:低温着火化学动力学反应期,向高温预混化学动力学反应过渡期,高温预混化学动力学反应期,向燃料和空气混合控制的扩散燃烧过渡期,火焰微元扩散燃烧期。认为主燃烧室的燃烧只有相关火焰微元的扩散燃烧阶段。用 Shell 着火模型、全局 Arrhenius 方程、相关火焰微元模型来分别模拟低温着火化学动力学反应阶段、高温预混化学动力学反应阶段和扩散燃烧阶段,用工质的局部温度作为低温着火向高温预混燃烧过渡的判据,用局部 Damkohler 数作为高温预混燃烧向扩散燃烧过渡的判据。该燃烧模型应用于准维模型当中,模型中不同位置处压力大小和放热率的预测值与实验很好地吻合。本文还对模型的结构参数和运转参数对压力的影响进行分析,结论为模型参数的最优化选取提供了相应的依据。

柴油机涡流室新的结构设计

田东波,杨文明,李德桃

[摘要]　本文基于对柴油机涡流室进行的多年基础性研究,设计了一种新型的涡流室结构,以改善柴油机的性能。试验结果表明,在负荷特性曲线上,采用新型涡流室结构的柴油机和原机相比标定点油耗略低,但部分负荷时油耗降低较多,排温也有所降低,50％负荷时油耗降低达18 g/(kW·h)。不仅如此,新燃烧室的冷起动性能也有较大改善。本文还通过放热规律研究分析了新燃烧室的燃烧特征。

　　涡流室式柴油机与直喷式柴油机相比存在着经济性和冷起动性较差的缺点,然而它也存在着排污少、噪声低和高速适应性好等优点。造成涡流室式柴油机经济性较差的主要原因有3个:(1)流动损失较大。连接通道导致了主、副室之间的节流损失。(2)传热损失较大。由于有涡流室和通道,其散热面积相对较大,传热损失增加。(3)该燃烧模式的循环效率较低[1]。针对这些原因,国内外内燃机工作者做了大量的研究工作,并且取得了一些令人满意的成果。如:日本三菱公司的涡流室连接通道由两段角度不同的通道组成[2],主燃室为楔形结构,这样既提高了通道流量系数又使燃气在主燃室的扩散与穿透得到强化,从而提高了柴油机的放热等容度,缩短了燃烧滞后期,最终改善了柴油机的性能;久保田公司的3旋流涡流室式燃烧系统(ETVSC)[3]通过镶块通道形状与活塞顶部凹坑的最佳匹配以及喷油特性的最佳化,使得采用这种燃烧方式的柴油机无论是经济性还是排放、噪声等都得到较大的改善。我们曾改进常柴Ⅱ号吊钟型涡流室并配以双楔形主燃室也取得了降低油耗、减少 NO_x 的效果[4]。

　　但是,现有的涡流燃烧室都未能克服该型燃烧室的基本缺点,要做到这一点,必须在结构上有所突破。为此,我们在对涡流室的温度场、浓度场、速度场以及燃烧过程作了多年基础研究的基础上,发展了一种新的涡流燃烧室结构。

1　涡流室内工质过程的若干基础研究的启示

　　我们的理论和试验研究已经表明[5],涡流室内气体的湍流运动对改善混合气质量和加速燃烧进程有着重要意义。最近,我们用激光多普勒测速仪 LDA(Laser Doppler Anemometer)对吊钟型涡流室内空气运动的流场分布进行了测量,部分结果如图1、图2所示。图1表明,在 BTDC 10°CA 和 TDC 时刻,涡流室中心附近的湍流强度相对较大。图2表示了上止点前不同时刻燃油喷射轴线(以下简称油线)上湍流强度的分布情况,它说明油线上方湍流强度较高,下方湍流强度则相对较低。文献[6]的高速摄影研究表明,涡流室式柴油机的混合气首先在油线下方靠近通道口附近着火,所以利用涡流室中心附近相对较高的湍流强度和提高油线下方的湍流强度对改善涡流室式柴油机的性能有着积极意义。

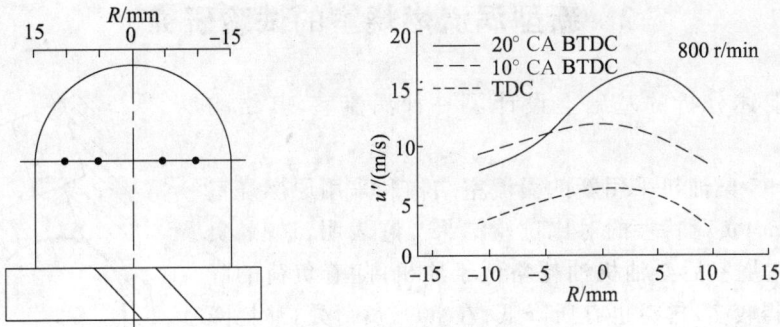

图 1　涡流室半径方向湍流强度分布

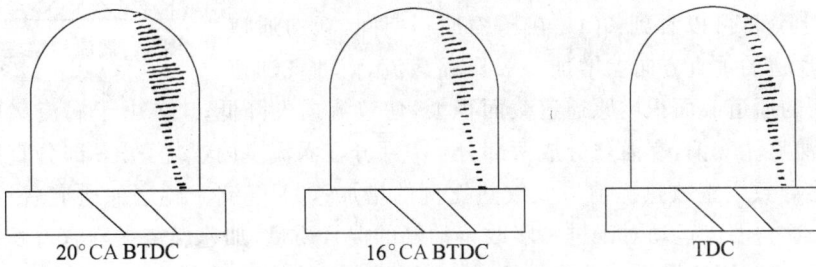

20°CA BTDC　　　　16°CA BTDC　　　　　TDC

图 2　燃油喷射轴线上湍流强度的分布($n=800$ r/min)

由图 3 可见,在压缩过程中涡流室的中部区域是高温区(此规律不随转速的变化而改变)[7]。如果让一部分燃油通过此高温区直接进入主燃室,将有利于燃油的蒸发、混合以及按直喷式柴油机的放热模型进行燃烧。

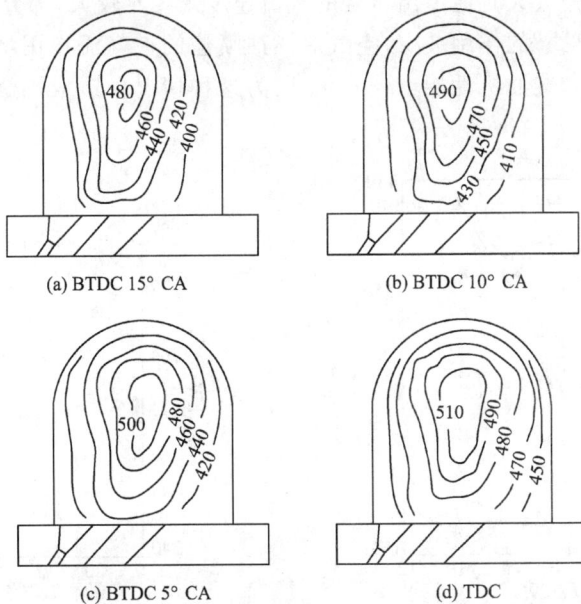

(a) BTDC 15° CA　　　　　　　(b) BTDC 10° CA

(c) BTDC 5° CA　　　　　　　　(d) TDC

图 3　压缩过程涡流室内温度分布($n=240$ r/min,不喷油)

171

2 新型涡流燃烧室的试验研究

作者在上述分析的基础上设计了一种涡流室,其结构如图 4 所示。

图 5 为 195 柴油机采用新的涡流室结构和采用原燃烧室在 2 000 r/min 时负荷特性的对比试验结果。它表明,100%负荷时采用新燃烧室后柴油机油耗略低于原机;随着负荷的降低,油耗降低得较多,排温也有所降低;在 50%负荷处,油耗降低了 18 g/(kW·h)。上述试验结果说明,新燃烧室的设计是合理的。

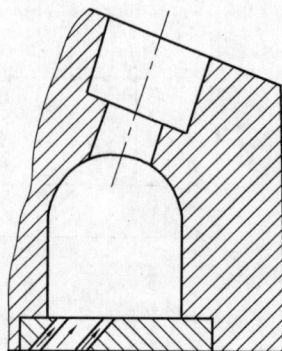

图 4　涡流燃烧室的新结构

从结构上分析可以看到:(1) 在压缩冲程期间,新的通道结构能在原通道的垂直方向上增加一股压缩涡流,使油气加速混合,而且新的通道截面积比原通道截面积大,使节流损失降低;(2) 由于涡流室内压缩涡流的运动将燃油喷注吹偏,燃油被分成两部分:一部分在涡流室内燃烧,另一部分直接进入主燃室燃烧,这将导致主室放热提前以及放热比例的增加;(3) 气流流经这种通道后,在通道口产生湍流运动,这一湍流运动对于进一步改善油气的混合质量、加速燃烧进程有着积极意义。

为了进一步分析新燃烧室的工作特性,进行了以下试验。

2.1　供油提前角试验

图 6 是采用新燃烧室进行供油提前角试验的结果。从图中可以看到,随着供油提前角的增加,柴油机的性能越来越差,而提前角为上止点前 12° CA 时性能最好。这可能是因为此时通道出口处的气流速度减弱,较多燃油能进入主燃室,空气对燃油喷注撞击的逆气流影响减轻所造成的。反之,提前角较大时通道出口处的气流速度则相对较大,部分燃油难以进入主燃室而留在涡流室内,"直喷式"的作用减少,空气对燃油喷注的逆气流作用增强,因而导致油耗增加。但供油提前角过小,柴油机性能也会恶化,因为这种情况会造成燃烧滞后,燃烧所作的有效功减少。

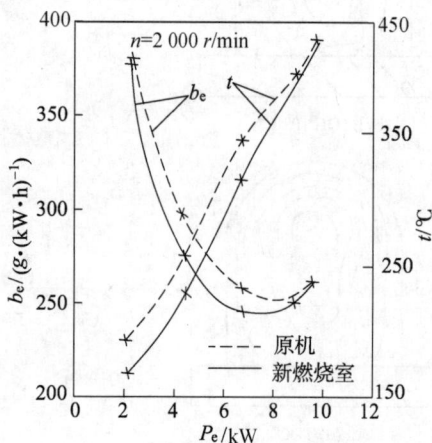

图 5　负荷特性试验对比结果

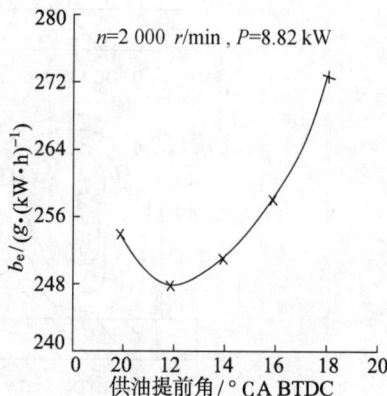

图 6　供油提前角的试验结果

2.2 通道角度试验

通道角度对柴油机性能的影响较大,在试验过程中作者分别选择了倾角 45°,40°,35°进行试验,试验结果如图 7 所示。从试验中可以看到,当倾角为 45°时柴油机油耗较高,当倾角为 40°和 35°时性能较好。倾角较大不利于燃气在主燃室的扩散与穿透,在涡流室中难以形成强涡流,所以性能较差;倾角过小时涡流室内的气流难以进入主燃室。因此,最好采用两段角度的通道,但工艺复杂。作为一个折中方案,我们选择通道倾角为 40°。

2.3 通道截面比试验

对于涡流室式柴油机而言,通道截面积 F_j 与活塞顶面积 F_p 之比 F_j/F_p 是影响柴油机性能的一个重要因素,最佳 F_j/F_p 值应根据发动机的用途和转速确定。图 8 是采用不同的 F_j/F_p 在 2 000 r/min 转速下进行对比试验的结果,从中我们可以看出,F_j/F_p 在 1.4%~1.6% 范围内,油耗有相对较低的值。

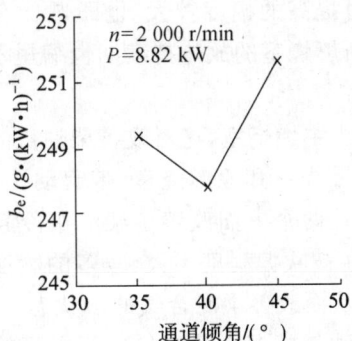

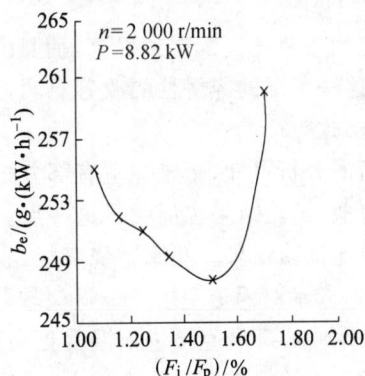

图 7　通道倾角的影响　　　　图 8　通道截面比的试验结果

2.4 放热分析

图 9 是原机分别在 8.82 kW(2 000 r/min)和 4.41 kW(2 000 r/min)时的放热率曲线图,图 10 则是采用新燃烧室后在上述两种工况下的放热率曲线图。通过放热率的对比可以看到:无论是在 8.82 kW 还是在 4.41 kW 时,原机的着火始点均为上止点前 5°CA,而新方案涡流室的着火始点较迟,为上止点前 3°CA;但主室的着火始点为上止点,较原机早 1°CA,这说明有燃油直接进入主燃室,在那里首先着火燃烧。从累积放热率的对比来看,在两种工况下,新

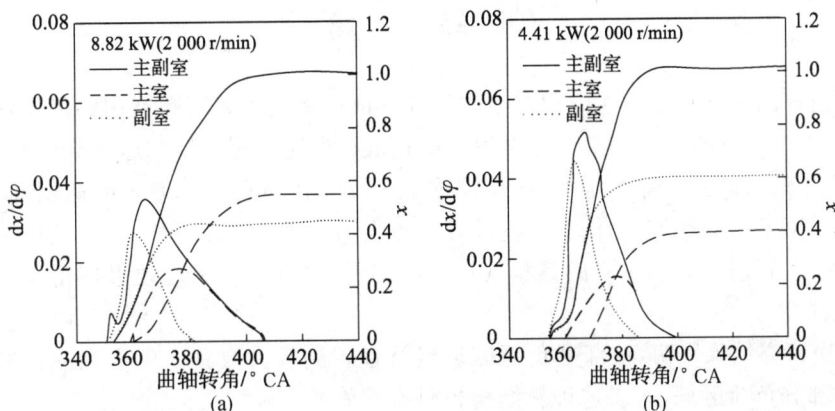

(a)　　　　　　　　　　　　(b)

图 9　原机放热率曲线图

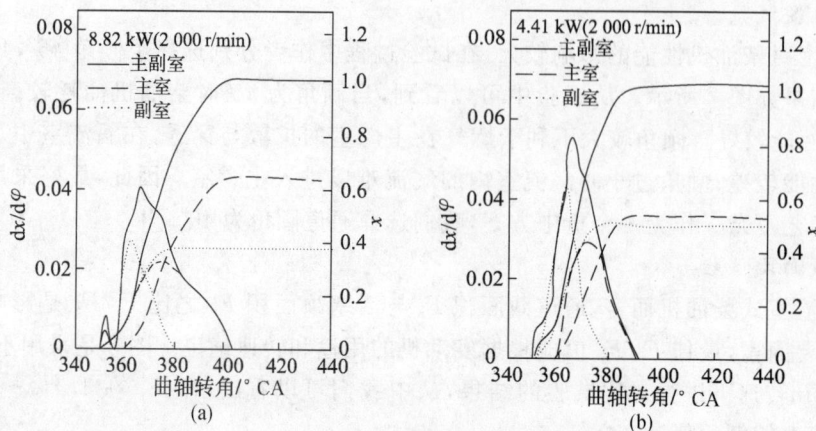

图 10　新结构燃烧室放热率曲线图

燃烧室主室的放热比例都得到提高,而且 50% 负荷时提高的幅度较大,也即燃烧有效功增加较多,因此这一工况的经济性的改善更明显些,同时新燃烧室的放热持续期也缩短了。

2.5　冷起动试验

由前面的分析可知,新燃烧室有部分燃油直接进入主燃烧室,这对改善柴油机的冷起动性能将大有裨益,试验结果也证明了这一点。表 1 为新燃烧室和原燃烧室(带起动孔)在冷起动条件下的对比试验结果。和原燃烧室相比,新燃烧室在两个方面改善了柴油机冷起动时的着火条件:(1)起动时涡流室内压缩涡流较弱,燃油喷注被吹偏程度不大,较多的燃油通过温度较高的涡流室中心区域直接进入温度更高的主燃室,从而改善了混合气的热状态;(2)新燃烧室中压缩气流能够促进燃油的离散雾化,从而改善了油气的混合状态。这两方面的改善导致涡流室式柴油机冷起动性能的较大提高。

表 1　冷起动试验结果

方　案	温度/℃			第一次起动		第二次起动	
	环　境	机　油	柴　油	状　况	时间/s	状　况	时间/s
新方案	3	3	3	成　功	4.5	成　功	4.5
原　机	3	3	3	成　功	5.5	成　功	5.5

3　结　论

(1)新设计的涡流室结构在应用直喷式燃烧室的一些设计原则和利用涡流室内气体的湍流运动方面取得了进展。台架试验表明,2 000 r/min 时的负荷特性和原燃烧室相比,新结构标定点油耗略低,部分负荷时油耗降低较多,排温也有所降低,50% 负荷时油耗降低达18 g/(kW·h)。不仅如此,新结构的冷起动性能也大为改善。放热分析表明,新结构主室的放热提前了,燃烧持续期缩短,主室的累积放热比例得到提高,这一综合效果使得柴油机实际热效率得到了提高。

(2)本次研究结果表明,在部分负荷区域经济性得到较大改善。由于柴油机实际运行时主要工作在部分负荷区域,所以这种新结构也具有较好的实际意义。

(3)新结构具有结构简单的特点,易于在生产中推广。

参 考 文 献

［1］李德桃.柴油机涡流燃烧室的研究与设计［M］.北京：机械工业出版社,1986.

［2］Tamura H，Hashimoto M，et al. New system of IDI diesel engine for passenger car application［J］. *SAE Paper*，890262，1989.

［3］寺下清司,山田喜一郎.小型通用柴油机的发展动向［J］.国外内燃机,1995(1).

［4］李德桃.低油耗、低污染、低爆压的柴油机涡流燃烧室.国家发明奖,1990.

［5］王谦.柴油机涡流室内空气运动的试验研究及数值模拟［D］.镇江：江苏理工大学,1996.

［6］李德桃,朱广圣.柴油机涡流燃烧室内喷雾和燃烧过程的研究［J］.内燃机学报,1989(1).

［7］熊锐.柴油机涡流室内温度场与密度场的研究［D］.镇江：江苏理工大学,1995.

（本文原载于《燃烧科学与技术》1997 年第 4 期）

New configuration design of the swirl chamber of diesel engines

Abstract：A new configuration of swirl chamber is developed to improve the performance of diesel engine based on the years' research on the swirl chamber diesel engine. The experimental results show that on the diagram of load character，the fuel consumption of the engine using new combustion chamber is a bit lower than that of original engine at the full load. With the load becoming lower，the fuel consumption becomes much lower，and the exhausting temperature is also lower. When at the 50％ load，the specific fuel consumption of new swirl chamber is 18 g/(kW・h) lower than that of the original. In adittion，the cold startability of new combustion chamber is improved a lot. The combustion character of new combustion chamber is also analysed by means of rate of heat release.

涡流室式柴油机燃油喷雾过程的
三维可视化研究

夏兴兰,李德桃,董　刚,吴志新

[摘要]　通过改进 KIVA-Ⅱ 程序中的油滴破碎和喷雾碰壁两个子模型,在微机上对涡流室式柴油机燃油喷雾过程进行可视化研究。画出油滴运动轨迹和燃料蒸汽度的三视图,研究涡流室内油滴的运动过程和喷油对涡流室内速度场和温度场以及平均温度和平均压力的影响,计算结果与高速摄影结果[1]基本一致。

　　燃油与空气的混合过程对柴油机的燃烧有着重要的影响。国内外有些学者进行过一些实验研究,但由于实验条件的限制,大部分实验研究都是在静态模拟容器中进行的,少部分是在柴油机低速运转条件下进行的。由于涡流室内存在高温高压环境和强烈的涡流运动,而涡流强度又与发动机转速密切相关,所以静态模拟容器和低速运转条件下的柴油机都与高速柴油机的实际运行状况有较大的区别。就现有技术而言,在高速柴油机中进行实验研究存在较大的困难。数值模拟可以模拟高速柴油机的实际运行状况,弥补实验研究的不足,且可以计算许多无法直接测量的结果。模拟结果的可视化研究可以形象地显示所得信息,使人们对诸如速度场、浓度场、油滴位置等有更清晰的认识。国外一些学者利用 KIVA-Ⅱ 程序对直喷式柴油机的燃油喷雾过程进行过模拟研究[2,3],但由于原始的 KIVA-Ⅱ 程序不适应涡流室式柴油机的工作过程,因而对涡流室式柴油机所进行的研究较少。

　　笔者在 KIVA-Ⅱ 程序的基础上,对燃油雾化和喷雾碰壁等子程序进行修改,使之适应涡流室式柴油机的条件,以 S195 柴油机为例,在微机上对燃油喷雾过程进行了三维可视化研究。

1　数学模型

　　在燃油喷雾过程中,由于油滴表面的微小扰动,产生不稳定波而使油滴破碎的现象,已为许多实验所证实[4-6]。在柴油机中,燃油从喷油器喷出后,形成图 1 所示的喷雾形态,喷雾可分为液核区和气液混合区[7]。不稳定性理论指出:射流表面的破碎过程可以用 K-H(Kelvin Helmotz)不稳定性理论来描述,而直接暴露于空气的微滴可以用 R-T (Rayleigh-Taylor)不稳定性理论来描述。所以,对涡流式柴油机

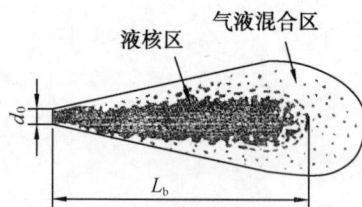

图 1　燃油喷雾形态

喷雾的液核区用 K-H 不稳定性来分析[8],对气液混合区用 R-T 不稳定性理论来分析[9]。

　　喷雾碰壁过程是一个非常复杂的过程,油滴在碰壁过程中伴随有粘附、反弹、扩散、飞溅和破碎等现象。本文将油滴碰壁过程分为粘附、反弹/粘附、飞溅/附壁射流等 3 种相互重叠的过程,以临界 Weber 数作为各方式转换的判据。当 $We \leqslant 5$ 时,为粘附过程;当 $5 < We \leqslant 80$ 时,为

反弹/粘附过程；当 $We > 80$ 时，为飞溅/附壁射流过程。

2 计算结果与分析

以 S195 柴油机为例，对涡流室式柴油机的喷雾过程进行模拟计算，得到可视化图形。柴油机在标定工况（8.82 kW/2 000 r·min^{-1}）下运行，每循环供油量为 0.037 4 g，喷油始于 10° CA BTDC，喷油持续角为 16° CA[1]，假定喷油规律为梯形波方式。此处暂不考虑燃料的燃烧。

燃油蒸汽质量分数图的截面位置见图 2，图 3 为燃油液滴轨迹的 3 个视图，燃油从喷油器顺气流射向涡流室，在喷雾初期，刚进入涡流室的油滴，由于其运动速度较大，运动轨迹几乎不受空气运动的影响。随后，在涡流室中强烈的有组织的涡流运动的作用下，进入涡流室的油滴和涡流室内的气体进行质量、动量和

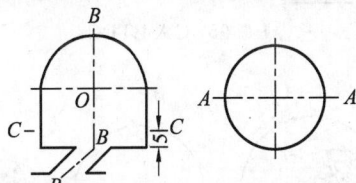

图 2 燃油蒸汽质量分数图

能量的交换后，其运动轨迹发生变化，转向偏左，并很快转移到空气运动的轨迹上。在喷雾的中后期，油滴运动轨迹与涡流运动方向相似，空气运动使油滴分散到更大的空间。在油滴运动的过程中，油滴几乎不与涡流室侧壁相碰，而在 4.76° CA BTDC，油滴与涡流室底面相碰，碰壁后，一部分油滴粘附在壁面上或沿壁面形成附壁射流，另一部分油滴反射回涡流空间。在 1.96° CA BTDC，油滴越过通道口，而后一小部分油滴沿涡流室右侧顺气流向顶部运行或进入主燃烧室，大部分油滴在越过通道口附近以前由于不断蒸发而消失。从图 3 和图 4 的侧视图和俯视图可看出：油滴主要集中在涡流室中心平面附近，而往前后方向扩散较少。涡流室喷油过程的高速摄影照片[1]证实了油滴的这种运动过程。在油滴的运动过程中，伴随有分裂、吸热、蒸发、气化和扩散等过程，逐步形成可燃混合气。图 4 为燃油蒸汽质量分数图，截面位置见图 2。图 3b、图 4b 表明，油滴与涡流室底部相碰后，由于碰壁产生的油滴分裂和壁面油膜的吸

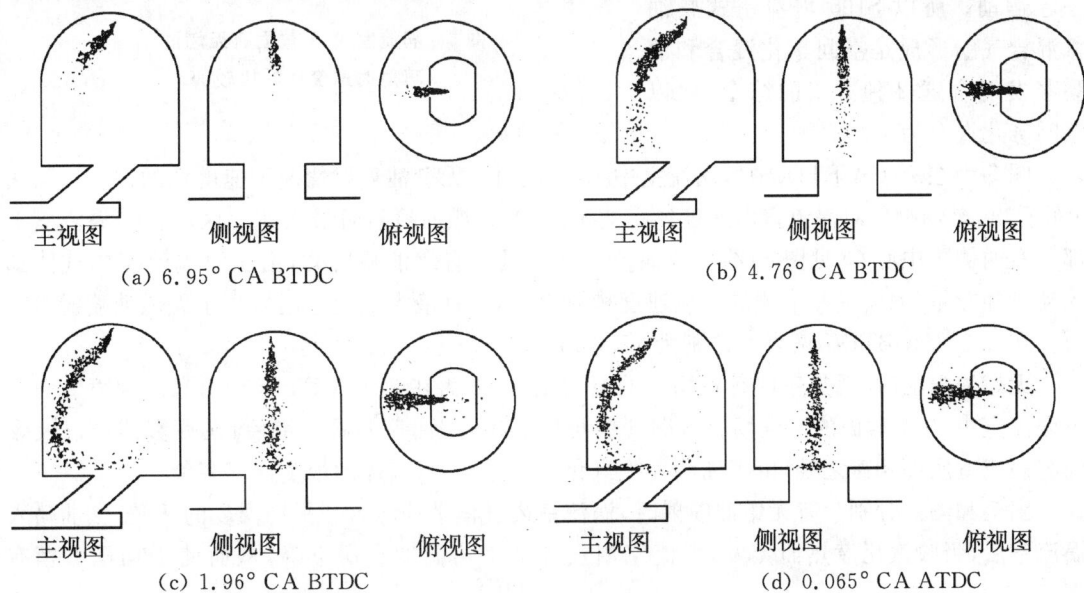

(a) 6.95° CA BTDC (b) 4.76° CA BTDC

(c) 1.96° CA BTDC (d) 0.065° CA ATDC

图 3 燃油液滴运动轨迹

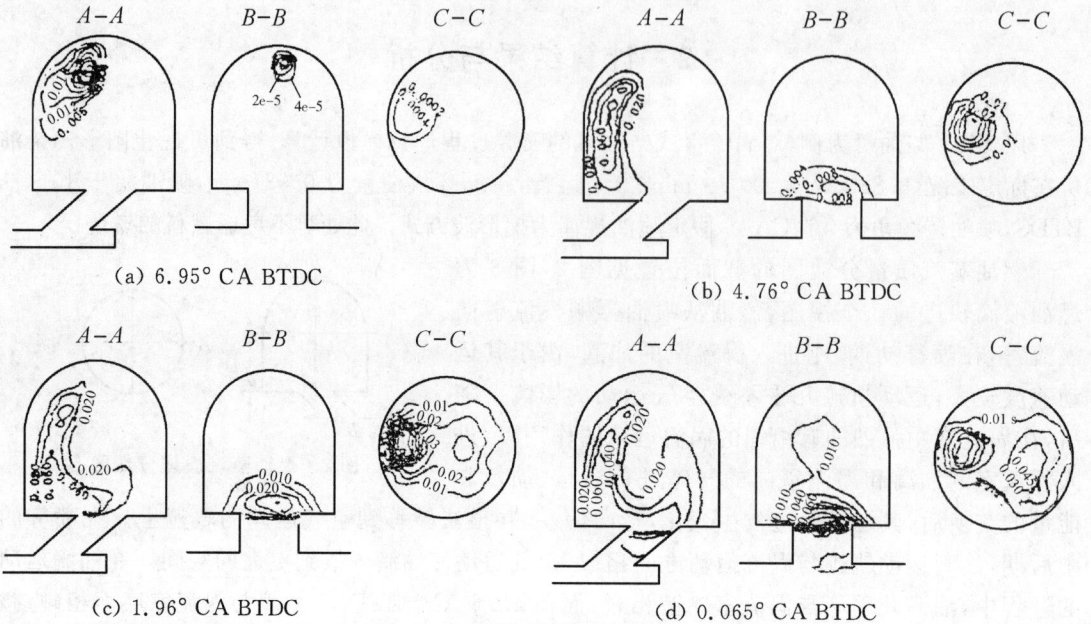

(a) 6.95° CA BTDC

(b) 4.76° CA BTDC

(c) 1.96° CA BTDC

(d) 0.065° CA BTDC

图 4　燃油蒸汽质量分数

热蒸发,在涡流室底部通道口左侧附近形成浓度适中的可燃混合气,此处最有可能首先着火。从图 3 可知,只有一小部分燃油在涡流室底部形成油膜,而大部分油滴在涡流室内随涡流一起运动。所以 S195 涡流室式柴油机混合气的形成是空间雾化混合和油膜蒸发混合这 2 种方式的综合,但以空间雾化混合为主。

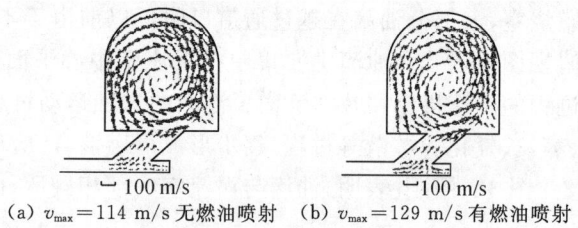

(a) $v_{max}=114$ m/s 无燃油喷射　　(b) $v_{max}=129$ m/s 有燃油喷射

图 5　涡流室 A-A 截面有无燃油喷射时速度场的比较

图 5 为在 2° CA BTDC 时,涡流室中 A-A 截面有无燃油喷射情况下速度场的比较。从图 5 可看出,燃油喷射对涡流室内速度场影响不大,有无燃油喷射时最大速度区均位于涡流室上部左右两侧距中心 O(见图 2)约 2/3 涡流室半径处。有燃油喷射时,由于对涡流室内气体加入动量和能量,涡流室左上侧的气流速度稍有增加,并且最大气流速度略高于无燃油喷射时的值。在 2.0° CA BTDC 两者的差别为 15.0 m/s。

无燃油喷射时,涡流室内有较均匀的温度场;图 6 为在 2° CA BTDC 时,有燃油喷射情况下涡流室中 A-A 截面的温度场。从图 6 中可以看出,燃油喷入涡流室后,在喷雾体内以及燃油与涡流室底面相碰之处,由于油滴吸收汽化潜热而蒸发,气体的温度明显降低。

图 7 和图 8 分别为有无燃油喷射时,涡流室内气体平均压力和平均温度的比较,燃油喷入涡流室后,吸收汽化潜热而蒸发,涡流室内气体的平均温度有所下降,从而使平均压力稍有下降。

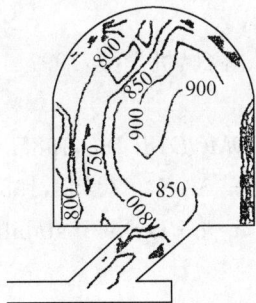

图 6　涡流室 *A-A* 截面有燃油喷射时温度场

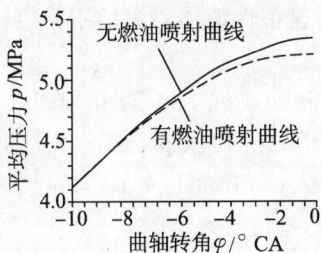

图 7　有无燃油喷射时涡流室平均压力比较

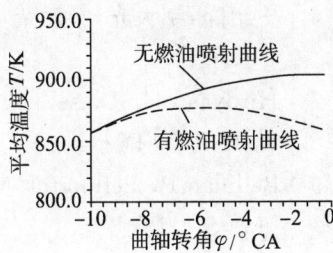

图 8　有无燃油喷射时涡流室平均温度比较

3　结　论

通过以 S195 柴油机为例的涡流室式柴油机喷雾过程的三维可视化研究,对涡流室式柴油机喷雾过程的油滴分布形态、燃料浓度分布状况、速度场、温度场以及平均温度和平均压力等有更清晰的认识,得到结论:

(1) 在喷雾初期,油滴运动受空气运动的影响较小;而在中后期,受空气运动的影响,油滴运动轨迹与涡流方向相似。

(2) 油滴几乎不与涡流室侧壁相碰,而于 $4.76°$ CA BTDC 和涡流室底面相碰,碰壁后大部分油滴反射回涡流室,一小部分粘附在壁面上或形成附壁射流。

(3) 大部分油滴在越过通道口少许以前由于不断蒸发而消失,一小部分油滴在涡流室右侧顺气流向顶部运动或进入主燃烧室。

(4) 在涡流室底部通道口左侧上方形成了浓度和温度适中的可燃混合气,此处有可能首先着火。

(5) 燃油喷雾对涡流室内速度场影响不大,但对温度场有较大的影响,在喷雾体内以及燃油与涡流室底面相碰之处,气体的温度明显降低,从而使平均温度和平均压力有所下降。

(6) 虽然本文计算的油滴运动轨迹与已发表的高速摄影结果[1]基本一致,但还需进一步开展实验研究,对诸如燃料浓度场等进行测量,提供更多可用以比较的实验数据。

参 考 文 献

[1] 李德桃,朱广圣.吊钟型涡流室内喷油和燃烧过程的研究[J].内燃机学报,1989(1):21-26.

[2] Tabata T, Ishii Y, Takatsuki T, et al. Numerical calculation of spray mixing process in a DI diesel engine and comparison with experiments[J]. *SAE*, 950853,1995.

[3] Hou Z X, Abraham J. Three-dimensional modeling of soot and NO in a direct-injection diesel engine[J]. *SAE*, 950608,1995.

[4] Liu A B, Mather D, Reitz R D. Modeling the effects of drop drag and breakup on fuel sprays[J]. *SAE*, 930072,1993.

[5] 史绍熙,郗大光,刘宁,等.高速液体射流的初始阶段的破碎[J].内燃机学报,1996

(4)：349-354.

［6］史绍熙,郗大光,秦建荣,等.高速粘性液体射流的不稳定模式[J].内燃机学报,1997(1)：
1-7.

［7］Hiroyasu H. Diesel engine combustion and its modeling[C]. *COMODIT '85*,1985.

［8］Reitz R D, Diwakar R. Structure of high-pressure fuel sprays[J]. *SAE*, 870598,1987.

［9］Bellman P, Pellington M. Effects of surface tension and viscosity on taylor instability
[J]. *SAME*, 1953,12(2)：103-111.

（本文原载于《中国机械工程》1998 年第 3 期）

Three-dimensional visual study of spray process in a swirl chamber diesel engine

Abstract：Visualized study on fuel spray process in a swirl chamber diesel on computer is carried out by using the updated KIVA-Ⅱ code, in which the submodels of fuel droplet breakup and spray impingement are improved. Three-dimensional graphs of fuel droplet motion locus and fuel vapor concentration are plotted. The influences of fuel droplet motion process and oil injection on the velocity field, temperature field, mean temperature and mean pressure in the swirl chamber are also investigated. The results predicted by the model agree with the result of the high-speed photo graphy basically.

柴油机涡流室内空气流动特性的
LDA 测试及数学模型[①]

王　谦，李德桃，许振忠

[摘要]　本文报道了首次在实机上采用激光多普勒测速仪(LDA)对柴油机涡流室内空气运动规律测试的研究结果。研究表明：涡流室内涡流在一定的半径范围内是刚体涡流，在涡流室周边区域可近似看作势涡流。涡流室连接通道处气流速度最高，涡流室上流侧区域气流速度较下流侧区域高，上部区域气流速度较涡流室下部高，涡流室中心附近气流速度最低。涡流室中心附近湍流强度较高，涡流室上流侧湍流强度明显高于下流侧，上部区域高于下部区域。在此基础上建立了吊钟型涡流室内空气运动的实用型简化数学模型。该模型考虑了曲轴转角、涡流室内不同区域、转速等因素对空气运动的影响，对进一步分析和模拟涡流室式柴油机的燃烧过程具有重要意义。

从我国柴油机实际发展情况来看，发展高喷射压力(70 MPa 以上)的直喷式柴油机以降低排气污染，工艺上面临很大困难。在此情况下，发展有害排气成分较少的涡流室式柴油机是一条适合国情之路。当然，涡流室式柴油机正面临不断增强的"高燃油经济性，低有害排放物"要求的压力。解决这一问题的途径之一就是深化对燃烧室内的气流运动、混合气形成和燃烧的基础研究，发展新型涡流燃烧系统。

柴油机涡流室内的空气运动与混合气形成的质量及燃烧速率密切相关。大量研究表明，强度适当的涡流运动能有效地促进燃油液滴的蒸发，有助于燃空混合。近年来，人们逐渐认识到涡流室内空气湍流运动在混合气形成和燃烧过程中的重要作用。研究者们利用多种方法对柴油机涡流室内空气运动进行了大量测试分析研究[1-4]，取得了很大进展。但由于其测试方法仅限于机械的、水模拟的接触式测量，测量范围和精度受到了很大限制。因此涡流室内的空气运动特性，尤其是湍流特性仍不十分清楚。

本文针对上述情况，采用先进的激光多普勒技术，利用二维氩离子激光多普勒测速仪(LDA)对柴油机涡流室内的空气流动特性进行了测试研究，并根据测量结果建立了涡流室内涡流特性的简化数学模型，为进一步研究燃烧过程打下了理论基础。

---

①　李德桃教授于 1992 年应邀到京都大学进行学术交流时，与日本著名的内燃机专家池上询教授讨论了 IDI 柴油机燃烧室内气流场的测量。此时池上询已用热线风速仪测量了球形涡流室内的流场，李德桃教授和他一直期望有人用激光测速仪作进一步的精确测量。李教授回国后，组织"跨学科、跨单位、跨地区"的人力和物力，在天津大学内燃机燃烧国家重点实验室，首次用激光多普勒测速仪对吊钟型涡流室的流场进行了测量，该文报道了测试和初步分析的结果。

——编者注

1 试验装置与方法

1.1 测试系统

（1）试验发动机。试验是在一台单缸涡流室式柴油机上进行的,其结构参数见表1。

表1 柴油机主要参数

主要参数	指　标
机　型	单缸、水冷、四冲程
缸数—缸径×行程/(mm×mm)	1—95×115
压缩比	19
气缸工作容积/L	0.815
涡流室型式	吊钟型
涡流室容积/L	0.025 7
标定功率/标定转速/(kW, r/min)	8.83 / 2 000

通过改装原涡流室式柴油机缸盖结构,并在涡流室两侧安装石英玻璃窗,成功地设置了涡流室内光学测量路径。图1为开设窗口后的涡流室形状及尺寸。观测窗口的安装不改变涡流室的压缩比和容积比,但对涡流室形状稍有改变,使涡流室由原吊钟型变成接近于扁型吊钟状。由于本次试验是测量涡流室主涡流平面内的流场,故涡流室形状的变动对测量产生的影响很小。图2所示为安装石英玻璃窗的观测套结构,这种结构便于安装和拆卸窗口。

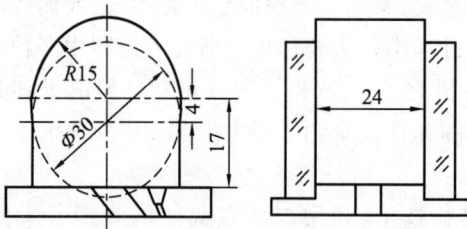

图1 涡流室形状(安装窗口后)

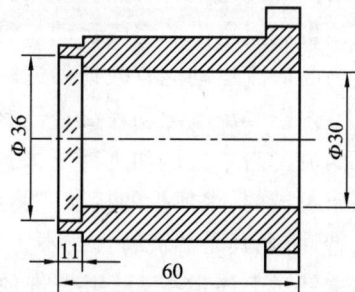

图2 观测套筒结构

（2）LDA设置与试验装置。试验所用LDA系统为DANTEC公司生产的3光束二维氩离子激光多普勒测速系统,整个LDA系统参数设定为激光器实用功率:1.2 W,氩离子;激光波长:蓝光4.88 nm,绿光514.5 nm;前透镜焦距:600 mm;接收透镜焦距:350 mm;频移方式:布拉格盒;接收方式:前向散射;信号处理器:频谱分析仪。

图3所示为LDA测量柴油机涡流室内流场的试验装置,主要包括:LDA系统;柴油机倒拖系统;示踪粒子加入装置;计算机控制系统。其中柴油机曲轴转角信号由轴编码器Model·TK-360产生,并用来触发LDA采样,整个试验由计算机控制。

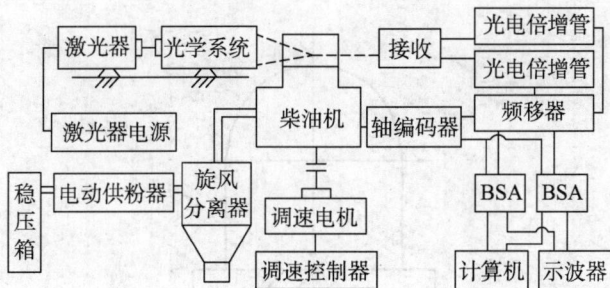

图 3　测量涡流室内流场的 LDA 测试装置

1.2　试验方案及流场测量数据处理

为研究涡流室内的涡流和湍流特性,将测量点布置在涡流室中心纵截面上(即测点处于主涡流平面内),如图 4 所示。在 x 方向上测点间距为 5 mm,在 y 方向测点间距为 4 mm。发动机转速分别为 400 r/min,600 r/min 和 800 r/min。

根据 LDA 所测数据的特点,本文以总体试验数据为依据,采用相平均法对试验数据进行处理,有关参量定义如下:

(1) 平均速度

$$u_m(\varphi) = \frac{1}{N_\varphi} \sum_{i=1}^{N_\varphi} u_i(\varphi)$$

式中 $u_i(\varphi)$ 为对应的曲轴转角下的瞬时速度;N_φ 为同一曲轴转角下所测得的瞬时速度的次数。

(2) 湍流强度

$$u'(\varphi) = \left\{ \frac{1}{N_t} \sum_{i=1}^{N_t} \left[u_i'(\varphi) - u_m(\varphi) \right]^2 \right\}^{\frac{1}{2}}$$

式中 N_t 为所测的总的瞬时速度次数。

(3) 相对湍流强度

$$\varepsilon = \frac{u'(\varphi)}{u_m(\varphi)} \times 100\%$$

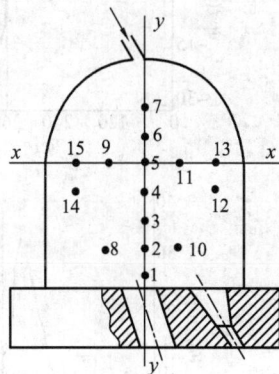

图 4　测点布置

2　试验结果与分析

2.1　涡流室连接通道处气体流动特性

连接通道附近的气体流动特性是影响涡流室内各处流动特性的重要因素之一,而它又直接地受到发动机结构参数和运转参数的影响。图 5 是连接通道口(测点 1)的 x 向、y 向及合成平均速度的变化曲线。图 6 是不同转速下平均速度随曲轴转角的变化规律。由图 5 可以看出,点 1 气流速度方向在压缩上止点前与 x 轴成 $30° \sim 60°$,随活塞下行,x 方向和 y 方向速度均呈现负值。由图 6 可知,在压缩冲程初期气流速度很低,从上止点前 $140°$ CA 左右起连接通道处气流速度明显升高,在上止点前 $24°$ CA 达到速度峰值,其后速度下降。在上止点时气流流动方向发生逆转。随着发动机转速升高,气流速度有所增大,但气流速度与曲轴转角变化规律十分相近(湍流强度的变化趋势亦与此相同)。另外,将该处的速度曲线与压缩过程主、副燃烧室的压力差曲线对比,两者具有相似变化规律。

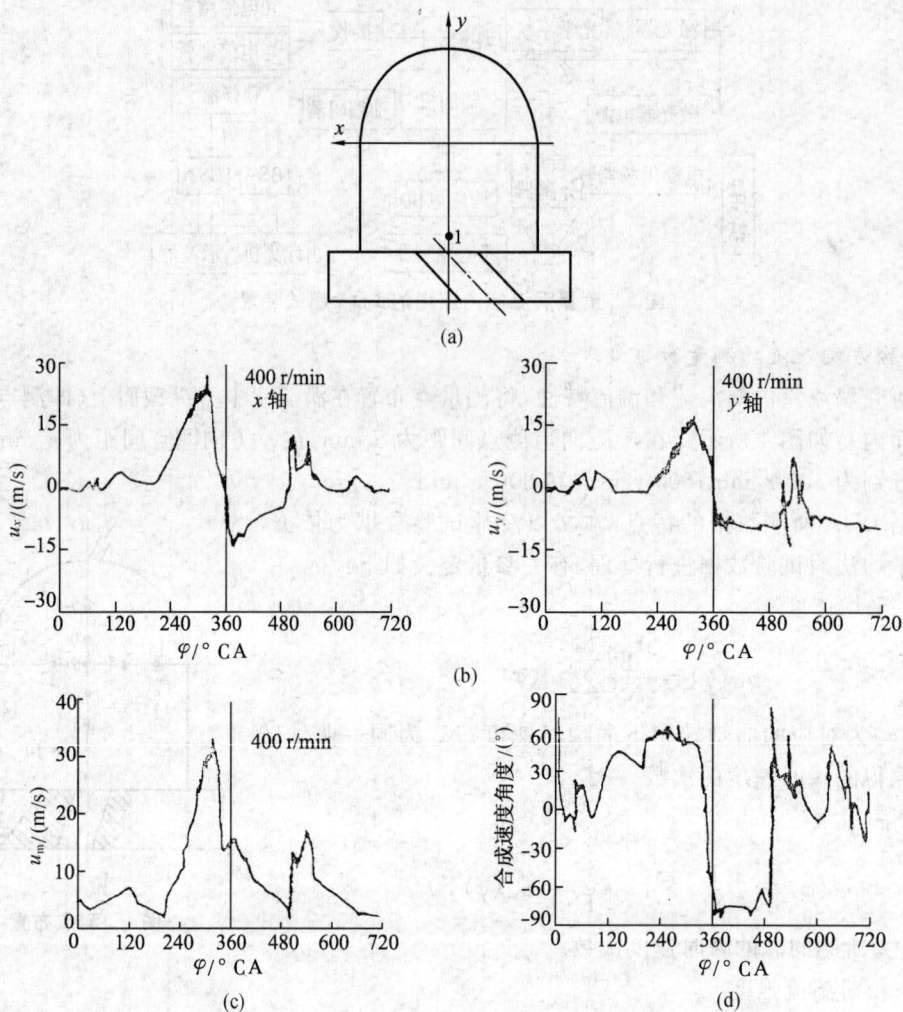

图 5　连接通道处 x 方向、y 方向速度矢量及合成速度大小、角度曲线

2.2　气流流动特性在涡流室内的空间分布

2.2.1　气体流动特性沿涡流室半径方向的分布

（1）气流平均速度沿涡流室半径方向的分布。

图 7 所示为在 3 种转速下，沿涡流室半径方向测点的速度变化曲线。由图可见，在涡流室中心处气流速度最低。沿着半径方向，在测量范围内（$-10\ \text{mm} \leqslant R \leqslant 10\ \text{mm}$）气流速度逐渐增加。涡流室上流侧气体流速比下流侧高，随活塞向上止点靠近，上、下流侧气体流速变得相对均匀。随着转速升高，上流侧速度峰

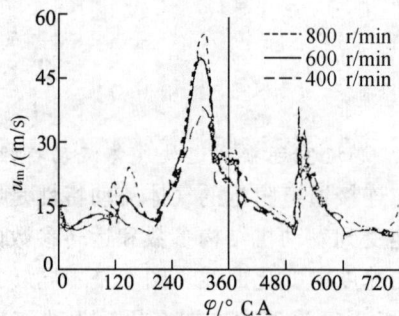

图 6　不同转速下合成速度曲线

值提前出现，而下流侧变化不明显。大量测试数据分析表明，在压缩冲程中其他曲轴转角下的速度分布具有与此相似的规律。

（2）气流切向速度在涡流室半径方向的分布。

图 8 示出了发动机转速在 800 r/min 时，涡流室内气流切向速度的变化。在压缩冲程中

后期,气流切向速度分布是从涡流室中心附近沿径向逐渐增加(到距壁面最近的测点为止),因此可以认为,在一定的半径范围内,涡流室内的涡流可近似为刚体涡流。另外在压缩冲程中,涡核位置不断变化。在 60° CA BTDC～TDC 之间,涡核偏离涡流室中心向左下方运动,最大偏离量为 5 mm。在涡流室中,切向速度占该点总动能的百分比由式 $\eta = (\overline{u}_\tau / \overline{U}^2) \times 100\%$ 计算可知,在压缩冲程中后期,涡流切向旋转动能占总动能的 $80\% \sim 99\%$,这进一步表明涡流室内的涡流运动在一定半径范围内为绕定轴旋转的圆周运动,即"刚体"涡流。

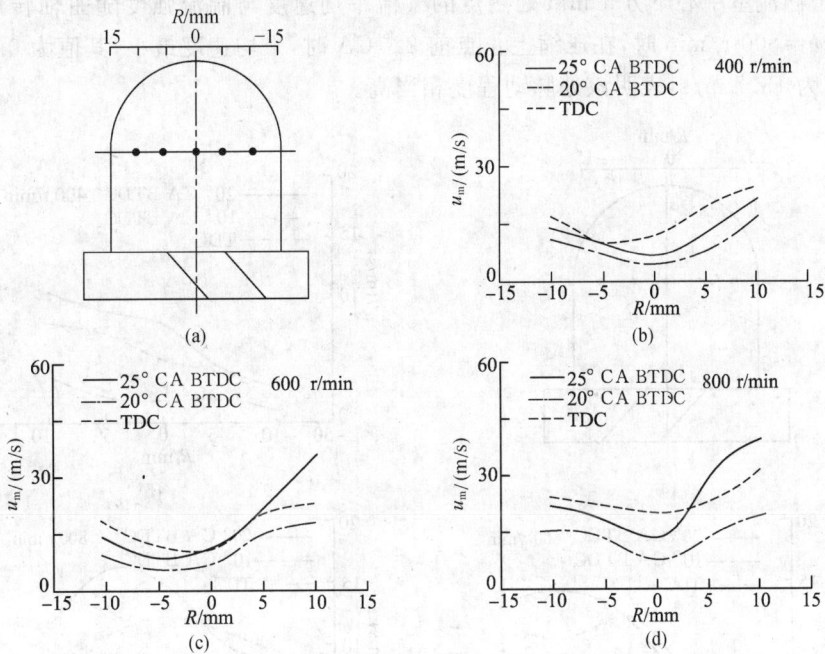

图 7　气流平均速度沿涡流室半径的分布

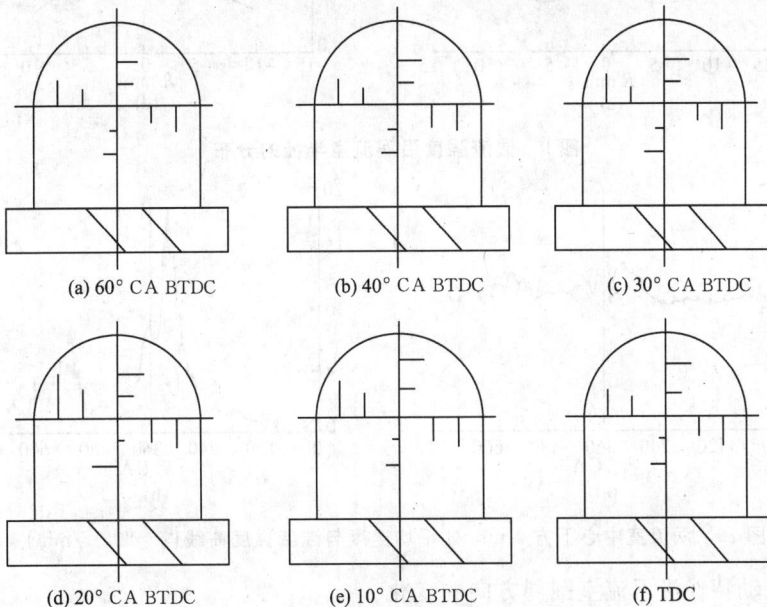

(a) 60° CA BTDC　　(b) 40° CA BTDC　　(c) 30° CA BTDC

(d) 20° CA BTDC　　(e) 10° CA BTDC　　(f) TDC

图 8　气流切向速度分布($n = 800$ r/min)

（3）湍流强度沿涡流室半径方向的分布。

图 9 是湍流强度沿涡流室半径方向的分布情况。由图可以看出，涡流室上流侧湍流强度明显高于下流一侧。转速较低时，涡流室上流侧湍流强度最高，中心部位次之，下流侧最低。发动机转速升高，湍流强度最大值逐渐向涡流室中心部位移动。随着活塞向上止点移动，湍流强度迅速衰减，在上止点附近，涡流室内湍流强度分布较均匀。但转速增大时，湍流强度最大值处在涡流室中心部位。湍流强度在涡流室中心附近较高是湍流强度分布的一个显著特点。图 10 示出了涡流室中心下方 4 mm 处测点的气流平均速度与湍流强度随曲轴转角的变化曲线。在转速 $n=800$ r/min 时，在压缩上止点前 $22°$ CA 时，平均速度最小，其值为 0.85 m/s，此时湍流强度为 13.7 m/s，表明该处脉动程度相当高。

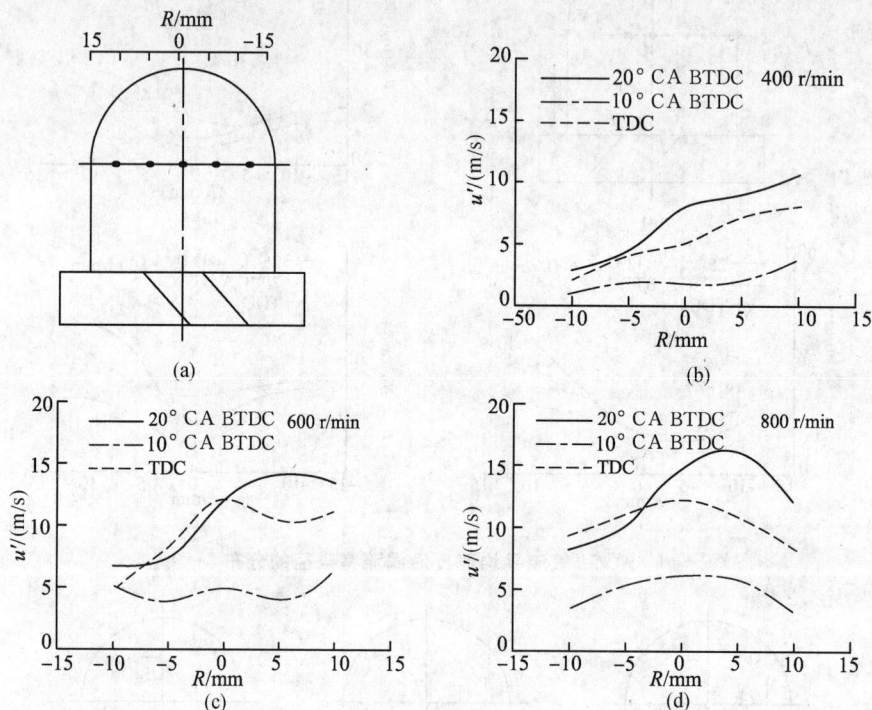

图 9　湍流强度沿涡流室半径的分布

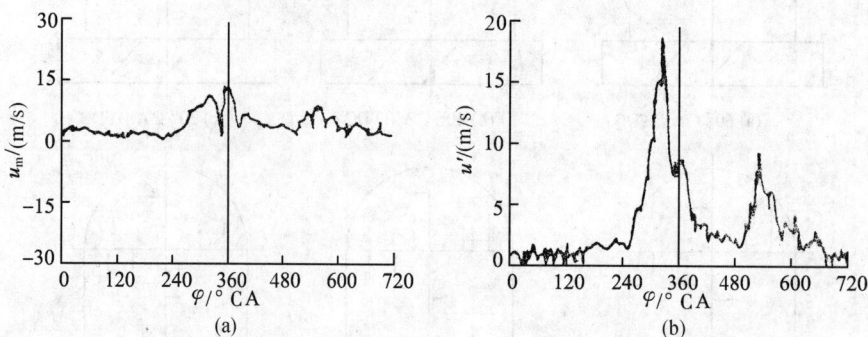

图 10　涡流室中心下方 4 mm 处平均速度与湍流强度曲线（$n=800$ r/min）

2.2.2　气体流动特性沿涡流室圆周方向的变化

（1）气流平均速度沿涡流室圆周方向的变化。

图 11 所示为气体平均速度沿涡流室圆周方向的变化曲线。其中 θ 是圆周角,研究点在 x 方向取距涡流室中心 10 mm 处测量点,在 y 方向取距涡流室中心 8 mm 处测量点。由图可知,在 $\theta=0°$ 时,气流速度最大;在 $0°<\theta<90°$ 范围内气流速度迅速下降;当 $\theta>90°$ 时,气流速度几乎保持一稳定值。在三种转速工况下,当活塞到达上止点时,气流速度在圆周方向的分布趋于均匀。

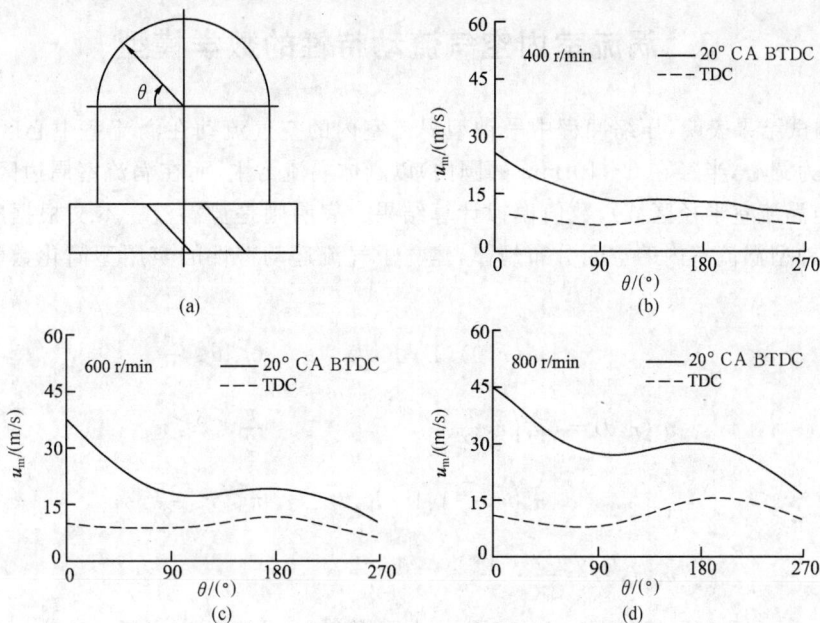

图 11　气流平均速度沿涡流室周向变化

(2) 湍流强度沿涡流室圆周方向变化。

图 12 所示为湍流强度沿涡流室圆周方向的变化。研究点与(1)中相同。当转速较低($n=$

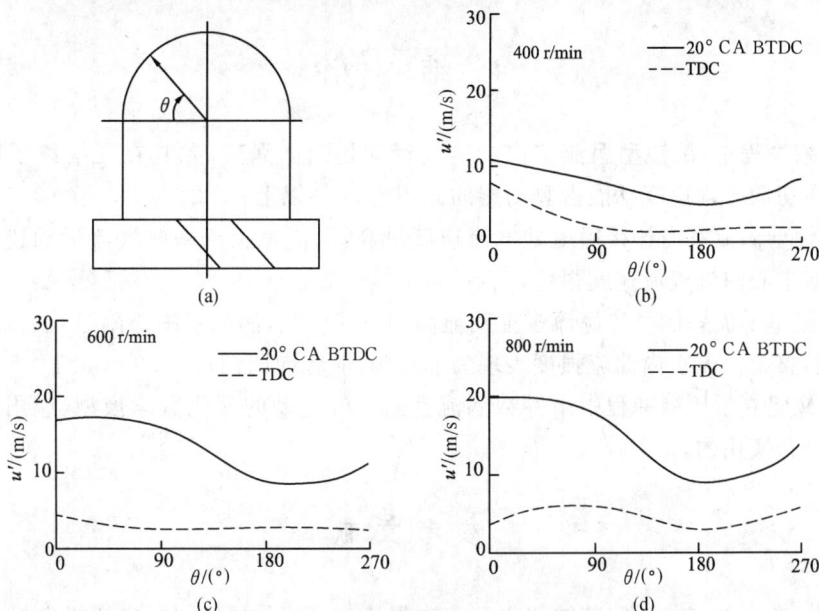

图 12　湍流强度沿涡流室周向的分布

400 r/min)时,湍流强度沿 θ 角逐渐衰减,转速增加($n=600$ r/min~800 r/min)后湍流强度最大值处在 $\theta=90°$ 处。在上止点前 20° CA 随转速升高各处湍流强度值增大,而在上止点时,湍流强度沿圆周方向的分布趋于均匀,不同转速下湍流强度的差值也减小,其他测点亦有类似规律。

3 涡流室内空气流动特性的数学模型

LDA 测试结果表明,压缩冲程中吊钟型涡流室内的空气流动在涡流室中心区域(以涡流室几何中心为圆心,半径在 0~10 mm 的圆内)以刚体涡流为主,而在涡流室周边区域(半径大于 10 mm 至涡流室半径区域),数值模拟计算结果[5]表明则是势涡流。本文根据所测得的气流特性在吊钟型涡流室内的空间分布规律,建立了气流运动特性的实用型简化数学模型。模型如下:

$$u_t(r,\theta)=\begin{cases}u_t(r,0)(1-K_1\theta) & 0\leqslant\theta\leqslant\dfrac{\pi}{2}\\[2mm]u_t\left(r,\dfrac{\pi}{2}\right) & \dfrac{\pi}{2}\leqslant\theta\leqslant\pi\\[2mm]u_t\left(r,\dfrac{\pi}{2}\right)(1-K_2\theta) & \pi\leqslant\theta\leqslant\dfrac{3\pi}{2}\end{cases}\qquad(r\leqslant r_0)$$

$$u_t(r,\theta)=\frac{\Gamma}{2\pi}\cdot\frac{1}{r}\qquad\qquad\qquad(r_0\leqslant r\leqslant R)$$

式中 $u_t(r,0)=r\omega,\theta=0°$ 时涡流的切向速度;θ 为圆周角;r_0 为刚体涡半径,取 $r_0=10$ mm;R 为涡流室半径;Γ 为速度环量;K_1,K_2 为系数。

此空气流动特性简化数学模型应用于涡流室式柴油机准维燃烧模拟时[6],模拟精度较好,证明了本模型的实用价值。

4 结 论

(1) 压缩冲程中,吊钟型涡流室内的空气运动是刚体涡流,尤其在压缩终了附近最为明显,此时气体切向平均速度动能占总动能的比例为 90% 以上;

(2) 吊钟型涡流室内连接通道处气流速度最高,涡流室上流一侧气流运动速度高于下流一侧,涡流室中心附近气流速度很低;

(3) 涡流室上流一侧气体湍流强度明显高于下流一侧,涡流室中心附近,湍流强度一直保持较高数值,整个涡流室内湍流强度表现为非均匀、非各向同性;

(4) 本文建立了压缩冲程吊钟型涡流室内空气流动的简化数学模型,应用结果表明此模型是合理的、实用的。

参 考 文 献

[1] 武善谋,杨海青,刘书亮.柴油机涡流室内涡流及紊流的研究[J].内燃机学报,1990(3).

[2] 李德桃.利用二维液流动模型模拟研究吊钟型涡流室内空气运动[J].内燃机工程,1983(1).

［3］小松源一,池上詢.デイーゼル機関のシミュレーション[J].内燃機関,86-0693A.

［4］王谦,罗惕乾,李德桃,等.柴油机涡流室内空气运动的实验和计算研究[J].内燃机学报,1994(2).

［5］王谦.柴油机涡流室内空气运动的实验研究及数值模拟[D].镇江:江苏理工大学,1996.

［6］吴志新.涡流室式柴油机准维燃烧模型的研究[D].镇江:江苏理工大学,1997.

(本文原载于《内燃机学报》1998 年第 3 期)

LDA measurements and mathematical model of air motion in swirl chamber of diesel engine

Abstract：Test investigations of air flow in the swirl chamber of diesel engine were successfully made by LDA for the first time. Velocity measurements at certain points locating on one plane in the swirl chamber in a motored condition are made for different engine speed. By using ensemble average method, the mean velocity and turbulence intensities are calculated. During compress stroke, the air motion in swirl chamber can regard as a solid vortex within a certain radius, and in the other zone can be approximate to a free vortex. The flow field in the swirl chamber is inhomogenous：the velocity at the connecting throat is very high. The velocity at up-stream side is higher than it at down-stream side. The velocity is lower near the central zone of the chamber. Regions of intensive turbulence are the up-stream side and the central zone of the chamber. A simplified and useful mathematical model for air motion in swirl chamber was developed for the first time considering the influences of the crank angle, different zones in chamber and engine speed, which is important to analyze and simulate combustion process of swirl chamber diesel engines.

涡流室式柴油机空气运动的三维数值模拟

夏兴兰,李德桃,吴志新,董　刚

[摘要]　通过对标准 k-ε 模型作适当修正来模拟气体的湍流运动,建立了涡流室式柴油机的空气运动的三维数值计算模型,开发了大型微机化三维数值模拟程序。首次对吊钟型涡流室式柴油机的流场进行了三维计算,计算结果与 LDV 实验结果具有良好的一致性。画出了涡流室纵截面、横截面和侧截面上的速度矢量图。分析了各截面上气体速度的变化过程和涡心运动轨迹,研究结果揭示了涡流室中空气运动的三维流动特性。

　　涡流室式柴油机有害排放物低,使用维修方便,在我国(尤其是广大的农村地区)仍具有广阔的市场,因此研究涡流室式柴油机,对我国的经济建设具有重要的意义。而空气运动对柴油机的混合气形成和燃烧有着重要的影响,数值模拟是研究柴油机中空气运动的重要手段之一,国内外已进行过许多研究,然而,这些研究大部分都是针对直喷式柴油机进行的。对涡流室式柴油机,由于其燃烧室结构复杂,在数值计算中存在较大的困难,所进行的研究较少。文献[1-4]将三维问题简化为二维问题,不能真正反映燃烧室和连接通道形状及位置的影响。文献[1,2]对涡流室式柴油机涡流室内的空气运动进行二维数值模拟计算时,需要实测通道入口处的边界条件。由于实测时受测点数目和转速的限制,很难测量入口处的速度分布及温度分布等边界条件,从而使其应用受到严格的限制。

　　美国 Los Alamos 国家实验室推出的 KIVA-II 程序[5]已被广泛应用于内燃机的流场计算,但原始的 KIVA-II 程序是针对 CRAY 巨型机编写的,且仅能直接应用于直喷式柴油机。本文在消化、吸收 KIVA-II 程序的基础上,开发了适合我国国情的大型微机化程序,对涡流室式柴油机的流场进行了三维数值模拟计算。

1　数学模型

1.1　气体运动控制方程

　　燃烧室内气流运动控制方程由一组守恒的偏微分方程、状态方程和湍流模型方程构成。

组成 m 的连续性方程

$$\frac{\partial \rho_m}{\partial t} + \nabla \cdot (\rho_m \boldsymbol{u}) = \nabla \cdot \left[\rho D \nabla \left(\frac{\rho_m}{\rho} \right) \right] \tag{1}$$

混合物的动量守恒方程

$$\frac{\partial (\rho \boldsymbol{u})}{\partial t} + \nabla \cdot (\rho \boldsymbol{uu}) = -\frac{1}{\alpha^2} \nabla p + \nabla \cdot \boldsymbol{\sigma} - \nabla \left(\frac{2}{3} \rho k \right) + \rho \boldsymbol{g} \tag{2}$$

能量守恒方程

$$\frac{\partial (\rho I)}{\partial t} + \nabla \cdot (\rho I \boldsymbol{u}) = -p \nabla \cdot \boldsymbol{u} - \nabla \cdot \boldsymbol{J} + \rho \varepsilon \tag{3}$$

状态方程

$$p = R_0 T \sum_m \left(\frac{\rho_m}{W_m} \right) \tag{4}$$

$$I(T) = \sum_m \left(\frac{\rho_m}{\rho}\right) I_m(T) \tag{5}$$

$$h_m(T) = I_m(T) + \frac{R_0 T}{W_m} \tag{6}$$

湍流模型方程是在标准的 k-ε 模型的基础上,作适当修正而得到的。

$$\frac{\partial(\rho k)}{\partial t} + \nabla \cdot (\rho k u) = -\frac{2}{3}\rho k \nabla \cdot u + \boldsymbol{\sigma} : \nabla u - \nabla \cdot \left[\left(\frac{\mu}{Pr_k}\right)\nabla k\right] - \rho\varepsilon \tag{7}$$

$$\frac{\partial(\rho\varepsilon)}{\partial t} + \nabla \cdot (\rho\varepsilon u) = -\frac{2}{3}(C_{\varepsilon1} - C_{\varepsilon3})\rho\varepsilon\nabla \cdot u + \nabla \cdot \left[\left(\frac{\mu}{Pr_\varepsilon}\right)\nabla\varepsilon\right] + \frac{\varepsilon}{k}\left[C_{\varepsilon1}\boldsymbol{\sigma} : \nabla u - C_{\varepsilon2}\rho\varepsilon\right] \tag{8}$$

k-ε 湍流模型中的常数为:$C_{\varepsilon1} = 1.44, C_{\varepsilon2} = 1.92, C_{\varepsilon3} = -1.0, Pr_k = 1.0, Pr_\varepsilon = 1.3$。以上各式中 ρ_m, ρ 分别为组分 m 的密度、总密度;D 为扩散系数;u 为气流速度;p, T, I, h 分别为气体的压力、温度、比内能、比焓;R_0 为通用气体常数;W_m 为组分 m 的分子量;k, ε 分别为湍流动能及湍流动能的耗散速率;g 为比体积力;α 为与压力梯度(PGS)有关的量;$\boldsymbol{\sigma}$ 为粘性应力张量,取为牛顿形式;$\boldsymbol{J}$ 为热流矢量,它是热传导和焓扩散的总和。

$$\sigma = \mu[\nabla u - (\nabla u)^{\mathrm{T}}] + \lambda(\nabla \cdot u)\boldsymbol{E} \tag{9}$$

$$\boldsymbol{J} = -K\nabla T - \rho D \sum_m h_m \nabla\left(\frac{\rho_m}{\rho}\right) \tag{10}$$

式中 μ, λ 为第 1 和第 2 粘性系数;K 为导热系数;$\boldsymbol{E}$ 为单位张量;上标 T 表示转置。

1.2 边界条件

固壁的速度边界条件可以是:(1) 自由滑移;(2) 无滑移;(3) 湍流壁面律。温度边界条件为绝热壁面或定温壁面。在以下的计算中取湍流壁面律和定温壁面为边界条件。

2 计算网格

合理地划分网格是进行数值计算的基础,本文以 S195 柴油机为例,对吊钟型涡流室式柴油机进行三维数值计算,发动机压缩比为 21,连接通道倾斜角为 40°,网格采用柱面直角坐标。图 1 为计算所用网格图。主燃烧室中网格为 $18 \times 16 \times 13$(在上止点附近为 $18 \times 15 \times 3$),连接通道中的网格为 $6 \times 16 \times 5$(为便于观察,连接通道示意图的比例已经放大,下同),涡流室中的网格为 $16 \times 16 \times 12$。

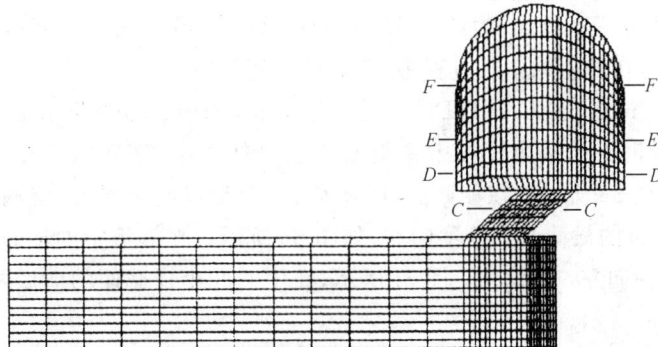

(a) 纵截面(A-A 截面)

(b) 连接通道横截面　　　　　　　　　　(c) 涡流室横截面

图1　涡流室式柴油机燃烧室的计算网格(40° CA BTDC)

3　计算结果及与实验结果的对比

3.1　速度场计算结果及分析

计算从压缩冲程下止点开始,假定初始速度为0,气缸内压力和温度均匀。计算结果表明:在2 000 r/min 时,在纵截面上于130° CA BTDC左右,开始在涡流室通道左侧边缘处形成初始涡流,随着压缩的进行,涡心向右上方向运动,运动到最高点时,在上止点附近和膨胀冲程,涡心向左下方向运动。文献[2]认为:受通道口入流的限制,涡心始终偏于涡流室左侧。本文计算的涡心位置位于连接通道在涡流室中延长线的左侧,而不是涡流室几何中心 $O(O$ 的位置见图2)的左侧。这可能是因为在二维计算中并不是把涡流室顶部当作半球形来处理,而是当作横的半圆柱形处理,从而未能真正反映涡流室的结构形状,不能计及涡流室前后部分对中心平面的影响;并且文献[2]中以实测涡流室通道入口处的速度为边界条件,而实测时只测了入口附近一个点的速度,并没有测得入口处的速度场,因此计算时还需假定一个入口处的速度场,此假定的速度场与实际的速度场会有一定的误差。在压缩过程中,活塞从下止点往上止点运动时,涡流室中的气流速度逐渐增加,约在20° CA BTDC 时,最大气流速度达到最大值,最大值为 179.43 m/s;随后,在上止点附近速度逐渐下降,大约于 2° CA ATDC 时,最大气流速度达到一最小值,最小值为 131.69 m/s;然后在膨胀初期,气流速度逐渐增加,大约于 15° CA ATDC 时最大气流速度再次达到一个峰值,其值为 194.47 m/s,在随后的膨胀过程中,气流速度逐渐减小。

图3为2 000 r/min、25° CA BTDC 时的速度分布,图中 V_{max} 为所示平面内速度矢量和其绝对值的最大值,此时涡流室内流场主要受通道入口处气流速度的影响。在纵横面(A-A 截面)上形成涡流,如图3a所示。在侧截面(B-B 截面)上形成两个涡流,但涡心位置较低,如图3b所示。在连接通道的横截面上速度向右,如图3c所示。在涡流室的横截面上,从下到上速度由向右逐渐转变为向左,如图3d～图3f所示,在中间几个横截面的前后两个半平面中各形成一个旋涡,如图3e所示。

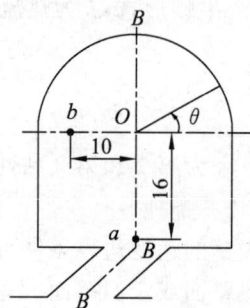

图2　纵截面上典型点位置

$V_{max} = 169.66$ m/s

(a) 纵截面（*A-A* 截面）

$V_{max} = 124.98$ m/s

(b) 涡流室侧截面（*B-B* 截面）

$V_{max} = 150.75$ m/s

(c) 连接通道横截面（*C-C* 截面）

$V_{max} = 125.33$ m/s

(d) 涡流室横截面（*D-D* 截面）

$V_{max} = 35.70$ m/s

(e) 涡流室横截面（*E-E* 截面）

$V_{max} = 58.78$ m/s

(f) 涡流室横截面（*F-F* 截面）

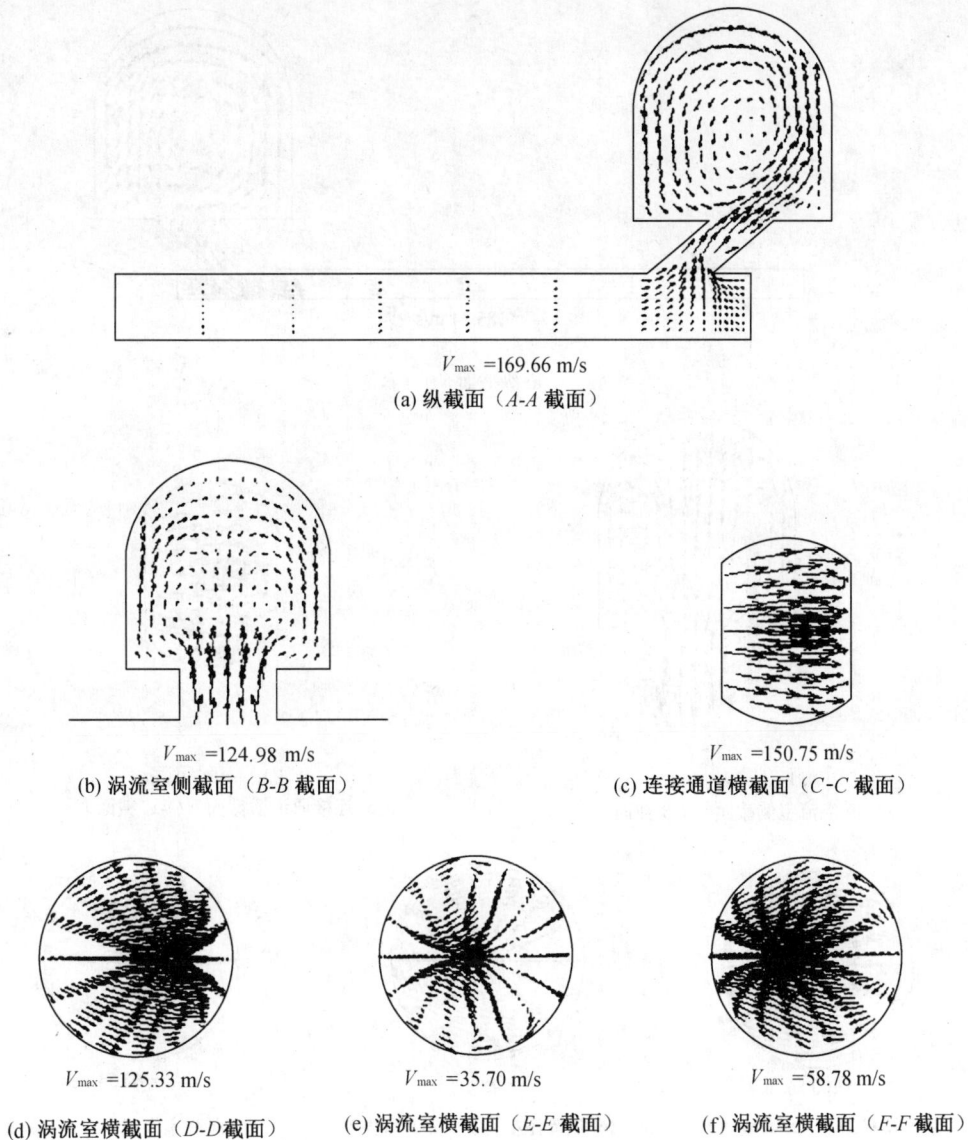

图 3　速度场(25° CA BTDC)

图 4 为 2 000 r/min,11° CA ATDC 时计算所得速度场。上止点后涡流室的纵截面上的涡流中心继续左移,如图 4a 所示。在侧截面上仍然形成两个涡流,但涡流方向与上止点前时的方向相反,如图 4b 所示。在连接通道的横截面上速度由向右逐渐改变为向左,如图 4c 所示。涡流室中的横截面上的速度矢量图与上止点前的速度矢量图类似,如图 4d～图 4f 所示。

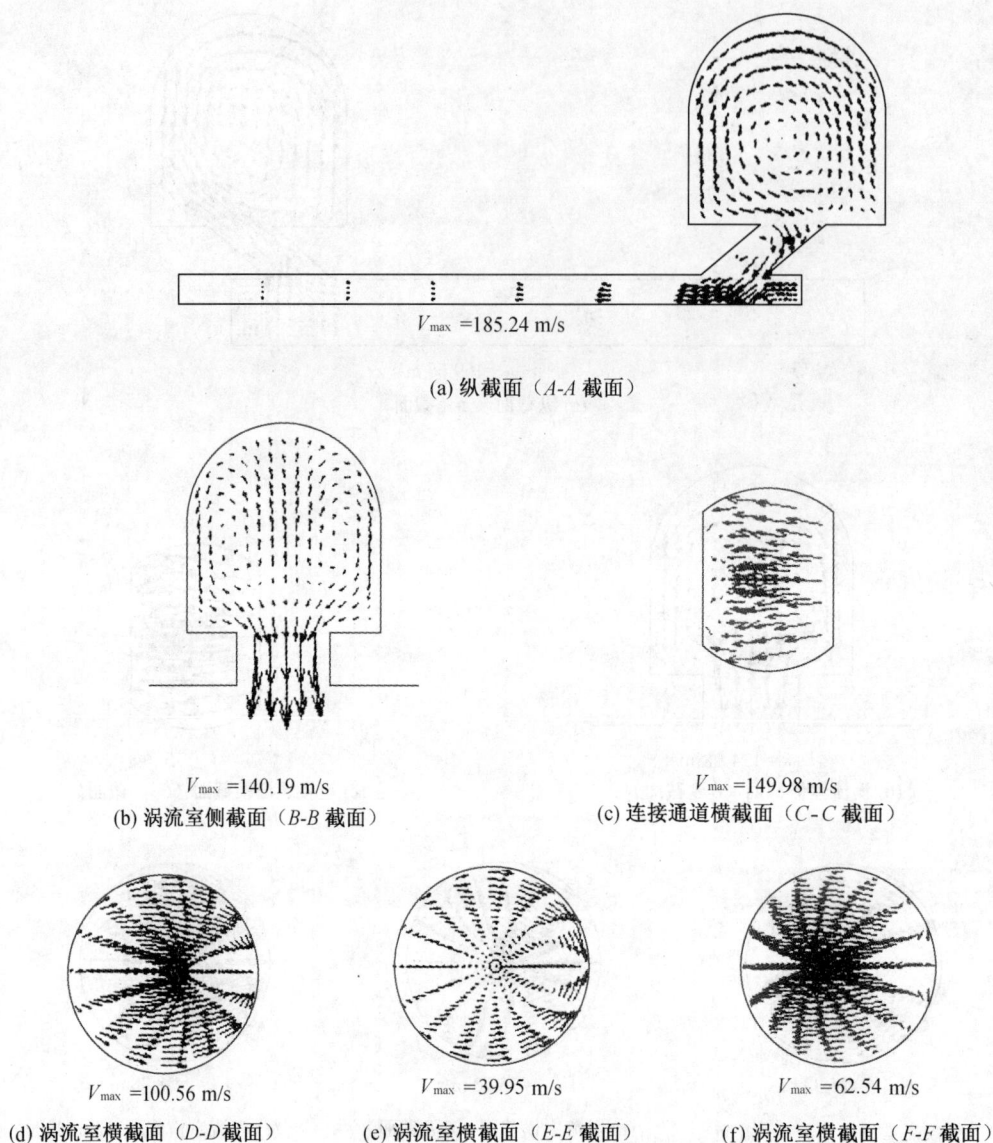

$V_{max} = 185.24$ m/s

(a) 纵截面（A-A 截面）

$V_{max} = 140.19$ m/s

(b) 涡流室侧截面（B-B 截面）

$V_{max} = 149.98$ m/s

(c) 连接通道横截面（C-C 截面）

$V_{max} = 100.56$ m/s

(d) 涡流室横截面（D-D截面）

$V_{max} = 39.95$ m/s

(e) 涡流室横截面（E-E 截面）

$V_{max} = 62.54$ m/s

(f) 涡流室横截面（F-F 截面）

图 4　速度场($11°$ CA ATDC)

3.2　计算结果与实验结果的对比

图 5a 为涡流室中心平面通道口附近(图 2 中点 a)气流速度的比较,计算结果与 LDA 实验结果[6]基本一致,约在 $25°$ CA BTDC 至 $30°$ CA BTDC 时,气流速度达到最大值。图 5b 为涡流室中心平面左侧(图 2 中点 b)气流速度的比较,在 400 r/min 时,点 b 的速度变化很小,转速升高到 800 r/min,约在 $12°$ CA BTDC 时速度出现最大值,计算和实验结果均反映了这一变化趋势。图 6 为涡流室中心平面圆周方向的速度比较,取距涡流室中心 10 mm 的点作为研究对象,θ 是与半径方向的夹角,计算结果与实验结果吻合良好,$\theta = 0°$ 时气流速度最大,$\theta = 270°$ 时气流速度最小,曲轴转角位于 TDC 时气流速度变化不大。

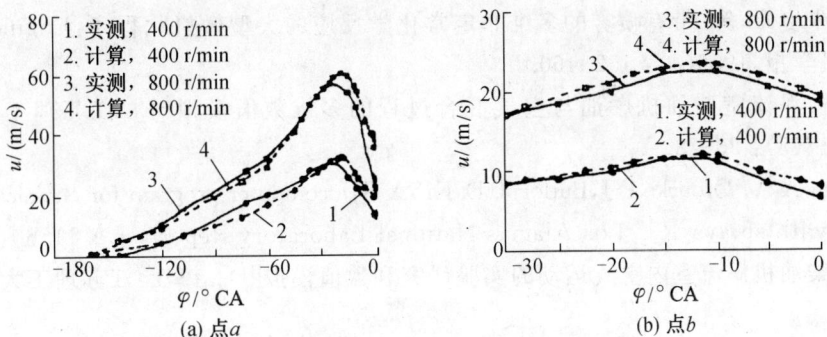

(a) 点a　　　　　　　　(b) 点b

图 5　涡流室中典型点气流速度的比较

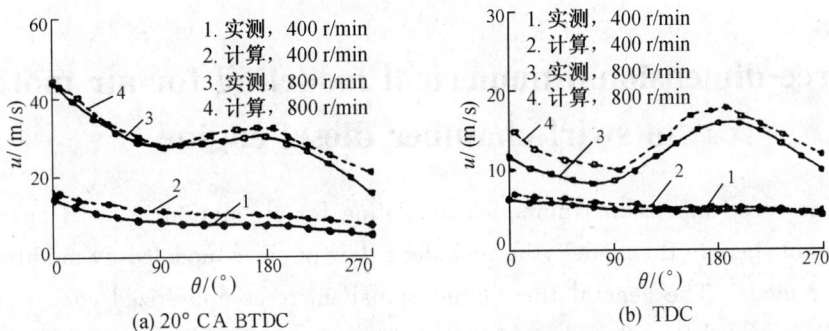

(a) 20° CA BTDC　　　　　　(b) TDC

图 6　涡流室圆周方向气流速度的比较

4　结　论

（1）涡流室式柴油机大型微机化三维数值模拟计算程序的开发,有助于国内开展涡流室式柴油机内流场的可视化研究。

（2）对吊钟形涡流室式柴油机的三维数值计算表明:气流速度关于中心平面（A-A 截面）完全对称,且在典型点的速度中,计算结果与实验结果吻合良好,这说明了程序的可靠性。

（3）涡流室式柴油机的纵截面、横截面和侧截面的速度矢量图有利于更清晰地认识此种柴油机的流场,涡流室的纵截面上于 130° CA BTDC 左右开始在连接通道左侧边缘形成涡流,随着压缩的进行,涡心向右上方移动,然后向左下方移动;侧截面上在压缩和膨胀过程中,均形成两个涡流,但涡心位置较低;横截面上,从下到上速度方向由向右逐渐改变为向左,在中间几个平面上形成两个涡流。连接通道的横截面上,在压缩过程中速度方向向右,在上止点附近逐渐改变为向左,在膨胀过程中速度方向向左。

参 考 文 献

［1］王谦,李德桃,等.柴油机涡流室内流场的实验和计算研究［J］.内然机学报,1994（2）:102-108.

［2］朱广圣,王谦,等.柴油机涡流室结构参数对涡流室内空气流动特性影响的研究［J］.燃烧

科学与技术,1996(4):322-328.

[3] 谭伟,朱埏章,等.带有喷雾的多维非定常化学反应的大型微机化程序:Engine CFD[J].
内燃机学报,1996(4):452-460.

[4] 杜家益.涡流室柴油机燃油与空气混合过程的多维数值模拟[J].江苏理工大学学报,
1996,17(5):20-23.

[5] Amsden A A, Orourke P J,Butlor T D. KIVA-Ⅱ:a computer program for chemically reacting
flows with sprays[R]. Los Alamas National Laboratory Report,LA-11560-MS,1989.

[6] 王谦.柴油机涡流室内空气运动的实验研究和数值模拟[D].镇江:江苏理工大学,1996.

（本文原载于《燃烧科学与技术》1998 年第 3 期）

Three-dimensional numerical modeling for air motion in swirl chamber diesel engine

Abstract: The three-dimensional numerical modeling for air motion in swirl chamber diesel engine is established. In the model ,the turbulent flow of air is modeled by modified standard k-ε turbulent model. The general three-dimensional micro-computerized code is developed. The three-dimensional flow field for the bell type swirl chamber diesel engine is first calculated. The calculated results and LDA measured data are compared which show fairly good agreement. The velocity profile in vertical section, cross section and side section is given. The change and development as well as the position of vertexes of air motion in swirl chamber are shown. The obtained results clearly show three-dimensional characteristics of air motion in the swirl chamber.

柴油机涡流室内湍流分布特性的研究

顾子良,王　谦,顾　滨,李德桃

[摘要]　报道了利用激光多普勒测速(LDA)技术测量柴油机涡流室内的湍流分布特性的方法以及 LDA 测试系统,提出了试验数据的处理方法,获得了倒拖工况下涡流室内湍流强度随时间和空间的变化规律,并分析了湍流特性对混合气形成及燃烧的影响。研究结果表明,涡流室内流场是非定常各向异性湍流场,其湍流强度分布特性对混合气形成及燃烧是有利的。

　　柴油机涡流室内的空气运动对混合气形成质量及燃烧速率有很大影响,是影响柴油机性能的重要因素。多年来国内外许多学者对涡流室内空气运动规律进行了测试分析和研究,取得了许多进展[1-5],使人们对涡流室内气流运动规律的认识逐步深化。一般认为:适当强度的涡流运动和适当的喷油方向相配合,能增强燃油与空气间的热交换,加速燃油蒸发,促使形成比较均匀的混合气。近年来,关于湍流运动对直喷式柴油机混合气的形成及燃烧的影响已有了较多的研究,但关于湍流对涡流室式柴油机混合气形成和燃烧的影响则研究得很少。

　　激光多普勒测速(LDA)技术是目前最先进的非接触式光学测量方法之一,它不仅对被测流场无干扰,而且具有高精度的时空分辨率,测试结果不需校正(热线风速仪测量前后均需校正),还具有良好的线性特征。本文用 LDA 技术获得在倒拖工况下的柴油机涡流室内的气流运动规律,着重对其湍流强度分布特性进行分析。

1　测试系统简介

　　试验是在一台改装的 195 型涡流室式柴油机上进行的,涡流室两侧开有用石英玻璃制作的观测窗,如图 1 所示。测试系统如图 2 所示,其中 LDA 为丹麦 DANTEC 公司的三光束二维氩离子激光多普勒测速系统,采用前向散射接收方式。

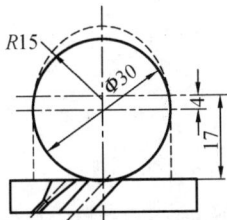

图 1　涡流室观测范围

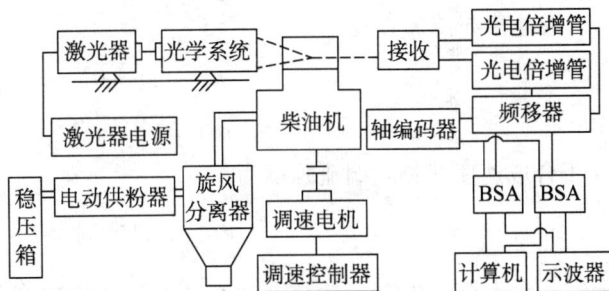

图 2　测试系统图

　　试验时,在进入燃烧室的空气中加入一定浓度的粒径为 1 μm 左右的二氧化锆微粒,涡流室内粒子运动所发生的多普勒频移信号及曲轴转角信号都经频谱分析仪(BSA)处理后输入计算机进行再处理,整个试验由计算机完成数据采集和控制。

根据吊钟型涡流室结构及空气运动特点,将测点布置在涡流室中心纵截面上,如图 3 所示。在 x 轴方向测点间距为 5 mm,y 轴方向测点间距为 4 mm,靠壁面最近的测点距壁面为 5 mm。测量过程中柴油机处于倒拖不喷油工况,倒拖转速分别为 800,600,400 r/min。

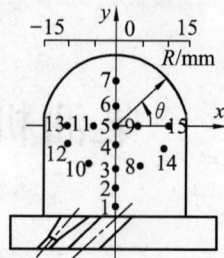

图 3　测点布置

2　测试数据的处理方法

对柴油机涡流室而言,由于进气过程的不稳定性、气缸内活塞运动及连接通道的节流作用等,使涡流室内流场很不稳定,因此涡流室内气流运动是非定常的涡流-湍流运动。对湍流流动参数常规处理方法是 Reynolds 时间平均法。对内燃机而言,发生在单个循环内的流动过程是随时变化的,各循环间的气流参数也会有所不同。根据这一特点,我们采用集时间平均法对试验数据进行处理,并把有关参数作如下定义:

(1) 平均速度

$$u(\varphi) = \frac{1}{N_\varphi} \sum_{i=1}^{N_\varphi} u_i(\varphi) \tag{1}$$

式中 $u_i(\varphi)$ 为对应于曲轴转角 φ 的第 i 个循环的瞬时速度;N_φ 为同一曲轴转角 φ 所测得的有效采样次数。

(2) 湍流强度

与式(1)对应的湍流强度为

$$u'(\varphi) = \left\{ \frac{1}{N_\varphi} \sum_{i=1}^{N_\varphi} [u_i(\varphi) - u(\varphi)]^2 \right\}^{1/2} = \left\{ \frac{1}{N_\varphi} \sum_{i=1}^{N_\varphi} [u_i'(\varphi)]^2 \right\}^{1/2} \tag{2}$$

式中 $u_i'(\varphi)$ 为对应于曲轴转角 φ 的第 i 个循环的脉动速度。

(3) 相对湍流强度

内燃机中习惯于用绝对值来表示湍流强度,而根据流体力学观点,应采用相对值来衡量其湍流强度,它反映了气流扰动的强弱,故定义相对湍流强度为

$$\varepsilon(\varphi) = \frac{u'(\varphi)}{u(\varphi)} \times 100\% \tag{3}$$

(4) 合成速度与湍流强度

$$u = \sqrt{u_x^2 + u_y^2} \tag{4}$$

$$u' = \sqrt{u_x'^2 + u_y'^2} \tag{5}$$

对应于合成速度的相对湍流强度为

$$\varepsilon = \frac{u'}{u} \times 100\% \tag{6}$$

式中 u_x, u_y, u_x', u_y' 分别表示 x 轴与 y 轴方向上的平均速度与湍流强度;u' 为对应于合成速度 u 的湍流强度。

涡流室内各测点 x, y 方向的瞬时速度由 LDA 测得。根据所编制的程序经计算机处理后可得上述各种流动参数的集时均值,并可绘制出相应的变化曲线。

3 涡流室内湍流的分布特性

3.1 连接通道处测点 1 的分布特性

测点 1 布置在涡流室底部连接通道出口处。图 4 为 800 r/min 时该点气流平均速度与湍流强度分布特性。最大速度值发生在压缩冲程上止点前（BTDC）26° CA，达 60 m/s，此时湍流强度绝对值及相对值分别为 8.1 m/s 及 13.5%。点 1 的最大湍流强度发生在 BTDC 17° CA，此时的数值分别为 10 m/s 及 20.5%。由此可见，气流从连接通道处以高速切入涡流室产生扰动，引起涡流运动和湍流，点 1 处的气流速度虽高，而湍流强度并不大。人们希望进入涡流室的气流能形成具有较大动量矩的涡流，正需要其进口有较高的气流速度，所以这种分布特性有利于混合气的形成。

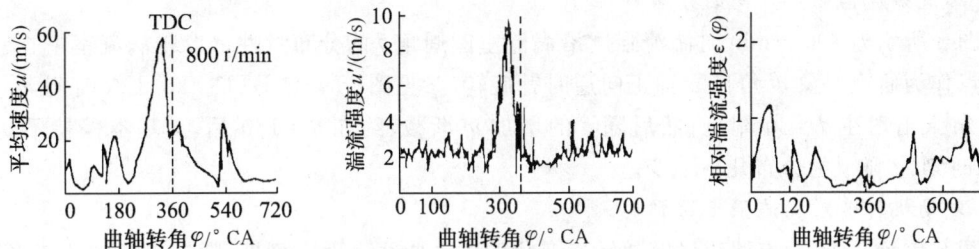

图 4　测点 1 的平均速度、湍流强度和相对湍流强度

3.2 涡流室两侧湍流强度分布特性

图 5 为 800 r/min 时涡流室两侧测点 9，11 的气流速度和湍流强度分布特性。由图可见，对处于上游侧的点 9，在 BTDC 18.5° CA 时湍流强度达到最大值 16 m/s，此时的气流速度为 43 m/s，相对湍流强度达 37.2%；对处于下游侧的点 11，湍流强度最大值 11 m/s 出现在 BTDC 14° CA，此时该点气流速度为 21 m/s，相对湍流强度达 52.4%。以上说明涡流室上下游两侧湍流强度都较大，亦即该处脉动速度均相当高，尤其是在喷油方向的下游侧，相对湍流强度更大，这对燃油与空气的快速均匀混合是十分有利的。

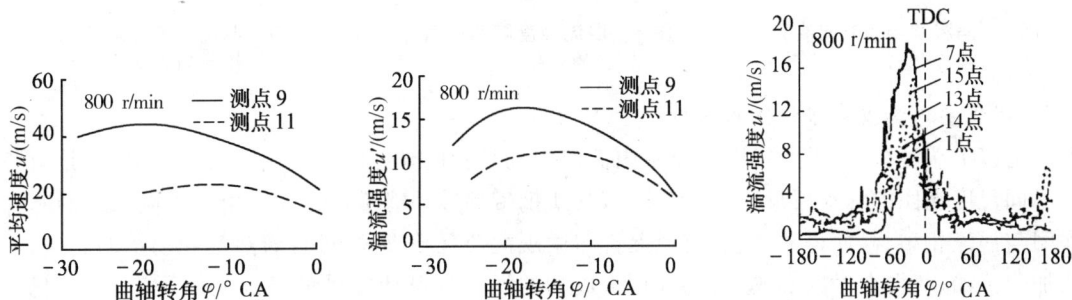

图 5　测点 9，11 的气流速度和湍流强度

图 6　湍流强度沿入流方向的变化

3.3 湍流强度沿入流方向的分布特性

图 6 为 800 r/min 时湍流强度沿入流运动方向的分布特性。由图可见，沿入流方向，在 BTDC 60° CA 到 TDC（上止点）附近，涡流室内气流的湍流强度分布是很不均匀的。如 3.1 节所述，连接通道点 1 处气流速度虽高，但气流脉动程度不大，绝对及相对湍流强度均较低；随着气流

流入涡流室,气流膨胀与扩散引起扰动,使湍流强度逐步增大,如图中的点14,15;气流继续向下游运动,使气流与涡流室壁发生碰撞并被迫转向,从而引起强烈的气流扰动,使涡流室上方点7处湍流强度达最高值;此后随着气流继续向下游运动,绝对湍流强度渐减,如图中点13所示。但在压缩冲程后期,各点的湍流强度均达最大值,这十分有利于改善柴油机的燃烧性能。

3.4 湍流强度沿涡流室半径 x 轴向的分布特性

图7为800 r/min时湍流强度沿涡流室半径 x 轴方向的分布特性。由图可见,在BTDC 20° CA时,湍流强度最大值在上游侧距涡心4 mm处,达 $u'_{max} = 16$ m/s,计算得相对湍流强度 $\varepsilon_{max} = 66.7\%$;而在BTDC 10° CA及TDC时涡流室中心附近的湍流强度最高,两侧均有所减弱;由图还可看到,随曲轴转角接近TDC,最大湍流强度值是逐渐减小的,而根据计算,最大相对湍流强度 ε_{max} 却由 66.7% 增大至 75%,说明此时气流脉动程度增强,这对混合气形成和燃烧是有利的。

3.5 湍流强度沿圆周方向的分布特性

图8所示为800 r/min时湍流强度沿涡流室圆周角 θ 的分布特性。取距涡流室中心5 mm处的点作为研究对象,θ 角自 x 轴正向逆时针旋转(参见图3)。在BTDC 20° CA时刻,$u'_{max} = 18.5$ m/s出现在 $\theta = 80°$ 附近;然后随 θ 的增加 u' 渐减,约在 $\theta = 180°$ 后,u' 基本保持不变;在TDC时刻,u' 随 θ 的分布几乎不变。

3.6 发动机转速对湍流强度的影响

图9表示9,11两点湍流强度随转速变化的关系曲线。由图可见,随转速增加湍流强度呈线性增加,涡流室内其他各点也呈这种变化关系。由此可见转速对涡流室内气流的湍流强度有着十分明显的影响。

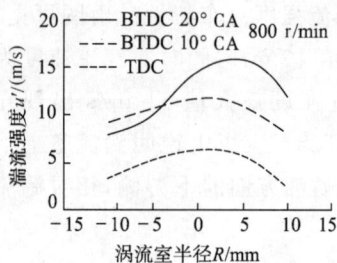

图7　涡流室半径方向湍流强度分布

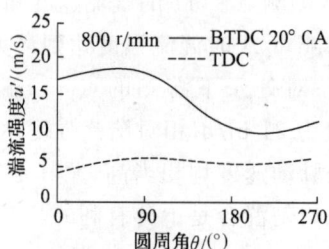

图8　湍流强度随圆周角 θ 的变化

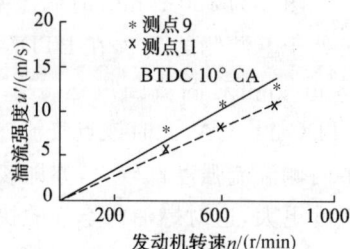

图9　湍流强度随发动机转速的变化

3.7 涡流室内湍流分布特性的分析

由以上湍流分布特性可见,涡流室内湍流流动是各向异性的。就湍流实质而言,它是由各种不同尺度涡团的脉动所形成,涡团越大,其所携带的脉动能量也越大。由于不同尺度的脉动涡团与涡流相结合,使得油、气的宏观混合与微观混合互相掺和,进行着热量和动量的交换,从而加速了油、气混合气的形成过程。因此,成功的涡流室设计应尽量形成人们所需要的湍流,尤其是在压缩冲程后期喷油开始时刻到TDC时刻之间,涡流室内要有较大的湍流强度,这对缩短滞燃期、加速火焰传播、改善燃烧性能、降低油耗和减少排放是十分有利的。由以上LDA所测得的湍流强度分布特性可见,由我国发展的吊钟型涡流室的设计是良好的。

4　结　论

(1) 用 LDA 技术测试涡流室内的气流运动规律是目前最为先进的测试手段之一,所得结果准确、可靠,所用数据处理方法合理;

(2) 涡流室内的流场是非定常的各向异性湍流流场,其湍流强度的空间分布特性及随时间的变化均有较大的差异;

(3) 涡流室两侧及顶部均有较大的湍流强度,这有利于油、气的均匀混合;

(4) 随发动机转速的增加,涡流室内湍流强度呈线性增加;

(5) 在压缩冲程后期至 TDC 之间,涡流室内湍流强度达到最大值,这对改善柴油机的燃烧性能十分有利。

参 考 文 献

[1] 李德桃.柴油机涡流燃烧室的研究与设计[M].北京:机械工业出版社,1986.

[2] 小松源一. デイーゼル機関の渦室内空気流動のシミュレーション(渦室形状および連絡孔配置の影響)[C]//日本機械学会論文集: B 编,1987,53(488).

[3] 宫本登.熱線法による渦室内のガス流動解析 [C].自動車技術会秋季学術講演会において発表,1989.

[4] Meintjes K, et al. An experimental and computational investigation of the flow in diesel prechambers[C]// Society of Automotive Engineers, 1993.

[5] 武善谋,等.柴油机涡流室内涡流及紊流的研究[J].内燃机学报,1990,8(3):209-216.

(本文原载于《农业机械学报》1998 年第 4 期)

Research on characteristics of turbulence distribution in the diesel swirl combustion chamber

Abstract: This paper presents the characteristics of turbulence distribution in the swirl combustion chamber of a motored diesel engine tested by LDA. The test system of LDA was summarized. Handling means of experimental data was raised. The law of changes of the turbulence intensity in swirl chamber to the changes of time and space was investigated. The influence of the turbulence features on the mixing of fuel oil-air and the combusting was also analyzed. It shows that the air-flow is unsteady and the turbulence structure is anisotropic. Its distribution characteristics of turbulence intensity is advantageous to the mixing of fuel oil-air and combusting.

用 LDA 研究柴油机涡流室内空气运动

王　谦,杨文明,李德桃,吴志新

[摘要]　本文介绍了采用二维氩离子 LDA(激光多普勒测速仪)对柴油机涡流室内空气运动特性测试的最新结果。测量结果表明:涡流室内气体平均速度及紊流强度是发动机曲轴转角的函数,在上止点附近随发动机转速升高,最大值出现时刻相对曲轴转角位置基本保持不变。这一点与使用 HWA(热线风速仪)对涡流室的测量结果不同。吊钟型涡流室内的空气涡流以刚体涡流为主。本文描述了吊钟型涡流室内涡流运动中涡核的运动轨迹,提出了进一步改进涡流燃烧系统的几点建议。

从我国柴油机现有技术来看,发展高喷射压力(70 MPa 以上)的直喷式柴油机以降低排气污染,在工艺上面临很大困难。在此情况下,发展有害排气成分较少的涡流室式柴油机是一条适合国情之路。目前涡流室式柴油机由于其高速适应性好、NO_x 排放低的突出优点已被用作轿车发动机。当然,涡流室式柴油机正面临着不断增强的"高燃油经济性、低有害排放物"要求的压力。解决这一问题的途径之一就是深化对涡流室内的气流运动、混合气形成和燃烧的基础研究,发展新型涡流燃烧系统[1]。

随着测试技术的不断发展和完善,以及测量仪器的更新,试验结果的可靠性、有效性和精度也得到了提高。文献[1]报道了使用 HWA(热线风速仪)测量涡流室内空气运动的情况。文中提到:随发动机转速升高,涡流室内气体温度、气流切向速度和紊流强度均有所增加,并且气体温度最大值出现时刻被推迟,而切向速度和紊流强度最大值出现时刻有所提前(在 900 r/min 时比 600 r/min 时提前 3° CA 左右)。对该现象的解释为:随发动机转速增大,涡流室通道的节流作用愈加明显,从而造成这一现象。

利用激光多普勒测速仪(LDA)测量涡流室气流运动,区别于热线风速仪(HWA)的测量结果,本文测量结果没有发现上述特征。比较 LDA 与 HWA 的测量特点(表1)可以看出:HWA 受测量条件的限制较多,LDA 的测量相对具有较大的优越性,可使测试结果更准确可靠。

表 1　LDA 与 HWA 测量方法比较

	LDA	HWA
测试范围	瞬时平均速度和紊流强度	同左
测量方式	光学点测量触发方式采样	线测量连续
数据处理	直接 BSA 微机	间接温度补偿微机
测量环境影响	无	温度、压力等
速度方向辨别	有	无
测量空间分辨率	高	较差
测量对流场干扰	无	有
测量精度及调整	精度高,调整方便	精度较低,调整复杂

如果 LDA 与 HWA 对涡流室气流运动的测量具有可比性,就存在对涡流室通道的节流效果和作用范围进一步探讨的必要性。表 2 是试验测量的有关参数,使用 LDA 对涡流室进行测量,可以更全面反映涡流室内空气运动的规律。

下面就根据 LDA 的测试结果对涡流室内的空气运动规律作进一步的分析研究。

表 2 试验发动机测量参数

参　　数		指　　标	参　　数	指　　标
激光波长/nm	氩离子绿光	514.5	示踪粒子	$1\mu mTiO_2$
	氩离子蓝光	488.0		
激光功率/W		标示 3,实际 1	记录长度/bit	32
光束直径/mm		1.25	信号处理器	DANTEC BSA
透镜焦距/mm		600	发动机	单缸 4 冲程
光束间距/mm		45.26	涡流室	平底吊钟型
光束夹角/(°)绿光(a/2)		2.20	缸径×行程/(mm×mm)	95×115
干涉条纹间距/μm 绿光		6.7	压缩比	19
探测体积/mm³		0.3×0.31×8.1	通道尺寸/(mm×mm)	6.5×14.5(角度 40°)
频移/MHz		9	发动机转速/(r/min)	400,600,800

1　试　　验

1.1　检测仪器

LDA 检测系统如图 1 所示,各部分仪器分别为 3 光束二维氩离子激光器(Model 2020)、频移器(55N11-12,丹麦)、触发信号频谱分析仪(57N10,BSA-Burst Spectrum Analyzer)、微机(IBM PC-XT)、轴编码器(Model TK-360)等,其中示踪粒子采用二氧化钛粉粒。

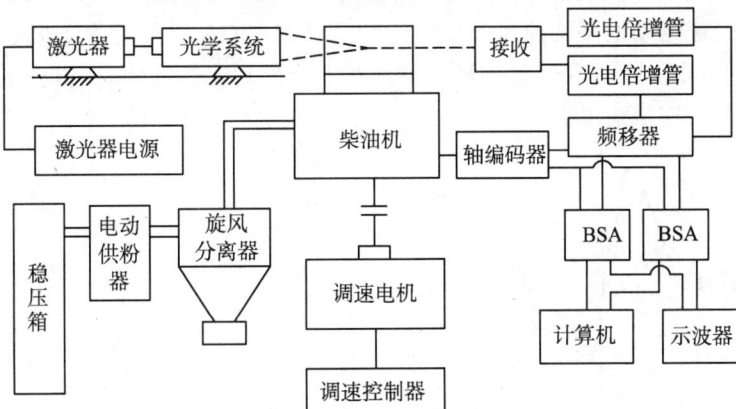

图 1　检测系统图

1.2　测点布置及数据采集

图 2 上各测量点位置均处于涡流室中心截面上,涡流室两侧安装石英窗口,采用后向散射方式采集数据,x,y 方向是设定的径向和切向标志,数据采集程序为 10 000 个有效数据(触发

方式采样)/单次记录,连同曲轴转角信号同步储存到计算机中。

1.3 数据处理

测量的平均速度、紊流强度以集总平均法处理,紊流结构参数如下:

(1)平均速度:

$$u(\overline{\varphi}, \Delta\varphi) = \frac{1}{N_t} \sum_i^{N_e} \sum_j^{N_d} u_{ij}\left(\overline{\varphi} \pm \frac{\Delta\varphi}{2}\right)$$

式中 N_d 为发动机第 i 个循环 $\overline{\varphi} \pm \Delta\varphi/2$ 转角内采集的个数;N_e 为测量的总循环数;N_t 为总的测量数,$N_t = N_d \times N_e$;$\Delta\varphi$ 为窗角(10° CA);u_{ij} 为瞬时速度。

(2)紊流强度:

$$u'(\overline{\varphi}, \Delta\varphi) = \left[\frac{1}{N_t} \sum_i^{N_e} \sum_j^{N_d} [u_{ij} - u(\overline{\varphi}, \Delta\varphi)]^2\right]^{1/2}$$

(3)相对紊流强度:

$$\varepsilon = \frac{u'(\varphi)}{u(\varphi)} \times 100\%$$

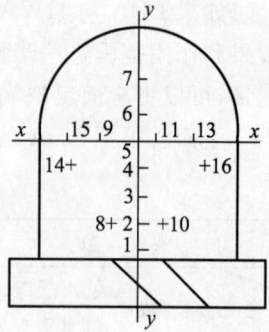

图 2 测点位置

2 试验结果及分析

2.1 连接通道口的流动特性

文献[2]中使用 HWA 的测量表明:涡流室连接通道口气体切向速度及紊流强度最大值出现时刻,随发动机转速增加有所提前(在 900 r/min 时比 600 r/min 提前 3° CA)。本文 LDA 测试结果与此有区别。

测量点 1 位于涡流室连接通道口处(如图 2),受通道口入射气流的作用较强。图 3 示出了其 x 方向和 y 方向的平均速度及合成平均速度变化曲线。从图中可以看出,在进气行程和压缩行程初期,气流运动的平均速度较低。到 BTDC 216° CA 时气流速度开始明显增加,且增加幅度较大,在 BTDC 24° CA 时达到最大值。在上止点后,x 方向和 y 方向的平均速度方向发生改变,气流流出连接通道。

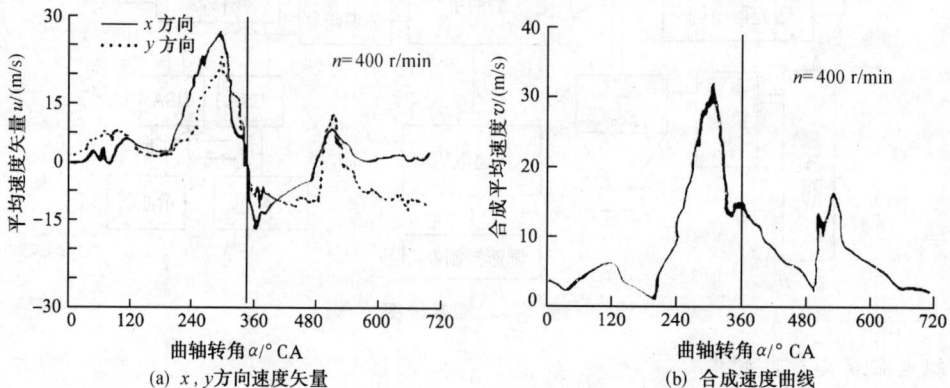

(a) x, y 方向速度矢量　　　　　(b) 合成速度曲线

图 3 连接通道处 x, y 方向平均速度及合成速度曲线

图 4 是在发动机不同转速下点 1 的平均速度的变化趋势。在压缩行程的中后期,随发动机转速升高,平均速度有所增大,但是变化的形态保持一致(紊流强度的变化趋势与此相同),并且平均速度(紊流强度)最大值出现时刻也保持不变。图 5 所示的合成平均速度方向随曲轴转角在垂直方向上的夹角变化曲线正说明了这一特点。显然,在压缩行程后期,不同的发动机转速下,平均速度的方向仍然保持相当的一致性。在速度峰值点附近,速度上升斜率一致,斜率转换时刻一致,则速度峰值出现时刻必然相同。因此,连接通道处平均速度最大值相对曲轴转角的出现时刻既不会超前也不会滞后。对其他各点的测量也得到了同样的结论,这说明了发动机转速的变化不会影响涡流室内的涡流形态。

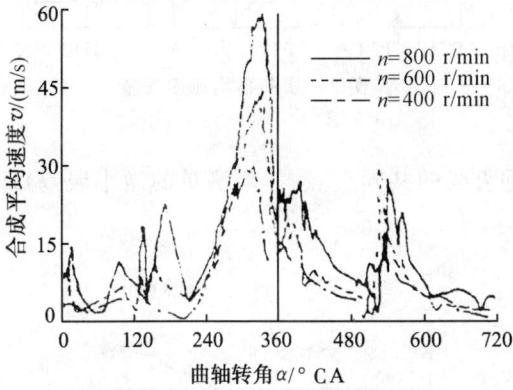

图 4 不同转速下的速度曲线(点 1)

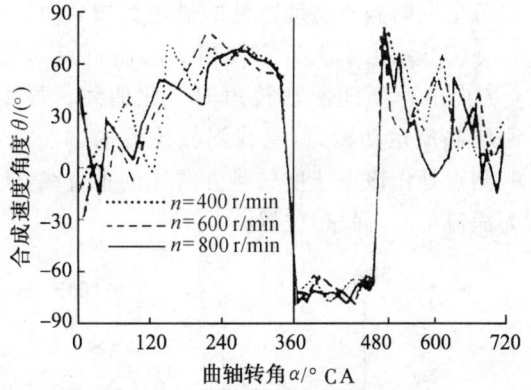

图 5 不同转速下合成速度角度曲线

2.2 涡流特性

对于涡流室内空气涡流特性,许多研究者采用不同的试验方法进行研究[3,4],得到比较一致的结论是:涡流室内的涡流在压缩行程初期是刚体涡(强制涡),然后逐步转变为势涡(自由涡)。以上结论主要是针对球型和彗星 V 号涡流室。由于涡流室形状对空气运动的影响很大[5],本文采用 LDA 的测量结果分析了国产吊钟型涡流室内的涡流特性,得到的结论与上述观点不完全一致。

图 6 是位于涡流室内 x 轴线上 9,11,13,15 点的切向平均速度曲线,图 7 是位于 y 轴线上测点 2,3,4,6,7 的切向平均速度曲线。图 8 是涡流室内各测点的切向速度矢量图。由图 8 可以直观地看出,在压缩行程后期(BTDC 30° CA ～ TDC),气流切向速度由涡流室几何中心位置沿半径方向逐渐增大,因此吊钟型涡流室内涡流可看作刚体涡流。

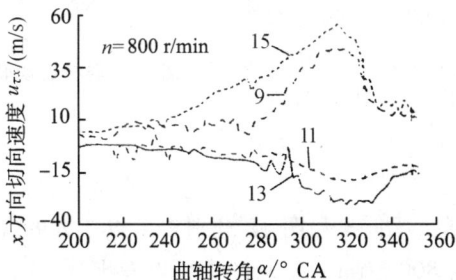

图 6 测点 9,11,13,15 切向平均速度曲线

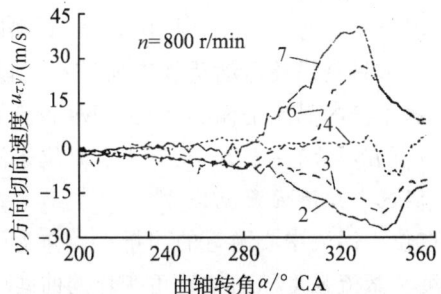

图 7 测点 2,3,4,6,7 切向平均速度曲线

2.3 涡流中心（涡核）的运动特性

以前的测试结果表明,涡流室内涡流在形成变化过程中,其涡流中心（涡核）的位置不是固定不变的,而是随着曲轴转角的变化不断运动[2,4,6]。但以前的研究由于测试方法的限制,对于涡核附近的气流速度的大小和方向不能准确测量,因此只能定性地指出涡核的运动范围（约在涡流室几何中心附近）[2,4,6]。本文的 LDA 测试结果能准确可靠地描述吊钟型涡流室内涡核的运动轨迹,从而对吊钟型涡流室内的涡流运动特性的描述更加全面。

图9示出了涡流室几何中心附近点 x 方向和 y 方向的速度随曲轴转角的变化曲线。根据前述刚体涡流运动特点,寻找涡核位置的方法是：在每一测点处作该点速度矢量方向线的垂直线,所有垂直线的共同交点（忽略测量的微小误差）即为涡流中心（涡核）位置。

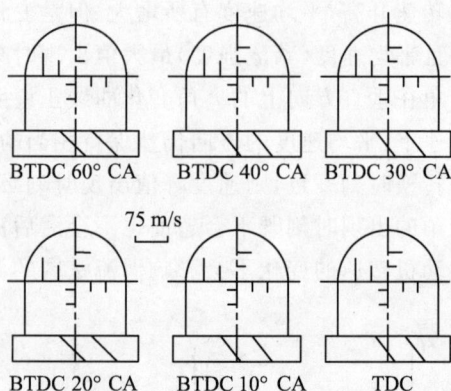

BTDC 60° CA   BTDC 40° CA   BTDC 30° CA

75 m/s

BTDC 20° CA   BTDC 10° CA   TDC

图 8   切向平均速度矢量

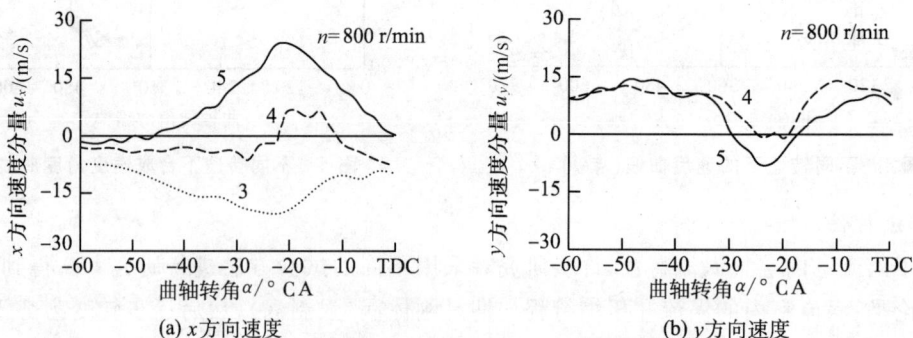

(a) x 方向速度   (b) y 方向速度

图 9   涡流室几何中心附近点的 x, y 方向速度曲线

图 10 是发动机转速为 800 r/min 时,压缩行程后期（BTDC 60° CA～TDC）吊钟型涡流室内涡核运动轨迹。由图可以看出：压缩行程中,在 BTDC 60° CA 时刻,涡核位于涡流室几何中心稍偏右处。随着活塞上行,涡核向右下方移动,直到 BTDC 20° CA 时刻,涡核在右方偏离涡流室中心 5 mm,下方偏离涡流室中心 3 mm 处。此后,随活塞向上止点移动,涡核迅速向左下方移动,在 TDC 时刻涡核位于涡流室中心正下方,距涡流室中心 5 mm。在 400 r/min 和 600 r/min 时,涡核运动轨迹与此类似,只是涡核运动范围有所缩小。以上表明由于受通道气流的影响,涡流室内涡流是偏心旋涡（相对涡流室几何中心）,其运动轨迹是先向右下方,再向左下方,在上止点时位于涡流室几何中心正下方。

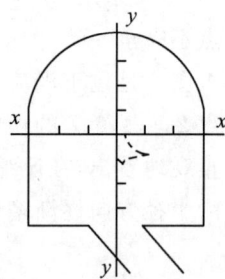

图 10   涡核运动轨迹

2.4 涡流中心附近紊流特性

取位于涡流中心附近的测量点 4 作为研究对象。图 11 和图 12 分别表示点 4 处的气流平均速度和紊流强度随曲轴转角变化的曲线（转速为 800 r/min）。从图中可以看出,尽管在压缩行程中,点 4 的气流平均速度很低,但其紊流强度很大。在压缩上止点前 22° CA 时,平均速度

值为0.85 m/s,此时紊流强度为13.72 m/s,脉动程度相当高。涡流中心附近其他各点的测试结果也显示了这一特点。因此,涡流室内涡流中心附近的紊流强度很高是涡流运动的重要特点。

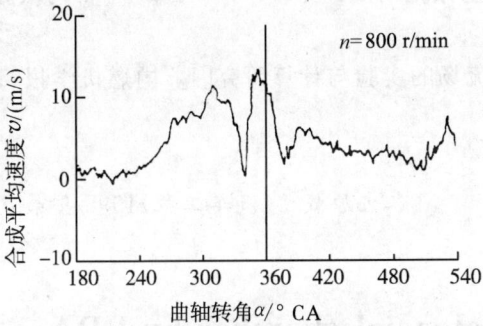

图11 点4的平均速度曲线

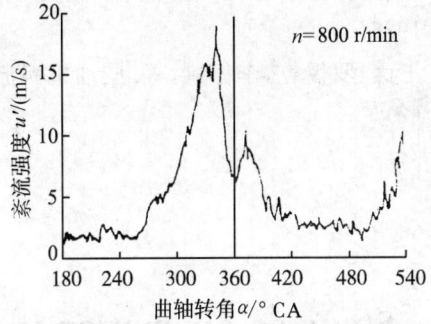

图12 点4的紊流强度曲线

2.5 改进涡流燃烧系统的几点建议

根据上述采用 LDA 对柴油机涡流室内空气运动的测试和分析结果,针对国产吊钟型涡流燃烧系统提出以下改进意见:

(1) 由于涡流室内涡流中心附近的紊流强度较高,应使喷油线与涡流室中心有适当错移,以利用涡流中心较高的紊流强度,促进油气的微混合,从而改善燃烧过程;

(2) 由于涡核运动范围偏离涡流室中心,使得涡流室下游一侧(喷油一侧)的气流速度小于上游一侧,这对燃油的蒸发和混合不利,在新型涡流室结构设计时应尽量提高下游一侧(喷油一侧)的气流速度;

(3) 改进镶块和连接通道形状,进一步增加通道口附近的紊流强度以及涡流室两拐角处的副涡强度(尤其是喷油一侧的副涡强度),从而改善涡流室的着火性能及涡流室柴油机的冷起动性能。

3 结 论

(1) 在压缩行程后期,随着发动机转速升高,平均速度、紊流强度增大,但其变化的基本形态不变,平均速度(紊流强度)最大值出现时刻相对曲轴转角保持不变;

(2) 在压缩行程中后期(BTDC 90° CA~TDC),吊钟型涡流室内涡流运动以刚体涡流为主,空气运动形态可用切向平均速度特征来描述;

(3) 吊钟型涡流室内涡流运动中涡核的位置是不断变化的,随曲轴转角变化,涡核先向涡流室中心右下方移动,再向左下方移动,在上止点附近处于涡流室中心正上方位置。

参 考 文 献

[1] 李德桃. 低油耗、低污染、低爆压的柴油机涡流燃烧室[P]. 86105755.4,1990.

[2] 武善谋,杨海青,刘书亮,等.柴油机涡流室内涡流及紊流的研究[J].内燃机学报,1990,
(3).

［3］Бондаренко Г П. *Исспедоьание Вихреьой Кашеры Дизиая М*［М］. Машгиз,1959.

［4］Nakajima K, et al. *An experimental investigation of the air swirl motion and combustion in the swirl chamber of diesel engines*［M］. FISITA, Barcelona, 1968.

［5］池上詢,宮本登. デイーゼル機関の渦室内空気流動のシミュレーション［J］. 内燃機関,1986(5).

［6］王谦,罗惕乾,李德桃,等. 柴油机涡流室内流场的实验与计算研究［J］. 内燃机学报,1994(2).

(本文原载于《汽车工程》1998 年第 6 期)

Study on air motion in diesel swirl chamber by LDA

Abstract：In this paper，a new measuring result of the air motion in swirl chamber of a diesel engine by the 2D Arion LDA was presented. It was found that the mean velocity and turbulence intensity are the function of the crank angle. Despite the engine speed increased，the peak values of mean velocity or turbulence intensity appeared at the same crank angle approximately. This result is different that of HWA test. The locus of vortex center in bell type swirl chamber the air motion of which can be regarded as a solid vortex was given in detail. Some suggestions for improving the swirl combustion system had also been given.

涡流室式柴油机相关火焰微元燃烧模型的三维数值模拟

夏兴兰,董　刚,李德桃,杨文明

[摘要]　把涡流室式柴油机不同区域与不同时期的燃烧过程分开处理,将涡流室的燃烧过程划分为 3 个阶段,即:低温着火化学动力学反应阶段、高温预混燃烧化学动力学反应阶段和相关火焰微元的扩散燃烧阶段,而认为主燃烧室的燃烧只有相关火焰微元的扩散燃烧阶段。用 Shell 着火模型、Arrhenius 方程和相关火焰微元模型来分别模拟低温着火、高温预混燃烧和扩散燃烧过程。开发了三维数值模拟计算程序并对燃烧过程进行计算,进而研究了涡流室中瞬态温度场的变化过程。模型预测的示功图和涡流室中的燃烧放热率与试验值吻合良好。

　　世界范围内日益高涨的降低发动机有害排放物的呼声,迫使人们更加重视内燃机燃烧过程的研究。Marble 和 Broadwell 针对剪切层中湍流扩散燃烧提出了相关火焰微元燃烧模型[1],将湍流反应区定义为火焰微元的集合,湍流流动使反应区扩展,而其局部仍保持层流火焰微元结构。由于人们对火花点火式发动机的火焰结构有了较深入的认识,Cheng[2],Boudier[3]和 Zhao[4]等人成功地应用相关火焰微元模型对火花点火式发动机均质混合气的预混燃烧进行了研究,模拟的气缸压力和燃烧放热率与实测值相符。Dillies 等人[5]尝试将相关火焰微元模型应用于直喷式柴油机燃烧过程的研究,对着火延迟、预混合燃烧和扩散燃烧分别建立了燃料与氧化剂接触表面总面积、预混合火焰表面积和扩散火焰表面积的输运方程,但该模型对预混合燃烧部分没有作出令人满意的描述。然而,在涡流室式柴油机中,由于其燃烧室结构的复杂性,主副燃烧室发生的燃料与空气混合、混合气的着火和火焰传播等现象都存在很大差别,同时两室的燃烧过程又通过连接通道而相互影响,从而使模型描述和数值计算都更加困难,多维相关火焰微元燃烧模型至今未能在这方面得到应用。本文对涡流室式柴油机的主副燃烧室进行不同处理,建立不同的燃烧模型,开发了三维数值计算程序并进行计算。

1　数学模型

　　燃烧过程的化学反应机理极其复杂,受计算机容量和计算速度的限制,要进行详细的化学反应计算是非常困难的,甚至是不可能的。相关火焰微元燃烧模型避开了对复杂化学反应机理的描述,认为化学反应发生在火焰面上,化学反应速率 W_i($i=$ox,f 分别代表氧气和燃料)可通过火焰面积密度 Σ(单位体积内的火焰面面积)求得:

$$W_i = \rho_{i,\infty} U_{D,i} \Sigma \tag{1}$$

式中 $\rho_{i,\infty}$ 为远离火焰面处的反应物密度;$U_{D,i}$ 为单位时间火焰面积内反应物的体积消耗量,它与火焰传播速度具有相同的物理意义,故也称为燃烧速度。为求解式(1),需要针对 Σ 和 $U_{D,i}$ 建立单独的子模型。

1.1 火焰面积密度的输运方程

火焰面在气流中传播时,会发生弯曲和变形。产生拉伸时其有效面积增加,使火焰面积密度增加;火焰面上的燃烧以及火焰面的过分拉伸、壁面激冷等均会使火焰面积密度减少。根据火焰面积密度的对流、扩散、生长和减少过程,按传统输运方程的形式,可写出火焰面积密度输运方程

$$\frac{\partial \overline{\rho}\,\overline{\Sigma}}{\partial t} + \nabla \cdot (\overline{\rho}\Sigma u) = \nabla \cdot \frac{\mu_t}{\sigma_{\Sigma}} \nabla \frac{\overline{\rho}\Sigma}{\overline{\rho}} + \alpha\,\overline{\rho}\varepsilon_s \Sigma - \beta\overline{\rho}\left(\frac{\rho_{f,\infty} U_{D,f}}{\overline{\rho}_{f,\infty}} + \frac{\rho_{ox,\infty} U_{D,ox}}{\overline{\rho}_{ox,\infty}}\right)\Sigma^2 -$$

$$2\alpha\,\overline{\rho}(\varepsilon_s - \varepsilon_{s,crit})h(\varepsilon_s - \varepsilon_{s,crit})\Sigma \tag{2}$$

式中"‾"表示单元体积平均;$\overline{\rho}_{ox,\infty}$、$\overline{\rho}_{f,\infty}$、$\overline{\rho}$ 分别为计算单元中氧气、燃料和总物质的平均密度;u 为单元中气流速度矢量;μ_t 为湍流动力粘性系数;σ_{Σ} 为火焰面的湍流 Schmidt 数;ε_s 为火焰面上流体的平均变形率;α 和 β 为模型常数,分别称为拉伸因子和耗散因子,本文取 $\alpha = 20.0, \beta = 5.0$;$\varepsilon_{s,crit}$ 为临界拉伸率,据试验观察[2],$\varepsilon_{s,crit} = 10^4 \sim 10^5$;$h$ 为阶跃函数。

上式中右边第 1 项为湍流扩散项,右边第 2 项为湍流运动引起的火焰面积密度增加项,右边第 3 项为反应物消耗引起的火焰面积密度减少项,右边第 4 项为过分拉伸引起的火焰面积密度减少项。

采用 k-ε 湍流模型预测流体的平均变形率,则

$$\varepsilon_s = \varepsilon/k \tag{3}$$

式中 k 为湍流脉动动能;ε 为湍流脉动动能的耗散率。

火焰面上燃料和氧气的体积消耗率 $U_{D,i}$ 为[1]

$$U_{D,i} = \Phi_0 Y_{i,\infty} \frac{\overline{\rho}}{\rho_{i,\infty}} \sqrt{\frac{\varepsilon_s \overline{D}}{2\pi}} \frac{(\Phi+1)}{\Phi} \exp\left\{-\left[rerf\left(\frac{\Phi-1}{\Phi+1}\right)\right]^2\right\} \tag{4}$$

式中 $\overline{D}$ 为单元中平均质量扩散系数;Φ 为燃空当量比;$rerf$ 为反误差函数;$Y_{i,\infty}$ 为离火焰面处物质的质量分数;Φ_0 为比例系数,当 $i = f$ 时,$\Phi_0 = 1$;当 $i = ox$ 时,$\Phi_0 = \Phi$。

1.2 燃烧过程描述和燃烧模型

涡流室式柴油机主副燃烧室的燃烧情况不同。本文将涡流室的燃烧过程划分为 3 个阶段 5 个时期来描述,即:着火延迟期,或称低温化学动力学反应期;向高温预混化学动力学反应过渡期;高温预混化学动力学反应期;向燃料和空气混合控制的扩散燃烧过渡期;火焰微元扩散燃烧期。主燃烧室的燃烧只有火焰微元扩散燃烧期。

1.2.1 低温化学动力学反应期

Halstead 提出的 Shell 着火模型是低温化学反应的多步 Arrhenius 动力学模型[6],它根据链式反应用 5 种普通物质建立了 8 个反应方程式。该模型在柴油机着火燃烧研究方面已有许多成功的应用[7],本文也采用此模型。

1.2.2 向高温预混化学动力学反应过渡期

涡流室内的低温氧化反应达到某一速度后,化学反应速度急剧增加,工质温度急剧上升,进入高温化学动力学反应期,本文以 $T_{crit} = 920$ K 为高温预混化学动力学反应期的始点,在过渡期内同时进行低温着火反应和高温化学动力学反应,此时燃料燃烧速率应为两者之和。

1.2.3 高温预混化学动力学反应期

高温预混化学动力学反应燃烧在着火延迟期内已形成的混合气,此时的化学反应迅速,局部温度急剧升高,燃烧主要由高温化学动力学控制。燃烧速率采用单步 Arrhenius 方程形式,燃烧的消耗率 W_f 为

$$W_{\mathrm{f}} = A\,\bar{\rho}^2\,\bar{Y}_{\mathrm{f}}^a\,\bar{Y}_{\mathrm{ox}}^b\exp\,(-E/T) \tag{5}$$

式中 $\bar{Y}_{\mathrm{f}}$，$\bar{Y}_{\mathrm{ox}}$ 分别为预混燃料和氧气的平均质量分数；A 为频率因子，$A=5\times10^{10}\ \mathrm{m}^3/(\mathrm{kg\cdot s})$；$E$ 为活化温度，$E=1.2\times10^4\ \mathrm{K}$；$T$ 为单元温度；a，b 为指数。

高温预混合燃烧只发生在扩散火焰的氧化剂一侧，因此必须将预混合并处于火焰氧化剂一侧的燃料与没有和氧气预混合并处于火焰燃料一侧的燃料区分开来，对预混合燃料密度 $\bar{\rho}_{\mathrm{f,p}}$ 采用新的输运方程

$$\frac{\partial\bar{\rho}_{\mathrm{f,p}}}{\partial t} + \nabla\cdot(\bar{\rho}_{\mathrm{f,p}}u) = \nabla\cdot\frac{\mu_t}{\sigma_\Sigma}\nabla\frac{\bar{\rho}_{\mathrm{f,p}}}{\bar{\rho}} + \dot{\rho}_{\mathrm{f,p}}^s + \dot{\rho}_{\mathrm{f,p}}^c \tag{6}$$

式中 $\dot{\rho}_{\mathrm{f,p}}^s$ 为液滴燃料蒸发的源项；$\dot{\rho}_{\mathrm{f,p}}^c$ 为预混合燃料化学反应的源项。

1.2.4　向扩散燃烧过渡期

当化学反应速度比燃料和空气混合的速度大得多时，燃烧前燃料不再有足够的时间与空气混合，燃烧由预混合燃烧开始向扩散燃烧过渡，利用临界 Damkohler 数 Da_{crit} 作为过渡点的判据

$$Da = \frac{A\exp\,(-E/T)}{\varepsilon/k} \tag{7}$$

当 $Da>Da_{\mathrm{crit}}$（本文取为 50）时，开始进行扩散燃烧，此时燃烧室内既有预混合燃烧又有扩散燃烧，但此时蒸发的燃料被限制在火焰面的燃料侧，不能越过火焰面进入氧化剂侧与氧气混合。当预混燃料全部被燃烧完后，预混燃烧结束，开始进入扩散燃烧阶段。

1.2.5　涡流室的相关火焰微元扩散燃烧期

火焰微元扩散燃烧的燃烧速率由方程(1)和(2)求得，预混燃烧转变为扩散燃烧后，扩散燃烧的火焰面面积必须有一个初始值，由于油滴是扩散燃烧的燃料源，因此，把此时燃料液滴的表面面积作为扩散燃烧火焰面的初始值。

柴油机的扩散燃烧与传统的扩散燃烧有较大的差别。在柴油机中，扩散火焰将每一计算单元分为两个区：火焰燃料侧的纯燃料区和火焰氧化剂侧的预混合燃料区。因此在每一单元内，混合物的组成是非均质的，在应用方程(4)时，必须确定火焰的相对位置以确定扩散火焰两侧的物质浓度。方程(6)描述了火焰氧化剂侧的预混合燃料量 $\bar{\rho}_{\mathrm{f,p}}$ 的输运关系，则火焰燃料侧的扩散燃烧燃料量 $\bar{\rho}_{\mathrm{f,D}}$ 为

$$\bar{\rho}_{\mathrm{f,D}} = \bar{\rho}_{\mathrm{f}} - \bar{\rho}_{\mathrm{f,p}} \tag{8}$$

氧气只在火焰的氧化剂侧，其他物质（包括燃烧产物和不参加反应的惰性物质）统称为惰性物质，可向火焰的两侧扩散。火焰燃料侧的惰性物质的输运方程可写为

$$\frac{\partial\bar{\rho}_{\mathrm{N,D}}}{\partial t} + \nabla\cdot(\bar{\rho}_{\mathrm{N,D}}u) = \nabla\cdot\frac{\mu_t}{\sigma_\Sigma}\nabla\frac{\bar{\rho}_{\mathrm{N,D}}}{\bar{\rho}} + \dot{\rho}_{\mathrm{N,Dcon}} + \dot{\rho}_{\mathrm{N,Ddiff}} \tag{9}$$

式中 $\bar{\rho}_{\mathrm{N,D}}$ 为火焰燃料侧的惰性物质的密度；$\dot{\rho}_{\mathrm{N,Dcon}}$，$\dot{\rho}_{\mathrm{N,Ddiff}}$ 分别为其对流源项和扩散源项。

1.2.6　主燃烧室的气体射流扩散燃烧

涡流室着火后，一部分燃料通过连接通道进入主燃烧室，与主燃烧室中的空气混合后再进行燃烧。由于此时主燃烧室的温度和压力较高，可认为主燃烧室中的燃烧为火焰微元的扩散燃烧，仍用方程(1)～(4)描述其燃烧过程。

1.3　其他相关模型

在整个燃烧模型中还包括空气运动、燃油喷雾、蒸发和混合等一系列子模型。

在空气运动模型中,采用标准的 k-ε 湍流模型,并作适当修正[8]。

在燃油喷雾模型中,将喷雾分为液核区和气液混合区,并分别采用 Kelvin-Helmhotz(K-H)和 Rayleigh-Tayhlor(R-T)不稳定性理论来模拟其中油滴的破碎过程;同时,根据 Weber 数,将喷雾碰壁分为粘附、反弹粘附、飞溅附壁射流 3 种相互重叠的过程来模拟喷雾碰壁[9]。

在油滴蒸发和与空气混合子模型中,采用油滴在高温高压下的蒸发模型,并考虑油膜蒸发过程[10]。

2 计算结果和分析

根据以上模型,以 195 柴油机为例,在标定工况(8.82 kW,2 000 r/min)下进行计算。柴油机每循环供油量为 0.037 4 g,喷油于 10° CA BTDC 开始,喷油持续角为 16° CA[11]。

为重点研究涡流室内燃烧过程,暂忽略主燃烧室中活塞顶凹坑的影响,即假定活塞顶为平顶。图 1 为涡流室中典型截面的位置。图 2 为计算所得的温度场随曲轴转角的变化。

从计算结果可知:在 4° CA BTDC 左右,涡流室底部通道口左侧附近首先着火;在 2° CA BTDC 时,燃烧火焰已达到通道口附近,在底部通道口处形成高温区,在涡流室左侧的喷雾外围由于喷射燃油的蒸发吸热,仍形成一狭长的低温区(如图 2a 所示)。随着燃烧的进行,涡流室中强烈的气流运动使一部分火焰向涡流室右上方扩展,另一部分火焰通过连接通道向主燃烧室扩展,在涡流室右侧和连接通道处形成高温区。在 6° CA BTDC 时,涡流室中的温度已经明显升高,说明火焰已扩展到整个涡流室,这与高速摄影照片结果相符[11],但此时喷油刚刚结束,涡流室左侧由于较多的燃油蒸发吸热,其温度仍然比右侧低,高温区位于右侧(如图 2b 所示)。受涡流室内旋转涡流的影响,燃烧后的高温产物由于其密度较小,被卷向涡流室中心,密度较大的油滴和燃油蒸气在由涡流室中心部分卷向外侧的同时继续燃烧。这样,在 10° CA ATDC 时,涡流室中的高温区开始向涡流室中心运动(如图 2c 所示)。到 20° CA ATDC 时,在涡流室中心附近形成高温区(如图 2d 所示),此时涡流室中的速燃期已经结束。随后,涡流室中燃烧较慢,温度趋向均匀,燃烧放热率计算也表明,在 20° CA ATDC 后,涡流室中的燃烧放热率极小。

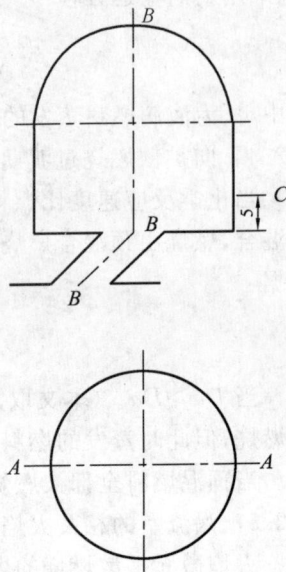

图 1 涡流室中典型截面位置

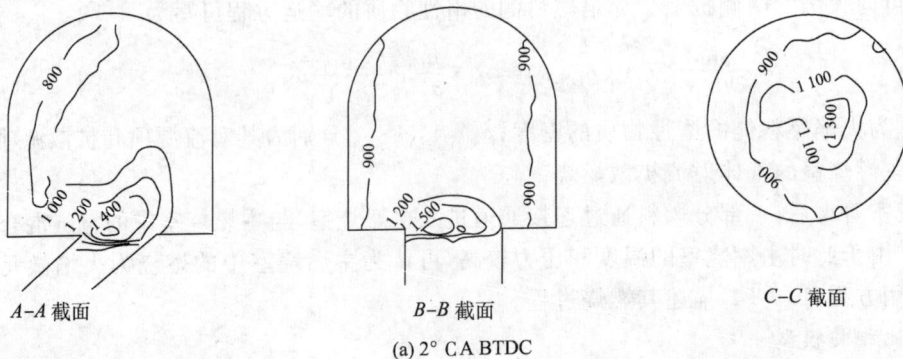

| A-A 截面 | B-B 截面 | C-C 截面 |

(a) 2° CA BTDC

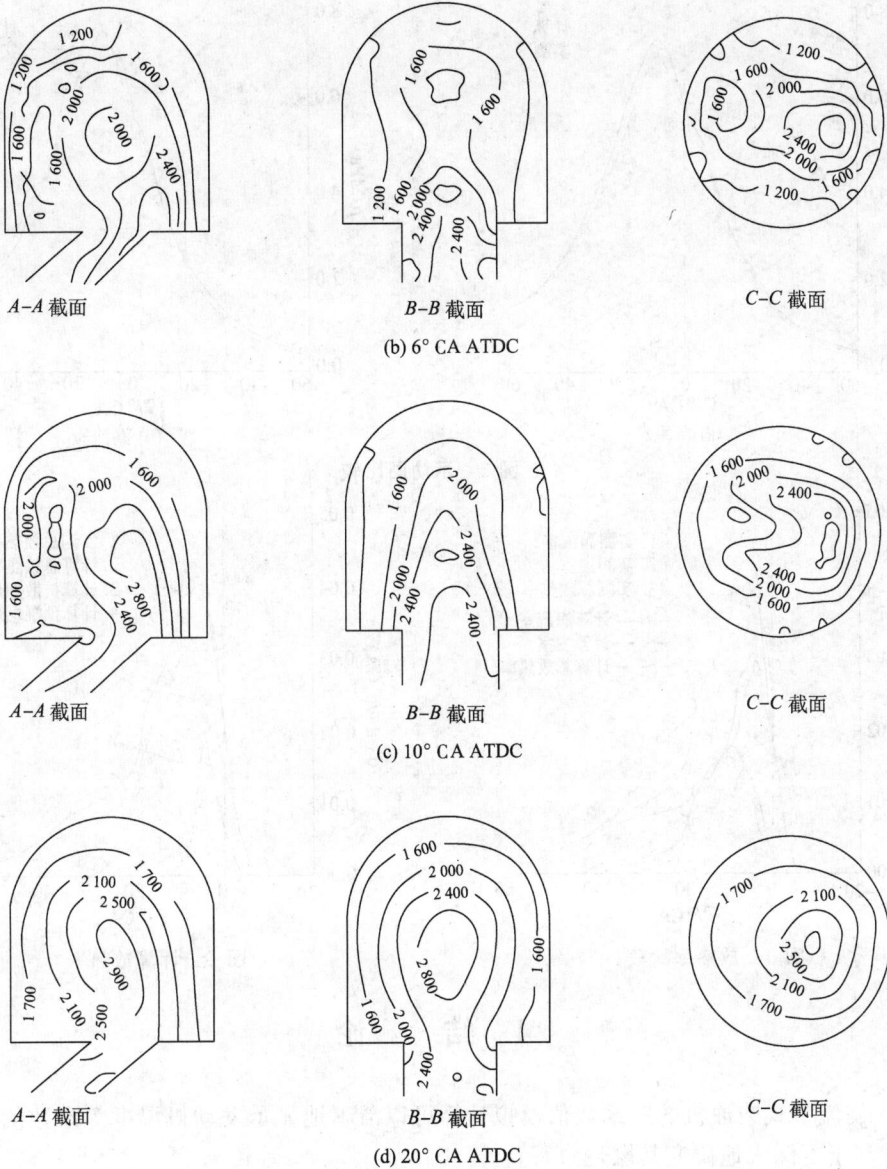

A-A 截面　　　　　　　　B-B 截面　　　　　　　　C-C 截面

(b) 6° CA ATDC

A-A 截面　　　　　　　　B-B 截面　　　　　　　　C-C 截面

(c) 10° CA ATDC

A-A 截面　　　　　　　　B-B 截面　　　　　　　　C-C 截面

(d) 20° CA ATDC

图 2　涡流室内温度场的变化

3　与试验结果的比较

在标定工况下实测 195 柴油机的示功图并进行放热率计算。图 3 为模拟计算与实测示功图的比较,图 4 为计算与实测(指由实测示功图计算而得)的放热率比较。由图可以看出:模型计算的示功图和涡流室中的燃烧放热率都与实测结果吻合良好。由于在模拟计算中,没有考虑活塞顶凹坑的作用,因此,计算所得的主燃室的放热率与实测值差别稍大,并且实测的主燃室燃烧放热更靠近上止点,计算所得主燃室的燃烧放热出现了第 2 个峰值。这也说明:主燃室中的凹坑具有促进燃料尽快燃烧,提高发动机热效率的作用。图 5 所示是计算所得预混燃烧和扩散燃烧的放热率。

(a) 主室

(b) 涡流室

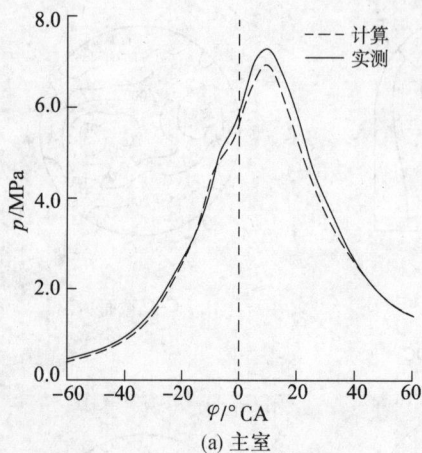

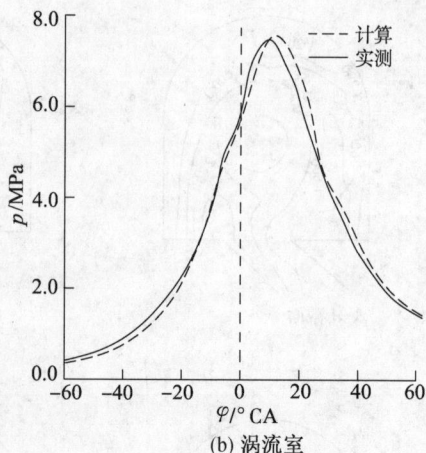

图 3　示功图比较

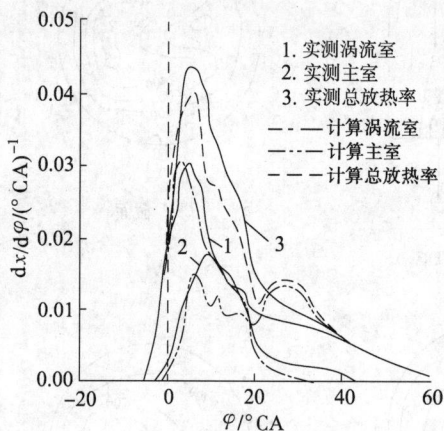

图 4　放热率比较

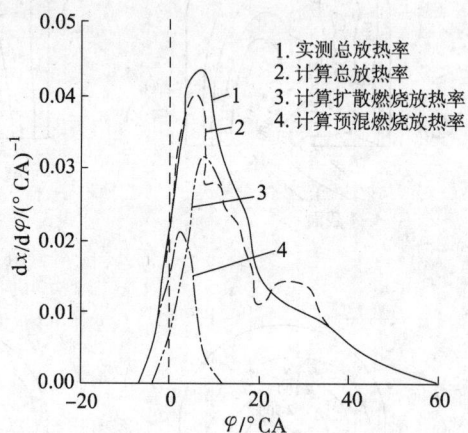

图 5　计算放热率

4　结　论

（1）涡流室式柴油机的三维数值模拟计算可以清晰地显示发动机温度等参数的瞬态分布规律，有利于更深入地研究其燃烧过程；

（2）相关火焰微元燃烧模型是一种新颖的燃烧模型，它对火焰结构进行了较好的描述，避免了对复杂化学反应机理的模拟，还可以定量区分柴油机的预混燃烧和扩散燃烧过程；

（3）模型预测的涡流室式柴油机的示功图和涡流室中的燃烧放热率与试验结果吻合良好。

柴油机燃烧过程数值模拟研究的主要目的之一就是对其有害排放物的生成进行预测，在现有工作基础上对涡流室柴油机有害排放物的生成进行三维数值模拟计算，是作者将进一步深入研究的课题。

参 考 文 献

［1］Marble F E，Broadwell. The coherent flamelet model for turbulent chemical reactions

[R]. Project Squid Report, TRW-9-PU, 1977.

[2] Cheng W K, Diringer J A. Numerical modeling of SI engine combustion with a flame sheet model[J]. *SAE Paper*, 910268, 1991.

[3] Bodier P, Henriot S, Poinsot T, et al. A model for turbulent flame ignition and propagation in spark ignition engines[C]// *24th Symposium (International) on Combustion*, 1992: 503–510.

[4] Zhao X W, Matthews R D, Ellzey J L. Three dimensional numerical simulation of flame propagation in spark ignition engines[J]. *SAE Paper*, 932713, 1993.

[5] Dillies B, Marx K, Dec J, et al. Diesel engine combustion modeling using the coherent flame model in KIVA-Ⅱ[J]. *SAE Paper*, 930074, 1993.

[6] Halstead M, Kirsh L, Quinn C. The autoignition of hydrocarbon fuels at high temperature and pressures-fitting of a mathematical model[J]. *Combustion and Flame*, 1997, 30: 45–60.

[7] Kong S C, Ayoab N, Reitz R D. Modeling combustion in compression ignition homogeneous charge engines[J]. *SAE Paper*, 920512, 1992.

[8] 夏兴兰, 李德桃, 吴志新, 等. 涡流室式柴油机空气运动的三维数值模拟[J]. 燃烧科学与技术, 1998, 4(3): 288–299.

[9] 夏兴兰, 李德桃, 董刚, 等. 涡流室式柴油机喷雾过程的三维可视化研究[J]. 中国机械工程, 1998, 9(3): 74–76.

[10] 夏兴兰. 涡流室式柴油机三维燃烧模拟计算与分析[D]. 镇江: 江苏理工大学, 1998.

[11] 李德桃, 朱广圣. 吊钟型涡流室内喷油和燃烧过程的研究[J]. 内燃机学报, 1989, 7(1): 21–26.

(本文原载于《内燃机学报》1999 年第 2 期)

Three-dimensional numerical simulation for coherent flamelet model in a swirl chamber diesel engine

Abstract: The combustion events of a swirl chamber diesel engine in distinct place and distinct stage are processed differently. The combustion in the swirl chamber is broken into three phases: low temperature ignition kinetics, high temperature premixed kinetics burn and flamelet diffusion burn phase. While the burn in the main chamber is considered as flamelet diffusion combustion. The Shell ignition model, global Arrhenius equation and coherent flamelet model are used to model low temperature ignition, high temperature premixed burn and diffusion burn respectively. Three-dimensional numerical calculation program is developed and the calculation of the combustion process has been carried out. The change of transient temperature fields in the swirl chamber is studied. The pressure and the heat release rate in the swirl chamber predicted by the model match well with the results derived from experiment.

Temperature measurement in the swirl chamber of an IDI engine using Moire deflectometry

Li Detao , Xiong Rui , Xue Hong

Abstract

The processes of ignition and combustion of a diesel engine are affected by temperature distribution formed during compression stroke. We present an applied study on the measurement of temperature distribution in the swirl chamber of an IDI engine using the Moire deflection technique. The method for determining the quantitative temperature distribution from detected Moire fringe shift has been established. From the measured temperature distribution, the evaporation and air mixing processes in the swirl chamber during compression stroke is analysed and discussed. The temperatures mapped by Moire deflectometry are compared and validated with the temperatures measured by hot-wire anemometry at several representative points. The experimental investigation opens up a new means for measuring the temperature distribution in the combustion chamber of IC engines.

1 Nomenclature

h	deviation of Moire fringe
l	length of the objective flow field along the transmitting direction of the light
n	refractive index
r	cylindrical co-ordinate, $r = \sqrt{x^2 + y^2}$
t	temperature
x, y, z	Cartesian co-ordinates as defined in Fig. 1
y^+	non-dimensional distance from wall
Δ	distance between two gratings
α	deflection angle

θ rotating angle of two gratings

λ wavelength of light

φ crank angle

2   Introduction

Measurement of temperature distribution in the combustion chamber of an IC engine is extremely important for understanding the combustion process of the engine. However, difficulties come from various aspects such as constrained structure, high temperature and high pressure in the chamber, and instability of the process. Over the years, contact measurements have been the dominant means to determine the temperature field[1], but the results of these methods are not satisfactory. The thermal inertia of contact sensor and the interference with measured flow field constantly affect accuracy and certainty of the measurement. On the other hand, when the non-contact optical method is used to measure the temperature distribution, the deflection effect of the light beam in the temperature field is usually neglected. Vest[2] suggested using an imagery lens for the correction of the deflection, but the process was rather tedious. In 1980, Kafrif[3] proposed for the first time that Moire deflectometry could be used for transient temperature measurement. Afterwards, Keren[4] used the method to measure one-dimensional(1D) temperature distribution of an axisymmetric steady flame. Recently, Almanasreh and Abushagur[5] used Moire deflectometry to measure the density distribution of a candle flame as a phase object. Toker et al[6] indicated that Moire deflectometry yielded a more accurate measurement of the temperature and density distribution in complex compressive flow fields, compared with Schlieren and shadowgraphy.

Moire deflactometry has a simple optical system, a moderate requirement on coherency of the light source, and is easy for operation. When it is used for measurement of an axisymmetric temperature field, mathematical formation is straightforward. As a result of these advantages, we are motivated to apply the Moire deflectometry to the swirl chamber of an IDI engine. In spite of the success by Stricker et al[7−9], who measured the distribution of density and temperature in the jet ejecting field of a rocket engine, the high temperature, high flow speed, and high pressure environment in the swirl chamber, set up many barriers for the application. The geometric constraints of the swirl chamber prevent observers from obtaining sufficient Moire deflection patterns at different directions.

Furthermore, in the swirl chamber, there is no surrounding area which can be used as a reference point to determine temperature quantitatively.

In this work, we developed a new method to measure the transient temperature distribution using Moire deflectometry in a swirl chamber of an IDI engine. The results of the measurement are validated by a hot-wire anemometry. Satisfactory results have been achieved. Possible improvements for design of the fuel evaporation and mixing processes in combustion chambers of IDI engines are suggested.

3   Principle

According to the Fermat principle, when a beam of light passes
through a flow field, the change in the
index of fluid refraction will produce a
deflection effect on the light beam.
The deflection effect is exactly em-

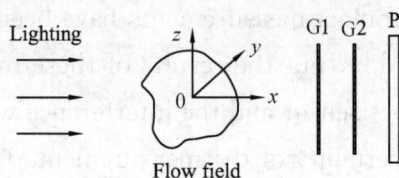

Fig. 1   Schematic of a simple Moire deflectometry

ployed by Moire deflectometry to measure the refractive index. Fig. 1 is a schematic of a simple Moire deflectometry. A sheet of light passes through an objective flow field, enters a pair of gratings G1 and G2, which are tilted with an angle $+\theta/2$ and $-\theta/2$ relative to the z axis, and finally reaches a receiving screen P where overlapped Moire fringes can be detected. As shown in Fig. 2, if the refractive index of the flow medium is uniform, a straight and unperturbed Moire pattern is obtained. In cases where changes of temperature and density distribution in the field occur, that is, the presence of a phase object with a gradient of refractive index, Moire fringes are then distorted on the receiving screen. The deflection can be determined carefully from the shift of the fringes.

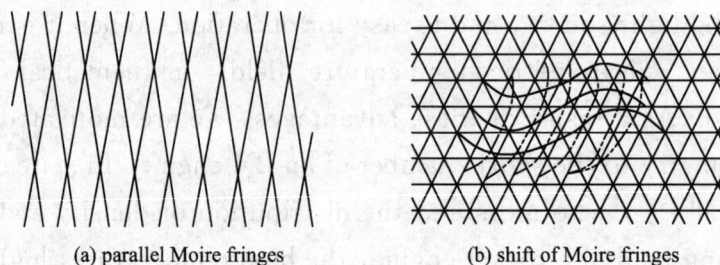

(a) parallel Moire fringes   (b) shift of Moire fringes

Fig. 2   Patterns of Moire fringes

If the objective flow field is axisymmetric, the deflection of the light can be easily related to the refractive index, which is a function of the location r in the cylindrical co-ordinate system shown in Fig. 3, by the following equation:

$$h(y) = \frac{1}{\lambda_0} \int_0^l [n(r) - n_0] \mathrm{d}x \qquad (1)$$

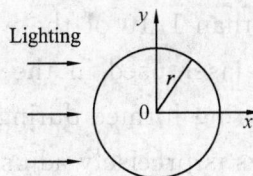

Lighting

Fig. 3   A sheet of light passing through an axisymmetry flow field

where l is the length of the objective flow field along the transmitting direction of the light, λ_0 is the wavelength of the light, $n(r)$ is the refractive index, and n_0 is the reference index.

Applying an Abel transform to the above equation, we obtain:

$$n(r) - n_0 = \frac{\lambda_0}{\pi} \int_r^{r_0} \frac{\mathrm{d}h(y)/\mathrm{d}y}{\sqrt{y^2 - r^2}} \mathrm{d}y \qquad (2)$$

From Eq. (2), the distribution of the refractive index can be determined once a reference index of medium refraction is chosen. For a laser light with a wavelength of 632.8 nm, assuming the medium in the measured flow field is an ideal gas, the relationship between the local refractive index and the temperature can be expressed precisely as[8]:

$$n - 1 = \frac{0.292\,015 \times 10^{-3}}{1 + 0.368\,184 \times 10^{-2} t} \qquad (3)$$

where t is the local temperature(℃).

Compared with shearing interferometry, Moire deflectometry provides an exact measure of the deflection angle and is much more flexible in terms of sensitivity. These are of immediate concerns of measurement in studying the combustion process of IC engines and related areas.

4   Temperature measurement

The temperature measurements using Moire deflectometry were carried out in the swirl chamber of an IDI testing engine. The main specification of the testing engine is listed in Tab. 1. The shape of the swirl chamber looks like a bell, as shown in Fig. 4. Circular viewing windows with a diameter of 30 mm were opened on both sides of the swirl chamber. The size of the window covers over 80% of the cross-sectional area to provide a comprehensive observation and measurement in the swirl chamber. The dimension of the window is carefully designed so that the opening of the window would not alter the original volume of the swirl chamber, so as to keep the compression ratio unchanged as the prototype engine. The viewing window is made of quartz, and the roughness of the surface is kept at less than 1/10 of the wavelength of the laser light, which is 632.8 nm in the He-Ne laser used in the experiments, in order to eliminate the Newton interference ring formed during measurement. The included angle between two quartz windows is precisely adjusted within a range of $10\sim20°$ to dispel the effect of the quartz background stripe on Moire fringes.

Tab. 1　Main specification of the testing engine

Type of engine	S195
Bore×stroke	95 mm×110 mm
Volume of swirl chamber	25.7 cm³
Displacement	850 cm³
Compression ratio	19.2

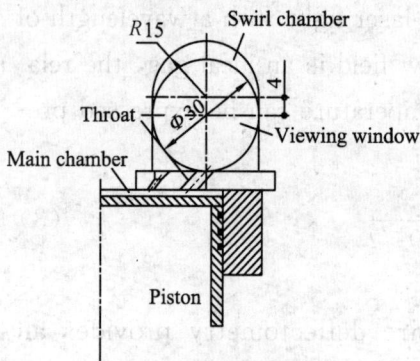

Fig. 4　Combustion chambers of the testing IDI engine

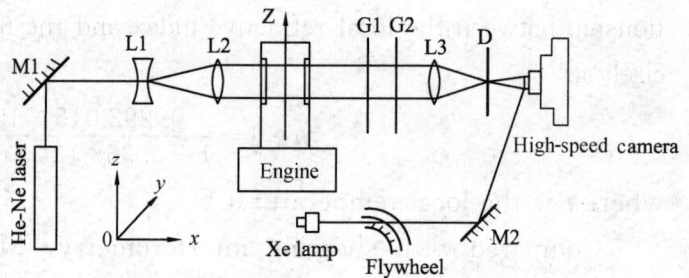

Fig. 5　Schematic of the experimental set-up

Fig. 5 shows the Moire deflectometry system. A 50 mW He-Ne laser is used as the lighting source. The laser beam passes mirror M1 and lenses L1 and L2 to form a light sheet shining through the two quartz windows on the swirl chamber of the testing engine. When the light sheet passes through a pair of Ronchi gratings (20 lines/mm, or grating constant $d=0.05$), which are tilted with an angle θ, Moire fringes are formed on the lens L3. The focusing lens L3 transmits the Moire deflection strips in $+1$(or -1) grade to the negative of a high-speed camera, after passing through a flitting hole D. The corresponding crank angles are recorded simultaneously by the high-speed camera.

Based on the principle of the Moire deflectometry and the coordinate system shown in Fig. 5, the deflection angle is related to the refractive index by:

$$\alpha = \frac{1}{n_0}\int_0^{x_0} \frac{\partial n}{\partial y}\mathrm{d}x \tag{4}$$

The deflection angle may also be expressed as:

220

$$\alpha = \frac{h\theta}{\Delta} \tag{5}$$

where Δ is the distance between the two gratings and h is fringe deviation.

Since the bell-shape swirl chamber has an axisymmetric structure. It is reasonable to assume that the flow field is also approximately in axisymmetry. Thus, Eq. (4) becomes:

$$\alpha(y,z) = 2y \int_y^{r_0} \frac{\partial n(r,z)}{\partial r} \frac{1}{\sqrt{r^2 - y^2}} dr \tag{6}$$

Applying Abel transformation and considering Eq. (5), Eq. (6) becomes:

$$n(y,z) = n_0 - \frac{\theta}{2\pi\Delta} \int_y^{r_0} \frac{h(r,z)}{\sqrt{r^2 - y^2}} dr \tag{7}$$

Provided that the reference index of refraction n_0 and the fringe deviation $h(r,z)$ are known, the refractive index can be easily obtained by Eq. (7). As a result, temperature field can be mapped by Eq. (3).

The reference index of refraction is usually determined under atmospheric conditions such as Lian and Trolinger[9,10], who measured the exhaust jet field of a rocket engine. However, such atmospheric reference field does not exist in the swirl chamber of IDI engines. In a confined combustion chamber, according to out experience, the temperature distribution in the turbulence viscous sub-layer ($y^+ < 5$) at the near wall region is relatively stable and uniform. Therefore, the temperature at the turbulence viscous sub-layer is taken as the reference temperature and is measured simultaneously by a surface thermocouple attached to the wall surface.

The experiments were carried out in pure compression condition while the engine speed was set at 240 r/min. The engine throttle was cut off to keep the quartz window free of injection fuel and improve the quality of the Moire fringe recording. The evolution of the Moire fringe during the entire compression stroke was recorded by the high-speed camera at a speed of 1 150 frames per second.

5　Results and discussion

Fig. 6 shows a series of Moire fringe photos at four representative crank angles during a compression stroke. The deviations of Moire stripes were read by overlapping two Moire fringe negatives. One negative of Moire fringes, which was taken before operating the testing engine, was overlapped on the objective negative of a distorted Moire deflection picture. The deviations on the overlapped pictures were then precisely determined by a special reading instrument. The data of deviation were fit to a polynomial of order 3 using the least-square algorithm. The distribution of the refractive index in the swirl chamber was obtained by integrating Eq. (7). Finally, the temperature mapping were obtained from Eq. (3), as shown in Fig. 7.

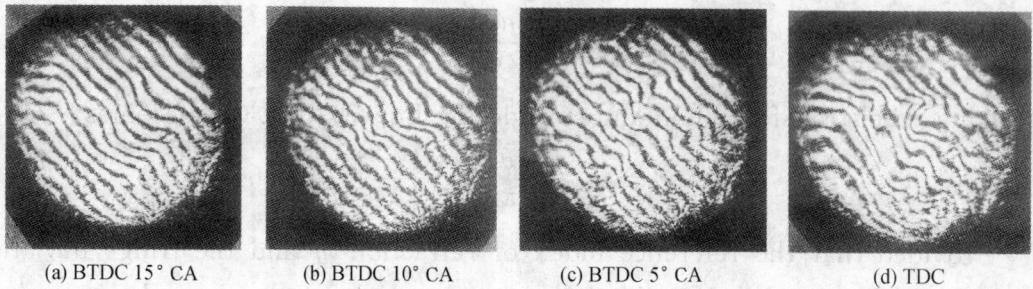

(a) BTDC 15° CA   (b) BTDC 10° CA   (c) BTDC 5° CA   (d) TDC

Fig. 6   Moire fringe photos at different crank angles

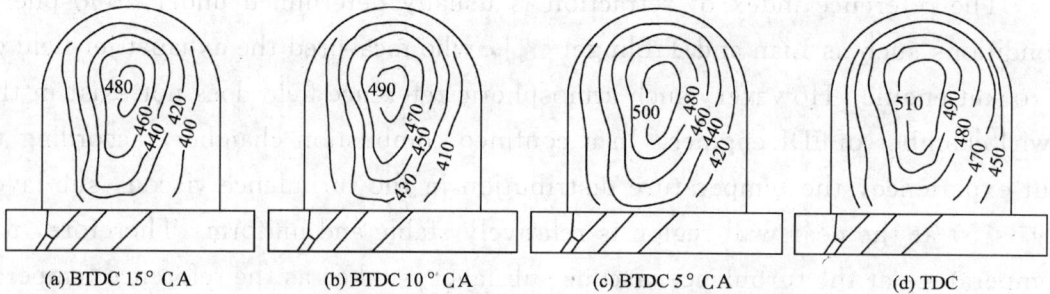

(a) BTDC 15° CA   (b) BTDC 10° CA   (c) BTDC 5° CA   (d) TDC

Fig. 7   Temperature distribution in the swirl chamber

It is found that the gas temperatures at the central region of the swirl chamber are higher than those at the near-wall region during a compression stroke, as a result of the cooling effect of the walls. With the progress of the compression stroke, the total temperature of the swirl chamber rises and the high temperature area at the up-central region gradually shifts down. There is a significant temperature difference between the left and right sides of the swirl chamber. However, the difference is getting smaller as the piston is close to TDC.

The above results imply that at the end of the intake stroke, cool and fresh air is not able to enter the swirl chamber because of structural constraints. Before a compression stroke takes place, the swirl chamber is filled with exhaust gases left over from the last cycle, and is continuously heated by the high temperature walls of the chamber. This results in the significant temperature difference between the main and swirl chambers at the early and middle stage of compression stroke. Cool air in the main chamber is initially ejected into the upright region of the swirl chamber through the inclined throat connecting the main combustion chamber and swirl chamber, as indicated in Fig. 4. Consequently, the temperature at the right side of the swirl chamber tends to be lower than that at the left. Another effect of this jet flow is that, when it impinges on the top of the chamber wall, it creates a non-uniform pressure field in the swirl chamber, that is, pressure is higher at the upper central region and is lower at the lower central region. However, the strength of the jet flow retards as the piston moves close to TDC, because the pressure difference between the main and swirl chamber decreases. This explains the reason why the high temperature region shifts down during the compression stroke.

From Fig. 7, it is obvious that the high temperature region is located at the central area of the swirl chamber at the late stage of the compression stroke. Therefore, the angle of engine injection can be optimized to allow most of the injected fuel to pass through the central region. The effect might be twofold. Promoting the evaporation and mixing of the fuel is definitely desirable. Injecting part of the fuel directly into the main combustion chamber after passing through the high temperature central region is again very important, since the combustion can be followed through the heat release model of DI engines.

In order to validate the temperature measured by Moire deflectometry, five representative temperature points were selected and measured by a hot-wire anemometry. The positions measured by the hot-wire anemometry are indicated in Fig. 8. The results are compared with the temperature mapped from Moire fringes at the five different locations over a complete compression stroke, as shown in Fig. 9. The results show that the measured temperatures by

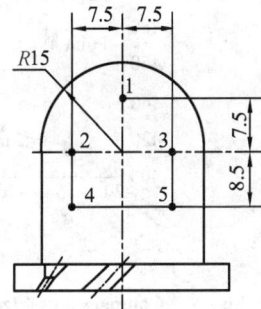

Fig. 8 **Measurement points for hot-wire anemometry**

hot-wire anemometry at all five points are slightly lower than that by Moire deflectrometry. The discrepancies increase slightly as the temperature is getting higher. Considering the effects of the thermal inertia of the hot-wire probe and the radiative heat exchange between the probe and the surroundings, the temperature distribution obtained by Moire deflectometry are remarkable.

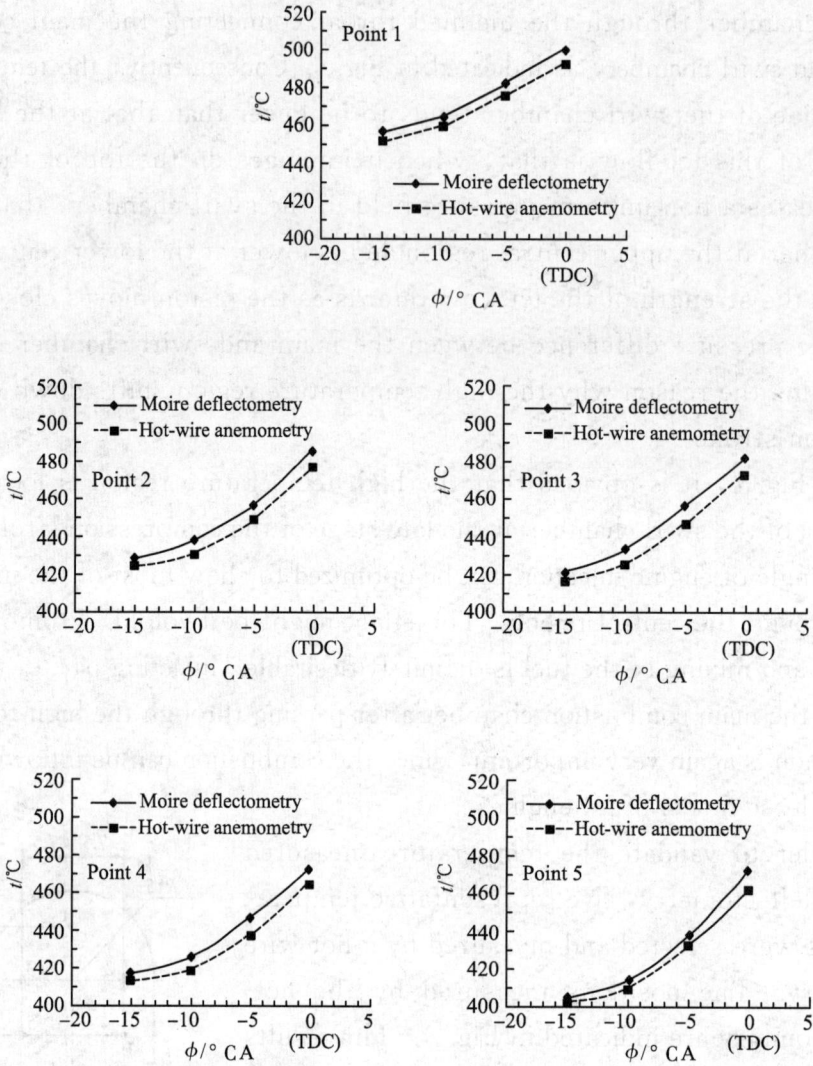

Fig. 9 Comparison of temperatures measured by Moire deflectometry and hot-wire anemometry

6 Conclusions

Moire deflectometry has been successfully used to measure the transient temperature distribution in the swirl chamber of an IDI engine. The technique has great promise for the temperature measurement in confined combustion chamber of IC engines for its advantages of field measurement, fast response and non-disruption. The method to determine the temperature distribution from the detected Moire fringes is established. It is suggested that the transient wall surface temperature of the swirl chamber can be used as the refe-rence point in the Moire fringe mapping. Measured temperatures in the swirl chamber are validated using hot-wire anemometry at the different positions, and the results are satisfactory.

The temperatures measured in the experiment provide useful information for the analysis of the fuel evaporation and mixing process in the swirl chamber during compression stroke. The engine performance could be improved by optimizing the design of the combustion system.

Acknowledgements

We are grateful to the National Science Foundation of China for their support in this research work, and to Professor Yan Dapeng and Professor He Anzi of Nanjing University of Science & Technology for their helpful instruction.

References

[1] Li D T. *Study and Design on the Swirl Chamber of IDI Engine*[M]. Beijing: Machinery Industry Press, 1986: 127. (in Chinese)

[2] Vest C M. Interferometry of strong refracting axisymmetric phase object[J]. *Appl Opt*, 1975, 14(7): 1601-1607.

[3] Kafrif O. Moire deflectometry: a ray deflection approach to optical testing[J]. *Optics Letters*, 1980, 5(12): 444-464.

[4] Keren E, et al. Measurements of temperature distribution of flame by Moire deflectometry[J]. *Appl Opt*, 1981, 20(12): 189-204.

[5] Stricker J. Analysis of 3-D phase object by Moire deflectometry[J]. *Appl Opt*, 1984, 23: 3657-3659.

[6] Almanasreh A, Abushagur M A G. Moire deflectometry in aero-optics analysis[C]// *Proceedings of the IEEE Southeastcon'92*, Birmingham, USA, 1992: 570-572.

[7] Toker G, Levin D, Stricker J. Experimental investigation of supersonic/hypersonic flow fields based on a new approach using holographic optical techniques[C]// *Proceeding of the 16th IEEE International Congress on Instrumentation in Aerospace Simulation Facilities*, Dayton, USA, 1995.

[8] Bar Ziv E, Sgulim S, Kafri O, Keren E. Temperature mapping in flames by morie deflectometry[J]. *Appl Opt*, 1983, 22(5): 543-549.

[9] Lian W Y, et al. The structure and internal properties of underexpanded exhaust jets [C]// *International Symposium on Refined Flow Modeling and Turbulence Measurements*, Iowa, USA, 1985: 53-57.

[10] Trolinger J D. Holography for aerodynamics[J]. *Astronautics Aeronautics*，1982,10(8)：56-61.

（From：*Applied Thermal Engineering*，1999,19)

应用莫尔偏折法测量 IDI 柴油机涡流室内的温度

[**摘要**] 柴油机的着火和燃烧过程受压缩冲程形成的温度分布的影响，作者采用激光莫尔偏折技术，首次成功地测量出 IDI 柴油机压缩过程涡流室内的温度场；还通过对比莫尔偏折法与热线风速仪的测量结果，分析了前者的测量精度，根据测得的温度场，讨论了涡流室由压缩冲程中的蒸发和混合过程。该项研究为内燃机燃烧室内温度场的测试，开发了一种新方法。

Numerical simulation of oil droplet breakup and spray impingement in a swirl chamber diesel engine

Li Detao, *Xia Xinglan*, *Xue Hong*

Abstract

The numerical simulation of oil droplet breakup and spray impingement is carried out in a combustion chamber of an IDI engine. The process of oil droplet breakup in the combustion chamber is divided into Kelvin-Helmhotz(K-H) instability breakup and Rayleigh-Taylor(R-T) instability breakup. Based on Weber number, the process of spray impingement is divided into three independent models: stick, rebound/stick, splash/tangential jet. The mathematical models of the oil droplet breakup and the spray impingement are established. The spray penetration and SMD calculated from the oil droplet breakup model agree with experimental data. The proposed spray impingement model predicts spray impingement patterns which agree with the results shown by high-speed photography. Furthermore, the models were applied to the real spray process in a swirl chamber diesel engine.

1 Introduction

The mixing process of fuel and air plays an important role on the combustion in a diesel engine. Oil droplet breakup and spray impingement are two key processes affecting the mixing. Most of the previous studies on these two processes were carried out in an enclosure with constant volume. The investigation of the impact of the two processes on a combustion chamber in a real diesel engine is rare especially in a swirl chamber diesel engine.

For the oil droplet breakup process, O'Rouke and Amsden[1] proposed a Taylor analogy breakup model(TAB). The model is based on an analogy between the vibrationally deformed droplet and a spring mass system. The governing equation is established based on an analogy between the air dynamic pressure, droplet surface tension and viscous stress in an oil droplet and the external force, recovery force and damping force in a spring-mass system. Reitz[2] applied Kelvin-Helmhotz(K-H) instability theory to the spray surface to describe the breakup process of the oil droplet. However, the K-H instability theory is only suitable for the surface breakup process of a jet, and is difficult to extend to the oil droplet which is directly exposed to the air. The description of the oil breakup process is then not complete.

In a study of spray impingement, Naber and Reitz[3] proposed a stick, rebound and tangential jet model to replace the simple stick model in the KIVA-II Code. But the model did not consider the energy loss of the impinged droplet. Tabata[4] proposed an impingement model with rebound and breakup processes. The results do not agree with experimental data since the effects of stick and tangential jet are not considered in his model.

In the present study, according to instability theory, the oil droplet breakup process into K-H and Rayleigh-Taylor(R-T) instability breakup, and spray impingement is divided into three overlapped processes: stick, rebound/stick, and splash/tangential jet. The energy loss of impinging droplets is also considered. The proposed model is validated through experiments carried out in a specially designed device. Furthermore the model is applied to a swirl chamber diesel engine to calculate the oil droplet spray development process. The results show reasonable agreement with high-speed photographs taken from a real working engine.

2　Mathematical model

2.1　Oil droplet breakup model

In a swirl chamber of a diesel engine, the spray is shown in Fig. 1 after the oil is injected from a high-pressure nozzle. The spray can be divided into two regions: a core region and a mixing region. There is no air involved in the core region. Based on experimental observations, the length of the core region L_b can be estimated by

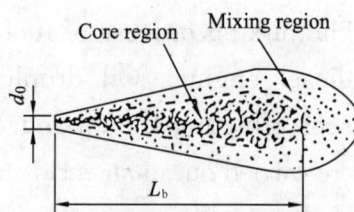

Fig. 1　Fuel spray in a diesel engine

$$L_b = 15.8 d_0 (\rho_l / \rho_g)^{0.5} \tag{1}$$

where d_0 is the diameter of nozzle and ρ_l, ρ_g, are the density of oil and surrounding air.

At the surface of the oil, when a perturbation $\eta = \eta_0 e^{ikZ + \omega t}$ is introduced, where η_0 is the initial amplitude, k is wave number, ω is the frequency of the wave, and the wave length is $\lambda = 2\pi k$. With increasing time, due to the breakup of the surface wave, droplets appear at the boundary area of the core region. Since the K-H instability is known to be suitable for the surface breakup of the jet, the oil breakup at the core region is described by the K-H instability. When the wavelength reaches most unstable wavelength Λ_{KH}, the droplet starts breaking up. The time of the breakup is[2]

$$\tau_{bKH} = (B_1 a / U_r) \sqrt{\rho_1 / \rho_g} \tag{2}$$

where a is the radius of the droplet, U_r is the relative speed between oil and air, and $B_1 = 10.00$ is a model constant. The unstable wavelength Λ_{KH} and the corresponding velocity Ω_{KH} can be obtained by using linear stability analysis on the jet[2].

The newly separated droplet from the core region due to K-H instability is usually located on the surface of the core region, and its radius r is equal to

$$r = B_0 \Lambda_{KH} \tag{3}$$

where the constant $B_0 = 0.61$.

If the breakup takes place at the mid-point of the amplitude, the volume of the droplet separated from the jet surface is

$$V_{2KH} = \frac{1}{6} \pi^2 La \eta_{bKH} \tag{4}$$

where η_{bKH} is the amplitude at breakup, La is the Laplace number, $La = \dfrac{\rho_1 d_1 \sigma}{\mu_1^2}$. Here σ is the surface tension of the oil droplet, μ_1 is the dynamic viscosity of the oil, and d_1 is the diameter of the droplet.

In the mixing region, the oil droplet is explicitly exposed to air, and R-T instability theory is used to describe the breakup process of the oil droplet. The breakup time is given by[7,8]

$$\tau_{bRT} = 1.66 Bo^{-0.25} \frac{1}{\Lambda_{RT}} \tag{5}$$

The droplet deformtion observed in a shock wave experiment shows that after the droplet reaches its maximum deformation, the dimension of the droplet remain constant for a period of time. The time period is about 4 to 7 times τ_{bRT}[9]. If 5 is used, the R-T instability breakup time is

$$\tau_{bRT} = 8.83 Bo^{-0.25} \frac{1}{\Lambda_{RT}} \tag{6}$$

where Bo is Bond number, $Bo = ga^2 \rho_l / \sigma$, and Λ_{RT} is the most unstable frequency of the wave. It is defined as

$$\Lambda_{RT}^2 = \frac{2}{3} \frac{1}{\sqrt{3\sigma}} \frac{[(g_0 + g)(\rho_l - \rho_g)]^{1.5}}{\rho_l + \rho_g} \tag{7}$$

where g_0 is the acceleration due to gravity, and g is the vertical acceleration, which can be obtained from the drag coefficient C_D

$$g = \frac{3}{8} C_D \frac{\rho_g U_r^2}{\rho_l a} \tag{8}$$

For a spherical droplet, the drag coefficient $C_{D, sphere}$ can be calculated based on the droplet Reynolds number Re_d

$$C_{D, sphere} = \begin{cases} \dfrac{24}{Re_d} \left(1 + \dfrac{1}{6} Re_d^{2/3}\right), & Re_d \leqslant 1\,000 \\ 0.424, & Re_d > 1\,000 \end{cases} \tag{9}$$

where $Re_d = 2\rho_g U_r a / \mu_g$.

A real oil droplet may not be spherically symmetric therefore a deformation parameter y is introduced to modify the drag coefficient. Based on the spring-mass analogy system in the TAB model[1,11], the modified drag coefficient is expressed as

$$C_D = C_{D, sphere} (1 + 2.632 y) \tag{10}$$

where y is in an range between 0 and 1.

When R-T breakup takes place, to simplify the calculation, an oil droplet is assumed to be randomly broken down into two smaller oil droplets with same speed but opposite directions (momentum conservation). The velocity component u_i after breakup can be calculated according to the initial velocity component u_{oi} before the breakup

$$u_i = u_{oi} \exp\left[\frac{-3 C_D U_r^2 t \rho_g / \rho_l}{16a}\right] \quad (i = x, y, z) \tag{11}$$

where t is the time period when the oil droplet exists in the flow field, and is of the same order as the breakup time.

2.2 Spray impingement model

Spray impingement is a very complicated process. It is accompanied by various phenomena such as stick, rebound, splash and breakup. Besides these, the approaching velocity of the oil droplet and the wall characteristics will also affect the impingement. According to experimental observation[5], we divide the spray impingement process into three modes: stick, rebound/stick, and splash/tangential jet. The Weber number of the oil droplet is used to classify the status of the oil droplet.

The Weber number is defined as

$$We = \frac{\rho U^2 d_I}{\sigma} \qquad (12)$$

where U is the normal velocity and d_I is the diameter of the approaching oil droplet.

The Weber number criteria are:

(1) If $We \leqslant 5$, then the spray impingement is in stick mode. The oil droplets are stuck on the wall after impingement.

(2) If $5 < We \leqslant 80$, the spray impingement is in rebound/stick mode. Part of the oil droplets are rebounded, while the rest are stuck on the wall. If the ratio of the rebounded oil droplets volume to impinged oil droplets volume is $V_R/V_I = 0.50$, considering the dynamic loss of the impinged oil droplets, the ratio of the rebounded oil droplet velocity to impinged oil droplet velocity becomes

$$\frac{U_R}{U_I} = \varepsilon \qquad (13)$$

where ε is the energy loss coefficient, $\varepsilon = [1 - 0.95\sin^2(\theta_I)]^{0.5\,[12]}$. When the oil droplet rebounds, the angle of impingement equals the angle of rebound, and the size and the quantity of the oil droplets are unchanged.

(3) If $We > 80$, the spray impingement is in splash/tangential jet mode. If we assume that the ratio of the splashed oil droplet volume to impinged oil droplet volume is $V_R/V_I = 0.50$, and that the ratio of the splashed oil droplet radius to impinged oil droplet radius $a_R/a_I = f$ (f is a random number between 0 and 1), then the velocity can still be calculated by Equation(13).

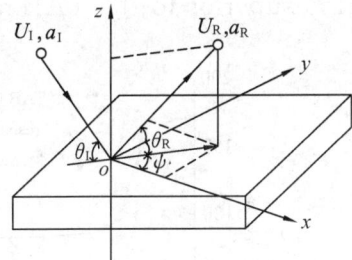

Fig. 2 Spray impingement on a wall

Based on the random number f, the rebound angle and azimuthal angle(as shown in Fig. 2) can be obtained from the following equations

$$\theta_R = \pi \exp(-cf) \qquad (14)$$

$$\Psi = 2\pi f \qquad (15)$$

where $c = 1.0$ is a constant. The three velocity components of the tangential jet are

$$u_{Jx} = u_{Ix}\varepsilon \tag{16}$$

$$u_{Jy} = u_{Iy}\varepsilon \tag{17}$$

$$u_{Jz} = 0 \tag{18}$$

3 Model validation

The above sub-models for oil droplet breakup and spray impingement are added to the KIVA-II code[10], and computational results are compared with the experimental data. In static air, a comparison of the calculated spray penetration length S and experimental data[13] is made in Fig. 3, where lines A, B, C correspond to different environmental pressures of 1.0 MPa, 3.0 MPa, and 5.0 MPa respectively. Fig. 4 shows the calculated value of SMD and experi-

Fig. 3 Calculated and measured spray penetration length

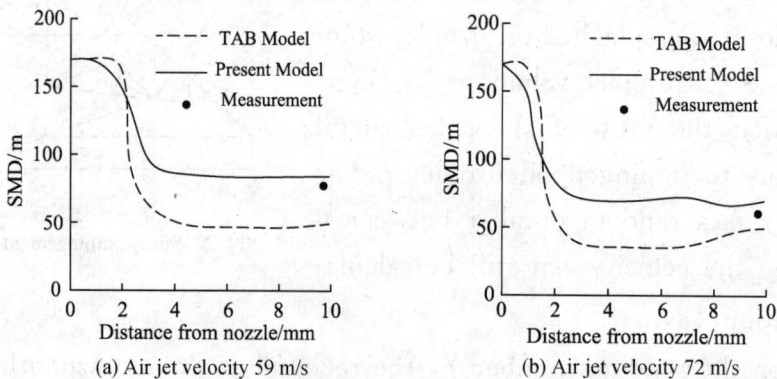

mental data, in which 4a and 4b correspond the air jet velocity of 59 m/s and 72 m/s respectively. The result obtained using TAB model is also shown in the figures. It can be seen that the reasonable agreement is achieved between the numerical simulation and experimental data. Especially for the value of SMD, our model is superior to the TAB model when compared with the experiment.

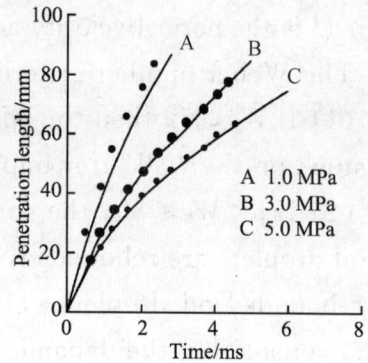

(a) Air jet velocity 59 m/s   **(b) Air jet velocity 72 m/s**

Fig. 4 Comparison of SMD

Fig. 5 shows results predicted by the spray impingement model and measurements obtained from high-speed photography[14]. The pattern of oil droplet distribution between them is very similar.

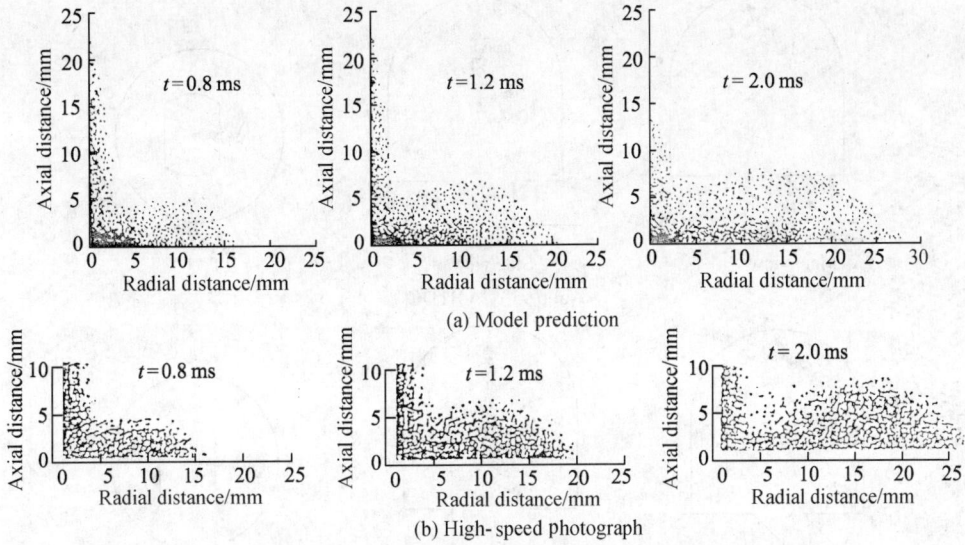

(a) Model prediction

(b) High-speed photograph

Fig. 5 Spray impingement development

4 Application to a swirl chamber diesel engine

The validated numerical model is applied to the real spray process in a swirl chamber of a diesel engine[15]. Under the normal working condition (8. 84 kW, 2 000 r/min), the fuel injection starts at 10° CA BTDC and the injection lasts 16° CA; the fuel consumption per cycle is 0. 037 4 g. To concentrate on the injection process, the combustion is not considered at the present stage.

Fig. 6 shows oil droplet distribution in three different views. After the fuel is injected from the nozzle to the swirl chamber, due to strong swirl existing in the chamber, the locus of the fuel will be immediately bent to follow the air movement. At the early stage, the front edge of the spray is close to the side wall of the chamber, and the strong momentum of the air flow deforms the spray right after it appears. The spray approaches the bottom of the chamber at 4. 76° CA BTDC. After the impingement, part of the oil droplets are stuck on the wall to form a thin oil film, while the rest are rebounded into the chamber. The calculated results agree with our previous high-speed photography taken under the same conditions.

Most of the oil droplets rebounded from the wall are evaporated or go into the main chamber through the swirl chamber air passage. Only small portion of oil droplets are continuously moved along the right wall of the chamber toward the top area.

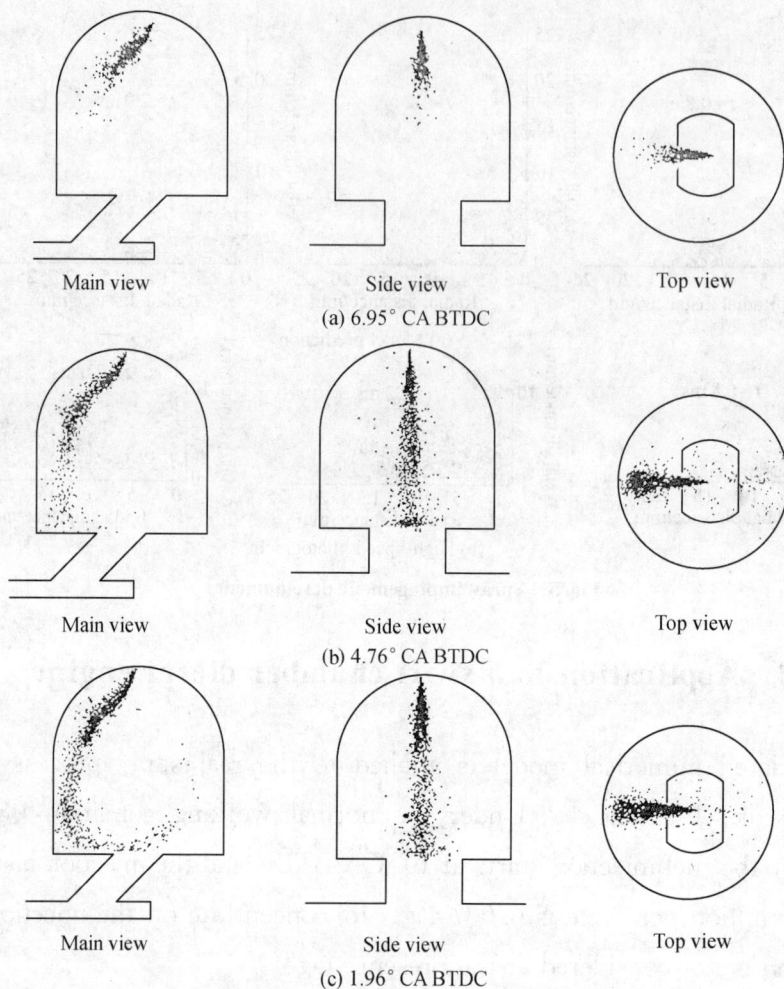

Main view   Side view   Top view

(a) 6.95° CA BTDC

Main view   Side view   Top view

(b) 4.76° CA BTDC

Main view   Side view   Top view

(c) 1.96° CA BTDC

Fig. 6 Oil droplet distribution at different crank angles

5 Conclusions

The oil droplet breakup and spray impingement are two very important but complicated processes happening during fuel injection in a diesel engine. Mathematical models are developed to predict the processes. The predicted results show reasonable agreement with experimental data. The application of the models to a practical swirl chamber diesel engine also results in a qualitative agreement with our previous high-speed photographs taken in an experiment. The following conclusions may be drawn.

（1）At the initial stage of injection，the oil spray is significantly deformed due to the strong air movement in the swirl chamber.

（2）The oil droplet impinged the bottom of the chamber at 4. 76° CA BTDC.

After the impingement, most of the oil droplets are rebounded back to the swirl chamber while a small portion of the oil droplets are stuck on the wall to form a tangential jet.

(3) Most of the oil droplets rebounded from the wall will disappear slight before coming across the air passage of the swirl chamber due to evaporation. Only a small portion of the oil droplets will continue to move along the right wall of the chamber and reach the top or directly go into the main combustion chamber.

References

[1] O Rourke P J, Amsden A A. The TAB method for numerical calculation of spray droplet breakup[J]. *SAE*, 872089,1987.

[2] Reitz R D. Modeling atomization processes in high pressure vaporizing sprays[J]. *Atomization and Spray Tech*, 1987(3): 309–337.

[3] Reitz R D. Modeling engine spray/wall impingement[J]. *SAE*, 880107,1988.

[4] Tabata T, Ishii Y, et al. Numerical calculation of spray mixing process in a DI diesel engine and comparison with experiments[J]. *SAE*, 950853,1995.

[5] Werlberger P, Cartellieri P. Fuel injection and combustion phenomena in a high speed DI diesel engine observed by means of endoscopic high speed photography[J]. *SAE*, 870079,1987.

[6] Hiroyasu H. Diesel engine combustion and its modeling[J]. *COMODIA*'85,1985: 53–75.

[7] Reyleigh L. On the instability of jets[C]// *Proc London*, *Math Soc*, 1979, 10(4).

[8] Hoyt J W, Taylor J J. Waves on water jets[J]. *J Fluid Mech*, 1977,83: 119–125.

[9] Simpkins L, Bales K. Water-drop response to sudden acceleration[J]. *J Fluid Mech*, 1972,55.

[10] Amsden A A, O Rouke P J, Butler T D. KIVA-Ⅱ: a computer program for chemically reactive flows with sprays[R]. Los Alamos National Laboratory Report, No LA-560-MS, 1989.

[11] Liu A B, et al. Modeling the effects of drop spray and breakup on fuel spray[J]. *SAE*, 930072,1993.

[12] Jayarantne D W, Mason B J. The coalescence and bounding of water drops at an air/water interface[C]// *Proc R Soc*, *Lond-A*,1964, 280: 545–565.

[13] Hiroyasu H, Kadeta T. Fuel droplet size distribution in diesel combustion chamber[J]. *SAE*, 740715,1974.

[14] Katsura N, Saito M, et al. Charateristics of a diesel spray impinging on a flat wall[J]. *SAE*, 890264,1989.

[15] Li D T, Zhu G S. Study on the injection and combustion process in a swirl chamber

diesel engine[J]. *Transaction of CSICE* (in Chinese)，1989 7(1)：21-26.

(From：*Proceeding of the Second Asia-Pacific Conference on Combustion*，Taiwan，1999)

涡流室式柴油机空油滴破碎和喷雾碰壁的三维数值模拟

[摘要]　根据不稳定性理论，将燃烧室中油滴破碎过程分为 K-H 不稳定破碎和 R-T 不稳定破碎；根据 Weber 数将喷雾碰壁过程分为粘附、反弹粘附和飞溅附壁射流 3 种模式，建立了油滴破碎和喷雾碰壁的数学模型。根据油滴破碎模型计算所得的喷雾贯穿距离和 SMD 值同试验结果相一致，喷雾碰壁模型计算所得的油滴位置图形与高速摄影结果一致。将已被试验验证后的模型应用于涡流室式柴油机中，计算了涡流室内喷雾的发展过程，计算结果与高速摄影结果吻合良好。

涡流室式柴油机燃油蒸发过程的三维数值模拟

夏兴兰,董 刚,李德桃

[摘要] 建立了高温高压条件下的油滴蒸发模型,并在不同的工质压力和温度下对模型进行了验证,模型计算结果和实验结果吻合良好。将此油滴蒸发模型和油膜蒸发模型相结合,应用到涡流室式柴油机的燃油蒸发过程中,通过三维数值模拟计算,研究了燃油蒸发过程中燃料蒸汽浓度和工质温度场的变化历程以及燃油蒸发对涡流室平均温度和平均压力的影响。

燃料蒸发对燃烧过程有着非常重要的影响。在柴油机中,液体燃料被高压喷射到燃烧室后,形成尺寸大小不等的大量液滴悬浮于空气中;同时,也有一部分燃料喷射到燃烧室壁面上,形成油膜。因此,柴油机燃烧室内燃料的蒸发包括油滴蒸发和油膜蒸发两种情况。一些学者对常温常压下液滴的蒸发规律进行了理论和实验研究[1],提出液滴直径 D 的计算式为:$D^2 = D_0^2 - Kt$(D_0 为初始直径,K 为蒸发常数)。在柴油机燃烧室中,由于高温高压环境的影响,其油滴蒸发规律不同于常温常压下。但许多研究者仍使用以上液滴直径 D^2 变化规律的计算式,这显然是不合适的。

对直喷式柴油机燃料蒸发过程的三维数值模拟,已有不少研究[2,3]。对涡流室式柴油机,由于其燃烧室结构复杂,三维模拟计算极其困难,人们所进行的研究较少。本文根据柴油机中高温高压条件,建立了油滴蒸发和油膜蒸发模型。将模型应用于涡流室式柴油机中,对其燃料蒸发过程进行三维数值模拟。

1 数学模型

1.1 油滴蒸发模型

假定油滴是球对称的,油滴内部温度均匀,则根据油滴与周围工质之间的能量平衡可得油滴温度 T_d 随时间 t 的变化率为

$$\frac{dT_d}{dt} = \frac{6}{\pi \rho_d D_d^3 C_{pd}} \left[Q_d + L(T_d) \frac{dm_d}{dt} \right] \tag{1}$$

式中 D_d,ρ_d,C_{pd} 分别为油滴的直径、密度和定压比热;$L(T_d)$ 为在温度 T_d 下的汽化潜热;Q_d 为向油滴的传热率;m_d 为油滴质量。

向油滴的传热率 Q_d 可表示为

$$Q_d = D_d^2 h^* (T - T_d) \tag{2}$$

式中 T 为工质温度;h^* 是传热系数,它受传质速率的影响,可表示为[4]

$$h^* = \frac{-dm_d/dt(C_{pf} + \beta C_{pa})}{\pi D_d^2 \left\{ \exp\left[\frac{-dm_d/dt(C_{pf} + \beta C_{pa})}{\pi D_d \lambda} \cdot \frac{1}{Nu} \right] - 1 \right\}} \tag{3}$$

式中 C_{pf}，C_{pa} 分别为燃油蒸汽和空气的定压比热；λ 为燃油的导热系数；β 为油滴表面上空气和燃料蒸汽的流量比；Nu 为 Nusselt 数，其计算公式[5]

$$Nu = (2.0 + 0.6Re_d^{1/2} Pr_d^{1/3}) \frac{\ln(1+B_d)}{B_d} \tag{4}$$

式中 Re_d，Pr_d，B_d 分别为油滴的 Reynolds 数、Prandtl 数、Spalding 传质数。

根据质量守恒，可得油滴直径 D_d 的变化率为

$$\frac{dD_d}{dt} = \frac{2}{\pi D_d^2 \rho_d} \left(\frac{dm_d}{dt} - \frac{\pi D_d^3}{6} \cdot \frac{d\rho_d}{dT_d} \cdot \frac{dT_d}{dt} \right) \tag{5}$$

在油滴表面上，燃料的质量传递速率为[4,5]

$$\frac{dm_d}{dt} = -\pi D_d^2 k^* \frac{Y_{f0}}{1-(1+\beta)Y_{f0}} \tag{6}$$

$$Y_{f0} = \frac{W_1}{W_1 + W_0 \left[\dfrac{p_0}{p_v(T_d)} - 1 \right]} \tag{7}$$

式中 Y_{f0} 为油滴表面燃料蒸汽的质量分数；W_1 为燃料蒸汽的分子量；W_0 为除燃料蒸汽外，所有其他物质的局部平均分子量；p_0 为工质总压力；$p_v(T_d)$ 为在温度 T_d 下燃料蒸汽的分压力。

传质系数 k^* 为

$$k^* = \frac{\rho_{air} D}{D_d} \cdot \frac{1-(1+\beta)Y_{f0}}{(1+\beta)Y_{f0}} Sh \cdot \ln \frac{1}{1-(1+\beta)Y_{f0}} \tag{8}$$

式中 ρ_{air} 为周围空气的密度；D 为燃料蒸汽在空气中的扩散系数；Sh 为 Shrwood 数[5]：

$$Sh = (2.0 + 0.6Re_d^{1/2} Sc_d^{1/3}) \frac{\ln(1+B_d)}{B_d} \tag{9}$$

$$B_d = \frac{Y_{f0} - Y_f}{1 - Y_{f0}} \tag{10}$$

式中 Sc 为油滴的 Schmidt 数；Y_f 为燃烧蒸汽的质量分数；在油滴表面上，空气和燃料蒸汽的流量比 β 为

$$\beta = -\frac{\rho_{air}(1-Y_{f0})}{\rho_d + \dfrac{D_d}{6} \cdot \dfrac{d\rho_d}{dt} / \dfrac{dD_d}{dt}} \tag{11}$$

由于柴油机燃烧室中工质压力较高，如果仍用理想气体状态方程式来描述工质状态，则存在较大误差，故应用 Redlich 和 Kwong 提出的状态方程[6]来描述：

$$\rho = \frac{RT}{v-b} - \frac{a}{T^{1/2} v(v+b)} \tag{12}$$

式中 v 为气体的比容；R 为气体常数；a，b 为与气体临界状态有关的两常数。

为了验证油滴蒸发模型，将模型计算结果和以正庚烷作燃料在定容容器中的实验结果进行对比。图 1 是工质温度为 573 K，不同压力下液滴粒度随时间变化规律与实验结果[4]的比较，图 2 是工质压力为 1 MPa，不同温度下液滴粒度随时间变化规律与实验结果[7]的比较（图中实线为计算结果）。可以看出，模型计算结果和实验结果吻合良好。

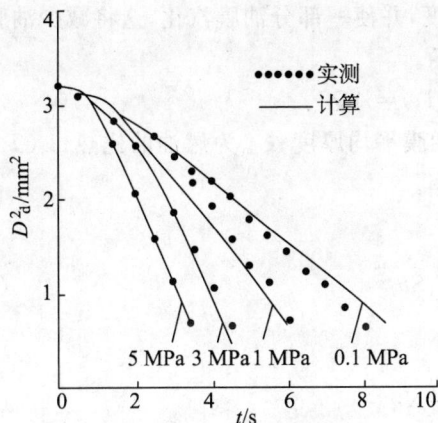

图 1　不同压力下液滴粒度比较

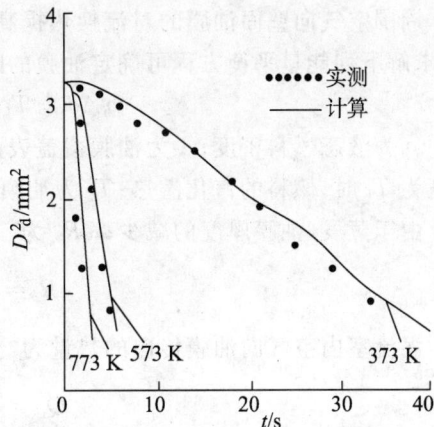

图 2　不同温度下液滴粒度比较

1.2　油膜蒸发模型

在涡流室式柴油机的喷雾过程中,一部分燃油会碰到涡流室底部并形成油膜,油膜厚度与喷油压力、喷油量、喷油持续期等因素有关。假定油膜厚度小于油膜的边界厚度和气流的湍流边界层厚度[8],则在油膜层中,流体的运动类似于原始碰壁点开始的短距离射流,并且随后由于边界层的粘滞力而滞止。因为涡流室壁面通过热传导传给油膜的热量比空气对流传给油膜的热量少得多,所以忽略其影响。Colburn 比拟适应于平板上湍流对流换热,且它与 Prandtl 数接近 1 的流体的实验数据较吻合,而涡流室底面与平板的情况接近,所以通过层流受迫对流分析和 Colburn 比拟来建立油膜与涡流室内空气间的对流换热及传质模型。

根据流动状况,用下列关系式来表示平均 Nussult 数和平均 Sherwood 数。

对于层流($Re_L < 5 \times 10^5$)

$$Nu_L = 0.664 Re_L^{1/2} Pr_L^{1/3} \frac{\ln(1+B_d)}{B_d} \tag{13}$$

$$Sh_L = 0.664 Re_L^{1/2} Sc_L^{1/3} \frac{\ln(1+B_d)}{B_d} \tag{14}$$

对于湍流区间 I($5 \times 10^5 < Re_L < 1 \times 10^7$)

$$Nu_L = (0.037 Re_L^{0.8} - 872) Rr_L^{1/3} \frac{\ln(1+B_d)}{B_d} \tag{15}$$

$$Sh_L = (0.037 Re_L^{0.8} - 872) Sc_L^{1/3} \frac{\ln(1+B_d)}{B_d} \tag{16}$$

对于湍流区间 II($Re_L > 1 \times 10^7$)

$$Nu_L = [0.228 Re_L (\lg Re_L)^{-2.584} - 872] Pr_L^{1/3} \frac{\ln(1+B_d)}{B_d} \tag{17}$$

$$Sh_L = [0.228 Re_L (\lg Re_L)^{-2.584} - 872] Sc_L^{1/3} \frac{\ln(1+B_d)}{B_d} \tag{18}$$

式中 $Nu_L = hL/K_{air}$,$Sh_L = h_m L/D_{air}$,$Re_L = \rho_{air} VL/\mu_{air}$,$Pr_L = \mu_{air} C_{p,air}/K_{air}$,$Sc_L = \mu_{air}/(\rho D)_{air}$,分别为油膜平均 Nusselt 数、平均 Sherwood 数、平均 Reynolds 数、平均 Prandtl 数、平均 Schmidt 数;L 为特征长度;V 为流过液体层的气流速度;h 和 h_m 是热量和质量传递数;ρ_{air},μ_{air},$C_{p,air}$,K_{air} 和 D_{air} 分别为空气的密度、动力粘性系数、定压比热、导热系统和质量扩散系统;$[\ln(1+B_d)]/B_d$ 项考虑了由于油膜表面附近燃油蒸汽的存在对传热及传质的影响。

周围空气向壁面油膜的对流换热提高了油膜温度,并使一部分油膜汽化,这将减少油膜厚度,求解下列能量平衡方程可确定油膜的瞬时温度 T_f。

$$\rho_f A \delta C_{liq} T_f - \rho_f A R_\delta L(T_f) = A Q_L \qquad (19)$$

式中 ρ_f 为液态燃料密度;A 为油膜覆盖表面积;δ 为油膜平均厚度;C_{liq} 为燃油的比热;$L(T_f)$ 是温度为 T_f 时,燃料的汽化潜热;T_f 为油膜温度变化率。

由于蒸发,油膜厚度的减少率 R_δ 为

$$R_\delta = -\frac{(\rho D)_{air}}{\rho_f L} B_d Sh_L \qquad (20)$$

涡流室内空气向油膜传递的热量为

$$Q_L = \frac{K_{air}(T - T_f)}{L} Nu_L \qquad (21)$$

式中 T 为油膜附近单元的空气温度。

2 计算结果及分析

将上述模型用于 195 涡流室式柴油机的燃油蒸发过程研究中[9]。柴油机在标定工况 (8.82 kW,2 000 r/min) 下工作,喷油提前角为 10° CA BTDC,喷油持续角为 16° CA[10],每循环供油量为 0.037 4 g。假定喷油规律为梯形波形式,为重点研究燃油蒸发过程,暂不考虑燃油的燃烧过程。

图 3 为涡流室中典型截面位置,图 4 为蒸发过程中燃料蒸汽质量分数的变化历程。可以看出,燃油喷入涡流室后,由于油滴蒸发,在喷雾体周围具有较浓的燃料蒸汽;由于一部分燃油喷到涡流室底部并形成油膜,而油膜蒸发使涡流室底部也形成较浓的燃料蒸汽。在涡流室中,随着空气的运动,燃料蒸汽顺涡流运动,上止点后,燃料蒸汽开始进入主燃烧室。

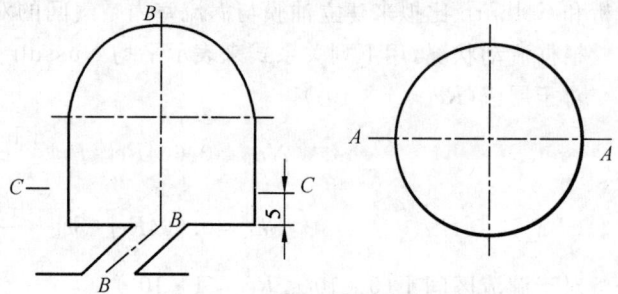

图 3 涡流室中典型截面位置

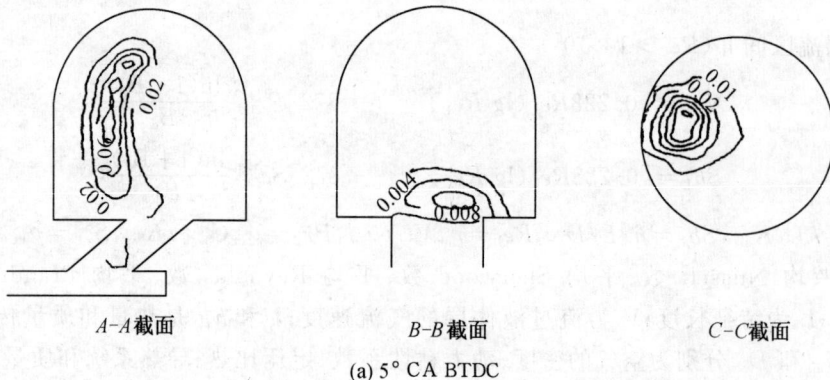

A-A 截面　　　　　　B-B 截面　　　　　　C-C 截面

(a) 5° CA BTDC

A-A 截面　　　　　　　　B-B 截面　　　　　　　　C-C 截面

(b) TDC

A-A 截面　　　　　　　　B-B 截面　　　　　　　　C-C 截面

(c) 5° CA BTDC

A-A 截面　　　　　　　　B-B 截面　　　　　　　　C-C 截面

(d) 15° CA BTDC

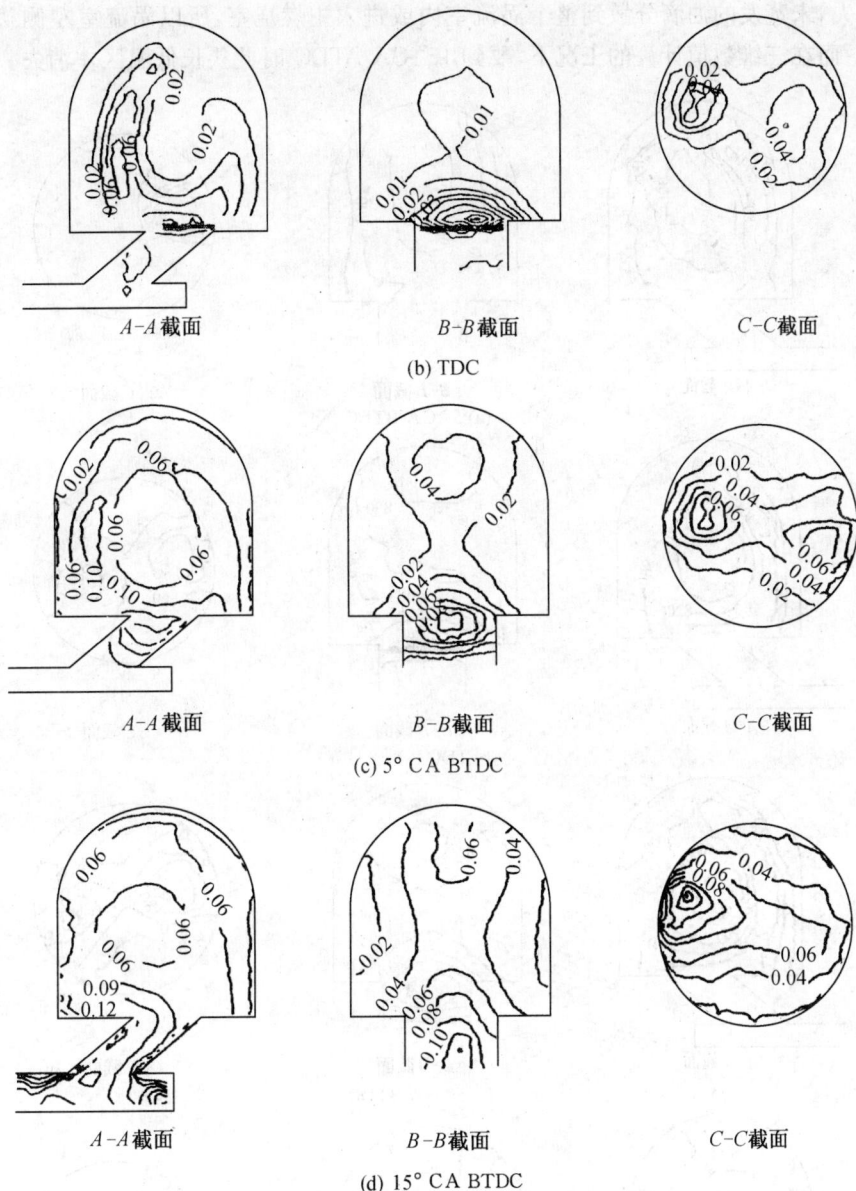

图 4　燃烧蒸汽质量分数变化历程

图 5 为燃料蒸发过程中涡流室内温度场变化情况。在纯压缩过程中,涡流室内有较均匀的温度场[9];燃油喷入涡流室后,由于油滴吸收汽化潜热而蒸发,涡流室内气体温度与纯压缩过程相比有明显差别。在 5°CA BTDC 时,温度较低的喷雾燃油已在涡流室左侧形成一个狭长的低温区,此低温区顺气流偏转,同时高温区左移。此时,在涡流室底部左侧上方,形成了温度和浓度适中的混合气,有利于着火。在 10°CA ATDC 后,由于较多油滴的分散和流向主燃烧室,左侧狭长的低温区开始消失,同时,在涡流室底部和连接通道处形成低温区,到 15°CA ATDC 时,此狭长的低温区已经消失,涡流室内形成较均匀的温度场。将图 5 与文献[11]所测得的燃油蒸发过程温度场对比可知,三维计算结果与实测结果具有相同的温度分布趋势。由于实测时发动机的转速较低且喷油量较少,因此燃油蒸发较快,在 10°CA ATDC 时,大部分燃油已经通过涡流室底

部,油束消失,未蒸发的油滴分散到整个涡流室内或进入主燃烧室,所以涡流室左侧狭长的低温区也消失。而在三维数值计算的工况下,要到 15° CA ATDC 时此狭长低温区才消失。

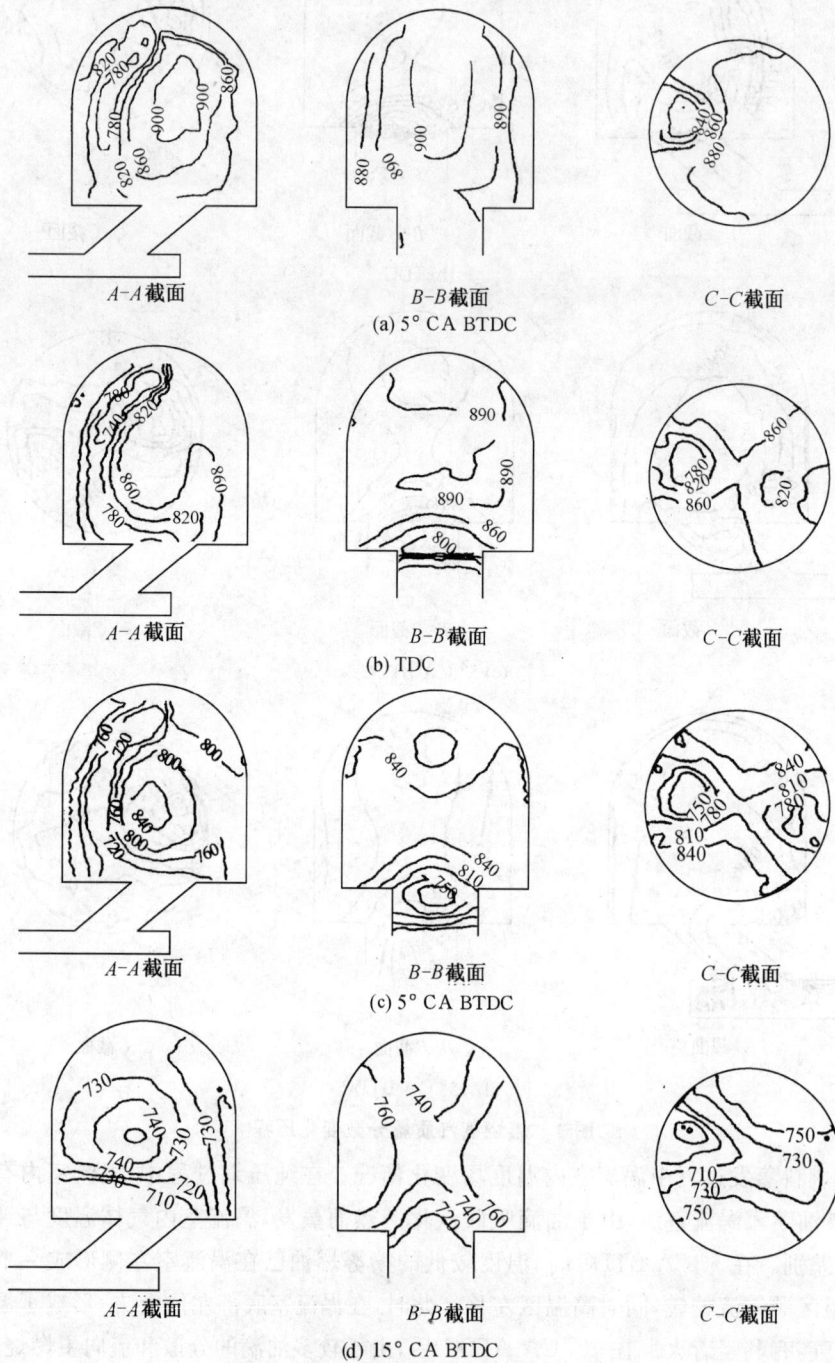

A-A 截面　　　　　　B-B 截面　　　　　　C-C 截面
(a) 5° CA BTDC

A-A 截面　　　　　　B-B 截面　　　　　　C-C 截面
(b) TDC

A-A 截面　　　　　　B-B 截面　　　　　　C-C 截面
(c) 5° CA BTDC

A-A 截面　　　　　　B-B 截面　　　　　　C-C 截面
(d) 15° CA BTDC

图 5　涡流室温度变化历程(单位:K)

图 6 为燃料蒸发过程和纯压缩过程涡流室中气体平均温度的比较。可见,由于燃料蒸发吸热,涡流室内的平均温度下降,并且从喷雾开始,随着曲轴的运动,燃油蒸发量增加,两者差别越来越大。涡流室内平均温度下降使平均压力也有所下降,如图 7 所示。

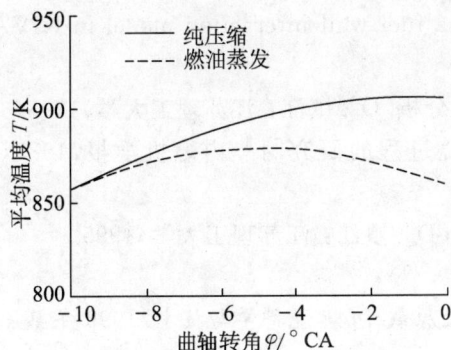

图 6　燃油蒸发对涡流室平均温度的影响

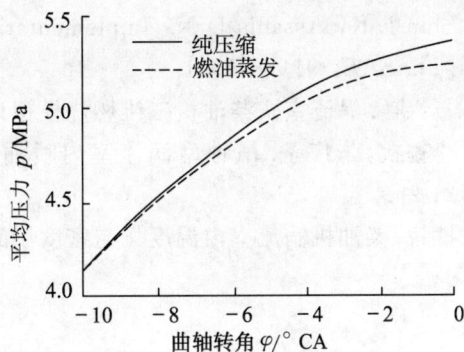

图 7　燃油蒸发对涡流室平均压力的影响

3　结　论

燃料蒸发是柴油机混合气形成过程中的一个重要环节。本文建立了高温高压条件下的油滴蒸发模型和油膜蒸发模型,并以 195 柴油机为例,对涡流室式柴油机中燃料蒸发过程进行了三维数值模拟计算,结果表明:

(1) 在不同的工质温度和压力下,所建立的高温高压条件下油滴蒸发模型的计算结果与实验结果吻合良好。

(2) 在喷雾体外围和涡流室底部左侧上方,具有较浓的燃料蒸汽,燃料蒸汽在涡流室中顺涡流运动,并于上止点后开始进入主燃烧室。

(3) 在涡流室底部左侧上方,形成了温度和浓度适中的混合气,此处有可能首先着火。

(4) 燃料蒸发对涡流室温度场有较大影响,在喷雾体内以及涡流室底部,气体温度明显降低,并且涡流室内的平均温度和平均压力也有所下降。

参　考　文　献

[1] 陈家骅,万俊华,魏象仪,等. 内燃机燃烧[M]. 哈尔滨:哈尔滨船舶工程学院出版社,1986.

[2] Gonzalez M A, Lian Z W, Reitz R D. Modeling diesel engines spray vaporization and combustion[J]. *SAE*, 920579,1992.

[3] Quoc H X, Brun M. Study on atomization and fuel drop size distribution in direct injection diesel spray[J]. *SAE*, 941019,1994.

[4] Hiroyasu H, Kadota, Arai M. 柴油机中燃料喷雾体的特性及其模拟[C]∥ Mattavi J N, Amann C A. 内燃机燃烧模拟论文集. 刘巽俊,译. 北京:机械工业出版社,1987.

[5] Amsden A A, O′Rourke P J, Butlor T D. KIVA-Ⅱ:A computer program for chemically reacting flows with sprays[R]. Los Alamos National Laboratory Report, LA-11560-MS, 1989.

[6] 沈维道,郑佩芝,蒋淡安. 工程热力学[M].2 版. 北京:高等教育出版社,1983.

[7] Rocco V. Results of quasi-steady evapporation model applied to multi-dimensional DI diesel combustion simulation[J]. *SAE*,930071,1993.

［8］Shin L K，Assanis D N． Implementation of a fuel wall interaction model in KIVA-Ⅱ ［J］． *SAE*，91178，1991．

［9］夏兴兰．涡流室式柴油机三维燃烧模拟计算与分析［D］．镇江：江苏理工大学，1998．

［10］李德桃，朱广圣.吊钟型涡流室内喷油和燃烧过程的研究［J］.内燃机学报，1989（1）： 21-26．

［11］熊锐.柴油机涡流室内温度场与密度场的研究［D］.镇江：江苏理工大学，1995．

（本文原载于《燃烧科学与技术》1999 年第 4 期）

Three-dimensional numerical simulation of fuel evaporation in a swirl chamber diesel engine

Abstract：A fuel droplet evaporation model at high temperatures and high pressures is developed. The model is verified at different pressures and temperatures of the working fluid，and the results predicted by the model agree with the experiment data fairly well. The fuel droplet evaporation model combined with the fuel film evaporation model is used in a swirl chamber diesel engine，and three-dimensional numerical simulation is carried out. The courses of fuel vapor concentration and temperature field during fuel evaporation in the swirl chamber are studied. The influences of fuel evaporation on the mean temperature and mean pressure in the swirl chamber are also investigated.

涡流室式柴油机燃油与空气混合过程的三维数值计算和试验验证

夏兴兰,熊　锐,李德桃

[摘要]　通过改进 KIVA-Ⅱ 程序中的油滴破碎和喷雾碰壁两个子模型,对涡流室式柴油机燃油与空气混合过程进行三维数值计算。获得了涡流室中典型截面的油滴运动轨迹和燃料蒸汽质量分数以及温度场的分布图。用激光莫尔偏折技术实测涡流室内的温度场,测试结果表明:三维数值计算与实测数据取得了定性上的一致。

柴油机中燃油与空气的混合过程对其燃烧起着重要的影响。国内外一些学者进行过一些试验研究,由于试验条件的限制,大部分试验研究都是在静态模拟容器中进行,少部分是在柴油机低速运转条件下进行的。因为涡流室内存在高温高压环境和强烈的涡流运动,涡流强度又与发动机转速密切相关,所以静态模拟容器和低速运转条件下的柴油机都与高速柴油机的实际运行状况有一定的差别。如要在高速柴油机中进行试验研究目前尚存在较大的困难。数值计算既可以模拟高速柴油机的实际运行状况,弥补试验研究的不足,又可以计算出许多无法直接测量的数据结果。国外一些学者利用 KIVA-Ⅱ 程序对直喷式柴油机的燃油与空气混合过程进行过数值计算[1,2]。由于涡流室式柴油机结构复杂,主副燃烧室中燃油与空气的混合过程存在较大的差别,同时两室通过连接通道而相互影响,使得模拟计算更加困难,并且 KIVA-Ⅱ 程序不完全适合涡流室式柴油机的工作过程,因此对涡流室式柴油机所进行的研究较少。

作者在 KIVA-Ⅱ 程序的基础上,对诸如燃油雾化和喷雾碰壁等子程序进行修改,使之适合涡流室式柴油机的条件,并以 195 柴油机为例,对燃油与空气的混合过程进行三维数值计算,并利用激光莫尔偏折技术对涡流室内燃油与空气混合过程的温度场进行实机测试,以验证数值计算结果的正确性。

1　数学模型简介

在燃油与空气混合过程中,由于油滴表面的微小扰动,产生不稳定波使油滴破碎的现象,这已为许多试验所证实[3-5]。在柴油机中,燃油从喷油嘴喷出后,形成图 1 所示的喷雾形态,喷雾可分为液核区和气液混合区[6]。不稳定性理论指出,射流表面的破碎过程可以用 K-H (Kelvin-Helmhotz)不稳定性理论来描述,直接暴露于空气的微滴可以用 R-T(Rayleigh-Taylor) 不稳定性理论来描述。所以,对涡流室式柴油机喷雾的液核区用 K-H 不稳定性来分析[7],对气液混合区用 R-T 不稳定性理论来分析[8]。

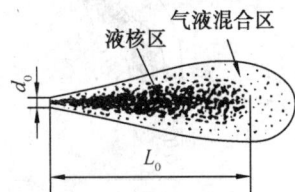

图 1　燃油喷雾形态

喷雾碰壁是一个非常复杂的过程,油滴在碰壁过程中伴随有粘附、反弹、扩散、飞溅和破碎

245

等现象。作者将油滴碰壁过程分为粘附、反弹/粘附、飞溅/附壁射流三种相互重叠的过程,以临界韦伯数 We($We=\rho_e dV^2/\sigma$,其中 ρ_e,d,σ 分别为油滴的密度、直径、表面张力,V 为入射油滴的法向速度)作为各方式转换的判据。当 $We\leqslant5$ 时,为粘附过程;当 $5<We\leqslant80$ 时,为反弹/粘附过程;当 $We>80$ 时,为飞溅/附壁射流过程。

2 计算结果与分析

以 195 柴油机为例,对涡流室式柴油机的燃油与空气的混合过程进行了三维数值计算。柴油机在标定工况(8.84 kW,2 000 r/min)下运行,每循环供油量为 32 mm³,喷油于 10° CA BTDC 开始,喷油持续角为 16° CA[9],假定喷油规律为梯形波方式。因重点研究燃油与空气的混合过程,暂不考虑燃料的燃烧。

图 3 为燃油液滴轨迹的三视图和燃料质量分数在涡流室三个典型截面(典型截面位置见图 2)的分布图,图 4 为其温度分布图。当燃油从喷油嘴顺气流射向涡流室时,涡流室的空气运动为强烈的有组织的涡流运动。在涡流运动的作用下,进入涡流室的油滴和涡流室内的气体进行质量、动量和能量的交换后,其运动轨迹发生变化,转向偏左,并很快转移到空气运动的轨迹上,空气运动使油滴分散到更大的空间里。在油滴运动的过程中,油滴几乎不与涡流室侧壁相碰,且到 5° CA BTDC 时,油束还未与涡流室底部相碰。温度较低的喷雾在涡流室左侧形成一个狭长的低温区,涡流室内的高温区处于涡流室的中间部分(如图 3a 和图 4a 所示)。在 4.76° CA BTDC 时,油滴与涡流室底面相碰,碰壁后,一部分油滴粘附在壁面上或沿壁面形成附壁射流,另一部分油滴反射回涡流室空间,并加入到空气的涡流运动中。在 1.96° CA BTDC 时,油滴越过通道口,一小部分油滴沿涡流室右侧顺气流向顶部运行或进入主燃烧室,大部分油滴在越过通道口附近之前由于不断蒸发而消

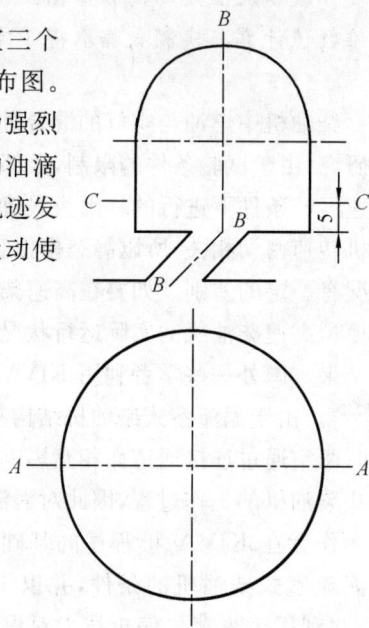

图 2 涡流室中典型截面位置

失,以往的高速摄影照片也证实了油滴的这种运动过程[9]。油滴在运动过程中,伴随有分裂、吸热、蒸发、汽化和扩散等过程,逐步形成可燃混合气。由于油滴吸收汽化潜热而蒸发,涡流室内气体的温度与压缩过程相比明显下降,随着油滴向涡流室底部运动,涡流室内高温区逐渐左移。

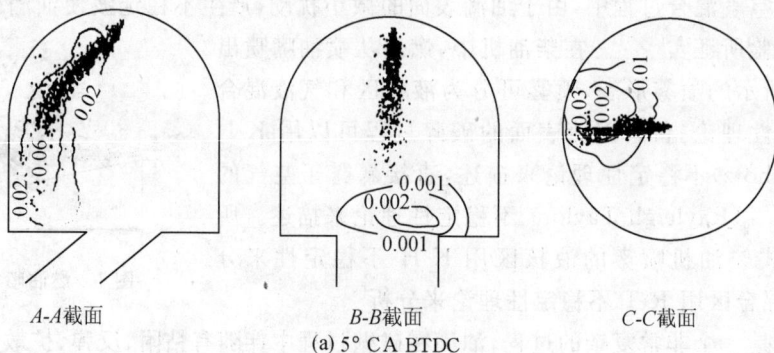

*A-A*截面   *B-B*截面   *C-C*截面

(a) 5° CA BTDC

A-A截面　　　　　　　B-B截面　　　　　　　C-C截面

(b) TDC

A-A截面　　　　　　　B-B截面　　　　　　　C-C截面

(c) 5° CA ATDC

A-A截面　　　　　　　B-B截面　　　　　　　C-C截面

(d) 15° CA ATDC

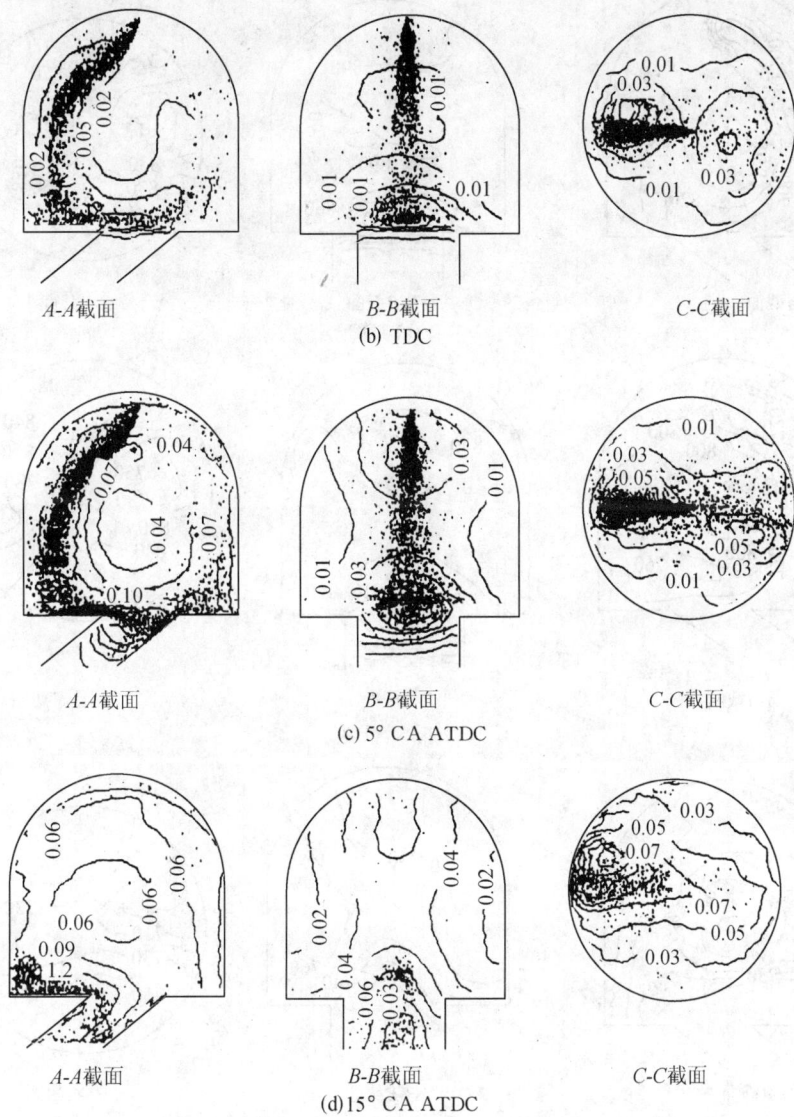

图3　燃油液滴运动轨迹和燃料蒸气质量分数(燃料蒸气质量与混合气总质量之比)

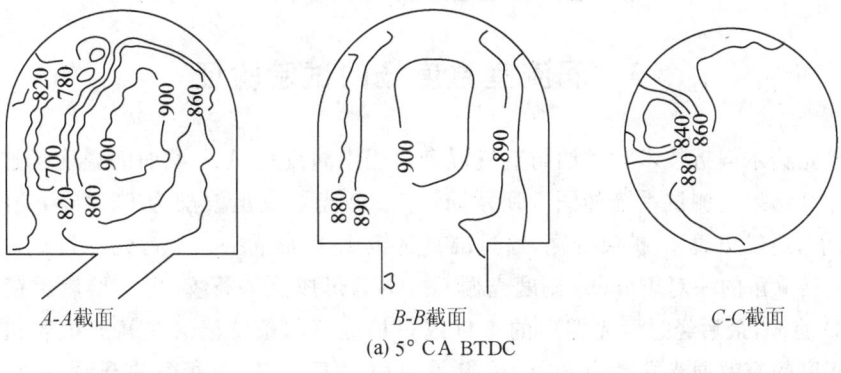

A-A截面　　　　　　　B-B截面　　　　　　　C-C截面

(a) 5° CA BTDC

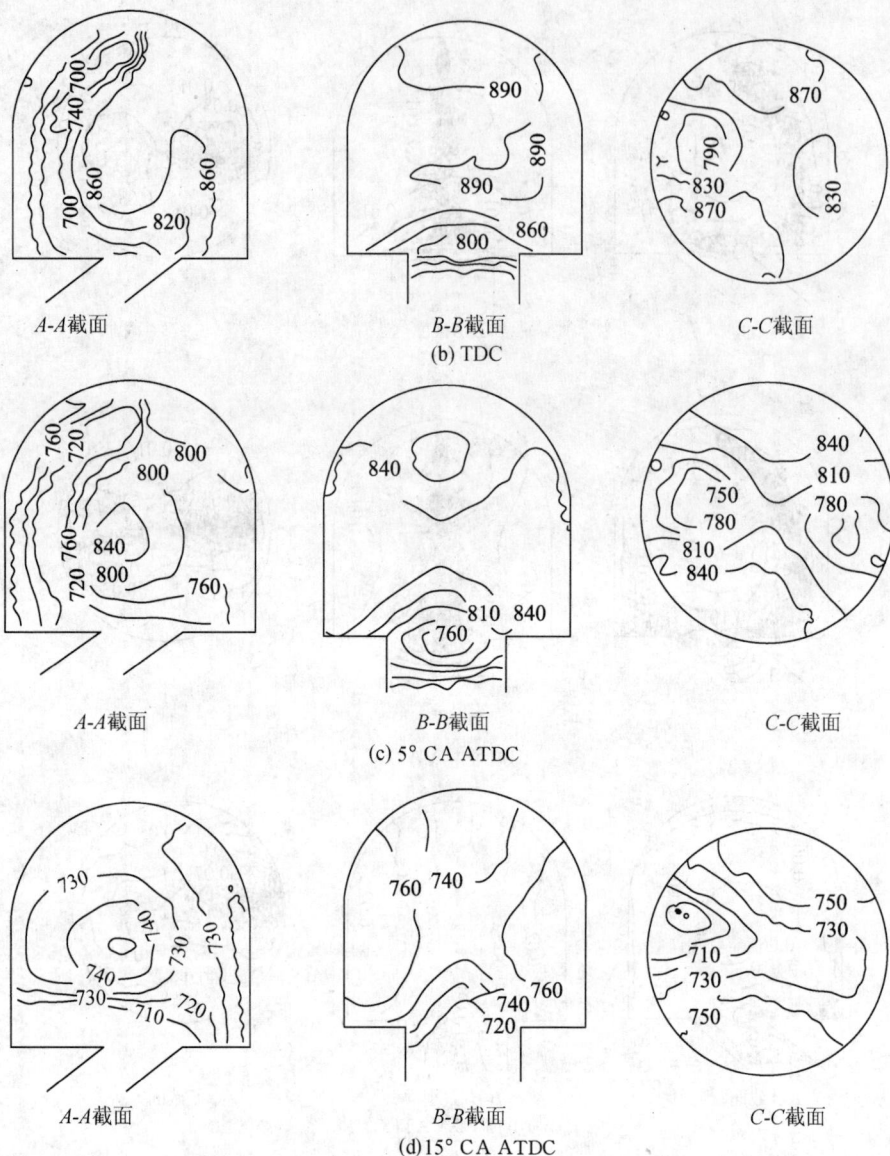

A-A截面　　　　　　　　　B-B截面　　　　　　　　　C-C截面

(b) TDC

A-A截面　　　　　　　　　B-B截面　　　　　　　　　C-C截面

(c) 5° CA ATDC

A-A截面　　　　　　　　　B-B截面　　　　　　　　　C-C截面

(d)15° CA ATDC

图4　涡流室温度分布(单位：K)

3　涡流室温度场的试验验证

利用激光莫尔偏折技术对燃油与空气混合过程中涡流室 A-A 截面的温度场进行测试，以验证数值计算结果。测试系统如图5所示，He-Ne 激光器发出波长为 623.8 nm 的激光由平面镜 M_1 反射，经扩束镜 L_1 扩束变宽，通过准直透镜 L_2 后形成一束平行光穿过涡流室流场，然后在相互偏转 θ 角的一对 Ronchi 刻度光栅 G_1，G_2 后形成莫尔条纹，再由聚焦透镜 L_3 成像及小孔光阑 D 滤波，最后经过双光栅后的＋1(或－1)级莫尔条纹记录在高速摄影机的胶片上。测试时观察窗的有效通光孔径为 30 mm，覆盖面积大于 80％，观察窗的开设不改变涡流室的容积，因此保证了压缩比与原机相同。观察窗用石英玻璃制作，其表面粗糙度小于测试时激光波长的 1/10,以避免平行光束经过观察窗时形成的牛顿干涉环;石英玻璃两平面间的夹角加工成 10′～20′,以消除玻璃背景条纹对莫尔条纹的影响。

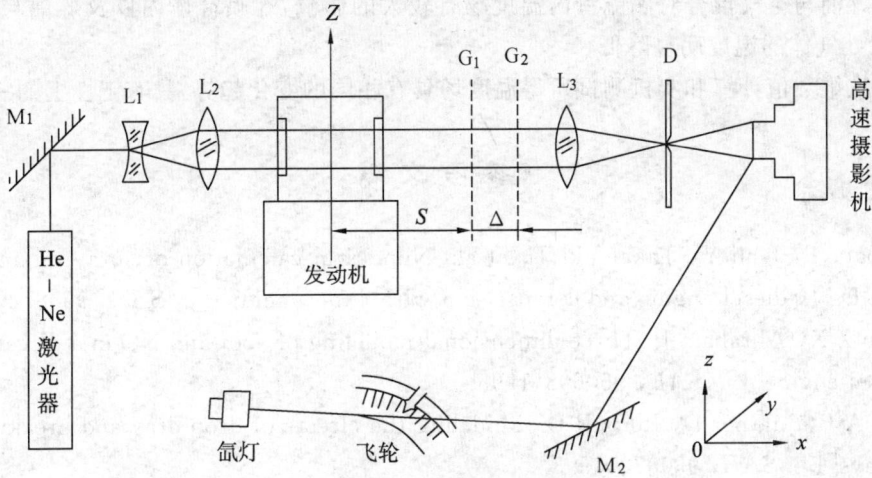

图 5　温度场测试系统示意图

因高转速时测量比较困难,故选择较低的发动机转速进行测试。测试时发动机的转速为 240 r/min,发动机上止点信号由 nac(E－10)高速摄影机记录在胶片上,高速摄影机以 1 150 帧/秒的速度拍摄了整个燃油与空气混合过程中各瞬间莫尔条纹的变化。

图 6 为测试所得温度场。对照图 4 可知:三维计算结果与实测结果具有相同的温度分布趋势,取得了定性上的一致。由于实测时发动机的转速较低且喷油量较少,因此燃油蒸发较快,在 10° CA ATDC 时,大部分燃油已经通过涡流室底部,油束已经消失,未蒸发的油滴分散到整个涡流室内或进入主燃烧室,涡流室左侧狭长的低温区消失。而在三维数值计算的工况下,则要到 15° CA ATDC 时此狭长低温区才消失。

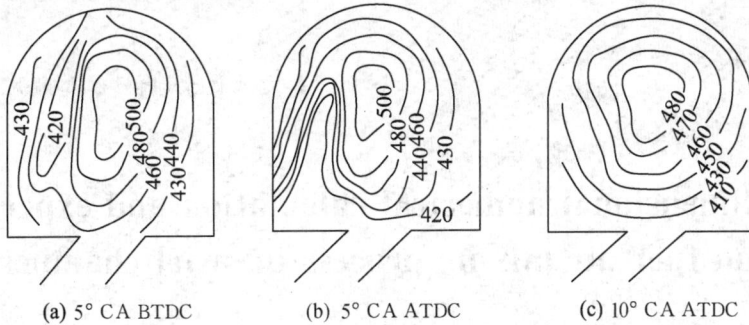

(a) 5° CA BTDC　　　　(b) 5° CA ATDC　　　　(c) 10° CA ATDC

图 6　涡流室内燃油与空气混合过程温度场测试结果(单位:K)
(转速:240 r/min,喷油量:23 mm³/循环)

4　结　　论

(1) 油滴几乎不与涡流室侧壁相碰,在 4.760° CA BTDC 时才和涡流室底面相碰,碰壁后大部分油滴反射回涡流室,一小部分粘附在壁面上或形成附壁射流;

    （2）大部分油滴在越过通道口以前由于不断蒸发而消失，一小部分油滴在涡流室右侧顺气流向顶部运行或进入主燃烧室；

    （3）燃油与空气混合对涡流室内温度场有较大的影响，在喷雾体内以及燃油与涡流室底部相碰之处气体的温度明显降低；

    （4）三维数值计算和实机测试所得温度场具有相同的变化趋势，具有定性上的一致。

参 考 文 献

［1］Tabata T，Ishii Y，Takatsuki T，et al. Numerical calculation of spray mixing process in a D. I. diesel engine and comparision with experiments[J]. *SAE*，950853，1995.

［2］Hou Z X，Abraham J. Three-dimensional modeling of soot and NO in a direct-injection diesel engine[J]. *SAE*，950608，1995.

［3］Liu A B，Mather D，Reitz R D. Modeling the effects of drop dray and breakup on fuel sprays[J]. *SAE*，930072，1993.

［4］史绍熙，郗大光，刘宁，等. 高速液体射流的初始阶段的破碎[J]. 内燃机学报，1996（4）：349-354.

［5］史绍熙，郗大光，秦建荣，等. 高速粘性液体射流的不稳定模式[J]. 内燃机学报，1997（1）：1-7.

［6］Hiroyasu H. Diesel engine combustion and its modeling[C]. *COMODIA*'85，1985：53-75.

［7］Reitz R D，Diwakar R. Structure of high-pressure fuel sprays[J]. *SAE*，870598，1987.

［8］Bellman P，Pellington M. Effects of surface tension and viscosity on Taylor instability [J]. *SAME*，1953，12（2）.

［9］李德桃，朱广圣. 吊钟型涡流室内喷油和燃烧过程的研究[J]. 内燃机学报，1989（1）：21-26.

（本文原载于《内燃机工程》2000 年第 1 期）

Three-dimensional numerical calculation and experimental study on fuel/air mixing process in swirl chamber diesel

Abstract：Three-dimensional numerical calculation on fuel/air mixing process in a swirl chamber diesel is carried out by using the updated KIVA-II code，in which the submodels of fuel droplet breakup and spray impingement are improved. The distribution of fuel droplet motion locus，fuel isoconcentraion contours and temperature field in the typical sections of the swirl chamber is plotted. The temperature fields in the swirl chamber are measured by using laser Moire deflectometry. The results show that the calculated data are agreement with the measured ones in quality.

改善涡流室式柴油机燃烧室结构降低排放研究

董　刚,李德桃,夏兴兰,杨文明

[摘要]　系统报道了通过改进燃烧室结构设计降低排放的实验方法和手段。结果表明,通过对涡流室柴油机涡流室、主燃室以及通道的形状和大小进行适当的改进,可以较好地改善柴油机混合气质量和燃烧过程,进而有效地抑制排放污染。

发动机的排气污染问题日益受到世界各国的重视,并相继制定了严格的排放法规来限制发动机排放[1]。我国产量最大的内燃机——涡流室式柴油机不仅作为农业机械和工程机械的动力,而且也广泛用于轻型车和轿车中,其排放对大气造成的污染是不容忽视的。为此,在分析国内外有关研究成果的基础上,结合国产机型的燃烧室结构设计进行了系统的试验研究和分析,为控制我国涡流室式柴油机排放找到了一些切实有效的途径。

1　改善涡流室结构设计降低排放

1.1　涡流室形状

图1为3种容积相同但形状不同的涡流室结构示意图,最左端的涡流室为传统形状,类型1具有曲线型壁面,类型2在对准喷嘴处安装了一档板。在由双弹簧控制双启喷压力(分别为15/20 MPa和25/30 MPa)下,类型1和类型2的烟度和NO_x的排放量与传统型的排放对比结果见图2[2]。

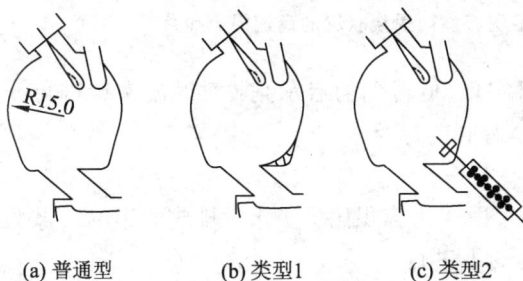

图1　不同形状的涡流室结构示意图

(a) 普通型　　(b) 类型1　　(c) 类型2

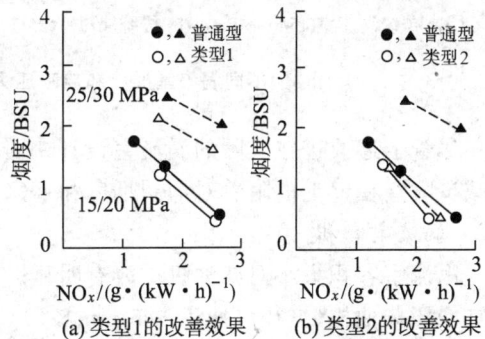

图2　涡流室形状变化对NO_x和烟度的改善

(a) 类型1的改善效果　　(b) 类型2的改善效果

由图2可以看出,在传统形状上改进的涡流室可以降低柴油机的烟度和NO_x排放,其中类型2比类型1的效果更好。图3给出了上述3种形状下,涡流室内混合气燃烧过程的高速摄影照片[2]。由图中右边三列可以看到,在相同喷油压力(25/30 MPa)下,具有曲线型壁面的涡流室(类型1)能够防止喷雾沉积在壁面上,这就使得混合气较好地流向主室,而在涡流室内对准壁面处加有一档板(类型2)后的涡流室由于喷雾沉积现象减少,混合气流动得以改善,因而碳烟的排放量得到了明显抑制。

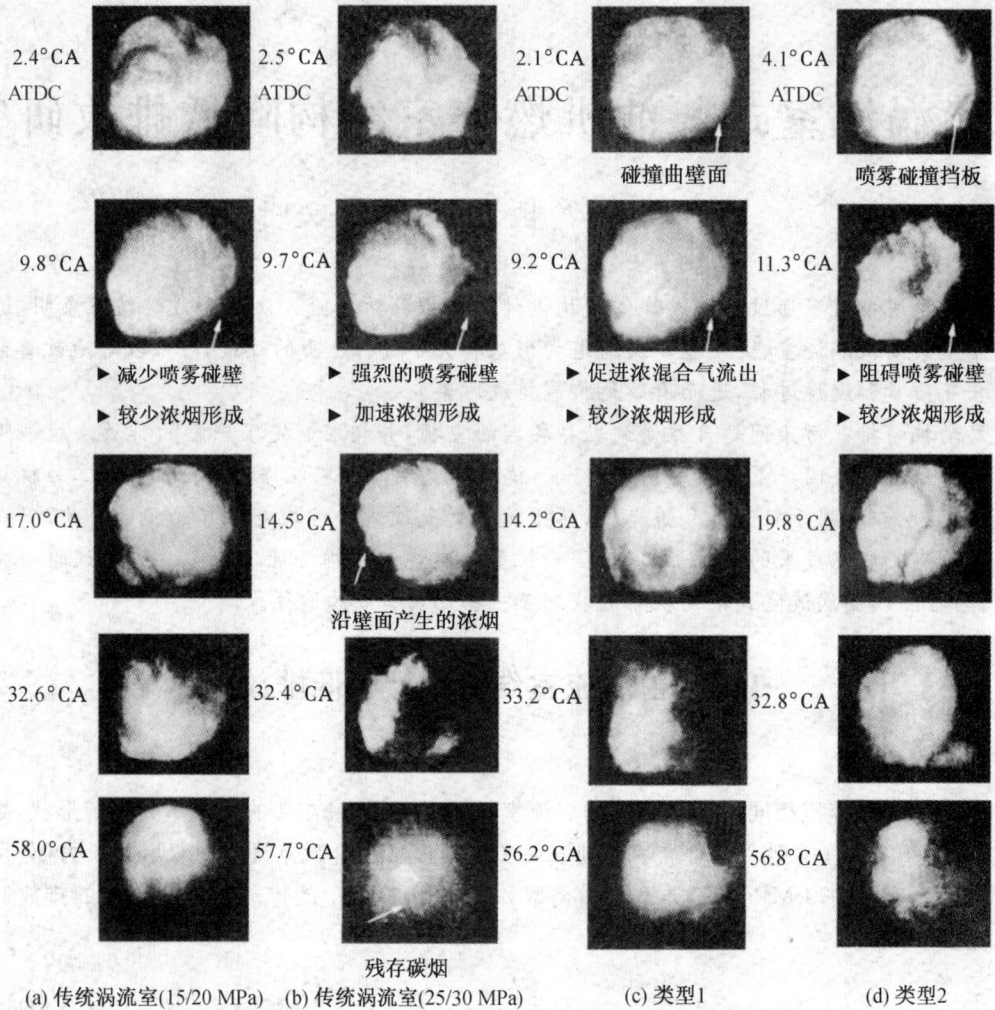

2.4°CA ATDC   2.5°CA ATDC   2.1°CA ATDC   4.1°CA ATDC

碰撞曲壁面   喷雾碰撞挡板

9.8°CA   9.7°CA   9.2°CA   11.3°CA

▶ 减少喷雾碰壁   ▶ 强烈的喷雾碰壁   ▶ 促进浓混合气流出   ▶ 阻碍喷雾碰壁

▶ 较少浓烟形成   ▶ 加速浓烟形成   ▶ 较少浓烟形成   ▶ 较少浓烟形成

17.0°CA   14.5°CA   14.2°CA   19.8°CA

沿壁面产生的浓烟

32.6°CA   32.4°CA   33.2°CA   32.8°CA

58.0°CA   57.7°CA   56.2°CA   56.8°CA

残存碳烟

(a) 传统涡流室(15/20 MPa)   (b) 传统涡流室(25/30 MPa)   (c) 类型1   (d) 类型2

图 3 不同涡流室形状和启喷压力下涡流室内燃烧状况的高速摄影照片

尽管上述改进在实际的涡流室设计中难以实现,但得到的启示是:对涡流室壁进行改进以减少喷雾在壁上的附着对抑制烟度和 NO_x 是有利的。

1.2 涡流室容积

涡流室容积大小对柴油机排放有明显影响。图 4 为容积比 β 变化对排放的影响。试验在四缸增压柴油机(缸径×冲程:96 mm×103 mm)上进行。

图 4 表明,随着容积比 β 的增加,HC、烟度和 PM 均有所下降,尤其在中高负荷时,烟度和 PM 下降十分明显,但 NO_x 的排放却略有增加。295 型柴油机的排放测试表明[4],随着容积比的增加,烟度和 CO 的排放也明显下降,而 NO_x 的排放量则增加较多,见图 5。

涡流室容积比的增加,使得进入涡流室的空气量增加,在高负荷时,混合气质量得到改善,使更多的油量在涡流室中就燃烧掉,因而 HC 和 CO 会有所下降,由于燃烧的改善,涡流室内燃烧温度有所提高,因而 NO_x 排放有所增加。文献[2]和[5]的高速摄影结果则表明,烟度和 PM 的增加主要是由于涡流室内喷雾碰到壁上形成过多的沉积引起的。随着容积比的增大,

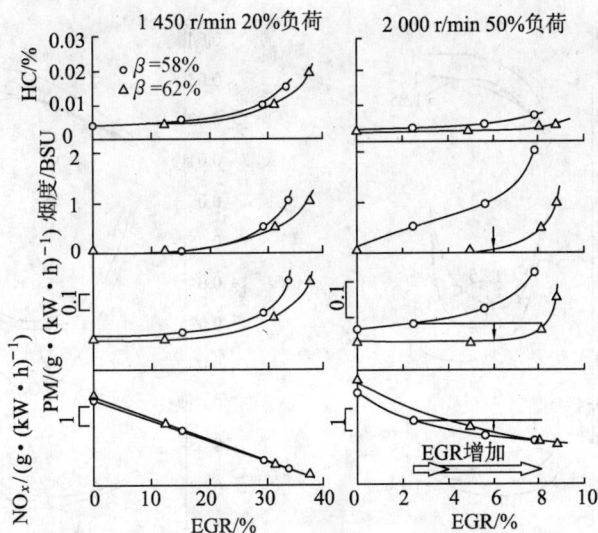

图 4　增压柴油机不同EGR时容积比β变化对排放的影响

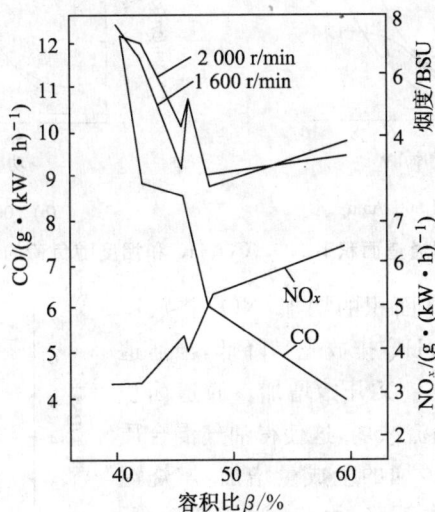

图 5　非增压柴油机不同容积比时排放的变化

喷嘴到壁面的距离也变大,这使得喷雾动量有所减小,喷雾也更加分散,同时喷雾的空气卷入量有所增加,这既有效地阻止了燃油在壁面上的附着,又稀释了过浓混合气,因而烟度和PM排放明显下降。

根据以上分析,涡流室容积适当扩大,对降低除 NO_x 以外的其他排放是有利的。

2　改善燃烧室通道设计降低排放

2.1　通道面积

在容积比一定的条件下,改变通道面积会影响柴油机的排放。图 6 为作者在天津大学内燃机燃烧学国家重点实验室一台 195 柴油机上测试的三种不同通道面积(通道面积比为 1∶1.14∶1.25)对排放的影响结果。

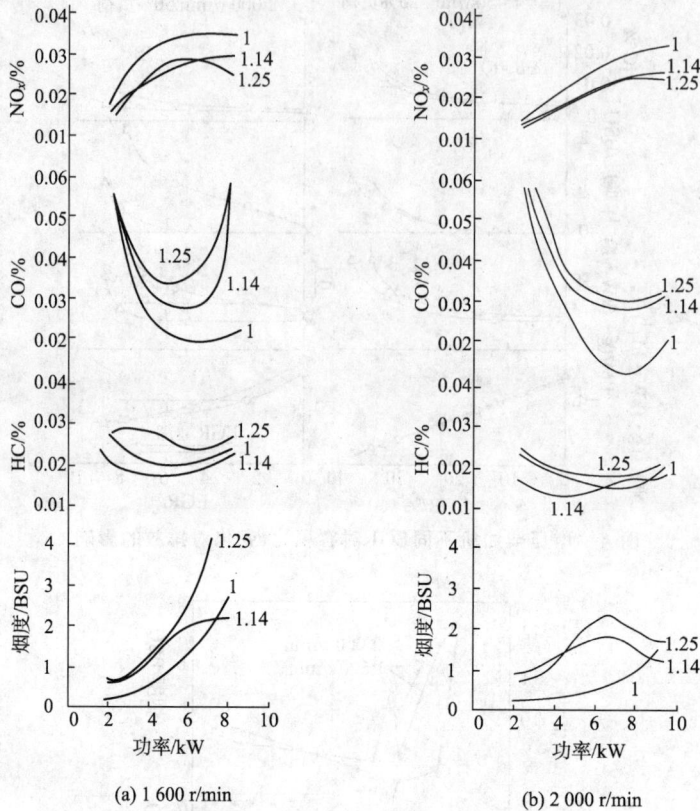

(a) 1 600 r/min　　(b) 2 000 r/min

图 6　不同通道面积下 NO_x,CO,HC 和烟度随负荷的变化关系

试验结果表明,随着通道面积的增加,NO_x 排放有所下降;CO 和烟度有所增加;HC 排放先是降低,当通道面积进一步增加时,HC 排放量又开始增加。通道面积增加,气体压缩涡流和燃烧涡流减弱,这使得油气混合质量变差,燃烧不完全,CO 和碳烟的生成量增加,燃烧温度下降,NO_x 的排放量就随之下降;由于燃烧温度降低,燃油的高温裂解现象减少,HC 的排放量有所下降,但通道面积进一步增加,相应的气流运动更弱,混合气质量更差,局部混合气过稀或过浓的现象增加,因而 HC 排放又增加。对 295 型涡流室式柴油机的排放测试也表明了 NO_x,CO 和烟度随通道面积变化所表现出的上述规律,见图 7[4]。

2.2　通道形状

通道截面形状对涡流室式柴油机的排放也有显著影响。图 8 为作者测试的三种不同形状(通道面积相同)的横截面。图 9 为这三种截面形状的通道对 195 型柴油机排放的影响。

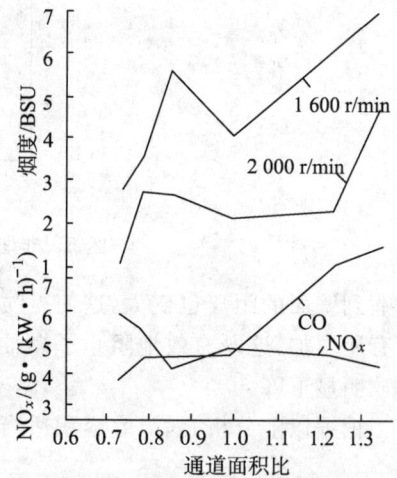

图 7　295 型柴油机不同通道面积时排放量的变化

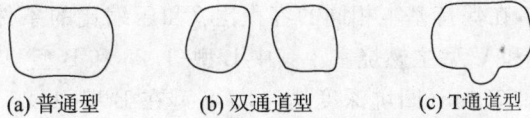

(a) 普通型　　　　(b) 双通道型　　　　(c) T通道型

图 8　通道的不同截面形状(通道面积相同)

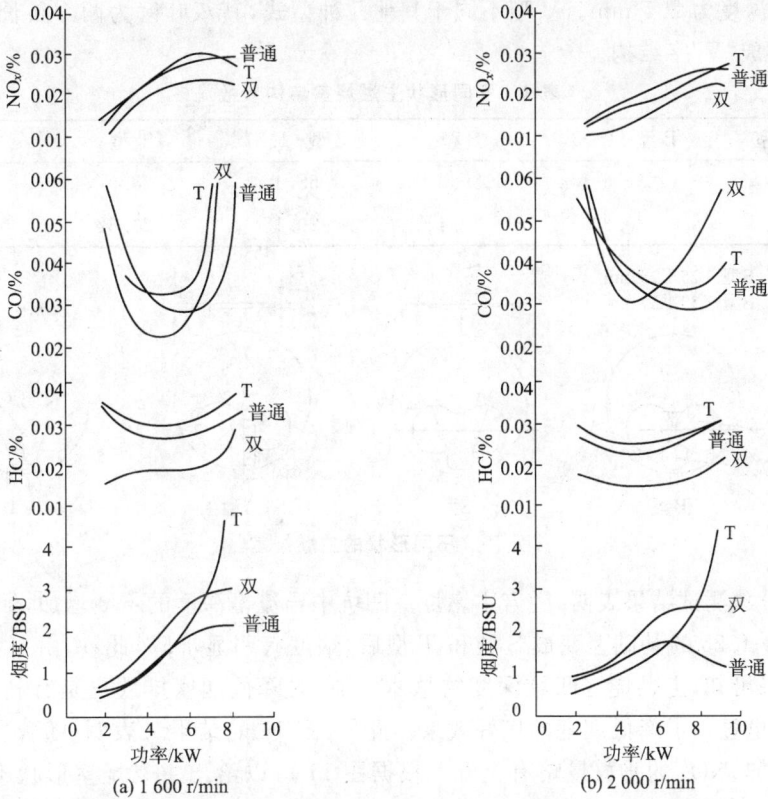

(a) 1 600 r/min　　　　(b) 2 000 r/min

图 9　不同通道形状下 NO_x,CO,HC 和烟度随负荷的变化关系

　　试验结果表明,在通道面积相同的条件下,双通道型的 NO_x 和 HC 的排放量要比其他两种形状的低,而 CO 和烟度在低负荷时也较低,但在高负荷时,不同的转速下双通道对 CO 和烟度的影响则不尽相同:在较低转速下(1 600 r/min),双通道型的 CO 和烟度均高于普通通道而低于 T 型通道;在较高转速下(2 000 r/min),双通道型的 CO 最高,而烟度则高于普通通道但却比 T 型通道要低。产生这一变化的原因在于,通道形状的改变会影响气体涡流运动的强度和方向,双通道型还会使气体产生节流现象,这些变化将最终影响混合气质量,从而影响到涡流室式柴油机各种有害排放的变化。

　　从这一试验结果来看,双通道型对降低涡流室式柴油机的总排放量似乎有更好的效果。

3　改善主燃烧室形状降低排放

　　主燃烧室中,活塞顶部的凹坑形状和深度对涡流室式柴油机的排放有影响。图 10 和表 1 给出了几种不同凹坑形状和深度的主燃烧室形状[5],图 11 为这些不同形状主燃烧室的放热规律及烟度、NO_x 的排放特性。

由图 10 和表 1 可见,在保持基本相同的容积比 β 和压缩比的条件下,在 B 型主燃烧室的基础上发展了 S 型、T 型和 V 型主燃烧室。其中 B 型-1.25 和 B 型-1.57 形状相同,其凹坑深度分别为 1.25 mm 和 1.57 mm(凹坑深度均匀)。S 型在 B 型-1.25 的基础上开了一个梯形导流槽,其深度为 1.57 mm。T 型则在 B 型-1.25 的基础上开了一个矩形导流槽,其接近混合气喷出部位的深度为 2.5 mm。V 型不同于其他几种型式,其双叶区为两个狭长的槽,深度均为 2.5 mm,构成"V"字结构。

表 1　不同形状主燃烧室结构参数

主燃烧室	B 型-1.25	S 型	B 型-1.57	T 型	V 型
容积比 β/%	59.6	59.3	58.4	58.9	59.3
压缩比	20.80	20.44	20.15	20.32	20.46

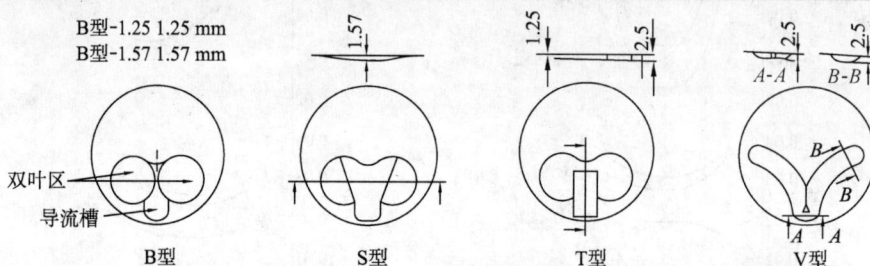

图 10　不同形状的主燃烧室

图 11 的排放测试结果表明,随着主燃烧室凹坑中导流槽深度的不断增加,烟度不断减小,尤其是在 B 型-1.25 的基础上变成 S 型和 T 型后,烟度有明显下降,此外,由 B 型-1.25 和 B 型-1.57 的结果可知,主燃烧室凹坑深度的总体增加,对降低烟度排放也是有利的,V 型结构(凹坑也很深)也显示了降低烟度的良好效果。此外,图 11 的结果又表明,随着主燃烧室凹坑导流槽深度增加,NO_x 的排放量略有上升。根据图 11 可以给出主燃烧室形状不同时两燃烧室(主燃烧室和涡流室)在不同时期的放热率,由图 12 可知[5],涡流室形状和容积在大体相同

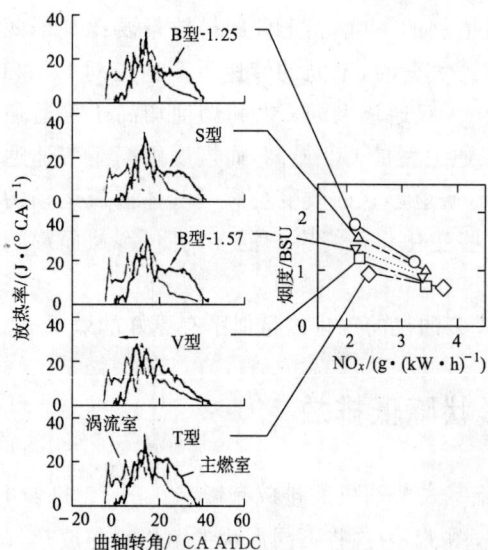

图 11　不同形状主燃烧室的放热规律和排放特性

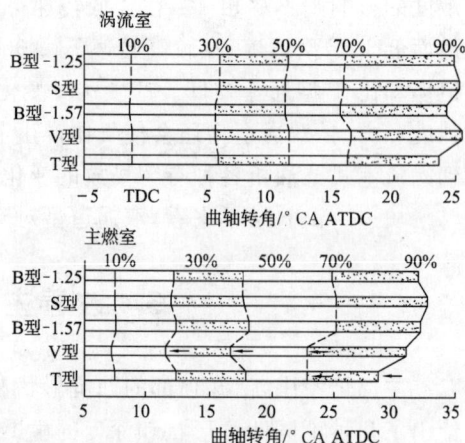

图 12　涡流室和主燃烧室放热率阶段比较

的条件下,其放热率的变化也大体相同,但是由于主燃烧室形状变化较大,对主燃烧室内放热率影响较大。T 型主燃烧室与 B 型、S 型相比,前 50% 的放热时期大致相同,在后 50% 里,T 型主燃烧室在更短的时间内完成了放热过程。对 V 型主燃烧室来讲,整个放热过程时间都较短。这一结果表明,主燃烧室活塞顶部凹坑(尤其是导流槽)深度的增加,对保证主燃烧室内更加充分的混合是必要的。混合气质量的提高,有利于主燃烧室保持充分的燃烧,因而缩短了燃烧期,使得烟度明显下降;反之,充分燃烧导致了燃烧温度增加,因而 NO_x 的排放有所上升。图 13 为上述几种形状主燃烧室内火焰发展的高速摄影照片,这些照片表明,凹坑深度(尤其是其中导流槽深度)的增加,可以有效地防止由涡流室喷出的火焰向活塞顶部凹坑以外的区域扩散,从而有效地使火焰更加集中于主燃烧室中。

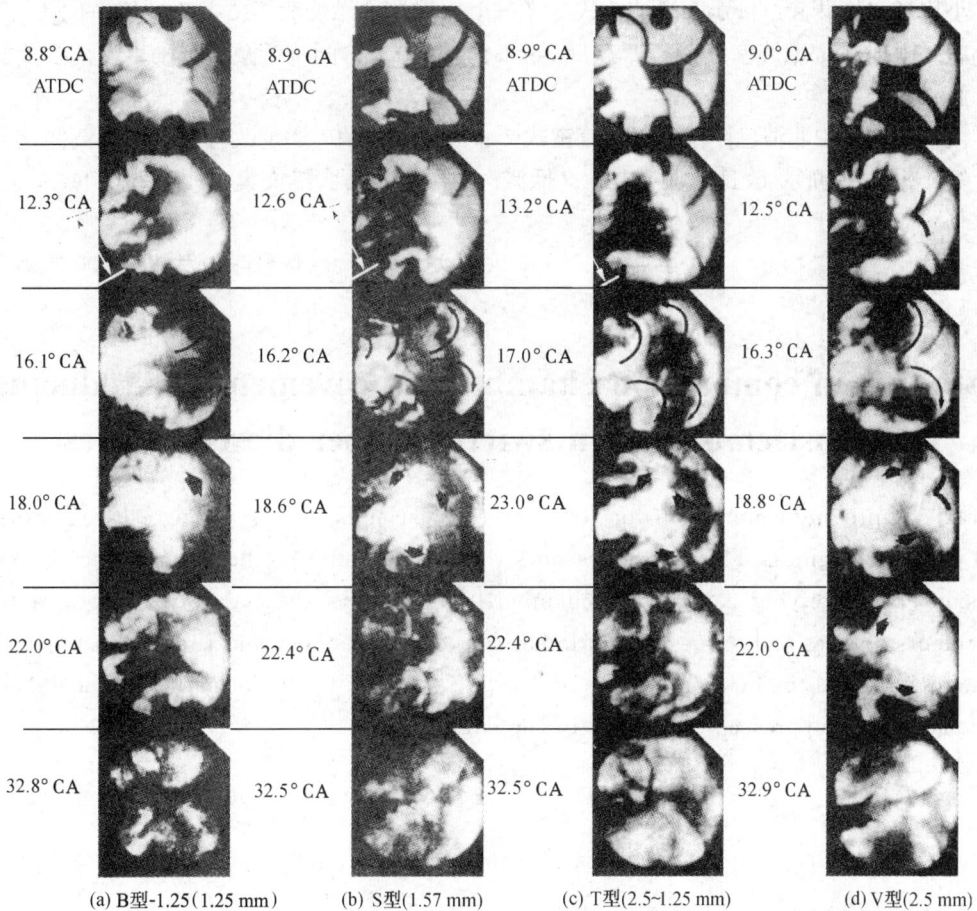

(a) B型-1.25(1.25 mm)   (b) S型(1.57 mm)   (c) T型(2.5~1.25 mm)   (d) V型(2.5 mm)

图 13　不同形状主燃烧室内火焰发展的高速摄影照片

4　结　　论

　　改进涡流室式柴油机燃烧室结构设计来降低其有害排放,是具有较高实用价值和理论意义的研究方向。本文的大量试验表明,通过对这类柴油机涡流室形状的大小、主副室通道面积和形状以及主燃烧室形状的适当改进,可以较好地改善涡流室式柴油机的混合气质量和燃烧过程,从而使其排放得到抑制。

参 考 文 献

[1] 史绍熙,李德桃,等. 建立我国车辆排放法规若干问题的研究[J]. 内燃机学报,1996,14
(2):111-118.

[2] 史北清已,稲吉三七二,ほか. 渦流室式デイーゼル機関の燃焼改善による排気浄化
[C]// 第3報:中負荷時のスモタ生成要因とその低減. 自動車技術会論文集,1996,27
(4):45-51.

[3] 小川孝,佐藤武,ほか. 渦流室式デイーゼル機関の燃焼改善による排気浄化[C]// 第1
報:噴射系・燃焼室改良による NO バライキエレトの同時低減. 自動車技術会論文集,
1996,27(4).

[4] 堀田義博,中北清已,ほか. 渦流室式デイーゼル機関の燃焼改善による排気浄化 [C]//
第4報:主室形状改良によるスモタ低減. 自動車技術会論文集,1997,28(20).

(本文原载于《燃烧科学与技术》2000 年第 1 期)

Studies on combustion chamber improvement for reducing exhaust emissions in swirl chamber diesel engines

Abstract: Improving combustion for reducing exhaust emissions is a top research project in swirl chamber engines. This paper presents the experimental methods for reducing exhaust emissions by modifying combustion chamber. The results show that the appropriate improvements of sizes and shapes of swirl chamber, main chamber and throat can improve the mixing quality and combustion process of charge, as a result, the formation of the exhaust emissions can be suppressed effectively in the engines.

改善涡流室式柴油机供油系统和利用
废气再循环降低排放的研究

董　　刚,李德桃,夏兴兰,杨文明

[摘要]　通过改进柴油机供油系统和利用废气再循环来改善混合气质量和燃烧过程,进而降低其有害排放。结果表明,推迟喷油提前角、适当降低喷油压力、抑制初期喷油率、减小高压油管直径、利用废气再循环,以及这几种方法的结合,能够有效地降低排放。

　　改善柴油机供油系统和进气系统对降低有害排放有重要影响,本文在分析国内外已有研究成果的基础上,从改进供油系统参数和利用废气再循环的角度,对降低涡流室式柴油机的排放进行了系统研究,以获得规律性结果。

1　改进供油系统参数以降低排放

1.1　喷油提前角

　　在 195 型柴油机上的排放测试表明,喷油提前角对各种有害排放有明显影响(见图 1)。

　　结果表明,随着喷油提前角的减小,在两种转速下 NO_x 排放量明显下降,CO,HC 和烟度的排放量有所增加。其中在低负荷时,CO 和 HC 排放增加明显;而在高负荷时,烟度增加明显。

　　喷油提前角的变化直接影响柴油机预混合燃烧量与扩散燃烧量的比例,这对整个燃烧过程以及排放的生成产生重大影响。喷油提前角减小,着火前的油气预混合比例变小,混合气浓度低,混合时间短,着火后燃烧不剧烈,燃烧温度低,因而 NO_x 排放量下降;而这种工况条件又造成了相对缺氧和燃烧不充分的环境,因而使得 CO,HC 和烟度的排放量又有所上

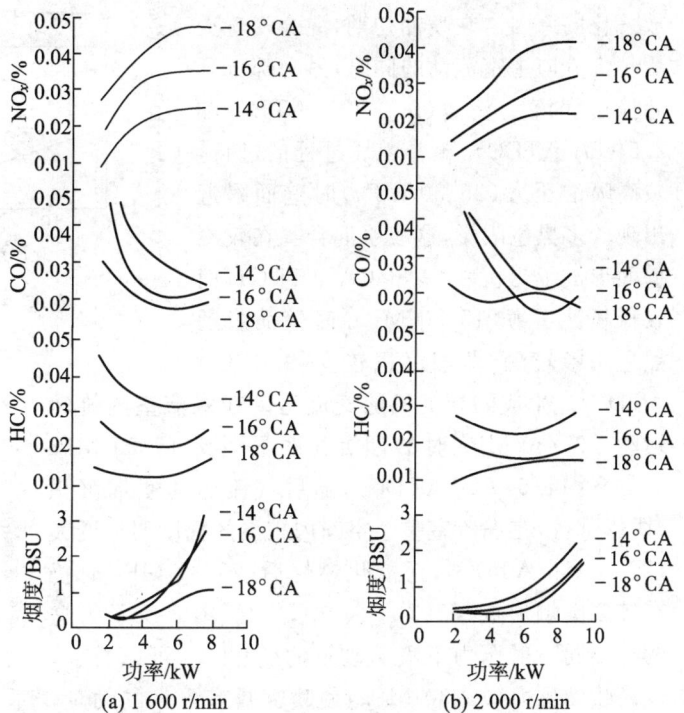

图 1　不同喷油提前角下 NO_x,CO,HC 和烟度随负荷的变化关系

升。由此可知,喷油提前角减小有利于减少 NO_x 排放,但其他有害排放又有所上升。文献[2]对 295 型涡流室式柴油机排放随喷油提前角变化的测试结果也基本反映了这一规律,只是 CO 排放量从喷油提前角 17° CA BTDC 到 15° CA BTDC 时呈现下降这一较为特殊的现象,其作者认为这是由于 295 型柴油机喷油持续期较短,未引起后燃造成的。

1.2 喷油压力

不同于直喷式柴油机,涡流室式柴油机能组织较强的空气涡流运动来促进混合气形成和燃烧,因而对喷油压力要求并不十分严格。这说明喷油压力不是影响涡流室式柴油机排放的主要因素。然而,近期有关增压四缸涡流室式柴油机的排放测试表明[3],提高喷射压力,在中等负荷下会使烟度和颗粒物的排放有所恶化,但对于 HC 和 NO_x 以及对低负荷下的烟度、微粒、HC 和 NO_x 均没有明显影响(见图 2)[3]。图 2 中喷射压力用 2 个弹簧的喷油器和 2 种启喷压力的变化来控制。

文献[1]中图 3 高速摄影的结果可以解释喷油压力增加导致烟度和 PM 排放恶化的原因。在喷油初期,混合气着火后一部分向主室流动的同时,另一部分开始撞击涡流室壁面(2.4° CA ATDC 和 2.5° CA ATDC)。随着燃烧过程的进行,较高喷油压力(25/30 MPa)时壁面附近出现较多黑色阴影,这表明混合气在较冷壁面形成了烟沉积(9.7° CA ATDC);而较低喷油压力(15/20 MPa)时壁面附近黑色阴影较少,即烟沉积较少(9.8° CA ATDC)。生成的烟沉积随气流运动在涡流室内旋转运动。25/30 MPa 时的烟沉积明显要比 15/20 MPa 时的多(14.5° CA ATDC),随后气流运动使烟沉积(阴影部分)在涡流室整个空间内扩展的幅度明显变大(32.4° CA ATDC),至氧化燃烧后,25/30 MPa 时仍有较多碳烟形成(57.7° CA ATDC)。图 3 所示为这两种不同喷射压力下喷雾动量的变化[4]。可以看到,较高喷油压力(25/30 MPa)使喷雾具有更大的动量,喷雾喷溅到燃烧室壁的燃油量增加。这使得文献[1]中图 3 左端两列所示的燃烧过程产生更多的碳烟和 PM。

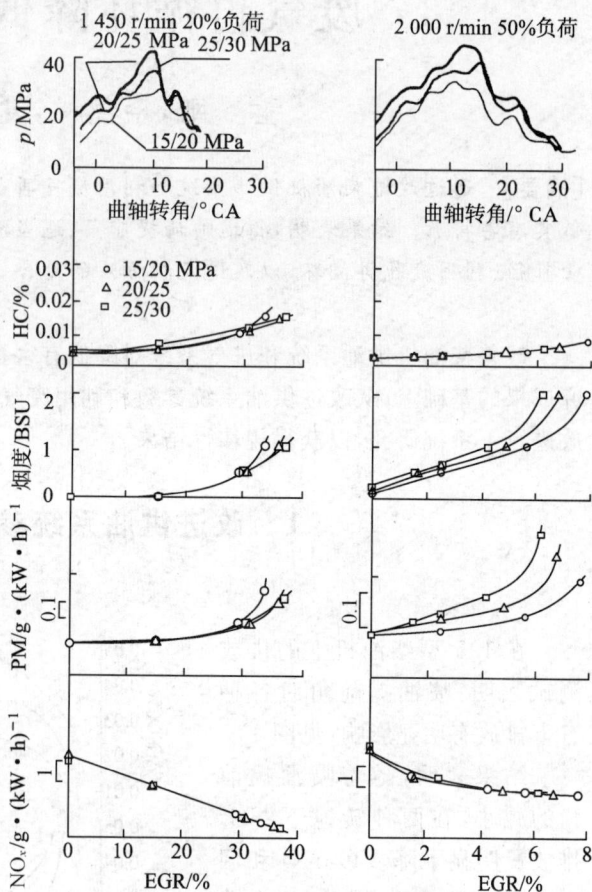

图 2　增压柴油机不同的 EGR 时不同喷油压力对排放的影响

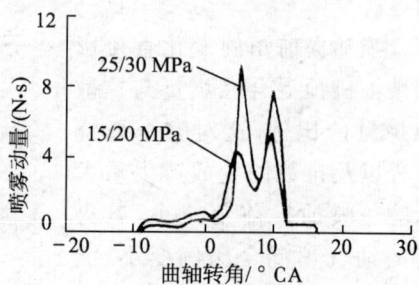

图 3　不同喷油压力下的喷雾动量

根据以上分析,适当降低涡流室式柴油机的喷油压力,对降低碳烟和 PM 是有利的。

1.3 初期喷油速率

初期喷油速率的变化对涡流室式柴油机的排放有着重要的影响。图 4 所示为初期喷油速率的变化对柴油机 HC、烟度、PM 和 NO_x 排放量的影响[3],图中初期喷油速率的变化分别由 1 个弹簧的喷油器(启喷压力 p_0 为 15 MPa)和 2 个弹簧的喷油器(启喷压力分别为 15/20 MPa 和 15/25 MPa)来控制。

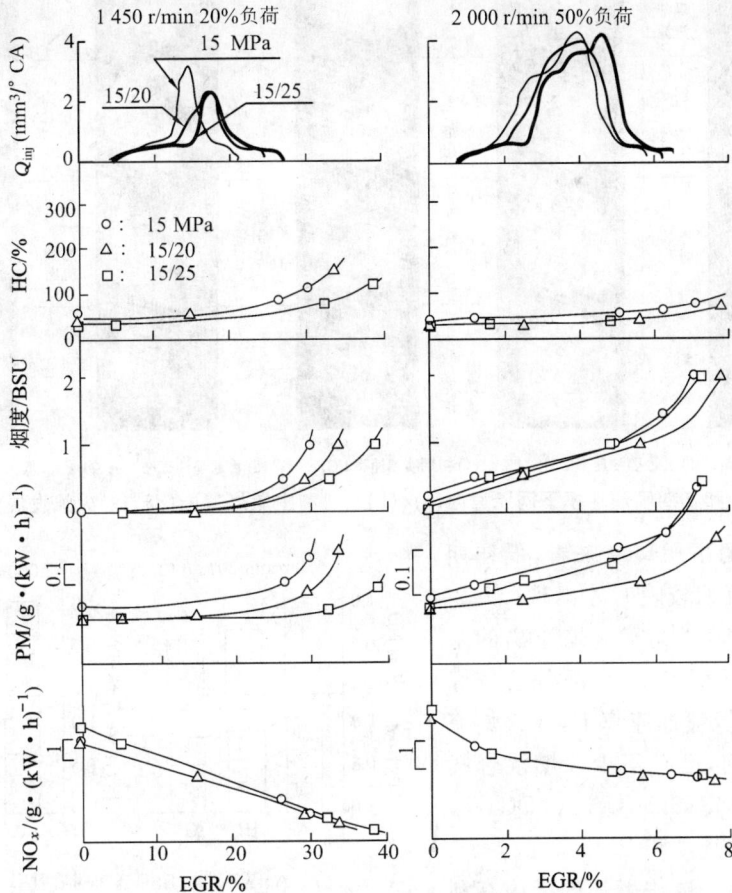

图 4 增压柴油机不同 EGR 时不同初期喷油速率对排放的影响

由图可见,随着初期喷油率的不断抑制(即 p_0 从 15→15/20→15/25 MPa),PM 和烟度得到了明显改善,这一状况在低负荷时尤为显著;而 NO_x 和 HC 的排放量没有明显变化。文献[5]对这一结果进行了研究,以解释造成这一变化的原因。图 5b、图 5c 为 EGR(废气再循环)为 24%时,两种不同初期喷油速率下,涡流室内混合气着火和火焰发展的高速摄影照片,图 5d、图 5e 为主燃烧室内火焰发展的高速摄影照片。

由图 5b 和 5c 可知,在传统喷油速率下,着火发生在喷嘴附近(1.3° CA ATDC);而在抑制初期喷油速率的条件下,着火点移至喷雾顶端(1.8° CA ATDC)。由于喷雾与壁面冲击处先着火,从而抑制了因冷却沉积所形成的烟。这表明,对初期喷油速率进行抑制可防止预混合气中过浓的燃油量附着于涡流室壁面而形成积碳。随后,在传统喷油速率下,5.0° CA ATDC 时便有烟沉积进入主燃烧室;而在抑制初期喷油率时,在大致相同时刻(5.1° CA ATDC),在喷雾顶端先着火的火焰便开始流向主燃烧室。烟沉积在 16.3° CA ATDC 时才流入主燃烧室。

261

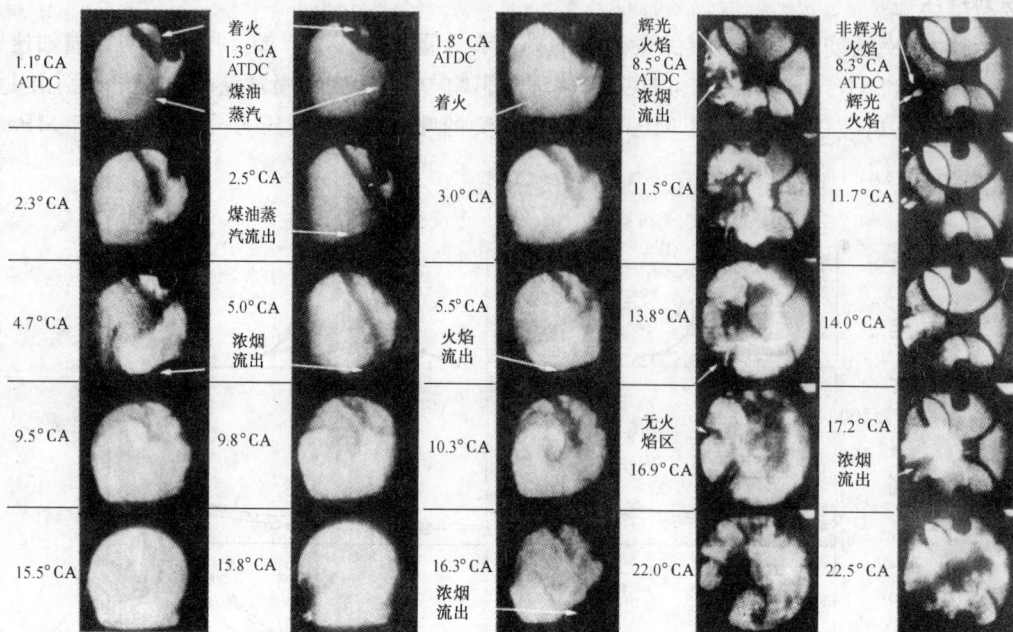

(a) 普通喷油速率 EGR 0%　(b) 普通喷油速率 EGR 24% (c) 低初期喷油速率 EGR 24% (d) 普通喷油速率 EGR 24% (e) 低初期喷油速率 EGR 24%

图 5　不同 EGR 和初期喷油速率下涡流室内(a),(b),(c)和主燃烧室内(d),(e)火焰发展的高速摄影照片

从图 5d 和 5e 的主燃烧室来看,低初期喷油速率的烟沉积较晚时才出现(17.2° CA ATDC),这表明先于火焰流出的未燃混合气和烟沉积较少,大部分已较好地燃烧;而在传统喷油速率下,在火焰流进主燃烧室之前,已有不少未燃混合气和碳烟附着于主室(8.5° CA ATDC),因而形成了大量的碳烟和 PM。

尽管采取单一改进燃烧的措施有利于减少部分有害排放,但另一部分排放指标却可能出现恶化的情况,因此综合利用各种改进措施可以明显降低涡流室式柴油机的各种有害排放物。图 6 为抑制初期喷油率和增大涡流室容积同时使用时,对涡流室式柴油机 NO_x 和 PM 排放的影响。

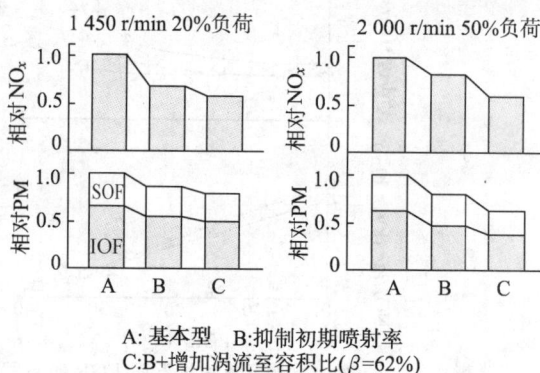

A: 基本型　B: 抑制初期喷射率
C:B+增加涡流室容积比(β=62%)

图 6　抑制初期喷油速率和增大涡流室容积同时使用时降低 NO_x 和 PM 的效果

在一般情况下,NO_x 和 PM 排放的形成因素是相互矛盾的,而由图 6 可见[3],在上述两种方法同时使用时,NO_x 和 PM(包括 SOF 和 IOF)均有所下降。

1.4　高压油管

高压油管变化会影响供油规律,从而对柴油机燃烧和排放物的形成产生一定的影响。图 7 为不同高压油管内径对 NO_x,CO 和烟度影响的变化规律[2]。

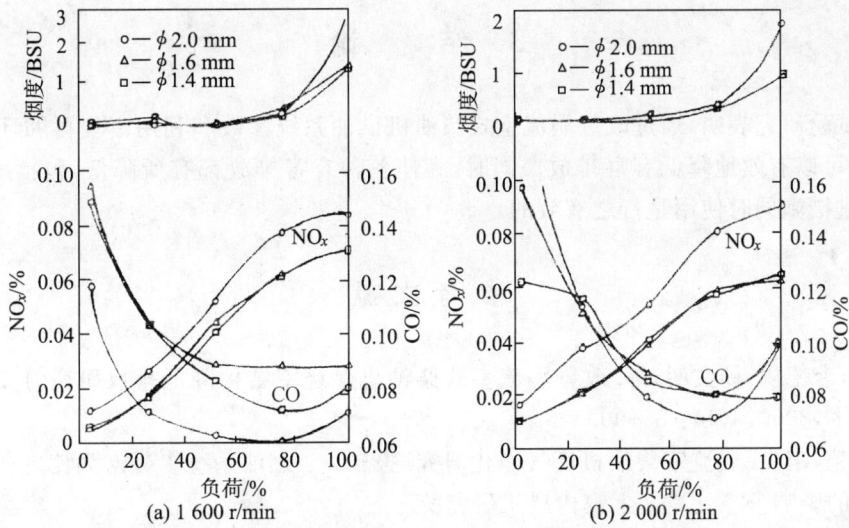

图 7　不同管径的高压油管对 NO_x,CO 和烟度的影响

由此可见,随着高压油管直径减小,NO_x 的排放和烟度有所改善,尤其是在高负荷时更为明显。因此对高压油管进行适当选择,对降低排放也是十分必要的。油管内径小,容积也小,这样,喷油时压力波动的影响就小,针阀开启灵敏,不易出现"滴油"等不正常的现象。但油管过细,也会使燃油流动阻力增大。

2　利用废气再循环降低排放

废气再循环(Exhaust Gas Recirculation,EGR)是指让一部分排气回到进气管,与新鲜空气混合后进入气缸作为工质参加气缸内的热循环。由文献[1]、图 2 和图 4 可以看出,无论改变喷油系统参数还是改变涡流室结构,随着 EGR 率的增加,NO_x 的排放量明显下降,而烟度、PM 和 HC 的排放量却有所增加。EGR 的作用原因可由图 5a、5b 两组高速摄影照片中有无 EGR 时涡流室内火焰发展的情况得到说明。由图 5b 可见,在 EGR 率为 24％时,火焰在喷雾根部着火,喷雾前端有大量的未燃混合气(1.3° CA ATDC);随后,火焰前的未燃混合气一部分先流入主燃烧室(2.5° CA ATDC)。而 EGR 率为 0％时,火焰发展较快,未燃混合气尚未进入主燃烧室(1.1° CA ATDC 和 2.3° CA ATDC)。在 5.0° CA ATDC 时,EGR 率为 24％的喷雾的烟沉积大量形成。由于有 EGR 时,火焰发展较慢,燃烧温度下降,因此 NO_x 生成量减少;未混合气燃烧不充分,PM、烟度以及燃烧中间产物 HC 的生成量大大增加。由于 EGR 对排放物的影响具有上述规律,因此,在使用 EGR 控制有害排放时有一定限度,以保证降低 NO_x 的同时,PM 和烟度等排放不致于恶化。

此外,文献[1]、图 2 和图 4 的结果还说明,同时使用 EGR 和其他几种措施中的一种(如增加涡流室容积,抑制初期喷油速率或减小喷油压力)是同时降低涡流室式柴油机各种有害排放的有效措施。

3 结 论

大量试验研究表明,通过改进涡流室式柴油机供油系统参数和利用废气再循环(EGR)来改善燃烧,可以有效地降低有害排放。同时,为使各种有害排放都有所降低,包括 EGR 在内的几种改进措施同时使用是行之有效的。

参 考 文 献

［1］董刚,李德桃,杨文明,等.改善涡流室式柴油机燃烧室结构降低排放研究［J］.燃烧科学与技术,2000,6(1)：85-90.

［2］邱培基,李贞.涡流室柴油机废气净化研究.报告一：燃油系统参数对涡流室柴油机有害排放的影响［R］.上海：上海内燃机研究所.

［3］小川孝,佐藤武,ほか.渦流室式デイーゼル機関の燃焼改善による排気浄化［C］∥ 第 1 報：噴射系・燃焼室改良による NO バライキエレトの同時低減.自動車技術会論文集,1996,27(4).

［4］史北清已,稲吉三七二,ほか.渦流室式デイーゼル機関の燃焼改善による排気浄化［C］∥第 3 報：中負荷時のスモタ生成要因とその低減.自動車技術会論文集,1996,27(4)：45-51.

［5］堀田義博,中北清已,ほか.渦流室式デイーゼル機関の燃焼改善による排気浄化［C］∥第 2 報：初期噴射率抑制による負荷時のパテイキユレート低減.自動車技術会論文集,1996,27(4).

(本文原载于《燃烧科学与技术》2000 年第 1 期)

Studies on fuel injection improvement and EGR for reducing exhaust emissions in swirl chamber diesel engines

Abstract：The quality of mixture has important influence on exhaust emissions in swirl chamber diesel engines. In this paper, it can be improved by improving fuel injection system and using EGR, the results show that the postponing injection timing, lowering injection pressure, suppressing initial injection rate, using EGR and the combination of the aboved methods can decrease exhaust emissions effectively.

利用混合破碎模型对涡流室内喷雾过程的三维数值模拟

杜爱民,朱埏章,姜树李,李德桃

[摘要] 利用已开发的大型流体计算软件包 EngineCFD-Ⅱ 对涡流室式柴油机内的喷雾过程进行了三维数值模拟。借鉴 WAVE 破碎模型的优点,进一步完善了 TAB 破碎模型,使用混合模型来描述不同阶段的油滴破碎过程。模拟结果与试验数据基本一致。

随着计算机技术和有关领域科学技术的飞速发展,内燃机工作过程的多维数值模拟得到了迅猛发展,并已经在内燃机的研究和开发过程中起到了非常重要的作用。EngineCFD-Ⅱ[1] 为最新开发的适用于涡流室式柴油机内工作过程三维数值模拟的大型软件包,该程序采用随机质点方法建立喷雾动态方程,考虑燃油的蒸发、碰撞/聚合、振荡与破碎、碰壁(包括反弹、粘附及飞溅)等基本物理过程,较真实地模拟燃油与空气混合的全过程。在研究过程中,发现正确合理的燃油破碎模型对模拟结果有重要影响。借鉴 WAVE[2-4] 破碎模型的优点,进一步完善了 KIVA-Ⅱ[5] 中 TAB[6] 破碎模型,使用混合模型来描述不同阶段油滴破碎过程。

1 破碎模型

目前通常使用两种破碎模型来模拟燃油破碎过程,一种是 TAB[6] 模型,由 O'Rourke 和 Amsden 在 Taylor 比拟的基础上提出来的;另一种是 WAVE[2-4,8] 模型,由 Reitz 和 Diwakar 提出。

1.1 TAB 破碎模型

TAB 模型是在 Taylor 将液滴振动及变形与弹性质量系统比拟(Taylor's analog)的基础上得到的。作用在液滴 m 上的空气动力 F 对应外力,弹性反应比拟液滴壁面张力 k,阻尼力比拟液滴粘性力 d,因此

$$m\ddot{x}=F-kx-d\dot{x} \tag{1}$$

式中 x 为液滴直径从平衡位置的位移。比拟为

$$\frac{F}{m}=C_F\frac{\rho_g u^2}{\rho_1 r}, \frac{k}{m}=C_k\frac{\sigma}{\rho_1 r^3}, \frac{d}{m}=C_d\frac{\mu_1}{\rho_1 r^2}$$

式中 ρ_g, ρ_1 为气体、液体密度;u 为气体与液滴的相对速度;r 为液滴半径;σ, μ_1 为液体表面张力、粘性。

若 $x>C_b r$ 时液滴破碎,其中 C_b 为另外一无量纲常数的 1/2。使 $y=x/C_b r$,方程变为

$$\ddot{y}=\frac{C_F}{C_b}\frac{\rho_g u^2}{\rho_1 r^2}-\frac{C_k\sigma}{\rho_1 r^3}y-\frac{C_d\mu_1}{\rho_1 r^2}\dot{y} \tag{2}$$

仅当 $y>1$ 时发生破碎。u 为常数时,方程的解为

$$y(t) = \frac{C_F}{C_k C_b} We + \mathrm{e}^{-t/t_d} \cdot \left[\left(y_0 - \frac{C_F}{C_k C_b} \right) \cos(\omega t) + \frac{1}{\omega} \left[\dot{y}_0 - \frac{y_0 - \frac{C_F}{C_k C_b} We}{t_d} \right] \sin(\omega t) \right] \quad (3)$$

式中 $\frac{1}{t_d} = C_d \dfrac{\mu_1}{2\rho_1 r^2}$; $\omega^2 = C_k \dfrac{\sigma}{\rho_1 r^3} - \dfrac{1}{t_d^2}$; $We = \dfrac{\rho_g u^2 r}{\sigma}$ 。

无量纲量 C_F, C_k 和 C_d 为模型常数,与试验数据相一致。在一定的物理条件下,$C_F = 1/3$,$C_k = 8$,$C_d = 5$(参见文献[6])。这些数据是在与基本振动模型匹配的临界 Weber 数大约为 6 时从振动试验得到的。

1.2 WAVE 破碎模型

WAVE 模型认为液滴的破碎是由于液滴表面的不稳定增长。无限小的轴对称位移形式的扰动波幅为

$$\eta = \eta_0 \mathrm{e}^{ikz + \omega t} \quad (4)$$

式中 η_0 为开始扰动值;ω 为波的增长率。对于这些波动可以写出并求解其运动方程和连续方程,得到一耗散方程求解波增长率和波长。最大波增长率 Ω 和相应的波长 Λ 与液体、气体的物理性质有关:

$$\begin{cases} \dfrac{\Lambda}{a} = 9.02 \dfrac{(1 + 0.45 Z^{0.5})(1 + 0.4 T^{0.7})}{(1 + 0.87 We_2^{1.67})^{0.6}} \\ \Omega = \left[\dfrac{\rho_1 a^3}{\sigma} \right]^{0.5} = \dfrac{0.34 + 0.38 We_2^{1.5}}{(1 + Z)(1 + 1.4 T^{0.6})} \end{cases} \quad (5)$$

式中 We_2 为气体的 Weber 数;Z 为 Ohnesorge 参数($= We_1^{0.5}/Re_1$),即液体 Weber 数和 Reynolds 数的比率;$T = Z/We_2^{0.5}$;a 为父液滴的半径。该模型假设半径为 a 的父液滴形成半径为 r 的新液滴:

$$\begin{cases} r = B_0 \Lambda, B_0 \Lambda \leqslant a \\ r = \min((3\pi a^2 u / 2\Omega)^{0.33}, (3a^2 \Lambda / 4)^{0.33}), B_0 \Lambda > a \end{cases} \quad (6)$$

父液滴半径按照下式变化

$$\frac{\mathrm{d}a}{\mathrm{d}t} = \frac{-(a - r)}{\tau}, \tau = \frac{3.726 B_1 a}{\Lambda \Omega} \quad (7)$$

式中 B_0 和 B_1 为模型常数;Reitz 分别为 30.61, 1.73 或 10[4,8]。

1.3 Hybrid(混合)模型

TAB 和 WAVE 两种模型都需要一些经验常数。Beatrice C 等[9]人在一定范围的喷射条件下进行了计算和试验比较,并就 TAB 和 WAVE 模型对经验常数的敏感性进行分析。分析表明,TAB 模型和 WAVE 模型预测结果有所不同,TAB 模型过低地预测贯穿度,因为对于非常小的液滴其破碎趋势太快;WAVE 模型则过高地预测贯穿度,特别是在喷射的早期。为了更加真实地进行燃油破碎模拟,这里采用 Beatrice C 等人所建立的混合模型("hybrid" model)。考虑到喷射液滴直径较大,混合模型在喷油开始时使用 WAVE 模型,直到液滴直径小于喷射最大直径的 95%,然后调用 TAB 模型。

两模型的选择从数值观点看是可以接受的,同时也有物理意义:对于喷油嘴处的大直径液滴,不稳定波在液滴表面传播模型更合适;相反,对于小直径液滴,把液滴比作质量弹性系统更合适。因此本文采用了此混合模型来模拟液滴破碎过程。

2 发动机参数与计算网格

本文主要对 S1100 涡流室式柴油机进行数值计算,其主要技术参数如表 1。

表 1 发动机主要技术参数

参　数	指　标	参　数	指　标
型号	S1100 柴油机	型式	卧式、单缸、四冲程、涡流室
气缸直径	100 mm	活塞行程	115 mm
12 小时功率,标定转速	10.3 kW,2 000 r/min	超负荷功率	11.32 kW
活塞排量	0.903 L	压缩比	20
平均有效压力	697 kPa	燃油消耗率	≤250.2 g/(kW·h)

S1100 涡流室式柴油机喷油嘴选用轴针式喷油嘴 ZS4S1。喷嘴处的流通面积为 0.785 mm²,喷油期间其平均流通面积约为 0.5 mm²,喷油器的安装角(喷油器中心线与涡流室纵向中心线的夹角)为 −20°。发动机在标定工况下转速为 2 000 r/min,每循环供油量约为 44.3 mg,喷油压力为 12 MPa,在上止点前 10° CA 开始喷油,喷油持续 16° CA[10]。

该发动机采用吊钟型涡流室,容积比 V_k/V_c(V_k 为涡流室及通道的容积;V_c 为整个压缩室容积)约为 54%。通道截面形状为圆矩形,通道倾角(通道中心线与气缸盖底面或镶块底面的夹角)为 40°,通道的长宽为长×宽=1.75 cm×0.65 cm,主燃室活塞顶部具有涡流槽。计算采用直角坐标系下整体网格。如果考虑涡流槽,网格的生成非常复杂,故暂不考虑此涡流槽,同时也不考虑起动孔的存在。为保证压缩比不变,上止点时活塞顶部余隙高度可通过压缩比的计算公式求得。此时根据所开发的网格自动生成技术所划分的网格如图 1 所示。其网格数为 $i×j×k$=25×25×30,涡流室内的网格为 25×25×18,通道内的网格为 5×9×4,且通道在 I,J 方向上的起始网格分别为 11 和 8。

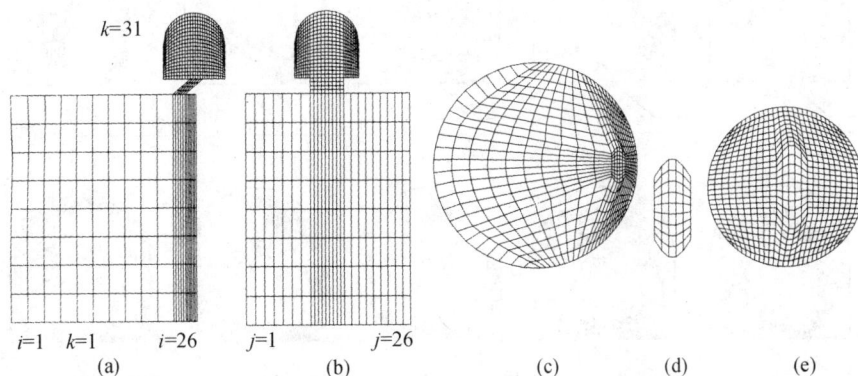

图 1 25×25×30 网格图

3 计算结果分析与验证

3.1 计算与试验结果对比

图2~图4分别为喷雾形态的主视图、侧视图和俯视图。由这些图可以看出,燃油喷入高压高速流动的涡流室内,由于气流的阻尼,加上油滴的破碎等作用,部分小液滴很快被"吹偏"而随气流一起运动。虽然油滴以一定的初始速度喷入涡流室,但进入高压气体后,由于气体的阻尼作用,液滴和气体间的相对速度迅速减小,很快小液滴与气体间的相对速度降低为零而随气流一起运动。大液滴虽然仍沿液滴速度方向运动,但很快大液滴就破碎成小液滴而被卷入气流运动中去。

由主视图可以看出,燃油虽然被气流"吹向"壁面,但随气流运动的小液滴几乎不与侧壁相碰。随气流运动的液滴有少部分与涡流室底面相碰,碰壁后的部分燃油反弹后随气体运动,很少一部分气体粘附在壁面形成油膜。燃油随气体运动到通道口附近,被由主燃室经通道压入的气流吹向涡流室的上部,但仍有少部分燃油随气流进入主视图中右下角的微涡流处,大部分燃油随主涡流一起运动,很快分布在整个涡流室区域。

由侧视图和俯视图可知,燃油随主涡流运动,并非只在一定的 J 截面内分布。如文献[1]所述,涡流室内不仅有主涡流和副涡流存在,从各 K 截面图可以看出,涡流室内还存在复杂的气流运动,正是在此复杂的气流运动下,燃油才得以在涡流室内分布的尽量均匀。

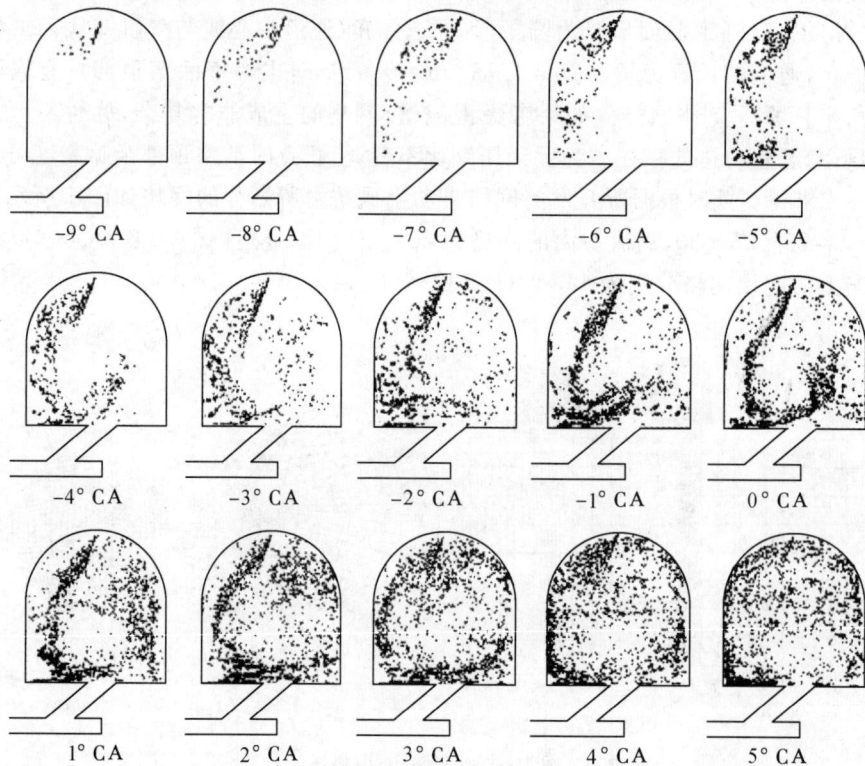

图 2 喷雾形态的主视图

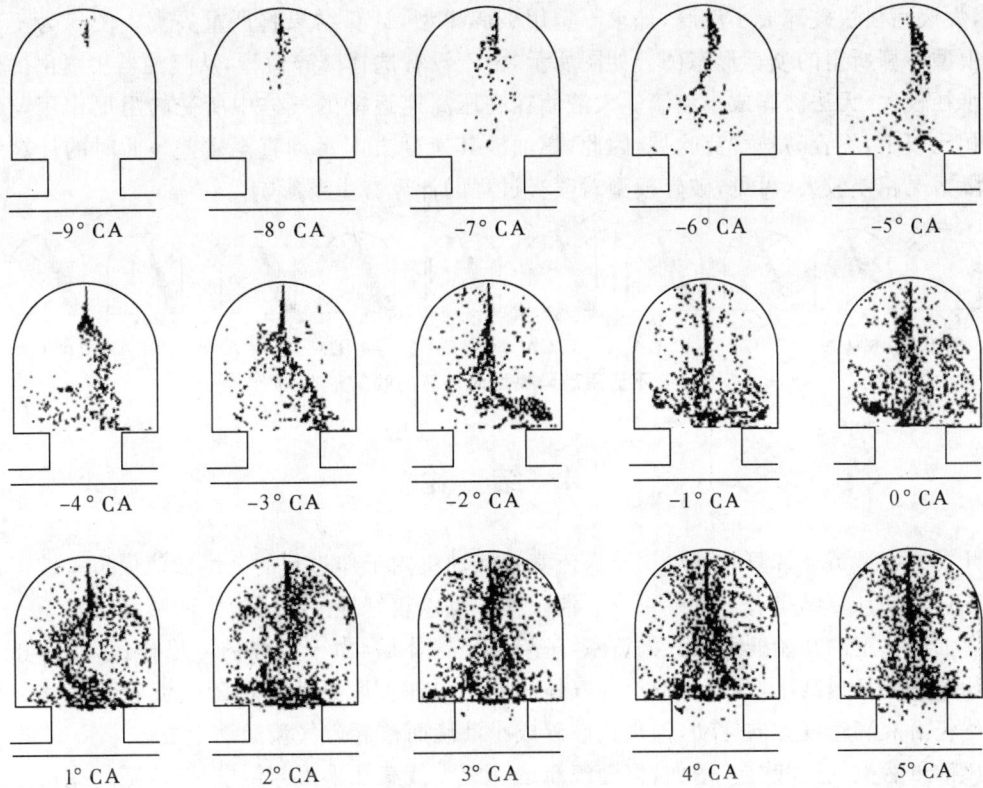

图 3 喷雾形态的侧视图

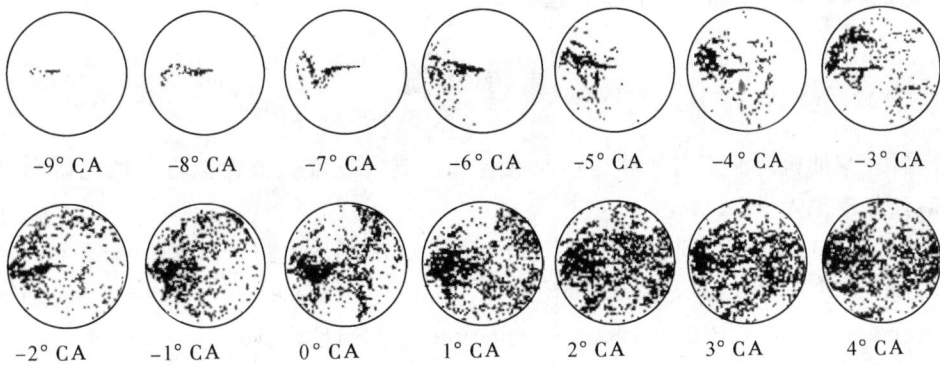

图 4 喷雾形态的俯视图

计算结果与文献[10]中高速摄影所得的结果是一致的,从而证明本研究数学模型和计算方法的正确性。与文献[10]的不同之处在于,本文计算结果表明气流运动对燃油的影响较强烈,而文献[10]的试验结果表明喷油的中间阶段受气流运动的影响较小。产生此差别的原因是由于试验样机与实际发动机并不完全相同,试验样机的压缩比降低到 12.8,转速也降低到 1 230 r/min,在该样机下,涡流室内的气流运动比实际发动机要弱得多,因此对喷油的影响也相对较弱。

3.2 破碎模型的影响

破碎模型的选择对计算结果有重要影响。试验表明,油滴在高压高速运动中的涡流室内运动,大液滴很快破碎成小液滴,如果不选用破碎模型,计算结果将严重失真。图5为不选用破碎模型计算所得的喷雾形态图。如图所示,由于没有选择破碎模型,从喷油器中喷出的燃油初始半径较大,无法破碎成小液滴。大液滴在高压高速运动的气体中所受的阻尼作用与惯性作用相比较小,不容易被气流吹偏,因此,燃油沿其速度方向运动直至碰壁。此时的计算结果与实际情况相差较大,可见,破碎模型对喷雾过程的计算有重要影响。

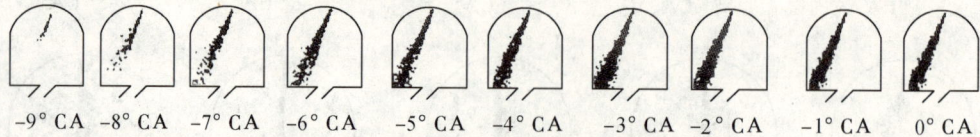

| −9° CA | −8° CA | −7° CA | −6° CA | −5° CA | −4° CA | −3° CA | −2° CA | −1° CA | 0° CA |

图 5　不选用破碎模型时计算的喷雾形态图

4　结　论

开发的大型流体计算软件可用于涡流室式柴油机内工作过程的三维数值模拟。破碎模型对喷雾过程的计算结果有着重要影响。混合模型可以较好地模拟涡流室的喷雾过程。

燃油以一定的初始速度喷入涡流室,进入高压气体后,由于气体的阻尼作用,液滴和气体间的相对速度迅速减小,很快小液滴与气体间的相对速度降低为零而随气流一起运动。大液滴虽然仍沿液滴速度方向运动,但很快破碎成小液滴而被卷入气流运动中去。

燃油虽然被气流"吹向"壁面,但随气流运动的小液滴几乎不与侧壁相碰,随气流运动的液滴有少部分与涡流室底面相碰,碰壁后的部分燃油反弹后随气体运动,很少一部分气体粘附在壁面形成油膜,大部分燃油随主涡流一起运动。

参 考 文 献

［1］杜爱民. 柴油机涡流室内气体运动及混合气形成的三维数值模拟与可视化研究［D］. 镇江：江苏理工大学,1998.

［2］Reitz R D, Diwakar R. Structure of high-pressure fuel sprays[J]. *SAE Paper*, 870589,1987.

［3］Gonzales M A, Borman G L, Reitz R D. A study of diesel cold-starting using both cycle analysis and multidimensional calculation[J]. *SAE Paper*, 910180,1991.

［4］Gonzales M A, Lian Z W, Reitz R D. Modeling diesel engine spray vaporization and combustio[J]. *SAE Paper*, 920579,1992.

［5］Amsden A A. KIVA-Ⅲ: a KIVA program with block-structed mesh for complex geometries[R]. Los Alamos National Laboratory Report, LA-12503-MS,1993.

［6］O' Rourke P J, Amsden A A. The TAB method for numerical calculation of spray droplet break up[J]. *SAE Paper*, 872089,1987.

［7］何学良,李疏松. 内燃机燃烧学［M］. 北京：机械工业出版社,1990.

［8］Liu A B, Mather D, Reitz R D. Modeling the effects of drop drag and break-up on fuel sprays[J]. *SAE Paper*, 930072,1993.

［9］Beatrice C, Belardini P, Bertoli C, et al. Fuel jet models for multidimensional diesel combustion calculation：an update[J]. *SAE Paper*，950086，1995.

[10] 李德桃，朱广圣. 吊钟型涡流室内喷油和燃烧过程的研究[J]. 内燃机学报，1989，7(1)：21-26.

[11] Amsden A A，O' Rourke P J，Butler T D. KIVA-Ⅱ：a computer program for chemically reacting flows with sprays[R]. Los Alamos National Laboratory Report，LA-11560-MS，1989.

[12] 杜爱民. 直喷式柴油机缸内气体流动和喷雾过程的三维数值模拟[D]. 镇江：江苏理工大学，1996.

（本文原载于《内燃机学报》2000 年第 1 期）

Three-dimensional numerical simulation of the spray process in swirl chamber with a hybrid breakup model

Abstract：With the newly developed software EngineCFD-Ⅱ，three-dimensional numerical simulation of the spray process in swirl chamber was carried out. According to WAVE model，the original TAB model in KIVA-Ⅱ is modified，the droplet's breakup process is simulated with a hybrid model which uses both TAB and WAVE model in different period. The numerical result is accordant with testing result.

LDA measurement and 3-D modeling of air-motion in swirl chamber of diesel engines

Li Detao, *Wang Qian*, *Xia Xinglan*, *Liu Jingping*

Abstract

Test investigations of air flow in the swirl chamber of a diesel engine were successfully conducted using Laser Doppler Anemometry (LDA). Air velocities at certain points in the swirl chamber were measured under motored condition for several engine speeds. The mean velocity and turbulence intensity was calculated using the ensemble average method. A general three-dimensional micro-computerized code was modified from KIVA II and developed to simulate the air-motion in the swirl chamber. In the model, the turbulent flow of air is modeled using the modified k-ε turbulent model. The model was applied on to a bell type swirl chamber diesel engine and fairly good agreement was shown between the LDA measurement data and the simulation results. The air velocity profile in the vertical section, cross section and side sections are given. The obtained results clearly demonstrate the three-dimensional characteristics of air-motion in the swirl chamber, including the change and development of vortices.

1 Introduction

Wide application of high injection pressure ($>$70 MPa) direct injection (DI) diesel engines are facing challenges in China from both the technical and economy sides. Swirl chamber indirect injection (IDI) diesel engines, due to their less harmful emissions and lower combustion noise, still take a significant market share in China. The more and more strict requirements of high fuel economy and low exhaust emissions put pressure on the improvement of swirl chamber diesels. One way to solve the problem is to strengthen the fundamental research efforts on the air-motion in combustion chamber and on the mixture formation and then combustion process of the mixture, and based on that to develop new swirl chamber combustion system.

The quality of air/fuel mixture and combustion process are strongly associated with the air-motion in the swirl chamber. It was shown by many previous researches that the swirl motion with proper intensity can effectively enhance the evaporation process of fuel droplets and help the mixture of fuel and air[1-4]. In recent years, it has been gradually realized that the turbulence motion of air plays an important role in the formation and combustion of fuel/air mixture. Although measurements of air-motion in the swirl chamber were conducted by previous researchers[5-7] with various methods and much progress was made, the experimental approach was limited to contact measurement with mechanical methods and water simulation rigs in which the measuring range and accuracy were limited. Therefore the air-motion in the swirl chamber, especially the turbulent flow, is still not very clearly understood.

On the other hand, numerical simulation of air-motion in the diesel engine combustion chambers has been widely used. Many researchers have applied multidimensional simulation on DI Diesel engines and very good results have been published. However, in swirl chambers of IDI engines, the complexity of the combustion chamber made it more difficulty to model and less modeling effort has been made[8-14]. Some previous researchers[15-18] simplified the 3-diemensional problem into 2-dimensional models. In the 2-dimensional models the influence of the position and orientation of the connecting channel can not be correctly studied, although it is critical to the air-motion in the swirl chamber. Besides, the 2-dimensional model needs the measured boundary condition at the connecting channel exit. Limited by the available measurement space, the air velocity and temperature distribution profiles are very difficult to determine under engine operational condition. Among the 3-dimensional modeling codes of air-motion in combustion chambers, KIVA II[19] had been widely applied on IC engines, but non application in "bell" type swirl chamber IDI engines had been found.

In this study, an experimental investigation of air-motion in the bell type swirl chamber was conducted with advanced Laser Doppler (LD) technology and 2-dimensional argon ion Laser Doppler Anemometry (LDA). Modified from the KIVA II Code, a 3-dimensional mathematical model and solution scheme were then developed for swirl chamber IDI diesel engines. Modification to the KIVA II code included: (1) automatic mesh generation; (2) modified boundary layer model (k-ε model); (3) modified fuel droplet impingement model to account for the swirl chamber geometry; (4) modified solution scheme suitable for running on a PC. The modified code runs much faster on micro-computers than the original KIVA II code.

2　LDA measurement of air-motion in swirl chamber

2. 1　Experimental apparatus

Fig. 1 shows the experimental measurement apparatus of the flow field in the swirl chamber, it mainly consists of four parts: (1) LDA system; (2) mechanical system driving the engine; (3) tracing particle injector; (4) computerized control system. The diesel engine crankangle signals are generated by a shaft angle encoder (Model TK-360) and are used to trigger the LDA sampling. The entire process is controlled by a computer.

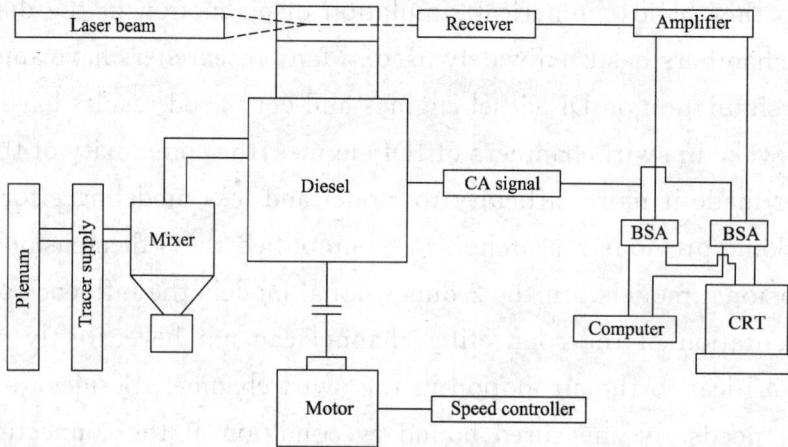

Fig. 1　LDA measurement setup of air-motion in the swirl chamber

The LDA system used in the experiment was manufactured by DANTEC Corporation and it was a 3-beam, 2-dimensional argon ion LDA system. Its parameters were set as follows:

Power: 1. 2 W, argon ion;

Wavelength of laser: blue light, 488 μm;

green light, 514. 5 μm;

Focus distance of launching lens: 600 mm;

Focus distance of receiving lens: 350 mm;

Receiving mode: forward scattering;

Signal processor: frequency spectrum analyzer.

The laser beam access to the swirl chamber was realized by changing the structure of cylinder head and installing quartz windows on both sides of the swirl chamber. The installation of quartz windows did not change the compression ratio of the engine nor the volume ratio of the swirl chamber to the main chamber. But the shape of swirl chamber changed a little, from bell type to flat bell type. Since the experiment was to measure the flow field on the main turbulence plane in the swirl chamber, the change of shape should have little influence to the velocity field.

The testing engine was a single cylinder bell type swirl chamber diesel engine, the major design parameters are listed in Tab. 1.

Tab. 1　Major design parameters of the testing engine

Parameters	Norm
Type	Single cylinder, water cooling, four stokes
No. of cylinders-Bore×Stroke(mm×mm)	1-95×115
Compression ratio	19. 2
Displacement(L)	0. 815
Type of swirl chamber	Bell type
Volume of swirl chamber(L)	0. 025 7
Rated power/speed(kW, r/min)	8. 83/2 000
Measurement speeds(r/min)	400,600&800

2. 2　Data processing

In order to investigate the characteristics of vortex and turbulence in the swirl chamber, the measuring points were arranged in the central vertical section(i. e. the points are distributed in the main vortex plane), which is shown in Fig. 2. The distance between measurement points was 5 mm in abscissa and 4 mm in ordinate. The motoring speeds of engine were 400 r/min, 600 r/min and 800 r/min respectively.

Because of the characteristics of data acquired from the LDA, ensemble average method must be employed to process

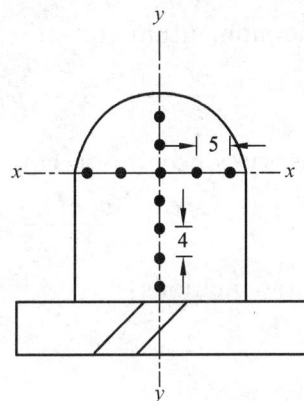

Fig. 2　Typical arrangement of measurement points

the experiment results. The interested parameters are defined as follows:

(1) Average velocity

$$u_{\mathrm{m}}(\varphi) = \frac{1}{N_\varphi} \sum_{i=1}^{N_\varphi} u_i(\varphi) \tag{1}$$

where $u_i(\varphi)$ is the instantaneous velocity at the corresponding crankangle; N_φ is the number of times of the instantaneous velocity measured at the same crankangle.

(2) Turbulence intensity

$$u'(\varphi) = \left\{ \frac{1}{N_{\mathrm{t}}} \sum_{i=1}^{N_{\mathrm{t}}} [u_i'(\varphi) - u_{\mathrm{m}}(\varphi)]^2 \right\}^{1/2} \tag{2}$$

where N_{t} is the total number of measurement being made.

(3) Relative turbulence intensity

$$\varepsilon = \frac{u'(\varphi)}{u_{\mathrm{m}}(\varphi)} \times 100\% \tag{3}$$

3　Mathematical modeling

3.1　Controlling equations of air-motion in the combustion chamber

The air-motion in the combustion chamber is described by a system of partial differential equations, state equation and turbulence equations:

Continuity equation of component m:

$$\frac{\partial \rho_m}{\partial t} + \nabla \cdot (\rho_m u) = \nabla \cdot \left[\rho D \nabla \left(\frac{\rho_m}{\rho} \right) \right] \tag{4}$$

The momentum conservation equation of a multi-composite mixture:

$$\frac{\partial (\rho u)}{\partial t} + \nabla \cdot (\rho u u) = -\frac{1}{\alpha^2} \nabla p + \nabla \cdot \sigma - \nabla \left(\frac{2}{3} \rho k \right) + \rho g \tag{5}$$

Energy conservation equation:

$$\frac{\partial (\rho I)}{\partial t} + \nabla \cdot (\rho I u) = p \nabla u + \nabla \cdot J + \rho \varepsilon \tag{6}$$

State equations:

$$p = R_0 T \sum \frac{\rho_m}{W_m} \tag{7}$$

$$I(T) = \sum \frac{\rho_m}{\rho} I_m(T) \tag{8}$$

$$h_m(T) = I_m(T) + \frac{R_0 T}{W_m} \tag{9}$$

The turbulence model used in the boundary layer was the modified k-ε

model:

$$\frac{\partial(\rho k)}{\partial t} + \nabla \cdot (\rho k u) = -\frac{2}{3}\rho k \nabla \cdot u + \sigma \nabla \cdot u - \nabla \cdot \left(\frac{\mu}{Pr_k}\nabla k\right) - \rho\varepsilon \qquad (10)$$

$$\frac{\partial(\rho\varepsilon)}{\partial t} + \nabla \cdot (\rho\varepsilon u) = -\frac{2}{3}(C_{\varepsilon 1} - C_{\varepsilon 3})\rho\varepsilon \nabla \cdot u +$$

$$\nabla \cdot \left(\frac{\mu}{Pr_\varepsilon}\nabla\varepsilon\right) + \frac{\varepsilon}{k}[C_{\varepsilon 1}\sigma\nabla u - C_{\varepsilon 2}\rho\varepsilon] \qquad (11)$$

In the above equations:

ρ_m, ρ: Density of component m and the overall mixture

D: diffusion coefficient

u: air-motion velocity

p, T, I, h: pressure, temperature, specific internal energy and enthalpy

R_0: gas constant

W_m: molecular weight of component m

k, ε: specific kinematic energy and dissipation rate

g: gravity

α: a variable associated with the pressure gradient scale(PGS)

σ: Newton viscosity tension

J: heat flux, J is the total of the thermal conductivity and enthalpy diffu-

sion:

$$\sigma = \mu[\nabla u - (\nabla u)^{\mathrm{T}}] + \lambda(\nabla \cdot u)E \qquad (12)$$

$$J = -K\nabla T - \rho Dm \sum h_m \nabla \left(\frac{\rho_m}{\rho}\right) \qquad (13)$$

Where:

μ, λ: The first and second viscosity coefficient

K: coefficient of thermal conductivity

E: unit tension

The constants in the k-ε model (10) and (11) are chosen as :

$C_{\varepsilon 1} = 1.44$

$C_{\varepsilon 2} = 1.92$

$C_{\varepsilon 3} = -1.0$

$Pr_k = 1.0$

$Pr_\varepsilon = 1.3$

3.2 Boundary conditions

Air-motion velocity next to the combustion chamber wall surface could be chosen as (1) free sliding, (2) zero velocity; or (3) to apply the turbulent boundary layer theory and solve the velocity gradient; The combustion chamber wall temperature boundary conditions could be (1) adiabatic or (2) constant temperature.

The turbulent boundary layer theory and constant wall temperature boundary conditions were applied to this study.

3.3　Computational grid and meshing

Meshing the combustion chamber is critical for the accuracy of the numerical simulation results. The meshing of the bell type combustion chamber of the S195 IDI diesel engine in this study is shown in Fig. 3. A cylindrical orthogonal grid was chosen in the modeling. The main combustion chamber was divided into $18 \times 16 \times 13$ meshes (The meshes become $18 \times 15 \times 13$ near TDC); Meshing in the connecting channel (it is enlarged in Fig. 3 to show the details) was divided into $6 \times 16 \times 5$ grid. The grid in the swirl chamber was $16 \times 16 \times 12$.

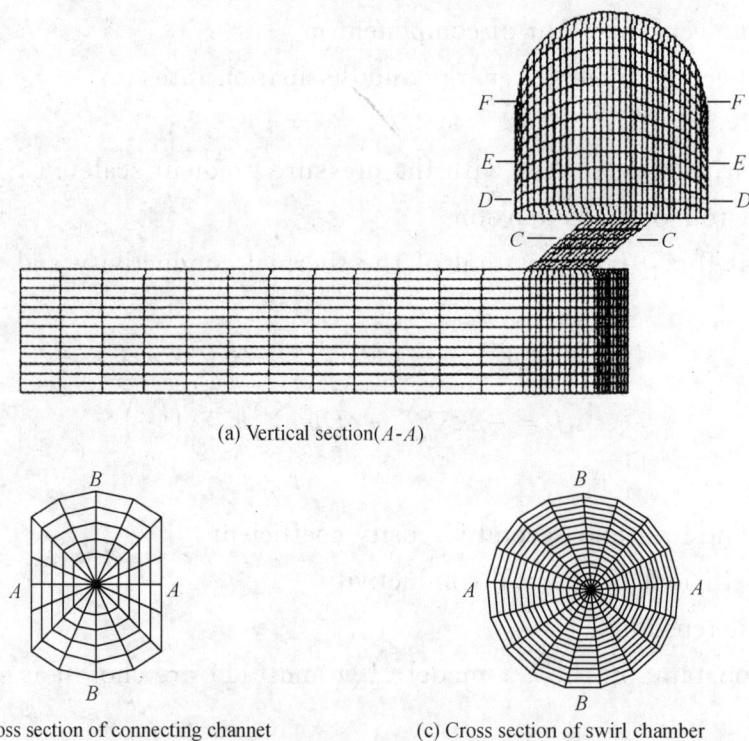

(a) Vertical section(A-A)

(b) Cross section of connecting channel　　　(c) Cross section of swirl chamber

Fig. 3　Meshing of the combustion chambers and computational grids ($40°$ CA BTDC)

4　Simulation results and comparison with experimental data

4.1　Comparison between simulation results and experimental data

The simulation started at the compression BTDC. The initial air velocity in the combustion chamber was assumed 0 and pressure and temperature were assumed uniform in the combustion chamber. Compared in Fig. 5a are the air velocity vs. crankangle at point a near the connection channel exit shown in Fig. 4. It is shown that the simulation results agree quite well with measurement data, and the trends match extremely well: both the measured and simulated air velocity reaches the peak at 25~30° BTDC.

The air velocity vs. crankangle histories at the left hand side of the swirl chamber, point b in Fig. 4 are compared in Fig. 5b. As shown, the air velocity at this location experienced less variation. At the engine speed of 800 r/min the air velocity peaks at around 12° BTDC.

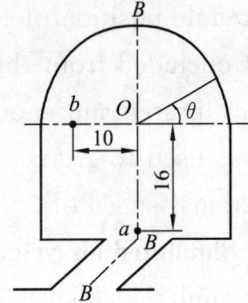

Fig. 4 **Typical points in the vertical section**

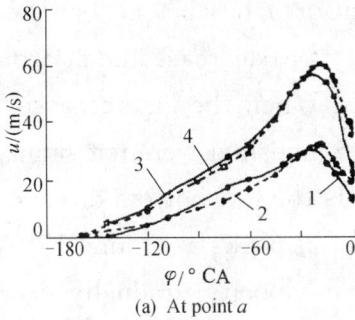

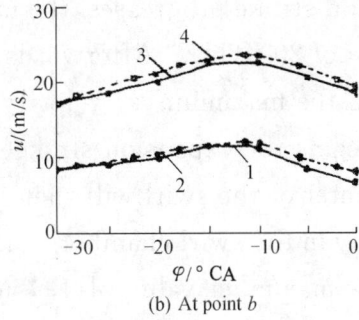

(a) At point a

(b) At point b

1—Measurement: 400 r/min 2—Simulation: 400 r/min 3—Measurement: 800 r/min 4—Simulation: 800 r/min

Fig. 5 **Comparison of measured and simulated air velocity vs. crankangle at typical points shown in Fig. 4**

Depicted in Fig. 6 are the comparison of air velocity along a circle of 10 mm radius from the swirl chamber center. In Fig. 6 the X axis θ is the angle shown in Fig. 4. Reasonable agreement is observed between the simulation results and measurement data, and the trend is very well captured in the simulation: i. e. the air velocity peaks at a θ angle of 0° and minimum value occurs at a θ angle of 270°. The velocity variation along the θ angle is less significant at TDC, compared to the

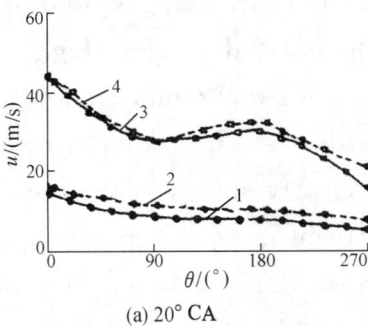

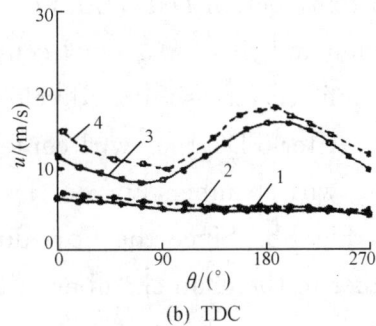

(a) 20° CA

(b) TDC

1—Measurement: 400 r/min 2—Simulation: 400 r/min 3—Measurement: 800 r/min 4—Simulation: 800 r/min

Fig. 6 **Comparison of measured and simulated air velocity along the circle of 10 mm radius from the swirl chamber center, θ angle is indicated in Fig. 4**

crankangle position of 12° BTDC.

Concluded from the above, the simulation model set up for this study was well calibrated and enough confidence had been gained. Then the simulation model was used to study the air-motion in the swirl chamber for the normal engine operational speed of 2 000 r/min.

4.2　Simulated air velocity field in the swirl chamber at 2 000 r/min

Simulation results indicated that at an engine speed of 2 000 r/min, the swirl starts to develop at a crankangle of 130° CA BTDC near the connection channel exit. Along with the compression process, the center of the swirl moves towards the upper right direction. In the mean time the air velocity increases as the compression stroke progresses, the maximum air velocity reaches 179 m/s at the crankangle of 20° BTDC. Afterwards the air velocity start to decrease and at around 2° ATDC the maximum air velocity drops to 132 m/s. When the piston passes TDC and begins the expansion stroke(fuel injection and combustion were not simulated), the center of the swirl will then move down towards the left bottom direction. Air velocity in the swirl chamber is accelerated again and it peaks at around 15° ATDC with a maximum value of 194 m/s, afterwards air velocity gradually decreases when the piston moves further down towards BTDC.

Fig. 7 depicts the velocity field in several sections of the swirl chamber at a typical crankangle position of 25° BTDC at an engine speed of 2 000 r/min.

V_{max} in each plot indicates the amplitude of the maximum velocity vector in this section. The swirl center of in section A-A is clearly shown in Fig. 7a. Two swirls centers are observed in section B-B, Fig. 7b, but they are located at lower positions. The velocity profile in the cross section of the connecting channel is shown in Fig. 7c. Along the horizontal planes of the swirl chamber from the top to bottom, section D-D, E-E, F-F, Fig. 7d to Fig. 7f, the swirl changes rotational direction and there is a swirl center at each side of the two half circles, Fig. 7e.

Similarly, the simulated velocity field at 11° ATDC of 2 000 r/min is shown in Fig. 8. After TDC the swirl center moves down towards the left bottom direction of the swirl chamber, Fig. 8a. Two swirl centers are clearly shown in the section B-B, Fig. 8b. Since the flow direction of the air stream is now from the swirl chamber to the main chamber, the velocity vectors in the cross section of the connecting channel change direction, Fig. 8c, compared to what before TDC shown in Fig. 7c. However, the most interesting point is that during the expansion process the velocity fields along the horizontal planes of the chamber is very similar to those during the compression process.

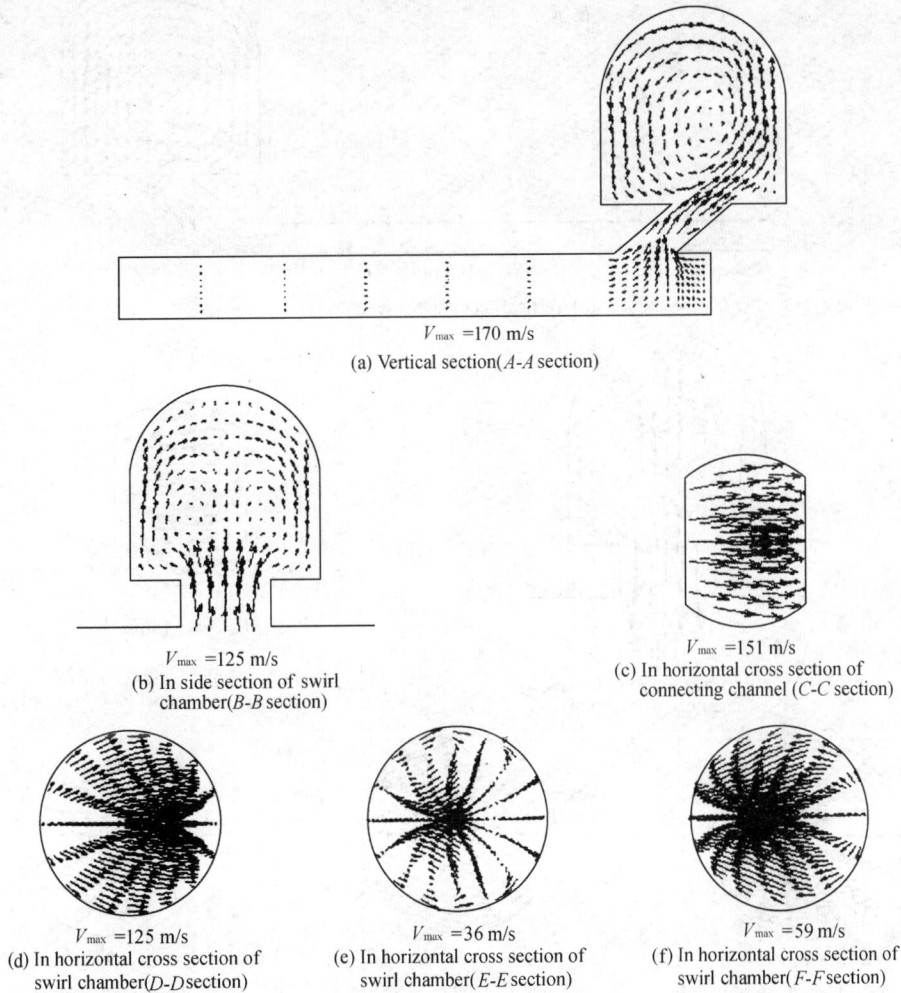

$V_{max} = 170$ m/s

(a) Vertical section(A-A section)

$V_{max} = 125$ m/s

(b) In side section of swirl chamber(B-B section)

$V_{max} = 151$ m/s

(c) In horizontal cross section of connecting channel (C-C section)

$V_{max} = 125$ m/s

(d) In horizontal cross section of swirl chamber(D-D section)

$V_{max} = 36$ m/s

(e) In horizontal cross section of swirl chamber(E-E section)

$V_{max} = 59$ m/s

(f) In horizontal cross section of swirl chamber(F-F section)

Fig. 7 Simulated typical air velocity field (25° CA BTDC), motored 2 000 r/min

The center of the swirl is always located at the left hand side of the extended center line BB' of the connecting channel during both the compression and expansion strokes, as shown in Fig. 9, but not at the left hand side of the geometric center O, as indicated in the 2-dimensional simulation results[16]. The possible reasons are: (1) the 2-dimensional simulation approach treated the top of the swirl chamber as half a circle instead of a hemisphere, by default a uniform velocity was assumed along the swirl chamber depth and the interference of the air-motion in that direction was neglected; (2) the 2-dimensional treatment required the velocity field at the connecting channel exit as the boundary condition. However, due to the limitation to the measurement space available, air velocity was only measured at one location near the exit, while a velocity field was assumed based on the measured velocity. The assumed velocity profile could be off from the reality.

$V_{max} =185$ m/s

(a) Vertical section(A-A section)

$V_{max} =140$ m/s

(b) In side section of swirl chamber(B-B section)

$V_{max} =150$ m/s

(c) In horizontal cross section of connecting channel (C-C section)

$V_{max} =101$ m/s

(d) In horizontal cross section of swirl chamber(D-D section)

$V_{max} =40$ m/s

(e) In horizontal cross section of swirl chamber(E-E section)

$V_{max} =63$ m/s

(f) In horizontal cross section of swirl chamber(F-F section)

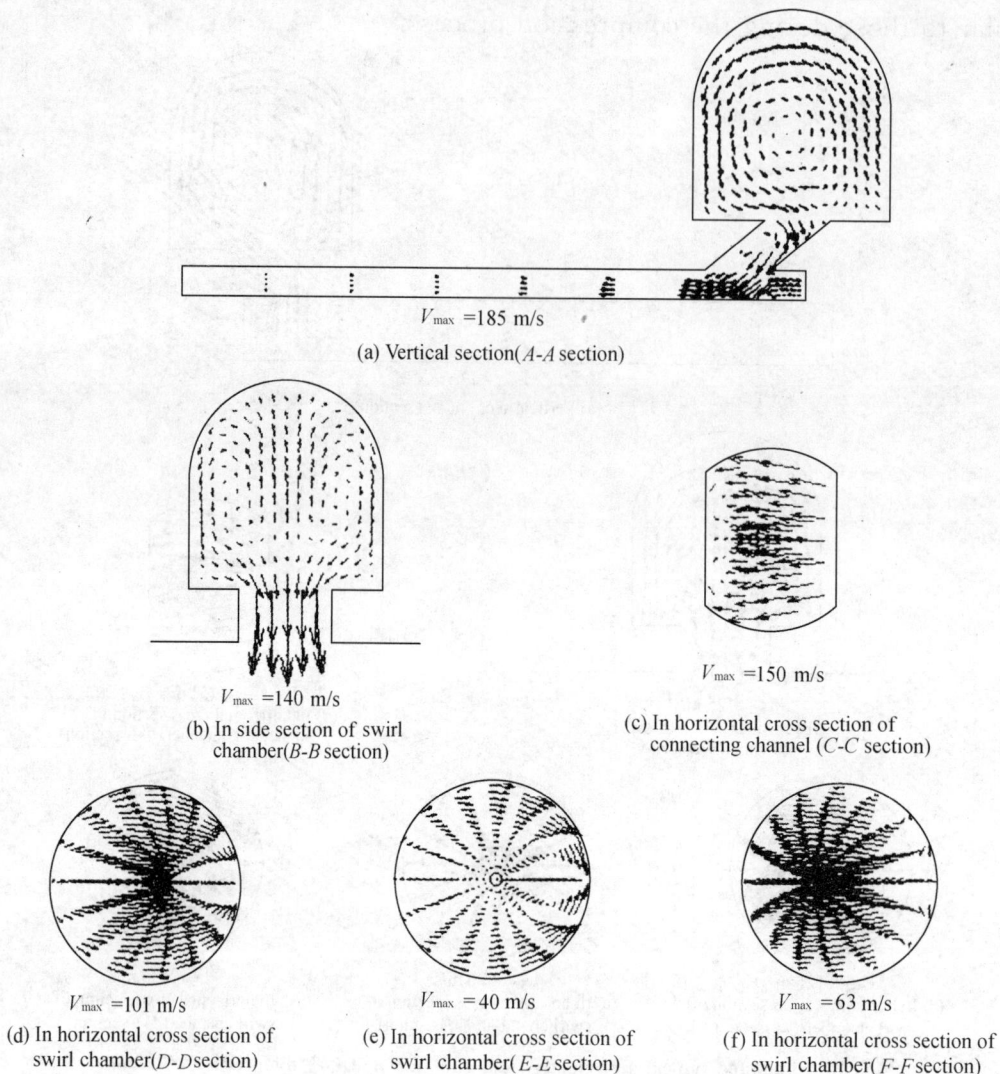

Fig. 8　Simulated typical air velocity field（11° CA ATDC）

5　Conclusions

（1）The air-motion characteristics in the swirl chamber of an IDI diesel engine were successfully revealed vs. crankangle positions and locations in the swirl chamber.

（2）A 3-D simulation code was developed based on the KIVA II code and successfully applied to study the formation and development of air velocity field in the swirl chamber vs. crankangle position.

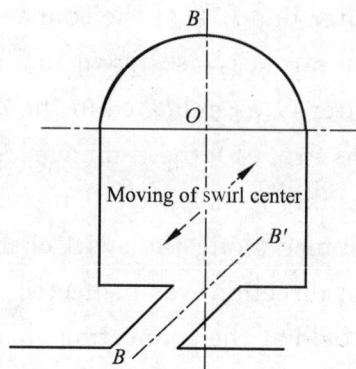

B

O

Moving of swirl center

B'

B

Fig. 9　Moving direction of the swirl center

(3) The modeling results agrees reasonably well with measurement data at typical measurement locations.

(4) The swirl starts to develop at around 130° BTDC during the compression stroke. With the progress of the compression stroke, the swirl center moves upwards towards the right of the swirl chamber, while during the expansion stroke it moves downwards to the left bottom. On the side vertical section of the chamber, there are always two swirl centers located at lower position, the swirling direction does not change from the compression to the expansion process; on the horizontal sections along the vertical position, the swirling experiences a change of direction from the top position to the bottom, but there are always two swirl centers, one at each side of the circle. For the obvious reason, the flow direction in the connecting channel changes from compression stroke to expansion stroke.

6 Acknowledgement

The authors acknowledge professor Su Wanhua and Professor Liu Shuliang of Tianjin Unversity for their support and instructions on the experimental measurement work, also like to thank the National Committee for Natural Science Research Funding for the financial support.

References

[1] Davis G C, et al. The effect of in-cylinder flow processes(swirl, squish and turbulence intensity) on engine efficiency: model prediction[J]. *SAE*, 820054, 1982.

[2] Arconmanis C, Whitelaw J H. Flow mechanics of internal combustion engine: a review [C]// *International Symposium of Flows in Internal Combustion Engine*-III, edited by Uzkan T, Tiederman W G, Novak J M, Nov. 17–22, 1985.

[3] Patterson M A, et al. Modeling the effects of fuel injection characteristics on diesel engine soot and NO_x emission[J]. *SAE*, 940523, 1994.

[4] Strauss T S, Schweimer G W, Ritscher U. Combustion in a swirl chamber diesel engine simulation by computation of fluid dynamics[J]. *SAE*, 950280, 1995.

[5] Nakajima K. An experimental research on the air swirl motion and combustion in swirl chamber of diesel engines[C]// *XII FISITA Congress*, Barcelona, 1968.

[6] Cole J B, Swords M D. Laser doppler anemometry measurements in an engine[J]. *Applied Optics*, 1979, 18(10).

[7] Rodney B, Rask. Laser doppler anemometry measurements in an internal combustion engine[J]. *SAE*, 790096, 1979.

[8] Koremenos D A,Rakopoulos C D. Thermodynamic analysis of indirect injection diesel engine by two-zone modeling of combustion[J]. *Transition of the ASME*,1990,112(1)：138-149.

[9] Princhon P,Guillot B. Thermodynamic and flow analysis of an indirect injection diesel combustion chamber by modeling[J]. *SAE*, 852686,1985.

[10] Hirt C W, et al. An arbitrary Lagrangian-Eulerian computing method for all flow speed [J]. *J Comp Phys*, 1974,14：227-253.

[11] Pinchon P. Three dimensional modeling of combustion in a prechamber diesel engine [J]. *SAE*, 890666,1989.

[12] Zellet M, et al. Three dimensional modeling of combustion and soot formation in an indirect injection diesel engine[J]. *SAE*, 900254,1990.

[13] Wakisaka T ,Shimoto Y, Isshiki Y. Three-dimensional numerical analysis of in-cylinder flow in reciprocating engines[J]. *SAE*, 860464,1986.

[14] Amato Cl, Bertoli C, Corcione F E, Petrillo F, Valentino G. Turbulent models validation by LDV in an internal combustion engine[C]// *COMODIA* 90,1990.

[15] Wang Qian, Li Detao, et al. The experimental and computational investigation of air-motion in swirl chamber diesel engines[J]. *Transaction of CSICE*,1994(2).

[16] Zhu Guangsheng, Wang Qian, et al. Influences on air-motion characteristics of swirl chamber design[J]. *Journal of Combustion Science and Technology*,1996(4).

[17] Tan Wei, Zhu Yanzhang, et al. Micro-computerized multi-dimensional engine simulation code with injection and chemical reaction models：engine CFD[J]. *Transaction of CSICE*,1996(6).

[18] Du Jiayi. Multi-dimensional simulation of fuel spray and mixing with air in swirl chamber diesel engines[J]. *Journal of JSUST*,1996(5).

[19] Amsden A A, O'Rourke P J,Butler T D. KIVA II:a computer program for chemically reacting flows with sprays[R]. Los Alamas National Laboratory Report, LA-11560-MS,1989.

[20] Wang Qian. Numerical simulation and experimental investigation of air-motion in swirl chamber of diesel engine [D]. Zhenjiang：JSUST,1996.

(From：*SAE Technical Paper Series*, 2002)

涡流室式柴油机空气运动的 LDA 测量与三维模拟计算

[摘要] 介绍了采用激光多普勒测速仪(LDA)对柴油机涡流室内空气运动规律测试的研究结果。测出了倒拖工况不同转速下,涡流室一些测点的空气流速,并采用相平均法计算了各点的平均速度和湍流强度。通过对标准 k-ε 模型作适当修正来模拟气体的湍流运动,建立了涡流室式柴油机的空气运动的三维数值计算模型,开发了三维数值模拟程序。运用该程序对吊钟型涡流室式柴油机的流场进行了三维计算,计算结果与 LDA 试验结果具有良好的一致性。作出了涡流室纵截面、横截面和侧截面上的速度矢量图,分析了各截面上气体速度的变化过程和涡心运动轨迹,研究结果揭示了涡流室中空气运动的三维流动特性。

十六烷值改进剂对涡流室式
柴油机排放特性的影响

董　刚,陈义良,吴志新,李德桃

[摘要]　将自行研制的有机硝酸酯类十六烷值改进剂以不同浓度加到柴油中,对影响95型涡流室式柴油机的排放特性进行了实验研究。结果表明:该十六烷值改进剂可明显降低柴油机 NO_x 的排放;在适宜浓度下,对 HC 和 CO 排放没有影响;但使排气烟度有所恶化。实验还测试了该剂对燃油经济性的影响,结果表明:该剂可略微降低柴油机的有效燃油耗。

柴油机的排放污染问题越来越受到世界各国政府和人民的高度重视。自 20 世纪 80 年代以来,大量的实验表明,柴油机的物理和化学性质对柴油机有害排放具有直接影响,可以预测,当改进发动机设计使有害排放降低到一定程度后,燃油品质将成为影响发动机有害排放的关键因素[1]。因此,从改善柴油品质的角度来降低柴油机有害排放具有重要实际意义。

满足低排放标准的柴油要具有低含芳烃量、低含硫量、低比重、低馏出温度以及高十六烷值的特性,其中提高柴油的十六烷值是控制柴油机有害排放的一条重要技术途径。Ullman 等人通过使用高十六烷值柴油或在柴油机中加入十六烷值改进剂的方法,在重型柴油机上进行了较为系统的排放测试[2,3],他们的结果表明,高十六烷值的柴油(包括加入十六烷值改进剂)可使重型柴油机 NO_x 和 CO 明显下降,同时 HC 和颗粒物质也会有所降低,但他们同时指出,在不同类型的柴油机以及不同的实验过程和条件下,要澄清柴油十六烷值对发动机排放的影响规律还需做进一步的研究。国内在这方面开展的工作还不足,尤其是针对我国产量最大、用量最广的 95 型涡流室式柴油机的研究还未见报道。鉴于此,本文通过使用自行研制的有机硝酸酯类十六烷值改进剂,以提高柴油十六烷值,进而改善柴油机着火燃烧特性的方法,研究了改进剂对柴油机有害排放的影响规律,同时还考察了这种改进剂对柴油机经济性的影响。

1　实　　验

1.1　实验装置和测试用油

实验用机为 S195 型涡流室式柴油机;利用天津大学内燃机重点实验室开发的 SINGAL-900 型发动机排放系统对该机的 NO_x,HC 和 CO 的排放进行了测试;为计算这 3 种物质的比排放量,还利用 LCQ-100 型气体尾流流量计测定了柴油机进气管内的空气流量。此外,还测试了该柴油机的烟度和油耗。

实验中使用了普通 0 号柴油(称基础油),自行研制的有机硝酸酯类十六烷值改进剂已在文献[4]中加以描述。将该剂按质量比为(下同)0.1%,0.3%,0.5% 和 1.0% 加到基础油中制成不同的加剂油,考虑到十六烷值改进剂的成本和添加效果等因素,一般添加量不宜大于0.5%,但为了测试该剂对柴油机排放的影响趋势,本实验也考虑了添加量为 1.0% 的情况。

1.2　测试过程

参照 GB 6456—86《柴油机排放实验方法》对基础油和 4 种添加量不同的加剂油进行了 13 个工况排放测试,并计算了 NO_x,HC 和 CO 的比排放量。实验从基础油做起,测度 13 个工况下柴油的 NO_x,HC 和 CO 的排放量以及烟度和燃油耗的值,再由进油管上的三通阀切换成加剂油,按添加量由低到高的次序重复上述过程。每一种油测试完成后,弃去滤清器中的剩余燃油,在 2 000 r/min 转速和 50% 负荷的工况条件下,先将柴油机运转约 5 min,以使柴油机油管和喷嘴中不含有上一次燃油,然后再进行下一次测量。

2　测试结果与分析

2.1　测试结果

图 1~图 5 是根据 13 个工况测试结果绘制的十六烷值改进剂添加量(以质量比计)、负荷(kW)和 NO_x,HC,CO,烟度(RB)及有效燃油耗的三维变化关系图,其中(a)(b)分别代表 2 000 r/min 和 1 600 r/min 转速下的测试结果。表 1 为不同添加量下 NO_x,HC 和 CO 通过干湿基浓度修正后的比排放量的计算结果。

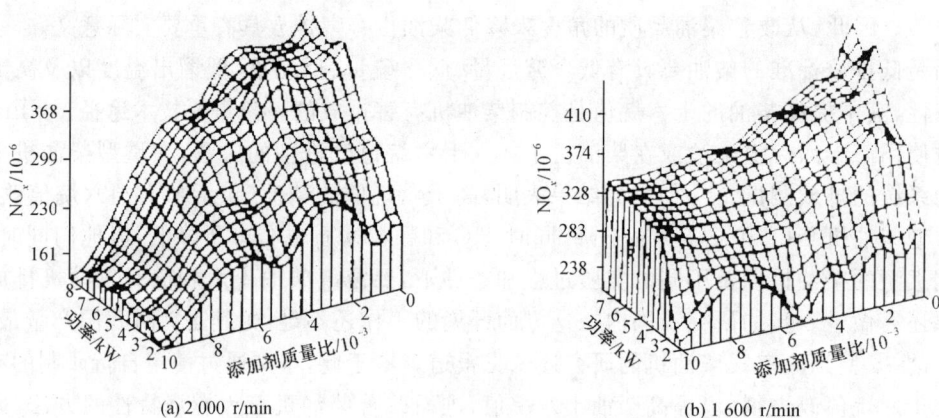

(a) 2 000 r/min　　　　　　　(b) 1 600 r/min

图 1　不同转速下,NO_x 随添加剂和负荷的变化关系

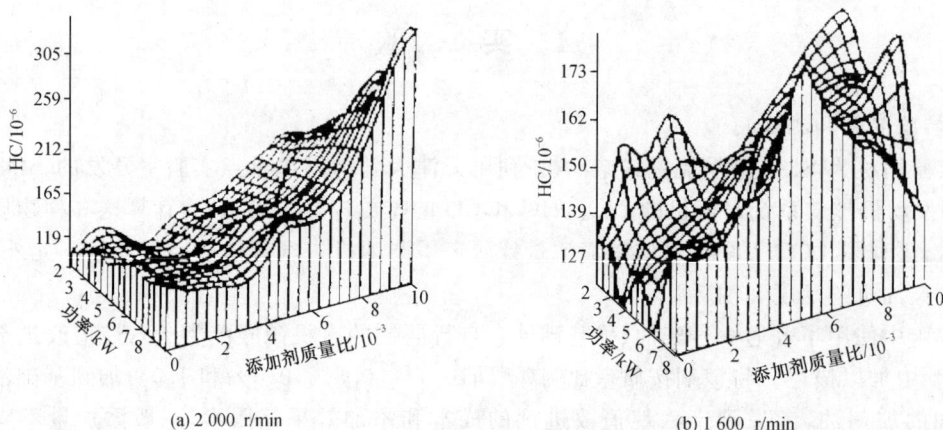

(a) 2 000 r/min　　　　　　　(b) 1 600 r/min

图 2　不同转速下,HC 随添加剂和负荷的变化关系

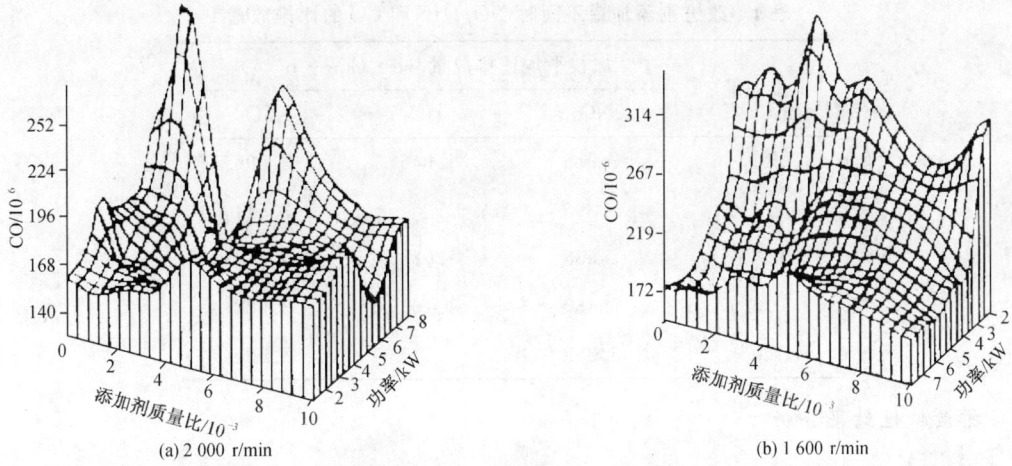

(a) 2 000 r/min (b) 1 600 r/min

图 3　不同转速下,CO 随添加剂和负荷的变化关系

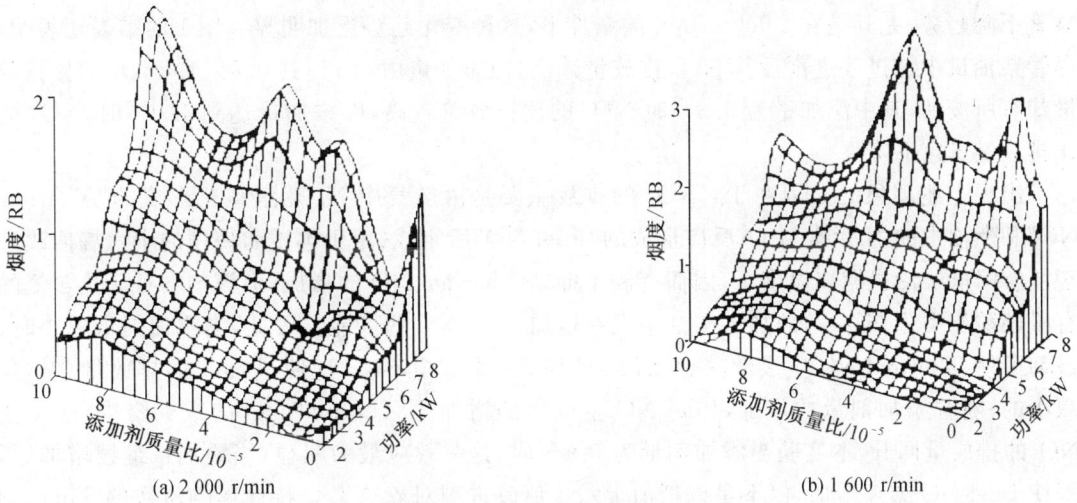

(a) 2 000 r/min (b) 1 600 r/min

图 4　不同转速下,烟度随添加剂和负荷的变化关系

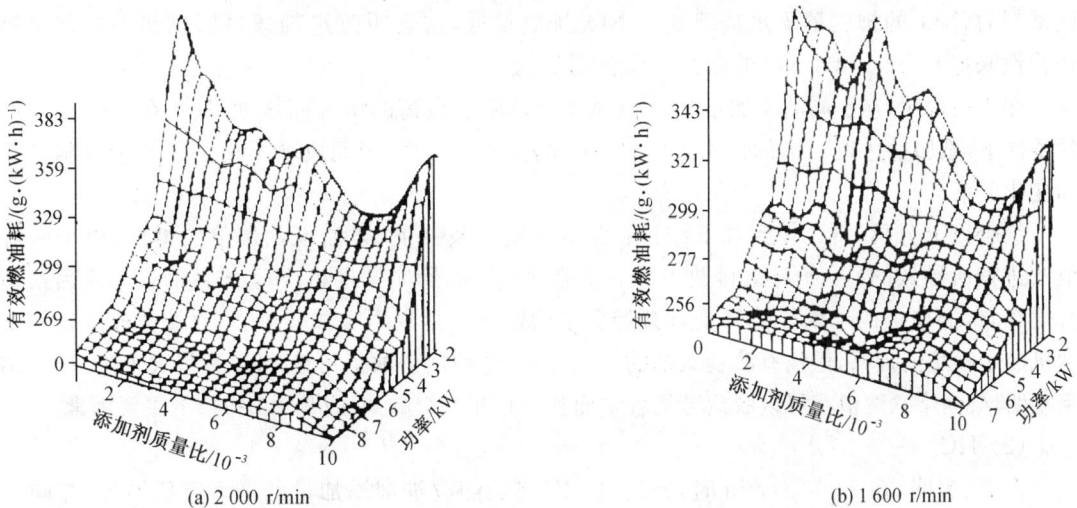

(a) 2 000 r/min (b) 1 600 r/min

图 5　不同转速下,有效燃油耗随添加剂和负荷的变化关系

表 1　改进剂添加量不同时 NO_x，HC 和 CO 的比排放量

添加量	加权平均比排放量/(g·(kW·h^{-1}))		
	NO_x	HC	CO
0	7.986	1.046	3.596
0.1%	7.365	1.054	3.611
0.3%	7.638	1.033	3.272
0.5%	7.480	1.160	3.582
1.0%	5.889	1.566	3.604

2.2　排放特性结果分析

（1）NO_x

由图 1 可以看出，在整个测试的工况范围内，NO_x 的排放量随改进剂添加量的增加而呈显著下降趋势，尤其是在 2 000 r/min 的条件下，这种变化趋势更加明显。图 1 的结果还表明，尽管柴油机中加进改进剂后其 NO_x 排放量减小，但添加量由 0.1% 到 0.5% 时，NO_x 的比排放量却有所波动，其中添加量为 0.3% 时 NO_x 的比排放量较高，而添加量达到 1.0% 时，NO_x 的比排放量迅速下降。

出现上述现象的原因在于：NO_x 的排放主要是由空气中 N_2 热解氧化形成的"Thermal NO"和燃油中的 N 元素与 O_2 反应形成的"Fuel NO"所组成；由于本试验使用的十六烷值改进剂能够降低柴油机燃烧温度[5]，因而有利于抑制 Thermal NO 的形成；但该剂本身又是含氮的有机硝酸酯类化合物[4]，故加入柴油中又有增加 Fuel NO 的可能。因此，当添加剂量较小时，该剂可使 Thermal NO 生成量减少，而 Fuel NO 的生成量不多，这使得 NO_x 的排放量总体是减小的；随着添加剂量的增加，Fuel NO 生成量的增加比 Thermal NO 的减少要快，这又使 NO_x 的排放量回升，本实验当添加剂量为 0.3% 时，这一效应最为明显；当添加量继续增加后，即使 Fuel NO 继续增加，但不足以抵消十六烷值改进剂对着火燃烧特性影响所致的 Thermal NO 的明显下降。因此可知，尽管含 N 的十六烷值改进剂本身对 NO_x 的排放有影响，但加入该试剂后，NO_x 的排放量要比基础油的 NO_x 排放量低，这表明改进剂通过影响柴油机着火燃烧特性来抑制 Thermal NO 的效果是十分明显的。

图 1 还表明，在不同的添加量时，NO_x 基本上均随着负荷的增大而增加，只是在 2 000 r/min 的条件下，添加量达到 1.0% 时，由于此时 NO_x 的排放量很低，因而反映不出 NO_x 排放量随负荷的变化关系。

综合以上分析，可以得到如下结论：在本实验的条件下，柴油十六烷值的增加（由十六烷值改进剂所导致的）可使 NO_x 排放下降。文献[6]认为，影响柴油机 NO_x 排放的最本质因素是芳烃含量，而柴油的十六烷值又受其芳烃含量的影响。但在本文试验中，十六烷值改进剂使柴油机的十六烷值提高，但没有改变柴油的芳烃含量，这种加剂油仍可使 NO_x 有很大下降，这一结果表明，柴油十六烷值是可以脱离芳烃含量而独立存在的影响柴油机 NO_x 排放的主要因素。

（2）HC

图 2a 表明，在 2 000 r/min 时，柴油机 HC 排放随改进剂添加量的增多有所增加，在同一添加浓度下，HC 的排放随负荷的增大而增加；图 2b 表明，在 1 600 r/min 时，柴油机 HC 排放

随改进剂添加量的变化有比较复杂的关系,但总体上亦呈增加的趋势,在同一浓度下,HC 排放呈现出在低负荷和高负荷较高,中间负荷较低的变化趋势。

影响柴油机 HC 排放的因素较多,其中在着火滞燃期中形成过稀混合气所导致生成的 HC 是比较主要的一部分[6]。在本实验中,由于十六烷值改进剂缩短了着火滞燃期[4],使这一过程中燃油蒸发量减少,因而形成了过多超过稀燃极限的混合气体过燃区,这导致了最终 HC 排放的增加。

从表 1 的结果来看,当十六烷值改进剂添加量不超过 0.5% 时,该剂对 HC 的比排放量没有很大影响,由于柴油机在 HC 的排放量相对较少,因此使用低于 0.5% 的添加量是可以满足柴油机 HC 排放要求的。

(3) CO

图 3a 表明,在 2 000 r/min 时,除低负荷下改进剂添加量在 0.3% 时 CO 排放量有明显减少外,在其他负荷和添加量下,该剂对 CO 的排放没有影响;图 3b 表明,在 1 600 r/min 时,CO 排放也未受到十六烷值改进剂的影响。在上述两种转速下,CO 的排放量随负荷的增加而逐渐减少,在 1 600 r/min 时,这种趋势更加明显。

CO 是烃类燃油燃烧的中间产物和不完全燃烧产物之一。在本实验条件下,尽管十六烷值改进剂对柴油机的着火特性有影响,但却不能改善整个工作过程中燃烧的完全程度,因此对 CO 的排放也不会有明显影响。

(4) 烟度

图 4 表明,在两种转速下,十六烷值改进剂的加入会使柴油机排气烟度有所恶化;在同一添加量下,柴油机排气烟度随负荷的增加也逐渐增大,在中高负荷时,排气烟度的增加速度较大。柴油中加入十六烷值改进剂后,其着火性能改善,着火滞燃期缩短,参与预混合燃烧的柴油量变少,大部分柴油以扩散燃烧的方式进行,这是造成柴油机排气烟度增加的主要原因。

2.3 燃油经济性分析

图 5 给出了 13 个工况中 2 000 r/min 和 1 600 r/min 时,不同负荷、不同十六烷值改进剂添加量下,柴油机有效燃油耗的变化情况。由图可见,在两种转速的低负荷下,随着添加量的增加,柴油机油耗逐渐降低,当添加量达到 1.0% 时,油耗略有上升;在中等负荷下,随着添加量的增加,柴油机油耗略有下降;在高负荷下,十六烷值改进剂对柴油机油耗没有明显影响。此外,在同一添加量下,两种转速下的柴油机油耗均随负荷变大而急剧减小,这与柴油机一般的负荷特性是相同的。由此可以得出:在本实验条件下,十六烷值改进剂对柴油机经济性没有负面影响,在适宜的添加量(一般不超过 0.5%)下,有些工况的有效燃油耗可以减少,即经济性有所提高。

3 结 论

本文通过使用自行研制的有机硝酸酯类柴油机十六烷值改进剂,对影响 95 型涡流室式柴油机的排放特性进行了实验研究,在本实验条件下,可得出如下结论:

(1) 本十六烷值改进剂可明显降低柴油机 NO_x 排放,且随着添加浓度的增大,降低 NO_x 排放的效果更加明显。实验结果还表明,在影响柴油机 NO_x 排放的诸燃油参数中,十六烷值是重要的独立表征量之一。

（2）该剂可使柴油机 HC 排放有所增加，但在适宜的添加浓度（不超过 0.5%）以内，HC 排放升高的趋势不明显。

（3）该剂对柴油机 CO 排放基本没有影响，但在少数工况下，随改进剂添加量的增加，CO 的排放也会明显下降。

（4）该剂可使柴油机排气烟度恶化。

（5）该剂不会使柴油机油耗增加，对柴油机经济性没有负面影响。

参 考 文 献

［1］Hutcheson R C. Diesel fuel quality—the future challenge fuels for automotive and industrial diesel engines[R]. The Institute of Mechanical Engineers, 1990：151-158.

［2］Ullman T L, Spreen K B, Mason R L. Effects of cetane number, cetane improver, aromatics, and oxygenates on 1994 heavy-duty diesel engines enmissions [J]. *SAE Paper*, 1994：682-691.

［3］Ullman T L, Spreen K B, Mason R L. Effects of cetane number on emissions from a prototype 1998 heavy-duty diesel engine [J]. *SAE Paper*,1995：255-271.

［4］董刚,李德桃,吴志新,等.有机硝酸酯类柴油十六烷值改进剂的研究[J].内燃机工程，1996,17(4)：21-27.

［5］董刚,陈义良,李德桃.利用放热规律研究十六烷值改进剂对柴油机着火燃烧特性的影响[J].内燃机工程,1999,20(4)：60-64.

［6］何学良,李疏松.内燃机燃烧学[M].北京：机械工业出版社,1992.

（本文原载于《燃烧科学与技术》2001 年第 1 期）

Effects of cetane number improver on emissions of swirl chamber diesel engine

Abstract：An experimental study of the effects of an organic nitrate cetane number improver which was made by the authors on the emission of the 95-type swirl chamber diesel engine was performed by adding to the improver diesel fuel in different concentrations. The results show that the cetane number improver decreases apparently the NO_x emission, and has no effect on the HC and CO emissions in proper concentrations, but increases the smoke. The effect of the improver on fuel economy is also investigated, and the result shows that the improver can decrease the effective fuel consumption slightly.

II. 柴油机冷起动的基础研究和改善措施

Primary Research and Improvements in a Swirl Chamber Diesel Engine in Cold-starting

Investigation on ignition and combustion at starting in a swirl chamber diesel engine with a starting throat[①]

Zhu Xiaoguang , Li Detao

Abstract

An investigation is made on transient ignition and combustion at starting in a swirl chamber diesel engine by high speed photography. It is found that the starting throat serves a dual purpose of direct ignition in main chamber and easier ignition in swirl chamber, and thus helps to shorten the starting process. The advantageous effects of the starting throat on the performance are also shown by examining the two starting processes in diesel engines with and without a starting throat. The flame motion pattern in cylinder is analyzed and the principle for the rational design of the main chamber is discussed.

1 Introduction

It has been shown, as the results of research work and practical application both in China and abroad, that the starting of the swirl chamber diesel engines can be improved and the starting fuel consumption can be reduced when a starting throat and adequate design of the main chamber are used[1-3]. But as far as the authors know, it is not clear yet what is the function of the starting throat and how the burning gas moves and the

---

① 本文和第Ⅱ部分前三篇等文揭示了涡流室式柴油机起动孔的作用和机理,结束了学术界对此问题进行了近30年的争论。

该文在实验研究的基础上,首次对涡流室式柴油机起动孔的作用机理,提出了科学合理的解释。在 IPC—5会上宣读后,经整理补充,发表于《内燃机学报》,1990,8(1)。此后又经多年的进一步的试验和理论分析,得到更全面系统的结果。先后在 SAE 会上宣读,在《内燃机学报》、《内燃机工程》上发表。有关系列论文,均收入本论文选集。

——编者注

flame spreads in the main chamber. To understand these, authors made direct observations and recordings by high speed photograghy of the ignition, combustion and flame spread and made detailed analysis.

2　Experimental apparatus and methods

Provided for the test is a modified S195 single cylinder swirl chamber diesel engine. Fig. 1 and Fig. 2 show respectively the lengthened cylinder body and modified head. The swirl chamber of S195 diesel engine with a starting throat is shown in Fig. 3. The fuel supplying and cooling as well as gas exchanging systems of original engine were not changed. The cy-linder body and the piston were lengthened so that photographs of the main chamber could be taken from the bottom of the piston. On both sides of the swirl chamber in the head and at the top of the piston were installed quartz windows. The main specifications of the engine are listed in Tab. 1.

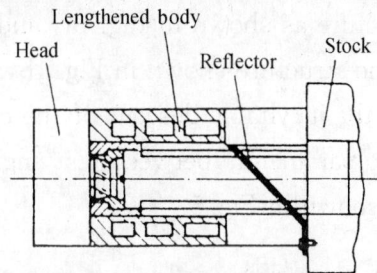

Fig. 1　Lengthened cylinder body

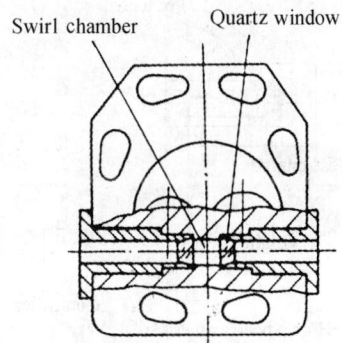

Fig. 2　Modified cylinder head

Tab. 1　Specifications of test engine

Parameters	Norm
Engine	S195
Type	indirect injecion
Bore	95 mm
Stroke	115 mm
Displacement	815 cm³
Compression ratio	20. 0
Rated output	10. 6 kW
	(2 000 r/min)
Injection pump	0—Type
Plunger diameter	8. 0 mm
Injection nozzle	throttle

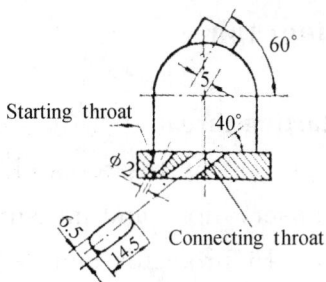

Fig. 3　Swirl chamber of S195 diesel engine with a starting throat

To keep the reliability and hermetic seal of the quartz window at the top of the piston and the original compression ratio unchanged, the top structure of the piston was specially designed (shown in Fig. 4c). Compared with some others currently used in

China and abroad, as seen in Fig. 4a and Fig. 4b, it has a larger supporting surface to be evenly pressed and the window is easier to clean. The structure surely avoids easily breaking from being unevenly pressed, as commonly happening in the

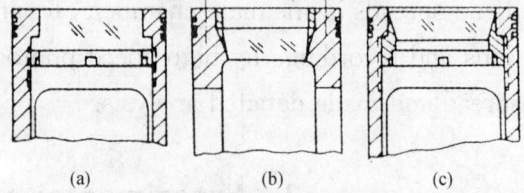

Fig. 4　Top structures of the lengthened piston

structure as shown in Fig. 4a, and makes up for the follwing defects, possibly existing in the structure shown in Fig. 4b, too much change in the volume of the main chamber and the inevitable descent of the compression ratio which causes greater differences of basic parameters between test engine and product one. The optical arrangements used are shown in Fig. 5.

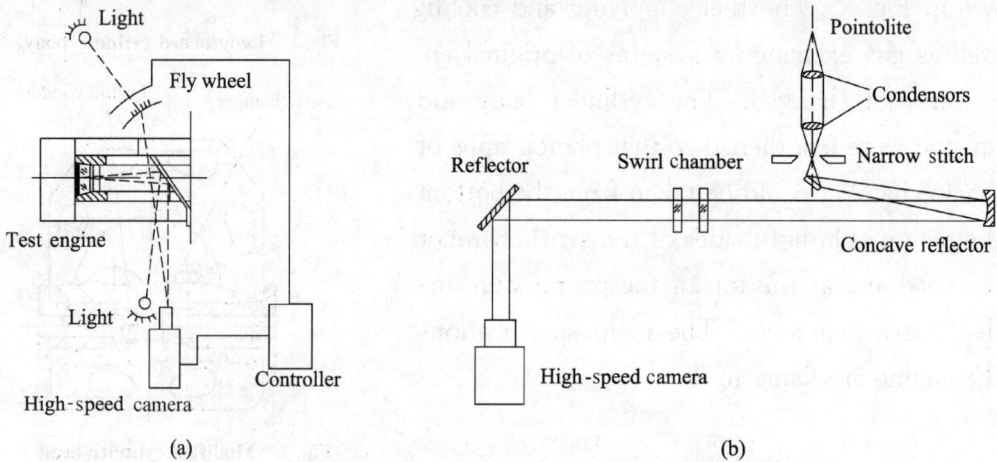

Fig. 5　Optical arrangements for swirl and main chambers

3　Analysis on high speed photograph

3.1　Ignition mechanism of the swirl chamber with a starting throat

The starting process was confirmed as a transient process by the research of scholars both in China and abroad[4,5]. After repeated observations and measurements on the large number of photographs using NAC 16 projector and NAC 160B motion analyzer, we found that each cycle in the starting process differs greatly. Fuel supplying quantities and speeds, ignition locations and instants, the instants at which flame blows out from connecting throat and the flame spreads are not the same for almost every cycle.

The starting throat functions in two aspects as shown in the two sets of photographs in Fig. 6.

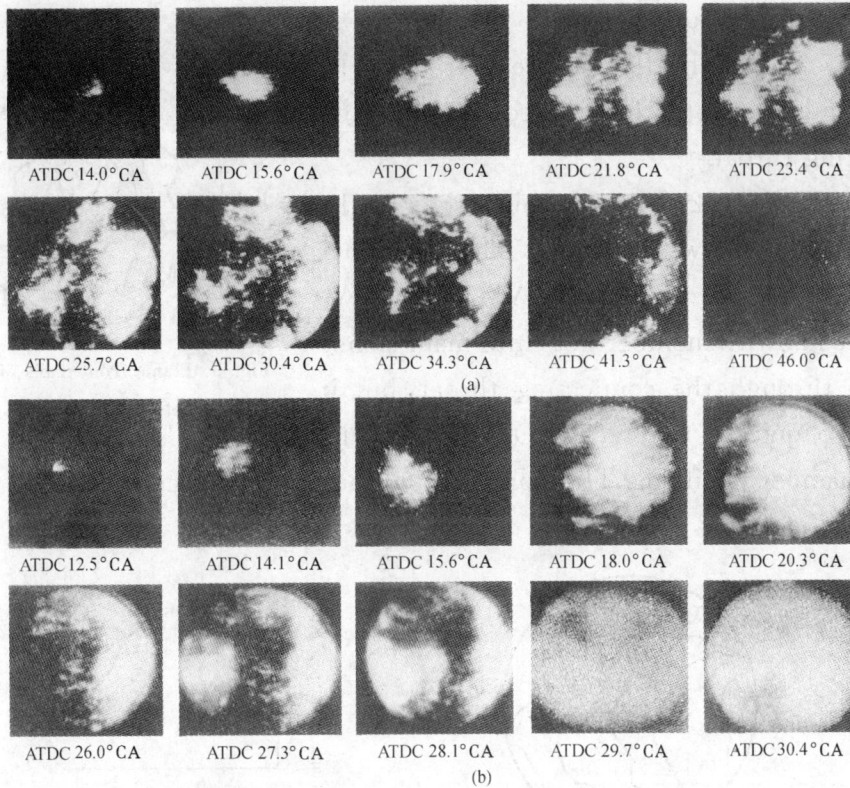

ATDC 14.0° CA	ATDC 15.6° CA	ATDC 17.9° CA	ATDC 21.8° CA	ATDC 23.4° CA
ATDC 25.7° CA	ATDC 30.4° CA	ATDC 34.3° CA	ATDC 41.3° CA	ATDC 46.0° CA

(a)

ATDC 12.5° CA	ATDC 14.1° CA	ATDC 15.6° CA	ATDC 18.0° CA	ATDC 20.3° CA
ATDC 26.0° CA	ATDC 27.3° CA	ATDC 28.1° CA	ATDC 29.7° CA	ATDC 30.4° CA

(b)

Fig. 6　Ignition and flame spreads in main chamber at starting speed 500 r/min

(1) Effects on ignition and combustion in the main chamber

Fig. 6a and b show two typically continous cycles of ignition and flame spread in the main chamber.

As shown in Fig. 6a, the fuel coming from the starting throat mixes up with the air in main chamber, and ignites near the starting throat outlet in the main chamber. The flame then propagates forward but not all over the whole chamber, and at ATDC 46° CA, it extinguishes.

It was found that, as shown in the photographs, no fuel or flame could be seen spurting out from the swirl chamber to the main chamber through the connecting throat. But at ATDC 30.4° CA, burnt gas rushing out from the connecting throat can be seen.

As seen in Fig. 6b, the fuel spray blown into the main chamber through the starting throat ignites at ATDC 12.5° CA, and at ATDC 20.3° CA, the flame spreads to most parts of the main chamber. Yet, no flame is seen this instant near the connecting throat outlet, which probably arises from the small pressure difference between main and swirl chambers that prevents both the flame in swirl chamber from blowing out and the flame in main chamber from spreading to the connecting throat outlet.

Fig. 7 shows the flame spread trail in the main chamber due to the fuel coming through the starting throat.

At ATDC 24. 2° CA, the burnt gas in the swirl chamber blows into the main chamber through connecting throat. At ATDC 26° CA, the flame in swirl chamber gets into main chamber also through the connecting throat, but it spreads so quickly that at ATDC 30° CA the whole chamber is filled. The spread trail and its speed curve of the flame are shown in Fig. 8.

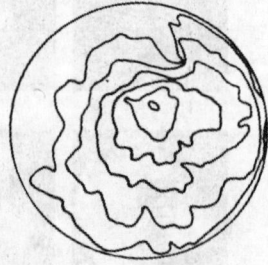

Fig. 7　Flame front trails, interval between two trails 5×10^{-4} s

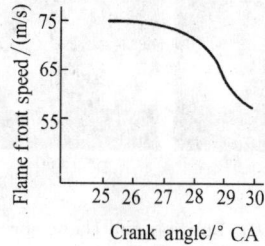

(a)　　　　　　　　(b)

Fig. 8　Flame front spread trails and speed curve, interval between two trails 2.5×10^{-4} s

(2) Functions in ignition and combustion in the swirl chamber

With the starting throat, the airflow produced in the opposite direction to the injected fuel accelerates dispersion and atomization of the fuel jet in swirl chamber, which shortens the ignition delay. This was proved by the holographs of Toshiaki Tanaka[1] and the high speed photographs provided in Fig. 9.

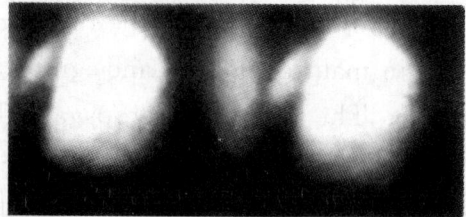

Fig. 9　Dispersion and atomization of the fuel jet in swirl chamber

According to the photographs of the main chamber shown in Fig. 10, another function of the starting throat——promoting the fuel in swirl chamber to ignite earlier——is further proved.

Fig. 10, provided as one of the series of photographs, shows a cycle of the starting process under another condition and is explained as following.

The fuel and flame go into the main chamber only through the connecting throat but no burning flame of the fuel is seen passing through the starting throat. It is explained that in this type of cycles, the counter-fuel spray airflow from the starting throat causes the fuel in swirl chamber to ignite earlier and the flame enters into the main chamber through the connecting throat earlier too.

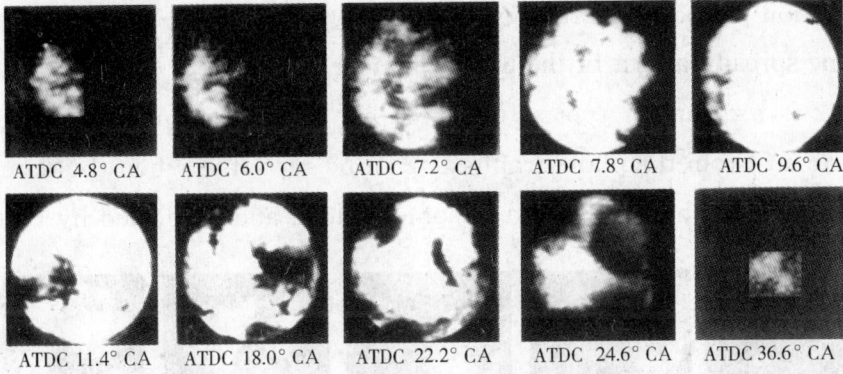

| ATDC 4.8° CA | ATDC 6.0° CA | ATDC 7.2° CA | ATDC 7.8° CA | ATDC 9.6° CA |

| ATDC 11.4° CA | ATDC 18.0° CA | ATDC 22.2° CA | ATDC 24.6° CA | ATDC 36.6° CA |

Fig. 10 Example of the combustion in main chamber at starting speed 350 **r/min**

Fig. 11 describes the instantaneous positions of the flame front and the corresponding front speed, which represent what is shown in Fig. 10.

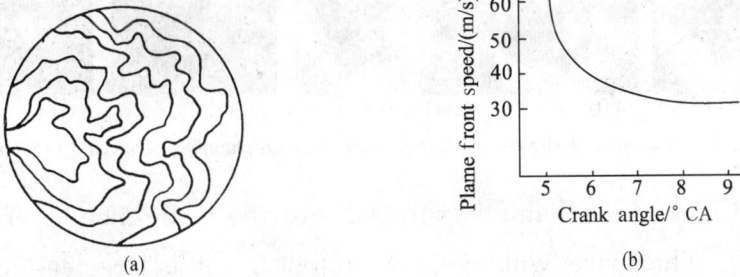

(a) (b)

Fig. 11 Flame front trails and spread speed curve, interval between two trails 2.5×10^{-4} **s**

The following is the summary from the figures above.

1) The starting is a transient process which consists of many cycles. In the starting process, the gas constituents and states, wall temperature, quantity of injected fuel and injecting speed both in swirl and main chambers are changing from one cycle to another. The enviroment and friction may be different. So under some conditions ignition takes place in the main chamber first; under some others in the swirl chamber first, or under some conditions, in both chambers at the same time. In our opinion, it is not worth to argue about in which chamber ignition occurs first. What is the most important is to find out the function of the starting throat to shorten the starting process.

2) In the starting process, the starting throat makes the ignition happen in main chamber directly, and ignition in swirl chamber easier. So it can be seen that the

starting throat serves a dual purpose of shortening the starting process.

3) Great changes and differences exist in cycles of the starting process. So it is unreasonable to make a judgement on the whole transient process by only one cycle.

4) In the starting process, the ignition delay is relatively longer, the delay and the combustion period vary from cycle to cycle.

3.2 Flame spread pattern in the main chamber

Fig. 12 is a group of typical photographs of the combustion process in the main chamber. From the photographs, it can be seen that when the flame spreads to whole chamber, the flame front becomes folded and deformed by turbulence.

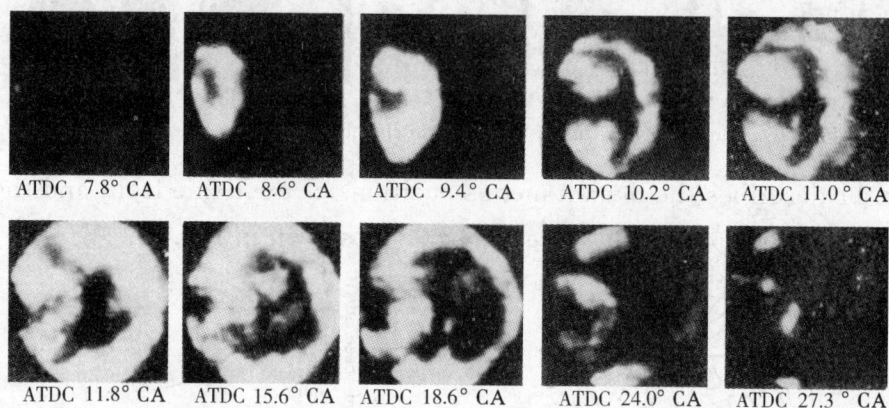

ATDC 7.8° CA   ATDC 8.6° CA   ATDC 9.4° CA   ATDC 10.2° CA   ATDC 11.0° CA

ATDC 11.8° CA   ATDC 15.6° CA   ATDC 18.6° CA   ATDC 24.0° CA   ATDC 27.3 ° CA

Fig. 12   Example of the combustion process in main chamber at speed 700 r/min

At ATDC 7.8° CA, flame is sprayed into the main chamber from the connecting throat. The quartz window at the top of piston is large enough so that the flame just blown out of the connecting throat can be seen.

At ATDC 11.8° CA, the flame has spreaded all over the chamber. Meanwhile, as shown in the corresponding p-ϕ diagram(Fig. 13), the pressure in the main chamber reaches its maximum.

At ATDC 9.4° CA, a puff of thick carbon smoke is seen blowing out through the connecting throat, following the flame front to spread forward. The puff of smoke possibly comes from the burning fuel which gathered at the bottom of the swirl chamber. Most of the smoke stays in the middle of the main chamber. It is considered that after the flame passes through the middle, there is too little fresh air in this area to enable the smoke to burn in time. Bright yellow rim at the edge of carbon smoke indicates that the unburnt smoke there is still burning. At ATDC 24° CA, the smoke burning stops.

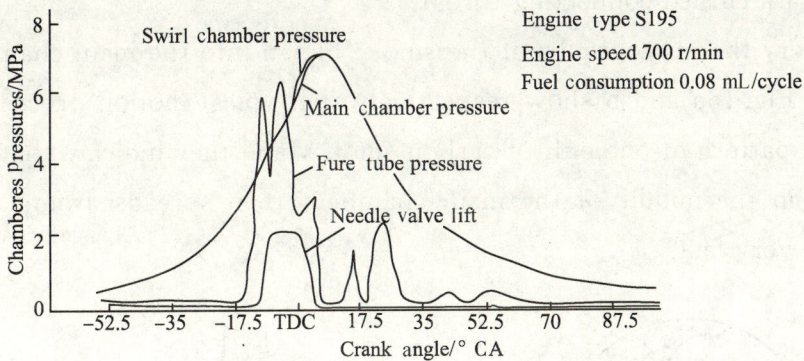

Fig. 13　**Pressures versus crank angle in swirl and main chambers during combustion process**

From what is mentioned above, it is concluded that the bisphenoid and dipper shapes of the main chamber, which lead the flame and smoke blowing out of the connecting throat to the surrounding area in the main chamber, are designed rationally. With such kinds of structure, the smoke will easily flow around the chamber and come in contact with the air quickly, then the combustion completes earlier.

At ATDC 15.6° CA, the smoke begins to rotate and spread all over the chamber. At ATDC 19.6° CA, it approaches the wall, and at ATDC 39.0° CA, the combustion is over. Quantitative analyses of the photographs under various conditions were made on NAC 160B film motion analyzer. As an example, according to Fig. 13, the transient flame fronts are shown in Fig. 14a and relevant flame speed curve in Fig. 14b. Based on the photographs and their quantitative analyses, the follows are concluded.

(1) The flame fronts are uneven and the local flame speeds differ greatly.

(2) When an engine runs at a lower speed, the speed of the flame front is also lower and relatively even.

(3) With a straight connecting throat, the speed angle of the flame is smaller

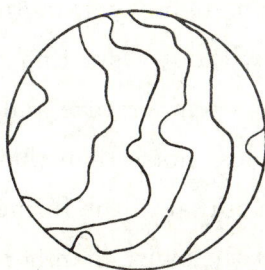

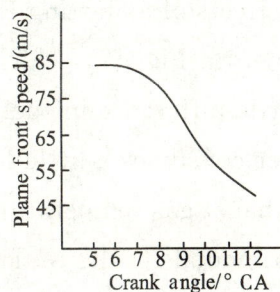

(a)　　　　　　　　　　　　　　　　　　　　(b)

Fig. 14　**Flame front trails and spread speed curve in main chamber, interval between two trails** 2.5×10^{-4} **s**

299

than that of a curved connecting throat.

Further, the photographs of the smoke blown into the main chamber are interpreted. Fig. 15a and b show respectively the typical motion process and relevant speed pattern of one puff of carbon smoke, and they indicate that the smoke stays first in the middle of the main chamber, then spreads around. Its speed goes down very fast.

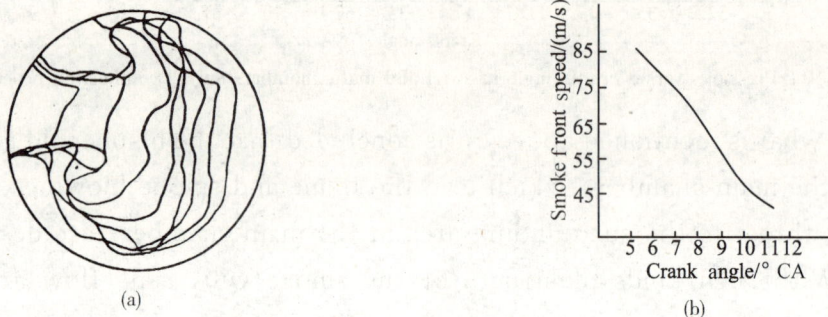

(a)   (b)

Fig. 15   Smoke front trails and its spread speed curve, interval between two trails 2.5×10^{-4} s

4   Practical effects of the starting throat

Fig. 16 shows two starting processes of an engine with and without the starting throat. With a starting throat, under 5 ℃ of the ambient temperature, once the engine was cranked, it started successfully in $2 \sim 3$ s. But without a starting throat, the starting began only when the inlet temperature was raised, the compression ratio or the cranking speed increased. Without starting aids, even if the ambient temperature rised to 20 ℃, it failed to be started, so it is known that the starting throat has significant effects on improving starting process.

The effects of a starting throat on performance of an engine working normally are shown in Fig. 17, which indicates that with or without a starting throat, there is little difference in fuel consumption, exhaust gas temperature and noise. The former consumes a little less fuel and gives a little less noise than the latter, but the exhaust gas temperature of the latter is lower than that of the former. Recent experimental results of Toshiaki Tanaka show that the swirl chamber with a dual-throat is able to bring about greater output and reduction of noise. His results are consistent with ours.

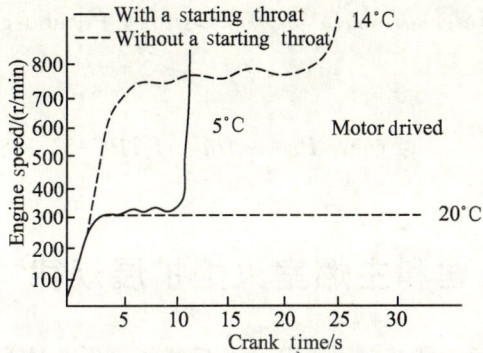

Fig. 16　Starting processes of the engine with and without a starting throat

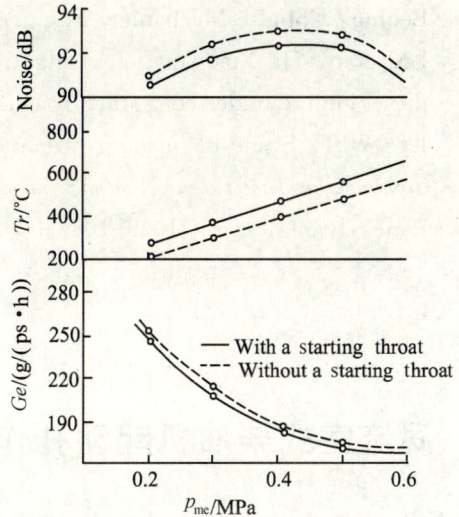

Fig. 17　Effects of a starting throat on performance

5　Conclusions

(1) The starting throat helps shorten the starting process in swirl chamber diesel engines. It causes direct ignition in the main chamber and easier ignition in the swirl chamber.

(2) The bisphenoid and dipper shapes of the main chamber have advantageous effects on burning in the main chamber.

(3) The flame spread angle of a straight connecting throat is smaller than that of a curved connecting throat. This is one of the reasons why the engine with a curved connecting throat performs better than one with a straight connecting throat.

(4) An engine with or without a starting throat has little change to its fuel consumption, exhaust gas temperature and noise.

References

[1] Toshiaki Tanaka, et al. Improvement in IDI diesel engine performance through dual-throat jet combustion[J]. *Transactions of the Society of Automotive Engineers of Japan*, 1986(32).

[2] Mataji Taleishi. New combustion system of the IDI diesel engine[J]. *SAE paper*, 841081,1984.

[3] Li Detao. *Investigation and Design of the Swirl Chamber Diesel Engines* [M].

Beijing：Chinese Mechanical Engineering Press，1986.

［4］Li Detao，He Xiaoyang. Investigation on unstable state combustion in swirl chamber diesel engine under cold-starting condition［J］. *Transactions of CSICE*，870020，1987.

［5］Lyn W T. Study of burning rate and nature of combustion in diesel engines［C］// *The Combustion Institute*，*Ninth Symposium（International）on Combustion*，Pittsburgh，Pennsylvania，1962：1069-1082.

（From：*Proceeding of IPC-5*，1989）

涡流室式柴油机起动孔作用机理和主燃室火焰扩展规律

［摘要］ 关于涡流室式柴油机起动孔的作用机理至今没有一种公认的合理解释。作者利用高速摄影对活塞顶部的着火和燃烧情况进行了观测，发现起动孔有使主燃室直接着火和使涡流室容易着火的双重作用，从而找到了起动孔加速起动过程的作用机理。本文讨论了起动孔对发动机性能的影响。文章还分析了主燃室内的火焰运动规律，为设计合理的主燃室结构型式提供了依据。

涡流室式柴油机冷起动过程的若干特征

李德桃，朱章宏，贾大锄，范永忠

[摘要]　作者运用自行建立的高速数据采集与分析系统，通过实测两种形式起动孔起动过程的连续示功图和瞬时转速，对涡流室式柴油机的冷起动过程进行了研究。文中分析了不同形式起动孔起动过程的转速变化特性，讨论了起动初期及滞速期循环的压力变化特性。研究结果为改善此类柴油机的冷起动性能提供了依据。

涡流室式柴油机是我国面广量大的一类柴油机，由于本身固有的特点，其起动性能与直喷式柴油机还有一定差距。不少研究者曾对起动孔改善涡流室式柴油机起动性能的作用机理进行过研究，但冷起动过程的基本规律尚不很清楚。作者曾多次运用高速摄影技术研究过涡流室式柴油机的冷起动过程[1,2]，发现柴油机的冷起动过程是一个非稳态工作过程。冷起动过程的这一特点，使得有关参数的测录非常困难。为此我们利用自行建立的高速数据采集和处理系统，编制了相应的软件，通过测录涡流室式柴油机冷起动过程连续的示功图及瞬时转速，对涡流室式柴油机的冷起动过程的基本特征作了进一步的分析。

1　试验装置与试验方法

由于柴油机的冷起动过程为典型的非稳态工作过程，不能用单个循环的或者多个循环平均的压力变化数据或示功图来全面而正确地描述，而必须研究冷起动的全过程。据我们所知，国内现有的（包括进口的）燃烧分析仪都不能满足涡流室式柴油机冷起动过程测试和分析的需要。

为此，我们利用磁带记录仪记录冷起动过程的压力变化数据和曲轴转角信号，然后利用自行研制的 A/D 转换装置，编制相应的数据采集和处理程序，将磁带记录仪记录的数据转存入计算机软盘，并通过测试曲轴转角信号间的时间间隔，得到了发动机的瞬时转速，从而在计算机上实现了对测试数据的处理。图 1 为测试装置和回放装置示意图，测试前后对传感器、电荷放大器和整个装置进行了标定。试验是在冷起动室进行的，试验用发动机为 S195 型发动机，试验中采用了图 2 所示的两种起动孔方案。

2　测试结果与分析

2.1　转速变化特性

图 3 显示了 S195 型柴油机起动后转速 n 随循环数 N 的变化情况。使用原机起动孔在环境温度－1 ℃时一次就起动成功（曲线 a），反映出较好的起动性能；使用直形起动孔在环境温度 0 ℃时前 3 次起动均失败，第 4 次方成功（曲线 b），此时已不再是绝对冷车，说明该起动孔起

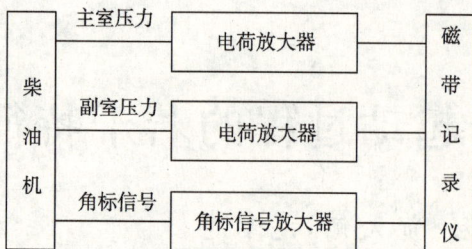

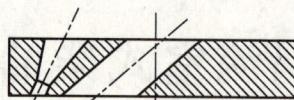

(a) 原机起动孔

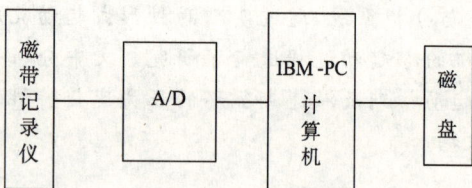

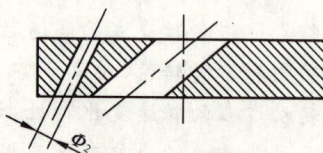

(b) 直形起动孔

图1　测试装置和回放装置示意图　　图2　试验用起动孔方案

动性能很差。两者起动后转速随循环数的变化情况表明,柴油机的冷起动过程都要经历初始加速期、滞速期和再次加速期。滞速期可占几十循环甚至上百循环。使用原机起动孔时,滞速发生在转速较高的时候,约 700 r/min,飞轮得到足够的惯性,因此,即使持续时间长一些,也不会停机;使用直形起动孔时,滞速发生在 400 r/min 左右,易熄火停机。因此,采用直形起动孔时,起动困难。

对于环境温度较高时的起动过程,情况有所不同。图4是使用直形起动孔在室温 21 ℃ 时测取的。该图表明,起动成功后转速即以较快的速度上升,至第 20 个循环时,转速已超过 1 300 r/min,没有明显的滞速期。

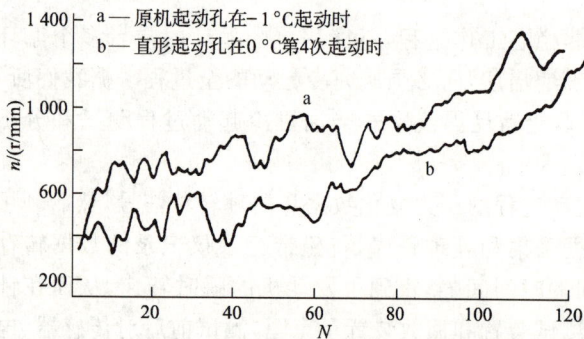

图3　S195型柴油机起动过程转速变化情况

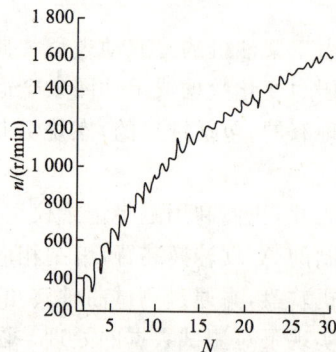

图4　S195型柴油机使用直形起动孔
在 21 ℃ 起动后转速变化情况

由此看来,形成滞速期的根本原因仍在于环境温度低,着火延迟期长。一定的着火延迟期应与一定的转速相适应。转速升高,则使以曲轴转角计的着火延迟期延长;超过可用延迟期就可能导致燃烧不良、转速下降。

因此,若使柴油机在低温起动时供油提前角适当提前,就可使发动机冷起动在较高的转速时发生滞速,增加起动成功的可能性。在文献[3]中也曾提到了采用手动控制装置或石蜡节温器将起动阶段的喷油时间提早的办法。

2.2 初期循环的压力变化

图 5 是采用原机起动孔在环境温度 −1 ℃时手摇起动测录的三维示功图和瞬时转速。图中显示了松开减压阀后 10 个循环的燃烧压力变化情况。从图可见，第 1 个循环在上止点前就着火，在上止点后 4° CA 达到最大值 9.65 MPa，爆发压力之高，上升速度之快，均显示了缸内较好的着火和燃烧条件。这一方面是由于前面多个循环内燃烧室中积累的油雾对该循环的着火有促进作用，另一方面是由于初始转速较低，可用延迟较长，使燃油能够有充分的时间雾化。因此第 1 个循环爆发压力较高，即使在使用起动性能很差的直形起动孔时，第 1 个循环也能够着火（见图 6）。

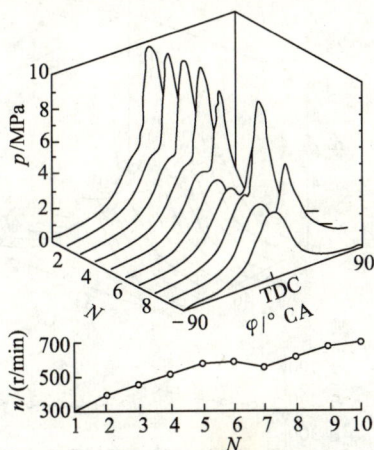

图 5　原机起动孔 −1 ℃手摇起动后
10 个循环的示功图和转速

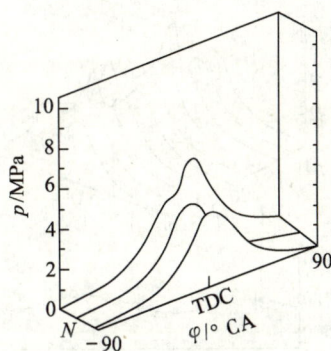

图 6　直形起动孔在室温 0 ℃时
起动失败的示功图

第 1 个循环爆发后，紧接着是 4 个循环的爆发，转速不断上升，使得以曲轴转角计的着火延迟期延长，图中清楚地显示了着火点不断推迟，第 5 个循环在上止点后 7° CA 着火，第 6 个循环则没有爆发。此时转速已近 600 r/min，进入了燃烧不稳定的时期——滞速期。

从转速变化看，初期循环表现为加速期，其着火和燃烧性能对起动的成功与否有重要的作用。直形起动孔起动性能差也表现在初期循环上。图 6 中第 1 循环在上止点后 2° CA 开始着火，至 17° CA 达最大压力值，仅有 5.02 MPa。爆发压力低，燃烧速度慢，说明燃气呈不良的混合状态。转速有所上升后，第 2 和第 3 循环着火而不自持，因而起动失败。

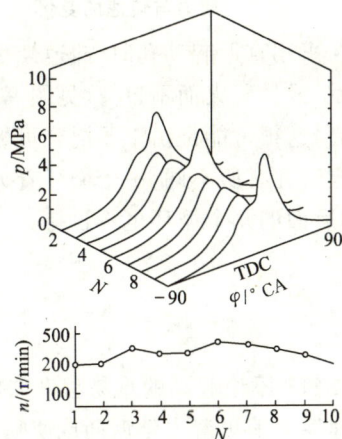

图 7　直形起动孔在室温 0 ℃时起动
成功的前 10 个循环

图 7 为直形起动孔在室温 0 ℃起动成功时的前 10 个循环。第 1 个循环减压阀开着，第 2 个循环才出现爆发，但爆发压力不高，紧接着是 2 个着火而没有自持的循环，第 5 个循环又出现爆发，其后的 4 个循环又均未自持。因此初期循环的加速性只表现在松开减压阀后的第 1 个循环，接着就出现了滞速现象。

2.3 滞速期的压力变化

图 8 是原机起动孔在室温－1 ℃起动时第 15 至 24 循环的示功图和转速变化情况。由图可见,前一个循环的爆发将导致后一循环以曲轴转角计的着火的推迟或熄火,该过程中转速基本不变。强化滞速期的着火和燃烧,使发动机转速上升后,着火点并不推迟,这对缩短起动过程是有利的。

文献[4]介绍了采用后加热电热塞改善滞速期燃烧的方法,文献[3]则介绍了分配式喷油泵上的"液压冷起动加速器(HKSB)",它能在发动机加速后将喷油时间提早,使发动机燃烧不再中断,转速持续上升。

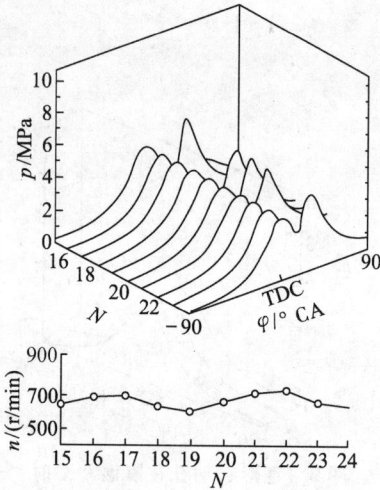

图 8 原机起动孔滞速期不同循环的
压力与转速的变化

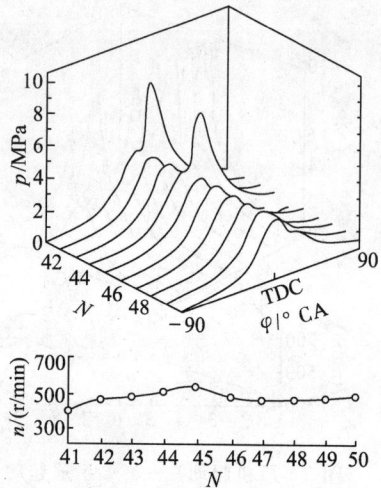

图 9 直形起动孔滞速期不同循环的
压力和转速的变化

采用直形起动孔时,滞速期更长,发生的转速也更低。滞速时 10 个循环中只有 2 个循环爆发,或虽着火而不自持(见图 9)。着火一般都在上止点附近发生,但燃烧速度慢,爆发压力不高,这很可能是由于预混合可燃气的量较少所造成的。直形起动孔滞速时的燃烧情况与原机起动孔有所不同,一个循环着火后,即使转速升高不多,以曲轴转角计着火延迟并不明显增加,后面的几个循环也不自持。

3 结 论

（1）不论起动的难易程度如何,柴油机的冷起动过程都要经历初始加速期、滞速期和再次加速期。缩短着火延迟期或使喷油定时提前,可使滞速发生在较高的转速水平,从而增加起动成功的可能性。

（2）起动初期循环的着火和燃烧特性对柴油机冷起动的成功与否有重要意义。柴油机涡流室镶块使用原机起动孔时,在冷起动过程中初始循环出现连续的旺盛燃烧,因而具有较好的起动性能;直形起动孔在第一循环着火后,便出现不着火或着火而不自持的循环,从而出现滞速,因而起动性能很差。

（3）滞速期在冷起动过程中占有一定地位。欲缩短滞速期,重要的是在转速上升后,着火并不因此推迟,这样转速才能够持续上升。

（4）直形起动孔滞速期的形成与原机起动孔有所不同，某一循环的着火要依靠前面多个不着火循环油雾的累积，因而形成爆发的间断性，外部表现为滞速期加长。

参 考 文 献

［1］李德桃，何晓阳.涡流室式柴油机在冷起动条件下非稳态燃烧过程的研究［J］.内燃机学报，1987，5(3)：193-204.

［2］李德桃，朱晓光.涡流室式柴油机起动孔作用机理和主燃烧室火焰扩展规律［J］.内燃机学报，1990，8(1)：27-32.

［3］Schmidt G F，et al. Optimierung des startverhaltens von wirbelkammer dieselmotoren［J］. *MTZ*，1984(2).

［4］Akio Kobayashi，et al. Analysis of cold start combustion in a direct injection diesel engine［J］. *SAE Paper*，840106，1984.

（本文原载于《内燃机学报》1992 年第 4 期）

Cold start behavior of a swirl chamber diesel engine

Abstract：Cold start process of a swirl chamber diesel engine was investigated with a high speed data acquisition and analysis system developed by authors. Engine speed variation and in-cylinder pressure in early and faltering stages during cold start were measured for two types of starting throat in swirl chamber. The results opened the way for improvement of cold startability of swirl chamber diesel engine.

涡流室式柴油机起动孔作用
机理的进一步探索

朱章宏，李德桃，贾大锄，范永忠

[摘要] 本文对于利用起动孔以改善涡流室式柴油机起动性能的作用机理，在前人研究基础上，作了进一步的探讨。通过试验研究进一步发现，虽然起动孔有使涡流室容易着火和使主燃室直接着火的双重作用，但前一种作用是主要的，即经过起动孔的逆气流促进涡流室中燃油的雾化混合对起动性能有决定意义；而经过起动孔进入主室的部分燃油对起动性能影响较小。从而进一步揭示了起动孔的作用机理，并指明了改善涡流室式柴油机起动性能的途径。

自从采用起动孔作为改善涡流室式柴油机冷起动性能的措施以来，由于效果显著，已在我国普遍应用。但是，对于起动孔改善涡流室式柴油机冷起动性能的作用机理仍有待进一步探索。一些研究者先后对不同结构的涡流室镶块起动孔进行了试验研究，认为起动孔的作用机理在于部分燃油通过起动孔进入温度较高的主燃烧室，并在主燃烧室中首先着火，改善了起动性能；而另一些研究者认为，起动时燃油将不能通过起动孔进入主燃烧室，起动孔的作用机理在于经过起动孔的逆气流将油束吹向温度较高的涡流室中心区域，改善了涡流室内的混合条件。文献[1]通过两种不同方案的起动孔镶块的对比试验，证实了逆气流对改善起动性能的作用，但燃油经过起动孔进入主燃烧室对起动性能的影响程度，该试验尚未作出令人信服的说明。

近年来，高速摄影技术的采用，使得直接观察涡流室式柴油机起动时燃油的雾化、着火情况变得可能。文献[2]的试验表明，不论转速高低，燃油均或多或少地经过起动孔进入主燃烧室。文献[3]还在发动机上直接观察和拍摄到了起动条件下燃油有时在主燃烧室中首先着火的情况。

前人的研究工作证明了：（1）起动时，燃油可以通过起动孔进入主燃烧室；（2）进入主燃烧室的燃油能够首先着火；（3）逆气流对起动性能有改善作用。但是，起动孔改善起动性能的作用机理仍有必要作进一步的阐明。起动孔的设置带来了双重作用：一方面使经起动孔进入副室的逆气流促进了副室中燃油的雾化混合；另一方面又使部分燃油进入主室甚至直接着火[3]。但是，究竟起动孔的哪一种作用对改善起动性能有决定意义，至今尚未得到严格的试验证明。为此，作者在 95 型单缸机上进行了试验研究，以寻求这一问题的解答。

1　试验设计

分析前人的工作，作者认为在对起动孔作用机理的研究中必须解决以下问题：由于起动孔存在两种作用，因此，要说明何种作用为主要的或决定性的，就必须以保持一种作用的强度或效果基本不变作为前提，通过改变另一种作用的强度，来观察对起动性能的影响。

为此，设计了三种起动孔方案（见图 1），包括原机起动孔，一共四种，进行了对比试验。

三种起动孔均由两部分组成，上半部分与原机起动孔相同，以保证燃油进入起动孔的条件不变，下半部为小直孔，直径与原机一致，但方向不同。通过小直孔不同的方向，改变燃油进入主室的难易程度，改变逆气流的作用效果。

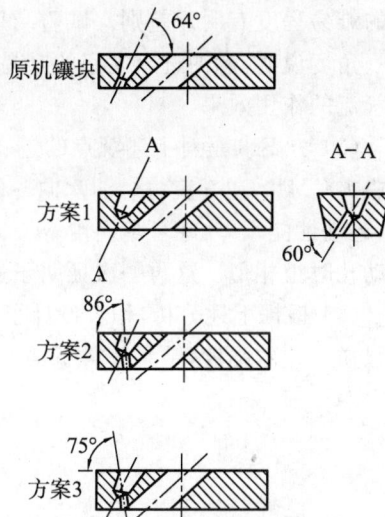

图 1　起动孔试验方案

2　试验结果及其分析

2.1　试验结果

起动试验是在低温室中进行的，所用机油为 11 号，柴油为 0 号，由同一人进行手摇起动，每次起动间隔 5 min，试验结果见表 1。

表 1　起动试验结果

镶块类型	温度/℃			第 1 次起动		第 2 次起动		第 3 次起动		备　注
	环境	机油	柴油	状况	时间/s	状况	时间/s	状况	时间/s	
原机	−2	−2	−2.5	成功	4～5	成功	4～5	成功	4～5	—
方案 1	−2	−2	−2	失败	＞5	成功	4～5	成功	4～5	第 1 次起动失败，但响了一声
方案 2	6	6	6	失败	＞5	失败	＞5	失败	＞5	后经多次泵油，第 4 次成功，但爆发不连续
方案 3	7	7	7	失败	＞5	失败	＞5	失败	＞5	后用电机以 420 r/min 拖动 20 s 后方成功

2.2　结果分析

（1）方案 1 和方案 2 两种起动孔，小直孔中心线与上半部锥孔中心线夹角均为 30°，两者燃油通过起动孔进入主燃烧室的难易程度是相似的。由于小直孔直径一致，因此，逆气流的强度是相同的，但方向不一样，方案 1 逆气流仍吹向涡流室中间区域，方案 2 的逆气流偏向壁面。前人的工作证明，气流将燃油吹向冷的壁面是不利于起动的。因此，如果两种方案的起动性能相似，这就说明起动孔改善起动性能的作用机理在于部分燃油进入主室首先着火，而与逆气流方向关系不大。但试验结果表明，两种方案起动性能相差较远，这就说明部分燃油进入主室甚至直接着火与起动性能关系不大，相反，逆气流作用对起动性能影响很显著。

（2）原机起动孔和方案 1，两者逆气流强度基本一致；效果上，原机起动孔逆气流方向处在涡流室中心，而方案 1 的逆气流方向要偏一些，因此起动性能应略差一些，但两者燃油进入主室的难易程度有明显差别。试验结果表明,两种起动孔的起动性能类似,而方案1略差。这从另一角度说明,燃油进入主室的难易程度对起动性能无影响或影响较小,起动性能主要取决于逆气流的作用效果。

(3)方案3是对上述观点的进一步验证,小直孔的角度继续偏向壁面,与上半部锥孔中心线成夹角41°,逆气流将燃油吹向冷的壁面,同时燃油进入主室的困难性也增大。试验结果表明,起动性能很差,在环境温度7℃时,用电动机以420 r/min的转速拖动20 s方能起动,与无起动孔时很相似。这再一次证明了逆气流的作用效果对起动性能的决定作用。

(4)根据上述分析,作者设计了与文献[4]中相似的起动孔方案,将小直孔改为倒锥体,以加强逆气流的强度,结果获得了类似的改进效果。

3　结　　论

(1)经过起动孔,由主燃烧室进入涡流室的逆气流,将燃油吹向涡流室中间区域,增加了涡流室中燃油的空间混合成分,是起动孔改善涡流室式柴油机冷起动性能的决定因素。

(2)经过起动孔,由涡流室进入主燃烧室的部分燃油,对于起动性能影响较小。

(3)加强由主燃烧室经起动孔进入涡流室的逆气流,进一步增加涡流室中燃油的空间混合成分,是涡流室式柴油机改善冷起动性能的重要途径。

参 考 文 献

[1] 李力. X195型柴油机冷起动性能的试验研究[J]. 内燃机工程,1989(1).
[2] 钱龙文,等. 涡流燃烧室柴油机起动时喷油与着火过程的研究[J]. 内燃机工程,1990(2).
[3] 李德桃,朱晓光. 涡流室式柴油机起动孔作用机理和主燃烧室火焰扩展规律[J]. 内燃机学报,1990(1).
[4] 刘铸民,李建伟. 涡流室式柴油机冷起动性能的改进试验[J]. 内燃机工程,1990(4).

(本文原载于《内燃机工程》1992年第2期)

A further study on the functions of starting throat in swirl chamber diesel engine

Abstract: A further study on the function of starting throat for improving the cold startability of swirl chamber diesel engines was made. With experimental investigation, it was found that the starting throat has both functions of causing direct ignition in main chamber and promoting the ignition of mixture in swirl chamber. The later function is considered the controlling factor.

The experimental results explained the function of starting throat and demonstrated the way to improve the cold startability of swirl chamber diesel engines.

柴油机冷起动时热力参数
计算模型的建立与应用

李德桃，朱章宏，朱晓光，贾大锄

[摘要]　依据热力学定律和试验结果，建立了一个较精确的柴油机冷起动条件下压缩终点温度和压缩压力的计算模型，并对漏气损失和传热损失进行了较深入的分析。该模型还为柴油机冷起动时传热损失的计算提供了一种简便而可靠的方法。

　　柴油机的冷起动性能取决于冷起动时的热力状态和混合状态[1]。柴油机冷起动时，由于活塞平均速度和壁温都较低，导致漏气损失和传热损失都较大[2,3]。如果不考虑这些损失，仅基于绝热压缩的简单热力学方法得到的压缩温度，误差可达 200～300 ℃。文献[4]虽然介绍了利用平均压缩多变指数的方法计算压缩温度，但由于此法不严格，且对漏气损失估计偏大，并不适用于大多数柴油机的起动过程。Fritz 和 Abata[3]虽然应用热力学方法，考虑了漏气损失和传热损失，建立了一个压缩温度的计算模型，但此模型尚不精确，漏气损失和传热损失的测试结果和计算方法也有待进一步研究。为此，我们在前人工作的基础上，根据热力学定律，建立了一个较精确的柴油机冷起动条件下压缩温度和压缩压力的计算模型，并对漏气损失和传热损失进行了深入的分析。

1　模型的建立

　　压缩终点温度的计算，针对的是进气门关闭时刻到压缩上止点这一过程。考虑到漏气的存在，将研究对象作为开式系统处理（见图1）。

　　假设工质为理想气体，系统内压力、温度均匀分布。于是，根据热力学第一定律可得

$$\delta Q = dU_{c.v} + \delta W + h \cdot dm_b \qquad (1)$$

式中 δQ 为系统与外界的热量交换，假定系统吸热为正，放热为负；$dU_{c.v}$ 为系统内能变化；δW 为系统与外界的功量交换，规定系统对外界作功为正，反之为负；m_b 为漏气质量；h 为焓；$h \cdot dm_b$ 为漏气带走的能量。

　　设传热损失率

$$\phi = \delta Q / \delta W, (0 < \phi < 1)$$

由
$$dU_{c.v} = d(mC_V T) = mC_V dT + C_V T dm$$

$$\delta W = PdV$$

$$h \cdot dm_b = C_P T dm_b$$

$$dm = -dm_b$$

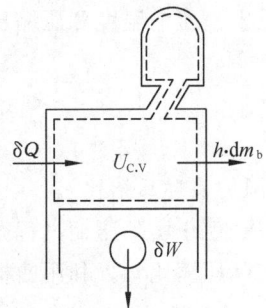

图1　发动机热力学模型

311

可得

$$mC_V dT + (1-\phi)P dV - (C_P - C_V)T \cdot dm = 0 \tag{2}$$

两边同除以 $mC_V T$,得

$$\frac{dT}{T} + (1-\phi)(k-1)\frac{dV}{V} - (K-1)\frac{dm}{m} = 0 \tag{3}$$

两边积分得到

$$\frac{T_2}{T_1} = \left(\frac{V_1}{V_2}\right)^{(1-\phi)(K-1)} \cdot \left(\frac{m_2}{m_1}\right)^{K-1} \tag{4}$$

式中 T_1, T_2 为压缩始点温度(进气温度)和压缩终点温度;V_1, V_2 为压缩始点和终点容积;m_1,m_2 为压缩始点和终点质量;K 为比热比。

令实际压缩比 $\varepsilon_a = V_1/V_2$,漏气率 $\beta = (m_1 - m_2)/m_1$

故

$$\frac{T_2}{T_1} = \varepsilon_a^{(1-\phi)(K-1)} \cdot (1-\beta)^{K-1} \tag{5}$$

也即

$$\frac{T_2}{T_1} = \varepsilon_a^{K-1} \cdot \varepsilon_a^{\phi(1-K)} \cdot (1-\beta)^{K-1} \tag{6}$$

式(6)清楚地显示了传热损失和漏气损失对压缩终点温度的影响。

利用理想气体状态方程式

$$P_1 V_1 = m_1 R T_1 \tag{7}$$

$$P_2 V_2 = m_2 R T_2 \tag{8}$$

得

$$\frac{P_2}{P_1} = \frac{V_1}{V_2} \cdot \frac{m_2}{m_1} \cdot \frac{T_2}{T_1}$$

$$= \varepsilon_a \cdot (1-\beta) \cdot \frac{T_2}{T_1} \tag{9}$$

$$= \varepsilon_a^{K-\phi(K-1)} \cdot (1-\beta)^K \tag{10}$$

$$= \varepsilon_a^K \cdot \varepsilon_a^{\phi(1-K)} \cdot (1-\beta)^K \tag{11}$$

利用实际压缩比 ε_a、漏气率 β、传热损失率 ϕ 的数值,按式(5)可计算出压缩终点温度,按式(10)可计算压缩终点压力。

由式(6)和式(11)还可看到:当漏气率 $\beta=0$,传热损失率 $\phi=100\%$ 时,式(6)和式(11)则变为等温压缩时的表达式;当 $\beta=0$,$\phi=0$ 时,式(6)和式(11)则又表现为绝热压缩时的情况。可见上述压缩温度和压缩压力的计算模型是具有普遍意义的。

2 模型参数分析

2.1 实际压缩比 ε_a

由于实际压缩过程开始于进气门关闭的时刻,故实际计算中不应取几何压缩比 ε,而应取实际压缩比 ε_a。利用曲柄连杆机构的运动学关系和进气门迟闭角 θ,可得

$$\varepsilon_a = 1 + \frac{\varepsilon-1}{2}\left[1 + \cos\theta + \frac{1}{\lambda}\left(1 - \sqrt{1-\lambda^2 \sin^2\theta}\right)\right] \tag{12}$$

式中 λ 为曲柄连杆比,即 r/l。

将 $\sqrt{1-\lambda^2 \sin^2\theta}$ 按泰勒级数展开并舍去高次项,代入式(12)可得

$$\varepsilon_a = 1 + \frac{\varepsilon-1}{2}\left(1 + \cos\theta + \frac{1}{2}\lambda \sin^2\theta\right) \tag{13}$$

对于 S195 型柴油机，$\varepsilon=20$，$\theta=43°$，$\lambda=1/3.65$，代入上式可算得 $\varepsilon_a=18$。

2.2 漏气率 β

柴油机的漏气是一项直接的能量损失。在正常运转条件下，漏气率一般在 1% 左右[5]，通常可被忽略不计；而在冷起动时，漏气率可达到甚至超过 10%[2]，它就是影响起动性能的一个重要因素。

柴油机的漏气，虽然可通过漏气测量仪器测出，但是仪器只能提供循环内的平均值，并且是具体发动机在具体条件下的数值，局限性很大，至于由此总结得到的计算漏气的经验公式，同样忽略了发动机加工精度、使用程度对漏气的影响，也是不严格的。为此，我们应用目前国外较常用的由 Ting 和 Mayer 提出的活塞环漏气计算模型[6]，根据实测的柴油机冷起动条件下拖动示功图和有关结构参数，可计算得到柴油机冷起动条件下的漏气率。

在漏气的计算模型中，假设活塞环的闭口间隙是唯一的泄漏途径，油环不密封气体。因此漏气通道便是一系列由活塞环开口处连接的腔室所组成。活塞环的漏气计算模型见图 2。气体的流动假设为理想气体的一元准稳态绝热流动。对于一个三道气环的发动机，考虑到压缩比不同和气体倒流的情况，我们得到如下公式：

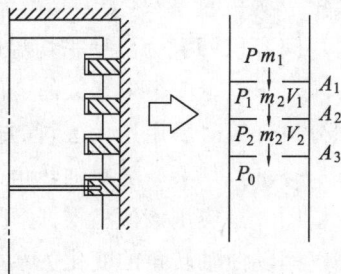

图 2　活塞环漏气模型

$$\frac{\mathrm{d}m_1}{\mathrm{d}\theta}=\frac{\mu A_1 P}{\omega \sqrt{RT_w}}\sqrt{\frac{2K}{K-1}}\cdot\left(\frac{2}{K+1}\right)^{\frac{1}{K-1}}, \quad \frac{P_1}{P}\leqslant\left(\frac{2}{K+1}\right)^{\frac{K}{K-1}} \tag{14}$$

$$=\frac{\mu A_1 P}{\omega \sqrt{RT_w}}\sqrt{\frac{2K}{K-1}\left[\left(\frac{P_1}{P}\right)^{\frac{2}{K}}-\left(\frac{P_1}{P}\right)^{\frac{K+1}{K}}\right]}, \quad 1\geqslant\frac{P_1}{P}>\left(\frac{2}{K+1}\right)^{\frac{K}{K-1}} \tag{14a}$$

$$=-\frac{\mu A_1 P_1}{\omega \sqrt{RT_w}}\sqrt{\frac{2K}{K-1}\left[\left(\frac{P}{P_1}\right)^{\frac{2}{K}}-\left(\frac{P}{P_1}\right)^{\frac{K+1}{K}}\right]}, \quad \left(\frac{K+1}{2}\right)^{\frac{K}{K-1}}\geqslant\frac{P_1}{P}>1 \tag{14b}$$

$$=-\frac{\mu A_1 P_1}{\omega \sqrt{RT_w}}\sqrt{\frac{2K}{K-1}}\left(\frac{2}{K+1}\right)^{\frac{K}{K-1}}, \quad \frac{P_1}{P}>\left(\frac{K+1}{2}\right)^{\frac{K}{K-1}} \tag{14c}$$

$$\frac{\mathrm{d}m_2}{\mathrm{d}\theta}=\frac{\mu A_2 P_1}{\omega \sqrt{RT_w}}\cdot\sqrt{\frac{2K}{K-1}}\cdot\left(\frac{2}{K+1}\right)^{\frac{1}{K-1}}, \quad \frac{P_2}{P_1}\leqslant\left(\frac{2}{K+1}\right)^{\frac{K}{K-1}} \tag{15}$$

$$=\frac{\mu A_2 P_1}{\omega \sqrt{RT_w}}\cdot\sqrt{\frac{2K}{K-1}\left[\left(\frac{P_2}{P_1}\right)^{\frac{2}{K}}-\left(\frac{P_2}{P_1}\right)^{\frac{K+1}{K}}\right]}, \quad 1\geqslant\frac{P_2}{P_1}>\left(\frac{2}{K+1}\right)^{\frac{K}{K-1}} \tag{15a}$$

$$=-\frac{\mu A_2 P_2}{\omega \sqrt{RT_w}}\sqrt{\frac{2K}{K-1}\left[\left(\frac{P_1}{P_2}\right)^{\frac{2}{K}}-\left(\frac{P_1}{P_2}\right)^{\frac{K+1}{K}}\right]}, \quad \left(\frac{K+1}{2}\right)^{\frac{K}{K-1}}\geqslant\frac{P_2}{P_1}>1 \tag{15b}$$

$$=-\frac{\mu A_2 P_2}{\omega \sqrt{RT_w}}\sqrt{\frac{2K}{K-1}}\cdot\left(\frac{2}{K+1}\right)^{\frac{1}{K-1}}, \quad \frac{P_2}{P_1}>\left(\frac{K+1}{2}\right)^{\frac{K}{K-1}} \tag{15c}$$

$$\frac{\mathrm{d}m_3}{\mathrm{d}\theta}=\frac{\mu A_3 P_2}{\omega \sqrt{RT_w}}\sqrt{\frac{2K}{K-1}}\left(\frac{2}{K+1}\right)^{\frac{1}{K-1}}, \quad \frac{P_0}{P_2}\leqslant\left(\frac{2}{K+1}\right)^{\frac{K}{K-1}} \tag{16}$$

$$=\frac{\mu A_3 P_2}{\omega \sqrt{RT_w}}\sqrt{\frac{2K}{K-1}\left[\left(\frac{P_0}{P_2}\right)^{\frac{2}{K}}-\left(\frac{P_0}{P_2}\right)^{\frac{K+1}{K}}\right]}, \quad 1\geqslant\frac{P_0}{P_2}>\left(\frac{2}{K+1}\right)^{\frac{K}{K-1}} \tag{16a}$$

$$= -\frac{\mu A_3 P_0}{\omega \sqrt{RT_w}} \sqrt{\frac{2K}{K-1}\left[\left(\frac{P_2}{P_0}\right)^{\frac{2}{K}} - \left(\frac{P_2}{P_0}\right)^{\frac{K+1}{K}}\right]}, \quad \left(\frac{K+1}{2}\right)^{\frac{K}{K-1}} \geqslant \frac{P_0}{P_2} > 1 \qquad (16b)$$

$$= -\frac{\mu A_3 P_0}{\omega \sqrt{RT_w}} \sqrt{\frac{2K}{K-1}} \cdot \left(\frac{2}{K+1}\right)^{\frac{1}{K-1}}, \quad \frac{P_0}{P_2} > \left(\frac{K+1}{2}\right)^{\frac{K}{K-1}} \qquad (16c)$$

$$\frac{\mathrm{d}P_1}{\mathrm{d}\theta} = \frac{RT_w}{V_1}\left(\frac{\mathrm{d}m_1}{\mathrm{d}\theta} - \frac{\mathrm{d}m_2}{\mathrm{d}\theta}\right) \qquad (17)$$

$$\frac{\mathrm{d}P_2}{\mathrm{d}\theta} = \frac{RT_w}{V_2}\left(\frac{\mathrm{d}m_2}{\mathrm{d}\theta} - \frac{\mathrm{d}m_3}{\mathrm{d}\theta}\right) \qquad (18)$$

式中 μ 为流量系数(为方孔的流量系数,$\mu = 0.65$);A_1, A_2, A_3 为各道气环的泄漏面积;V_1, V_2 为环间容积;T_w 为缸壁平均温度;ω 为角速度。

对于以上 5 个未知数的一阶常微分方程组,可应用龙格-库塔(Runge-Kutta)法进行数值积分。计算时采用迭代法,首先预估 P_1, P_2 值,在循环始点、终点的 P_1, P_2 值一致时,便求得 m_1, m_2, m_3, P_1, P_2 随曲轴转角的变化规律。

图 3 是应用该模型根据 S195 型柴油机冷起动条件下拖动示功图计算得到的各道气环的漏气率随曲轴转角的变化关系。图 4 表示了漏气率与拖动转速的变化关系。

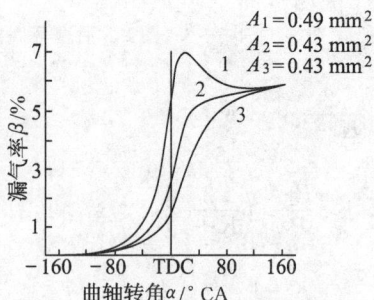

图 3 各道气环的漏气特性

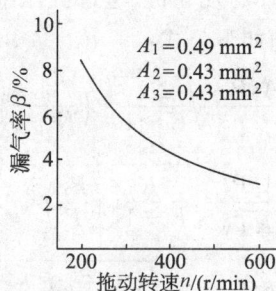

图 4 漏气率与拖动转速的关系

2.3 参数的敏感性分析

敏感性分析可以帮助我们深入准确地了解漏气率、传热损失率对压缩温度和压缩压力的影响程度。

由式(5)可得

$$\frac{\partial T_2}{\partial \beta} = T_1 \cdot \varepsilon_\alpha^{(K-1)(1-\phi)} \cdot (K-1)(1-\beta)^{K-2} \cdot (-1)$$

$$\Delta T_2 \approx -(K-1) \cdot T_1 \varepsilon_\alpha^{(K-1)(1-\phi)} \cdot (1-\beta)^{K-2} \Delta\beta$$

故

$$\frac{\Delta T_2}{T_2} \approx -\frac{K-1}{1-\beta} \Delta\beta \qquad (19)$$

同理可得

$$\frac{\Delta P_2}{P_2} \approx -\frac{K}{1-\beta} \Delta\beta \qquad (20)$$

由式(19)和(20)可知,当 β 由 0 增到 0.10 时,压缩温度下降 4.3%,而压缩压力下降近 15%。因此,漏气对压缩终点的温度影响较小,一般情况下,漏气造成的压缩温度降低不超过 30 ℃,而漏气对压缩终点压力的影响较大(见图 5)。

传热损失率 ϕ 对压缩终点温度和压力的影响,可由式(5)和式(10)导出

$$\frac{\Delta T_2}{T_2} \approx -(K-1) \cdot \ln \varepsilon_a \cdot \Delta \phi \tag{21}$$

$$\frac{\Delta P_2}{P_2} \approx -(K-1) \cdot \ln \varepsilon_a \cdot \Delta \phi \tag{22}$$

可见,传热损失率对压缩终点温度和压力的影响程度是相同的。当 $\varepsilon_a = 18$,ϕ 增加 0.01,则压缩温度和压力下降近 1.2%(见图6)。

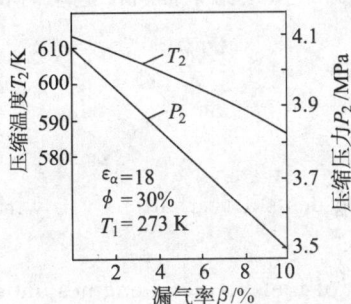

图5　压缩温度、压缩压力与漏气率的关系

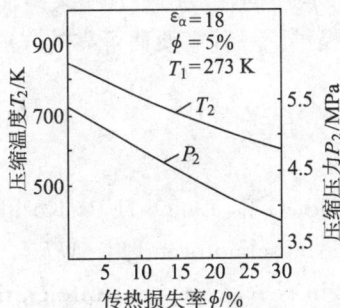

图6　压缩温度、压缩压力与传热损失率的关系

3　模型应用

在上述模型的实际应用中,由于冷起动过程中传热损失的计算较为复杂,目前尚无可借鉴的计算模型,不少研究者在对冷起动过程进行热力计算时都避开了这一问题,因此直接由式(6)和式(11)计算压缩温度和压缩压力,还存在一定的困难。但是,由于示功图的测录比较容易,因此,该模型可在利用压缩压力数据计算传热损失、压缩温度方面获得应用。

由式(11)两边取对数可得

$$\ln \frac{P_2}{P_1} = \ln \left[\varepsilon_a^K \cdot (1-\beta)^K \right] + \phi(1-K) \ln \varepsilon_a$$

$$\phi = \frac{\ln \left[\varepsilon_a^K (1-\beta)^K \cdot \dfrac{P_1}{P_2} \right]}{\ln \varepsilon_a^{K-1}} \tag{23}$$

利用式(23)计算传热损失简单、可靠,为冷起动时传热损失的计算提供了一条新的途径。

我们测得 S195 型柴油机在环境温度 0 ℃时,以 300 r/min 拖动时的示功图,$P_2 = 3.72$ MPa,由漏气计算模型可算得 $\beta = 5.8\%$。于是由式(23)可得传热损失率 $\phi \approx 30\%$。

如进一步根据示功图计算出压缩功,便可得到传热损失的绝对数值,即

$$传热损失 = \phi \int P\mathrm{d}V$$

压缩温度可由式(9)推算出

$$T_2 = T_1 \cdot \frac{P_2}{P_1} \cdot \frac{1}{\varepsilon_a(1-\beta)} = 599 \text{ K}$$

4 结 论

(1) 依据热力学定律和试验结果,可以建立一个较精确的柴油机冷起动条件下压缩温度和压缩压力的计算模型;

(2) 漏气损失与活塞环数目、泄漏面积、气缸压力、转速均有密切关系,冷起动时,起动转速越低,漏气损失越大;

(3) 该模型同样适用于直喷式柴油机和分隔式柴油机,模型还为柴油机冷起动时传热损失的计算提供了一种简便而可靠的方法。

参 考 文 献

[1] Schmidt G F, Lange H P, Kornher H. Optimierung des startverhaltens von wirbelka-mmer-dieselmotoren [J]. *MTZ*, 1984(2): 65-69.

[2] Henein N A, Lee C S. Autoignition and combustion of fuels in diesel engines under low ambient temperatures[J]. *SAE Transactions*, 861230, 1986(4): 986-999.

[3] Steven G Fritz, Duane L Abata. A photographic study of cold start characteristic of a spark assisted diesel engine operating on broad cut diesel fuels[J]. *SAE Transations*, 871674, 1987.

[4] Дробышевский Ч Б. Оценка термодинамических параметров заряда дизеля на нусковых режимах и возможности пуска[J]. *Двигателестроеиие*, 1984(11): 9-11.

[5] 钱耀义. 汽车发动机漏气性能的研究[J]. 汽车技术, 1985(5): 4-9.

[6] Ting L L, Mayer Jr J E. Piston ring lubrication and cylinder bore wear analysis [J]. *Journal of Lubrication Technology*(*Part 1-Theory*), 1974: 305-314.

(本文原载于《农业机械学报》1992 年第 4 期)

Development and application of a model for calculatiing thermodynamic parameters in a diesel engine under cold starting conditions

Abstract: Based on the first law of thermodynamics and experimental results, an even more accurate mathematical model for calculating compression temperatures and pressures in a diesel engine under cold starting conditions has been developed in this paper. A further analysis of the air blowby loss and the heat transfer loss has also been carried out. According to this model, a simple but trustworthy method is provided for calculating the heat transfer loss of diesel engine during cold start.

涡流室式柴油机冷起动过程燃烧室
瞬态壁温测量与分析[①]

单春贤,李德桃,吕兆华

[摘要]　作者采用具有良好瞬态测量特性的壁面热电偶,结合自行开发的计算机数据采集与处理系统,实测了涡流室式柴油机冷起动过程燃烧室的瞬态壁温,详细分析了冷起动过程燃烧室壁面瞬态温度的变化规律和特征,首次提出了冷起动过程瞬态壁温变化的四阶段模型。对有、无起动孔时,壁温的变化规律也进行了对比分析,揭示了起动孔在冷起动过程中所起的作用。

近年来,作者开展了燃烧室瞬态壁温测量技术的研究,分析了现有薄膜热电偶的性能和特点,设计了一种实用的壁面瞬态温度测量传感器,克服和解决了冷起动过程瞬态参数测量技术和数据处理技术存在的困难,建立了内燃机非稳态过程计算机数据采集与处理系统。作者运用此系统,测试了涡流室式柴油机冷起动过程燃烧室瞬态壁温的变化,分析了壁温的变化规律,首次提出了冷起动过程非稳态温度场的四阶段模型。为分析研究柴油机冷起动过程的机理,提供了一种有效的研究途径和一些具有参考价值的研究结果。

1　壁面瞬态温度传感器——壁面热电偶

常用的薄膜热电偶[1]是以热电偶端部的薄膜层作为感温面,时间常数与薄膜厚度密切相关。为满足动态特性的要求,薄膜层一般均在 5 μm 左右,因此抗振动、抗冲击性能较差,使用寿命仅 2～10 h。壁面热电偶两电极 a,b 的热接点温度相等时,第三种材料 c 的加入对输出热电势 E 没有任何影响,并且 E 与第三种材料的温度与冷接点温度之差成正比,即

---

① 开展燃烧室壁温变化对涡流室式柴油机冷起动性能影响的研究,是研究柴油机冷起动机理的一种有效途径。然而当时计算机数据采集技术刚刚发展起来,抗干扰技术与动态数据采集技术尚不完善,无法实现现场动态数据采集,只能利用模拟磁带记录仪现场记录,再通过 A/D 转换板进行后期数据采集与处理。在研究初期,首先需要解决的一个棘手问题就是瞬态温度的测量,常用的薄膜热电偶由于抗冲击能力、使用寿命和价格因素不宜使用,普通热电偶也由于时间常数大满足不了动态测量的要求。在已有的测试手段都不能很好的满足试验条件的情况下,只能寄希望于开发一种新的测试手段。课题组成员经过翻阅大量的资料、反复推演、实验验证,最终完成了"壁面热电偶"的研制,有效地解决了燃烧室瞬态壁温的测量。在顺利地测录了各种工况下柴油机冷起动过程瞬态温度场后,通过信号回放才发现:测录的温度信号发生严重畸变、完全"湮没"在各种干扰信号中。其间试验了各种滤波及数据处理方法均无济于事,此时时间已经是 5 月份,气温回升,在无冷起动实验室的情况下,要重做冷起动试验只得等来年。在连连受挫的逆境中,课题组成员发扬百折不饶、勇于探索的精神,在艰难的条件下,利用先前积累的频谱分析技术,经过反复试验,终于将完全"湮没"在干扰中的温度信号提取出来,为涡流室式柴油机冷起动过程燃烧室瞬态壁温的分析奠定了基础。

——编者注

$$E = \int_{t_{col}}^{t_{hot}} S_a \, \mathrm{d}t + \int_{t_{hot}}^{t_{hot}} S_c \, \mathrm{d}t + \int_{t_{hot}}^{t_{col}} S_b \, \mathrm{d}t = \int_{t_{col}}^{t_{hot}} S_{ab} \, \mathrm{d}t = S_{ab}(t_{hot} - t_{col}) \tag{1}$$

式中 S_a,S_b,S_c 分别为材料 a,b,c 所具有的热电能;S_{ab} 是热电极 a,b 的热电能之差;t_{hot},t_{col} 分别为热端与冷端温度。

壁面热电偶的结构示于图 1,将两根标准的热电偶丝用专用焊机分别焊在传感器座感温端面上,传感器座材料必须与被测物体材料一致,座端面即构成热电偶的第三种材料。焊点必须很小且尽可能靠近,以保证两接点处的温度基本相同,并都等于热端温度。根据式(1),此时热电偶输出电势将不受壁面材料的影响,但却与壁面温度有关。壁面热电偶利用壁面本身作为传感器的一部分,感温面即

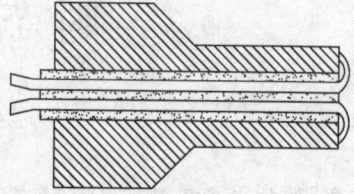

图 1　壁面热电偶的结构

被测面。因此,保证了被测部分的壁面温度与传感器表面温度转换时间一致,能动态地跟踪壁面温度变化,而无其他热电偶存在的响应滞后问题,测得的温度真实地反映了该时刻壁面的瞬时温度。经实验证明,壁面瞬态温度传感器使用效果好,寿命长,输出电压与温度之间的线性度、稳定性、重复性等均较好。

2　瞬态测试系统及实验条件

柴油机冷起动过程瞬态温度场的测录系统如图 2 所示。壁面热电偶输出的热电势分别经热电偶温度放大器放大 200 倍后,输入多通道模拟磁带记录仪,信号的模数转换由每度的曲轴转角信号作为转换开始的外触发信号,即采样间隔为 $1°$ CA,转换后的数字信号经 DMA 输入计算机,由计算机完成进一步的分析与计算,并由绘图仪和打印机输出测量和计算结果。

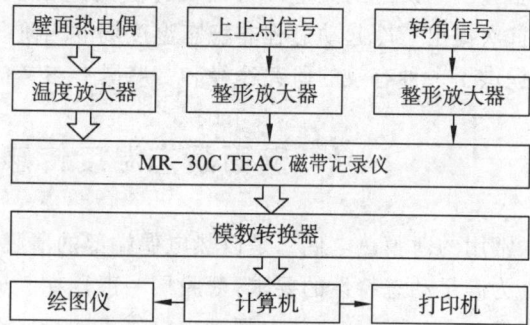

图 2　瞬态温度场测录系统

对于冷起动过程这种典型的非稳态过程,引起温度信号发生畸变的各种外界干扰信号,不宜采用稳态过程常用的多循环"均化"处理来消除。本文提出采用快速傅里叶变换技术(FFT),来滤除周期及高频随机干扰信号,显著地提高了测量数据的真实性及可靠性。

试验是在改装的 195 型柴油机上进行的。该机为单缸、水冷、四冲程涡流室式柴油机,标定功率 $8.8\,\mathrm{kW}$,$2\,000\,\mathrm{r/min}$。为了分析和研究涡流室式柴油机在冷起动过程中燃烧室内壁面温度的分布及变化规律,为冷起动过程传热的研究提供精确的边界条件,在燃烧室壁面布置了 5 支壁面热电偶。壁面热电偶的安装必须保证感温面与燃烧室壁面平齐,且与燃烧室壁可靠接触,以减少安装壁面热电偶对燃烧室内流场和温度场的干扰。

本次试验是在 5 ℃ 环境温度下进行的。为了考察涡流室式柴油机起动孔的作用机理,实验测录了两种不同起动孔方案。方案一为原机镶块,方案二为无起动孔镶块。

3 实验结果分析

3.1 冷起动过程燃烧室瞬态壁温的变化规律

图 3 是 195 型柴油机在冷起动过程中,燃烧室壁温随曲轴转角及经历时间变化的三维曲线。由图可见,就单个循环来说,温度波动的总规律与稳态过程基本相似。在上止点附近,温度变化率较大;在上止点后,温度值达到最大。但各循环之间存在着明显的差异。从随时间变化的

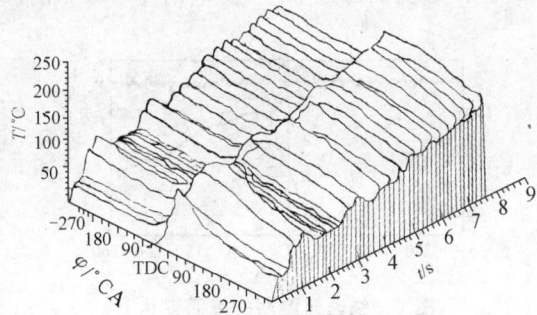

图 3　燃烧室壁面温度三维变化曲线

总趋势看,循环的最高温度和平均温度是逐渐上升的,但各循环之间的差异将随着过程的进行而逐渐缩小,如图 4 所示。然而,在冷起动过程中,燃烧室壁面的最高温度、平均温度和温度变化率等并不是呈线性升高的,不同时期存在着明显的差别,呈现出各自不同的变化规律。一系列测试结果表明,涡流室式柴油机冷起动过程燃烧室瞬态壁温的变化可以划分为四个阶段,即初始升温期、快速升温期、滞温期和缓慢升温期。图 5～图 8 分别示出了这四个阶段中有代表性的壁面温度和温度变化率的曲线。

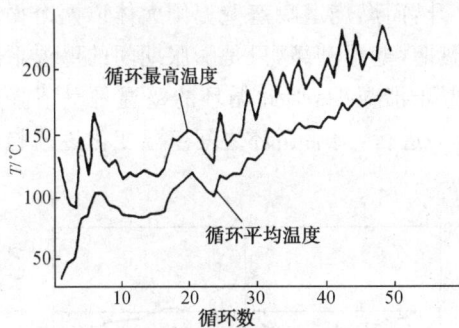

图 4　循环最高温度与平均温度变化

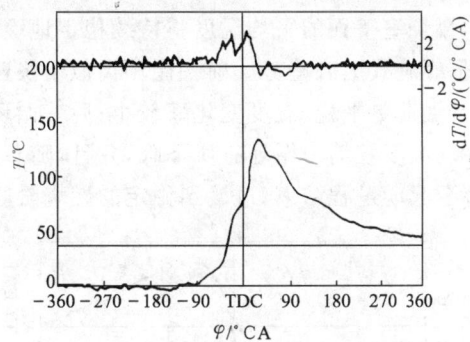

图 5　初始升温期的壁温变化

3.1.1 初始升温期(图 5)

一般均出现在减压阀关闭后的第一个循环。其特点是壁温变化迅速,壁温变化率较大,温度在上止点后达到最大值,并迅速下降。

初始升温期形成,是由于前面多个循环积聚了大量的油雾,形成了良好的可燃混合气浓度,且转速低,以时间计的着火延迟期较长,燃烧的物理化学过程准备充分,故易于着火燃烧,温度上升较快。但是,因燃烧室壁内层温度较低,从内表面向外表面的导热量大,使得燃烧室壁面温度在达到最大值后,迅速下降。

3.1.2 快速升温期(图 6)

出现在初始升温期后的几个循环,是冷起动成功与否的重要标志。其特点是壁温迅速增高,其平均温度亦提高,转速上升,壁温变化较初始升温期平缓。

经历初始升温期后,壁温上升,由工质传给燃烧室壁的热量减少,着火延迟期虽有所缩短,

但因转速升高,使着火可用延迟期大大缩短,着火困难,转速降低,为后续循环的着火和燃烧作好了准备,使下一循环燃烧充分,温度迅速升高。受初始升温期的影响,此时燃烧室壁平均温度较高,内外壁面温差缩小,在该循环的膨胀阶段,壁面温度下降趋于平缓。

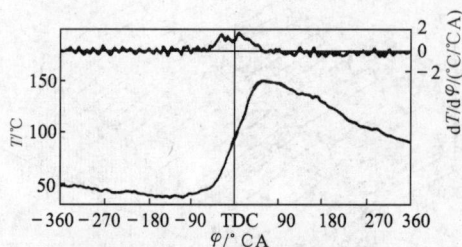

图 6　快速升温期的壁温变化　　　　图 7　滞温期的壁温变化

3.1.3　滞温期(图 7)

此期间为不稳定燃烧阶段,约可持续 10 个循环。此期间的特点是燃烧室壁面平均温度基本上维持不变,温度变化率较小,壁温在上止点后略有上升。

滞温期是由于快速升温期后,壁温和转速迅速升高,着火可用延迟期缩短,着火点后移造成的。燃烧室内无明显放热反应,燃烧相当不稳定,着火困难,着火后火焰不能自持而熄灭。此时微弱的放热与散热相平衡,使燃烧室壁面平均温度保持相对稳定,转速逐渐降低。

3.1.4　缓慢升温期(图 8)

此期间为过渡燃烧阶段,也是冷起动过程的最后阶段,可持续数十个至上百个循环。在此期间,燃烧室壁面的平均温度将持续稳定地缓慢上升,而燃烧室瞬态壁温则大体以两个循环为周期呈螺旋式上升趋势。前一循环类似于快速升温期,壁温升高,只是膨胀期间的壁温下降较快速升温期更平稳,温度变化率较小;后一循环类似于滞温期,但此循环的壁温是呈下降趋势的,平均温度较前一循环有所下降,并且,随着过程的进行,前后循环之间壁温变化逐渐趋于一致,直至起动过程结束,过渡到稳定的燃烧阶段。

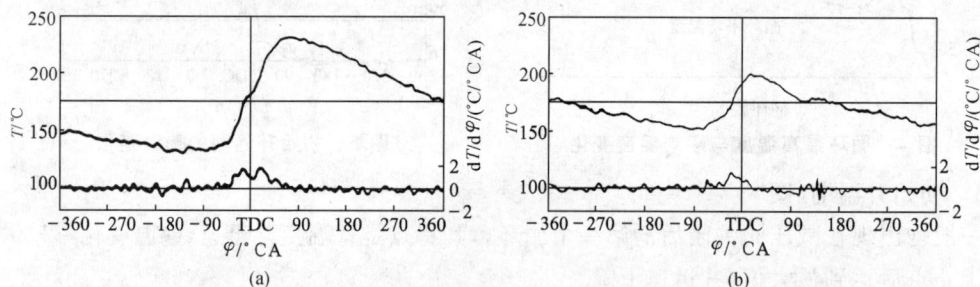

图 8　缓慢升温期的壁温变化曲线

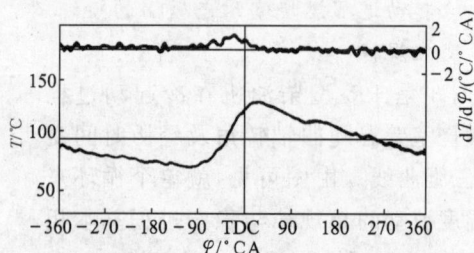

缓慢升温期壁温的变化特点,反映了燃烧室内燃烧的不连续性,前一个循环的燃烧将导致后一循环着火的推迟,使转速在某一值附近剧烈波动,历经滞温期后,燃烧室内外壁面的温差进一步减少,传热损失下降;且由于燃烧的不稳定,转速逐渐降低,着火可用延迟期延长,为缓慢升温期第一循环的到来,做好了必要的准备。经前一循环后,壁温上升,转速增加,着火可用延迟期缩短,保证正常着火所需的压缩温度及壁面温度必须很高,使下一循环着火困难,出现后一循环壁温的变化规律。

图 9 给出冷起动过程初期 10 个循环的壁温变化曲线。该图反映了初始升温期、快速升温

期及滞温期的壁温变化特点。图10为缓慢升温期连续10个循环的壁温变化曲线。从中可见缓慢升温期瞬态壁温周期变化规律。上述结果与冷起动过程的压力变化规律也相吻合。

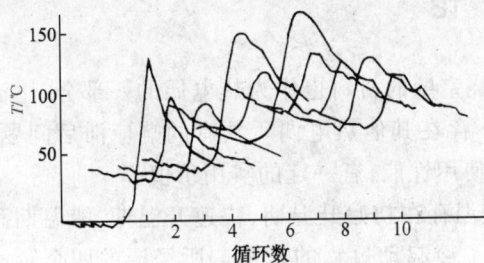

图 9　冷起动初期10个循环的壁温

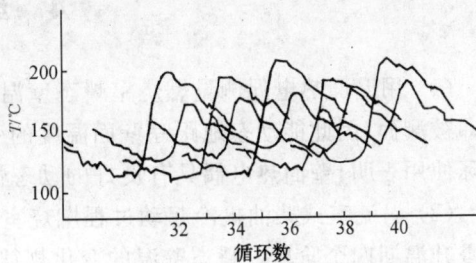

图 10　缓慢升温期连续10个循环的壁温

3.2　无起动孔冷起动过程瞬态壁温的变化特点

图11为无起动孔时的燃烧室瞬态壁温在冷起动过程中的三维变化图形。由图可见,燃烧室壁温始终处于较低的温度水平,增加幅度十分缓慢,壁温变化率较有起动孔时显著减少,这进一步证实了起动孔在涡流室式柴油机冷起动过程中所起的作用。取消了起动孔,改变了起动过程燃油与空气的混合方式,燃烧室内不能组织起良好的可燃混合气,着

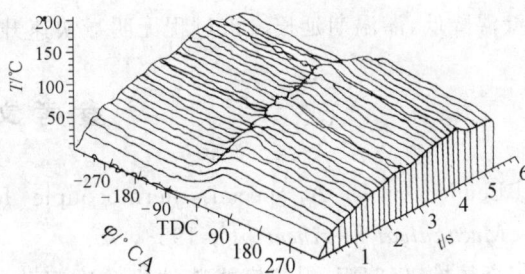

图 11　无起动孔燃烧室瞬态壁温三维曲线

火延迟期延长,着火点后移,着火后也难以自持[2-4],燃烧相当不稳定,燃烧室内没有强烈燃烧放热反应,放热与散热相平衡,使燃烧室壁温基本维持不变。

无起动孔冷起动过程瞬态壁温变化的各个阶段与有起动孔时相比,有明显的差异,初始升温期温升幅值较低,温度变化率较小,着火点滞后,没有出现旺盛燃烧(图12a)。初始升温期后,燃烧室壁温仍然处于较低的温度水平,着火延迟期长,而转速却略有上升,可用着火延迟期相对减少,使得后续循环的着火更加困难。该起动过程没有经历快速升温期,就进入了滞温期。由于滞温期的平均温度较低,着火不易,难以自持,膨胀阶段的壁温下降较快(图12b),使滞温期明显延长,燃烧室壁温的增加幅度十分缓慢,即使转速下降也满足不了该转速下正常着火所要求的温度水平,使缓慢升温期迟迟不能到达。

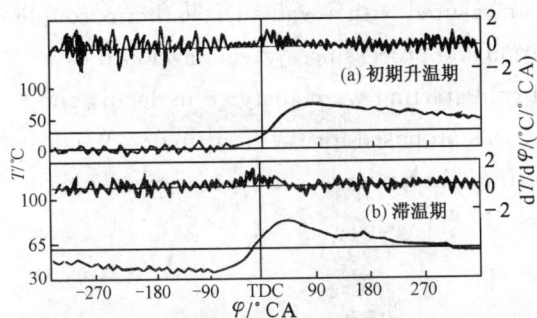

图 12　无起动孔燃烧室瞬态温升变化

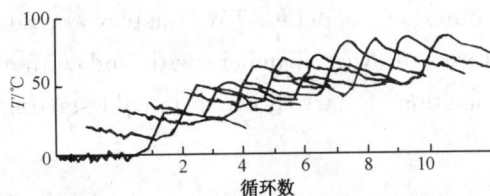

图 13　无起动孔冷起动初期壁温变化

图13给出了无起动孔冷起动过程初期的壁温变化曲线。当柴油机与起动电机脱离后,由

于不能保持稳定的着火与燃烧,转速将不断下降,燃烧室壁温也随之降低,最终导致起动失败。

4 结 论

(1)用壁面热电偶测量燃烧室瞬态壁温,将燃烧室壁的材料也作为热电偶的一部分,感温面即被测面,因此能动态地跟踪壁面温度的变化,不存在其他热电偶所共有的响应滞后问题。实际使用表明,壁面热电偶具有良好的动态性能和使用性能,有一定的实用价值。

(2)涡流室式柴油机冷起动过程燃烧室瞬态壁温存在初始升温期、快速升温期、滞温期和缓慢升温期四个阶段。瞬态壁温的变化规律,反映了冷起动过程的机理。所经历的四个阶段中,快速升温期是反映冷起动过程成功与否的重要标志。

(3)取消起动孔,改变了冷起动过程燃油与空气混合方式,使燃烧室壁温的变化幅度减小,壁温降低,滞温期延长。该过程无明显快速升温期,致使起动过程不能顺利进行。

参 考 文 献

[1] Bendersky D. A special thermocouple for measuring transient temperature [J]. *Mechanical Engineering*,1953,75.

[2] 李德桃,何晓阳.涡流室式柴油机在冷起动条件下非稳态燃烧过程的研究[J].内燃机学报,1987,5(3).

[3] 李德桃,朱晓光.涡流室式柴油机起动孔作用机理和主燃室火焰扩展规律[J].内燃机学报,1990,8(1).

[4] Fritz S G,Abata D L. A photographic study of cold start characteristics of a spark assisted diesel engine operation on broad cut diesel fuels[J]. *SAE Paper*,871674,1987.

(本文原载于《内燃机学报》1993 年第 1 期)

Transient chamber wall-temperature in a swirl chamber diesel engine at cold-starting

Abstract:Transient wall temperature(TWT) was measured with a wall surface thermocouple with a good dynamic response,and a data acquisition and processing system developed by authors. TWT histories and their characteristics at cold-starting were analyzed in detail,and a four-stage model for TWT history at cold-starting was proposed for the first time. TWT histories of swirl chambers with and without a starting throat were compared,revealing the function of starting throat in cold-starting process.

An analysis of the unsteady combustion in swirl chamber diesel engine in cold-starting

Li Detao , Shan Chunxian , Jia Dachu , Fan Yongzhong , Zhu Xiaoguang

Abstract

In recent years, by means of the technology of high speed photographing and the data collecting, the authors have proved that the cold-starting is a typical process of unsteady combustion. According to the results of photography, test and record achieved by authors, this paper analyses comprehensively the internal and external characteristics, mechanisms and regularities of the unsteady combustion in cold-starting. On this basis, this paper brings to light the direction and the way to improve starting performance.

1 Introduction

The reliability of cold-starting is one of the significant subjects for the development of the diesel engine at present. Although the development of the study on improving the cold-starting ability and that of diesel engine are in the same step, the research jobs having ever been done before are seen mainly at the angle of adding auxiliary devices, not of researching the starting process itself. It's recent years that the study on cold-starting process's ignition and combustion and their relationship with relative factors have been developed[1-4].

In recent six years, authors have been developing the study of unsteady combustion in cold-starting on basis of the jobs the former persons had done. During this processing, the difficulties about the test technology and the data processing in the fields have been overcome and solved. The new developments in the fields such as understanding the mechanism of the process and etc. have been achieved. The characteristics of this process have been revealed. This paper gives a comprehensive report to this research achievement.

2   Characteristics of unsteady combustion in starting process

2. 1   Characteristics of indicator diagram, flame forms and rate of heat release （ROHR）

We analyse the internal characteristics of unsteady combustion in starting according to different flame shapes, indicator diagrams, the ROHR and the change regularities of the transient wall temperature.

Fig. 1 shows parts of the 160 cycles in the indicator diagram measured and recorded continuously at $-7\ ℃$ in ambient temperature. In the initial, middle and late periods of the whole starting process, more than ten continuous representative cycles pressure are separately shown in each period(Fig. 1a, 1b, 1c). These figures show that the indicator diagrams with different shapes generally belong in three different typical types.

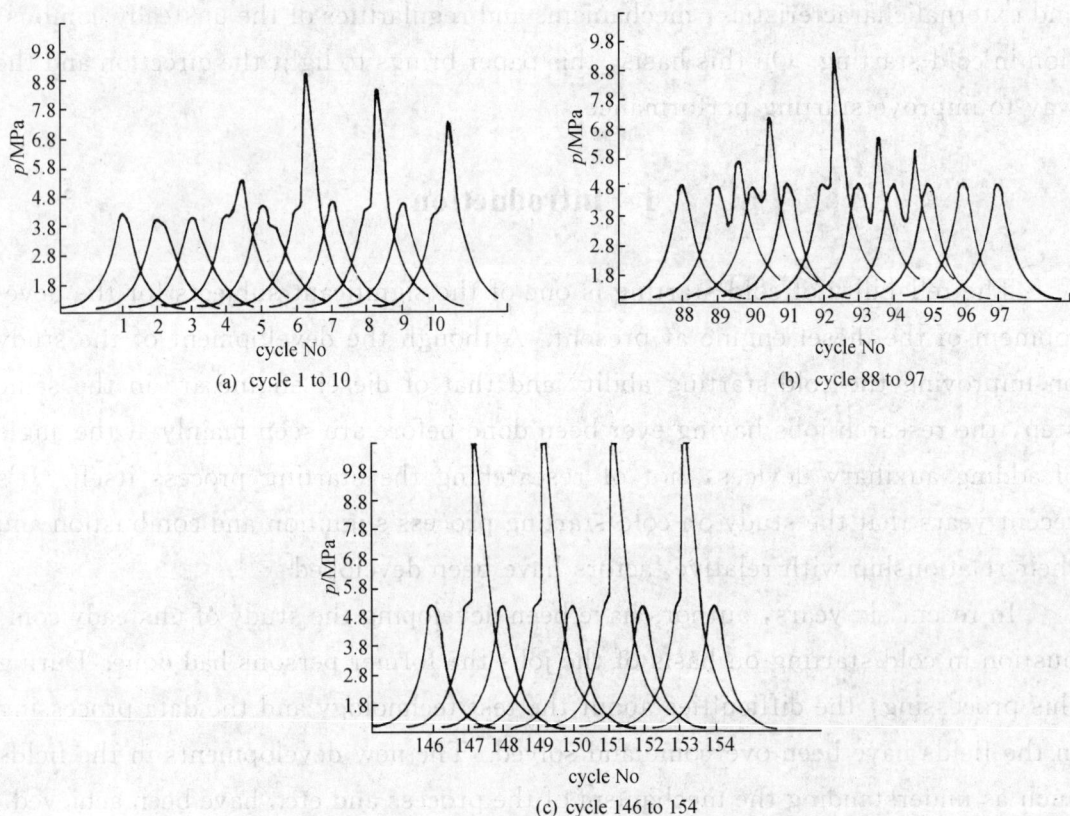

(a) cycle 1 to 10

(b) cycle 88 to 97

(c) cycle 146 to 154

Fig. 1   Parts of the pressure indicator diagrams

(1) Type V, that is, Type H in reference [2]. This type's characteristics are that its shape is similar to the indicator diagram without ignition; its combustion

pressure has no obvious increase; and its ignition delay period (IDP) is longer. The relevant flame shape in swirl chamber is in fibrous state. This type of combustion often takes place in the initial period of the cold-starting.

(2) Type W, that is, Type X in reference [2]. This type's characteristics are that there are two peaks in the whole process of pressure change; its IDP is longer. As the piston has moved down, then, the maximum combustion pressure and the pressure rise rate are not very high, the combustion often takes place in the middle period of the starting.

(3) Type Y, this type's characteristics are that they are similar to those shown in the indicator diagram of the normal combustion; its IDP in crankshaft angle is shorter. There are usually several cores of fire in swirl chamber. After ignition, the flame expends rapidly and covers the whole swirl chamber. The combustion of this type takes place in the whole starting process, but more often takes place in late period of starting process.

Because the invented images of these three indicator diagrams are separately similar to English alphabets 'V', 'W' and 'Y', we call them Type 'V', 'W' and 'Y' of combustion after their images. And the corresponding high speed photography are shown in Fig. 2.

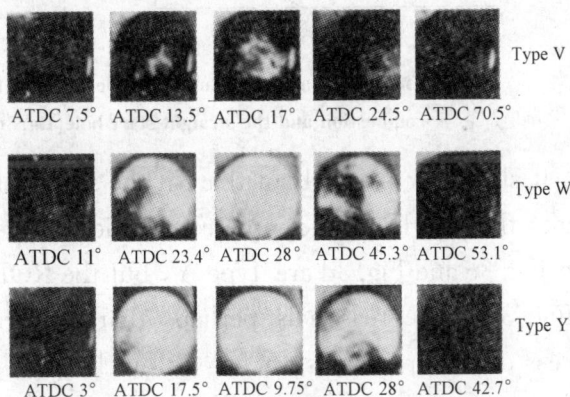

Fig. 2　Combustion flame figure

Fig. 3 is the indicator diagrams and the ROHR of Type V, W and Y combustion. From these figures, we can see that the ROHR of Type V is very low, but the duration of heat release is very long, the precentage of heat release is also smaller. The heat release beginning of Type W is comparatively late. In most cases, the ROHR of Type W is higher than that of Type V, but lower than that of Type Y. Type Y's heat release beginning is near the TDC, its ROHR is high, the percentage of heat release will reach to the maximum rapidly and the duration of heat release is short. The heat release cases as mentioned above coincide with the flame shape and expansion as shown in Fig. 2.

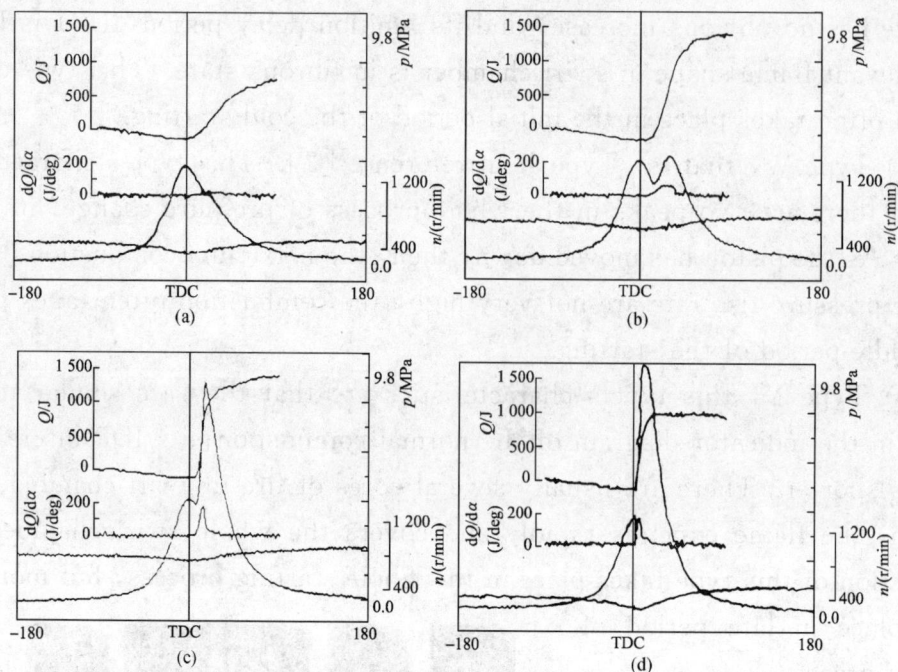

Fig. 3   The ROHR and the transient cycle speed of type V(fig. 3a) W(fig. 3b) Y(fig. 3c)
combustion and the straight start hole(fig. 3d)

It should be pointed out that, if the start holes are different or the periods are different, the ROHR of the same combustion shape also appear different. For example, both Fig. 3c and Fig. 3d are Type Y, but the ROHR of Fig. 3c is much higher than that of Fig. 3d. The reason is perhaps that the ignition's physicochemistry preparation process of the ordinary start hole is much more perfect than that of the straight start hole.

2.2 The variation characteristics of transient cycle speed

Now, we analyse the external characteristics of the unsteady combustion in starting according to the change of the transient speed.

As seen from the Fig. 3, in every cycle, whatever the combustion form is, the transient speed always changes. In compression process, the part of the kinetic energy of engine is transformed into compressing work, the remains is used to overcome the friction drag, the transient speed has been decreasing, it will usually reach the minimum near the TDC of the cycle. In the expansion process, the transient speed increases, the range of increase depends on the level of its cycle speed and the indicated work it gives out. On condition that the indicate work is fixed, the lower the cycle speed is, the larger the range of increase is. On condition that the levels of the cycle speed are nearly the same, the indicated work given out by Type V combustion is little, so the increase of cycle speed is small, but the indicated work given out by Type W or Y combustion is larger. This causes the cycle speed increase in larger range and then the level of the successive cycle speed is able to be enhanced.

The IDP in crankshaft angle is increasing, available IDP becomes short, ignition is caused to be retarded, even the cycle without ignition will occur. This phenomenon is particularly obvious in the initial period of the starting when the cycle speed is lower. Transient cycle speed and IDP act on each other, influence on each other. Acceleration and retardation occur in the whole starting process several times(Fig. 4).

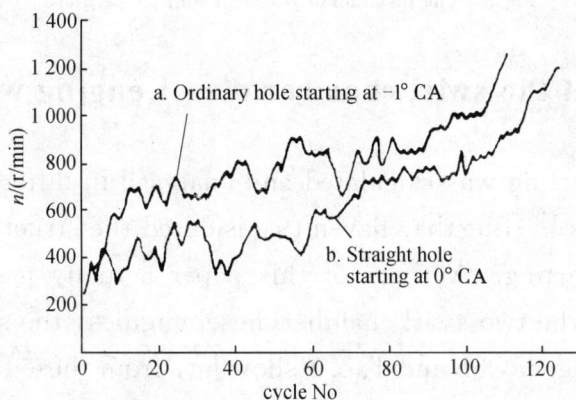

Fig. 4   The variation characteristics of transient speed

2.3   The variation characteristics of transient wall temperature

Now, through the change of the wall temperature in swirl chamber, We further analyze the internal characteristics of unsteady combustion.

Fig. 5 is the three dimension figure of the wall temperature's change of the whole starting process(5a) and the plane figure of the wall temperature change of the first ten cycles(5b) in the swirl chamber with the ordinary start hole at 7 ℃ in ambient temperature. These figures show:

(1) In cold-starting, similar to the different types of combustion and different forms of indicator diagrams, the different kinds of transient wall temperature occurs. This kind of change of transient wall temperature reflects the characteristics of unsteady heat transfer and unsteady combustion in cold-starting, too.

(2) As the process goes on, the average wall temperature is enhanced continuously. The wall temperature of the successive cycle usually will not reduce to the minimum of the last one. In another word, as the process goes on, the minimum wall temperature of the cycle is being enhanced gradually. This is one of the significant reasons that the IDP measured in time is shortening gradually.

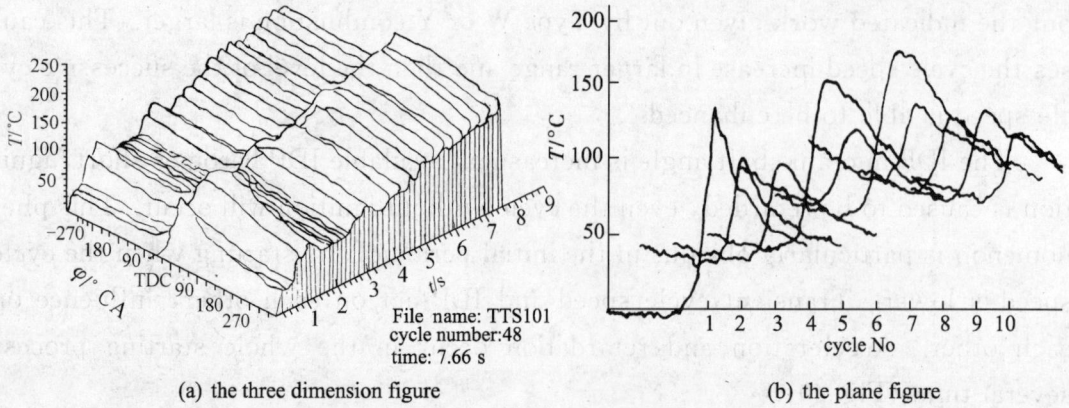

(a) the three dimension figure

File name: TTS101
cycle number:48
time: 7.66 s

(b) the plane figure

Fig. 5   The variation of transient wall temperature

3   The IDP of the swirl chamber diesel enging with start hole

The IDP of starting was calculated and analysed in different calculately formulas in References[1]. But they haven't considered the structure factors influencing the IDP in starting. Authors of this paper actually measured the IDP in starting process in the two swirl chamber diesel engines, the start holes of which are different. As Fig. 6a, 6b and Tab. 1 showing, from these figures and table we can see that:

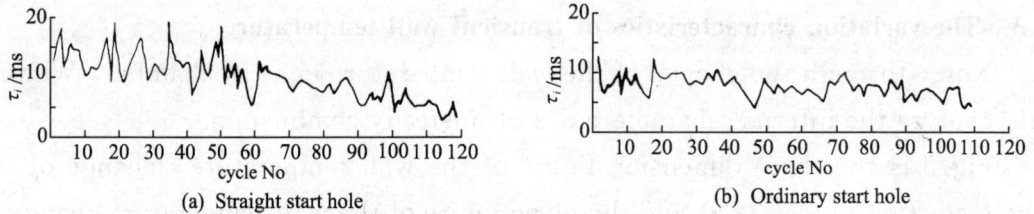

(a) Straight start hole

(b) Ordinary start hole

Fig. 6   The ignition delay period in starting

Tab. 1   The Cold start characteristics of different start hole at different conditions

H\P	C	1	2	3	4	5	6	7	8	9	10	11	12
Straight start hole	$n/(r/min)$	296	394	356	364	436	420	386	340	280	354	354	438
	τ_i/ms	12.056	15.619	17.916	11.085	12.868	15.224	14.351	—	14.084	9.857	9.743	11.203
	p_i/MPa	0.810	0.225	0.272	0.874	0.395	0.166	0.293	0	0.783	0.262	0.776	0.753
Ordinary start hole	$n/(r/min)$	288	390	442	506	574	588	554	626	680	696	670	692
	τ_i/ms	12.550	7.318	6.320	7.817	7.425	—	10.035	5.992	10.734	—	7.140	8.367
	p_i/MPa	0.802	0.749	0.855	1.0221	0.955	0	1.035	1.022	0.977	0	1.104	0

(H: Hole state, P: Parameter, C: Cycle No.)

（1）The change of IDP has close relationship with the form of unsteady combustion.

(2) As the starting process goes on, all the IDP measured in time are tending to shorten. But under ordinary start hole condition, the descending rate is faster than that under the straight hole condition.

(3) The large number of tests made by us indicate that the difference of the minimum start temperature will reach 30 ℃ around with or without the start hole. So we can see that the structure factors have significant influence on IDP and unsteady combustion in cold-starting process. In the calculation model of the IDP if this influence is expressed by related parameters, the calculation model is more perfect and the calculation result is more accurate.

4 Conclusion

(1) The cold-starting process of diesel engine is a typical unsteady ignition and combustion one. The combustion forms generally can be divided into Type V, Type W and Type Y according to the indicator diagram, flame shape and heat release regularity of every cycle.

(2) The ambient temperature has a significant influence on the mechanism of cold-starting. The lower the ambient temperature is, the more unsteady the process is and the larger the change range of the related process parameters such as the maximum pressure, IDP and etc. are.

(3) The transient speed and the IDP act on each other, and influence on each other, this causes the accelaration and retardation several times in whole starting.

(4) The structure factors of combustion system have significant influence on the IDP and the unsteady combustion process in cold-starting and even on the whole starting process. In the calculation model of the IDP this influence factor should be expressed by related parameters.

Reference

[1] Steven G F, Dvane L A. A photography study of cold start characteristic of a spark as-sisted diesel engine operating on broad cut diesel fuels [J]. *SAE Paper*, 871674,1987.

[2] Li Detao, He Xiaoyang. A study of unsteady combustion process of swirl chamber diesel en-gine on cold start condition[J]. *Internal Combustion Engine Journal*, 1987(3).

[3] Li Detao, Zu Xiaoguang. Start hole action mechanism and main chamber flame expan-ding regularity in swirl chamber diesel engine [J]. *Internal Combustion Engine*

Journal，1990(1).

[4] Chang Jiang, Zheng Hao, et al. A study of startability of small type direct injection diesel engine[J]. *Small Type Internal Combustion Engine*，1991(2).

（From：*2nd Asian-Pacific International Symposium on Combustion and Energy Utilization*，Beijing，1993）

涡流室式柴油机冷起动时非稳态燃烧过程分析

[摘要]　近年来，作者通过对柴油机冷起动过程进行高速摄影、高速采集缸内压力数据、温度数据和瞬态转速，证明该过程是一个典型的非稳态燃烧过程。本文根据作者拍摄、测试和记录所得结果，综合分析了冷起动过程非稳态燃烧的内外部特征、机制和规律。在此基础上，揭示了改善冷起动性能的方向和途径。

带有不同形式起动孔的涡流室式柴油机
冷起动全过程的研究

贾大锄,范永忠,李德桃,朱章宏

[摘要]　作者通过采集带有不同形式起动孔的涡流室式柴油机冷起动全过程的主、副室压力示功图,对起动过程的每循环的转速、平均指示压力、滞燃期、压力升高率及主副室压差等进行了详细分析。提出了三种不同形式的主副压差图,分析了冷起动条件下的三种典型的非稳态燃烧示功图,阐明了涡流室式柴油机冷起动过程的机制,为改善这类柴油机的冷起动性能指出了方向和途径。

涡流室式柴油机的冷起动过程是一种非稳态过程,涡流室镶块上的起动孔能改善起动性能已被国内外学者所证明[1-2]。但此非稳态过程的规律性及起动孔的机理尚需作更全面和合理的解释。为此,作者利用先进的测试仪器,采集了带有不同形式起动孔的涡流室式柴油机冷起动全过程的主、副室压力示功图,并利用自行研制成功的数据处理装置及技术对大量数据进行计算和分析。

1　试验装置与试验方法

试验是在一台 195 型柴油机上进行的。将两个压电石英传感器分别齐平安装在主燃烧室气缸盖底平面和涡流室纸眉孔处,图 1 为试验装置简图。作者分别采集了正常起动孔和直起动孔,在环境温度分别为 $-7\,^\circ\mathrm{C}$、$-1\,^\circ\mathrm{C}$ 和 $0\,^\circ\mathrm{C}$ 状态下起动过程的主、副室压力示功图。

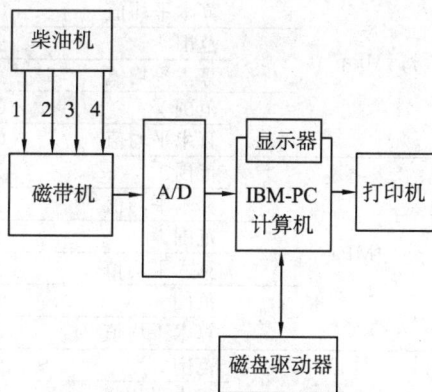

2　典型的非稳态燃烧过程

文献[1]对涡流室式柴油机在冷态条件下的起动过程进行了研究,提出了涡流室式柴油机的起动过程是非稳态过程,首先发现起动过程中分别定义为 H 型、X 型、Y 型三种燃烧过程。但由于当时的条件和测试设备所限,所得示功图只能进行定性分析。作者采用先进的测试手段和自行研制的数据处理装置及技术,精确地取得了这三种典型的非稳态燃烧示功图。图 2 是冷起动过程连续 10 个循环示功图的重叠显示。从中可见,每循环的着火状况都不尽相同。对两种起动孔、三种起动环境温度下起动全过程的测试结果见表 1。

1—主室压力信号　2—副室压力信号
3—角标信号　　　4—上止点信号

图 1　数据采集系统框图

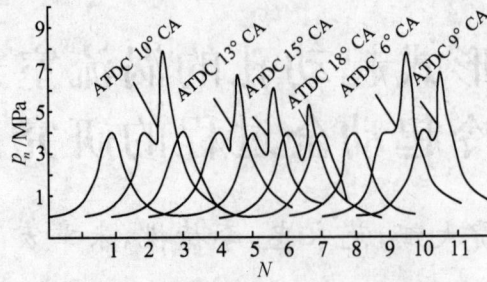

p_n——主室压力　　N——循环数

图 2　冷起动过程连续示功图

表 1　各种燃烧过程参数

参　数	序号	统计内容	燃　烧　过　程		
			H　型	X　型	Y　型
τ_i/ms	1	范围	8.58~17.46	5.22~15.8	2~11
		算术平均值	11.265	7.95	6.5
	2	范围	无	4.74~11.54	3.42~9.91
		算术平均值	无	7.02	6.14
	3	范围	8.97~21.77	74.85~10.85	2.39~11.99
		算术平均值	12.83	6.91	6.32
φ/° CA	1	范围	20~45	21~42	17~29
		算术平均值	31.636	32.67	20.65
	2	范围	无	28~45	16~33
		算术平均值	无	32.85	25.6
	3	范围	20~42	25~41	15~33
		算术平均值	29.4	32.28	22.48
p_i/MPa	1	范围	0.103~0.551	0.349~1.177	0.8~1.2
		算术平均值	0.424	1.013	1.013
	2	范围	无	0.782~1.164	0.749~1.207
		算术平均值	无	1.038	1.046
	3	范围	0.088~0.834	0.897~1.114	0.678~1.209
		算术平均值	0.39	1.015	0.972
p_{max}/MPa	1	范围	3.7~4.02	3.8~6.6	5.8~10.3
		算术平均值	3.818	4.788	8.169
	2	范围	无	4.076~6.696	6.376~9.78
		算术平均值	无	5.66	8.228
	3	范围	3.26~3.918	3.818~6.97	4.633~10.42
		算术平均值	3.743	5.371	8.24
$(dp/d\varphi)_{max}$/(MPa/° CA)	1	范围	0.17~0.243	0.2~0.6	0.282~5.03
		算术平均值	0.205	0.313	1.24
	2	范围	无	0.224~0.769	0.487~2.298
		算术平均值	无	0.478	1.144
	3	范围	0.171~0.312	0.218~0.801	0.284~5.12
		算术平均值	0.213	0.503	1.175
C 点的主、副室压差 $(p_n-p_s)_c$/MPa	1	范围	0.4 左右	0.35~-0.6	0.5~-3.2
	2	范围	无	0.202~-0.915	0.907~-2.66
	3	范围	0.165~0.411	0.31~-0.63	0.45~-2.78

参　数	序号	统计内容	燃　烧　过　程		
			H　型	X　型	Y　型
分布情况	1	范围（循环号）	2～86	26～115	1～123
		次数（百分比）	36(29.26)	26(21.11)	25(20.3)
	2	范围（循环号）	无	10～108	1～109
		次数（百分比）	无	34(31.2)	31(28.44)
	3	范围（循环号）	1～61	40～139	5～157
		次数（百分比）	23(14.6)	34(21.65)	48(30.57)

注：表中序号 1 为直起动孔，环境温度 0 ℃，起动转速 296 r/min；序号 2 为正常起动孔，环境温度 −1 ℃，
起动转速 286 r/min；序号 3 为正常起动孔，环境温度 −7 ℃，起动转速 272 r/min。
表中括号内数据为每种类型燃烧出现的百分比。

2.1　H 型燃烧过程

图 3 为典型的 H 型燃烧过程示功图。这种燃烧过程多发生在冷起动初期、转速较低的状况下，随着环境温度的降低，这种燃烧过程出现的可能性越大。不同形式的起动孔对这类燃烧过程出现的次数也有影响。

H 型燃烧过程的特征：

（1）以 ms 计的滞燃期 τ_i 很大，约为正常运转状况下的 6 倍，以曲轴转角计的滞燃期转角 φ 亦较长。

（2）平均指示压力 p_i 很小，但此时摩擦力矩较大，这种燃烧过程不会导致转速的升高，绝大多数状况下转速会下降，若转速降至不能依靠惯性越过上止点，则导致起动失败；其最高燃烧压力 p_{max} 和最大压力升高率 $(\mathrm{d}p/\mathrm{d}\varphi)_{max}$ 均很小。

（3）主、副室压力差 $(p_n - p_s)$ 在燃烧期间内始终大于零。

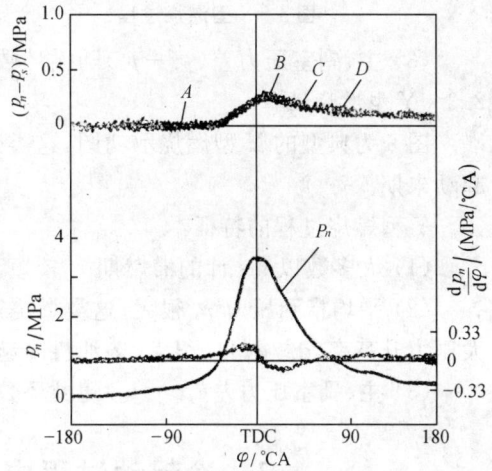

p_n——主室压力　　p_s——涡流室压力
$(p_n - p_s)$——主、副室压力差

图 3　H 型燃烧过程

2.2　X 型燃烧过程

图 4 为典型的 X 型燃烧过程示功图，这类燃烧过程大多发生在冷起动中期。

X 型燃烧过程的特征：

（1）绝大多数以 ms 计的滞燃期 τ_i 较长，约为正常运转状况下的 3 倍，以曲轴转角计的滞燃期转角 φ 较长。

（2）平均指示压力 p_i 较大，绝大多数这类燃烧过程能克服摩擦阻力使转速升高。其最高燃烧压力 p_{max} 及最大压力升高率 $(\mathrm{d}p/\mathrm{d}\varphi)_{max}$ 较小。

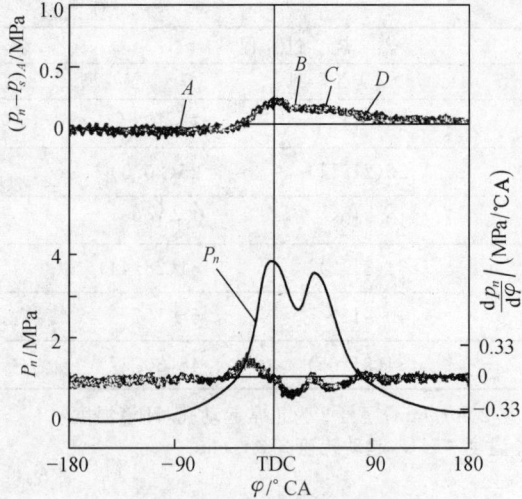

图中参量代号含义同图 3

图 4 X 型燃烧过程

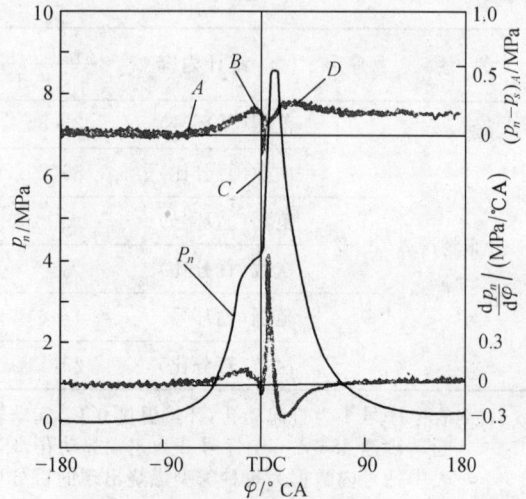

图中参量代号含义同图 3

图 5 Y 型燃烧过程

(3) 主、副室压力差 $(p_n - p_s)$ 随循环数的增加由正值变为负值,其变化幅度不大。

2.3 Y 型燃烧过程

图 5 为典型的 Y 型燃烧示功图,这类燃烧过程在整个起动过程都会出现,但大多发生在起动末期。

Y 型燃烧过程的特征:

(1) 大多数以 ms 计的滞燃期 τ_i 较短,以曲轴转角计的滞燃期转角 φ 亦较短。

(2) 平均指示压力 p_i 很大,这类燃烧过程致使转速明显升高。其最高燃烧压力 p_{max} 及最大压力升高率 $(dp/d\varphi)_{max}$ 很大,柴油机冷起动噪声主要是由这类燃烧引起的。

(3) 主、副室压力差 $(p_n - p_s)$ 随循环数增加由正值变为负值,其变化幅度较大。

3 冷起动过程中典型的主、副室压差形式

由于分隔式柴油机的主、副室存在着压力差,所以作者通过对表中所列三种起动条件下的主、副室压差进行如下分析。

3.1 不着火的主、副室压差图

不论是起动过程中未着火的示功图还是起动成功后停止供油的未着火示功图,其主、副室在上止点 $\pm 50°$ CA 范围内形成一凸起 0.3 MPa 左右的压力差(图 6)。这是由于在压缩过程中,流入涡流室的气体在通道处有节流,导致主室压力高于副室压力,膨胀过程随着活塞的下行主室压力随之下降,主、副室压力差逐渐减小,排气门开启后,主、副室压力差基本一致。

3.2 起动后期典型的主、副室压差图

如图 5、图 7 所示,这类压差相应的燃烧过程大多为 Y 型和 X 型,以 ms 计的滞燃期较短。

在滞燃期内主、副室压差变化与不着火的示功图类似,在上止点附近形成 A-B 段凸起。在急燃期 B-C 段,由于滞燃期较短,进入副室的燃油可能在未到达起动孔前就着火,使得涡流室压力高于主室。随着涡流室中未燃烧的燃料、空气及燃气经通道及起动孔进入主燃烧室中,与主燃烧室的空气进一步混合燃烧,主室压力又急剧升高,使 C-D 段主、副室压差又迅速上升。

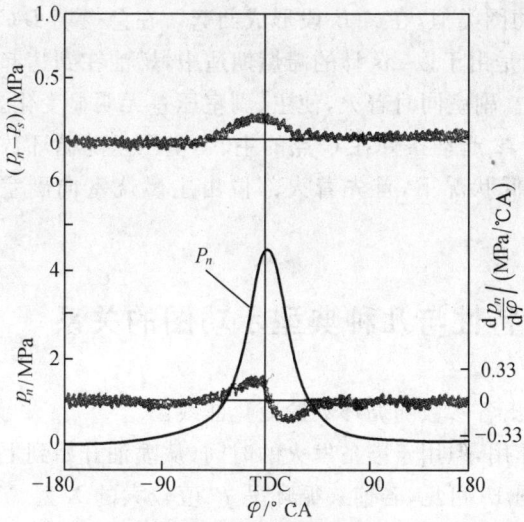

图中参量代号含义同图 3

图 6　不着火过程

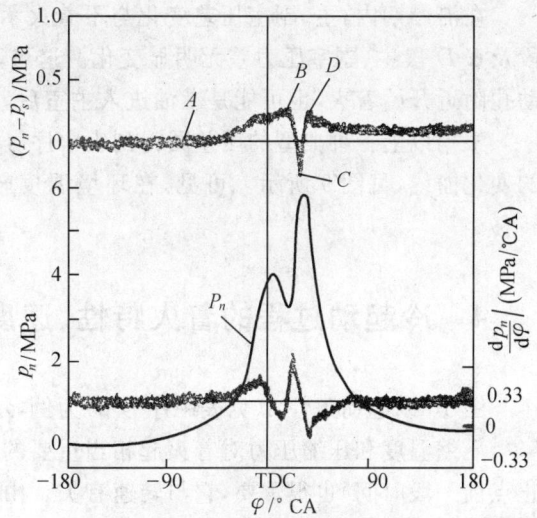

图中参量代号含义同图 3

图 7　起动后期的燃烧过程

3.3　起动初期典型的主、副室压差图

如图 3、图 8 所示，这类压差相应的燃烧过程多为 H 型和 Y 型，以 ms 计的滞燃期较长。

在滞燃期内，主、副室压差变化与不着火示功图类似，在 A-B 段形成凸起。在急燃期 B-C 段，由于滞燃期较长，燃油可经起动孔进入压力较高的主燃烧室，致使着火首先在主室发生，主燃烧室压力急剧上升。由于进入主室的燃油相对较少，主室着火后未燃烧的燃料较少，而涡流室中燃油多数落在壁温较低的镶块上，致使随后的副室着火较弱，C-D 段主、副室压差下降较平缓。

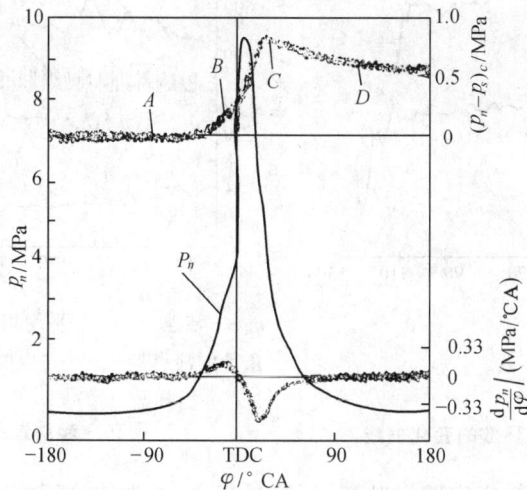

图中参量代号含义同图 3

图 8　起动初期燃烧过程

3.4　起动中期典型的主、副室压差图

如图 4 所示，这类压差相应的燃烧过程多为 X 型，以 ms 计的滞燃期适中。

在滞燃期内,主、副室压差变化与不着火示功图类似,在 A-B 段形成凸起。在急燃期 B-C 段及 C-D 段主、副室压力差无明显变化。这可能是由于以 ms 计的滞燃期适中,燃油在到达起动孔附近开始着火,也可能是燃油进入主室后,主、副室同时着火,使主、副室压差无明显变化。

根据以上三种典型的压差图得到表中序号 1,2 起动条件在 C 点的主、副室压差随循环数的变化曲线,见图 9 所示。可见,在环境温度较低状况下,首先着火部位由主燃烧室向副室转移。

4 冷起动过程的着火特性、速度特性与几种典型示功图的关系

图 10 以起动条件 3(见表中序号 3)为例,示出各参数随循环数的变化曲线图。

压缩温度和压缩压力对着火起相当重要的作用,但用于准备发火的时间(从喷油开始到上止点前一段时间)也很重要,它与转速有关。由图 10 可见,若前一循环呈 p_i 值较大的 X 型、Y 型燃烧过程,则发动机转速急剧上升,准备发火的时间下降较大。若准备发火的时间降至小于以 ms 计的滞燃期,这便导致下一循环不着火,从而发动机转速下降。此外,当前一循环呈平均指示压力较小的 X 型、Y 型燃烧过程时,发动机转速上升不大,准备发火的时间下降较小,但不致小于以 ms 计的滞燃期,导致下一循环以曲轴转角计的滞燃期增大;连续几个循环着火后,准备发火的时间又小于以 ms 计的滞燃期,使再下一个循环不能着火。这就是起动过程为何出现 X 型、Y 型、H 型和不着火的燃烧示功图的原因之一。

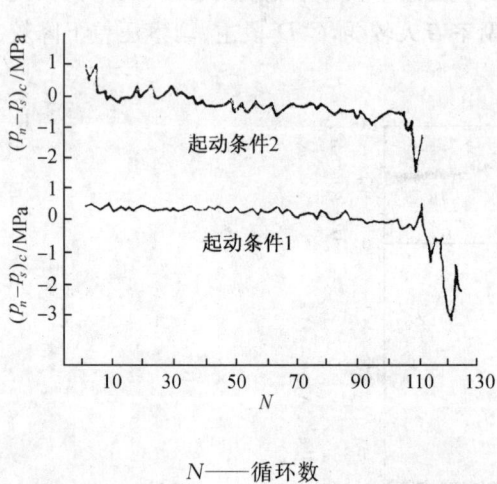

图 9 主、副室压差随循环数的变化曲线

N——循环数

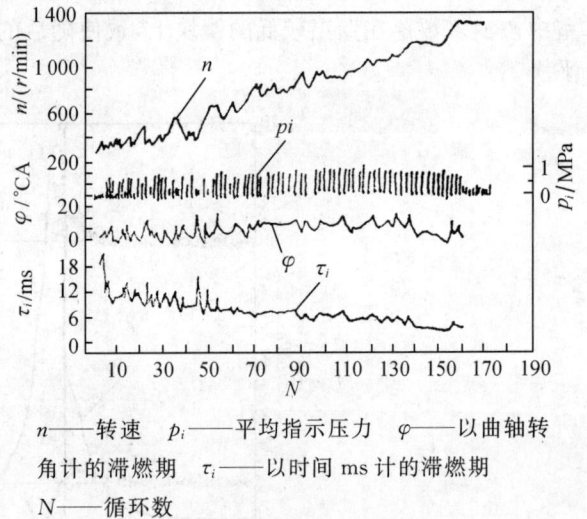

n——转速 p_i——平均指示压力 φ——以曲轴转角计的滞燃期 τ_i——以时间 ms 计的滞燃期

N——循环数

图10 起动过程的着火特性

综上所述,因着火使燃烧室壁温升高,压缩过程中,外界传给压缩空气的热量增加,压缩温度升高,致使以 ms 计的滞燃期降低,其相应的准备发火时间就随之减小,这就为柴油机不断加速创造了条件,以致最后达到稳定的燃烧。

5 结 论

(1) 柴油机的冷起动是一种非稳态过程,起动过程中含有 H 型、X 型、Y 型三种燃烧过程和不着火循环。

(2) 有起动孔的涡流室式柴油机冷起动时的主、副室压差图与正常运转状况下不同,起动燃烧过程存在初期、中期、后期三种不同的压差形式和不着火的压差形式。起动孔的作用是在起动初期使燃油进入压力温度较高的主燃烧室首先着火并由主室向副室转移。

(3) 转速和以 ms 计及曲轴转角计的滞燃期在起动过程中是相互影响的,它们之间的变化是导致起动过程中产生几种典型的燃烧过程的原因之一。降低以 ms 计的滞燃期对改善起动性能有重要作用。

(4) 随着环境温度的降低,起动初期燃烧呈着火特性较差的 H 型燃烧过程的趋势愈大,这类燃烧过程在几个循环范围内难以起动。不同形式的起动孔影响着这类燃烧过程的出现次数。

参 考 文 献

[1] 李德桃,等.涡流室式柴油机在冷起动条件非稳态燃烧的研究[J].内燃机学报,1987,5(1).

[2] 长江,正浩,等.小型直喷式柴油机起动性能的研究[J].小型内燃机,1992(2).

[3] Lyn W T.柴油机燃烧研究(2)[J].国外内燃机,1980(3).

[4] 石田明男,等.柴油机燃油喷射的新概念[J].油泵油嘴技术,1987(1).

[5] Akio Kobayashi,等.直喷式柴油机冷起动时的燃烧分析[J].*SAE Paper*,840106,1984.

(本文曾在中国内燃机学会第三届学术年会上宣读,原载于《内燃机工程》1994 年第 1 期)

A study on whole process of cold start of swirl chamber diesel engine with different forms of starting hole

Abstract: This paper analyses in detail the rotation of initial cycle, mean indicated pressure, ignition delay period, pressure increase rate, pressure difference between the main and the swirl chamber and so on, by means of collecting the main and the swirl chamber's pressure work-showing figures of the swirl chamber diesel engine with different forms of start hole. In this paper, the pressure difference figures of three different forms of main and swirl chamber are introduced, three typical unsteady state combustion work-showing figures in cold start condition are analysed further, the mechanism for the cold start process of swirl chamber diesel engine is elucidated. To improve the cold startability of this kind of diesel engine, the paper points out the direction and way.

Transient combustion process of an IDI diesel engine with dual-throat jet at cold-starting

Li Detao, *Zhu Xiaoguang*, *He Xiaoyang*, *Peng Lixin*

Abstract

The dual-throat jet technique has been successfully used to improve cold-starting of the swirl-type IDI diesel engines. It has been proven that, with the aid of the second throat connecting the swirl chamber and the main combustion chamber, the cold-starting process was more stable, quiet and clean. However, the understanding of the mechanism of this technique is far less than satisfied as regarding to the better control and further development.

An intensive fundamental experimental investigation of the transient process of the ignition and combustion at cold-starting had been conducted on a swirl-chamber IDI diesel engine with the help of high-speed photography. Based on the results of this investigation, the following conclusions have been approached: 1) There exist three types of heat release rate pattern at the cold-starting. Different pattern will result in different engine behavior. 2) The secondary throat has double effects on the engine starting characteristice: (a) improvement of the ignition conditions in the main chamber by directly delivering pilot fuel; (b) improvement of the fuel/air mixing process by intensifying turbulence in the swirl chamber, which depends on the geometrical shape and orientation of the throat.

1 Introduction

IDI diesel engine experiences difficulties during the cranking and starting, particularly under cold ambient conditions. With higher kinematic viscosity and density when it is cold, diesel fuel sprays into the combustion chamber with larger droplets[1]. These larger droplets results in less total fuel surface area, and as the consequence, produce a longer ignition delay. In addition to that, the greater momentum of the larger droplets results in greater fuel penetration. There will be an even longer ignition delay due to more fuel spread onto the cold wall. Plus the bigger heat loss through the cold wall, the combustion at cold-starting becomes a unstable and rough process.

Adding a secondary throat connecting the swirl and main chambers in addition to the primary throat had been proven significantly beneficial to the cold-starting characteristics of the swirl-type IDI diesel engines[2]. One of the successful examples of this dual-throat technique is shown in Fig. 1[3,4].

The secondary throat for this design was oriented opposite to the direction of the fuel spray in the swirl chamber. It had been confirmed by experimental test that, with the secondary throat, the cold-starting process of the engine was more stable, quiet and clean.

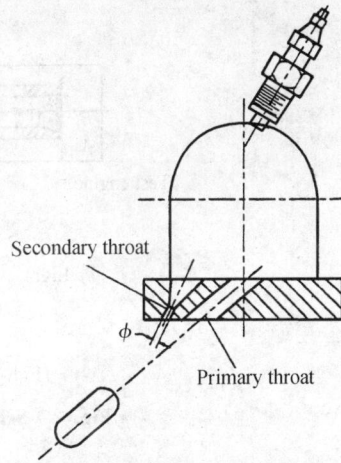

Fig. 1　Swirl chamber with dual-throat

This study was designed to identify the basic patterns of combustion process at cold-starting in an IDI diesel engine, and to uncover the underlying mechanism of how the cold-starting was improved by the introduction of the secondary throat. The study was based on an intensive experimental investigation of the combustion process employing high-speed photograph technique.

2　Experimental setup

The engine studied was a swirl type IDI diesel engine with the following specifications:

single cylinder, four stroke

bore/stroke: 95/115 mm

compression ratio: 19. 16 : 1

volume ratio of the swirl chamber: 57%

rated power: 8. 8 kW/2 000 r/min

injector opening pressure: 12. 75 MPa

A 16mm E-10 rotational high speed camera with $300 \sim 10\ 000$ frames/second was used in the study. Fig. 2 shows a schematic configuration of the experimental setup.

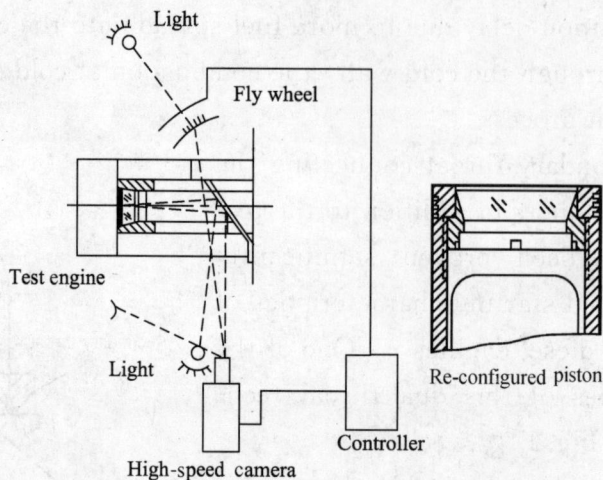

Fig. 2 Schematical diagram of the camera setup

3 Three typical transient combustion patterns

During the cranking and starting, particularly cold ambient conditions, IDI diesel engines experience a transient combustion process. By observing a large number of combustion cycles in both the swirl chamber and the main chamber, the authors found that the combustion process differs greatly from cycle to cycle during the course of the cold-starting. These differences include the distribution of fuel in the two chambers, location and timing of the ignition, as well as the instant at which flame thrusts into the main chamber. Whether this transient process could lead to a smooth engine start-up depends essentially on the combustion pattern produced during this process.

It had been found from this study that there exist three typical transient combustion patterns for IDI diesel engine at the cold-starting. Fig. 3 shows three groups of photographs taken from the swirl chamber demonstrating these different combustion processes, together with the corresponding cylinder and swirl chamber pressure diagrams. The shape of these diagrams are similar to the up-side-down letters V, W and Y. For convenience these three combustion patterns are named after these letters respectively by the authors.

（1） V-Pattern-Initial Combustion. The absolute ignition delay in microseconds(ms) is much longer, in the order of five times, than that of the normal working conditions. The ignition occurs at locations close to the bottom of the swirl chamber. Without inflaming all of the injected fuel, the combustion terminates at very early stage, which is too week to support the engine for speeding up.

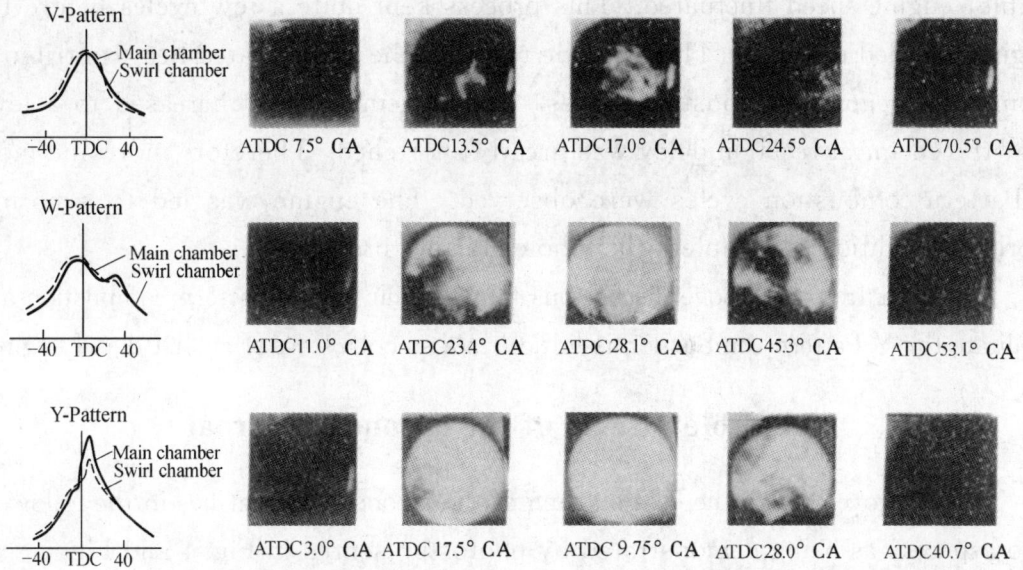

Fig. 3　Three typical combustion patterns, photos taken from the swirl chamber

(2) W-Pattern-Unstable Combustion. The relative ignition delay in crank angle (CA) is longer than that of the V-Pattern. Ignition appears at 11.0 ATDC for the observed cycle. Because of the improved combustion condition resulted from higher engine speed and warmer environment, the ignition occurs at locations closer to the center of the chamber. On the other hand, since the descending movement of the piston after the TDC the combustion is weak and unstable. Late-combustion dominates those cycles.

(3) Y-Pattern-Dynamic Combustion. The ignition delay in this stage is much shorter and similar to that of the normal combustion. Several ignition are observed in the swirl chamber. The flame sweeps throughout the chamber rapidly. The bright flame photo indicates a dynamic combustion process. The combustion completes within about 30° CA resulting in a positive engine power output.

For a typical cold-starting of an IDI diesel engine the following process of the combustion was observed. The process started from a few cycles of V-Pattern combustion at cranking. Then Y-Pattern combustion cycles were observed. At those cycles engine speed was very low, the relative ignition delay in CA was short. The heat released from the combustion helped speeding up the engine. As a consequence, a longer ignition delay in CA was resulted from the higher engine speed although an absolute ignition delay in ms was shorter due to the improved combustion conditions. The ignition took place later after TDC. As a result, there came a serials of typical W-Pattern combustion cycles. Combustion was un-

341

stable，engine speed fluctuated. This process kept quite a few cycles before the engine speeded up again. This was the result of the accumulated heat contributed from the foregoing combustion cycles. Engine temperature increased to a level that the chemical reaction delay was greatly shortened. Therefore more and more Y-Pattern combustion cycles were observed. The engine was led to a normal working condition to complete the whole starting process.

It is clear from the above discussions that to limit the W-Pattern combustion and facilitate the Y-Pattern combustion is critical to a smooth start of an IDI diesel engine.

4　Double effects of the secondary throat

The improvement to the cold-starting of the secondary throat lies in the following two aspects，as being demonstrated by photos presented in Fig. 4 and Fig. 5.

（1）To produce direct ignition in the main combustion chamber by direct pilot fuel supply. Fig. 4 shows two consequent cycles of direct ignition in the main chamber and the followed flame propagation. The first cycle is shown in Fig. 4a.

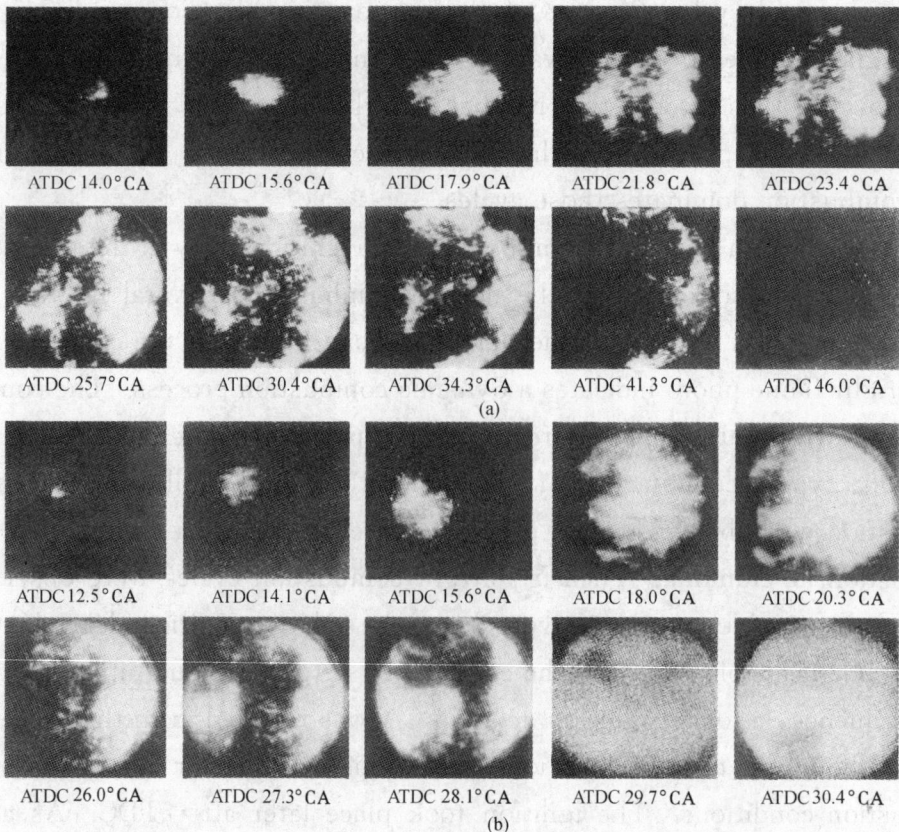

Fig. 4　**Two continuous combustion cycles with direct ignition at the main chamber**（$n=500$ r/min）

The fuel was directly injected into the main chamber through the secondary throat, ignition occurred near the port of the throat. The flame then propagated forward at a limited speed and terminated at 46° CA ATDC without completing the entire chamber. No clear evidence from the photos showed either the injection of fuel or flame into the main chamber through the primary throat.

The succeeding cycle shown in Fig. 4b indicates a direct ignition and more dynamic combustion in the main chamber. The ignition appeared at 12.5° CA ATDC and the flame spread most of the chamber at 20.3° CA ATDC.

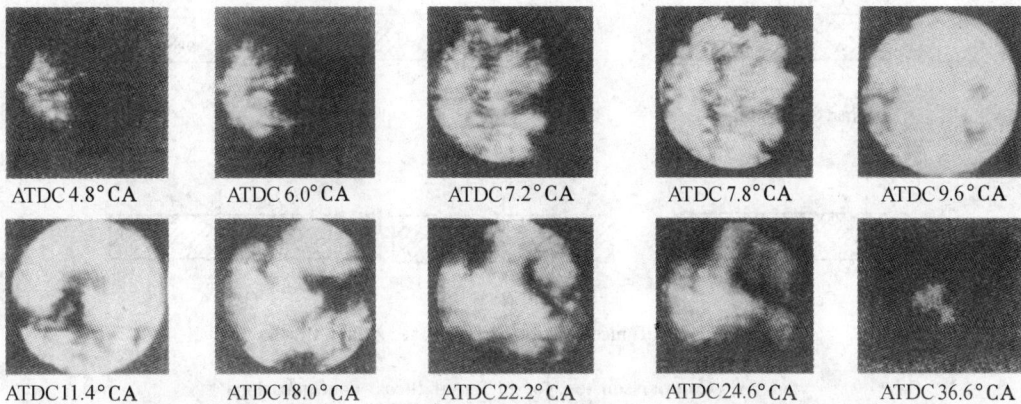

ATDC 4.8° CA   ATDC 6.0° CA   ATDC 7.2° CA   ATDC 7.8° CA   ATDC 9.6° CA

ATDC 11.4° CA   ATDC 18.0° CA   ATDC 22.2° CA   ATDC 24.6° CA   ATDC 36.6° CA

Fig. 5   **An example of the combustion cycle without direct ignition at the main chamber**($n = 350$ r/min)

(2) To assist the ignition and combustion in the swirl chamber. Since the secondary throat was orientated in the opposite direction to the fuel injector, a stream of air was generated in the compression stroke which penetrated onto the fuel spray and accelerated the dispersion and atomization of the fuel spray in the swirl chamber. As the result, the ignition delay was shortened.

Fig. 5 provides a group of photos showing such an effect. Ignition occurred first in the swirl chamber near TDC. Then, the burning mixture thrust through the primary throat into the main chamber at 4.8° CA ATDC. No ignition was observed in the main chamber resulted from the pilot fuel spray through the secondary throat. This small amount of fuel was burnt together with the flame from the swirl chamber through the primary throat.

Apparently the introduction of the secondary throat shortened the ignition delay and minimizes the W-Pattern combustion in the swirl chamber. The cold start of the IDI diesel engine is made smoother and quieter.

To identify which of the above discussed two effects of the secondary throat plays the most important role in improving the cold-starting, further investigation was conducted looking into the different configurations of the secondary throat.

With different configurations of the throat, either the amount of pilot fuel sprayed into the main chamber or the angle of the air stream towards the fuel spray in the swirl chamber could be varied. Five different throat configurations including the original throat had been investigated experimentally in this study based on the previously defined engine. These configurations are shown in Fig. 6. The test conditions and results are listed in Tab. 1. The engine was cranked by the same technician and there was a 5-minute interval between each cranking.

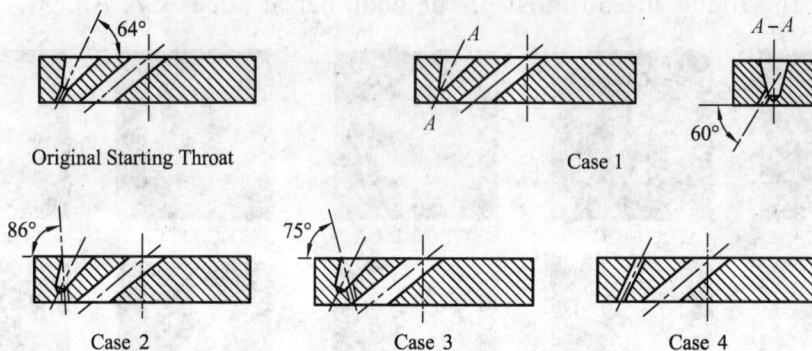

Fig. 6   Configurations of different secondary throat

Tab. 1   Comparison test for different throat configurations

Insert type	Temperature/℃			First start		Second start		Third start	
	Room	Oil	Fuel	State	Time/s	State	Time/s	State	Time/s
Original	−2	−2	−2.5	OK	4∼5	OK	4∼5	OK	4∼5
Case 1	−2	−2	−2	Fail	>5	OK	4∼5	OK	4∼5
Case 2	6	6	6	Fail	>5	Fail	>5	Fail	>5
Case 3	7	7	7	Fail	>5	Fail	>5	Fail	>5
Case 4	−2	−2	−2	Fail	>5	Fail	>5	OK	4∼5

The throat in case 1 and case 2 are different from each other at the lower part. In case 1 the air stream was directed to the center of the chamber, while in case 2 air blew the fuel spray towards the cold wall, which was obviously harmful to the cold-starting. Both cases had the same upper part of the throat. The same amount of pilot fuel should be directly delivered into the main chamber. If the ignition in the main chamber were the essential factor, those two throats would had resulted in more or less the same engine cold-starting behavior. However, from Tab. 1 one can see that was not the case. Case 1 was much better than case 2 as regarding to the cold-starting. This implies that the counter stream to the fuel spray in the swirl chamber, which helps to shorten the ignition delay in the swirl chamber, significantly improved the cold-starting process. Compared with the original configuration, the secondary throat for case 1 only tiled the stream angle by a few degrees towards the wall. The test result agrees with the conclusions made for the case 1 and case 2: case 1 was worse than the original throat as regarding the cold-starting because of the direction of the air stream generated.

Turning the direction of the air stream further towards the wall without changing the conditions of fuel delivered into the main chamber, as was configured in case 3, the engine could hardly be started.

In this case the engine cold-starting process was deteriorated by the introduction of the secondary throat. A simple straight throat with the same angle as the original throat was configured in case 4. The conditions for both of the two effects of the second throat were changed: little fuel could be delivered into the main chamber and more importantly, the range of the impingement of the counter air stream with the fuel spray in the swirl chamber was greatly narrowed. Consequently little improvement of the cold-starting was obtained(refer to Tab. 1).

5 Summary and conclusions

Experimental investigations had been conducted in an effort to reveal the mechanism of the improvement to the cold-starting of IDI diesel engines by introduction of the secondary throat. The following conclusions are suggested from those investigations:

(1) There exist three types of combustion patterns according to the heat release rate pattern at cold-starting. How to limit the W-pattern unstable combustion and to facilitate the Y-pattern dynamic combustion is critical to the smooth cold-starting.

(2) The secondary throat has double effects on the engine characteristics: (a) improvement of the ignition conditions inside the main chamber through direct pilot fuel supply; (b) improvement of the fuel/air mixing in the swirl chamber by intensifying the turbulence during the compression stroke, which effect is strongly influenced by the geometry and orientation of the second throat.

(3) The counter stream of air against the fuel spray accelerates dispersion and atomization of the fuel spray in the swirl chamber. It plays the primary role in assisting the cold-starting of the engine.

References

[1] Steven G F, Duane L A. A photographic study of cold start characteristics of a spark assisted diesel engine operating on broad cut diesel fuels [J]. *SAE Paper*, 871674, 1987.

[2] Toshiaki Tanaka, et al. Improvement in IDI diesel engine performance through dual-throat jet combustion[J]. *Transactions of the Society of Automotive Engineers of Japan*, 1996, 32.

[3] Li Detao, He XiaoYang. Investigation on unstable state combustion in swirl chamber diesel engine under cold-starting condition [J]. *Transactions of CSICE*, 870020, 1987.

[4] Zhu Xiaoguang, Li Detao. Investigation on ignition and combustion at starting in a swirl chamber diesel engine with a starting throat[J]. *Transactions of CSICE*, 891327, 1989.

(From: *SAE Paper*, 960029)

具有双气流通道的 IDI 柴油机冷起动时的瞬态燃烧过程

[摘要]　采用双气流通道的涡流室式柴油机,已成功地用于改善冷起动性能。其中一个小通道,即所谓"起动机",可使起动过程加速、排放污染物降低。然而,对于起动机的作用机理,却长期未得到合理阐明。

本文作者通过高速摄影拍摄了涡流室式柴油机的冷起动全过程,表明冷起动过程有三种不同的燃烧放热模式,并通过试验证明,起动机有使涡流室加速混合和着火,使主燃室直接着火的双重作用。在一般情况下,后一种作用是主要的。

改善483Q柴油机起动可靠性的研究

杨文明，李德桃，龚金科，何烈秋，胡明爱

[摘要]　483Q柴油机是一种高速国产柴油机，在其上增设起动孔，在不用电热塞的情况下，可改善起动可靠性。

483Q柴油机是目前国产转速最高的一种柴油机。它具有性能稳定、噪声低、排放较好及油耗较低等一系列优点，已作为轻型卡车、空调大巴客车乃至小轿车的配套动力。然而，由于该机型是国外60年代设计的，而且若干配件尚未国产化，因此，如何根据当前国内生产的实际需要，作进一步的改善，成了该机型发展中的主要课题，改善其起动可靠性就是这类课题之一。

1　基本思路和基础研究

483Q柴油机的原设计完全是依靠电热塞来保证冷起动性能的。由于电热塞插入涡流室的深度较大，且可能对电热塞的布置未进行优化，因此，电热塞使用寿命较短，即使使用进口电热塞亦是如此。如何在尽可能不影响现产品工艺性的条件下，改善该机型的起动可靠性就成了本课题的研究目标。

国产涡流室式柴油机的长期生产和使用经验表明，在涡流室开设起动孔有利于改善起动性能[1]。李德桃和Konishi Y等人的研究也表明，采用起动孔，有使涡流室容易着火和主燃室直接着火的功能[2,3]。但是在483Q柴油机上，开设何种形式的起动孔？起动孔布置在何处？开设起动孔后，发动机的基本性能如何？这些都是需要从理论和试验两方面加以分析和验证的问题。

为此，我们针对两种不同形式的起动孔，在单缸机上对涡流室式发动机冷起动过程中的非稳态燃烧放热特性进行了研究和分析；并以此为基础，对483Q柴油机的冷起动性能进行了改进。

图1和图2分别为带锥起动孔与直形起动孔在冷起动过程中的一组连续示功图[1,4]。从图中可以看出在几乎相同的环境温度下，图1和图2相比Y型燃烧过程多，V型燃烧过程少。下面对上述示功图的放热特性分别进行讨论。

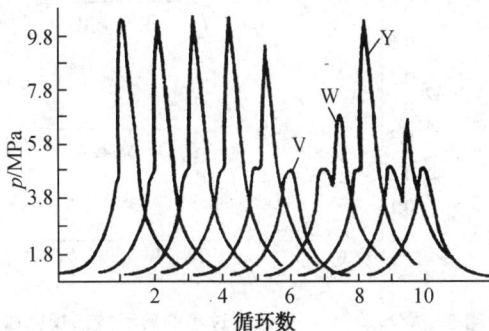

图1　环境温度－1℃、带锥起动孔的连续示功图

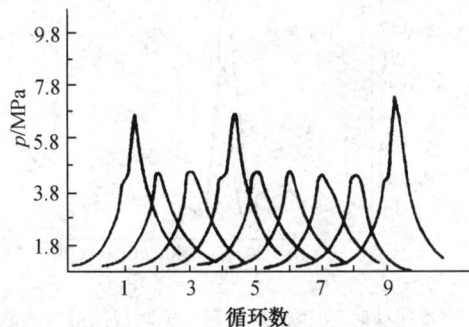

图2　环境温度0℃、直形起动孔的连续示功图

图 3 所示为主、副燃烧室的 V 型示功图。图 4 和图 5 分别为它们的瞬时放热规律图和累积放热规律图。从图上可以看出,V 型燃烧过程一般在上止点前 5° CA 附近开始着火,主、副燃烧室的着火始点相差不大,其燃烧持续时间为 25° CA 左右,涡流室放热量很少,只在燃烧开始时,涡流室中有一极短的放热过程,主燃烧室的最大瞬时放热率也较低,约为 0.02 kJ/° CA 左右。

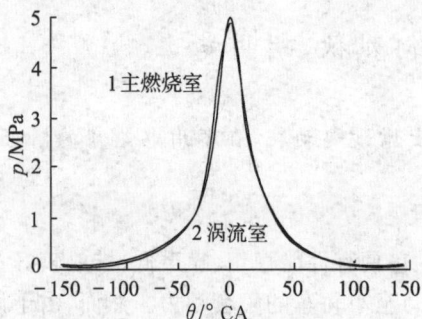

图 3　冷起动过程中的 V 型示功图

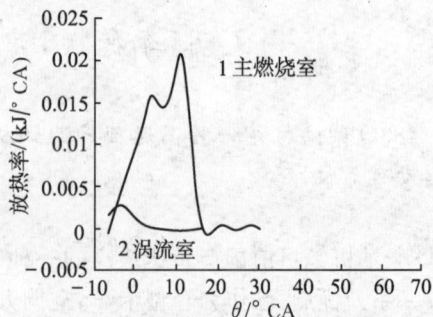

图 4　冷起动时 V 型燃烧过程的瞬时放热规律曲线

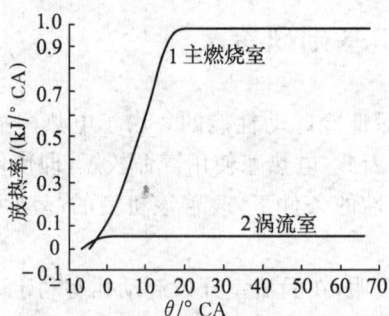

图 5　冷起动时 V 型燃烧过程的累积放热规律曲线

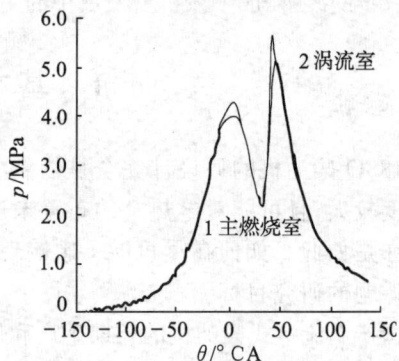

图 6　冷起动过程中的 W 型示功图

图 6 所示为主、副燃烧室的 W 型示功图。图 7 和图 8 分别为 W 型燃烧的瞬时放热规律图及累积放热规律图,从图中可以看出,涡流室的着火始点为上止点前 3° CA 左右,而主燃烧室的着火始点在上止点附近。一开始,主、副燃烧室中的放热率曲线比较平缓,这种状况一直延续到上止点后 23° CA 附近,首先是主燃烧室的瞬时放热率急剧增大,紧接着涡流室中的放热率也急剧增大,在上止点后 28° CA 附近达到最大值(主、副燃烧室的最大瞬时放热率分别为

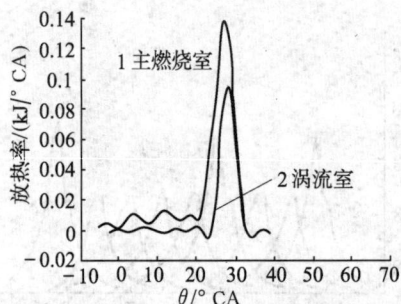

图 7　冷起动时 W 型燃烧过程的瞬时放热规律曲线

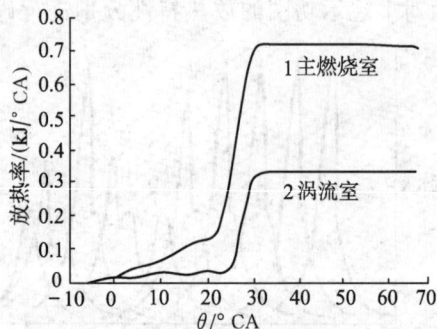

图 8　冷起动时 W 型燃烧过程的累积放热规律曲线

0.14 kJ/°CA,0.095 kJ/°CA),主、副燃烧室中的燃烧几乎同时于上止点后 35°CA 附近结束,整个燃烧持续时间约为 40°CA。

图 9 和图 10 分别为 Y 型燃烧示功图与标定工况下的示功图。图 11 和图 12 分别为 Y 型示功图的瞬时放热规律曲线和累积放热规律曲线。图 13 和图 14 分别为标定工况下示功图的瞬时放热规律图和累积放热规律图。对比以上有关图形可以看出,涡流室 Y 型燃烧的着火始点在上止点前 1°CA 左右,而主燃烧室的着火始点为上止点后 5°CA 左右,均比正常燃烧时的着火始点要迟。着火后,涡流室中有一段较短的缓慢放热期。随后其放热率迅速增大;主燃烧室在落后于涡流室 3°CA 的时候也开始迅速放热,其最大瞬时放热率分别达到 0.10 kJ/°CA 与 0.12 kJ/°CA,明显大于标定工况下的最大瞬时放热率。高的瞬时放热率导致高的压力升高率,从而引起冷起动时的振动和噪声加大,这是在发动机设计中所必须考虑的。涡流室中的燃烧于上止点后 12°CA 左右基本结束,而主燃烧室中的燃烧一直延续到上止点后 25°CA 左右才结束,整个燃烧过程约为 25°CA,明显较正常燃烧时短。

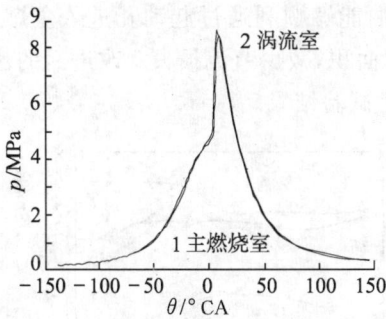

图 9　冷起动时 Y 型燃烧过程的示功图

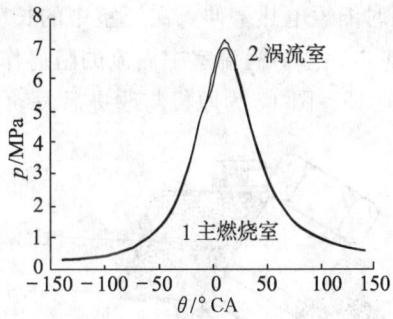

图 10　正常工作时标定工况下的示功图

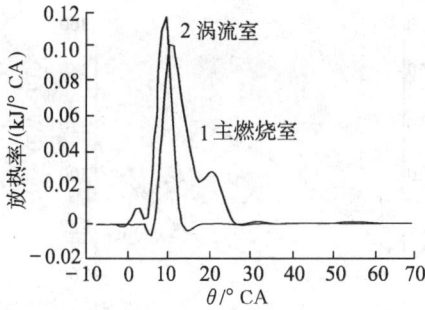

图 11　冷起动时 Y 型燃烧过程瞬时放热规律曲线

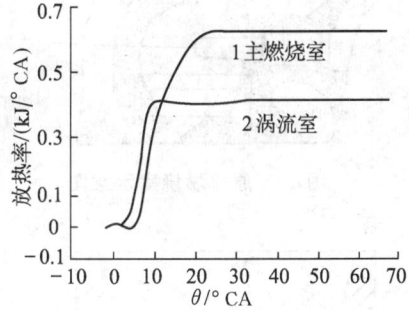

图 12　冷起动时 Y 型燃烧过程累积放热规律曲线

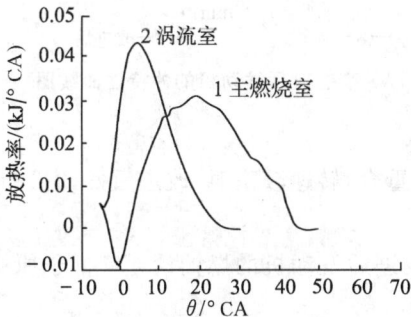

图 13　标定工况下的瞬时放热规律曲线

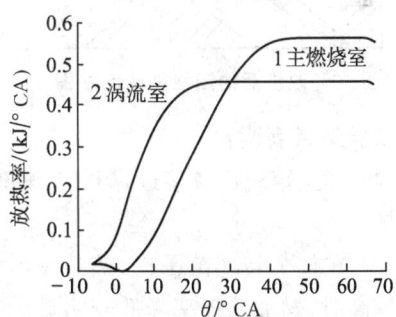

图 14　标定工况下的累积放热规律曲线

从上面的研究中,可以看出:直形起动孔与带锥起动孔相比,V型示功图多,Y型示功图少,因而起动性能较差。

从V型到W型,再到Y型燃烧过程以及正常燃烧,涡流室中的放热量逐渐增大,放热始点具有从主燃烧室向涡流室转移的特点。由此可见,在涡流室中采取一些措施,如设置电热塞和优化涡流室结构等,以改善冷起动时涡流室中的燃烧状况,使V型、W型尽快向Y型燃烧过程转变,对于改善发动机的冷起动性能是非常必要的。

2 改进方法与性能对比

图15为原机燃烧室结构示意图。从图中可以发现,该燃烧室没有起动孔,电热塞的布置和涡流室结构均有待进一步优化。为此我们提出了以下改进方案:舍弃原机镶块的凸台形式,增大涡流室容积,以改善涡流室中燃油的雾化;在稍微偏离油线位置的地方开一个带锥起动孔;同时缩短电热塞伸入涡流室中的长度,以使部分燃油能够顺利通过起动孔进入主燃烧室,并减少电热塞对涡流室中涡流的阻碍作用;增大通道截面积,减少节流损失。改进后的镶块结构见图16。图17为原机与改进后发动机的外特性对比试验结果。

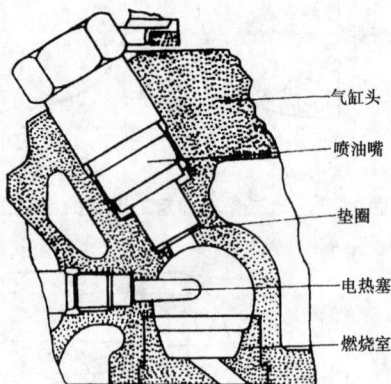

图15 原机燃烧室示意图

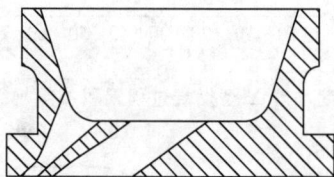

图16 改进后的镶块结构示意图

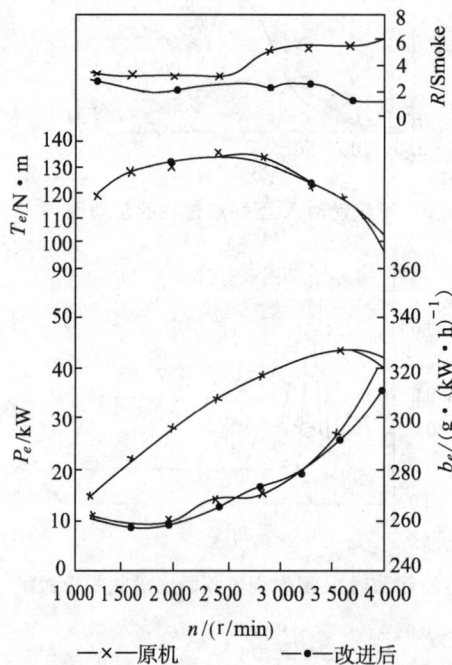

图17 改进前后发动机的外特性曲线图

试验结果表明:

(1) 其扭矩线比较接近,当转速较低时,两者几乎重合,转速较高时,改进后发动机的扭矩有所增加。

(2) 在整个转速范围内,其功率曲线几乎重合;改进后发动机的燃油消耗率低于原机,当转速较高时,燃油消耗率降低较明显。

(3) 改进后的烟度值降低非常明显,几乎只有原机的一半。

3 结 论

(1) 在 483Q 柴油机上增设起动孔,在不用电热塞的情况下,可使最低起动温度降低 15 ℃左右,而且可大大加速起动过程。

(2) 采用起动孔可减少冷起动过程中 V 型燃烧的数目,并使 V 型燃烧加速向 Y 型燃烧转变。

(3) 当采用电热塞时,其位置和深入涡流室长度,必须同燃烧室其他参数(如容积比、面积比)进行优化。

参 考 文 献

[1] 李德桃. 柴油机冷起动的基础研究和改善措施[M]. 北京:科学出版社,1997.

[2] Konishi Y. The development of the hight perfomance IDI and low emission diesel engine[J]. *IMech E*, 1992.

[3] Li Detao, Zhu Xiaoguang. Investigation on ignition and combustion at starting in a swirl chamber diesel engine with a starting throat[C]. *Proceeding of IPC*-5,1989.

(本文原载于《车用发动机》1998 年第 6 期)

A study of improved start reliability on 483Q diesel engine

Abstract:483Q diesel engine is a high speed homemade diesel engine. Experimental testing results indicate that the starting reliability of 483Q diesel engine without glow plug can be improved if a start hole is designed.

非直喷式柴油机低温起动时
非稳态燃烧过程分析[①]

杨文明,李德桃,夏兴兰,龚金科,王　谦

[摘要]　利用精确的零维模型和程序对实测涡流室式柴油机在低温起动时的非稳态燃烧过程进行了分析,分别计算和探讨了主、副燃烧室的放热规律,由此得出若干对改善柴油机低温起动性能具有理论价值和实际意义的结论。

非直喷式柴油机低温起动时的着火和燃烧过程非常复杂。这一过程除受到通常条件下的内燃机燃烧过程的诸多因素影响外,环境温度和燃烧室结构型式(例如,有无起动孔,采用何种形式的起动孔)也有很大的影响[1],这就导致对该过程进行数学分析十分困难。鉴于此,把实测的非直喷式柴油机(有起动孔、无电热塞)低温起动时的连续示功图,分成几种典型的循环(典型的燃烧过程),然后运用热力学方法,对不同形式的循环进行分析,由此得出了一些既有理论价值,也有实际意义的结论[1,2]。

1　理论模型

1.1　放热率

对于非直喷式柴油机燃烧系统,根据能量守恒方程、质量守恒方程及状态方程,并考虑到低温起动时的漏气损失,经推导和分析,得到主、副燃烧室的放热规律计算式[1-3]:

$$\frac{dQ_{bm}}{d\theta} = \frac{\gamma}{\gamma-1} p_m \frac{dV_m}{d\theta} + \frac{1}{\gamma-1} V_m \frac{dp_m}{d\theta} - c_p T_{k,m} \frac{dm_u}{d\theta} + c_p T_m' \frac{dm_t}{d\theta} + \frac{dQ_{wm}}{d\theta} \tag{1}$$

$$\frac{dQ_{bk}}{d\theta} = \frac{1}{\gamma-1} V_k \frac{dp_k}{d\theta} + c_p T_{k,m} \frac{dm_u}{d\theta} + \frac{dQ_{wk}}{d\theta} \tag{2}$$

式中 V_k, V_m 为涡流室(副燃烧室)容积和主燃烧室容积; $\dfrac{dQ_{bm}}{d\theta}$ 为主燃烧室的燃烧放热率; $\dfrac{dQ_{bk}}{d\theta}$ 为涡流室的燃烧放热率; $\dfrac{dm_u}{d\theta}$ 为通过连接通道主燃烧室和涡流室交换的工质质量; $\dfrac{dm_t}{d\theta}$ 为通过活塞环端口的漏气量; $\dfrac{dQ_{wm}}{d\theta}$, $\dfrac{dQ_{wk}}{d\theta}$ 为主、副燃烧室对壁面的传热率; T_m, T_k, p_m, p_k 为主、副燃

---

①　把柴油机非稳态过程的示功图划分成几种典型的形态进行分析,是李德桃教授创新的一种方法,由此得到的结果,对于认识发动机的非稳态过程的特征,寻求改善非稳态过程的经济性,降低排气污染,有着重要的学术意义和应用价值。该项研究在日本京都大学和上智大学进行学术交流时,曾得到池上询、五味努等教授的赞赏。

——编者注

烧室的工质温度和压力,当工质从涡流室流向主燃烧室时,$T_{k.m}=T_k$,反之 $T_{k.m}=T_m$;c_p 为工质的定压比热;γ 为比热比,$\gamma=c_p/c_v$。

1.2　气缸容积

可以认为涡流室的容积 V_k 为常量,即

$$V_k = 常量 \tag{3}$$

$$dV_k = 0 \tag{4}$$

主燃烧室容积 V_m 可用下式计算:

$$V_m = \frac{V_h}{2}\left[\frac{2}{\varepsilon-1}+1-\cos\theta+\frac{1}{\lambda_s}(1-\sqrt{1-\lambda_s^2\sin^2\theta})\right] \tag{5}$$

$$V_h = \frac{\pi D^2 r}{2} \tag{6}$$

式中 D 为气缸直径;r 为曲柄半径;λ_s 为曲柄连杆比;V_h 为活塞排量。

1.3　工质质量及成分计算

将柴油机的燃烧室分成两个相互关联的热力系统:主燃烧室和涡流室,并规定输入系统的能量和质量为正值,离开系统的能量和质量为负值。

设 m_{bk} 为涡流室已燃燃料的质量,m_{lk} 是涡流室中空气的质量,m_k 为涡流室中工质质量,L_0 为 1 kg 燃料完全燃烧所需的理论空气量,m_{bm} 和 m_{lm} 分别是主燃烧室中已燃燃料的质量和空气量,则涡流室内的质量平衡方程为

$$\frac{dm_k}{d\theta} = \frac{dm_u}{d\theta} + \frac{1}{H_u}\frac{dQ_{bk}}{d\theta} \tag{7}$$

$$\frac{dm_{bk}}{d\theta} = \frac{1}{H_u}\frac{dQ_{bk}}{d\theta} + \frac{m_{bi}}{m_i}\frac{dm_u}{d\theta} \tag{8}$$

$$\frac{dm_{lk}}{d\theta} = \frac{dm_u}{d\theta}\left(1-\frac{m_{bi}}{m_i}\right) \tag{9}$$

燃烧室内的质量平衡方程为

$$\frac{dm_m}{d\theta} = \frac{dm_u}{d\theta} + \frac{1}{H_u}\frac{dQ_{bm}}{d\theta} + \frac{dm_t}{d\theta} \tag{10}$$

$$\frac{dm_{bm}}{d\theta} = \frac{1}{H_u}\frac{dQ_{bm}}{d\theta} + \frac{m_{bi}}{m_i}\frac{dm_u}{d\theta} + \frac{m_{bm}}{m_i}\frac{dm_t}{d\theta} \tag{11}$$

$$\frac{dm_{lm}}{d\theta} = \frac{dm_m}{d\theta} - \frac{dm_{bm}}{d\theta} + \frac{1}{H_u}\frac{dQ_{bm}}{d\theta}L_0 \tag{12}$$

$$\lambda_i = \frac{m_{li}}{L_0 m_{bi}} \tag{13}$$

式中 H_u 为燃料的低热值;m_i,m_{bi},m_{li} 为主、副燃烧室工质质量、已燃燃料质量及空气量。当工质从主燃烧室流向涡流室时,$i=m$;当工质从涡流室流向主燃烧室时,$i=k$。

1.4　工质对气缸壁的热交换

气体与壁面间交换的热量可以由下式计算:

$$\frac{dQ_w}{d\theta} = \frac{h_c A_t}{\omega}(T_w - T_g) \tag{14}$$

式中 h_c 为气体对壁面的热交换系数;A_t 为参与热交换的瞬时气缸壁面面积;T_g 为工质温度;T_w 为气缸壁面平均温度;ω 为发动机角速度。

经过计算比较,选用下列公式计算热交换系数:

主燃烧室 $\qquad h_c = (Nu)_m = 0.048\,2(Re)_m^{0.792\,4}$ （15）

涡流室 $\qquad h_c = (Nu)_k = 0.035(Re)_k^{0.771\,2}$ （16）

式中 Nu 为努谢尔数; Re 为雷诺数。

1.5 通过活塞环的漏气量

对于柴油机某些非稳态过程,如冷起动过程,由于活塞平均速度和壁温都较低,导致漏气损失较大,所以必须考虑活塞环的漏气损失。

在漏气模型中,假设活塞环的端口间隙是唯一的泄漏途径,油环不密封气体。如图 1 所示,气体的流动假设为理想气体的一元准稳态绝热流动。对于一个 3 道气环的发动机,考虑到不同的压力比及气体倒流情况,得到如下公式:

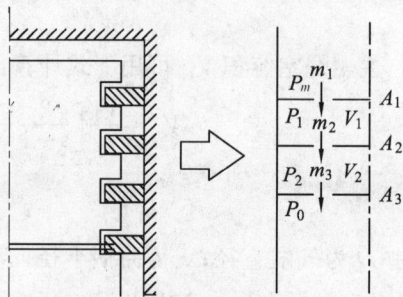

图 1　活塞环漏气模型

$$\frac{\mathrm{d}m_t}{\mathrm{d}t} = \frac{\mu A_1 p_m}{\omega} \frac{1}{\sqrt{RT_w}} \sqrt{\frac{2\gamma}{\gamma - 1}\left[\left(\frac{p_1}{p_m}\right)^{2/\gamma} - \left(\frac{p_1}{p_m}\right)^{(\gamma+1)/\gamma}\right]}$$

（17）

式中 μ 为流量系数(方形小孔的流量系数, $\mu = 0.65$); A_1 为第一道气环的泄漏面积; T_w 为缸壁平均温度; R 为气体常数。

1.6 主燃烧室和涡流室交换的工质质量

为了计算主燃烧室和涡流室之间通过连接通道交换的工质质量,采用一维、非稳态、可压缩、绝热流动的计算公式:

$$\frac{\mathrm{d}m_u}{\mathrm{d}\theta} = \frac{\mu A p_i}{\omega} \frac{1}{\sqrt{RT_w}} \sqrt{\frac{2\gamma}{\gamma - 1}\left[\left(\frac{p_j}{p_i}\right)^{2/\gamma} - \left(\frac{p_j}{p_i}\right)^{(\gamma+1)/\gamma}\right]}$$

（18）

式中 p_i 为系统上游的压力; p_j 为系统下游的压力; A 为连接通道截面积; μ 为连接通道的流量系数。

2　放热分析

图 2 为冷起动过程中的一组连续示功图。从图中可以看出,示功图根据其形状可以分为 3 种典型的型式,即 V 型、W 型和 Y 型,如图 3 所示。图 3～图 8 中的数字 1、2(包括下标数字)分别为主燃烧室和涡流室。图 4 所示为标定工况下的示功图。通过比较发现:V 型示功图的爆发压力最小,压力升高率也较低,主要集中于起动初期;W 型示功图的爆发压力有所增加,其压力升高率也较大,主要集中于起动中期;Y 型示功图有点类似于标定工况下的示功图,爆发压力最高,有时超过标定工况下的爆发压力,其压力升高率也较大,主要集中于起动后期。

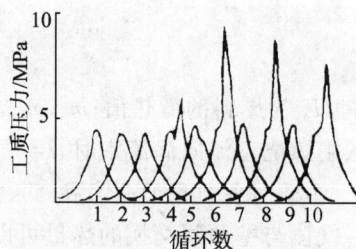

图 2　冷起动过程中的连续示功图

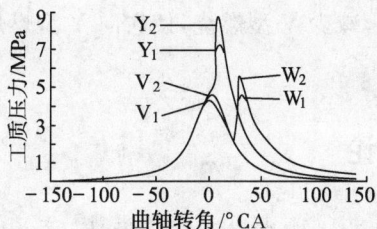

图 3　冷起动过程中的 V，W，Y 示功图

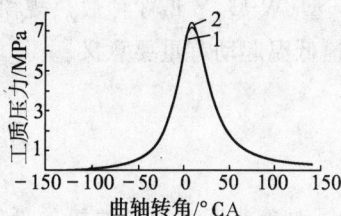

图 4　正常工作时标定工况下的示功图

图 5～图 8 所示分别为上述 4 种燃烧的瞬时放热图和累积放热图。从图上可以看出，V型燃烧过程一般在上止点前 5°CA 附近开始着火，主、副燃烧室的着火始点相差不大，其燃烧持续时间为 25°CA 左右，涡流室放热量很少，只在燃烧开始的时候，涡流室中有一极短的放热过程，主燃烧室的最大瞬时放热率也较低，约为 0.02 kJ/°CA。

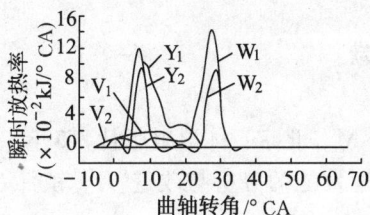

图 5　V，W，Y 型燃烧的瞬时放热率

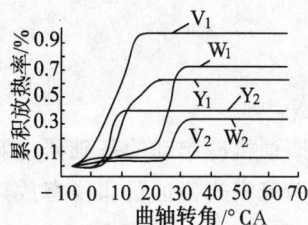

图 6　V，W，Y 型燃烧的累积放热率

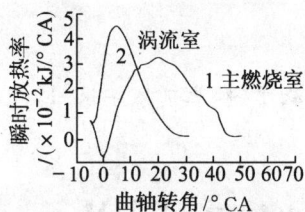

图 7　标定工况下的瞬时放热率

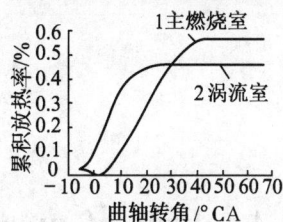

图 8　标定工况下的累积放热率

在 W 型燃烧过程中，涡流室的着火始点为上止点前 3°CA 左右，而主燃烧室的着火始点在上止点附近。一开始，主、副燃烧室中的放热率曲线比较平缓，这种状况一直延续到上止点后 23°CA 附近。首先是主燃烧室的瞬时放热率急剧增大，紧接着涡流室中的放热率也急剧增大，在上止点后 28°CA 附近达到最大值（主、副燃烧室的最大瞬时放热率分别为 0.14 kJ/°CA，0.095 kJ/°CA），主、副燃烧室中的燃烧几乎同时于上止点后 35°CA 附近结束。整个燃烧持续过程约为 40°CA。

在 Y 型燃烧过程中，涡流室的着火始点在上止点前 1°CA 左右，而主燃烧室的着火始点为上止点后 5°CA 左右，均比正常燃烧时的着火始点要迟。着火后，涡流室中有一段较短的缓慢放热期，随后其放热率迅速增大，主燃烧室在落后于涡流室 3°CA 的时候也开始迅速放热，其最大瞬时放热率分别达到 0.10 kJ/°CA 与 0.12 kJ/°CA，明显大于标定工况下的最大瞬时放热率。高的瞬时放热率导致高的压力升高率，从而引起冷起动时的振动和噪声增大，这是在发动机设计中必须考虑的问题。涡流室中的燃烧在上止点后 12°CA 左右基本结束，而主燃烧室中的燃烧一直延续到上止点后 25°CA 左右才结束，整个燃烧过程约为 25°CA，明显较正常燃烧时短。

对照 V 型、W 型、Y 型 3 种放热规律图可以看出,减少 V 型燃烧,使它向 Y 型燃烧过程转变,对于加速低温起动有重要意义。

3 结　论

(1) 为了改善非直喷式柴油机的低温起动,必须改进燃烧室的结构设计,或增加辅助装置,以减少 V 型燃烧过程所占比例,使 V 型燃烧过程加速向 Y 型燃烧过程转变。

(2) 从 V 型→W 型→Y 型→正常燃烧,放热量具有从主燃烧室向涡流室过渡的特征,因此,采取技术措施,加速副燃烧室的着火和燃烧过程,对改善起动过程有着更为重要的意义。

(3) 在冷起动过程中,Y 型燃烧过程的最高爆发压力和最大瞬时放热率及压力上升率有时大大超过正常燃烧时的相应值,这是发动机设计者必须注意的一个问题。

参 考 文 献

[1] 李德桃. 柴油机冷起动的基础研究和改善措施[M]. 北京:科学出版社,1998.

[2] 杨文明. 涡流室式柴油机非稳态放热特性和高速性能的分析与改进[D]. 镇江:江苏理工大学,1997.

[3] Heywood J B. *Internal Combustion Engine Fundamentals*[M]. New York:McGraw-Hill,1988.

(本文原载于《农业机械学报》1999 年第 1 期)

Analysis of unsteady combustion of IDI diesel engine in cold-starting stage

Abstract:A great attention has been paid to the cold-starting of swirl chamber diesel engine because of its complexity. This paper uses the precise zero-dimentional model and program to analyze the unsteady combustion process of IDI diesel engine by experiment,estimate the rules of heat release of three types of typical combustion(type V,type W and type Y) of main chamber and swirl chamber at cold-starting and also the rules of heat release of steady combustion at stated working. Some important conclusions for improving the performance of a diesel engine at cold-starting are obtained,which are valuable in both theory and practice.

涡流室式柴油机冷起动时的
准维燃烧模拟计算

李德桃,杨文明,姜树李,刘桃英,卢伯林

[摘要] 本文在实测涡流室式柴油机非稳态连续示功图的基础上,首次建立了该型柴油机的准维燃烧模型,并对该模型进行了详细分析,然后对其放热特性和排放生成进行了计算和探讨,为改善柴油机的冷起动性能提供了可靠的理论基础。

冷起动性能是柴油机的重要性能之一,低温下不能起动或不能可靠起动不仅降低了柴油机的利用率,而且还给使用者带来许多不便。同时,柴油机在低温起动时的有害排放物水平较高也给人们健康带来不良的影响。目前,国内外学者对涡流室式柴油机冷起动性能的研究还停留在试验与热力学模型的基础上,对涡流室式柴油机在冷起动时的排放生成情况没有进行模拟计算和分析,这是与涡流室式柴油机的发展不相适应的。正是基于这一考虑,本文在对试验结果进行分析研究的基础上,建立了涡流室式柴油机低温起动时非稳态燃烧的准维模型,研究了涡流室式柴油机在低温起动时的排放生成情况。

1　模型的建立

考虑到冷起动时非稳态燃烧的特殊性,本文对曾经建立的涡流室式柴油机稳态燃烧时的准维模型进行了一些改进:

(1) 以实测的燃烧室瞬态壁温取代原有的定值[1];

(2) 增加了活塞环漏气子模型;

(3) 增加了变曲轴转角的子模型,以取代原模型中给定的定转速值;

(4) 增加了喷雾碰壁子模型及油膜蒸发模型。

1.1　活塞环漏气子模型

柴油机在冷起动过程中,由于活塞平均速度和活塞环温度都较低,导致漏气损失较大,压缩终了的工质温度、压力较低,从而使燃油的雾化及着火条件变差,因此在冷起动非稳态燃烧模拟中必须考虑通过活塞环的漏气损失。在活塞环漏气模型中,假设活塞环开口间隙是泄露的唯一途径,认为气体的流动是一维准稳态绝热流动。考虑到不同压力比和气体倒流等情况,得到如下公式[1]:

$$\frac{\mathrm{d}m_t}{\mathrm{d}\varphi}=\frac{C_d A_c}{\omega}\sqrt{\frac{2k}{k-1}\cdot\frac{p_2}{v}\left[\left(\frac{p_2}{p_1}\right)^{\frac{2}{k}}-\left(\frac{p_2}{p_1}\right)^{\frac{k+1}{k}}\right]} \tag{1}$$

式中 C_d 为流量系数(方形小孔, $C_d=0.65$); A_c 为泄漏面积; φ 为曲轴转角; ω 为瞬时角速度; k 为比热比; p_1, p_2 分别为上、下流侧工质压力。依次研究每一对相邻容积,联立求解即可得到活塞环的漏气质量流量。

1.2 变曲轴转角子模型

有关资料进行的冷起动试验测量表明[2],压缩行程中起动转速减小。这说明在冷起动模拟研究中,有必要考虑这一现象。在本研究中,基于对其他发动机试验数据的分析,选用压缩期间变化幅度为 20% 的正弦波作为发动机的瞬时转速波形,计算公式为

$$n = n_i [1 + 0.2\cos(\varphi + 180)] \tag{2}$$

式中 n_i 为发动机在第 i 个循环的平均转速,r/min。

1.3 喷雾碰壁子模型及壁面油膜蒸发子模型

在冷起动过程中,由于压缩压力、温度较低,会有部分燃油喷到涡流室壁面形成油膜,影响涡流室式柴油机的冷起动性能,因此有必要建立喷雾碰壁子模型及壁面油膜蒸发子模型。

喷雾碰壁子模型采用 Leonard 和 Assanis Dennis N[3] 建立的模型,将液滴的 Weber 数作为确定液滴碰壁结果的判据:

$$We = \frac{\rho_L u^2 d_L}{\sigma} \tag{3}$$

式中 ρ_L 为液态燃油的密度;u 为油滴碰壁前一瞬间的法向速度;d_L 为入射油滴直径;σ 为油滴表面张力。

当 $We \leqslant 80$ 时,油滴碰壁后反弹,且在碰撞和反弹过程中不分裂,反弹后随空气作近似刚体运动。

当 $We > 80$ 时,油滴破碎,并粘附在壁面上,形成液态油膜。

由式(3)可见,高速大油滴碰壁后在壁面上形成液态油膜,油膜厚度与喷油压力、喷油量、喷油持续期有关。假定油膜厚度小于或接近油膜层流边界层厚度(约 0.1 mm)和湍流气体边界层厚度(0.01~0.10 mm)[3]。这样假设液态油膜像附壁射流一样从碰撞点开始运动一段距离,随后由于边界层的粘滞力而滞止。因为涡流室壁面通过热传导传给油膜的热量比空气对流传给油膜的热量少得多,所以,忽略其影响。Colburn 比拟适用于平板上的湍流对流换热,且它与 Prandtl 数接近 1 的流体的试验数据较吻合;而涡流室底面与平板的情况比较接近,所以通过层流受迫对流分析和 Colburn 比拟来建立油膜与涡流室内工质间的对流换热及传质模型[4]。

根据流动状况,用下列关系来表示平均 Nusselt 数和 Sherwood 数。

对于层流($Re_L < 5 \times 10^5$),有

$$Nu_L = 0.664 Re_L^{1/2} Pr_L^{1/3} \frac{\ln(1 + B_d)}{B_d} \tag{4}$$

$$Sh_L = 0.664 Re_L^{1/2} Sc_L^{1/3} \frac{\ln(1 + B_d)}{B_d} \tag{5}$$

对于湍流区间 I($5 \times 10^5 < Re_L < 1 \times 10^7$),有

$$Nu_L = (0.037 Re_L^{0.8} - 872) Pr_L^{1/3} \frac{\ln(1 + B_d)}{B_d} \tag{6}$$

$$Sh_L = (0.037 Re_L^{0.8} - 872) Sc_L^{1/3} \frac{\ln(1 + B_d)}{B_d} \tag{7}$$

对于湍流区间 II($Re_L > 1 \times 10^7$),有

$$Nu_L = [0.228 Re_L (\lg Re_L)^{-2.584} - 872] \cdot Pr_L^{1/3} \frac{\ln(1 + B_d)}{B_d} \tag{8}$$

$$Sh_L = [0.228Re_L(\lg Re_L)^{-2.584} - 872] \cdot Sc_L^{1/3} \frac{\ln(1+B_d)}{B_d} \tag{9}$$

式中 $Nu_L = \dfrac{hL}{k_{air}} =$ 平均 Nusselt 数,$Sh_L = \dfrac{h_m L}{D_{air}} =$ 平均 Sherwood 数;$Re_L = \dfrac{\rho_{air} VL}{\mu_{air}} =$ 平均 Reynolds数;$Pr_L^{1/3} = \dfrac{\mu_{air} c_{p,air}}{k_{air}} =$ Prandtl 数;$Sc_L^{1/3} = \dfrac{\mu_{air}}{\rho_{air} D_{air}} =$ Schmidt 数;V 为缸内气体流过液态油膜的速度;h 和 h_m 分别为热量和质量传递系数;L 为特征尺度;ρ_{air},μ_{air},$c_{p,air}$,k_{air} 和 D_{air} 分别为气体的密度、动力粘性系数、定压比热容、导热率和质量扩散系数。

$\dfrac{\ln(1+B_d)}{B_d}$ 项考虑了液态油膜表面附近燃油蒸气的存在对热量和质量传递的影响。B_d 为 Spanding 质量传递数,即

$$B_d = \frac{Y_1^* - Y_1}{1 - Y_1^*} \tag{10}$$

式中 Y_1 为所在单元内燃油蒸气质量百分数;Y_1^* 为液态油膜附近燃油蒸气的质量百分数。

从周围气体向壁面上液态油膜进行的对流换热把热量传递给液态油膜,用来提高其温度并蒸发部分积累的燃油,减少油膜厚度。下列能量平衡方程用来确定油膜瞬时温度:

$$\rho_f A \delta c_{liq} \dot{T}_f - \rho_f R \delta L(T_f) = AQ_L \tag{11}$$

式中 ρ_f 为液态燃油密度;A 为油膜覆盖面积;δ 为油膜厚度;c_{liq} 为液态燃油的比热容;$L(T_f)$ 为燃油在温度 T_f 时的蒸发潜热;$\dot{T}_f$ 为油膜温度的变化速率。

由于蒸发,油膜厚度的减少率 R_δ 为

$$R_\delta = \frac{(\rho D)_{air}}{\rho_f L} B_d Sh_L \tag{12}$$

从缸内气体向液态油膜进行的热传递速率 Q_L 为

$$Q_L = \frac{k_{air}(T - T_f)}{L} Nu_L \tag{13}$$

式中 T 为缸内气体温度。

计算从第 1 滴液滴在燃烧室壁面碰撞开始,根据未反弹油滴的质量、液滴湿润表面积和燃油密度,将油膜厚度初始化。假设一初始厚度,用显式替代方法求解能量方程直到在选定的时间间隔内得到收敛解。在时间间隔 Δt_{cvap} 内蒸发的燃油质量 Δm_f 可用下式计算:

$$\Delta m_f = AR_\delta \Delta t_{cvap} \tag{14}$$

燃油蒸气直接加入到经过油膜上空的单元内。为进行下一步长油膜温度的计算,油膜厚度需根据蒸发掉的燃油和碰壁增加的燃油来更新。

注意给定时间步长内从油膜内蒸发的最大燃油蒸气质量为

$$\frac{\Delta m_f}{\text{单元总质量}} \leq \frac{1}{2} \frac{\ln(1+B_d)}{Sc} \tag{15}$$

2 计算结果分析[5]

2.1 V 型燃烧过程

图 1 所示为 V 型燃烧放热规律的计算值与实测值(由实测示功图计算得到)对比。在 V

型燃烧过程中,由于主、副燃烧室中的放热量都很小,工质温度较低,因此几乎没有形成 NO_x 和微粒排放。

2.2 W 型燃烧过程

图 2 所示为 W 型燃烧过程的放热率对比。从该图可以看出,W 型燃烧过程的瞬时放热率较大,以曲轴转角计的燃烧持续时间较标定工况时短,大部分燃油在 25° CA 范围内燃烧完毕。

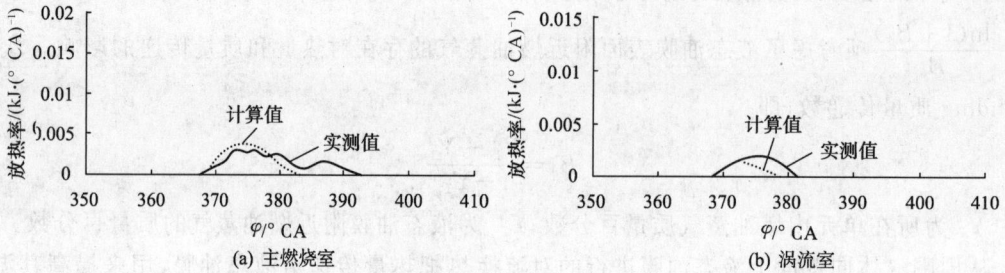

图 1　V 型燃烧过程的放热率对比

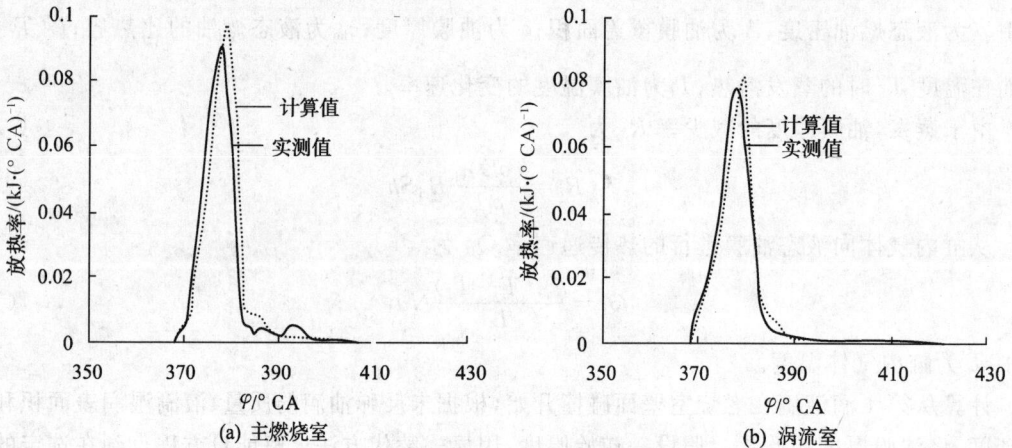

图 2　W 型燃烧过程的放热率对比

图 3、图 4 所示为 W 型燃烧过程的排放生成情况。从图 3 可以看出,W 型燃烧过程生成的 NO_x 浓度明显较标定工况时少。这主要是因为 W 型燃烧过程的着火延迟期较长,着火发生在活塞下行一定距离之后,降低了燃烧室中的工质温度,而工质温度是影响 NO_x 生成的主要因素,因此 W 型燃烧过程生成的 NO_x 浓度较低。从图 4 可以看出,W 型燃烧过程中生成的微粒也较标定工况时少,且微粒生成的峰值较晚。这主要是因为 W 型燃烧过程着火较晚,初期油雾单元混合较好造成的。

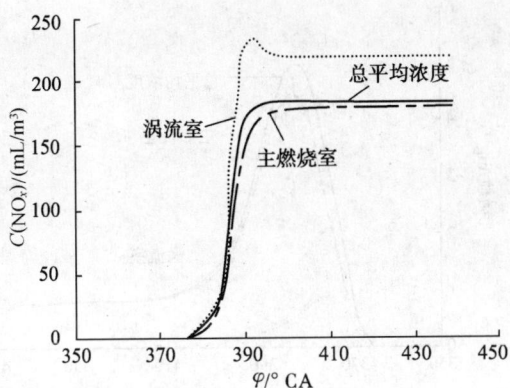

图3 W型燃烧过程中 NO_x 的生成历程

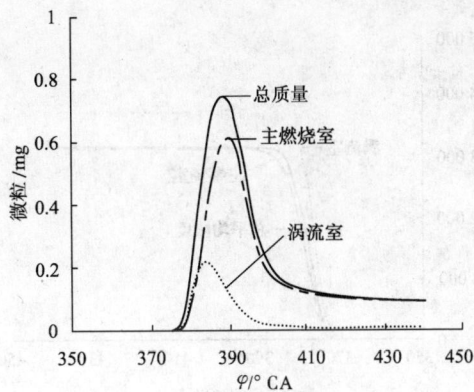

图4 W型燃烧过程中微粒的生成历程

2.3 Y型燃烧过程

图5所示为Y型燃烧过程的放热率对比。从该图可以看出,Y型燃烧的瞬时放热率很大,以曲轴转角计的燃烧持续时间短,燃烧集中在 $20°$ CA范围内完成。Y型燃烧在整个起动过程的初、中、后期都有可能发生,但以后期最多。在起动初期,虽然压缩温度较低,以时间计的着火延迟期较长,但因为转速低,当活塞运行到上止点附近时,喷雾已有较长时间进行着火前的准备工作,这样就有可能形成Y型燃烧过程。特别是当前一循环有部分燃油残留在燃烧室中时,随着起动过程的进行,燃烧室缸壁温度逐渐升高,通过活塞环的漏气损失降低,因此,Y型燃烧过程出现的比例也越大。

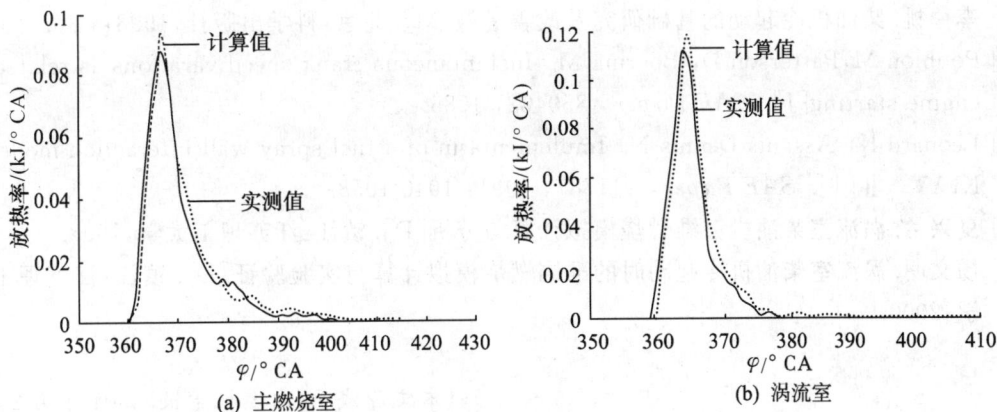

(a) 主燃烧室

(b) 涡流室

图5 Y型燃烧过程放热率的对比

图6、图7所示为Y型燃烧过程中 NO_x 和微粒排放的生成情况。由图可知,Y型燃烧过程产生的 NO_x 浓度和微粒量均较标定工况时高,这是由于在Y型燃烧过程中,燃油集中在较短时间内完成,使工质温度较标定工况时高造成的。

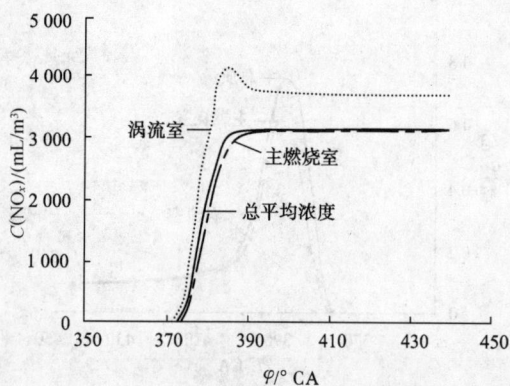

图 6 Y 型燃烧过程中 NO_x 的生成历程

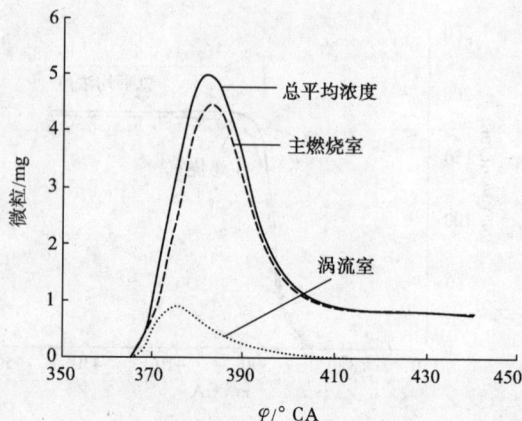

图 7 Y 型燃烧过程微粒的生成历程

4 结 论

（1）在冷起动过程中，3 种典型燃烧过程的放热规律及排放生成情况具有较大的差异。从 V 型至 W 型、Y 型，其放热率及排放生成逐渐增大。

（2）为了改善发动机的冷起动性能，采取一些辅助措施，如加装电热塞、进气管加热等方法，加速 V 型燃烧与 W 型燃烧向 Y 型燃烧转变是必要的。此外，还可以通过优化涡流室结构参数，如增加压缩比等方法来改善其冷起动性能。

参 考 文 献

［1］李德桃. 柴油机冷起动的基础研究及改善措施［M］. 北京：科学出版社，1998：4-30.

［2］Poublon M，Patterson D，Boerma M. Instantaneous crank speed varations as related to engine starting［J］. *SAE Paper*，850482，1985.

［3］Leonard K，Assanis Dennis N. Implementatin of a fuel spray wall interaction model in KIAV－Ⅱ［J］. *SAE Paper*，911787，1991：1040-1058.

［4］夏兴兰. 涡流室柴油机三维燃烧模拟计算与分析［D］. 镇江：江苏理工大学，1998.

［5］杨文明. 涡流室柴油机冷起动时的准维燃烧模拟计算与实验验证［D］. 镇江：江苏理工大学，2000.

（本文原载于《内燃机学报》2001 年第 2 期）

Qussi-dimensional combustion simulation of swirl chamber diesel engine at cold-starting

Abstract：Base on the actual measurement on the unsteady continuous *P-V* figure of swirl chamber diesel engine，a quasi-dimensional model is developed first and analyzed in detail，then the heat release rate and emission formation are calculated and discussed. The results offer the reliable theoretical basis for improving the cold-starting of diesel engine.

Combustion simulation of IDI engine at cold-starting

Li Detao, *Pan Jianfeng*, *Liang Fengbiao*, *Yang Wenming*

Abstract

According to the *P-V* figures measured by experiments, we know that there are three types of typical unsteady combustion process during cold-starting of swirl chamber diesel engine: V, W and Y type. Based on a series of experiment study, a quasi-dimensional model has been developed to simulate the unsteady combustion process at cold-starting. The heat release rate and emissions formation of two types combustion process both in main chamber and swirl chamber are calculated in this paper. It is instructive for us to improve the cold-starting performance of swirl chamber diesel engine.

1 Introduction

Cold-starting performance is one of the most important performances of swirl chamber diesel engine. It not only reduces the extent of diesel engine being used, and also arises so much inconvenience to user, if the diesel engine can't be started successfully at cold ambient. In addition, the high level of harmful emissions at cold-starting will affect the health of humanity. But the research to the cold-starting performance of swirl chamber diesel engine still retains on experiments and simulations by thermodynamic model. It is inconsistent with the rapid development of swirl chamber diesel engine. In order to improve the cold-starting performance of swirl chamber diesel engine more conveniently, based on a lot of experimental results, we develop the quasi-dimensional model of swirl chamber diesel engine at cold-starting, and study the formation of harmful emissions. The results of experiments have been described in reference[1,2], so we won't introduce it here again.

2　Mathmatical model

Considering the peculiarity of swirl chamber diesel engine at cold-starting, some reformations are done to the quasi-dimensional model of steady combustion:

(1) The leakage through clearance of piston ring is very little when engine is working steadily, so we often neglect it when simulates. But when engine works at cold-starting conditions, it becomes so much(about $4\% \sim 10\%$ of total mass, primarily caused by low rotating speed) that we can't neglect it again. Below is the leaking rate of mass through clearance of piston ring[1]:

$$\frac{dm}{d\varphi} = \frac{\mu A p_1}{\omega \sqrt{RT_w}} \sqrt{\frac{2k}{k-1} \frac{p_2}{p_1}^{\frac{2}{k}} - \frac{p_2}{p_1}^{\frac{k+1}{k}}} \qquad (1)$$

where ,μ is coefficient of flow, A is the area of flow passage ,p_1, p_2 is the pressure of downstream and upstream of flow passage, respectively, k is specific heat ratio, T_w is temperature of the wall of cylinder, ω is angular speed, and R is constant number of gas.

(2) Some experiments on performance of diesel engine at cold-starting indicate that the rotating speed change not only with cycle number but also with piston location within a cycle. It subsequently affect the performance of cold-starting of swirl chamber diesel engine. So we must consider the change of rotating speed[3]:

$$n_i = \bar{n}_i [1 + 0.2\cos(\varphi + 180)] \qquad (2)$$

where, n_i, $\bar{n}_i$ is the rotating and average rotating speed of the cycle i, respectively, φ is the angle of shaft.

(3) Because the wall temperature of swirl chamber and the inject pressure of oil at cold-starting is very low that a part of oil adheres to the wall and form oil film. It affects remarkably the cold-starting performance of swirl chamber diesel engine, so we develop here the model of spray impinging on wall and the model of evaporation of oil film.

According to reference[4], the Weber number of liquid oil drops is

$$We = \frac{\rho_1 u^2 d_1}{\sigma} \qquad (3)$$

where, ρ_1 is the density of liquid oil drops, d_1 is the diameter, u is the speed of oil drops, and σ is the surface stress.

When, $We \leqslant 80$, the oil drops bounce and don't break up after impinging on the wall; $We > 80$, the oil drops break up and adhere to the wall of swirl chamber forming oil film.

The reducing rate of thickness of oil film is[5]

$$R_\delta = -\frac{(\rho D)_{air}}{\rho_f L} B_d Sh_L \qquad (4)$$

The mass of oil vaporized within Δt is

$$\Delta m_f = A R_\delta \Delta t \qquad (5)$$

where, D is the mass diffusion coefficient, L is characteristic length, B_d is Spanding mass transfer number, Sh_L is Sherwood number, and A is the area of oil film.

3   Result and discussion

Because the heat released during V type combustion process is so little that there is almost no harmful emissions generating, and so we won't discuss it in this paper.

The heat release rate of W type combustion process at cold-starting is illustrated in Fig. 1, Fig. 2 and Fig. 3 illustrate the formation of NO and soot of W type combustion process respectively. We can know from Fig. 1 that the computational result is close to the test result, so we think that the model is accurate enough. We can also know from Fig. 1 that the heat release rate of W type combustion process is higher than that of rated power, most of the oil is burnt out within $30°$ CA. According to Fig. 2 and Fig. 3, we know that the mass fraction of NO and soot generated during W type combustion process are less thanthat of

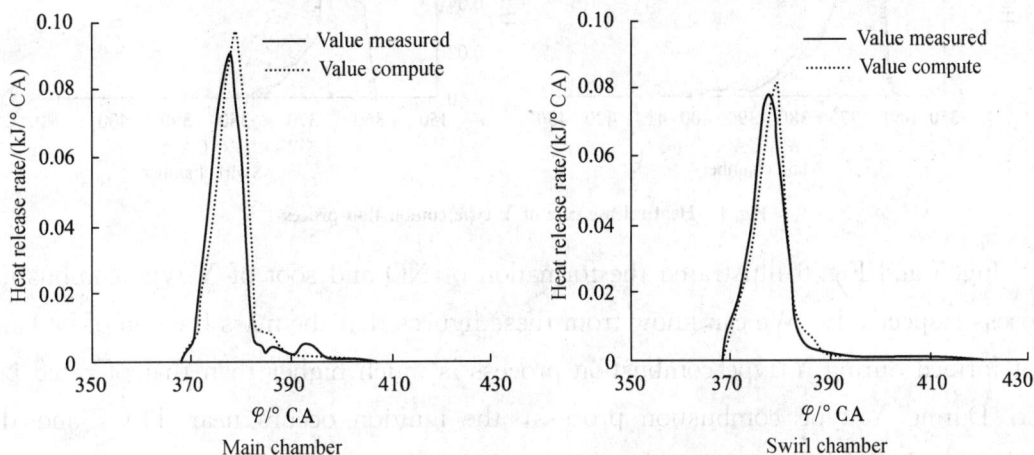

Fig. 1   Heat release rate of W type combustion process

rated power[2]. This is caused primarily by the long ignition delay time. When ignition occurs, the piston has been far from TDC, subsequently reduces the temperature and pressure of working mass in the cylinder. So less emissions are generated.

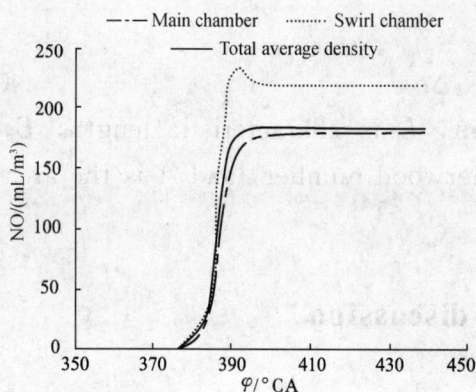

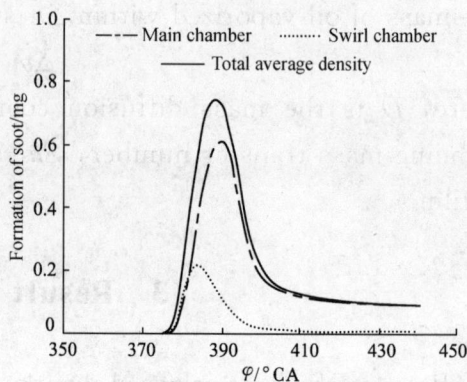

Fig. 2　Formation of NO of W type combustion process　**Fig. 3　Formation of soot of W type combustion process**

The heat release rate of Y type combustion process at cold-starting is illustrated in Fig. 4. According to the figure, we can know that the heat release rate of Y type combustion process is higher than that of rated power. The last time of combustion is also very short, and most of the oil is burnt out within 20° CA.

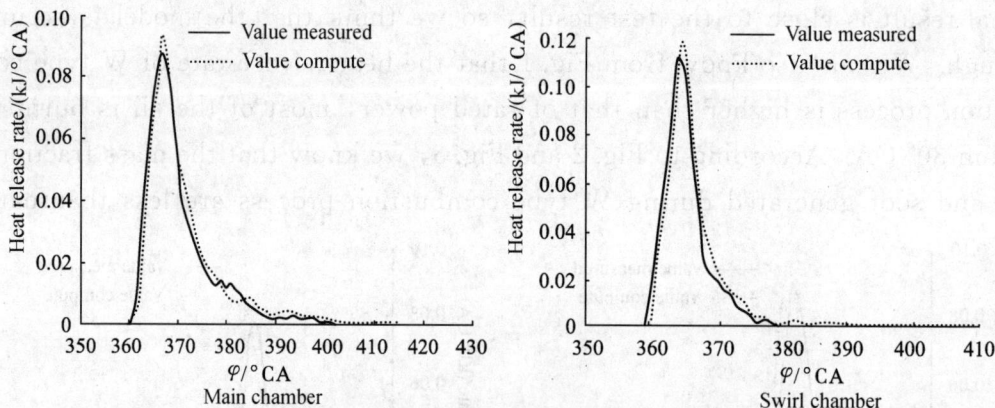

Fig. 4　Heat release rate of Y type combustion process

Fig. 5 and Fig. 6 illustrated the formation of NO and soot of Y type combustion process respectively. We can know from these figures that the mass fraction of NO and soot formed during Y type combustion process is much higher than that of rated power. During Y type combustion process, the ignition occurs near TDC, and the combustion is finished in such a short time, so the temperature of working mass is very high, subsequently generates more emissions.

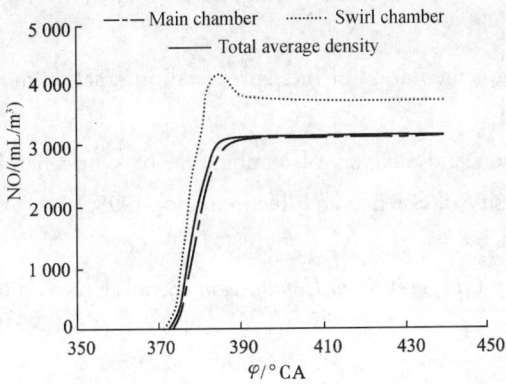

Fig. 5　Formation of NO of Y type combustion process

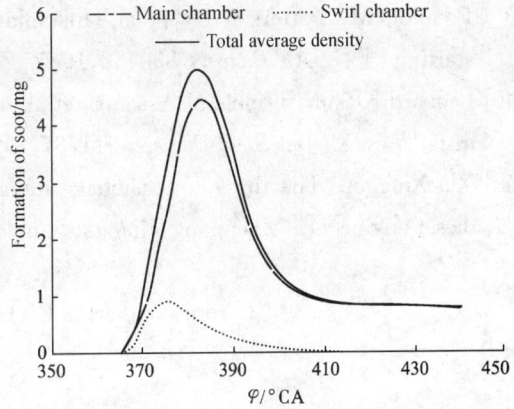

Fig. 6　Formation of NO of Y type combustion process

4　Conclusion

During cold-starting process, the heat release rate of three types of different combustion process is much different from each other, the mass fraction of emissions formed is also much different from each other, but grows from V to W to Y type combustion process. According to the experiments of swirl chamber diesel engine at cold-starting[1], we can also know that V type combustion process mainly occurs at initial stages of cold-starting, and W type mainly occurs at middle stages, while Y type occurs more frequently at last stages. In order to improve the performance of cold-starting of swirl chamber diesel engine, it is very important for us to take some means to accelerate the transform of V to W to Y type combustion process.

Acknowledgement

Dr. Zhu Xiaoguang and Dr. He Xiaoyang contributed their experience to the experiments on the performance of cold-starting of swirl chamber diesel engine. The work was largely supported by Chinese national fund of natural science.

References

[1] Li Detao. *The Foundational Study on Performance of Cold-Starting of Diesel Engine* [M]. Beijing: Science Press, 1998. (in Chinese)

[2] Yang Wenming. Quasi-dimensional combustion simulation and verification of swirl chamber diesel engine during cold-starting [D]. Zhenjiang: Jiangsu University of Science and Technology, 2000.

[3] Poublon M, Patterson D, et al. Instantaneous crank speed variations as related to engine starting [J]. *SAE Paper*, 850482, 1985.

[4] Leonard K Shih, Dennis N Assanis, et al. Implementation of fuel spray wall interaction model in KIVA—X [J]. *SAE Paper*, 911787, 1991.

[5] Xia Xinglan. The three-dimensional simulation and analysis of combustion of swirl chamber diesel engine[D]. Zhenjiang: Jiangsu University of Science and Technology, 1998.

(From: 3*rd Asia-Pacific Conference on Combustion*, Seoul, Korea, 2001)

IDI 柴油机冷起动燃烧过程模拟分析

[摘要]　从实测示功图的形状可知,涡流室式柴油机在冷起动过程中存在三种典型的非稳态燃烧过程:V型、W型和Y型。本文在试验研究的基础上,首次建立了涡流室式柴油机冷起动时非稳态燃烧过程的准维模型,分别计算和分析了三种不同燃烧过程主、副燃烧室中的放热特性及排放生成情况,以期为进一步改善涡流室式柴油机的冷起动性能打下基础。

Ⅲ. 柴油机共轨喷油系统的研究

Study on Common Rail Fuel Injection System for Diesel Engine

Investigation on new type of hydraulically intensified common-rail injection system[①]

Hu Linfeng , Liang Fengbiao , Li Detao , Wu Jian

Abstract

This paper analyses the deficiencies of accumulator-type common-rail injection system. By improving the injector structure, the new-style hydraulically intensified common-rail injection system realizes pilot injection and gets ideal injection rate. After establishing the calculation model and through the numerical simulations, the injection characteristics of the common-rail injection system have been investigated. The experiment investigations prove that the numerical simulations are accurate. The new-type hydraulically intensified common-rail injection system can meet the requirements of future engines.

1 Introduction

The increasing needs for higher fuel economy, less toxic, exhaust-gas emissions, together with never-ending demands for reductions in diesel engine noise can no longer be met by mechanically governed fuel-injection system[1]. The very high injection pressure coupled with a precise discharge rate and metered injected fuel quantities are needed to meet the above requirements.

The common-rail fuel-injection system is the most advanced and efficient

---

① 文中介绍的电控共轨喷油系统,早期为增压式的共轨喷油系统,后为高压共轨式喷油系统。作者在开发增压式共轨喷油系统时,曾得到美国著名的内燃机专家、两届 SAE 年会主席 John Beck 的帮助。此后,通过"产学研"对该系统进行了设计和理论分析,为下一代系统的开发积累了坚实的理论基础和丰富的实践经验。后来,第一作者领衔开发的高压共轨喷油系统,在系统结构上有多方面的突破,获得了十几项国家专利,项目先后得到国家发改委、科技部和江苏省科技厅以及一汽集团公司的大力支持,该系统现已中试生产准备和全面推广应用。

——编者注

technology for reducing emission, fuel consumption and noise levels. It has wide pressure controllability, high injection pressure and flexible injection rate controllability. A new-type hydraulically intensified common-rail injection system has been developed, based on an accumulator type injection system. Unlike the accumulator type, the new type has the advantages of adapting injection pressure independent of the fuel quantity and achieving desirable injection rate shaping for diesel engine combustion, enhancing the controllability over the injection process and better meeting the requirements of future diesel engines.

2 Deficiencies of accumulator type common-rail injection system

The BKM servojet injection system is a representative of diesel fuel injection system using a simple, medium-pressure, common-rail system, with pressure intensifier and accumulator type unit injectors. Accumulator type nozzle is controlled by an electronically controlled three-way solenoid. The injector diagram is shown in Fig. 1[2].

An electronic pressure regulator is used to control rail pressure by a digital pulse-width modulated signal from a electronic control module. The fuel of the common-rail system is supplied by a simple constant flow positive displacement pump. The pressure intensifier is a simple plunger and bushing with a area ratio of 10 : 1 between the high pressure and low pressure

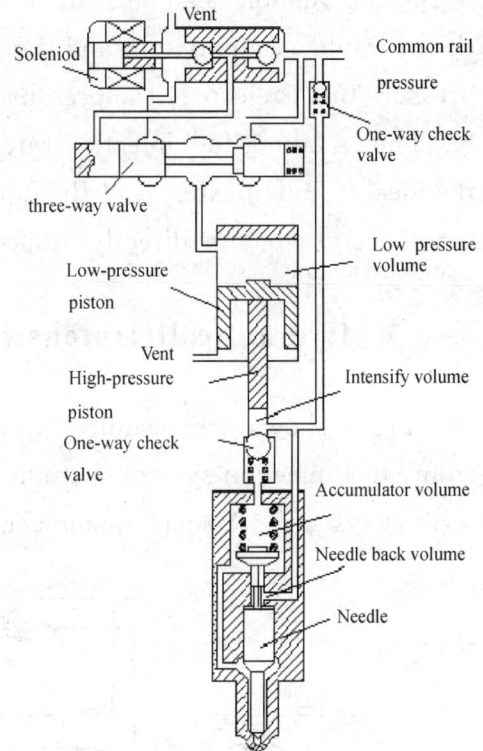

Fig. 1 Accumulator-type injector

piston. The solenoid valve operates under the pressure of 100 bar and then acts on the low-pressure piston which in turn intensifies the pressure in the accumulator nozzle to about 1 000 bar.

The injection process starts at the fuel supply module, in which the certain pressure fuel is generated and supplied to each unit injector via a common rail. To initiate the injection process, the electronic control module energizes the solenoid operated flow control valve. The rail pressure is then applied to the intensifier. The 100 bar common-rail pressure then can be intensified to 1 000 bar approximately. The high-pressure plunger injects the required fuel volume to the accumulator and compresses it to the desired maximum injection pressure. Needle valve doesn't lift during this period as the intensified inlet pressure is acted on the top of the needle valve system. The accumulator chamber is separated from the intensifier chamber by a one-way check valve. Injection is initiated when the solenoid is de-energized allowing the two intensifier pistons to retract, This action creates an unbalanced pressure acting on needle, witch lifts rapidly then. The check valve remains seated and fuel is ejected out of the nozzle spray holes and into the combustion chamber. In this system, the pressure is firstly stored inside the hydraulic accumulator, and then is released during the injection. As the high-pressure fuel is ejected, the pressure in the accumulator chamber drops accordingly. Thus a falling fuel injection rate is produced in this case, which is contrary to the ideal injection rate. And the injection quantity is controlled by the accumulator pressure which is directly proportional to the rail pressure.

3 Hydraulically intensified common-rail injection system

Fig. 2 shows a hydraulically intensified medium pressure common-rail electronic fuel injection system is studied in this paper[3]. It is improved from the BKM servojet system. A supply pump gets the fuel from the tank, and the high-pressure

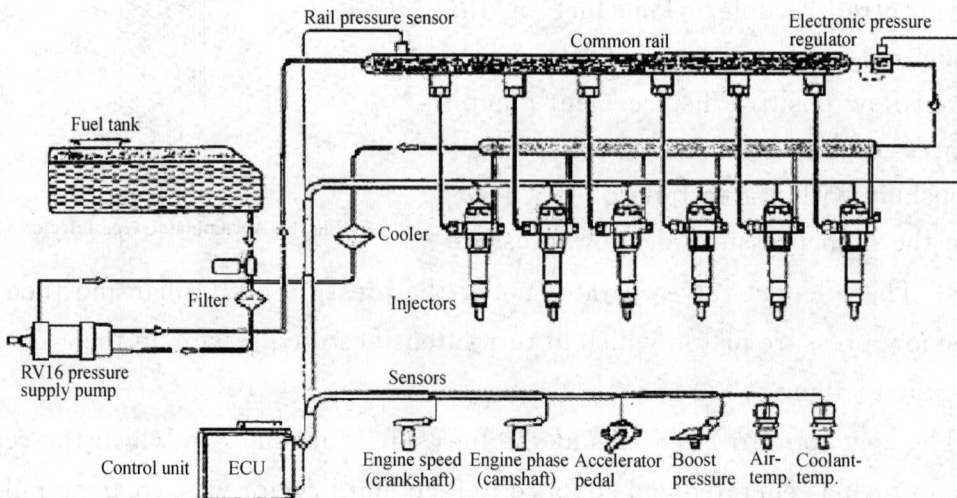

Fig. 2 System overview of the new-style injection system

pump delivers the fuel to the common-rail. The fuel volume in the rail dampens the oscillations caused by the high-pressure pump and the injection process. A rail pressure sensor measures the pressure in the rail and provides this value to the electronic control unit (ECU). The ECU compares the actual value of the rail pressure with the required value determined by engine speed and injection quantity. The ECU controls the fuel pressure delivered by the pump in order to achieve minimum pressure deviation in the rail. This new-style injection system is much more complicated than the accumulator-type and is different in the pressure generation and the related injection process. In the new type, the high-pressure generation and injection occur simultaneously, producing a near square injection rate profile. For the pressure is intensified in the injector, the high-pressure rail and high-pressure volume decrease greatly. In turn, reduced high-pressure volume improves the pressure intensifier efficiency. Though the flux of the high-pressure supply pump is high, the supply pressure is rather low. In such a low pressure, the fuel compressibility is very small, and the high-pressure pump has a very high volume efficiency.

The merits of the accumulator type injection system are taken and the shortcomings are overcome, then pilot injection and ideal fuel injection rate are realized on the basis of improving the injector structure. Fig. 3 shows the structure of the hydraulically intensified injector. The one-way check valve under the pressure intensifier is removed and a precise passage is added to the plunger bushing, which releases high pressure fuel to the needle back volume for a certain period during the injection, generating a lower pressure than in the intensifier chamber. Thus the needle valve lifts rapidly and pilot injection is realized. The length of the passage determines the injection timing and injection quantity of the pilot

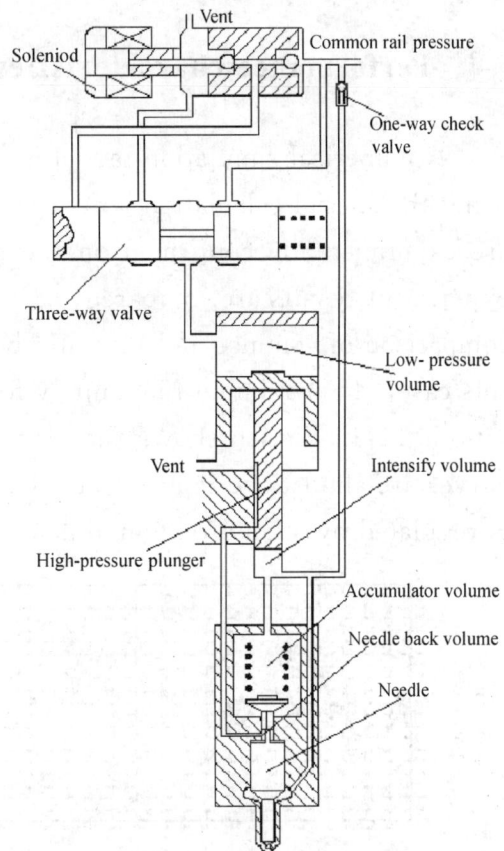

Fig. 3　Hydraulically intensified injector

injection.

In the accumulator-type injection system, the pressure is firstly stored and then released during injection as a subsequent event. In the intensifier-type, by improving the response velocity and precision of the solenoid valve, together with increasing the common-rail pressure and getting the appropriate compression ratio, the system can realize the generation and injection of much higher pressure simultaneously. With the common-rail pressure and compression ratio increased, the pressure in the intensifier chamber can be increased even during the injection process, and thus persistent injection is possible. The nozzle is charged and discharged by the opening and closing of the solenoid valve.

In the new-style injection system, the changes of pressure intensification and injection process make it possible to control injection quantity by changing common-rail pressure and the pulse length determined by the ECU. It overcomes the deadly shortcomings of accumulator-type injection system and realizes high pressure injection even for a small fuel injection quantity.

4 Performance characteristics of the new-style injection system

A numerical simulation model is established and the calculation is performed using the special calculation software of Austria AVL Co. The calculated results are experimentally confirmed on the test-bed[4]. The calculated results and the experiment results are compared, as shown in Fig. 4 and Fig. 5. Fig. 4a shows the comparison of common-rail pressure between the calculated and the measured. In this case, the maximum fuel supply for the injector is 80 mm³. The compression ratio is 14:1. Fig. 4b shows the pressure curves measured along the rail. These curves are almost overlapped. In this system, the pressure in the supply pump is regulated by a hand-regulated discharge valve. The flow in the supply pump is

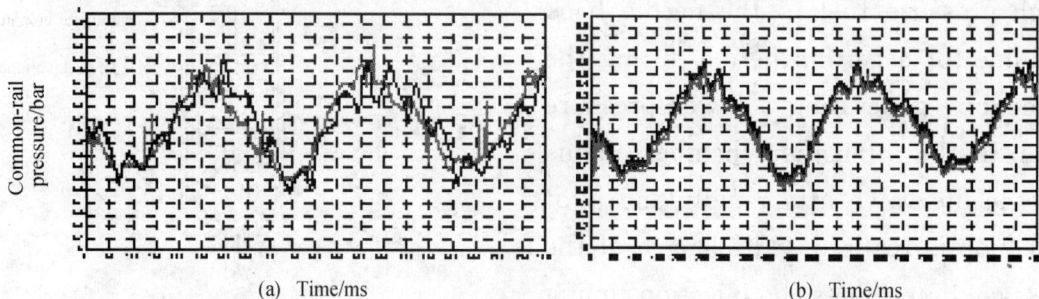

(a) Time/ms　　　　　　(b) Time/ms

Fig. 4 Calculated results of supply pump and pressure regulator

measured by measuring cup of the time-measuring system on the test-bed. The BOSH injection rate instrument transfers the injection rate to the digital wave display and prints it out.

Fig. 5 shows the com-

Fig. 5 Comparison of calculated and measured injection rate

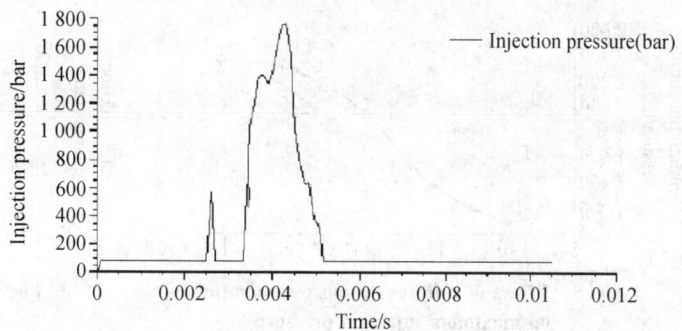

parison of calculated and measured injection rate. It can be seen that the calculated results and measured results are quite close.

With the common-rail pressure of 100 bar, a given pulse length from the ECU and the compression ratio of 18 : 1, Fig. 6 shows the injection pressure. It realizes pilot injection. The injection rate approaches rather to the ideal one. The maximum injection pressure reaches 1 700 bar. With the increasing of common-rail pressure and compression ratio, the injection system can produce even higher pressure.

With the compression ratio of 18 : 1 and the effective intensifier chamber vol-

Fig. 6 Calculated injection pressure

ume of 440 mm^3, the injection quantity curves under the different pulse length and varied common-rail pressure are shown in Fig. 7. The most sensitive pulse length to the injection quantity is from 3 ms to 7 ms. This system realizes exclusively flexible control of injection pressure and quantity. When the common-rail pressure required is very high, the pulse length can be adjusted for the certain injection quantity.

Fig. 8 shows the influence of varied compression ratio on the maximum injection pressure. In a certain area, the injection pressure increases with the compression ratio. Coupled with the appropriate high-pressure volume, the regulating area of injection quantity can be raised greatly. In this system, the high-pressure generation and injection events occur simultaneously, thus the increase of injection pressure is not proportionate to the increase of compression ratio when the

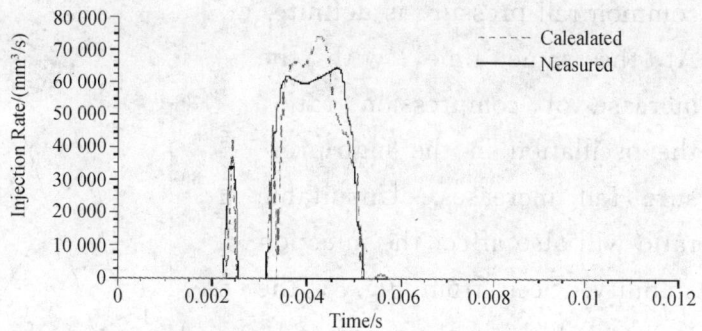

375

common-rail pressure is definite. At the same time, with the increase of compression ratio, the oscillation in the high-pressure rail increases. Unsuitable ratio will also affect the injection stability. Seen from Fig. 8, the ratio should be in the range of 10 to 20.

Fig. 9 shows the influence of the section area of the precise passage on the pilot injection quantity. Reasonable selection of the section area can obtain the best pilot fuel injection.

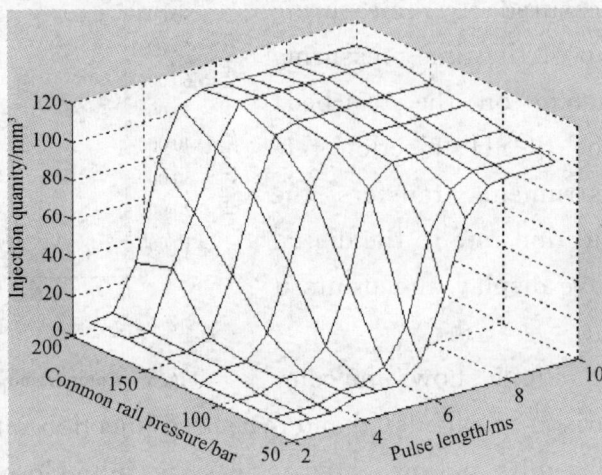

Fig. 7 Fuel supply characteristics of the system

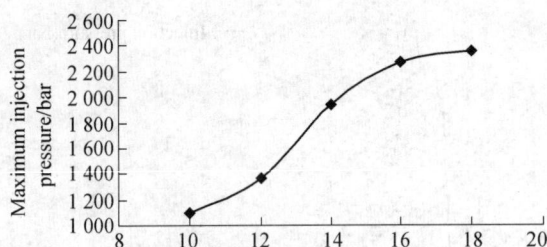

Fig. 8 Effect of different compression ratio on maximum injection pressure

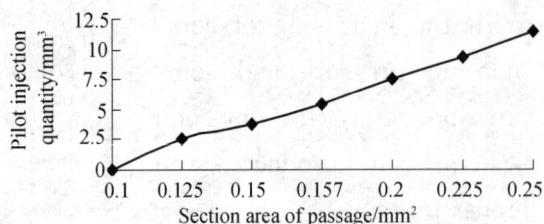

Fig. 9 Effect of section area of passage on pilot injection quantity

The numerical simulations show that the new-style injection system has the superiority over the accumulator type injection system.

Conclusions

The new-style injection system has the following advantages:

(1) Fuel injection quantity can be regulated by the pressure and the pulse length, and it would be possible to fully optimize the injection pressure.

(2) The injection rate shape is much closer to that of the ideal one, and the pilot injection can be realized.

(3) High injection pressure can be freely selected.

(4) With the improvement of the injector structure, higher working reliability can be obtained and higher requirements of future diesel engines can be met.

References

[1] Wu Jian, et al. Three-dimensional numerical simulation of multi-branch pipe in common rail system of diesel engine[J]. *Chinese Society for Internal Combustion Engines*, 2003, 5(20).

[2] John Beck N, et al. Direct digital control of electronic unit injectors[J]. *SAE Paper*, 840273, 1984.

[3] Hu Linfeng, et al. Design and investigation on new type of hydraulic intensified common-rail injection system[D]. Ph. D Dissertation.

[4] Hu Linfeng, et al. Simulation and analysis of pressure-intensified CR injection system [J]. *Chinese Society for Internal Combustion Engines*, 2002, 3(20).

(From: *The 6th Annul academia congress of CSICE*, 2003)

新型增压式柴油机共轨喷射系统的研发

[摘要]　作者研究开发的增压式共轨喷射系统,是一种在蓄压式共轨系统基础上改进而成的新型共轨系统。实验和研究结果表明,该系统可完全实现压力、油量和定时的独立控制,燃油喷射率较理想,达到了产品研发的目标,可满足不同柴油机排放和性能要求。

增压式共轨喷射系统的模拟计算和试验分析

胡林峰,吴　建,李德桃,梁凤标,张建新

[摘要]　通过对原蓄压式共轨喷射系统基本工作原理的重大改进,新型增压式共轨喷射系统实现了真正意义上的压力和时间的独立控制。同时,利用 Hydsim 软件模块化模拟技术,对该系统进行了较为全面的模拟分析。计算与试验结果表明,喷油器的结构参数对液力流动性能的影响很大。合理地匹配控制油道和喷孔面积、针阀升程和开启压力以及增压比等参数,可以进一步扩大系统的可控范围,为喷油性能的优化提供更好的前提条件。

符 号 说 明

m——运动件质量；

μ——液力推力系数(考虑滑动面之间的运动摩擦)；

p_1,p_2——运动件两端的液压力；

x_0——弹簧的预压行程；

K——流体体积弹性模量；

u——x 向流体的流动速度；

f——摩擦系数；

$\Delta p_{out}(t)$——t 时刻油管出口处的压力扰动值；

$\Delta p_{in}(t)$——t 时刻油管入口处的压力扰动值；

$\overline{p_{m,n}}$,$p_{m,n}$,$\overline{u_{m,n}}$,$u_{m,n}$——某点处的压力和速度的预估值和修正值,m 和 n 表示位置和时间的计算点；

x——x 轴向位移；

A——液力作用面积；

L——油管长度；

k——弹簧常数；

D_c——阻尼系数；

V——控制体或容腔体积；

a——声速；

D——油管内径或当量液力直径；

ρ_M——某点上的流体密度；

Δ_t,Δ_x——时间和位移的计算步长。

　　共轨式燃油喷射系统具有喷射压力高、控制柔性好和精度高等特点,是公认的下一代新型柴油机最有潜力的喷射系统[1,2]。在共轨喷射系统中,由于机械和液力参数对控制性能参数的可控范围和控制特性有着决定性的影响,所以,必须对系统参数的匹配进行模拟和优化。

　　本文所讨论的增压式共轨喷射系统,是在原蓄压式共轨喷射系统[3]基础上改进而来的新型共轨喷射系统,理论上可灵活实现多次喷射过程,但工作过程复杂。作者旨在通过该系统的液力模拟以及与试验对比分析,从液力流动的角度去理解和掌握共轨喷射系统中各种液力元件的特性参数对系统喷射性能的影响,消除高压喷射过程中可能出现的不稳定流动现象,最终通过系统主要参数的合理匹配和优化,满足不同发动机性能的要求。

1　增压式共轨喷射系统的组成及模拟方法

　　图 1 为增压式共轨喷射系统结构示意图,基本参数如表 1 所示。系统主要由高压供油泵、压力调节器、共轨部件和喷油器总成等组成。其中,高压供油泵是一种轴向柱塞泵,共有 7 个供油部件,安装在一个由凸轮轴驱动的转子中。柱塞的行程由斜盘的倾斜角决定。进油和供

油过程由与转子精密配合的配流盘控制,当某个压油部件正处于回行时其油道就会与配流盘的进油道相通开始进油过程,并一直进行到这个油道被配流盘完全盖上为止。当转子转动到该泵油部件开始上行时,其油道则与配流盘的出油道相接,即开始出油过程。共轨的压力通过一个比例电磁调压阀来控制。图1中所示的喷油器是一种全新的增压式喷油器。共轨压力作用于上部的控制电磁阀驱动增压活塞和柱塞向下运动,压缩柱塞下部的燃油使之压力提高,当柱塞上的回油道与针阀上部的油道接通时,针阀顶部的压力迅速下降,在针阀下部液压力的作用下,克服上部调压弹簧力使针阀开启,开始喷油。喷油过程一直进行到电磁阀开始换向后终止。如果柱塞上的回油道做成几段的结构,便可进行多次喷射,且其间隔时间长度可以通过调节共轨压力在一定范围内进行调整。为了对这些液力及机械部件的工作原理准确地进行模拟,计算程序中将这些部件按要求拆分为下列几种机械电子和液力单元:喷油器针阀阀体、针阀、流量孔组、泵油元件、阀组、执行元件组、容积、油路、刚体质量组、凸轮组和边界组等等,将这些模拟元件按照系统的组成以一定的方式互相连接起来(如液力连接或机械连接方式),就构成了系统的液力计算模型。图2为高压供油泵的液力模拟图。

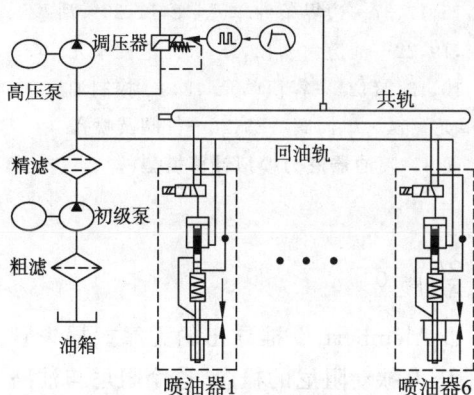

图 1　共轨喷射系统结构示意图

表 1　喷油系统的基本参数

参　　数	指　　标
供油泵柱塞直径×升程	ϕ 12 mm×10 mm
喷油器增压比	10~20
最大喷油量	120 mm³/str
喷油嘴开启压力	40~45 MPa
最高共轨压力	30 MPa
最高喷射压力	250 MPa
喷油嘴喷孔数及直径	5~6 个,0.18~0.25 mm
喷油嘴型号	S/P 系列

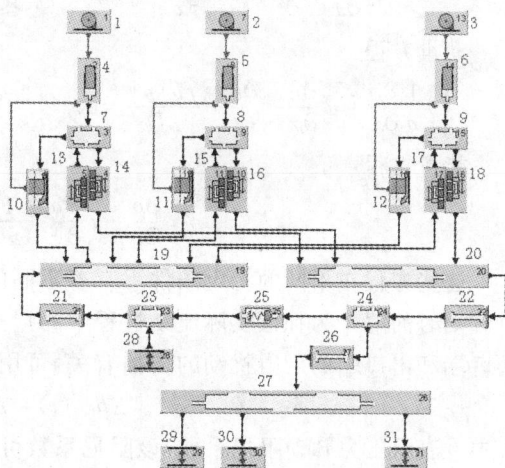

1,2,3—凸轮组　4,5,6—柱塞组

7,8,9,19,20,23,24,27—容积单元

10,11,12—泄漏单元　13,14,15,16,17,18—控制

油道单元　21,22,26—油道　25—溢流阀

28,29,30—边界条件单元

图 2　高压供油泵 Hydsim 计算模型图

共轨系统的压力调节阀是一种电子控制的比例电磁阀,通过调节比例电磁阀脉宽即占空比来调节压力值,占空比为 1 时压力最高,占空比为 0 时压力最低。模拟计算时,压力调节阀可用设定某一压力值的溢流阀来代替。

为了简化计算,将喷油器上部的二位三通滑阀用一个压力边界来代替,如图3所示。

2　模型及解法

本模拟计算采用了 AVL 公司的一维的 Hydsim 模拟计算软件。基本控制方程包括运动件的运动方程和流体的流动控制方程。本系统中的运动方程主要是针阀的运动方程：

$$m \frac{\mathrm{d}^2 x}{\mathrm{d}t^2} = \mu A(p_1 - p_2) - k(x + x_0) - D_c \frac{\mathrm{d}x}{\mathrm{d}t}$$

容积单元中的连续性方程为

$$\frac{\mathrm{d}p}{K} = \frac{\mathrm{d}m}{m} - \frac{\mathrm{d}V}{V}$$

高压油道或油管中一维可压缩流体的流动控制方程[4]为连续性方程：

$$u \frac{\partial \rho}{\partial x} + \frac{\partial \rho}{\partial t} + \rho a^2 \frac{\partial u}{\partial x} = 0$$

动量方程：

$$\frac{1}{\rho} \frac{\partial \rho}{\partial x} + u \frac{\partial u}{\partial x} + \frac{\partial u}{\partial t} + \frac{fu|u|}{2D} = 0$$

能量方程：

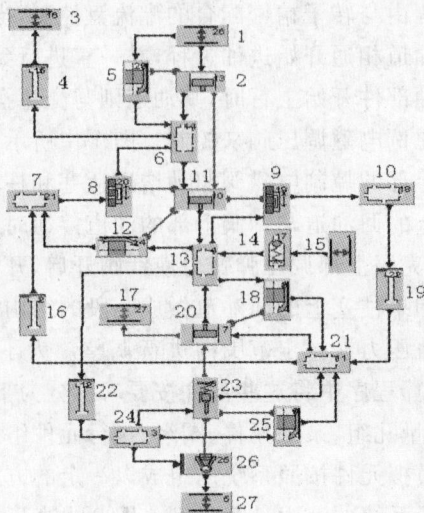

1,3,15,17,27—边界条件单元　2,11,20—活塞
4,16,19,22—油道　5,12,18,25—泄漏单元
6,7,10,13,21,24—容积单元　8,9—控制油道
单元　14—单向阀　23,26—针阀及喷孔

图 3　喷油器液力模拟计算模型

$$\frac{\partial \rho}{\partial t} + u \frac{\partial \rho}{\partial x} - \frac{1}{a^2} \left(\frac{\partial p}{\partial t} + u \frac{\partial p}{\partial x} \right) = 0$$

这些方程在不同的特定条件下有着不同的解法。如 Alembert D 推导出的无摩擦损失管内流动时的解。为补偿实际流动过程中的损失，提出了压力脉动阻尼的概念，这个阻尼与油路入口端和出口端的压力扰动的传播有关，可用一个经验公式来表示：

$$\Delta p_{\mathrm{out}}(t) = \Delta p_{\mathrm{in}}(t - \Delta t) \mathrm{e}^{\beta L}$$

式中 β 表示压力脉动阻尼系数，该阻尼系数可用下面的公式[5]来估计：

$$\beta = -\frac{6.6}{D} \sqrt{\frac{\nu}{aL}}$$

式中 ν 为流体的运动粘度。从物理学的观点来看，这个摩擦模型不是很准确，但在很多情况下仍具有足够的计算精度。

在拉普拉斯[6]的解法（又称 Kroller 法）中则考虑了摩擦力的平均值 R，

$$R = \frac{1}{A} \iint \nu \left(\frac{\partial^2 u}{\partial y^2} + \frac{\partial^2 u}{\partial z^2} \right) \mathrm{d}y \mathrm{d}z$$

在 Melcher 的著作[7]中给出了方程的解。当假定平均摩擦力仅取决于管壁上的剪切力时，Melcher 推导出了下面形式的摩擦函数：

$$R = \int_0^t \frac{\nu}{A} \frac{\partial u}{\partial t} D(t - \tau) \mathrm{d}\tau$$

式中 $D(t)$ 为阻尼函数，对于圆截面管可表示为如下的形式：

$$D(t) = -4\pi \sum_1^\infty \mathrm{e}^{\omega_n^2 \nu t / r^2}$$

式中 ω_n^2 为 0 阶贝塞尔函数的根,本程序中取前 20 个根;r 是油管的截面半径。

此外还有特征线法和 Godunov 法等求解方法。这些方法在理论上均是处理纯液体的流动特征。当在系统流动中出现气穴现象即二相流动时,其计算结果便会有较大的误差。因为流体的 ρ, K, a 等参数会出现较大的变化而降低计算精度。本程序中模拟油管中小空气分量的气液二相流动方程的求解用 MacCormack 线法。在这种方法中应用气泡的动态气穴模型并结合 FCT(Flux Corrected Transport)法进行求解。

MacCormack 线法分预估和修正两步进行计算。预估应用向前差分法,而修正计算则应用向后差分法,也可反过来应用。为了获得更好的精度,一般将这两个过程结合起来使用。在发生液力冲击的情况下,如果差分方向与解的不连续性传播方向相同,MacCormack 线法将获得更好的不连续解。如果预估应用向前差分法,修正计算用向后差分法,则预估的流动速度和压力的解分别为

$$\bar{u}_{m,n+1} = u_{m,n} - \frac{\Delta t}{\Delta x}\left[(u_{m+1,n}u_{m+1,n} - u_{m,n}u_{m,n}) + \left(\frac{p_{m+1,n}}{\rho_{M,m+1,n}} - \frac{p_{m,n}}{\rho_{M,m,n}}\right)\right] - \Delta t R_f$$

$$\bar{p}_{m,n+1} = p_{m,n} - \frac{\Delta t}{\Delta x}\left[(u_{m+1,n}p_{m+1,n} - u_{m,n}p_{m,n}) + (\rho_{M,m+1,n}a_{m+1,n}^2 u_{m+1,n} - \rho_{M,m,n}a_{m,n}^2 u_{m,n})\right]$$

修正的速度和压力计算公式为

$$u_{m,n+1} = \frac{1}{2}(u_{m,n} + \bar{u}_{m,n+1}) - \frac{1}{2}\frac{\Delta t}{\Delta x}\left[(u_{m,n}\bar{u}_{m,n+1} - u_{m-1,n}\bar{u}_{m-1,n+1}) + \right.$$

$$\left.\left(\frac{\bar{p}_{m,n+1}}{\rho_{M,m,n}} - \frac{\bar{p}_{m-1,n-1}}{\rho_{M,m-1,n}}\right)\right] - \frac{\Delta t}{2}\bar{R}_f$$

$$p_{m,n+1} = \frac{1}{2}(p_{m,n} + \bar{p}_{m,n+1}) - \frac{1}{2}\frac{\Delta t}{\Delta x}\left[(u_{m,n}\bar{p}_{m,n+1} - u_{m-1,n}\bar{p}_{m-1,n+1}) + \right.$$

$$\left.(\rho_{M,m+1,n}a_{m+1,n}^2 u_{m+1,n} - \rho_{M,m,n}a_{m,n}^2 u_{m,n})\right]$$

在计算中,计算步长用 Courant 等人定义的稳定性法则确定,即

$$n_x = \frac{L}{2a\Delta t}$$

式中 a 为声速;Δt 为时间步长;L 为油管长度。

$$\Delta t \leqslant \frac{\Delta x}{u+a} = \frac{L}{n_x(u+a)},$$

式中 $\Delta x = \dfrac{L}{n_x}$;$u$ 为计算单元中的流动速度。

3 计算结果及讨论

为了对模型的正确性以及计算的精度进行分析比较,作者建立了如图 4 所示的试验系统。在该系统中,供油泵的供油压力通过一个手动的溢流阀调节,压力值经压力变送器由专用仪器显示。供油泵的流量借助于试验台的计时系统通过量杯测量。而喷油器的喷油规律则用 BOSCH 的喷油规律仪将信号和逻辑控制波形一起传送给数字示波器显示并可由打印机打印。图 5a 为共轨压力的计算和测量值比较。此时,喷油器的最大供油量为 80 mm³,喷油器的

增压比为 14。由于共轨压力比较低，共轨中的燃油仍可作为不可压缩的流体进行处理。此外，尽管共轨的容积较大，长度较长，其基本特征仍可作为统一的容积处理。图 5b 是在共轨的不同位置上测得的压力变化曲线，它们几乎全部重叠在一起。

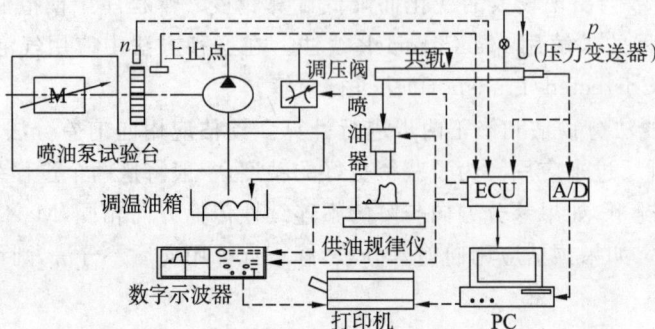

图 4　增压式共轨喷射系统性能试验系统

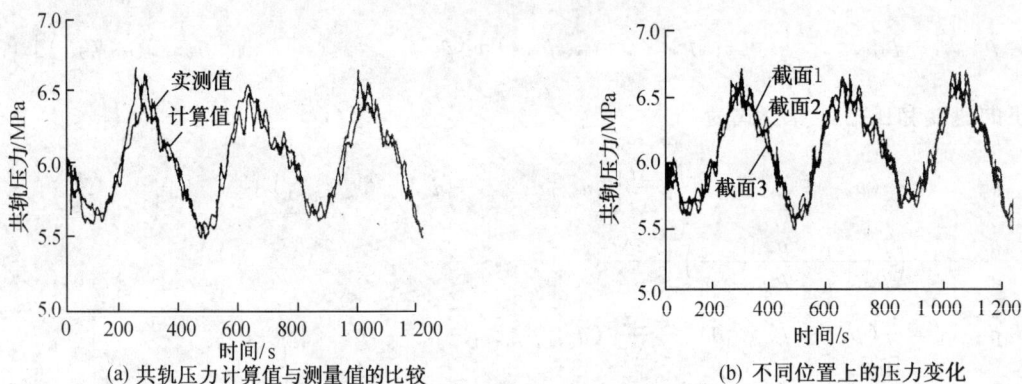

(a) 共轨压力计算值与测量值的比较

(b) 不同位置上的压力变化

图 5　共轨压力实测及模拟计算结果

　　图 6 是喷油器压力室内的压力和喷油规律的变化曲线，以及与实测喷油规律的比较曲线。此时喷油器的增压比为 18，最高计算共轨压力为 25 MPa，高压腔容积在 500 mm³ 以内。从计算结果与实测值的比较可以看到，喷油器结构及控制方法改进后，主喷射的喷射压力达到了 180 MPa 以上，预喷射时的压力则很好地维持在 60 MPa 左右。从喷油规律的曲线图形看，喷油规律的形状接近矩形，但喷油规律的初期相对比较平缓，符合性能和排放对喷油规律的要求。预测值的曲线变化趋势和绝对值与实测值非常吻合，可以满足工程计算的精度要求。

　　图 7 是增压比为 18，有效蓄压室容积为 420 mm³ 的喷油器在不同的电磁阀通电时间长度和共轨压力的条件下的供油量特性曲线。其中，电磁阀的通电时间对供油量最敏感的区域在 3～7 ms时间段。计算和试验结果表明，本系统在原来蓄压式共轨喷油系统的基础上，实现了真正意义上的喷射压力和供油量的独立控制。当发动机的转速和负荷较低时，所需要的喷油量相应较小，可以通过调节时间来改变供油量，仍然能够保证具有很高的喷射压力，以改善发动机的低速经济性和排放性能。

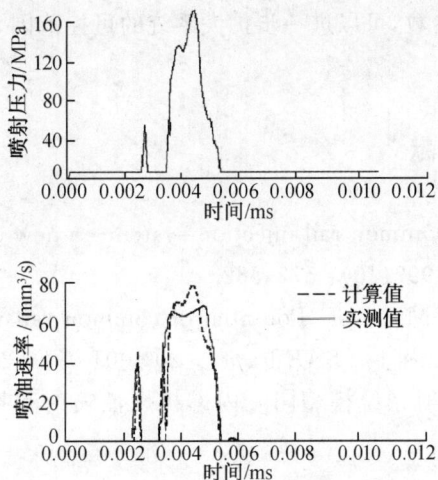

图 6　喷射压力和喷油规律曲线

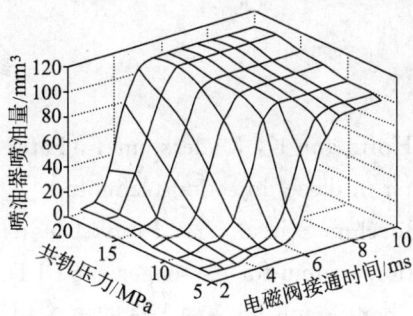

图 7　喷射系统供油特性曲线

图 8 表示不同增压比对最高喷射压力的影响。在一定范围内,增压比的提高会大大提高系统的喷射压力,与一定的高压腔容积相配合,可以使共轨系统的可调油量范围大大增加。但是,增压比的增加使增压液体流量迅速增加,这会造成共轨压力波动的加剧。对于整个喷油系统而言,若供油泵的能力贮备系数不大,增压比过大,喷射压力不仅不能有效地提高,还会影响系统喷油的稳定性。图 9 示出了控制油道截面的大小对预喷量的影响,合理地选择这个面积,在较大的共轨压力范围内可以获得最佳的预喷量。除此之外,喷油器的开启压力、喷油嘴的结构、喷孔的大小以及针阀升程的大小都会影响系统的喷油特性和控制特性。在系统油道中,如局部压差过大,还会造成流动不稳定性的加剧,产生空穴等现象。

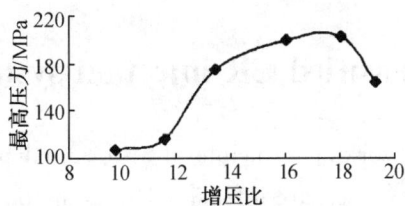

图 8　不同增压比对最高喷射压力的影响

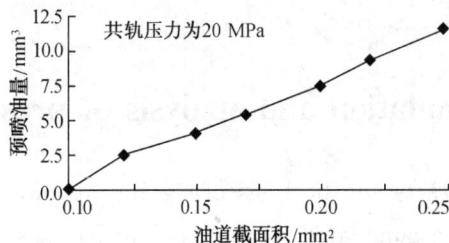

图 9　控制油道截面积对预喷量的影响

4　结 束 语

作者在对蓄压式共轨喷射系统多年试验研究的基础上,对原蓄压式共轨喷射系统的基本工作原理作了重大的改进。试验及计算结果表明,改进后的喷油系统实现了真正意义上压力和时间的独立控制。同时,作者利用 Hydsim 软件模拟技术,对该喷射系统进行了较为全面的模拟分析。计算结果与试验结果的比较表明,喷油系统的改进达到了预期的效果,模拟计算的精度也完全满足设计要求。

计算及试验结果表明,喷油器的结构参数对液力流动性能的影响很大,合理地匹配控制油道和喷孔面积、针阀升程、开启压力以及增压比等参数,可以进一步扩大系统的可控范围,为喷油性能的优化提供更好的前提条件。

参 考 文 献

[1] Hoffmann K，Maderstein T，Peter A. The common rail injection system—a new chapter in diesel injection technology[J]. *MTZ*,1997(10)：572-582.

[2] Joachim Schommers，Frank Duvinage，Stotz M，et al. Potential of common rail injection system for passenger car DI Diesel Engines[J]. *SAE Paper*, 2000-01-0944,2000.

[3] Chen Simon K. 一种先进的电控液压燃油喷射系统模型[J]. 内燃机燃油喷射和控制，1996(2)：27-32.

[4] Zhu Yuhua，Reitz Rolf D. Modeling fuel system performance and its effect on spray characteristics[J]. *SAE Paper*, 2000-01-1253,2000.

[5] Romig B E，Strunk R D，Weinert M S. Performance comparison of unit injector and pump-line-nozzle injection system[J]. *SAE Paper*, 840274,1984.

[6] Kroller M. *Efficient Computation of a Mathematical Model for the Damping of Pressure Waves in Tubes of Circular Form（Numerical Methods for Partial Differential Equations）* [M]. John Wiley & Sons Inc, 1995.

[7] Melcher K. Ein Reibungsmodell zur Berechnung von Instationaeren Stroemungen in Rhorleitungen an Brennkraftmaschinen[R]. Bosch Tech Berichte 4，1974.

(本文原载于《内燃机学报》2002 年第 3 期)

Simulation and analysis of pressure-intensified CR injection system

Abstract：On the basis of significant improvement of working principle of an original pressure-accumulated common rail injection system，a new pressure-intensified in common rail injection system can realize independent control of injection pressure and duration. At the same time，by the method of modularizing simulating computation technique of HYDSM program，this system has been analyzed systemically. The results have demonstrated that the configuration of injector have significant effects on the injection characteristics. After rational coupling control of line section and injection hole section，nozzle lift，opening pressure of injector and intensified ratio，the controllable range of injection parameters，which are the precondition for optimization of injection performance，can be further improved.

蓄压式电控喷油器燃油喷射过程的
模拟计算分析

吴　建，胡林峰，李德桃，张建新

[摘要]　建立了蓄压式喷油器的计算模型,并对其喷射过程进行了模拟计算,分析了共轨压力、增压活塞面积比、针阀弹簧预紧力、蓄压室容积以及电磁阀通电时间等因素对喷射过程的影响。所得结果可以为蓄压式喷油器的设计开发提供有价值的参考依据。

为了适应日益严格的排放法规的要求,减少柴油机的有害排放,要求提高柴油机燃油喷射压力并精确控制喷射时间,而机械式喷射系统已很难满足要求,必须采用时间控制式的电控喷射系统。

共轨电控系统就是一种具有代表性的压力时间控制式喷射系统[1]。在本文所研究的蓄压式电控喷射系统中,电控喷油器是实现高压喷射和喷油量控制的关键部件。它通过电磁阀控制增压活塞上腔回流卸压的时间来控制喷油始点,用压力控制方式将经过压缩获得的高压燃油储存在蓄压室内,当喷射开始时,蓄压室内的高压燃油从喷孔喷出。通过对蓄压式电控喷射系统进行模拟计算,可以分析系统的结构参数对燃油喷射过程的影响,为蓄压式电控喷射系统的设计提供参考依据,从而缩短喷射系统的设计周期,降低开发设计成本。本文利用 AVL 公司的 Hydsim 软件对蓄压式电控喷射系统的喷射过程进行了模拟计算,并分析了主要结构参数对喷射过程的影响。

1　计算模型及基本方程

蓄压式电控喷油器结构如图 1 所示。建立计算模型时,首先根据蓄压式电控喷油器的结构将电磁阀、滑阀、增压活塞、单向阀、针阀等基本运动部件以及增压腔、蓄压室等容积单元和油道等单独进行模型化,建立相应的基本计算模型单元;然后按蓄压式电控喷油器的工作原理将各基本计算模型单元连接起来,组合成蓄压式电控喷油器的整体计算模型(如图 2 所示)。

在模拟计算中,作如下假设:

(1) 在喷油过程中,燃油温度保持不变;

(2) 燃油密度随压力变化,用经验公式计算;

(3) 不考虑各部件的弹性变形;

(4) 喷油器中的油道较短,可不考虑压力波的传播,同时忽略油道中的流动阻力损失;

(5) 仅在增压活塞和针阀处考虑径向间隙所引起的泄漏。

计算模型可用粘性流体一维不稳定流动方程来表示。

连续性方程:

$$\frac{\partial \rho}{\partial t} + u \frac{\partial \rho}{\partial x} + \rho \frac{\partial u}{\partial x} = 0$$

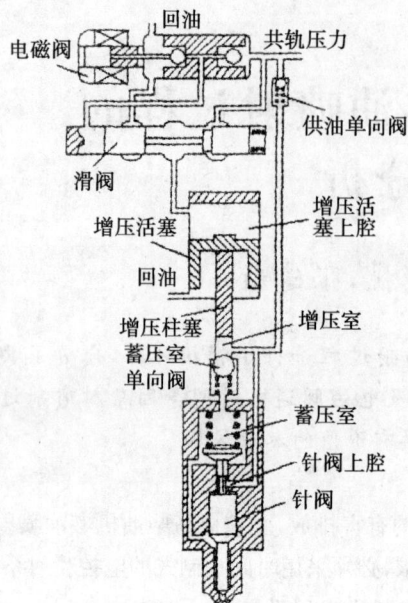

图 1　蓄压式电控喷油器结构示意

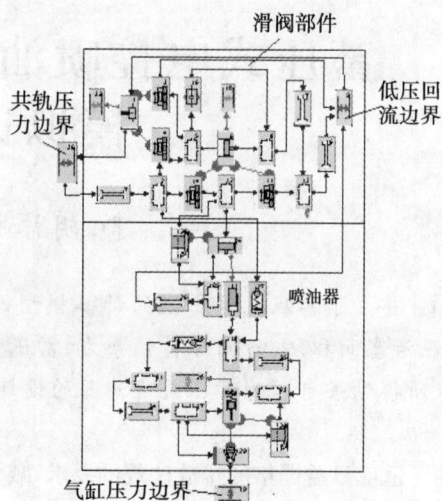

图 2　蓄压式电控喷油器的计算模型

动量方程：

$$\frac{\partial u}{\partial t}+u\frac{\partial u}{\partial x}+\frac{1}{\rho}\frac{\partial p}{\partial x}=\frac{4}{3}\mu\frac{\partial^2 u}{\partial x^2}$$

密度随压力的变化关系[2]：

$$\frac{\mathrm{d}\rho}{\mathrm{d}p}=\frac{0.69\times10^{-9}}{(1+3.23\times10^{-9}p)^2}\rho_0$$

牛顿公式：

$$m\frac{\mathrm{d}^2 x}{\mathrm{d}^2 t}=F_1$$

式中 ρ 为流体密度；ρ_0 为 820 kg/m³；u 为流体速度；p 为流体压力；μ 为流体动力粘度，取为 0.002 46 N·s/m²；F_1 为作用在运动件上的合力；m 为运动件质量。

根据基本方程，结合各基本计算模型单元的具体特点，可建立与基本计算模型单元相应的计算方程。

共轨压力边界可以采用实测的共轨压力，低压回流边界的压力取为 0.2 MPa，嘴端压力边界为 8 MPa。

各计算模型单元的初始条件根据经验确定，计算进行 2～3 个循环后，由于初始条件的偏差所引起的计算结果的误差基本可以消除。本文中的计算结果均为第 3 循环的计算结果。

2　计算结果及分析

2.1　蓄压式喷油器的喷射过程

图 3 所示为喷射过程的计算结果。计算时泵轴转速为 1 000 r/min，共轨压力为 7.5 MPa，其波动幅度为 0.8 MPa，增压活塞的面积比为 14，电磁阀通电时间为 20 ms（相当于泵轴转角

120°)。由图可见：当电磁阀有方波信号输入时，电磁阀的高压通道导通，滑阀在来自共轨的燃油压力作用下产生位移。当滑阀达到一定位移时，其控制的高压端通道开启，来自共轨的燃油进入活塞上部的低压腔，低压腔的压力逐渐升高到共轨压力。在燃油压力的作用下，活塞发生位移。活塞开始移动的时刻要滞后于滑阀，滞后的时间取决于滑阀的结构尺寸。随着活塞的移动，增压腔、蓄压室的压力上升。当活塞两端的作用力平衡时，活塞停止移动，这时低压腔、增压腔和蓄压室的压力稳定在一定的数值上。由于共轨压力的波动，低压腔的压力也会发生相应的波动，引起活塞上下移动，导致增压腔的压力也发生波动。由于蓄压室和增压腔之间有单向球阀，蓄压室的压力保持稳定。当方波结束时，电磁阀的低压通道导通，滑阀在弹簧力的作用下反向移动，使其控制的高压端通道关闭，低压通道开启，活塞上部低压腔的燃油迅速

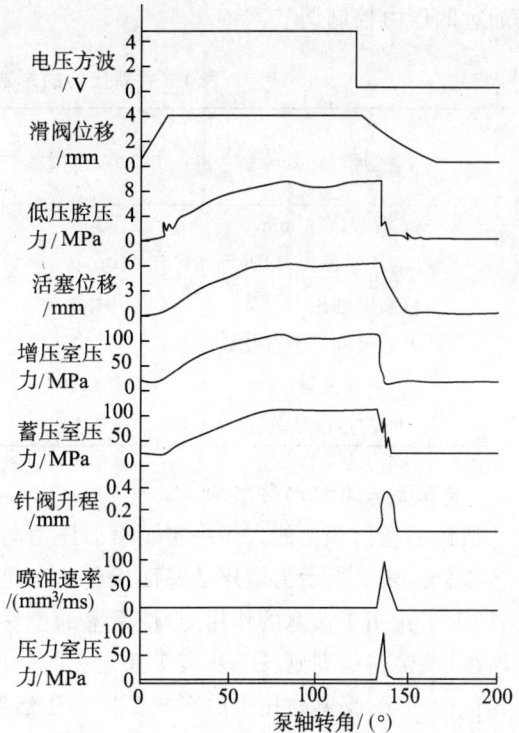

图 3 蓄压式喷油器喷射过程的计算结果

流出，压力急剧下降。低压腔的燃油流出的时刻也由于滑阀结构的原因而滞后于滑阀。由于低压腔的压力迅速下降，使活塞迅速上升，增压腔的压力也迅速下降，针阀背压也同时下降。由于针阀背压的迅速下降，油槽(蓄压室的一部分)压力和针阀背压的压差急剧增大，远大于针阀弹簧的预紧力，因而使针阀迅速抬起，喷油开始。随着喷油过程的进行，蓄压室压力下降，喷油压力降低，喷油速率降低。当油槽压力和针阀背压的压差降到小于针阀弹簧的作用力时，在针阀弹簧的作用下，针阀回落，直至针阀关闭，喷油结束。

从计算结果来看，在所给条件下，蓄压室压力可以达到 105 MPa，在喷油过程中，油嘴压力室压力最高也可以达到 92 MPa。但随着燃油的喷出，蓄压室压力下降较快，使压力室压力也降低较快，因此，平均有效喷油压力只有 35 MPa，喷油速率也呈前高后低的特点，这与蓄压室喷油器的结构有关，应设法加以改进。

2.2 共轨压力的影响

表 1 所示为共轨压力对蓄压室压力、喷油压力及循环喷油量的影响；图 4 为不同的共轨压力对喷油规律的影响。从表 1 和图 4 中可以看出：在一定的增压比下，随着共轨压力的提高，增压活塞的位移量加大，对蓄压室内的燃油压缩加强，使蓄压室压力提高，喷油压力相应提高；喷油开始的时刻基本不变，喷油结束时刻后延，喷油持续期延长，喷油量增加。可见喷油量与共轨压力关系密切，可以通过调节共轨压力来调节喷油量，实现对

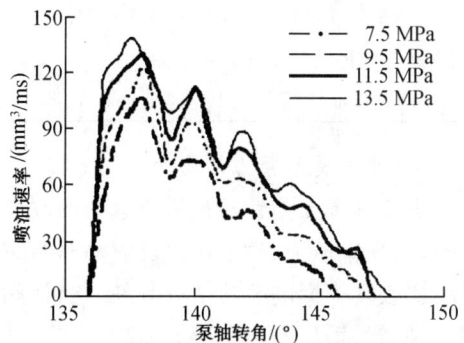

图 4 共轨压力对喷油规律的影响

喷油量的压力控制调节。

表 1　共轨压力的影响(增压活塞面积比 14)

指　标	共轨压力/MPa			
	7.5	9.5	11.5	13.5
活塞位移/mm	4.95	6.43	7.51	8.00
蓄压室压力/MPa	105.9	133.5	161.6	175.8
最高喷油压力/MPa	91.67	123.00	156.50	170.10
平均喷油压力/MPa	33.0	46.4	55.6	57.8
循环喷油量/mm³	78.46	101.00	122.00	128.28
喷油持续期/(°)	9.00	10.25	10.25	11.25

2.3　增压活塞面积比的影响

　　增压活塞面积比的大小决定了喷油压力与共轨压力的比值,是蓄压式喷油器的一个重要结构参数。表 2 所示为增压活塞面积比的变化对喷油过程的影响。随着增压活塞面积比的增大,增大了作用于活塞的作用力,使活塞的位移量增大,蓄压室压力、喷油压力相应提高,喷油量增加,喷油持续期延长。从表 1 和表 2 的对比来看,共轨压力和增压活塞面积比对喷油过程有相似的影响,但共轨压力的影响要比增压活塞面积比的影响更显著。在计算中,活塞的最大行程为 8 mm,但在 7.5 MPa 的共轨压力下,即使是增压活塞面积比为 18 时,也未达到活塞的最大行程。可见,活塞对蓄压室燃油的压缩还有潜力。因此,在确定增压活塞的面积比时,应考虑共轨压力的高低,应尽可能使活塞达到其最大行程,充分发挥活塞对燃油的压缩能力,提高喷油压力。

表 2　增压活塞面积比的影响(共轨压力 7.5 MPa)

指　标	面　积　比				
	10	12	14	16	18
活塞位移/mm	3.45	4.21	4.95	5.58	6.18
蓄压室压力/MPa	77.35	92.90	105.60	120.10	133.40
最高喷油压力/MPa	61.80	77.62	91.67	106.70	118.90
平均喷油压力/MPa	24.57	29.64	33.00	37.65	43.90
循环喷油量/mm³	52.74	65.68	78.46	88.87	98.46
喷油持续期/(°)	7.50	8.50	9.25	10.00	10.50

2.4　针阀弹簧预紧力的影响

　　表 3 和图 5 为针阀弹簧预紧力的改变对喷油过程影响的计算结果。当针阀弹簧预紧力改变时,蓄压室压力、喷油压力及喷油开始时刻基本不变。针阀弹簧预紧力的改变对喷油过程影响主要表现为影响喷油结束时刻。当针阀弹簧预紧力提高时,针阀关闭时刻提前,喷油持续期缩短,同时,蓄压室残余压力提高。如果共轨压力有较大的波动,而针阀弹簧预紧力较小,则可能会出现不正常喷射。图 6 就是当共轨压力为 9.5 MPa,压力波动幅度为 2.4 MPa 时的计算结果。弹簧预紧力从 650 N 到 1 250 N 变化时均未出现不正常喷射,而当弹簧预紧力降低为

450 N 时,在正常喷射前,针阀就开启喷油。其原因是:当共轨压力有较大波动时,活塞上部的低压腔压力也要随之波动,导致活塞在压缩过程中发生反向移动,使增压腔压力降低,而蓄压室压力并未降低(如图 7 所示)。当二者压差大于弹簧预紧力时,针阀就会开启,进行喷油。喷出一定数量燃油后,蓄压室压力也降低,当二者压差小于弹簧力时,针阀关闭。随着共轨压力的上升以及活塞再次开始压缩,蓄压室和增压腔的压力又开始升高,直至正常喷射。因此,要保证燃油喷射的正常进行,一方面要减小共轨压力的波动,同时也要适当提高针阀弹簧预紧力,既避免出现不正常喷射,又能使喷油结束迅速,减少低压喷射的油量。

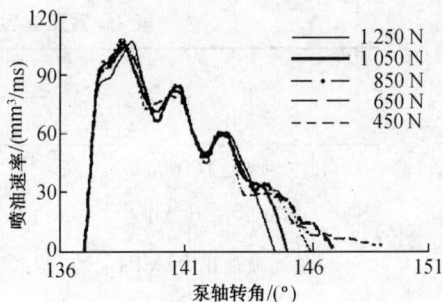

图 5　针阀弹簧预紧力对喷油规律的影响

表 3　针阀弹簧预紧力的影响(共轨压力 9.5 MPa,压力波动 2.4 MPa)

指　　标	弹簧预紧力/N				
	1 250	1 050	850	650	450
蓄压室压力/MPa	135.0	134.7	134.3	134.0	133.8
蓄压室残余压力/MPa	26.58	21.68	17.81	11.23	9.85
最高喷油压力/MPa	120.8	117.0	112.6	116.3	113.0
平均喷油压力/MPa	55.27	53.06	48.75	44.80	37.99
循环喷油量/mm³	91.66	95.54	98.78	100.20	102.80
喷油持续期/(°)	7.0	8.0	9.0	10.5	12.0

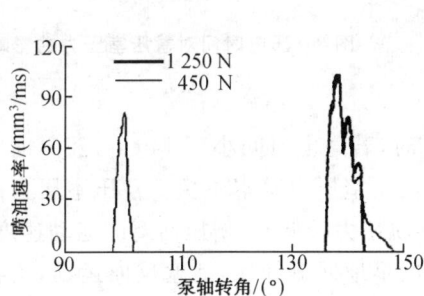

图 6　低预紧力时出现的不正常喷射

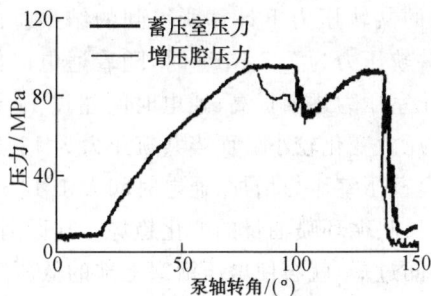

图 7　不正常喷射时的压力对比

2.5　蓄压室容积的影响

在蓄压式喷射系统的燃油喷射过程中,燃油的喷射实际上是蓄压室内高压燃油压力的释放过程,蓄压室容积的大小是影响燃油喷射过程的一个重要参数,它影响到喷油量和喷油持续期的长短。表 4 所示为蓄压室容积变化对燃油喷射过程的影响。由表 4 可见:当活塞增压面积比不变、蓄压室容积增大时,蓄压室燃油压力略有降低,喷油压力没有太大变化,而循环喷油量随之增大,喷油持续期延长。这是因为蓄压室容积增大后,在活塞压缩燃油过程中,进入蓄压室的燃油量增多,在喷油过程中,喷出的油量也随之增多,持续时间增加。

表4　蓄压室容积的影响(共轨压力9.5 MPa)

指　　标	蓄压室容积/mm³				
	900	1 000	1 100	1 200	1 300
活塞位移/mm	5.62	6.06	6.49	6.93	7.37
蓄压室压力/MPa	136.64	135.61	134.90	134.33	133.34
最高喷油压力/MPa	122.34	123.38	127.39	122.17	126.00
平均喷油压力/MPa	40.46	40.10	43.01	43.58	42.35
循环喷油量/mm³	83.84	92.90	101.10	110.30	118.76
喷油持续期/(°)	9.00	9.50	10.25	11.00	11.50

本文计算中,活塞没有达到计算设定的最大行程8 mm。虽然蓄压室容积增大,但活塞的位移量随之增大,活塞对燃油的压缩量增大,从而保证蓄压室内的燃油压力变化不大。如果活塞位移已经达到其最大行程,随着蓄压室容积的增大,由于燃油的压缩量不再增大,燃油的相对体积变化率减小,导致蓄压室内的燃油压力降低。因此,在确定蓄压室容积时,应综合考虑共轨压力、增压活塞的面积比以及行程等因素。

2.6　电磁阀通电时间的影响

蓄压室喷射系统喷油开始时刻主要是由电磁阀通电时间决定的,因此,电磁阀通电时间的长短将对燃油喷射过程有很大的影响。图8所示的结果是在增压活塞面积比为14,保持电磁阀通电开始时刻不变,改变通电结束时刻来改变通电时间,并在不同的共轨压力下计算所得到的结果。由图可见:当共轨压力为7.5 MPa时,随着通电时间的延长,蓄压室压力逐渐提高;通电时间超过17 ms时,

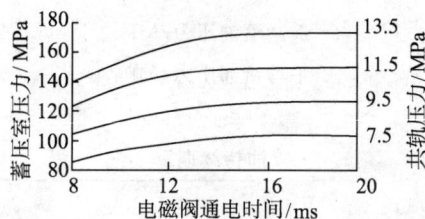

图8　通电时间对蓄压室压力的影响

蓄压室压力变化较小。而当共轨压力为13.5 MPa时,若通电时间小于14 ms,随着通电时间的延长,蓄压室压力增加;通电时间大于14 ms时,蓄压室压力基本不变。蓄压室压力的变化趋势也反映循环喷油量的变化趋势。其原因是当共轨压力较低时,增压活塞的运动速度慢,若通电时间过短,就会使增压活塞上部的低压腔燃油过早地回流卸压,活塞反向运动,不再对蓄压室燃油进行压缩,致使蓄压室压力较低,未达到由增压活塞面积比和共轨压力所决定的压缩压力。随着通电时间的延长,蓄压室燃油受到的压缩增加,其压力也相应提高。当通电时间达到某一临界值时,蓄压室压力达到由增压活塞面积比和共轨压力所决定的压缩压力,再延长通电时间,蓄压室压力不会发生大的变化。对于较高的共轨压力而言,由于增压活塞运动的速度较快,通电时间的临界值较小。因此,对于不同的共轨压力、增压活塞面积比和活塞最大行程的不同组合,存在不同的通电时间临界值,设计时要保证通电时间大于该临界值。另外,无论共轨压力高低,随着通电时间的延长,电磁阀通电结束时刻推迟,喷油开始时刻也随之后延。同时由于蓄压室压力提高,使喷油压力相应提高。

3 结 论

本文所建立的计算模型可以很方便地对蓄压式喷油器的燃油喷射过程进行模拟计算,并可分析各参数对喷射过程的影响。通过计算分析可知:电磁阀通电结束时刻决定了喷油开始时刻,而针阀弹簧预紧力和共轨压力决定喷油结束时刻;共轨压力的高低决定了循环喷油量的多少;在共轨压力波动较大时,应考虑适当提高针阀弹簧预紧力,既避免出现不正常喷射,又能使喷油迅速结束,减少低压喷射的油量;蓄压室容积也是影响喷油量的重要参数;增压活塞面积比的大小决定了蓄压室压力和喷油压力高低,是蓄压式喷油器的一个重要结构参数;对于共轨压力、增压活塞面积比和活塞最大行程的不同组合,存在不同的通电时间临界值,要保证通电时间大于该临界值。

参 考 文 献

[1] Keiki Tanabe, Susumu Kouketsu, Koji Mpri,等. 用计算机模拟分析共轨系统的喷射特性[J]. 国外内燃机, 2000(5): 36-39.
[2] 王钧效,等. 柴油机喷油过程模拟计算中的几个经验公式研究[J]. 车用发动机, 2001(5):6-11.

(本文原载于《内燃机工程》2002 年第 4 期)

Simulation and analysis of injection process of electronic-controlled fuel injector with accumulator booster

Abstract: The model of the electronic-controlled fuel injector with accumulator was presented in this paper. The injection process of the injector was simulated. The factors such as the common-rail pressure, the area ratio of boost piston, the preload of needle spring, the volume of the accumulator and the dwell time of the electromagnetic valve were simulated and their influences on injection process were analyzed. The obtained results can be used in the design of the electronic-controlled fuel injector with accumulator.

柴油机共轨系统中多分支共轨的
三维模拟计算和分析

吴 建,胡林峰,李德桃,龚金科,张建新

[摘要] 利用 PISO 算法对多分支共轨管内的非稳态流动进行了三维模拟计算,并对共轨喷射系统进行了测试,数值模拟计算结果和实测结果基本吻合,计算结果反映了共轨内压力场的基本情况。通过计算,分析了共轨容积和结构尺寸对共轨内压力波动及喷油的影响。

为了提高燃油经济性、降低排气污染和噪声,对柴油机燃油喷射系统进行细致的试验研究和数值模拟计算是非常必要的[1,2]。国内外许多学者对柴油机燃油喷射系统进行了大量模拟计算研究,如 Laforgia 的适用于直列泵系统和转子泵系统的计算程序,以及 Catania 开发的适用于直列泵系统的模拟计算程序等[3]。但新发展起来的柴油机电控共轨式喷射系统的模拟计算与传统的模拟计算有很大区别。在共轨系统中,多分支共轨是向各缸电控喷油器提供压力尽可能稳定的燃油的重要部件,共轨内燃油的流动及压力波动将影响到向各缸喷油器的供油。由于共轨容积相对于循环喷油量很大,而且由于各喷油器按一定时序喷油,导致轨内的流动产生脉动和压力波动,具有明显的非稳态流动特征。一维计算方法只能将共轨视为集中容积,无法分析共轨的几何参数对多分支共轨内压力场的影响。本文利用 PISO 算法[4](Pressure Implicit solution by Split Operator)对多分支共轨内的非稳态流动进行了三维模拟计算,并对共轨系统进行了测试,以验证其数值模拟计算的正确性,同时分析了共轨容积和结构尺寸对共轨压力波动的影响以及共轨压力对喷油的影响。

1 计算模型及计算方法

1.1 控制方程

本文所考虑的是非定常粘性流体的流动问题,同时还考虑流场的紊流特征,忽略了温度的影响。其控制方程如下[5]:

质量守恒方程:

$$\frac{\partial \rho}{\partial t} + \frac{\partial}{\partial x_j}(\rho u_j) = s_m \tag{1}$$

$$\frac{\partial}{\partial x_j}(\rho u_j) = \frac{\partial}{\partial x_1}(\rho u_1) + \frac{\partial}{\partial x_2}(\rho u_2) + \frac{\partial}{\partial x_3}(\rho u_3) \tag{2}$$

式中 ρ 为燃油密度;x_j 为坐标($j=1,2,3$);t 为时间;u_j 为流速在三个坐标方向上的分量;s_m 为质量源项。

动量守恒方程:

$$\frac{\partial(\rho u_i)}{\partial t} + \frac{\partial}{\partial x_j}(\rho u_j u_i - \tau_{ij}) = -\frac{\partial p}{\partial x_i} + s_i \tag{3}$$

$$\tau_{ij} = 2\mu s_{ij} - \frac{2}{3}\mu\frac{\partial u_k}{\partial x_k}\delta_{ij} - \rho\overline{u'_i u'_j} \tag{4}$$

$$s_{ij} = \frac{1}{2}\left(\frac{\partial u_i}{\partial x_j} + \frac{\partial u_j}{\partial x_i}\right) \tag{5}$$

$$-\rho\overline{u'_i u'_j} = 2\mu_t s_{ij} - \frac{2}{3}\left(\mu_t\frac{\partial u_k}{\partial x_k} + \rho k\right)\delta_{ij} \tag{6}$$

式中 p 为流体压力；μ 为动力粘性系数；s_i 为动量源项；τ_{ij} 为作用在与 i 方向相垂直的平面上的 j 方向上的应力；s_{ij} 为流体变形率张量；u' 为紊流脉动速度；δ_{ij} 为克罗内尔符号；μ_t 为紊流粘性系数；k 为紊流脉动动能；ε 为紊流脉动动能的耗散率。

密度与压力的关系可表示为 $\mathrm{d}p = E\mathrm{d}\rho/\rho$，$E$ 为流体的弹性模量。

1.2 紊流模型

在方程(4)出现了 $-\rho\overline{u_i u_j}$ 项，该项来源于非线性对流项的分解和平均，体现了紊流的输运作用，它的出现使体系中未知数的数目超过了独立方程的数目，方程变得不封闭，因此需要建立紊流模型，使控制方程封闭，从而获得控制方程的解。

本文采用标准的 $k\text{-}\varepsilon$ 模型[5]：

$$\frac{\partial(\rho k)}{\partial t} + \frac{\partial}{\partial x_j}\left(\rho u_j k - \frac{\mu_{\mathrm{eff}}}{\sigma_k}\frac{\partial k}{\partial x_j}\right) = \mu_t(P + P_B) - \rho\varepsilon - \frac{2}{3}\left(\mu_t\frac{\partial u_i}{\partial x_i} + \rho k\right)\frac{\partial u_i}{\partial x_i} \tag{7}$$

$$\frac{\partial(\rho\varepsilon)}{\partial t} + \frac{\partial}{\partial x_j}\left(\rho u_j\varepsilon - \frac{\mu_{\mathrm{eff}}}{\sigma_\varepsilon}\frac{\partial\varepsilon}{\partial x_j}\right) = \frac{\varepsilon}{k}\left[\begin{array}{l}(C_{\varepsilon1}P + C_{\varepsilon3}P_B)\mu_t - \\ \frac{2}{3}C_{\varepsilon1}\left(\mu_t\frac{\partial u_i}{\partial x_i} + \rho k\right)\frac{\partial u_i}{\partial x_i}\end{array}\right] - C_{\varepsilon2}\rho\frac{\varepsilon^2}{k} + C_{\varepsilon4}\rho\varepsilon\frac{\partial u_i}{\partial x_i} \tag{8}$$

标准的 $k\text{-}\varepsilon$ 模型的系数见表1。

表 1 标准的 $k\text{-}\varepsilon$ 模型的系数

C_μ	σ_k	σ_ε	$C_{\varepsilon1}$	$C_{\varepsilon2}$	$C_{\varepsilon3}$	$C_{\varepsilon4}$
0.09	1.0	1.22	1.44	1.92	0.0/1.44*	−0.33

注：当 $P_B > 0.0$ 时，$C_{\varepsilon3} = 1.44$；其他 $C_{\varepsilon3} = 0.0$。

1.3 网格生成及边界条件

多分支共轨结构如图1所示。各分支油管均布在共轨上。将分支油管和共轨分成两个子块，利用 STAR-CD 软件，采用分块耦合的方法生成三维贴体网格，如图2所示。由于共轨的网格尺度较分支油管的网格尺度大，所以在共轨和分支油管连接处做局部加密处理。

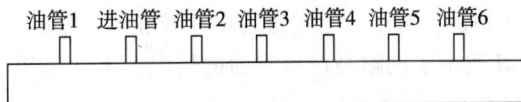

油管1 进油管 油管2 油管3 油管4 油管5 油管6

图1 多分支共轨管结构简图

图2 多分支共轨管贴体网格

出口边界一般可以采用压力边界或速度边界。在本文的计算中,如果采用压力边界,可利用动网格生成方法来模拟电磁阀的开闭,借助 STAR-CD 软件所提供的连接(Attach)边界实现油路的开闭。当电磁阀开启时,连接边界相邻两层的边界单元连接;当电磁阀关闭时,连接边界相邻两层的边界单元分离,边界转为固壁边界,从而实现分支油管的开闭,保证在模拟计算中各分支管内的流动按相应时序进行。边界的压力值可直接采用试验得到的测量值。这样处理的出口边界不能反映电磁阀关闭后仍然有燃油流出(对喷油器的蓄压室进行充油)的情况。如果采用速度边界,则可以克服上述的不足之处,但需要根据油管嘴端压力测量值计算出油管的出口流速。本文采用的是速度边界,利用作者建立的蓄压室式喷油器模型,根据实测得到的油管嘴端压力计算出油管的出口流速。

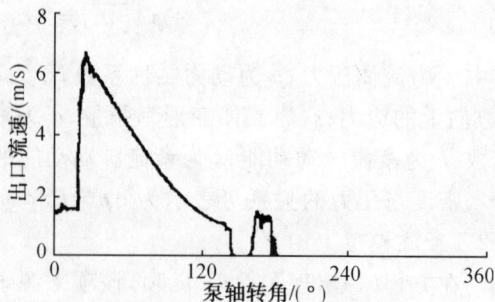

图 3　油管出口边界的流速

图 3 为油管 1 的出口流速。向喷油器提供的油量为每循环 2.14×10^{-6} m^3。

计算时,温度取为常数,固壁边界取无滑移速度边界;进口边界为压力边界,用实测的压力值代入。

1.4　数值求解

求解动量方程需预先知道压力场 p,但通常压力场是未知的。一般是采用迭代方法求解速度场,即假定一压力场 p^*,然后代入动量方程求解速度场 U^*。如果该速度场满足质量守恒方程,则流场求解结束。如果所得的速度场不满足质量守恒方程,则需对压力场 p^* 进行修正。反复此过程,直至由该压力场所得到的速度场能满足质量守恒方程为止。

根据压力场 p^* 的不同计算方法形成了不同的算法。常用的算法有 SMPLE 算法、PISO 算法、SMPLEPISO 算法。SMPLE 算法是求解压力耦合方程的半隐式法,该算法只对动量方程进行一次校正,尽管 SMPLE 算法也可用来求解非稳态流动问题,但由于在处理变量之间的耦合时,该方法过于依赖迭代,计算成本较大,所以适用于求解稳态流动问题。PISO 算法是利用分裂算子求解压力的隐式法,其基本思想是利用常微分方程中的预报—校正法,把每一时间步分裂为一个预报步和一个或多个校正步,在时间差分上仍是全隐式,故时间步长不受稳定性限制,即使将空间网格加密,也无需减少时间步长。与 SMPLE 算法相比,PISO 算法更适用于非稳态流动问题的求解。SMPLEPISO 算法是 SMPLE 算法和 PISO 算法的结合,它除了对动量方程进行一次校正外,还借用了 PISO 算法的做法,对压力方程进行多次校正,以减少压力梯度与网格表面不垂直而引起的误差。此方法尤其适用于网格被严重扭曲的情况。本文采用 PISO 算法。

2　压力测量

测试系统如图 4 所示。在共轨与供油泵之间的管路上设有压力调节阀,用来调节共轨平均压力(由数显压力表显示)。在共轨系统的进油管端和 6 个与喷油器相连的油管嘴端布置了压力传感器,以测量油管的嘴端压力以及共轨的进油压力。在共轨总管上布置了 2 个压力测点(油管 1 与进油管之间为测点 1,油管 4 与油管 5 之间为测点 2),测量共轨内压力的变化。由 ECU 控制 6 个喷油器按照发火顺序依次喷油,用数据采集计算机记录各被测处的压力变化。试验所用仪器设备见表2。试验工况为供油泵转速 1 000 r/min,共轨平均压力为90 MPa,电磁阀通电时间为 22 ms。

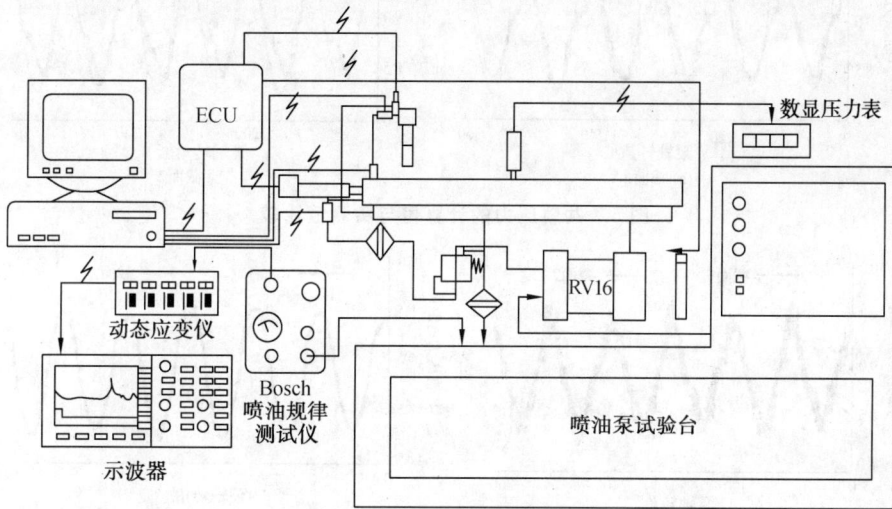

图4 试验设备示意图

表2 试验仪器

名　称	型　号	产　地
油泵试验台	12PSDW110-Ⅱ	成都八一机电厂
喷油规律仪	EFEP481	Bosch 公司
动态应变仪	CDV-110A	[日]KYOWA
数字示波器	TDS340A	Tektronix
数显压力表	YMK-1	上海仪表四厂
采样计算机	PR75	联想集团

3　计算结果和测试结果的对比分析

共轨管横断面上的压力为该断面处一单元层的压力的平均值。图5为七分支共轨管(ϕ36 mm×556 mm,容积为 $5.66×10^{-3}$ m³,油管长度为 300 mm)内非定常流动中压力的计算结果与测量结果的对比(截面1、2分别对应于测点1、2)。图中的曲线为压力随泵轴转角的变化曲线,从图5(a)、图5(b)的曲线对比来看,压力的计算值和实测值基本吻合。计算所得到的结果基本上可以反映共轨管内压力场的情况。图6为共轨内不同截面处压力之间的比较。由图6曲线对比可见:在轨内不同断面处曲线基本重合,压力基本相同。这表明在同一时刻,共轨内各处压力基本相同。

从计算和试验结果对比来看,本文所建立的计算模型和计算方法是正确的,可以用于共轨内压力的计算分析。

共轨内压力的波动主要是由于油泵供油压力的波动和按一定时序向各喷油器供油而产生的。共轨的结构尺寸及容积大小对轨内的压力波动有很大的影响。利用上述方法对不同容积的共轨在相同的边界条件下的轨内压力进行了模拟计算,计算结果如图7所示。

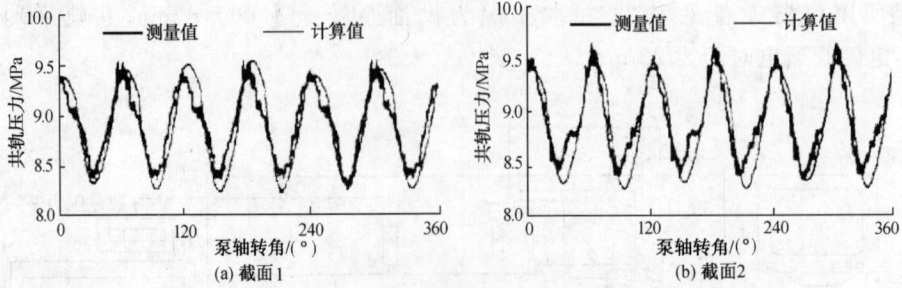

图 5　共轨压力的计算值与实测值比较

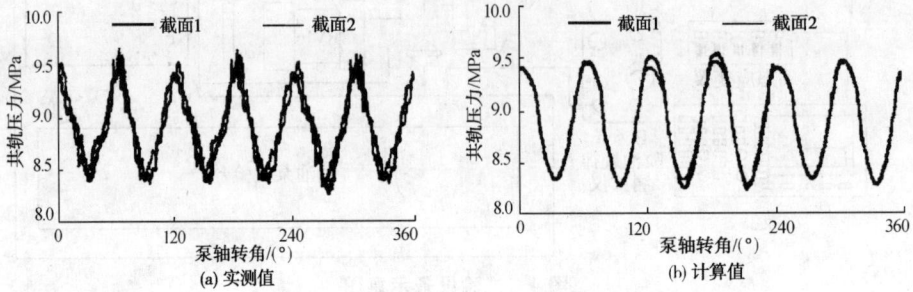

(a) 截面1

(b) 截面2

图 6　不同截面处共轨压力的比较

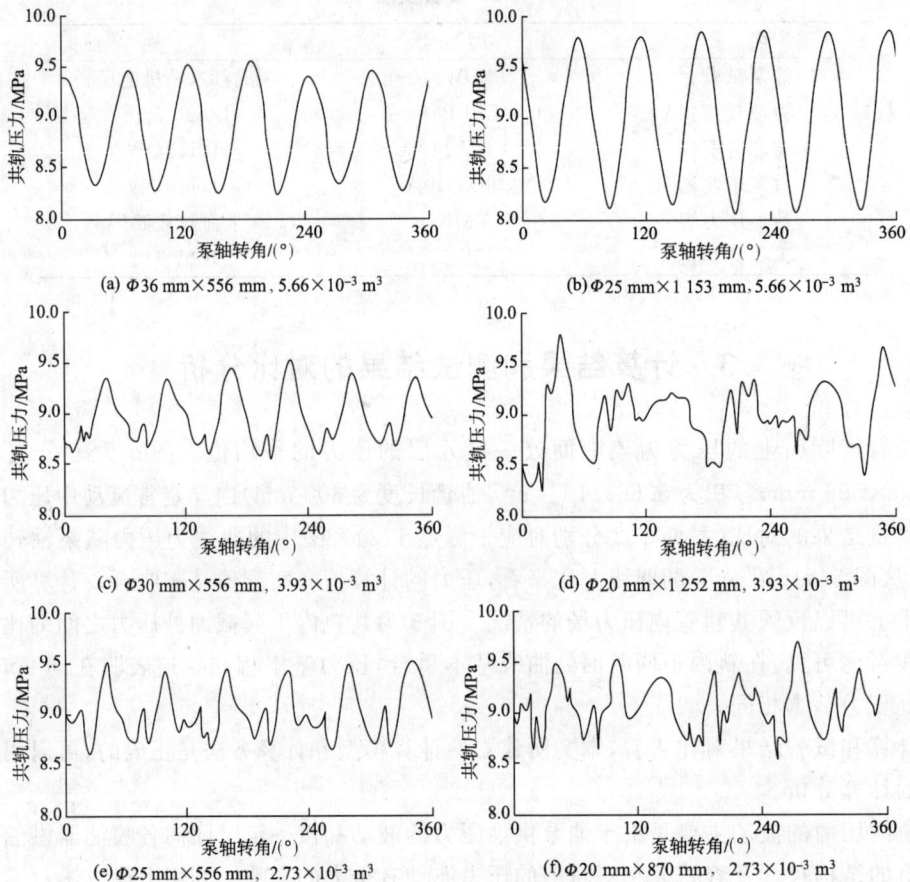

(a) 实测值

(b) 计算值

(a) Φ36 mm×556 mm, 5.66×10⁻³ m³

(b) Φ25 mm×1 153 mm, 5.66×10⁻³ m³

(c) Φ30 mm×556 mm, 3.93×10⁻³ m³

(d) Φ20 mm×1 252 mm, 3.93×10⁻³ m³

(e) Φ25 mm×556 mm, 2.73×10⁻³ m³

(f) Φ20 mm×870 mm, 2.73×10⁻³ m³

图 7　不同尺寸共轨压力波动的计算结果

由图 7(a)、图 7(b)可见：当共轨容积较大时，共轨压力的波动幅度较大，压力波动具有与喷油器工作时序相应的周期性；当直径减小、长度增加时，压力波动幅度增大。由图 7(c)～图 7(f)可知：当共轨容积较小时，共轨压力的波动幅度有所减小，但压力波动的频率有所增加，尤其是在共轨直径较小而长度较长时，压力变化没有明显的周期性。这是由于油管嘴端压力(见图 8)在停止向喷油器供油后会出现压力振荡，这个压力振

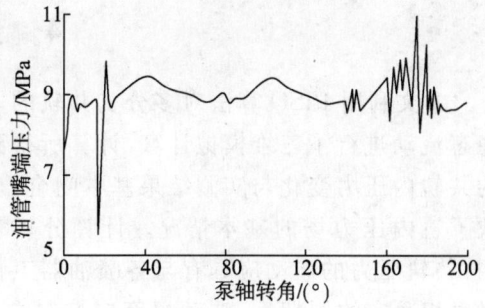

图 8　油管端压力计算结果

荡会波及到共轨。如果容积小、共轨直径小，则在油管入口附近的燃油相对较少，因而对由嘴端传来的压力振荡的吸收和衰减能力小。共轨压力受其影响产生振荡，各油管产生的振荡与喷油器供油所产生的压力变化叠加，就会使共轨压力产生如图 7(d)和图 7(f)所示的波动。共轨压力波动幅度较大会影响到喷油器的喷油量，而压力波动的不规则将影响各缸喷油的均匀性。

为了了解共轨压力波动对喷油的影响，利用作者建立的蓄压式喷油器的计算模型，以波动幅度不同的共轨压力为边界条件进行计算，结果如图 9 所示。图中喷油量的变化是相对于无压力波动时的喷油量的变化，时间为电磁阀通电时间。由图 9 可见：轨内压力波动幅度越大，则喷油量变化越大。变化量还与电磁阀通电时间有关。以图 7(c)和图 7(d)中的压力为边界条件对 6 个喷油器的喷油过程进行计算，结果如图 10 所示。图中喷油量的变化是相对于喷油器 1 的变化。由图 10 可见：轨内压力波动的不规则性使各缸喷油的均匀性变差。要保证柴油机稳定工作，各缸喷油均匀性要好，循环油量的变化不应大于 ±3 %。因此，共轨压力的波动幅度应控制在 1 MPa 以内，轨内压力变化应具有与喷油定时相应的周期性，以保证各缸喷油器工作时具有基本相同的压力变化。从图 7、图 9 和图 10 来看：共轨容积的大小要适当；在共轨容积一定时，其长径比不宜过大。在本文的计算条件下，直径为 30 mm、长度为 556 mm、容积为 3.93×10^{-3} m³ 的共轨最好。

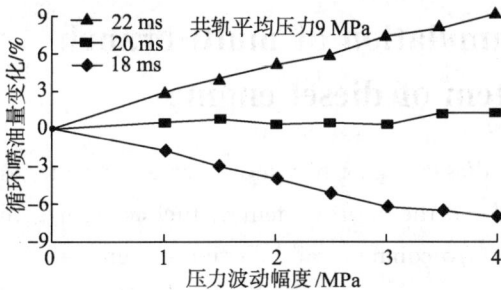

图 9　共轨压力波动对喷油量的影响

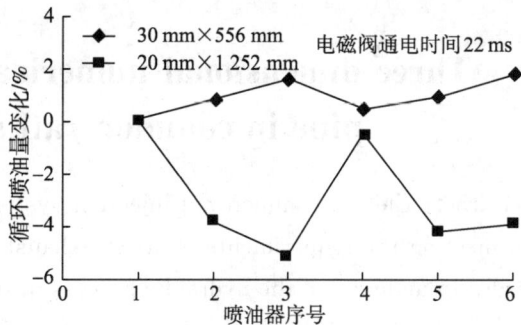

图 10　共轨压力波动对喷油均匀性的影响

在计算中发现：不论轨内压力如何变化，沿共轨轴线各截面的平均压力基本相同，如图 11 所示(ϕ30 mm×556 mm)。图中 1、3、5 截面分别为油管 1、3、5 所对应的截面，截面 2、4、6 分别为油管 1 和进油管之间、油管 2 和 3 之间、油管 4 和 5 之间的截面。由图可见：各截面的压力基本相同，压力差不大于 0.1 MPa。所以，进、出油管的布置不会对共轨内的压力波动以及喷油过程产生影响，可以根据柴油机整机布置的要求确定。

4 结 论

本文利用 PISO 算法对多分支共轨的非稳态流动进行了三维模拟计算,计算所得到的共轨内压力变化与实测结果基本吻合,反映了轨内压力场的基本情况。计算分析表明:共轨压力的波动应具有与各喷油器工作时序相适应的周期性,波动幅度应控制在 1 MPa 以内;共轨容积应适当,直径不应过小,共轨长径比不宜过大;沿共轨轴线各截面的平均压力基本相同,进、出油管的布置可以根据柴油机整机布置的要求确定。

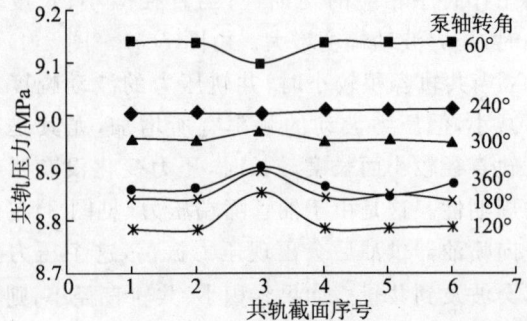

图 11 共轨各截面间压力的比较

参 考 文 献

[1] Kato T, Tsujimura K M, Minami T, et al. Spray characteristics and combustion improvement of DI diesel engine with high fuel injection[J]. *SAE Paper*, 890265, 1989.

[2] Digesu P, Ficarella A, Laforgia D. Diesel electro-injector: a numerical simulation code [J]. *SAE Paper*, 940193, 1994.

[3] Ficarella A, Laforgia D. Injection characteristics simulation and analysis in diesel engines[J]. *Meccanica*, *International Journal of AMETA*, 1993,28:239-248.

[4] Issa R I. Solution of the implicitly discredited fluid flow equations by operator-splitting [J]. *J Comp Phys*, 1986,62:40-65.

(本文原载于《内燃机工程》2002 年第 5 期)

Three-dimensional numerical simulation of multi-branch pipe in common rail system of diesel engine

Abstract: Electro-common rail injection system of diesel engine has shown great advantages in meeting the requirements of strict exhaust laws and the improvement of fuel economy. In order to understand the hydraulic process in the electro-common rail injection system and optimize the common rail injection system, it is necessary to make detailed experimental and numerical investigations on the electro-common rail injection system. In order to understand the pressure field, the three-dimensional numerical simulation of multi-branch pipe in common rail system was carried out by the use of PISO algo rithm, and the measurement of pressure in the system was carried out. The calculated results have a good agreement with the measured ones and demonstrate the basic characteristics of the pressure field in the multi-branch pipe in common rail system.

滑阀参数对蓄压式电控喷油器
喷射过程影响的计算分析

吴　建,卫　尧,李德桃,胡林峰,张建新

[摘要]　本文利用建立的蓄压式电控喷油器的计算模型,对喷射过程的模拟计算,分析了滑阀高压通道截面积、节流孔和滑阀弹簧预紧力对喷射过程的影响。计算结果对滑阀参数的确定和优化具有一定的参考价值。

在蓄压式共轨电控喷油系统中,电控喷油器是实现高压喷射和喷油量控制的关键部件(图1),而作为喷油器的控制单元,电磁阀部件对燃油喷射过程有很大的影响。电磁阀部件由先导阀和滑阀组成。由于燃油喷射所需的高压是经过增压活塞和增压柱塞对燃油进行压缩获得的,喷油始点又取决于增压活塞上腔泄流降压、增压活塞和增压柱塞复位的时刻,而滑阀直接控制着增压活塞上腔的进出口流通截面开闭,所以滑阀的运动过程对燃油喷射过程有很大的影响。因此,了解滑阀结构参数对喷射过程的影响,对于蓄压式电控喷油器的设计和结构参数的优化是非常重要的。

作者利用 AVL 公司的 Hydsim 软件,对蓄压式电控喷油器的喷射过程进行模拟计算,分析了滑阀主要结构参数对喷射过程的影响。

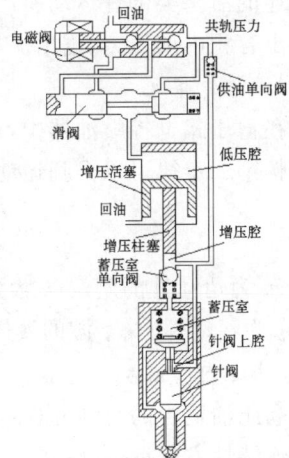

图1　蓄压式电控喷油器结构示意

1　蓄压式电控喷油器的数学描述

建立计算模型时,首先根据蓄压式电控喷油器的结构将电磁阀、滑阀、增压活塞、单向阀、针阀等基本运动部件以及增压活塞上腔、蓄压室等容积单元和油道等单独进行模型化,建立相应的基本计算模型单元;然后按蓄压式电控喷油器的工作原理将各基本计算模型单元连接起来,组合成蓄压式电控喷油器的整体计算模型[1]。

在模拟计算中,作如下假设:(1) 在喷油过程中,燃油温度保持不变;(2) 燃油密度随压力变化,用经验公式计算;(3) 不考虑各部件的弹性变形;(4) 喷油器中的油道较短,可不考虑压力波的传播,同时忽略油道中的流动阻力损失;(5) 仅在增压活塞和针阀处考虑径向间隙所引起的泄漏。

计算按粘性可压缩流体进行。根据流体力学的一维非稳态流动基本方程,结合各基本计算模型单元的具体特点,可建立与基本计算模型单元相应的计算方程[2]。

1.1　容积单元

考虑到与燃油的弹性模量相比,系统中的容积壁面在受液体压力的作用时变化极小,可以近似地将模型中的容积视为刚性容积。

容积的质量连续性方程为

$$\frac{\mathrm{d}p}{K}=\frac{\mathrm{d}m}{m}-\frac{\mathrm{d}V}{V} \tag{1}$$

密度与压力的关系

$$\frac{\mathrm{d}\rho}{\rho}=\frac{\mathrm{d}p}{K} \tag{2}$$

存在气穴时的燃油密度

$$\rho=\alpha\rho_v+(1-\alpha)\rho_1 \tag{3}$$

式中 K 为液体的弹性模量;ρ 为密度;p 为压力;ρ_v,ρ_1 分别为燃油的饱和蒸汽密度和液态密度;α 为气穴分量或气穴率。

方程(1)反映了容积中压力 $\mathrm{d}p$ 与质量 $\mathrm{d}m$ 和容积 $\mathrm{d}V$ 的关系,式中的容积变化是由于与容积相连的活塞类部件运动所产生的。该方程可用到系统中所有的容积中。利用上述三个方程可求出容积内的压力、进出流量以及容积体积等参数。

1.2 运动件单元

在蓄压式电控喷油器中,运动件单元主要有滑阀体、增压活塞和柱塞、单向阀和针阀等。由牛顿第二定律可以得到运动件的运动方程:

$$m\frac{\mathrm{d}^2x}{\mathrm{d}t}=\mu A(p_1-p_2)-k(x+x_0)-D\frac{\mathrm{d}x}{\mathrm{d}t} \tag{4}$$

式中 m 为运动件的质量;k 为弹簧弹性系数;x 为运动件的位移;x_0 为弹簧的初始预压缩量;p_1,p_2 为运动件上、下游的液体压力;D 为阻尼系数;μA 为有效承压面积。

1.3 油道内的流动

高压油道内的流动方程可由以下两式来表示。

连续性方程

$$\frac{\partial\rho}{\partial t}+u\frac{\partial\rho}{\partial x}+\rho\frac{\partial u}{\partial x}=0 \tag{5}$$

动量方程

$$\frac{\partial u}{\partial t}+u\frac{\partial u}{\partial x}+\frac{1}{\rho}\frac{\partial p}{\partial x}=\frac{4}{3}\mu\frac{\partial^2 u}{\partial x^2} \tag{6}$$

式中 ρ 为流体密度;u 为流体速度;p 为流体压力;μ 为流体动力粘度,取常数为 $0.002\,46\ \mathrm{N\cdot S/m^2}$。

2 计算结果及分析

图 2 为滑阀工作原理图,图 3 是滑阀运动规律的计算结果。当电磁线圈接通电流时,先导阀打开共轨与滑阀控制室的通道。滑阀在高压油的作用下,迅速向右运动。滑阀移动到 a 点时,关闭增压活塞上腔与回油道的通路;滑阀移动到 b 点时,开启增压活塞上腔与共轨的高压油通道,高压油进入增压活塞上腔,使增压活塞开始增压过程。当电磁阀断电时,先导阀关闭共轨与滑阀控制室的通道,并开启滑阀控制室与低压回油的通道,滑阀控制室内的压力迅速降低,在滑阀弹簧的作用下,滑阀向左运动,滑阀移动到 c 点时,关闭增压室活塞上腔与共轨的高压油通道;滑阀移动到 d 点时,开启增压活塞上腔与回油道的通路(回流通道),此时,增压活塞

上腔的燃油迅速从回油道回流,压力降低,使增压活塞在增压室压力的作用下向上运动。针阀背压降低,针阀升起,开始喷油。因此,滑阀的运动规律将直接影响增压过程和喷油时刻。同时,增压活塞上腔与共轨的高压油通道面积的大小也影响增压过程,进而影响喷油过程。

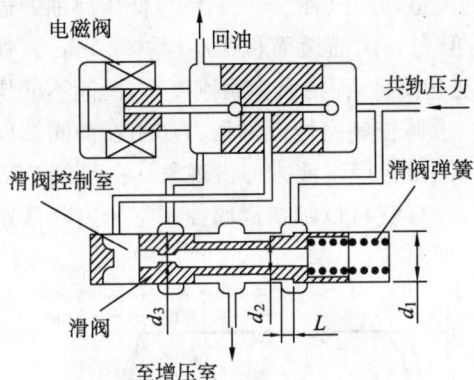

图 2 滑阀的工作原理示意图

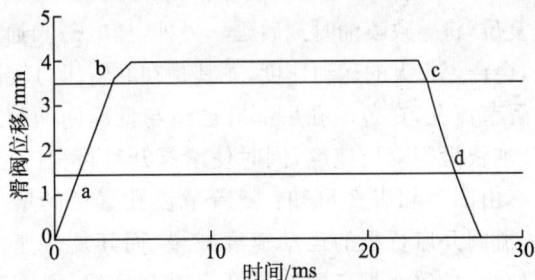

图 3 滑阀的运动规律

在计算中,电磁阀通电时间为 20 ms(计算始点即为电磁阀通电始点),泵轴转速为 1 000 r/min。

2.1 滑阀高压通道流通截面积的影响

滑阀高压通道流通截面是增压活塞上腔与共轨高压油通道的流通截面。从图 2 可知,流通截面位于 d_1 和 d_2 所组成的环形通道或者由相应的肩胛面处。流通截面积可由下式进行计算:

$$S = \min\{\frac{\pi}{4}(d_1^2 - d_2^2), \pi d_1 L\} \tag{7}$$

式中 S 为最小流通截面积,其他符号的意义见图 2。

由表 1 可知:当滑阀高压通道流通截面积改变时,高压通道开启时刻和回流通道开启时刻均未发生变化,说明通道面积的大小不会影响滑阀的运动规律。随着通道面积的增大,燃油流过通道截面的最大流量增大,这就使增压活塞的增压速度加快,蓄压室达到最高压力的时间缩短,最高压力提高。蓄压室燃油压力的提高,将会使喷油压力提高,喷油量增加;同时,通道面积增大,进入增压活塞上腔的燃油增加,油管嘴端的压力降也会随之增大。油管嘴端过大的压力降会波及共轨,影响了共轨压力的稳定。从表 1 中还可看出:当通道截面积大于 2.85 mm² 时,随截面积增大,最大流量和最高压力增加得比较平缓;而当通道截面积小于 2.85 mm² 时,截面积的变化对最大流量和最高压力的影响较大。

表 1 滑阀高压通道流通截面积的影响

指标	流通截面积/mm²				
	2.35	2.85	3.35	3.85	4.35
高压通道开启时刻/ms	3.13	3.13	3.13	3.13	3.13
回流通道开启时刻/ms	25.96	25.96	25.96	25.96	25.96
最大流量/(×10³ mm³/s)	189.30	221.40	241.30	255.70	265.80
达到最高压力时刻/ms	19.83	17.33	15.50	14.13	13.04
蓄压室最高压力/MPa	107.34	125.79	127.12	128.30	129.60
油管嘴端压力降/MPa	5.80	6.20	6.50	6.80	7.00

注:共轨压力 9.2 MPa,增压比 14。

因此,在确定滑阀高压通道流通截面积时,既要保证有足够的流量,以获得较高的增压压力,又要避免在油管嘴端产生过大的压力降低。

2.2 节流孔的影响

当电磁阀断电时,先导阀关闭共轨与滑阀控制室的通道,并开启滑阀控制室与低压回油的通道,滑阀控制室内的燃油通过先导阀回油,压力降低。但先导阀通道面积很小(约 $0.4\ mm^2$),如果仅仅依靠先导阀通道回油,则回油速度慢,压力不能迅速降低,而滑阀弹簧刚度小,滑阀不能快速复位,将导致喷油时刻后延。另外,当电磁阀通电时,滑阀控制室内燃油压力作用在滑阀端面上,会产生很大的运动速度,尤其是在共轨压力较高时,甚至会产生液力冲击现象。在滑阀芯部设置节流孔,在节流孔后部有通道与低压回油腔相通,这样就可以加快滑阀控制室压力降低速度,加快滑阀复位速度,同时使滑阀开启行程的速度有所减缓。

由图4和表2可知:随着节流孔直径的增大,滑阀开启行程的运动速度减慢,而其复位速度加快,导致滑阀高压通道开启时刻推迟,滑阀回流通道开启时刻提前。这意味着在相同的通电时间下,当节流孔直径变化时,滑阀的响应速度在两个运动过程中呈相反的变化。同时,在电磁阀通电时间不变的条件下,滑阀高压通道关闭时刻略有提前,因此,随着节流孔直径增大,高压通道开启的持续时间缩短,增压活塞进行增压的时间相应缩短,使得蓄压室燃油的最高压力降低,这将导致喷油压力降低。而滑阀回流通道开启时刻的提前,将导致喷油时刻的提前。节流孔直径增大,也会导致高压油的损失,造成油泵驱动功率的损失。因此,节流孔直径的确定,应综合考虑滑阀的响应速度、蓄压室燃油压力以及通电时间等因素,通过试验确定。

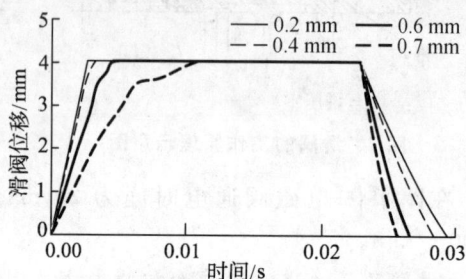

图4 节流孔直径 d_3 对滑阀运动规律的影响

表2 节流孔直径 d_3 的影响

指　　标	节流孔直径 d_3/mm				
	0.0	0.2	0.4	0.6	0.7
高压通道开启时刻/ms	2.37	2.46	2.75	3.70	6.70
高压通道关闭时刻/ms	23.96	23.87	23.83	23.62	23.54
回流通道开启时刻/ms	27.37	27.04	26.42	25.50	24.92
蓄压室最高压力/MPa	127.50	127.00	125.10	121.40	117.00

注:共轨压力9.2 MPa,增压比14。

2.3 滑阀弹簧预紧力的影响

滑阀弹簧的作用就是使滑阀在电磁阀断电时迅速复位。当弹簧预紧力较大时,滑阀开启行程的阻力大,滑阀的速度慢;而在复位行程中,弹簧预紧力较大,滑阀运动速度快。图5和表3所示为弹簧预紧力的影响:当预紧力增大时,高压通道开启的持续时间略有减少,使增压时间有所缩短,导致蓄压室最高压力略有降低,但影响较小。滑阀回流通道

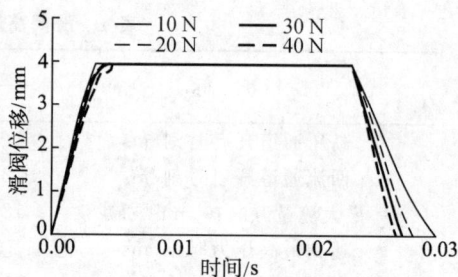

图5 弹簧预紧力对滑阀运动规律的影响

开启时刻随预紧力增大而提前,增压活塞的复位相应提前,这就意味着喷油时刻将随预紧力增

大而提前。随着预紧力增大,喷油时刻相对于电磁阀断电时刻的延迟缩短,但弹簧预紧力过大,会使滑阀在复位中对止口产生较大的冲击;而且在共轨压力较低时,过大的弹簧预紧力会使滑阀在开启行程中速度降低。

表3 弹簧预紧力的影响

指 标	弹簧预紧力/N			
	10	20	30	40
高压通道开启时刻/ms	2.88	3.08	3.38	3.58
高压通道关闭时刻/ms	23.88	23.79	23.63	23.58
回流通道开启时刻/ms	26.54	25.92	25.46	25.21
蓄压室最高压力/MPa	123.60	123.10	122.90	122.60

注:共轨压力9.2 MPa,增压比14。

3 结 论

(1) 增大滑阀高压通道流通截面积,可以提高增压活塞的增压速度,使蓄压室内燃油压力提高,喷油压力相应提高和喷油量增大。同时,较大的流通截面积会使油管嘴端产生较大的压力降低,会对共轨压力的稳定产生影响。确定滑阀高压通道流通截面积时,应保证足够流量,避免在油管嘴端产生过大的压力降低。

(2) 节流孔直径影响滑阀高压通道开启持续时间和滑阀复位速度,进而影响蓄压室压力和喷油时刻。节流孔直径的确定,应综合考虑滑阀的响应速度、蓄压室燃油压力以及通电时间等因素,通过试验确定。

(3) 预紧力增大,滑阀响应加快,喷油时刻提前,但弹簧预紧力不宜过大,应避免滑阀在复位中对止口产生较大的冲击。

参 考 文 献

[1] 吴建,胡林峰,李德桃,等. 蓄压式电控喷油器燃油喷射过程的计算分析报告[R]. 无锡:无锡油泵油嘴研究所,2002.
[2] AVL-Hydsim Reference Manual V4.2[R]. 2001:145–210.

(本文原载于《内燃机工程》2003年第2期)

Simulation and analysis of effects of slide-valve on injection process of electronic-controlled fuel injector with accumulator

Abstract:In this paper the model of the electronic-controlled fuel injector with accumulator was built and the effects of the construction of slide-valve parameters such as flow area of high pressure gallery, diameter of the throttle and spring preload were analyzed in injection process of electronic-controlled fuel injector with accumulator by simulating injecting process. The obtained results can be references for designing a slide-valve.

柴油机中压共轨式喷油系统的模拟计算与分析

潘剑锋,梁凤标,李德桃,胡林峰,吴　建

[摘要]　在改进喷油器结构的基础上,在中压共轨喷油系统中实现了预喷。建立了中压共轨喷油器的液力计算模型,模拟计算了喷油特性,并进行了试验验证,证明了模拟计算的准确性,得到了若干有价值的结论。

随着对柴油机在排放、噪声、排烟等方面的要求日趋严格,传统的机械式柴油机燃油喷射系统因其固有的缺点已无法满足要求[1]。高压喷射、可变喷油定时及喷油速率柔性控制成为柴油机燃油喷射装置的发展方向。共轨式燃油喷射系统能很好地满足高压喷射和对喷油定时、喷油速率的柔性控制,从而能有效地减少柴油机排放和噪声,改善燃油经济性,提高动力性和可靠性[2]。本文模拟计算了增压式共轨燃油喷射系统喷油器的喷油特性,获得了一些有参考价值的结论。

1　蓄压式共轨喷射系统

在中压共轨式喷油系统中,喷油器部件集中了对增压、喷油定时、喷油量以及喷油率曲线形状的控制等功能,是喷油系统最关键的部件。蓄压式电控中压共轨喷油系统结构如图1所示,该系统中蓄压过程和喷射过程是连续进行的两个步骤,即在完成蓄压过程后再进入喷射过程。在喷射过程中,除了喷射始点可控外,喷射终了是一个自然形成的过程。设定了针阀的关闭压力,喷油开始后,随着喷射过程的进行,喷油压力迅速降低,一直到接近喷油器关闭压力时终止喷油,喷油器的喷油量取决于喷油器蓄压室内燃油的体积。因此其喷油规律是先急后缓,断油不迅速,与理想的喷油规律(如图2[3])相反,很难满足新一代柴油机对喷油系统的要求。

喷油率是控制柴油机燃烧过程的重要参数之一。为了同时改善柴油机的动力性和经济性,降低噪声和污染物

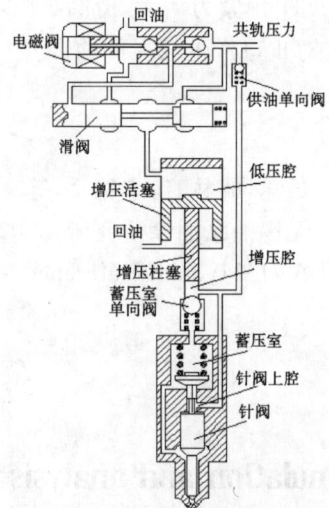

图1　蓄压式喷油器结构示意图

排放,应按图2所示理想的喷油率曲线喷油,可以形成最佳的混合气,实现理想的燃烧过程。应该注意到理想的喷油率曲线形状不是固定不变的,而应随柴油机转速和负荷的变化相应调整成最佳的形状(如图3所示),这种调整包括各个喷油阶段的油量比和喷油定时[4]。

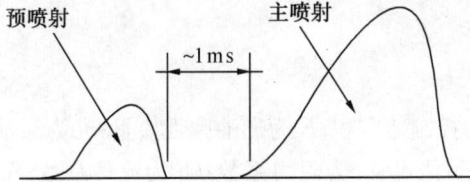

图 2　理想的喷油率曲线

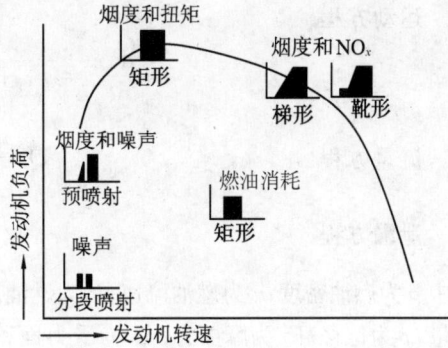

图 3　各运行条件最佳喷油率波形

原蓄压式共轨喷油系统存在着两个缺点：一是供油量与蓄压压力成正比,在喷油量较低时无法获得较高的喷射压力;二是喷油规律与理想的喷油规律相反。为此需对其结构进行改进[5]。

2　新型增压式共轨喷射系统

新型增压式共轨喷射系统(图 4)充分吸取了原蓄压式共轨喷射系统的长处,通过提高电磁阀的响应速度和精度,适当提高共轨压力和增压比,实现增压过程和喷射过程的重叠进行,并获得更高的喷射压力。当高压腔压力达到喷油器针阀的开启压力时,喷油器就开始进行喷射。由于共轨压力和增压比的提高,对增压活塞与增压柱塞的面积比和增压柱塞与针阀喷孔总面积之比进行合理选择,高压腔内的压力在喷射发生的同时也能进一步提高,持续进行喷射。喷射终了时间则由电磁阀控制,从而实现了喷射时间的全面控制。喷油器增压过程和喷射过程的改变,使喷油系统的喷油量可以通过改变共轨压力和增压时间来控制。这种控制方法突破了传统喷油系统无法控制喷油压力的缺点,在小喷油量的情况下也能实现较高的喷射压力。

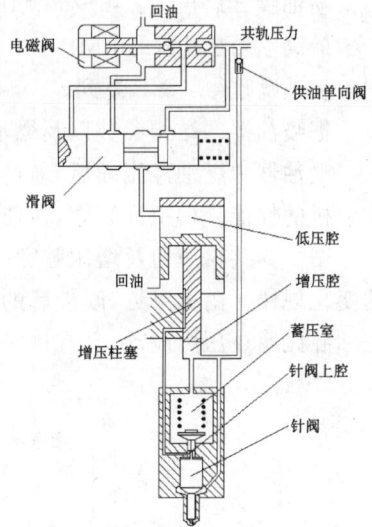

图 4　新型增压式共轨喷射系统

3　模拟计算

本文利用 AVL 公司的 Hydsim 软件对改进后的增压式共轨燃油喷射系统进行模拟计算。在建立计算模型时,把喷油器结构模块化,选取 Hydsim 中相应的功能元件,组合成液力计算模型,即可进行模拟计算。采用的基本方程[6]为：

流体连续性方程 $\qquad \oint \rho u \, dA = 0$

流体可压缩性方程 $\qquad dP = -(E/V)dV$

运动方程

$$m \frac{\mathrm{d}^2 x}{\mathrm{d}t^2} = F_0 + \sum p_i A_{\mathrm{p}i} - Kx - c_{\mathrm{r}} \frac{\mathrm{d}x}{\mathrm{d}t}$$

流量方程

$$Q_i = C_{\mathrm{d}} A_i \sqrt{\frac{2 \Delta P_i}{\rho_i}}$$

泄漏方程

$$q_i = \frac{\pi d_i \delta_i^3}{12 \mu L_i} \Delta P_i$$

式中 ρ 为燃油密度;u 为燃油流速;P 为燃油压力;V 为控制体容积;E 为燃油弹性模量;m 为运动件质量;x 为位移;F_0 为摩擦力;A_{p} 为受力面积;K 为弹簧刚度;c_{r} 为阻力系数;C_{d} 为流量系数;A_i 为流通面积;d_i 为偶件直径;δ_i 为偶件间隙;μ 为动力粘度;L_i 为偶件长度;Q_i 为流量;q_i 为泄漏流量。

喷油器的结构参数按照实际尺寸进行计算,其基本参数如下:

供油泵柱塞直径及升程:$\Phi 5 \text{ mm} \times 8 \sim 10 \text{ mm}$;

喷油嘴喷孔数及喷孔直径:$6 \times \Phi 0.20(0.22) \text{ mm}$;

喷油器增压比:$10 \sim 20$;

最大喷油量:$120 \text{ mm}^3 / \text{次}$;

最高共轨压力:20 MPa;

喷油嘴开启压力:$40 \sim 50 \text{ MPa}$;

最高喷射压力:220 MPa;

喷油嘴型号:S/P 系列。

把喷油器的结构参数以及燃油特性参数输入程序中即可进行模拟计算和分析。

喷油器中燃油经压缩后,通过喷孔雾化成细粒直接喷入发动机燃烧室,这一过程直接影响发动机的性能。模拟计算中,转速为 1 700 r/min,标定负荷。

图 5 为共轨压力及增压脉宽一定、增压比为 10 时,模拟计算所得增压室及喷嘴喷油压力的变化规律。由图可见,改进后的增压式共轨喷射系统能实现预喷,主喷结束时停油迅速,高压油腔无明显压力波动,具有良好的喷油率曲线形状。共轨压力为 15 MPa 时,喷油压力高达 150 MPa。

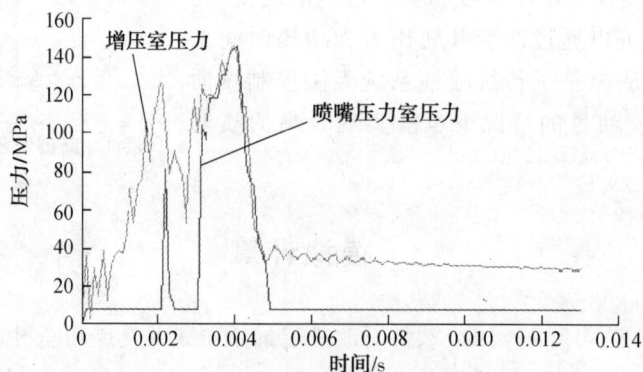

图 5　增压室及喷嘴压力的变化规律

喷油器每一循环的燃油喷射量取决于储存在喷油器增压室的燃油压力、针阀关闭压力和增压室容积。针阀的关闭压力由控制室弹簧的预紧力决定,通过改变弹簧调整垫可调整喷油器的关闭压力。对某一型号喷油器而言,由于其增压室容积、针阀关闭压力已经确定,所以此时喷射量仅取决于燃油压力。当增压室燃油压力高时,喷射量大;当增压室燃油压力低时,喷射量相应减小。

共轨式电控喷油系统的流量特性就是喷油量随增压脉宽、针阀调压弹簧预紧力及增压腔容积变化的规律。图 6 为共轨压力一定时每循环喷油量随增压时间变化的规律,图 7 为每循环喷油量随针阀调压弹簧预紧力变化的规律,图 8 为共轨压力一定时每循环喷油量随增压室容积变化的规律。

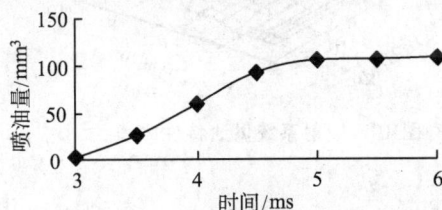

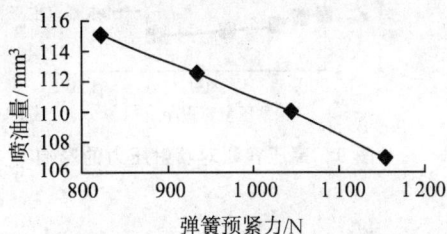

图 6　循环喷油量随增压时间变化的规律　　　图 7　循环喷油量随弹簧预紧力的变化规律

由图 6 可知,在喷油系统其他参数确定的情况下,电磁阀通电时间(即增压脉宽)有一个临界值。此时,喷油量和喷油压力刚好达到可能的最大值。当电磁阀通电时间大于其临界值时,喷油量和喷油压力维持在最大值不变;小于其临界值时,随电磁阀通电时间的增加喷油量呈增加趋势。

由图 7 知,针阀调压弹簧预紧力的值决定了喷油结束的时刻。预紧力大,喷油结束早,喷油持续时间短,因而喷油量小;预紧力小,则喷油结束晚,喷油持续时间长,喷油量大。在共轨系统中,喷油开始时刻几乎与弹簧预紧力无关,因此针阀开启压力由弹簧预紧力和针阀顶端与高压腔的压力差共同决定。喷油开始前弹簧预紧力与压力差相比相差甚远,可忽略。

由图 8 知,当共轨压力一定时,随着增压室容积 V 的增大,增压室内被压缩的燃油也增多,则喷油量增加。弹性模量和密度之间的关系为:

$$\frac{1}{E}=\frac{1}{V}\frac{\mathrm{d}\rho}{\mathrm{d}p}$$

即

$$\mathrm{d}p=\frac{K}{V}\mathrm{d}\rho$$

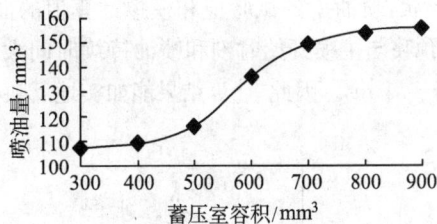

图 8　循环喷油量随增压室容积的变化规律

从中可以看到,增压腔内压力的升高值与液体的弹性模量和密度的增长成正比,而与容积的大小成反比。即当其他结构参数一定时,如果高压腔的容积越小,增压时高压腔的容积变化率就越大,相对应的密度的变化就越大,所以,最终的压力变化也越大。在系统的设计中,最小的高压容积可以设计小于 400 mm^3,高压容积的减小还可以提高喷油器的增压效率。因此,增压腔容积的减小,无论对于提高喷射压力(图 9 所示),还是提高系统的工作效率都是十分有利的。

在新型增压式共轨喷射系统中,喷油量可根据柴油机的转速和负荷灵活控制。如图 10 所示,可通过改变电磁阀通电时间和共轨压力灵活控制喷油量和喷油压力,从而实现了真正意义上的喷射压力和喷油量的柔性控制。作者还模拟计算了其他参数对喷油器喷油特性的影响,计算表明共轨压力、电磁阀通电时间、高压腔容积和增压比对喷油特性的影响较大,通过合理调节系统参数,能满足发动机不同工况的性能要求。

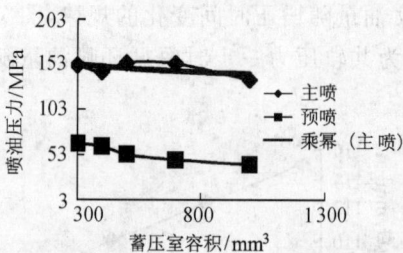

图 9　高压容积对喷射压力的影响

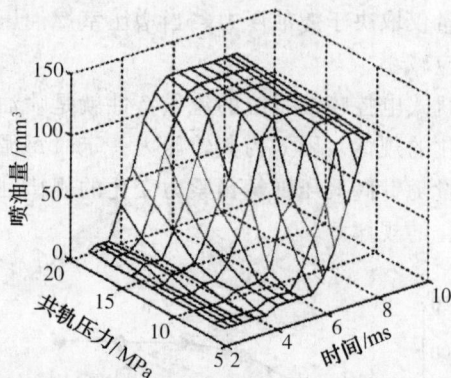

图 10　喷射系统供油特性曲线

4　试验验证

在增压比为 10 时检测增压式喷油器在特定工况下的参数,以验证喷油规律、时间和喷油量与计算值是否相吻合。对增压式喷油器的喷油率进行测量,结果如图 11 所示。由图可见,有明显的预喷,同时喷油率曲线形状也较为理想。

为了验证模拟结果的正确性,在改变发动机转速和共轨压力的条件下,比较了试验台实测值和计算值,结果如图 12 所示。在不同共轨压力和转速条件下,有较理想的喷油率形状,喷油率的试验值和模拟计算值的曲线规律大体一致;预喷与主喷开始时刻和喷油持续时间大致相同,其误差小于 0.1 ms。因此,计算结果能如实地反映喷油规律。

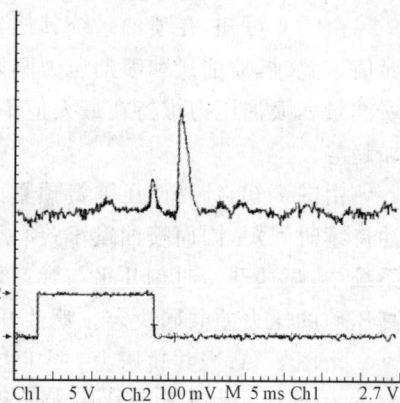

图 11　试验台架上测量所得的喷油率

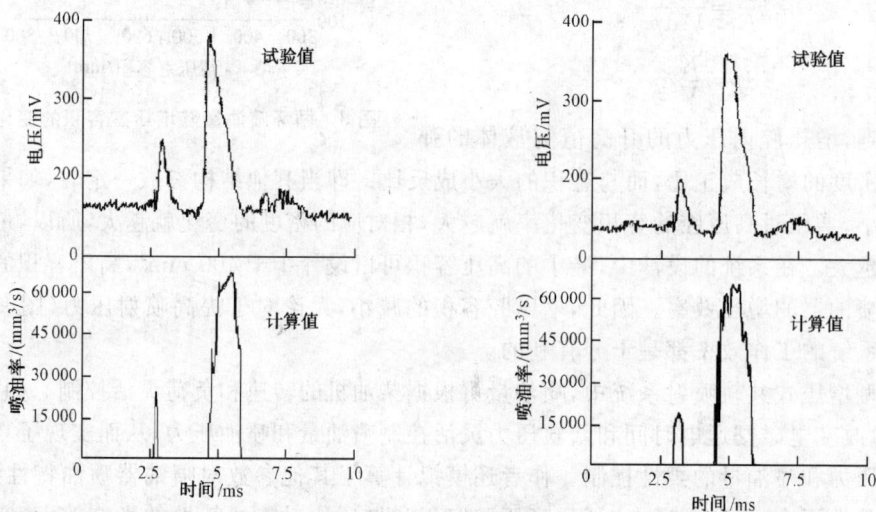

(a) 转速 1 000 r/min,共轨压力 10 MPa,每次喷油量 100 mm³　(b) 转速 1 600 r/min,共轨压力 15 MPa,每次喷油量 115 mm³

图 12　不同转速和共轨压力下试验值与计算值的比较

在该试验台上进行了不同转速条件下喷油量、喷油正时与共轨压力之间关系的测定,测试结果见表1、表2。表1、表2中括号内数据为预先设定的油量、正时MAP图中的计算值,另一数据为试验值(取10次的平均值)。从表1、表2中可以看出,喷油量、喷油正时与共轨压力和转速具有所给定的关系,其误差一般不超过5 mm³(5％)和0.2 ms,基本符合要求,这说明该电子控制燃油喷射系统具有良好的可控制性和可靠性,其调节范围广,控制灵活,系统工作稳定。持续时间上存在的误差,主要与电磁阀的响应速度有关,尤其是闭阀响应速度。

表 1   不同压力、转速下喷油量试验值和计算值的比较　　　　　　mm³

转　速	共　轨　压　力		
	10 MPa	12 MPa	15 MPa
1 500 r/min	113(116)	123(125)	137(140)
1 200 r/min	105(110)	117(120)	124(128)
1 000 r/min	100(105)	108(113)	117(120)

表 2   不同压力、转速下喷油持续时间的试验值和计算值的比较　　　ms

转　速	共　轨　压　力		
	10 MPa	12 MPa	15 MPa
1 500 r/min	2.2(2.1)	2.4(2.4)	2.6(2.5)
1 200 r/min	2.0(1.8)	2.2(2.1)	2.5(2.3)
1 000 r/min	1.9(1.7)	2.0(2.0)	2.1(2.0)

由试验验证可知模拟计算结果是准确的,它可以用于分析中压共轨式喷油器的结构参数对喷油特性的影响。

5　结　束　语

研究结果表明,原蓄压式共轨喷油系统存在不足,如供油量与共轨压力直接相关,喷油规律与理想的喷油规律相反等。采用改进后的新型增压式共轨喷射系统,能实现预喷和高压喷射;喷油规律与理想喷油规律较为接近;可根据共轨压力和电磁阀通电时间全面控制喷油量。这样的喷油系统,才能实现真正意义上的喷油量和喷油压力的柔性控制。

参 考 文 献

[1]宓浩祥,等.国内外柴油机电控技术发展动态及我国今后发展的方向[J].内燃机燃油喷射和控制,1996(6):3-23.

[2]皇甫坚,等.现代汽车电子技术与装置[M].北京:北京理工大学出版社,1999.

[3]董尧清.柴油机共轨式电控喷射系统的进展[J].国外内燃机,2000(5):26-35.

[4]王钧效,等.柴油机共轨式喷油系统喷油率控制技术分析[J].内燃机燃油喷射和控制,2001(3):9-13.

［5］胡林峰,等.新型增压式柴油机共轨喷射系统的研究与开发[J].内燃机工程,2002,23(1):23-27.

［6］AVL-Hydsim Reference Manual V4.2[R]. 2001：145-210.

（本文原载于《汽车工程》2003 年第 4 期）

Simulation and analysis of medium pressure common rail electronic diesel injection system

Abstract：The pilot injection is realized in medium pressure common rail electronic diesel injection system on the basis of injector structure modification. A model for hydraulic calculation is built and the injection characteristics of the common-rail system are simulated. The experiment proves the correctness of numeral simulations.

柴油机高压共轨喷油系统内的瞬变流动研究

何志霞,李德桃,胡林峰,袁建平,王　谦,唐维新

[摘要]　在分析柴油机高压共轨喷油系统的工作过程及系统特性的基础上,研究了该种喷油系统内部的瞬变流动现象,建立了高压共轨喷油器内部流动的数学模型,并进行了讨论和分析。

　　近10年来,柴油机电控共轨喷油系统在国外得到了很大发展[1,2],而国内的研究尚处于起步阶段。该喷油系统的模拟研究集中于电磁阀的响应特性和ECU电控单元,基本出发点都是希望喷油器能更快响应电压输入信号,阀的特性更接近开关型。但电控及电磁阀本身的特性只是影响喷油系统工作的一个方面,而喷油器内部不稳定流动现象的存在也是另一不容忽视的重要方面,所以有必要对后者作进一步的研究。本文对高压共轨喷油系统内部的瞬变流动现象进行研究,用瞬变流动理论建立系统内部的数学模型,并进行数值模拟。

1　基本过程及特性

　　柴油机高压共轨喷油系统主要由高压供油泵、带调压阀的共轨管(CR)、带电磁阀的喷油器和ECU电控单元四部分组成,图1为其示意图。

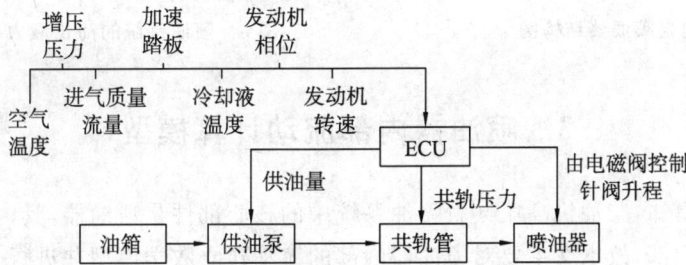

图1　高压共轨喷油系统示意图

　　由图1可看出共轨系统基本工作过程:高压油泵将加压后的高压燃油送到共轨管中,并在其中保持恒定高压;ECU电控单元得到由安装在车辆和发动机上的发动机转速、发动机相位、冷却液温度等传感器所输出的相关信息,控制供油泵供油量和共轨管油压;当控制喷油器上部的电磁阀通电时,喷油器针阀运动,共轨管中的高压燃油便通过高压油管送到喷油器喷射;电磁阀断电时,喷油过程结束。

2　瞬变流动现象

　　经过对高压共轨喷油系统中流体动力学的研究,得知整个系统内部的流动为瞬变流动。图2为一种高压共轨喷油器的结构图。在电磁阀未通电时,A节流孔被关闭,Z节流孔打

开,共轨管中的高压燃油作用于控制活塞 P 上部,使针阀 D 关闭喷油嘴。而当电磁阀通电后,A 节流孔被打开,控制活塞上部压力迅速下降,针阀向上运动,开始喷油;电磁阀关闭时,控制室压力上升,针阀向下运动,关闭喷油嘴,停止喷油,完成喷油过程。由分析知其内部的不稳定现象由三方面引起[3]:(1)衔铁的残余自由运动和电磁阀的再次打开引起的不稳定流动;(2)针阀的残余运动引起的不稳定流动;(3)控制活塞的自由振动产生控制室节流,从而导致喷油管和各室的残余压力波动引起的不稳定流动。

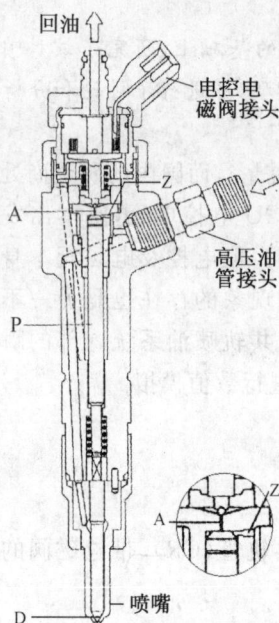

图 2　BOSCH 电控喷油器结构图　　　　　图 3　喷油系统的简化液力控制模型

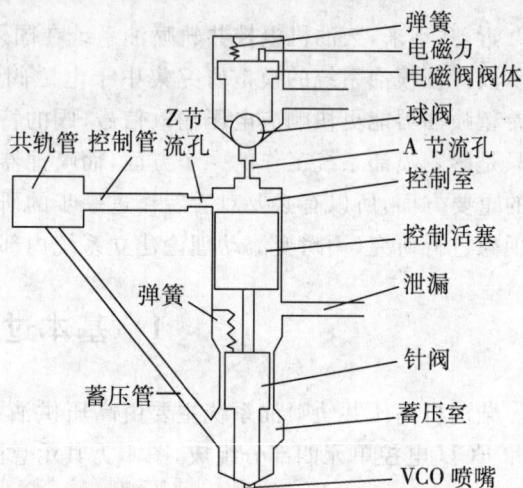

3　喷油器内部流动计算模型

由上述分析得知,柴油机高压共轨喷油系统中的核心部件是喷油器,其内部不稳定现象主要产生于喷油器内部,故本文主要对喷油器内部的流动建立数学模型并进行分析。

3.1　喷油器简化液力模型的建立

在对整个共轨喷油系统结构以及该系统工作原理、工作过程分析的基础上,对高压燃油从共轨管出来通过高压油管进入喷油器内,进而从喷油器喷嘴喷入气缸这一流体动力学过程进行简化,建立图 3 所示的喷油系统简化液力模型。它如实反映了高压共轨喷油系统的工作过程,同时大大简化了对问题的求解。根据图 3 可对高压共轨喷油器各个单元模块分别建立流动控制方程。

3.2　流动控制方程的建立

3.2.1　高压油管

高压油管包括控制管和蓄压管。高压油管内的流动可看作一维可压缩、等截面、有阻力的瞬变流动。为简化模型进行如下假设[4]:和流体压缩性相比,忽略管壁弹性;温度变化很小,粘度恒定,无热量传递。

· 运动方程为

$$\frac{\partial u}{\partial t} + u\frac{\partial u}{\partial x} + \frac{1}{\rho}\frac{\partial p}{\partial x} + \frac{fu|u|}{2D} = 0 \tag{1}$$

连续性方程为

$$\frac{\partial \rho}{\partial t} + u\frac{\partial \rho}{\partial x} + \rho\frac{\partial u}{\partial x} = 0 \tag{2}$$

能量守恒方程为

$$\frac{\partial \rho}{\partial t} + u\frac{\partial \rho}{\partial x} - \frac{1}{a^2}\left(\frac{\partial p}{\partial t} + u\frac{\partial p}{\partial x}\right) = 0 \tag{3}$$

式中 ρ 为燃油密度；a 为压力波传播速度；f 为 Darcy-Weisbach 摩擦系数[5]；D 为油管直径。

式(3)减去式(2)得

$$\frac{\partial u}{\partial x} + \frac{u}{a^2\rho}\frac{\partial p}{\partial x} + \frac{1}{a^2\rho}\frac{\partial p}{\partial t} = 0 \tag{4}$$

可用特征线法求解式(1)和式(4)，但非常繁琐，且耗时多。试将 $u\dfrac{\partial u}{\partial x}$ 和 $u\dfrac{\partial p}{\partial x}$ 略去，使方程变为线性偏微分方程。经计算对比发现简化对计算结果的影响很小，又大大节省时间，故方程(1)和(4)简化为

$$\rho\frac{\partial u}{\partial t} + \frac{\partial p}{\partial x} + \frac{\rho fu|u|}{2D} = 0 \tag{5}$$

$$\frac{\partial u}{\partial x} + \frac{1}{a^2\rho}\frac{\partial p}{\partial t} = 0 \tag{6}$$

式(5)和(6)对 x 求导，并整理得

$$\frac{\partial^2 u}{\partial x^2} - \frac{1}{a^2}\frac{\partial^2 u}{\partial t^2} - \frac{f|u|}{2a^2D}\frac{\partial u}{\partial t} = 0 \tag{7}$$

$$\frac{\partial^2 p}{\partial x^2} - \frac{1}{a^2}\frac{\partial^2 p}{\partial t^2} - \frac{f\rho|u|}{2D}\frac{\partial u}{\partial t} = 0 \tag{8}$$

采用有限差分法由式(7)和(8)可求得近似数值解。

3.2.2 控制室

将文献[5]中对一般空腔、容积所建的基本控制方程用于建立控制室控制方程，即

$$\frac{\mathrm{d}p_c}{\mathrm{d}t} = \frac{E}{V_c(h_p)}\left(Q_{in} - Q_{out} - A_p\frac{\mathrm{d}(h_p)}{\mathrm{d}t} - Q_{leak}\right) \tag{9}$$

其中 $Q_{in} = C_1 A_{in}\sqrt{\dfrac{2(p_r - p_c)}{\rho}}$；$Q_{out} = C_2 A_{out}\sqrt{\dfrac{2(p_c - p_0)}{\rho}}$；$Q_{leak} = \dfrac{\pi \cdot d_p \delta^3}{12\mu l_p}\Delta p^{[3]}$；$V_c(h_p) = V_{c0} - A_p h_p$。

式中 p_c 为控制室压力；p_r 为共轨压力；p_0 为外界压力；E 为燃油弹性模量；V_c 为控制室容积；h_p 为控制活塞升程；Q_{in} 为由 Z 节流孔流入控制室的流量；Q_{out} 为经 A 节流孔流出控制室的流量；Q_{leak} 为泄漏量；C_1、C_2 为流量系数；A_p 为控制活塞横截面积；A_{in} 为 Z 节流孔横截面积；A_{out} 为 A 节流孔横截面积；V_{c0} 为控制室初始容积；δ 为控制活塞与控制套间隙；l_p 为控制活塞长度；d_p 为控制活塞直径；μ 为动力粘度；Δp 为间隙两端压力差。

由工作过程和上述方程可看出：当电磁阀通电，球阀开启后，控制室压力下降；当下降至低于燃油气化压力 p_{cav} 时，不再满足上述方程。故压力降至 p_{cav} 后，可假定维持该压力不变，即

不再下降,然后可引入气化容积 V_{cav} 对上述方程进行修正。

3.2.3 控制活塞

控制活塞的受力平衡方程为

$$m_p \frac{d^2 h_p}{dt^2} = p_c A_p - p_r A_n - f_c \tag{10}$$

式中 m_p 为控制活塞质量;A_n 为针阀横截面积;f_c 为活塞运动阻力,$f_c = c' \frac{dh_p}{dt}$,c' 为阻尼系数。

3.2.4 电磁阀

图 4 为电磁阀结构示意图。阀所受力有电磁力 F_m、弹簧预紧力 F_0、液压力 F_{hyd}、摩擦力 F_{frict} 以及当球阀碰到阀座和超过最大行程 x_{lift} 时所受的附加力 F_1,F_2。故其运动方程为

图 4 电磁阀结构示意图

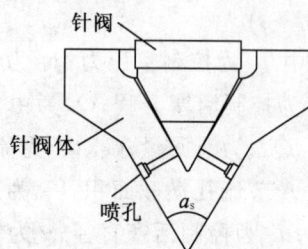

$$m_b \frac{d^2 x}{dt^2} + c \frac{dx}{dt} + kx = F_m - F_0 - F_{hyd} - F_{frict} - F_1 - F_2 \tag{11}$$

其中 $\quad F_0 = p_{open} \frac{\pi}{4} (d_b \cos \alpha_s)^2$

$$F_{hyd} = (p_{out} - p_{in}) \frac{\pi}{4} (d_b \cos \alpha_s)^2$$

式中 c,k 为球阀弹簧的阻尼系数、弹性系数;m_b 为球阀质量;p_{open} 为开阀所需压力;α_s 为阀座锥角。

球阀碰到阀座($x < 0$)时,

$$F_1 = c_1 \frac{dx}{dt} + k_1 x$$

球阀超过最大行程($x > x_{lift}$)时,

$$F_2 = c_2 \frac{dx}{dt} + k_2 (x - x_{lift})$$

而球阀流量则为

$$Q = \sqrt{\frac{1}{\zeta_{eq}} \frac{2(p_{in} - p_{out})}{\rho}} A_{flow} \tag{12}$$

其中

$$\zeta_{eq} = \frac{\zeta_{in}}{A_{in}^2} + \frac{\zeta_{seat}}{A_{seat}^2} + \frac{\zeta_{out}}{A_{out}^2}$$

$$A_{seat} = \pi x \sin \alpha_s \cos \alpha_s (d_b + x \sin \alpha_s)$$

式中 ζ_{eq} 为总局部阻力系数;A_{flow} 为球阀流通面积。

3.2.5 VCO 喷嘴喷孔

该喷油器选用了 VCO(Valve Closed Orifice)喷嘴,即针阀密封喷孔式喷嘴,也可称为无压力室喷嘴(如图 5 所示),根据伯努利方程,喷嘴流量为

$$Q_{vco} = C_{vco} A \sqrt{\frac{2}{\rho} | p_{in} - p_{out} |} \tag{13}$$

只有当 $p_{in} > p_{out}$ 时,才有燃油从喷嘴流出,此时 Q_{vco} 取正值。N_{holes} 为喷孔数,d_{hole} 为喷孔直径,作如下定义:

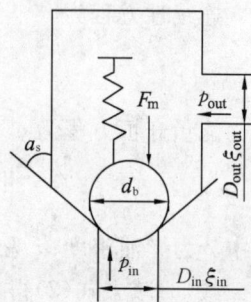

图 5 VCO 喷嘴结构简图

柴油机高压共轨喷油系统内的瞬变流动研究

$$A_{holes} = \frac{\pi}{4} N_{holes} d_{hole}^2$$

$$A_{nsh} = N_{holes} d_{hole} \pi x_{lift} \sin \frac{\alpha_s}{2}$$

此时，A 可取 A_{nsh}，A_{holes} 的最小值。

上述方程为不考虑空穴时的基本控制方程，但实际上在喷孔处一定存在着空穴现象，而且研究表明正是产生于喷孔内的空穴对液流造成的紊乱使液体喷束雾化得以改善。

3.2.6 泄漏

在高压共轨系统中，泄漏是系统主要的能量损失源。喷油器各配合偶件总处于很高的燃油压力作用下，配合面间的泄漏时间较常规系统大大增加，故总泄漏量是十分重要的参数。由于喷油器各配合偶件间隙均很小，故泄漏可视为环形狭窄间隙中的层流边界层流动。图 6 为配合偶件泄漏流动模型，对 N-S 方程简化后有

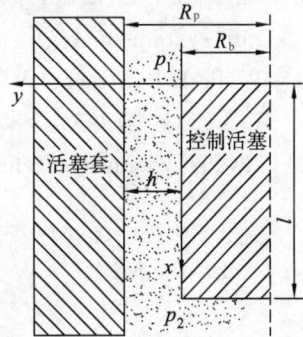

图 6　泄漏流动模型

$$\frac{\mathrm{d}p}{\mathrm{d}x} = \mu \frac{\mathrm{d}^2 u}{\mathrm{d}y^2}$$

$$\frac{\mathrm{d}p}{\mathrm{d}x} = \frac{p_1 - p_2}{l} = \mathrm{const}$$

轴向速度 u 的边界条件为

$$\begin{cases} u = u_b, & y = 0 \\ u = 0, & y = h \end{cases}$$

故

$$u(y) = \frac{1}{2\mu} \frac{\mathrm{d}p}{\mathrm{d}x}(hy - y^2) + \frac{y}{h} u_b - u_b$$

其中 $h = R_p - R_b$。

由于 h 非常小，故偶件间隙的泄漏量为

$$Q_{leak} = 2\pi R_b \int_0^h u(y)\mathrm{d}y = 2\pi R_b \left(\frac{1}{12\mu} \frac{p_1 - p_2}{l} h^3 - \frac{h u_b}{2} \right) \tag{14}$$

式中 u_b 为控制活塞速度。

虽然对喷油器各模块单元建立起数学模型，但实际情况远比上述计算复杂。如压力极高时，流体温度不再保持恒定，会有较大范围的变化，流体粘度也相应发生变化；A 孔、Z 孔中一定会有空穴现象存在；针阀、压力杆或多或少会发生变形；阀的流量特性、燃油特性受各种因素的影响等，这都会使计算精度、准确性受到较大影响。故在上述方程的基础上，还需作更进一步的分析，从而不断修正和完善各控制方程及相关的边界条件。

4　结　束　语

对高压共轨喷油系统中的喷油器的结构进行液力简化，并进行了模块化，针对各个具体单元分别进行分析，建立数学模型。通过对内部流动现象的模拟仿真分析，可以进一步改进系统的各种结构参数，进一步优化喷油系统的喷油性能。

参 考 文 献

[1] Hlousek J. *Common rail fuel injection system for high speed large diesel engine*[M]. Copenhagen：CIMAC Congress，1998：25-32.

[2] Glassey S F，Stockner A R，Flinn M A. HEUI——a new direction for diesel engine fuel systems[J]. *SAE*，930270，1993.

[3] Laforgia F D，Landriscina V. Evaluation of instability phenomena in a common rail injection system for high speed diesel engines[J]. *SAE Paper*，1999-01-0192，1999.

[4] Wylie E B，Streeter V L. *Fluid Transients in Systems*[M]. New Jersey：Prentice-Hall Inc，1993：37-95.

[5] AVL-Hydsim Reference Manual V4. 2[R]. 2001：145-210.

（本文原载于《农业机械学报》2004 年第 1 期）

Study of transient flow in a common rail injection system of diesel engine

Abstract：A high-pressure，common-rail injection system can increase the power capability of a diesel engine and decrease the fuel consumption，noises and emissions. The working process and the system characteristics of this injection system were introduced in the paper. The transient flow phenomenon of the injection system was analyzed. The progress of the fluid dynamics from the common-rail tube to the cylinder was simplified and a calculation model of flow in the common-rail injection system was established and analyzed.

喷油嘴喷孔内部空穴两相流动数值模拟分析

何志霞,李德桃,胡林峰,袁建平,王　谦,袁文华

[摘要]　在对柴油机喷油嘴喷孔内部空穴流动现象分析的基础上,建立了完全发展了的空穴流动的二维气液两相流空穴模型,并进行了喷孔内部气液两相流数值模拟。计算结果与国外已有试验数据的对比表明:所建模型是正确的。在此计算模型框架下分析了喷孔上下游压差和喷孔入口圆角半径、喷孔直径和长度等几何特征参数对喷孔内空穴区域分布的影响,进而分析了对喷雾特性的影响。

　　柴油机喷油嘴是柴油机最关键的部件之一。近年来国外大量研究表明:燃油喷雾的形成、雾化过程及雾化质量除了受其与周围空气摩擦的影响外,更受到喷油嘴内部流动情况,即湍流和空穴的影响[1]。故控制和优化喷孔内部流动对良好的喷雾特性就异常重要,它影响到柴油机的燃烧及排放性能。但喷孔内部流动的高度湍动以及气液两相流动的复杂性使得我们对这种影响机理和影响程度尚不很清楚。通常喷孔尺寸相当小($10^{-4} \sim 10^{-3}$ m),内部流动流速很高,进行试验存在相当的难度,因而对喷孔内部流动进行数值模拟研究非常有必要。同时,这也可为模拟发动机气缸内喷雾发展的 CFD 喷雾模型提供必要的边界条件。Chaves[2],Roosen[3]等人用试验对喷孔内气泡的形成过程进行了照相记录,Schimidt[4]则对喷孔内流动情况进行较为详尽的研究,而国内的相关研究则很少。

1　喷油嘴喷孔内部流动空穴现象分析

　　掌握喷油嘴喷孔内部流动特性非常有助于对喷雾破裂模型的处理。喷孔非常小,典型喷孔通常大约只有 1 mm 长,直径也只有零点几毫米,通过喷孔的流速又非常高,接近每秒几百米的数量级,而且流动是瞬态的,喷油持续期只有几毫秒,这使直接观察喷孔内部流动非常困难。喷孔内部流动是剧烈的湍流(雷诺数在 50 000 个数量级以上),且为两相流,因此这种喷油嘴内流体动力机理的复杂性,已使其自身成为一研究领域。对喷油嘴喷孔内部流动的研究对象包括湍流、发展了的管路流动和空穴。尽管湍流仍然是影响喷孔内部流动的一个重要因素,但空穴现象更是造成喷孔出口燃油雾化的一个重要原因,于是空穴成为近来研究的一个重点。

　　图 1 所示为喷孔空穴简化模型。根据不可压缩流体伯努利方程,沿流线速度越高,压力越低。在高速流动的喷孔入口处,由于拐角的存在,产生局部流动分离以及孔口收缩,引起横截面面积的减小。根据质量守恒方程,此截面上的流速大大

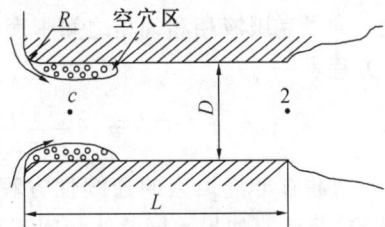

图 1　喷孔流动空穴模型

增加,从而引起压力降低,当局部压力降至低于该液体饱和蒸气压时,该区域就会出现空穴现

象,形成空穴区。而其中气泡的存在使液相流体的连续性遭到破坏,在喷孔内形成蒸气泡、液体中的未溶气体和液体的两相流动。照片显示空穴仅仅发生在喷孔入口处,但通常都会向下游延伸而形成长而薄的空穴层。气泡的形成很大程度上会依赖于喷孔几何特征(R 与 L 等)、喷孔壁面上的毛刺、几何缺陷以及上下游压差。这种空穴是一种高速现象,其产生、发展和溃灭过程均十分复杂和短暂,需要建立起简化、有效的喷油嘴空穴流动数学模型,进行数值模拟研究。

2   喷孔内空穴流动的数学模型

本文对喷孔内部完全发展了的空穴流动进行了依赖时间的二维气液两相数值模拟。采用了混合(Mixture)多相流模型,附加以空穴模型(气泡的生长和溃灭模型),求解依赖于时间的空穴区在喷孔内的分布情况。对流体的湍动,采用了 k-ε 模型。对方程的离散采用有限容积法。

2.1   混合多相流空穴模型

在喷油嘴喷孔内部的空穴流动中,气液两相被认为是均匀混合体,所以也可理解为单一流体。密度、粘度均采用混合密度和混合粘度,则连续性方程和动量方程可采用与单相流体相同形式的方程描述。同时假设两相以相同速度运动,混合模型也就简化为均匀两相流模型。

此时混合相的连续性方程和动量方程分别为:

$$\frac{\partial \rho}{\partial t} + \nabla \cdot (\rho \boldsymbol{v}) = 0 \tag{1}$$

$$\frac{\partial (\rho \boldsymbol{v})}{\partial t} + \nabla \cdot (\rho \boldsymbol{v} \boldsymbol{v}) = -\nabla p + \nabla [\mu (\nabla \cdot \boldsymbol{v} + \nabla \cdot \boldsymbol{v}^{\mathrm{T}})] + \rho \boldsymbol{g} + \boldsymbol{f} \tag{2}$$

式中 $\boldsymbol{v}$ 为流速;p 为压力;$\boldsymbol{f}$ 为体积力;ρ, μ 分别为混合密度和混合粘度。

$$\rho = (1 - \alpha) \rho_\mathrm{l} + \alpha \rho_\mathrm{v} \tag{3}$$

$$\mu = (1 - \alpha) \mu_\mathrm{l} + \alpha \mu_\mathrm{v} \tag{4}$$

式中 α 为气相体积分数(蒸气在整个气液混合体中所占比例);$\rho_\mathrm{l}, \rho_\mathrm{v}, \mu_\mathrm{l}, \mu_\mathrm{v}$ 分别为纯液相和纯气相的密度与粘度,可认为是常数。假定空穴区中的气相是由许多很小的球状气泡组成,气泡半径为 r 时,气液比 α 可按下式计算:

$$\alpha = \frac{n_0 \cdot 4\pi r^3 / 3}{1 + (n_0 \cdot 4\pi r^3 / 3)} \tag{5}$$

式中 n_0 为单位体积纯液体中所含的气泡数,是预先给出的。于是空穴形成发展过程中气相体积分数的变化率为

$$\frac{\mathrm{d}\alpha}{\mathrm{d}t} = (1 - \alpha) \frac{4\pi n_0 r^2}{1 + (n_0 \cdot 4\pi r^3 / 3)} \cdot \frac{\mathrm{d}r}{\mathrm{d}t} \tag{6}$$

为了克服液相和气相之间密度存在很大差异所带来的数值求解困难,采用非守恒型连续性方程

$$\nabla \cdot \boldsymbol{v} = -\frac{1}{\rho} \left(\frac{\partial \rho}{\partial t} + \boldsymbol{v} \cdot \nabla \rho \right) = -\frac{1}{\rho} \cdot \frac{\mathrm{d}\rho}{\mathrm{d}t} = \frac{\rho_\mathrm{l} - \rho_\mathrm{v}}{\rho} \cdot \frac{\mathrm{d}\alpha}{\mathrm{d}t} \tag{7}$$

可将该非守恒型的连续性方程和动量方程(2)联立求解,对压力的修正方法采取 SIMPLEC 算法。如果不同流体相间不存在质量输运,此时附加项 $\mathrm{d}\alpha/\mathrm{d}t$ 即为 0。

对气相体积分数 α,结合式(6)、式(7)就可导出完全不同于连续性方程(1)的一个输运方程(体积分数方程):

$$\frac{\partial \alpha}{\partial t} + \nabla \cdot (\alpha \boldsymbol{v}) = \frac{\mathrm{d}\alpha}{\mathrm{d}t} + \alpha \nabla \cdot \boldsymbol{v} = \frac{(1-\alpha)\rho_l}{(1-\alpha)\rho_l + \alpha\rho_v} \cdot \frac{4\pi n_0 r^2}{1 + n_0 \cdot 4\pi r^3/3} \cdot \frac{\mathrm{d}r}{\mathrm{d}t} \tag{8}$$

对单个气泡的成长破裂过程,在等温下,忽略蒸发潜热,由 Rayleigh-Plesset 方程有:

$$r \frac{\mathrm{d}^2 r}{\mathrm{d}t^2} + \frac{3}{2}\left(\frac{\mathrm{d}r}{\mathrm{d}t}\right)^2 = \frac{p_b - p}{\rho_l} - \frac{2\sigma}{\rho_l r} - 4\frac{\mu_l}{\rho_l r} \cdot \frac{\mathrm{d}r}{\mathrm{d}t} \tag{9}$$

此处 p_b 表示气泡内的压力,为蒸气部分压力 p_v 和不凝性气体部分压力 p 之和,σ 是表面张力系数。对上述方程作进一步简化后有:

$$\frac{\mathrm{d}r}{\mathrm{d}t} = \begin{cases} \sqrt{\dfrac{2(p_b - p)}{3\rho_l}}, & p_v > p \\[3mm] -\sqrt{\dfrac{2(p_b - p)}{3\rho_l}}, & p_v < p \end{cases} \tag{10}$$

2.2 湍流模型

与单相流动相比,多相流动动量方程中所模拟的项数增加了,使得多相流模拟中的湍流模型非常复杂,而当前尚没有非常合适的湍流模型。作为初步的研究,笔者采用了用于单相流动的标准 k-ε 模型来模拟气液混合相的流动。该混合相湍动能 k 方程和耗散率 ε 方程[6]分别为:

$$\frac{\partial}{\partial t}(\rho k) + \nabla \cdot (\rho \boldsymbol{v} k) = \nabla \cdot \left(\frac{\mu_t}{\sigma_k}\nabla k\right) + G_k - \rho\varepsilon \tag{11}$$

$$\frac{\partial}{\partial t}(\rho\varepsilon) + \nabla \cdot (\rho \boldsymbol{v} \varepsilon) = \nabla \cdot \left(\frac{\mu_t}{\sigma_\varepsilon}\nabla\varepsilon\right) + \frac{\varepsilon}{k}(C_{1\varepsilon}G_k - C_{2\varepsilon}\rho\varepsilon) \tag{12}$$

式中 $\mu_t = \rho C_\mu \dfrac{k^2}{\varepsilon}$;$G_k = \mu_t \nabla \cdot \boldsymbol{v}[\nabla \cdot \boldsymbol{v} + (\nabla \cdot \boldsymbol{v})^{\mathrm{T}}]$;模型常量 $C_{1\varepsilon} = 1.44, C_{2\varepsilon} = 1.92, \sigma_k = 1.0, \sigma_\varepsilon = 1.3, C_\mu = 0.09$。

k-ε 模型通过对粘性支层的积分可获得适当的准确解。但是在高雷诺数下,与壁面邻接的粘性支层非常薄,很难采用足够多的网格点来求解。为了减少计算时间,节省内存,笔者采用了壁面函数法来处理。在壁面函数方法中,与壁面剪切应力有关的切向速度服从对数分布规律:

$$u^+ = \frac{u}{u_\tau} = \frac{1}{\kappa}\ln\left(\frac{yu_\tau}{\upsilon}\right) = \frac{1}{\kappa}\ln(y^+) + B \tag{13}$$

式中 $u_\tau = \sqrt{\tau_{\mathrm{wall}}/\rho}$ 为切应力速度;y^+ 为距壁面的无量纲距离;u 为速度沿壁面的切向分量;y 为壁面距离;υ 为运动粘度;冯卡门常数 $\kappa = 0.42, B = 5.44$。

通常当 y^+ 在 $30 \sim 500$ 时,上述对数法则是有效的。而本研究中,近壁区域被分成两部分,对应有如下的关系式:

$$u^+ = \begin{cases} y^+, & y^+ \leqslant 11.225 \\[2mm] \dfrac{1}{\kappa}\ln(y^+) + B, & y^+ > 11.225 \end{cases} \tag{14}$$

3 数值模拟验证

本文以 Roosen 等人[3]试验所用的单孔轴对称喷嘴为例,喷孔直径为 0.28 mm,长度为 1 mm,入口圆角半径 $R = 28~\mu m$。该试验选用的流体介质为 20 ℃ 的水。利用上述所建模型进行二维气液两相流的数值模拟,基于内节点有限容积法对 $u, p, k, \varepsilon, \alpha$ 等值进行了计算。首先在新的时步下,求解气相输运方程(8),解出 α 值,然后以该新计算的 α 和混合密度 ρ 通过一定的迭代再求解动量方程。

计算的初始及边界条件为:喷油压力 $p_{in}=8$ MPa,喷油嘴出口压力 $p_{out}=2.1$ MPa。进行依赖时间的瞬态求解,时间步长 $\Delta t=3\times10^{-8}$ s,相应的 CFL 数则在 $0.1\sim0.5$ 之间取值。对空穴模型中气泡初始直径 R_0 和单位体积纯液体中所含的气泡数 n_0 的取值,参考相关资料在一定范围内进行了反复的试算,最终确定:当 $n_0=1.5\times10^{14}$, $0.1\ \mu m\leqslant R_0\leqslant1.2\ \mu m$ 时,计算结果中密度分布图所反映出的空穴区域分布情况和 Roosen 等人的试验所拍摄图片(图2)均能很好地保持一致。故取 $n_0=1.5\times10^{14}$, $R_0=0.5\ \mu m$,入口湍流强度 $I=0.16\times(Re)^{-1/8}$ (Re 为雷诺数),湍流长度尺度 $l=0.07D$,壁面为无渗透、无滑移壁面,压力的法向分量为 0,采用上述的标准壁面函数法,最终数值模拟结果如图3所示。

图 2　试验得出的喷孔内密度分布

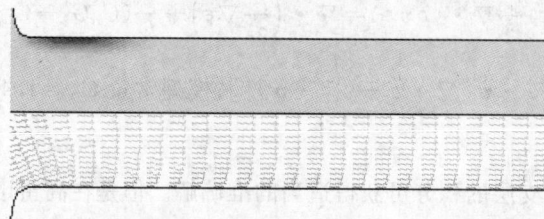

图 3　数值模拟计算所得到的喷孔内密度和速度分布

由图2试验拍摄得到的图片和图3数值计算出的喷孔内密度分布图所反映出的空穴区域分布情况,可看出空穴层延伸长度均大致为 200 μm,这验证了本文数值计算模型的正确性。于是在该计算模型框架下,可分析喷油嘴几何特征参数以及上下游压差对内部所形成的空穴区域的影响,进而分析对喷雾的影响。

4　数值模拟的计算结果对比分析

利用本文计算模型,以上述喷油嘴为参考,分别单独改变上下游喷油压差、喷油嘴入口圆角半径、喷油嘴喷孔直径及长度进行数值模拟。图4为下游压力依次由 2.1 MPa 减小为 1.5 MPa,1.1 MPa,0.5 MPa,其他条件及上游压力不变时所计算出的喷嘴孔内密度和速度分布图。图5为入口圆角半径依次由 28 μm 变为 56 μm,14 μm,0 μm 时的密度和速度分布图。图6、图7则分别为保持喷油嘴喷孔直径不变而改变喷孔长度、保持喷孔长度不变而改变喷孔直径时计算出的喷孔内密度和速度分布图,这两图实际上反映了不同长径比 L/D 时喷孔内部空穴区域的分布情况。对图4~图7的对比分析得出以下结论:

(1)随着上下游压差的增加,喷孔内部空穴层延伸长度不断增加。当压差达到一定程度,空穴延伸超出喷嘴出口,出现所谓的"超空穴"现象。而下游压力相对较高时,空穴区较短,不会延伸至出口(图4)。

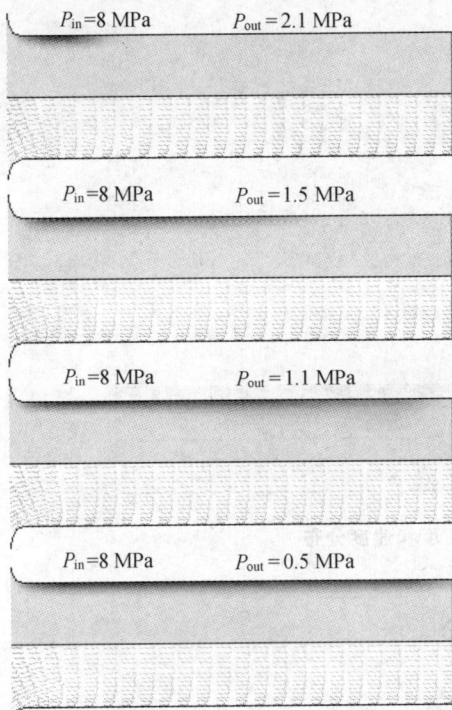

图 4　不同喷油压差下喷孔内密度和速度分布

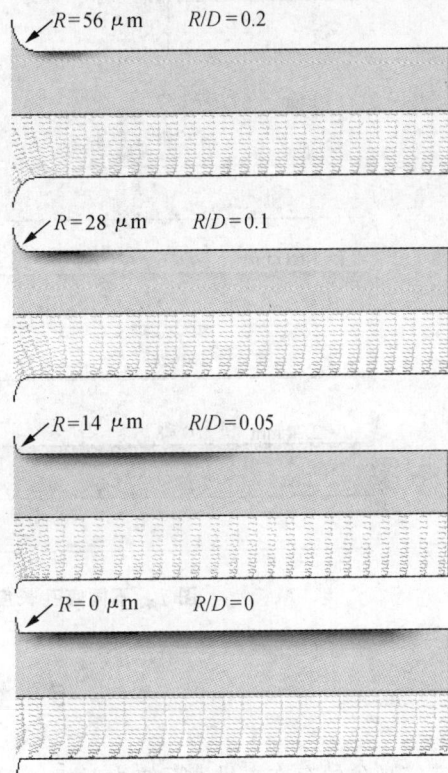

图 5　不同入口圆角半径下喷孔内密度和速度分布

(2)喷孔入口圆角半径 R 越大,即 R/D 值越大,入口圆角边越圆,流量系数会越大。通常喷孔轴线液相核心区速度接近于伯努利速度,故该液相核心区相应也越大,即空穴层越薄,同时延伸长度也越短。图 5 的计算结果证实了这一点。

(3)由图 6 可看出在喷孔长度 L 不变的情况下,随着孔径 D 的减小,空穴层延伸长度增加。图 7 则反映出保持 D 不变,长度 L 增加,空穴层延伸长度相应地增加。结合图 6 和图 7 可看出,在其他参数不变情况下空穴层延伸长度关键由长径比 L/D 决定,即 L/D 值越大,则空穴层延伸长度越长,但同时空穴层厚度几乎没什么变化。L/D 变化对流量系数的影响没有 R/D 变化对流量系数的影响显著。对于较长的孔,接近壁面的空穴层在出口处会离开壁面,短孔则不会出现该现象。

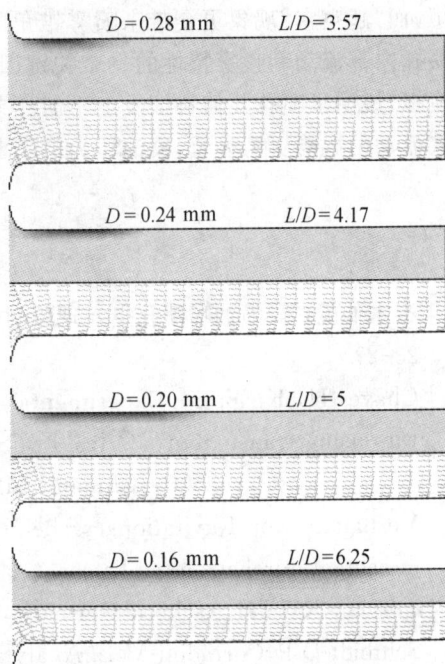

图 6　不同喷孔直径下喷孔内密度和速度分布

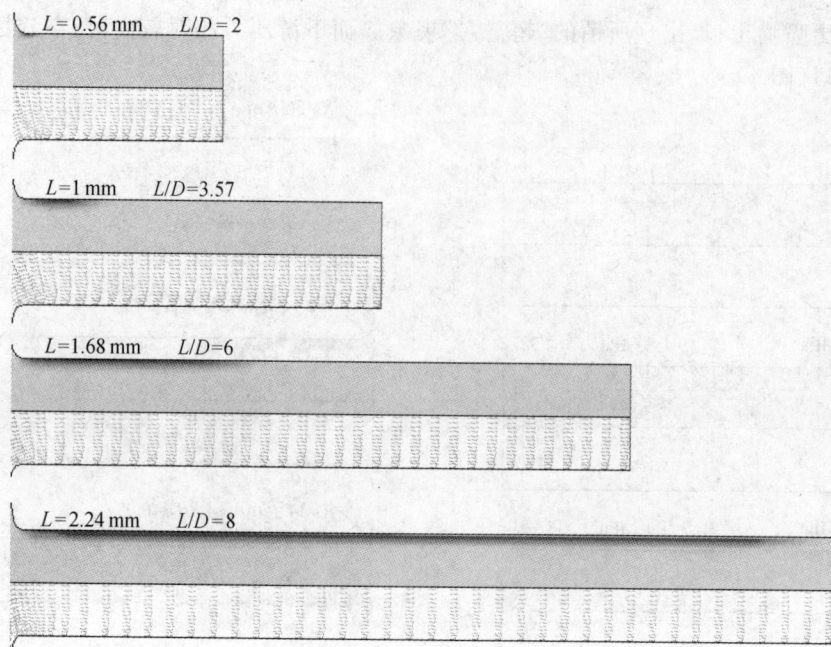

图 7　不同喷孔长度下喷孔内密度和速度分布

5　结　　论

　　空穴现象的存在在供油系统中对机械零件有负面影响,而对喷雾分裂过程则起到积极作用。目前国外相当一部分试验均观察到了出现空穴时,喷雾锥角会显著增加,喷雾破裂长度会缩短,而"超空穴"现象更会带来喷雾锥角的显著增大,从而优化喷雾特性。空穴现象已成为联系喷孔内部流动和喷雾特性的一个关键因素。本文对单孔轴对称喷嘴喷孔内部空穴流动的数值模拟,说明了喷孔几何特征及上下游压差对空穴现象的影响程度,喷孔入口圆角半径对内部空穴流动的影响更为显著,喷油压力的不断提高在很大程度上改善了喷雾特性。

参 考 文 献

[1] 玉木伸茂,等.喷油嘴喷孔空穴现象对液体喷束雾化的影响[J].国外内燃机,1998,338(8): 23-29.

[2] Chaves H, Knapp M. Experimental study of cavitation in the nozzle hole of diesel injectors using transparent nozzles[J]. *SAE Paper*, 950290,1995.

[3] Roosen P, Unruh O, Behmann M. Untersuchung und Modellierung des transienten Verhaltens von Kavitationserscheinungen bei ein-und mehrkomponentigen Kraftstoffen in schnell durchstromten Düsen[C]// *Report of the Institute for Technical Thermodynamics*, RWTH Aachen, Germany,1996.

[4] Schmidt D P,Corradini M L. Analytical prediction of the exit flow of cavitating orifices [J]. *Atomization and Sprays*, 1997,7(6): 54-65.

［5］Schmidt D P，Corradini M L，Rutland C J. A two-dimensional non-equilibrium model of flashing nozzle flow［C］// *3rd ASME / JSME Joint Fluids Engineering Conference*，1999.

［6］陶文铨. 数值传热学［M］. 西安：西安交通大学出版社，1988.

（本文原载于《内燃机学报》2004 年第 5 期）

Numerical simulation and analysis of two-phase cavitating flow in injection nozzles

Abstract：Based on the analysis of the phenomenon of cavitating flow in the nozzles of the diesel injector，a two-dimension and two-phase model of cavitating flow in nozzles was developed for a well-developed cavitating flow and the simulation on gas-liquid two-phase flow was performed. The calculated results from the model have good agreement with the experimental results available in the literatures. Using this model，The effect of pressure difference between upstream and downstream，the corner radius of the nozzle inlet，the nozzle hole diameter and length on the cavitation location in injection nozzle and the spray characteristics were analyzed.

垂直多孔喷嘴内部空穴两相流动的
三维数值模拟分析

何志霞,李德桃,王　谦,袁建平,胡林峰

[摘要]　对完全发展了的空穴流动建立起多维空穴两相流动数学模型,对多孔垂直喷嘴进行了喷孔内部空穴两相流动的三维数值模拟。首次提出将计算区域的出口边界延伸至气缸内部,以减小出口对喷孔内部求解区域的影响。对针阀偏心时各喷孔内部空穴流动特性进行了分析研究,阐明了喷孔内部空穴流动特性及其对喷孔出口流动分布的影响以及对喷雾油束雾化所产生的作用。

柴油机燃油喷射系统利用高喷油压力和极高的喷油速度在喷孔出口形成由很多细小液滴组成的燃油喷雾。长期以来,燃油雾化机理普遍被认为是基于气动雾化理论[1],但对喷射雾化进一步的理论和试验研究[2,3]表明,喷射发展过程不存在一个完整的分裂长度,而是包括两个主要过程:喷孔内部空穴与湍动所引起的初次雾化过程和上述的气动雾化理论所解释的二次雾化过程。因此不考虑喷嘴所带来的影响而建立的预测喷雾形成的理论是不完善的,但喷嘴内部流动的高度湍动和气液两相的存在等燃油流动的复杂性因素使得对喷嘴内部流动物理机理的认识还很不充分。通常喷孔直径只有零点几毫米,喷油持续期只有几毫秒,这使得直接观察喷孔内部流动非常困难,于是对喷嘴头部空穴流动开展气液两相数值模拟研究就显得非常有必要。

1　数学模型

从喷嘴一维模型要获取尽可能多的信息受到很大限制。为了更多地了解空穴在喷嘴内部流动情况,必须发展多维空穴流动数学模型。当前单相的 CFD 已经比较成熟,而对两相空穴流动模拟仍非常有限。多维空穴流动数学模型的建立以 Rayleigh 所发展的单气泡溃灭模型为基础[4]。

应用一套量级分析法的分析[5]表明,目前对于喷嘴空穴流动,可不考虑两相间的相对运动。假定气液两相无差别,是均匀混合体,也可理解为单一流体,即采用单流体法。此时基本的运动方程、连续性方程和单相流体运动有相同的表达形式,只是喷嘴空穴两相流动中的密度、粘度均采用混合密度 ρ 和混合粘度 η 来表示:

$$\rho=(1-\alpha)\rho_l+\alpha\rho_v \tag{1}$$

$$\eta=(1-\alpha)\eta_l+\alpha\eta_v \tag{2}$$

式中 $\rho_l,\rho_v,\eta_l,\eta_v$ 为纯液相和纯气相的密度和粘度,视为常数;α 为气相体积分数,即空穴在气液混合体中所占比例。

假定空穴区中的气相由许多很小的球状气泡组成,气泡既不创造也不被破坏,气泡半径为

r 时，α 作如下计算：

$$\alpha = \frac{V_v}{V_1 + V_v} = \frac{n_0 4\pi r^3/3}{1 + n_0 4\pi r^3/3} \tag{3}$$

于是可得喷嘴内部三维空穴两相湍流流动数学模型基本控制方程[6-8]。

（1）连续性方程：

$$\frac{\partial \rho}{\partial t} + \nabla \cdot (\rho \boldsymbol{v}) = 0 \tag{4}$$

（2）动量方程：

$$\frac{\partial(\rho \boldsymbol{v})}{\partial t} + \nabla \cdot (\rho vv) = -\nabla p + \nabla[\eta(\nabla \cdot \boldsymbol{v} + \nabla \cdot \boldsymbol{v}^{\mathrm{T}})] + \rho \boldsymbol{g} + \boldsymbol{f} \tag{5}$$

（3）气相输运方程：

$$\frac{\partial \alpha}{\partial t} + \nabla \cdot (\alpha \boldsymbol{v}) = \frac{\mathrm{d}\alpha}{\mathrm{d}t} + \alpha \nabla \cdot \boldsymbol{v} = \frac{(1-\alpha)\rho_1}{(1-\alpha)\rho_1 + \alpha\rho_v} \cdot \frac{4\pi n_0 r^2}{1 + n_0 4\pi r^3/3} \cdot \frac{\mathrm{d}r}{\mathrm{d}t} \tag{6}$$

（4）对单个气泡的成长破裂过程，根据 Rayleigh-Plesset 方程[7]有：

$$r\frac{\mathrm{d}^2 r}{\mathrm{d}t^2} + \frac{3}{2}\left(\frac{\mathrm{d}r}{\mathrm{d}t}\right)^2 = \frac{p_b - p}{\rho_1} - \frac{2\sigma}{\rho_1 r} - 4\frac{\eta_1}{\rho_1 r}\frac{\mathrm{d}r}{\mathrm{d}t} \tag{7}$$

（5）k-ε 方程

$$\frac{\partial}{\partial t}(\rho k) + \nabla \cdot (\rho \boldsymbol{v} k) = \nabla \cdot \left(\frac{\eta_t}{\sigma_k}\nabla k\right) + G_k - \rho\varepsilon \tag{8}$$

$$\frac{\partial}{\partial t}(\rho\varepsilon) + \nabla \cdot (\rho \boldsymbol{v} \varepsilon) = \nabla \cdot \left(\frac{\eta_t}{\sigma_\varepsilon}\nabla\varepsilon\right) + \frac{\varepsilon}{k}(C_{1\varepsilon}G_k - C_{2\varepsilon}\rho\varepsilon) \tag{9}$$

$$\eta_t = \rho C_\mu \frac{k^2}{\varepsilon} \tag{10}$$

$$G_k = \eta_t \nabla \cdot \boldsymbol{v}[\nabla \cdot \boldsymbol{v} + (\nabla \cdot \boldsymbol{v})^{\mathrm{T}}] \tag{11}$$

（6）壁面律方程

$$u^+ = \frac{u}{u_\tau} = \frac{1}{\kappa}\ln\left(\frac{yu_\tau}{\upsilon}\right) = \frac{1}{\kappa}\ln(y^+) + B \tag{12}$$

$$k = \frac{(u^*)}{(C_t)^{1/2}}, \quad \varepsilon = \frac{(u^*)^3}{\kappa y} \tag{13}$$

$C_{1\varepsilon} = 1.44, C_{2\varepsilon} = 1.92, \sigma_k = 1.0, \sigma_\varepsilon = 1.3, C_\mu = 0.09, \kappa = 0.42, B = 5.44$。

式中 $\boldsymbol{v}$ 为流速；p 为压力；$\boldsymbol{f}$ 为体积力；t 为时间；ρ 为混合密度；η 为混合粘度；n_0 为单位体积纯液体中所含气泡数；k 为湍动能；ε 为耗散率；p_b 为气泡内压力；σ 为表面张力系数；y 为壁面距离；y^+ 为距壁面的无量纲距离；u 为速度沿壁面的切向分量；u_τ 为切应力速度；υ 为运动粘度。

2 三维数值模拟

针对有压力室的多孔垂直喷嘴内部流动，采用 Fluent 6.0 软件进行了三维两相流数值模拟，下面为其基本几何参数。

喷孔数：$n = 4$；

喷孔直径：$d = 0.32$ mm；

喷孔长度：$L = 1.3$ mm；

喷孔与针阀轴线间夹角：$\alpha=75\,^\circ$；

喷孔入口圆角半径：$r=0$ mm。

2.1 求解区域

多孔垂直喷嘴的各喷孔中心线与针阀轴线间夹角相同,各喷孔沿圆周均匀分布。所以,除了进行针阀偏心时的流动分析外,其他分析均针对 $1/n(n$ 为喷孔数)区域进行。本计算取整个流动区域的 1/4 作为计算区域,采用分块耦合的方法在 Gambit 2.0 中生成三维结构化网格,对喷孔处进行局部加密处理,见图 1。当前国外所进行的为数不多的三维计算的流动区域均是直接以喷孔出口作为出口边界,设为压力边界。但由于喷孔内部空穴的出现,喷孔出口压力往往已不同于气缸压力。当出口压力设为气缸压力时,势必影响喷孔内部流动计算的准确性,所以提出将出口边界延伸至气缸内部,建立图 1 所示的计算求解区域,这使得出口边界取为气缸压力时,更贴近实际,也大大减小了出口对喷孔内部求解区域的影响。

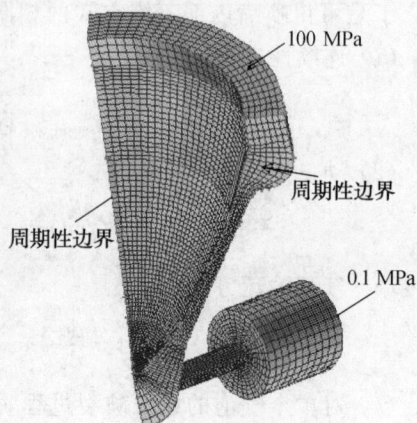

图 1 喷嘴计算网格

2.2 边界条件

表 1 列出了喷嘴计算的边界条件。进出口均为压力边界,由于计算区域内湍流很强烈,所以进出口边界上 k,ε 取值对计算结果影响不大,可据表 1 所列公式计算。在固壁边界上流速无滑移,均取为 0,而压力则取第二类边界条件。截取 1/4 作为计算区域时所产生的两个截面同为旋转周期性边界条件,可认为通过该周期性边界的平面不存在压力降。

表 1 喷嘴边界条件

边 界	速 度	压 力	$k\text{-}\varepsilon$
入 口	$\partial u/\partial n=0$	$p=100$ MPa	$I=0.16\times(Re)^{-1/8}$ $l=0.07D$
出 口	$\partial u/\partial n=0$	$p=0.1$ MPa	$I=0.16\times(Re)^{-1/8}$ $l=0.07D$
壁 面	$u=0$	$\partial p/\partial n=0$	$k=(u^*)^2/(C_t)^{1/2}$ $\varepsilon=(u^*)^3/\kappa y$
旋转周期性边界	—	$\partial p/\partial n=0$	—

2.3 方程的离散和求解

对方程的离散均采用基于内节点的有限容积法。气相输运方程式(6)中的对流项采用结合了中心差分和迎风差分的混合方法：

$$\alpha_c=\beta\alpha_u+(1-\beta)\alpha_d \tag{14}$$

式中 $\beta=0.75$。

动量方程式(5)的对流项的空间差分采用一阶迎风格式,其他则采用二阶中心差分,时间采用隐格式。

计算过程为:首先在新的时步下求解气相输运方程式(6),解出 α 值,然后以该新计算的 α 和由此得到的混合密度 ρ,采用 SIMPLEC 算法,将动量方程和连续性方程耦合迭代求解。在具体计算过程中,为更快地获得解,可在开始计算时不求解气相输运方程,而快速获得单相收敛的初始流场,之后再增加气相输运方程,起动混合多相流模型,进行混合相的计算,计算流程如图 2 所示。

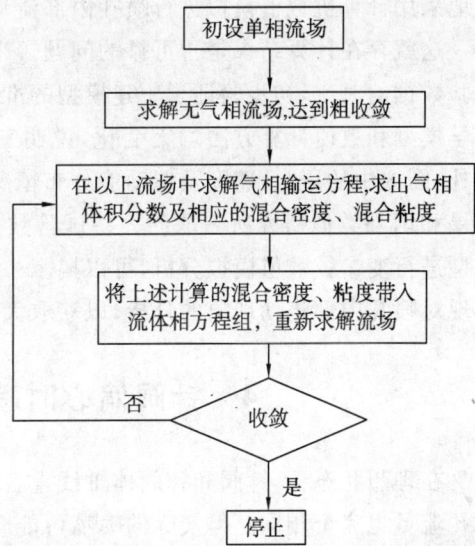

图 2　计算流程

3　数值计算结果

图 3 即为采用上述计算模型、边界条件及求解方法所得的数值计算结果。图 3a 为压力分布,图 3b 为湍动能分布,图 3c 为气相体积分数分布,也可认为是空穴分布,图 3d 为速度分布。

由计算结果可看出,喷嘴内部燃油流动的压力降基本发生在喷孔入口处,而在喷孔入口上部存在最低的局部压力,由于喷孔入口圆角半径为 0,锐边过渡,使得该处必定出现空穴现象,而且会延伸至喷孔出口,从而在喷孔出口形成一定的涡旋,增加液流的紊乱,引起燃油在喷孔出口的初次雾化。

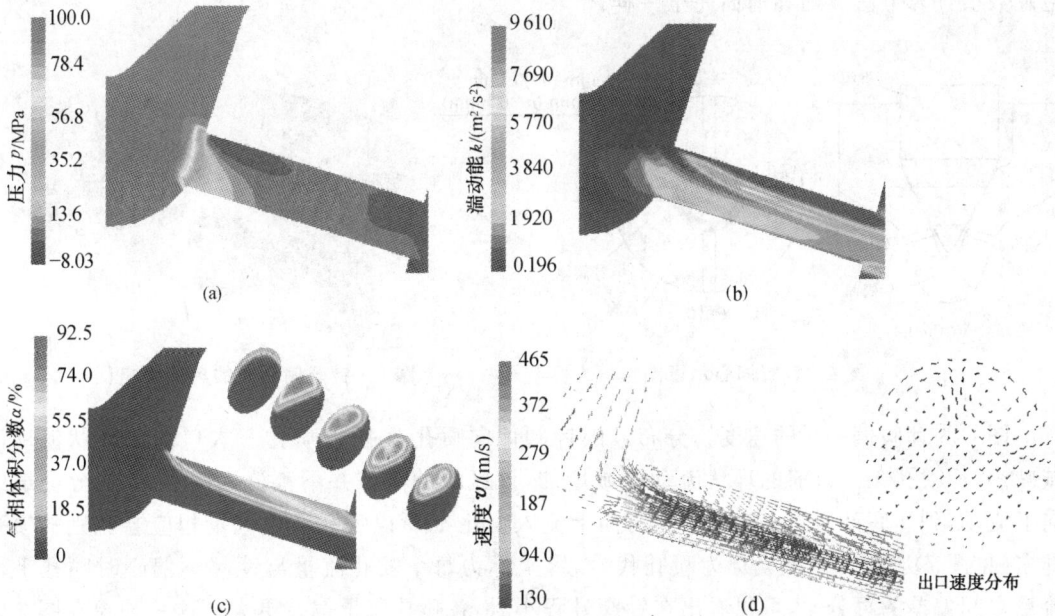

图 3　数值计算结果

当前,对喷孔如此小的尺寸,国内尚未能进行喷孔内部流动的可视化试验研究,所以尝试性地采用计算机数值模拟进行喷孔内部流动分析,获取有价值的信息和资料就显得非常有必要。这就存在计算结果是否可靠的问题。采用上述数学模型进行二维计算,通过和国外已有试验数据的对比,初步验证了所建模型的准确性[8]。三维计算是在二维计算基础上进行,基本数学模型和数值计算方法均无变化,说明三维数值计算具备了可行性和一定程度的可靠性。另外,图3中的压力、速度及湍动能分布情况,特别是空穴区域在喷孔内部的分布情况的计算结果和国外类似研究所获取的一些试验数据及数值计算结果[9,10]很接近,这均说明利用所建模型进行变参数数值模拟分析,可以揭示一些喷孔流动现象及规律,并能全面、深入和高效地实现对喷孔内部流动现象的分析,以获取大量有价值的信息资料。

4 针阀偏心时喷孔内部空穴流动分析

在理想状态下,针阀和针阀体轴线重合,针阀和针阀体间隙区域内的流动完全轴对称,各喷孔流量也完全相同。但实际的喷嘴可能会由于喷孔尺寸加工偏差、喷孔粗糙度的不同、喷孔入口形状的不同以及针阀的偏心等原因,或多或少出现不对称流动现象。特别是对多喷孔的VCO(无压力室)喷嘴,该现象更为典型,内部流动会明显不同于对称结构喷嘴,各喷孔流量不再相同。同时,这种微小且无法控制的不对称性对喷孔内部空穴分布会带来很大影响,出现对称喷嘴中所没有的具有复杂特征的流场,所以对不对称喷嘴流动的研究更具有实际意义。

当VCO喷嘴针阀升程$h=0.1$ mm和$h=0.3$ mm时,分别进行针阀偏心时的空穴流动数值计算。图4为喷孔偏心示意图,针阀升程$h=0.1$ mm时的偏心量取0.031 mm,$h=0.3$ mm时的偏心量取0.1 mm,针阀均偏向喷孔3,远离喷孔1。图5则为$h=0.3$ mm时的计算网格,网格数为134 180个。除了计算边界中少了旋转周期性边界外,其他的计算边界条件、数学模型及数值计算方法等均和前面完全一样。

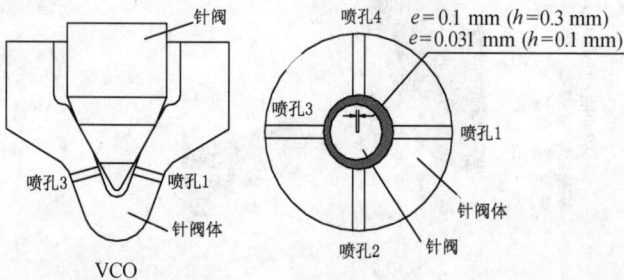

图4 针阀偏心示意图

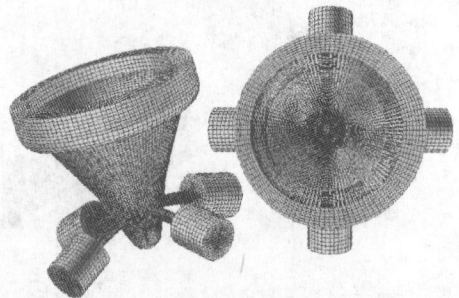

图5 针阀偏心时的计算网格($h=0.3$ mm)

图6为针阀偏心时的速度场分布。因针阀偏向喷孔3,所以喷孔3入口前的环状流动区域减小,而喷孔1入口前的环状流动区域增加,该区域的流体并不都流入喷孔1,部分会继续向下流动,到达底部压力室容积中后再向上流入喷孔3,所以喷孔3中流量相应会比别的喷孔都多(见图7)。图7中横坐标为喷孔代号,纵坐标为每个喷孔流量与针阀不偏心时喷孔平均流量之间偏差的百分比,可以看出在针阀升程小时,各喷孔流量偏差更大。这一现象在图6中同样有所反映,即流体由底部流入喷孔3的流动趋势在针阀小升程时更为显著,喷孔内流速比起高针阀升程来说也更为均匀。

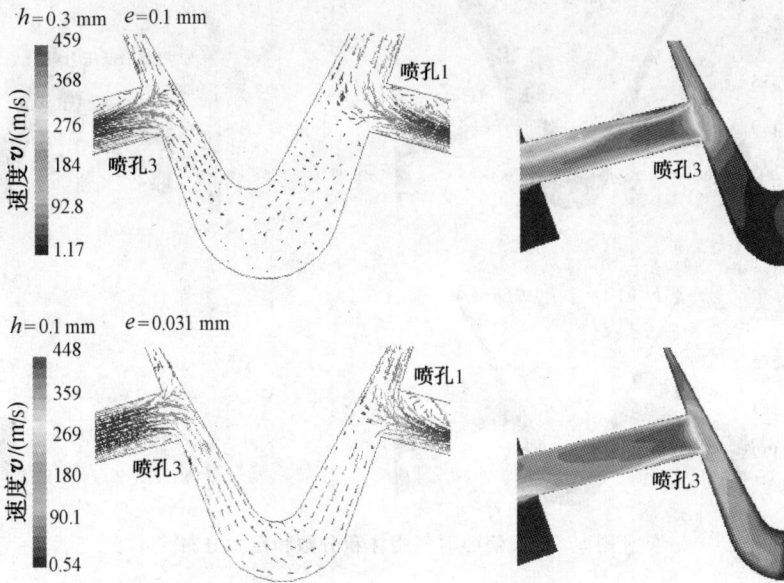

图 6　针阀偏心时的速度分布

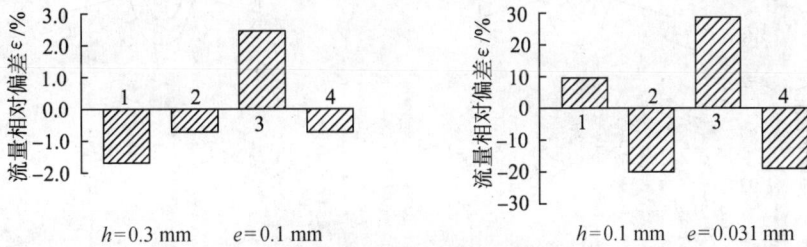

图 7　各喷孔流量相对偏差百分比

图 8 为空穴和压力分布图。从图 8 中发现,在高针阀升程时,喷孔 3 的局部最低压力出现在入口上部,而在低针阀升程时则出现在入口下部。所以对应高针阀升程时空穴主要由上部向喷孔出口延伸,而低针阀升程空穴则主要发生在喷孔下部。这主要由于在针阀升程低时,喷孔 3 内的流体绝大部分都是由针阀体底部压力室容积中自下而上流入的。

图 9 则为喷孔入口处,沿 A-A 方向所作截面上的速度分布。明显看出,流动从喷孔 1 向喷孔 3,2 和 4 的流动,而在低针阀升程情况下,还出现了流动从喷孔 3 处向喷孔 2 和 4 的流动趋势。

图 10a 和图 10b 分别为针阀升程 0.3 mm 和 0.1 mm 时各喷孔出口速度、空穴和湍动能的分布,其中喷孔 2 和 4 是相似的,故此处没有显示喷孔 2。计算结果表明,各喷孔空穴分布有所不同:针阀小升程时差别更大,空穴现象整体更为严重;随针阀升程的增加,喷孔出口空穴的出现也由中心向喷孔上方偏斜。由速度分布得知,针阀偏心引起的涡旋运动被传入喷孔内部,并延伸至喷孔出口,特别是喷孔 2 和 4,涡旋更为强烈,针阀升程小时,更为显著。文献[11]中对 VCO 喷嘴做喷雾试验所获得的图片显示,喷雾锥角的形成正是由喷孔出口处这种涡旋运动及由此产生的离心力所致。各喷孔出口湍动能更是有很大差别,由计算结果的对比分析看出,当涡旋运动强烈并且空穴现象严重时,对应的湍动能相应较大,出口液流的这种极度紊乱正是燃油雾化的原因所在。所有这些均显示出针阀偏心所产生的影响在针阀小升程时更为显著。

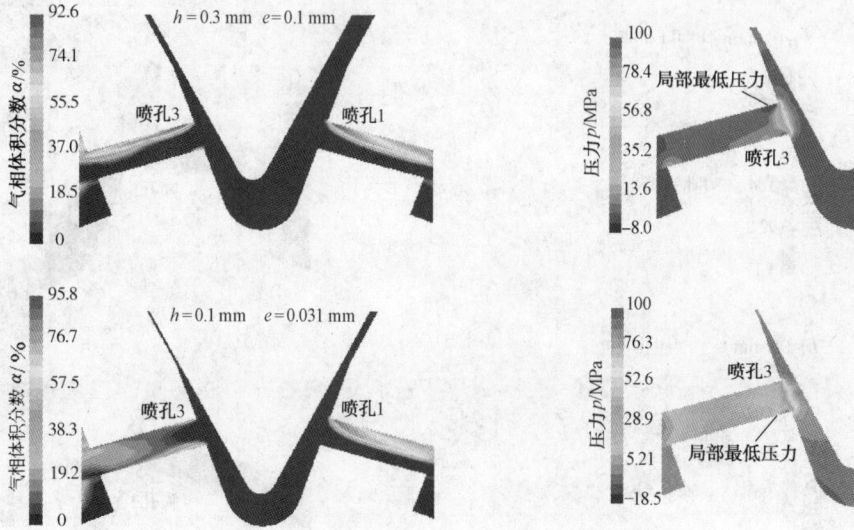

图 8　针阀偏心时气相体积分数和压力分布

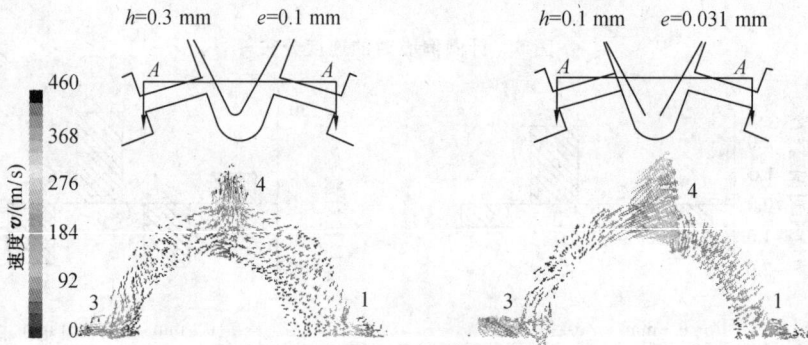

图 9　针阀偏心时 A-A 方向速度分布

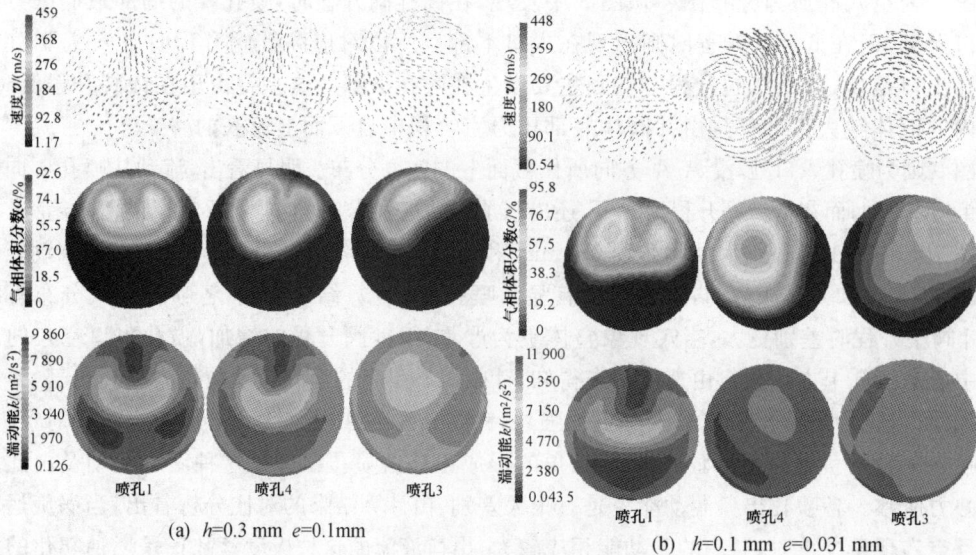

(a) h=0.3 mm e=0.1mm

(b) h=0.1 mm e=0.031 mm

图 10　不同喷孔的出口速度、空穴和湍动能分布

5 结 论

采用混合多相流空穴模型进行了多孔垂直喷嘴三维空穴两相湍流流动的数值模拟。首次提出了将计算区域的出口边界延伸至气缸内部，从而减小了出口对喷孔内部求解区域的影响。对 VCO 型喷嘴分别在针阀升程为 0.1 mm，0.3 mm 时进行针阀偏心后的数值模拟分析。针阀偏心时会在压力室产生一种强烈的涡旋运动，被传入喷孔内部，延伸至出口，使得各个喷孔出口流动形态不同，喷孔流量有很大差异；在针阀小升程时，空穴主要出现在喷孔入口下部，这与针阀高升程时空穴主要出现在喷孔入口上部有所不同；针阀偏心对流动所产生的影响在针阀小升程时更为显著。所有分析表明正是喷孔入口处所存在的低压回流区强烈影响着喷孔出口的流动分布。

参 考 文 献

［1］玉木伸茂,等. 喷油嘴喷孔空穴现象对液体喷束雾化的影响[J]. 国外内燃机,1998,338(8)：23-29.

［2］Chaves H, Knapp M. Experimental study of cavitation in the nozzle hole of diesel injectors using transparent nozzles[J]. *SAE Paper*, 950290,1995：199-211.

［3］Schmidt D P,Corradini M L. Analytical prediction of the exit flow of cavitating orifices [J]. *Atomization and Sprays*, 1997,7(6)：54-65.

［4］Rayleigh Lord. On the pressure developed in a liquid during the collapse of a spherical cavity[J]. *Phil Mag*, 1917,34：94-98.

［5］Brennen C E. *Cavitation and Bubble Dynamics*[M]. New York：Oxford University Press,1995.

［6］郭烈锦. 两相与多相流动力学[M]. 西安:西安交通大学出版社,2002.

［7］陶文铨. 数值传热学[M]. 西安:西安交通大学出版社,1988.

［8］何志霞,李德桃,胡林峰,等. 喷油器喷嘴孔内部空穴两相流动数值模拟分析[J]. 内燃机学报,2004,22(5)：433-438.

［9］Arcoumanis C,Gavaises M,Nouri J M,et al. Analysis of the flow in the nozzle of a vertical multi-hole diesel engine injector[J]. *SAE Paper*, 980811,1998：1245-1259.

［10］Arcoumanis C,Flora H,Gavalses M, et al. Cavitation in real-size multi-hole diesel injector nozzles[J]. *SAE Paper*, 2000-01-1249,2000：1458-1499.

［11］Soteriou C E, Andrews R J, Smith M. Direct injection diesel sprays and the effect of cavitation and hydraulic flip on atomization[J]. *SAE Paper*, 950080,1995：27-52.

（本文原载于《机械工程学报》2005 年第 3 期）

Three-dimensional numerical simulation and analysis of cavitating two-phase flow in a veritical mulit-hole nozzle

Abstract: The multi-dimensional two-phase mathematical model of cavitating nozzle flow was developed for the developed cavitating flow. Thus the three-dimensional numerical simulation of cavitating flow in nozzle holes of a vertical multi-hole injector is dealt with. The hole exit boundary of the computational area was extended to outside of the nozzle holes for the first time to decrease the effect of the nozzle holes exit on the inside flow of the computational area. At the same time the cavitating flow in the various nozzle holes under needle eccentricity was analyzed and studied in detail. Thus it was pointed out clearly that the characteristics of the cavitating flow in nozzle holes and the effect of the cavitating flow on the flow of the nozzle holes exit and on the fuel spray atomization.

柴油机喷嘴结构优化的数值模拟分析

何志霞,袁建平,李德桃,梁凤标

[摘要] 柴油机喷嘴头部细微结构的变化对内部流动有显著的影响,进而影响到喷雾油束雾化性能。利用混合多相流空穴模型对 STD(Standard)标准型、VCO(Valve Closed Orifice)无压力室型及 IMPROVED 改进型的垂直多孔喷嘴内部完全发展了的空穴流动进行了不同针阀升程条件下的三维数值模拟,结合试验对比分析证实:改进型喷嘴能在低喷油速率下获得相对良好的喷雾性能,综合特性优于 STD 标准型和 VCO 无压力室型喷嘴。

改善柴油机燃油喷射系统对喷雾油束的控制是优化燃油雾化及提高空气利用率的关键,特别是喷嘴头部细微结构的变化对喷孔出口喷雾特性有很大的影响。对喷嘴而言,通常考虑尽可能采用多喷孔、小孔径及小的压力室,以产生最优的喷雾油束。燃油喷雾雾化过程被分成两个主要部分:初次破裂和二次破裂,这一观点已逐步被内燃机界所认可,即发生于接近喷孔处的初次雾化除了由液相和气相相互作用所决定之外,也受喷嘴内部流动的湍动及空穴现象的影响[1]。所以对喷嘴头部结构不断作出改进后,对喷嘴内部流动及喷雾特性的研究是不可或缺的。当前国内由于各种客观原因尚未对喷嘴内部流动进行可视化试验研究,本文结合国外所做试验[2-3],对有压力室(通常所说的标准型 STD)喷嘴、无压力室(VCO)喷嘴及在这两者基础上加以改进的喷嘴——IMPROVED 喷嘴内部流动进行了计算机三维数值模拟分析。

1 计算模型和数值计算网格

本文对图 1 所示的三种结构的垂直多孔喷嘴(4×0.32 mm),在喷油压力为 100 MPa,气缸背压为 0.1 MPa 时,分别按不同的针阀升程(0.1 mm,0.2 mm 和 0.3 mm)进行计算机数值模拟分析。

由图 1 可看出,三种结构喷嘴只有头部细微结构的不同。STD 喷嘴在喷孔之前有相对较大的压力室。VCO 喷嘴则基本无压力室,燃油通过狭窄的环形空间进入喷孔。IM-PROVED 喷嘴对上述两种结构进行了融合,有压力室,但要小于 STD 结构,喷孔入口的位置接近 VCO 结构,位置偏上,位于燃油流经通道开始扩散处。

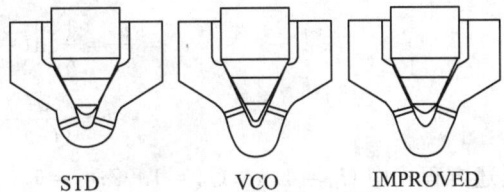

图 1 三种喷嘴结构

对多孔垂直喷嘴内部完全发展了的空穴流动进行依赖时间的三维气液两相湍流数值模拟。数学模型以 Rayleigh 所发展的单气泡溃灭模型[4]为基础而建立,采用混合多相流模型。量级分析法[5]表明,对目前的喷嘴空穴流动,可不考虑两相间的相对运动,将气液两相看做是均匀混合体,采用单流体法。此时基本的运动方程和连续性方程等与单相流体运动采用同一套表达形式,只是喷嘴空穴两相流动中的密度、粘度均采用混合密度 ρ 和混合粘度 μ 来表示:

$$\rho=(1-\alpha)\rho_l+\alpha\rho_v \tag{1}$$

$$\mu=(1-\alpha)\mu_l+\alpha\mu_v \tag{2}$$

式中 $\rho_l,\rho_v,\mu_l,\mu_v$ 分别为纯液相和纯气相的密度和粘度,为常数;α 为气相体积分数,即空穴在气液混合体中所占比例。假定空穴区中的气相由许多很小的球状气泡组成,α 由下式计算[6]:

$$\alpha=\frac{V_v}{V_l+V_v}=\frac{4\pi r^3 n_0/3}{1+4\pi r^3 n_0/3} \tag{3}$$

于是喷嘴三维空穴两相湍流模型基本控制方程如下[6-9]:

(1) 连续性方程

$$\frac{\partial \rho}{\partial t}+\nabla \cdot (\rho \boldsymbol{v})=0 \tag{4}$$

(2) 动量方程

$$\frac{\partial (\rho \boldsymbol{v})}{\partial t}+\nabla \cdot (\rho \boldsymbol{v}\boldsymbol{v})=-\nabla p+\nabla[\mu(\nabla \cdot \boldsymbol{v}+\nabla \cdot \boldsymbol{v}^{\mathrm{T}})]+\rho \boldsymbol{g}+\boldsymbol{f} \tag{5}$$

(3) 气相输运方程

$$\frac{\partial \alpha}{\partial t}+\nabla \cdot (\alpha \boldsymbol{v})=\frac{\mathrm{d}\alpha}{\mathrm{d}t}+\alpha \nabla \cdot \boldsymbol{v}=\frac{(1-\alpha)\rho_l}{(1-\alpha)\rho_l+\alpha\rho_v} \cdot \frac{4\pi n_0 r^2}{1+n_0 \cdot 4\pi r^3/3} \cdot \frac{\mathrm{d}r}{\mathrm{d}t} \tag{6}$$

(4) 单个气泡的成长破裂过程,根据 Rayleigh-Plesset 方程[7]有

$$r\frac{\mathrm{d}^2 r}{\mathrm{d}t^2}+\frac{3}{2}\left(\frac{\mathrm{d}r}{\mathrm{d}t}\right)^2=\frac{p_b-p}{\rho_l}-\frac{2\sigma}{\rho_l r}-4\frac{\mu_l}{\rho_l r} \cdot \frac{\mathrm{d}r}{\mathrm{d}t} \tag{7}$$

(5) k-ε 方程

$$\frac{\partial}{\partial t}(\rho k)+\nabla \cdot (\rho \boldsymbol{v}k)=\nabla \cdot \left(\frac{\mu_t}{\sigma_k}\nabla k\right)+G_k-\rho\varepsilon \tag{8}$$

$$\frac{\partial}{\partial t}(\rho\varepsilon)+\nabla \cdot (\rho \boldsymbol{v}\varepsilon)=\nabla \cdot \left(\frac{\mu_t}{\sigma_\varepsilon}\nabla \varepsilon\right)+\frac{\varepsilon}{k}(C_{1\varepsilon}G_k-C_{2\varepsilon}\rho\varepsilon) \tag{9}$$

$$\mu_t=\rho C_\mu\frac{k^2}{\varepsilon} \tag{10}$$

$$G_k=\mu_t \nabla \cdot \boldsymbol{v}[\nabla \cdot \boldsymbol{v}+(\nabla \cdot \boldsymbol{v})^{\mathrm{T}}] \tag{11}$$

(6) 壁面律方程

$$u^+=\frac{u}{u_\tau}=\frac{1}{\kappa}\ln\left(\frac{yu_\tau}{\upsilon}\right)=\frac{1}{\kappa}\ln(y^+)+B \tag{12}$$

$$k=\frac{(u^*)}{(C_t)^{1/2}}, \quad \varepsilon=\frac{(u^*)^3}{\kappa y} \tag{13}$$

上述方程式中 $C_{1\varepsilon}=1.44$;$C_{2\varepsilon}=1.92$;$\sigma_k=1.0$;$\sigma_\varepsilon=1.3$;$C_\mu=0.09$;$k=0.42$;$B=5.44$;$\boldsymbol{v}$ 为流速;p 为压力;$\boldsymbol{f}$ 为体积力(不包括重力);t 为时间;n_0 为单位体积纯液体中所含气泡数;k 为湍动能;ε 为耗散率;p_B 为气泡内压力;σ 为表面张力系数;y 为壁面距离;y^+ 为距壁面的无量纲距离;u 为速度沿壁面的切向分量;u_τ 为切应力速度;υ 为运动粘度。

方程的离散均采用基于内节点的有限容积法,压力修正采用 SIMPLEC 算法。对流项的空间差分采用一阶迎风格式,其他则采用二阶中心差分,时间采用隐式法。

进出口均采用压力边界,但由于喷孔内部空穴的出现,喷孔出口压力往往已不同于气缸压

力。当出口压力设为气缸压力时势必影响喷孔内部流动计算的准确性,所以本文将出口边界延伸至气缸内部,建立图2所示的计算求解区域,使得出口边界取为气缸压力时,更贴近实际,也大大减少了出口对喷孔内部求解区域的影响。表1列出了喷嘴计算的边界条件。由于计算区域内湍流运动很强烈,进出口边界上 k,ε 取值对计算结果影响不大,可据公式(4)计算。对固壁边界,流速无滑移,取为0,压力取第二类边界条件。对垂直多孔喷嘴,喷孔沿圆周均匀分布,所以取喷嘴流动区域的 $1/n(n$ 为喷孔数)进行数值模拟,采用分块耦合的方法生成三维结构化网格,对喷孔处做局部加密处理。

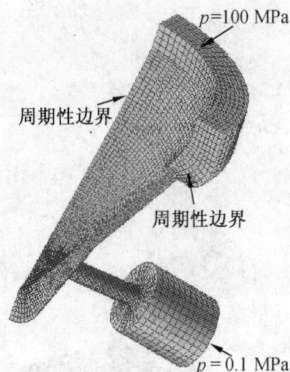

图2　喷嘴计算网格

表1　喷嘴边界条件

边界	速度	压力	k-ε
入口	$\partial u/\partial n=0$	$p=100$ MPa	$I=0.16\times(Re)^{-1/8}$ $l=0.07D$
出口	$\partial u/\partial n=0$	$p=0.1$ MPa	$I=0.16\times(Re)^{-1/8}$ $l=0.07D$
壁面	$u=0$	$\partial p/\partial n=0$	$k=(u^*)^2/(C_t)^{1/2}$ $\varepsilon=(u^*)^3/\kappa y$
旋转周期性边界	—	$\partial p/\partial n=0$	—

2　STD 喷嘴不同针阀升程时的流动对比分析

图3、图4分别为不同针阀升程时的压力 p、喷油速度 v 和气相体积分数 α 分布的计算结果,其中气相体积分数分布反映了空穴在喷孔内部的分布情况。从中看出,对 STD 喷嘴,针阀升程小时($h=0.1$ mm),喷孔入口处压力和喷油速率的值均较小,喷孔出口处喷油速率同样也较小,即通过控制针阀升程可以控制喷油速率。但针阀升程小时,喷孔入口前的喷油压力也相应降低,喷雾特性会变得很差。所以采用控制针阀升程的座面节流,很难同时控制喷油速率及获得好的油束雾化,由此对结构作出改进,就有了随后 VCO 和 IMPROVED 喷嘴的出现。

由速度分布图可以大致作以下判断:针阀升程小时,其喷孔有效流通面积 A_{He}(定义为速度矢量指向喷孔出口时喷孔中的最小面积[10])也略有减小。而文献[10]中提到:对标准喷嘴,由于针阀到喷孔之间的距离 L 不会跟 VCO 和 IMPROVED 一样随针阀升程的变化而变化,而是基本保持常数,压力室中的燃油流动不随针阀升程的变化而变化,燃油流通的有效喷孔面积也不随针阀升程而变。此处出现这种矛盾的主要原因在于文献[10]中所做的计算没有考虑空穴的存在,而我们从图4(不同针阀升程时的空穴分布情况)看出,针阀升程小时,空穴层相应地会厚很多,从而肯定会使喷孔有效流通截面有所减小。由于空穴现象的存在,喷孔的流量系数也有很大的不同。

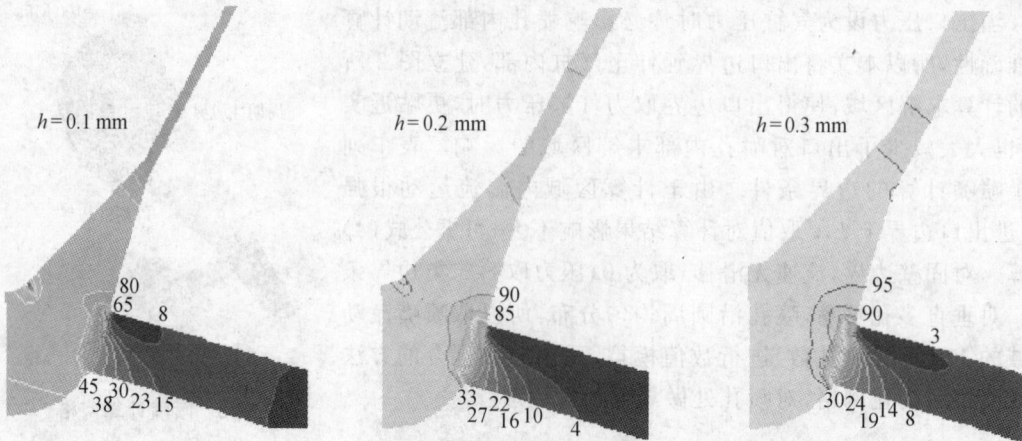

图 3　STD 喷嘴不同针阀升程时压力分布(MPa)

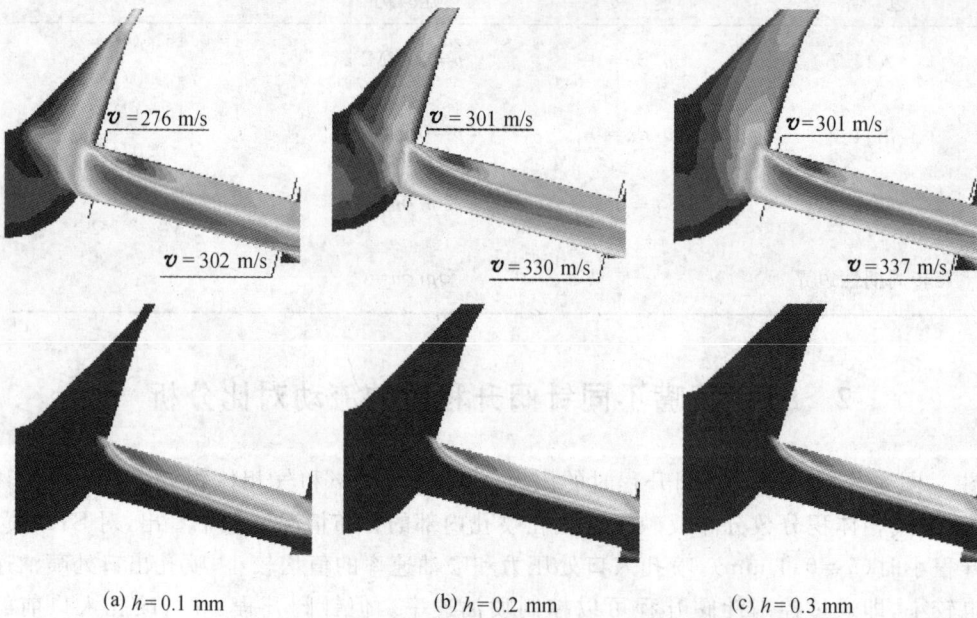

(a) $h=0.1$ mm　　　　(b) $h=0.2$ mm　　　　(c) $h=0.3$ mm

图 4　STD 喷嘴不同针阀升程时的喷油速度和空穴分布

3　不同类型喷嘴流动对比分析

喷嘴结构由 STD 标准型、VCO 无压力室型发展到 IMPROVED 改进型,结构的不同对流动的影响不同,对低针阀升程来说影响更为显著。故本文在针阀升程 $h=0.1$ mm 的情况下,对 3 种结构喷嘴进行数值模拟对比。图 5 中的 a,b 和 c 分别为 3 种喷嘴的压力、速度和空穴分布的数值计算结果。

由图 5 看出,在 VCO 喷嘴和 IMPROVED 喷嘴中,在低针阀升程时,喷孔入口处的喷射压力和速度均比标准型的有所提高。经燃油流动分析,可以解释为当针阀升程减小时,针阀座面流通面积明显减小,对 VCO 喷嘴和 IMPROVED 喷嘴而言,这会限制喷孔入口燃油流通,相反 STD 喷嘴则不存在这种情况。所以在小针阀升程下,与 STD 喷嘴相比,由于燃油是通过受限制的喷孔流通面积,VCO 喷嘴和 IMPROVED 喷嘴喷孔入口前的实际喷射压力和速度均自然要高于 STD。对于由湍动造成的喷孔出口液流紊乱,VCO 和 IMPROVED 喷嘴比 STD 严重,从这个角度我们预测 VCO 喷嘴和 IMPROVED 喷嘴在喷射开始(低针阀升程)时产生的油雾雾滴比标准 STD 喷嘴要小。

图 5   不同结构喷嘴的计算结果($h=0.1$ mm)

就 VCO 无压力室喷嘴和 IMPROVED 改进型喷嘴而言,低针阀升程时的喷孔内部流动的压力损失系数不同于高针阀升程时。所以在低针阀升程时,需对在正常喷油状态(高针阀升程)下分析获得的压力损失进行修正,损失修正系数为 ζ_c,有如下的经验公式[10]:

$$\zeta_c = C_1/L^2 - C_2$$

式中 C_1,C_2 为常数;L 为针阀到喷孔间的垂直距离。显然损失修正系数 ζ_c 受 L 的影响,当 L 减小到一定程度后,ζ_c 开始显著增加,喷孔流量系数就开始降低。对 VCO 喷嘴,L 在针阀升程变化的整个过程中始终比较小,喷油速率也就被限制在低值上。而对 IMPROVED 来说,针阀的头部锥面与喷嘴体锥面的不平行性增加,接近 STD 型,L 值也因而增加,ζ_c 减小,从而使喷孔流量系数会比 VCO 型的有所增加,流动的综合特性好于 VCO 型。

从图 5 空穴分布发现,VCO 喷嘴内部空穴层最薄,而 STD 最厚,即空穴造成的液流紊乱在 VCO 喷嘴中最微弱,IMPROVED 结构喷嘴则介于 STD 和 VCO 之间,优于 VCO 喷嘴,所以从该角度考虑,同样认为 IMPROVED 喷嘴雾化特性好于 VCO 喷嘴。

图 6 和图 7 分别是 VCO 喷嘴和 IMPROVED 喷嘴在不同针阀升程时的速度和空穴分布图。

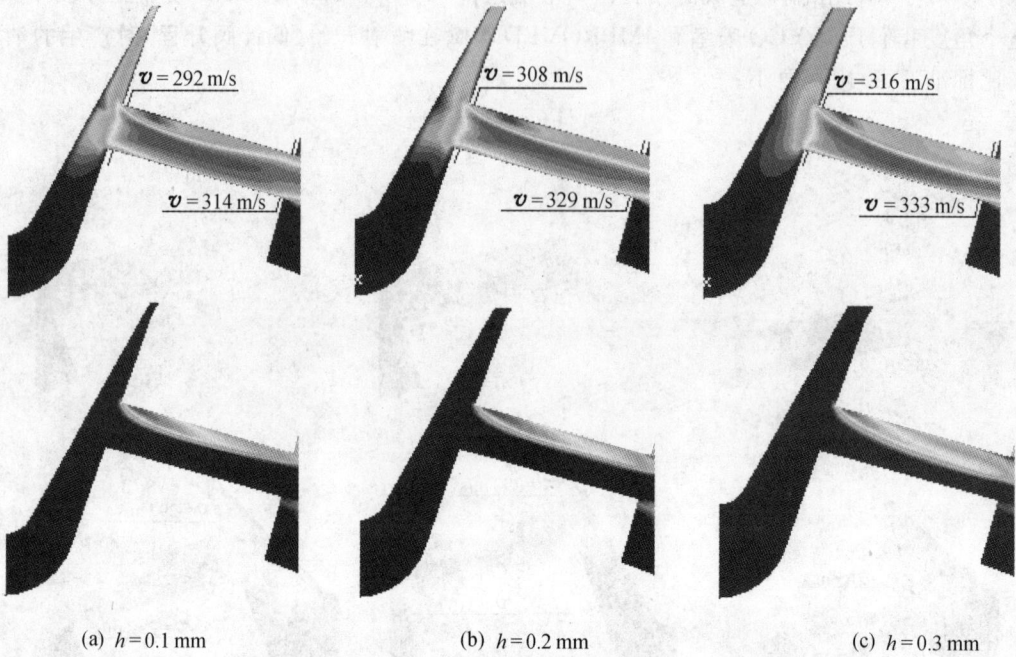

(a) $h = 0.1$ mm   (b) $h = 0.2$ mm   (c) $h = 0.3$ mm

图 6   VCO喷嘴不同针阀升程时的喷油速度与空穴分布

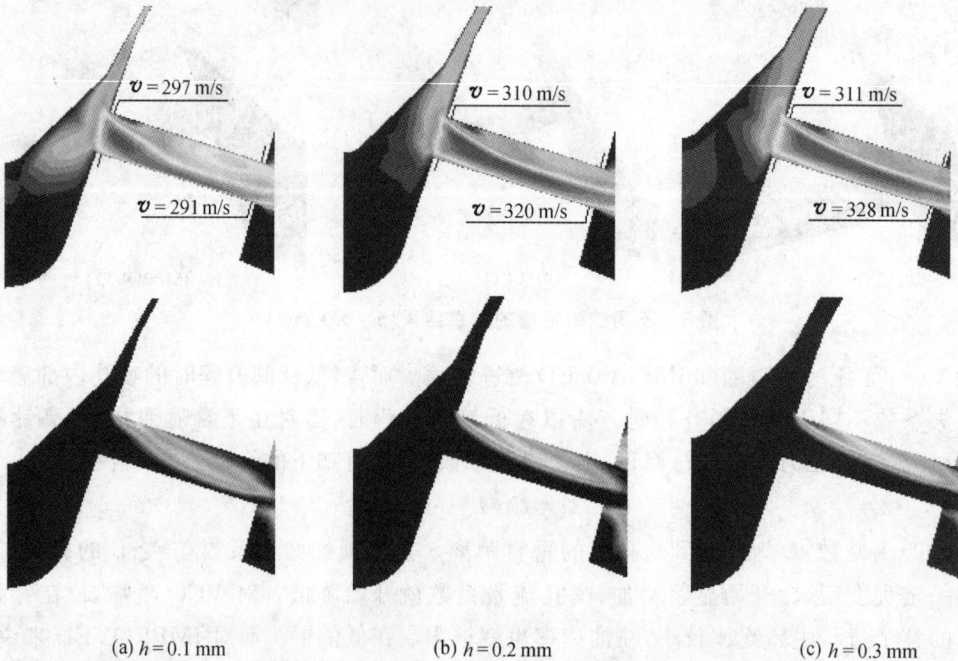

(a) $h = 0.1$ mm   (b) $h = 0.2$ mm   (c) $h = 0.3$ mm

图 7   IMPROVED喷嘴不同针阀升程时的喷油速度与空穴分布

结合图 4,可以看出这 3 种类型喷嘴,随针阀升程的增加,喷孔内平均速度均会增加,特别是在小针阀升程时,由于座面流通面积小,喷孔入口处,速度均不能达到较高值。喷孔有效流通面积也均与喷孔内部空穴区域的分布有很密切的关系。喷孔入口和喷孔出口处平均速度的差异正是喷孔内部空穴现象的直接反映,差异的大小也直接与喷孔空穴分布区域的大小有关。

4   试验对比

在自行研制的喷雾燃烧过程多功能动态可视化试验装置上,进行了不同喷嘴结构喷雾发展过程的可视化试验,并对比分析试验结果。

图 8 为该试验装置的光路系统示意图[11]。激光器发出的激光点光源 5 经扩束镜 4 和准直镜 3 转换为准直平行光,经半透半反镜 2,透射后到达平面反射镜 7(与水平方向呈 45°放置),被反射后直接穿过石英玻璃将燃烧室内照亮。当出现喷雾和燃烧时,相应的喷雾像和燃烧火焰像再经石英玻璃和 45°的平面反射镜到达半透半反镜,此时利用其反射功能将像反射进入高速数码摄像机,完成喷雾和燃烧全过程记录。

利用上述试验装置,在喷油压力为 18 MPa 时,对 STD 型、VCO 型和 IMPROVED 型 3 种喷嘴结构分别进行试验。其中,高速数码摄像机的拍照速

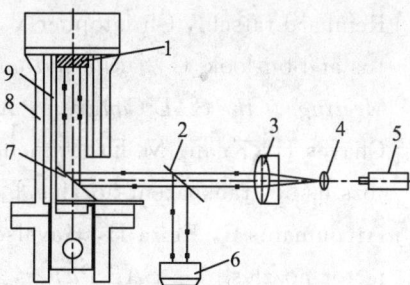

1—石英玻璃(镀银薄膜) 2—半透半反镜
3—准直镜 4—扩束镜 5—激光光源
6—高速摄影机 7—平面反射镜
8—加长缸套 9—加长活塞
图 8   光路系统示意图

度为 5 000 幅/秒,此时所拍摄的每两张照片间的时间间隔 Δt 为 0.2 ms。随后,对通过高速数码摄像机拍摄得到的每幅试验照片进行简化处理,得到的喷雾发展过程投影图见图 9[9]。

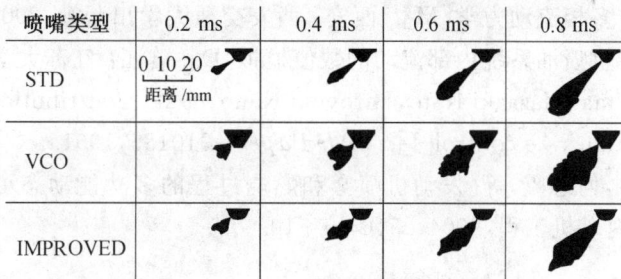

图 9   不同类型喷油嘴的喷雾发展过程投影图($p_{in}=18$ MPa)

由试验数据所获得的 3 种不同结构类型喷嘴的喷雾发展过程投影图可以清楚地看出:STD 喷嘴喷雾锥角小,VCO 喷嘴喷雾锥角大大增加,但同时其喷雾贯穿距减小,喷雾形状有了较大变化。将这两种喷嘴的结构特点进行综合和改进后出现的 IMPROVED 结构喷嘴在喷雾锥角增加的同时,具备了类似于 STD 喷嘴的优异性能,即 VCO 喷嘴所存在的喷雾贯穿距减小和喷雾变化问题得以改进。通过试验的对比分析得到了和数值模拟分析同样的结果:IMPROVED 喷嘴综合性能优于 STD 喷嘴和 VCO 喷嘴。

5   结   论

不同结构喷嘴(STD,VCO 和 IMPROVED)在不同针阀升程时内部气液两相流动的三维数值模拟的可视化计算结果表明:喷嘴头部压力室体积的大小及细微结构的变化显著影响喷嘴内部流动特性,相应也会显著影响到随后的喷雾特性。从内部流动数值模拟和随后喷雾发

展过程可视化试验对比均证实：IMPROVED 改进型喷嘴能在低喷油速率下获得相对良好的喷雾性能，综合特性优于 STD 标准型和 VCO 无压力室型喷嘴。

参 考 文 献

[1] Reinhard tatschl, Christopher V, Künsberg sarre, et al. IC-engine spray modeling status and outlook[C]// *International Multi-dimensional Engine Modeling User's Group Meeting at the SAE Congress*, Austria, 2002.

[2] Chaves H, Knapp M. Experimental study of cavitation in the nozzle hole of oiesel injectors using transparent nozzles[J]. *SAE Paper*, 950290, 1995.

[3] Arcoumanis C, Flora H, Gavalses M, et al. Cavitation in real-size multi-hole diesel injector nozzles [J]. *SAE Paper*, 2000-01-1249, 2000.

[4] Rayleigh Lord. On the pressure developed in a liquid during the collapse of a spherical cavity[J]. *Phil Mag*, 1971, 34: 94–98.

[5] Brennen C E. *Cavitation and Bubble Dynamics* [M]. New York: Oxford University Press, 1995: 416–483.

[6] 何志霞, 李德桃, 胡林峰, 等. 喷油器喷嘴孔内部空穴两相流动数值模拟分析[J]. 内燃机学报, 2004, 22(5): 433–438.

[7] 陶文铨. 数值传热学[M]. 西安: 西安交通大学出版社, 1988: 433–438.

[8] 郭烈锦. 两相与多相流动力学[M]. 西安: 西安交通大学出版社, 2002: 21–65.

[9] 何志霞. 高压共轨喷油系统内部流动的数值模拟[D]. 镇江: 江苏大学, 2004.

[10] Masahiro Okajima, Masaaki Kato, Hiroyuki Kano, et al. Contribution of optimum nozzle design to injection rate control[J]. *SAE Paper*, 910185, 1991.

[11] 潘剑锋, 李德桃, 何志霞, 等. 发动机喷雾和燃烧过程的多功能动态可视化测试装置的研制和应用[J]. 内燃机工程, 2004, 25(6): 7–10.

（本文原载于《内燃机学报》2006 年第 1 期）

Numerical simulation on optimization of diesel nozzle

Abstract: The changing of detailed structure of diesel nozzle tip greatly affect the flow characteristics in the nozzle hole and the performance of the subsequent breakup of spray. This paper dealt with three-dimensional numerical simulation of the cavitating flow fully developed in the three types of nozzles, which are the standard(STD), the valve closed orifice (VCO) and the improved nozzle(IMPROVED) of the vertical multi-hole injector, under different needle lifts with mixing multi-phase cavitating flow model. The study showed that the IMPROVED nozzle has better spray characteristics at low injection velocity and the general performance of IMPROVED nozzle is better than that of the STD nozzle and the VCO nozzle.

垂直多孔喷嘴内部流动空穴现象数值模拟分析

何志霞,袁建平,李德桃

[摘要] 柴油机喷嘴内部空穴流动是影响喷雾特性的重要因素。在对柴油机喷嘴喷孔内部空穴流动现象分析的基础上,利用混合多相流空穴模型对垂直多孔喷嘴完全发展了的空穴流动进行了三维数值模拟,分析了喷油压力、气缸背压及喷孔入口圆角半径与喷孔倾角对喷孔内部空穴分布的影响。验证了低空穴参数对应的空穴层延伸长度较长,喷孔入口锐边过渡会增加喷孔出口液流紊乱,从而加速雾化。

为适应日益严格的排放法规,柴油机的燃油喷射系统要求提供越来越高的喷油压力,以获得更高的雾化质量及更为充分的空气利用效果,然而高喷油压力势必增加油泵的驱动力矩及尺寸,从而增加成本。分析喷嘴内部燃油流动,优化喷孔内部流动在一定程度上能起到提高雾化质量的效果。近年来,众多的研究成果已使人们认识到,由产生在喷嘴喷孔内的空穴给液流造成的紊乱对液体喷束雾化产生的影响,远远大于由喷束与周围空气摩擦所产生的影响[1]。国外的大量研究将空穴现象看做将喷嘴流动和喷雾行为联系起来的一种关键因素[2,3]。所以深入分析和了解喷嘴内部流动,特别是深入研究空穴现象和机理会有助于对喷雾破裂模型的处理,以及对柴油机喷雾传播特性的认识。

1 喷嘴喷孔内部流动空穴现象分析

图 1 所示为喷孔内部流动简化模型[4]。根据不可压缩流体伯努利方程,沿流线速度越高,压力越低。在高速流动的喷孔入口处,由于拐角的存在,产生局部流动分离及孔口收缩,引起横截面面积的减小。根据质量守恒定律,此截面上的流速会大大增加,从而引起压力降低。当局部压力降至低于该液体饱和蒸汽压时,该区域就会出现空穴现象,形成空穴区。气泡的存在使液相流体的连续性遭到破坏,在喷孔内形成蒸汽泡、液体中的未溶气体和液体的两相流动。国外由原型喷嘴试验所得照片[5]显示空穴仅仅发生在喷孔入口处,但通常都会向下游延伸形成长而薄的空穴层。气泡的形成很大程度上依赖于喷孔几何特征(r,L,D 等)与喷孔壁面上的毛刺、几何缺陷,而空穴层的延伸长度则由上下游压差决定。某些条件下,流动完全脱离喷

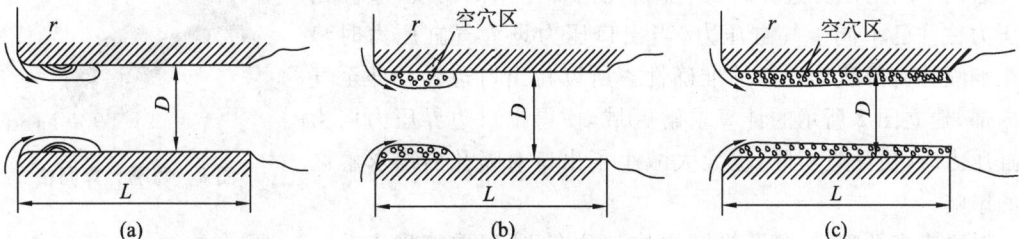

图 1 喷孔流动简化模型

孔壁面,喷孔流量与喷孔入口处的压力状态有关,而与喷孔出口压力无关,从而出现所谓的"水击"现象。一旦出现"水击",喷孔内部空穴不会再发生,喷孔出口流动会比较平滑、稳定。可见实际的喷嘴空穴流动过程非常复杂。

与流体粘性运动存在动力相似的无量纲参数雷诺数 Re 一样,对空穴流动同样定义相应的相似无量纲参数,即空穴参数

$$K = \frac{p_1 - p_c}{p_1 - p_2}$$

式中 p_1 为无限远处上游侧压力(喷油压力);p_2 为下游侧压力或周围空气压力(背压);p_c 为孔口收缩处的压力。

K 值会随喷油压力和背压压差的减小而显著增大。上式可看成导致空穴溃灭的可用压力($p_1 - p_c$)与有助于空穴形成和发展的可用压力($p_1 - p_2$)之比,故可将其理解为空穴参数 K 越小,空穴越容易产生。于是存在一临界空穴参数

$$K_{crit} = \frac{p_1 - p_v}{p_1 - p_2}$$

式中 p_v 为液体的饱和蒸气压。

当 K 大于 K_{crit} 时,不会有空穴现象发生,只是湍流射流,湍流涡旋使喷束表面形成波皱,向中心扩散,成为二次雾化的扰动源(见图1a)。而当 K 值一旦小于 K_{crit},喷孔入口就会出现空穴流动(见图1b)。随上游压力的增加,K 值进一步减小,空穴区延伸超出喷孔,形成所谓的"超空穴"现象(见图1c)。此时喷孔出口处,蒸汽泡在内外压差的突变作用下,克服表面张力而破裂,形成明显的浓密油滴,不会再有油束核的存在,从而雾化分裂。因此喷孔内部空穴现象是使喷孔出口燃油雾化的一个重要因素。

当前国内尚未进行喷孔内部流动的可视化试验研究,所以有必要尝试性地采用计算机数值模拟技术进行喷孔内部空穴现象分析。

2　CFD 模型和数值计算方法

2.1　数学模型和网格生成

对多孔垂直喷嘴内部完全发展了的空穴流动进行了依赖时间的三维气液两相数值模拟。采用混合多相流模型,并假定两相以相同速度运动,附加以空穴模型(气泡的生长和溃灭模型),求解空穴区在喷孔内的分布情况;对流湍,采用标准 k-ϵ 模型。基本数学模型方程在文献[6]中详细列出。

进出口均为压力边界,但由于喷孔内部空穴的出现,喷孔出口压力往往已不同于气缸压力,当出口压力设为气缸压力时势必影响喷孔内部流动计算的准确性。所以将出口边界延伸至气缸内部,建立图2所示的计算求解区域,使得出口边界压力取为气缸压力时,更贴近实际,也大大减小了出口对喷孔内部求解区域的影响。

对垂直多孔喷嘴,喷孔沿圆周均匀分布,所以取喷嘴流动区

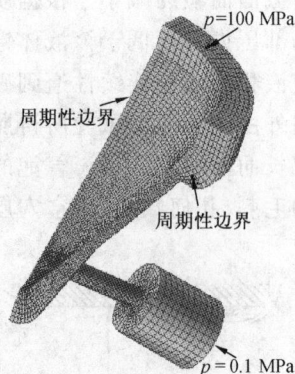

图 2　喷嘴计算网格

域的 $1/n$(n 为喷孔数)进行数值模拟,采用分块耦合的方法生成三维结构化网格,对喷孔处进行局部加密处理(图 2),网格总数为 32 565 个。

2.2 边界条件

表 1 列出了喷嘴计算的边界条件。进出口均为压力边界,由于计算区域内湍流很强烈,所以进出口边界上 k,ε 取值对计算结果影响不大,可据表列公式计算。在固壁边界上流速无滑移,均取为 0,而压力则取第二类边界条件。截取 1/4 作为计算区域时所产生的两个截面同为旋转周期性边界条件,认为通过该周期性边界的平面不存在压力降。

表 1　喷嘴边界条件

边　界	速　度	压力/MPa	$k\text{-}\varepsilon$
入　口	$\partial u/\partial n=0$	100	$I=0.16Re^{-1/8}$ $l=0.07D$
出　口	$\partial u/\partial n=0$	0.1	$I=0.16Re^{-1/8}$ $l=0.07D$
壁　面	$u=0$	$\partial p/\partial n=0$	$k=(u^*)^2/(C_t)^{1/2}$ $\varepsilon=(u^*)^3/(\kappa y)$
旋转周期性边界	—	$\partial p/\partial n=0$	—

2.3 方程的离散和求解

对方程的离散均采用基于内节点的有限容积法。动量方程的对流项的空间差分采用一阶迎风格式,其他则采用二阶中心差分,时间采用隐格式。计算时首先在新的时步下求解气相输运方程,解出气相体积分数 α 值,然后以该新计算的 α 和由此得到的混合密度 ρ,采用 SIMPLEC 算法,将动量方程和连续性方程耦合迭代求解。在具体的计算过程中,为更快地获得解,可在开始计算时不求解气相输运方程,而快速获得单相收敛的初始流场,之后再增加气相输运方程,起动混合多相流模型,进行混合相的计算,计算流程如图 3 所示。

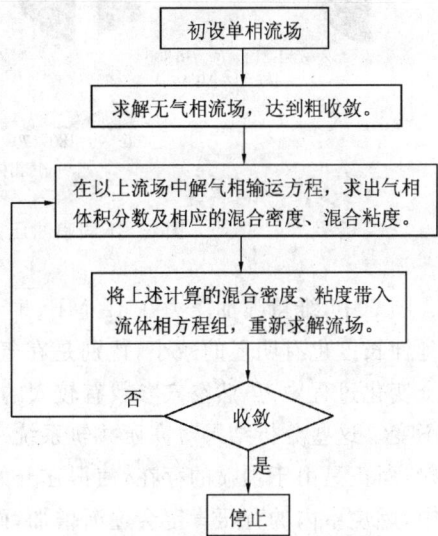

3　不同喷油压差下的空穴流动分析

由空穴参数的定义得知,空穴的发生及发展主要受喷油压力及下游气缸压力影响。故针对 VCO 型喷嘴,在针阀升程 $h=0.3$ mm 条件下,分别改变喷油压力和下游气缸压力,进行了空穴流动分析。图 4 为下游气缸压力保持 0.1 MPa,喷油压力分别由 100 MPa 改变至 140 MPa,50 MPa,12 MPa,5 MPa 和 3 MPa 时喷孔内部空穴分布情况。图 5 为上游喷油压力恒定在 12 MPa,改变下游气缸压力(0.1 MPa,1 MPa,1.5 MPa,2 MPa,2.5 MPa 和 3 MPa)时的计算结果。其中气相体积分数分布即反映了空穴

图 3　计算流程图

443

的分布情况。

由图 4 看出在很高的喷油压力(140 MPa)时,空穴层紧贴壁面延伸出喷孔,而随着喷油压力的减小,空穴层在喷孔出口处开始脱离壁面,而当压力减小至 10 MPa 以下(5 MPa 和 3 MPa)时,空穴层延伸长度明显开始缩短。在如此小的气缸压力下,上游喷油压力的大幅变化,使空穴参数的增加并不是很多,空穴层也基本都可延伸至喷孔出口,长度变化总的来说不是很明显。这也说明对锐边过渡喷嘴,通常喷孔内部都会出现比较明显的空穴现象。同时注意到在改变喷油压力时,空穴参数和雷诺数都在发生变化。

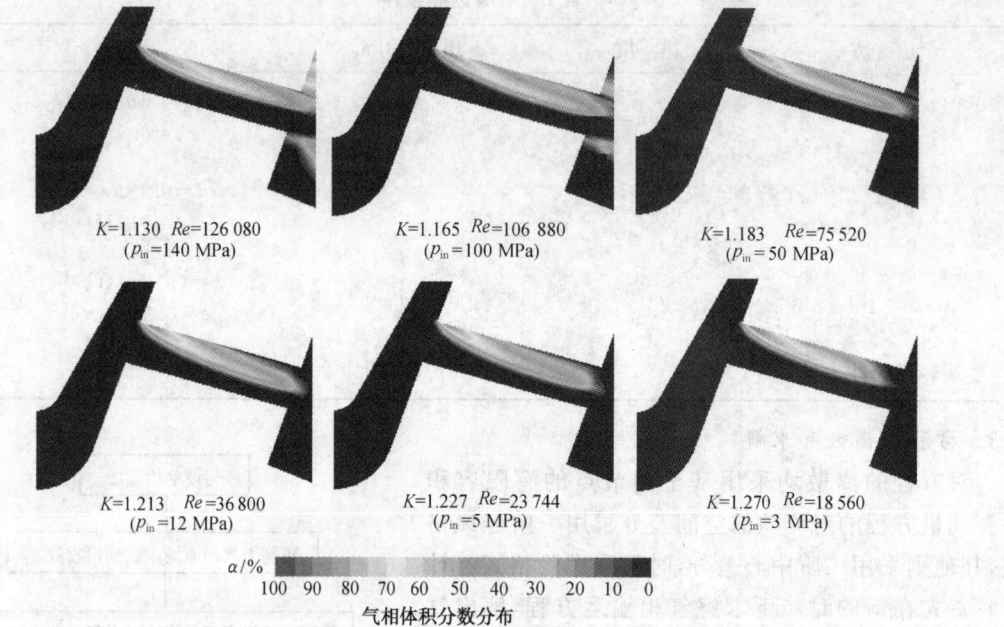

K=1.130 Re=126 080
(p_{in}=140 MPa)

K=1.165 Re=106 880
(p_{in}=100 MPa)

K=1.183 Re=75 520
(p_{in}=50 MPa)

K=1.213 Re=36 800
(p_{in}=12 MPa)

K=1.227 Re=23 744
(p_{in}=5 MPa)

K=1.270 Re=18 560
(p_{in}=3 MPa)

α/%
100 90 80 70 60 50 40 30 20 10 0
气相体积分数分布

图 4   不同喷油压力下的空穴分布(p_{out} = 0.1 MPa)

图 5 中,维持喷油压力在 12 MPa 时,随着下游气缸压力的增加,空穴参数增加显著,空穴层延伸长度也有明显的减小,特别是在气缸压力增至 3 MPa 时,空穴区域几乎快要消失。在这个变化过程中,虽然空穴参数有较大的变化,但通过喷孔的流速基本不变,雷诺数也基本保持不变。这些分析结果对高压共轨系统来说十分重要:高压共轨系统通常需要能够提供多次喷射,而正是由于预喷的存在,使得在开始主喷之前,燃烧室内已经开始燃烧。因此,在喷油过程中,燃烧室内的背压肯定会逐渐增加,而喷孔内部流动的空穴参数会随之减小。

同时改变上下游压力,而保持空穴参数基本相同情况下,喷孔内部空穴流动的计算结果如图 6 所示。可以看出,虽然上下游压力及压差不同,喷孔内部平均流速不同,相应的雷诺数也不同,但由于空穴参数相同,喷孔内部空穴层厚度、延伸长度等分布情况还是相似的,雷诺数的变化可以说对喷孔内部流动结构没有太大的影响。本文计算结果与文献[7]中试验结果均完全一致,即空穴参数基本相同的情况下,随着雷诺数的增加,由试验获得的流动图像基本相同,喷孔内部流动结构并没显示出有多大的差异,各种形式的空穴分布情况也均相同。本文数值模拟再次证明:空穴参数是影响喷孔内部流动结构最为重要的参数。

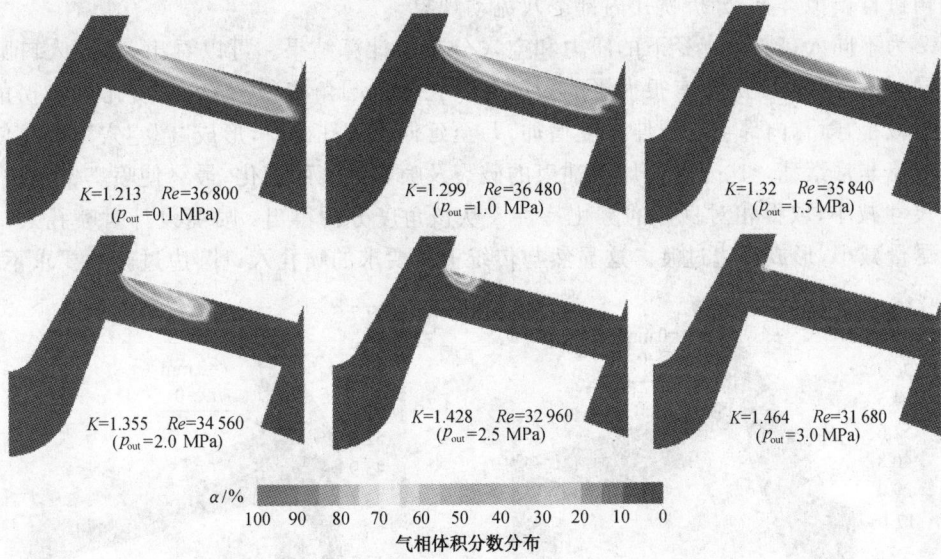

$K=1.213$ $Re=36\,800$
$(p_{out}=0.1\ \text{MPa})$

$K=1.299$ $Re=36\,480$
$(p_{out}=1.0\ \text{MPa})$

$K=1.32$ $Re=35\,840$
$(p_{out}=1.5\ \text{MPa})$

$K=1.355$ $Re=34\,560$
$(p_{out}=2.0\ \text{MPa})$

$K=1.428$ $Re=32\,960$
$(p_{out}=2.5\ \text{MPa})$

$K=1.464$ $Re=31\,680$
$(p_{out}=3.0\ \text{MPa})$

$\alpha/\%$

100 90 80 70 60 50 40 30 20 10 0
气相体积分数分布

图 5　不同气缸背压下的空穴分布($p_{in}=12\ \text{MPa}$)

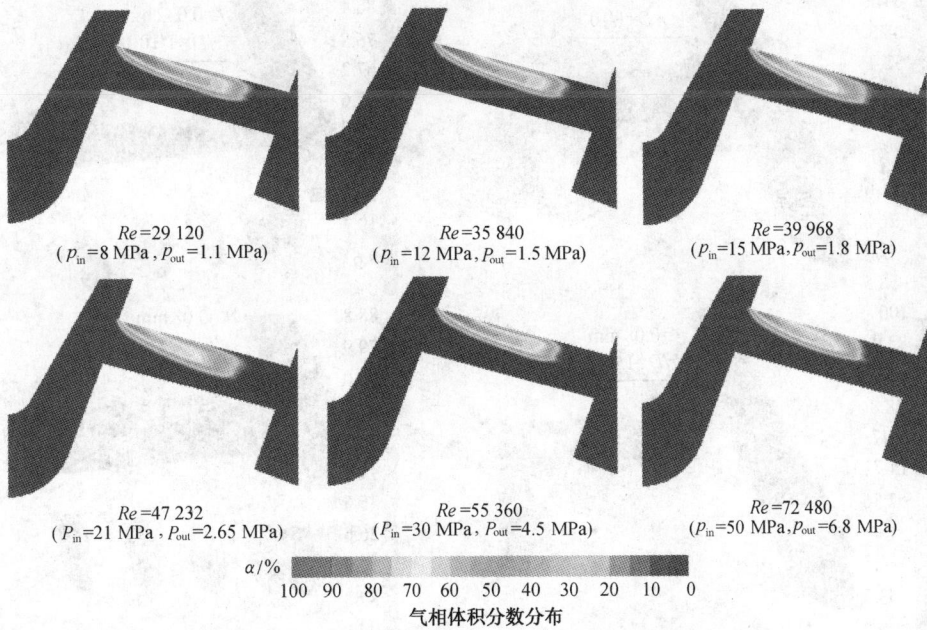

$Re=29\,120$
($p_{in}=8\ \text{MPa}$, $p_{out}=1.1\ \text{MPa}$)

$Re=35\,840$
($p_{in}=12\ \text{MPa}$, $p_{out}=1.5\ \text{MPa}$)

$Re=39\,968$
($p_{in}=15\ \text{MPa}$, $p_{out}=1.8\ \text{MPa}$)

$Re=47\,232$
($P_{in}=21\ \text{MPa}$, $p_{out}=2.65\ \text{MPa}$)

$Re=55\,360$
($P_{in}=30\ \text{MPa}$, $p_{out}=4.5\ \text{MPa}$)

$Re=72\,480$
($p_{in}=50\ \text{MPa}$, $p_{out}=6.8\ \text{MPa}$)

$\alpha/\%$

100 90 80 70 60 50 40 30 20 10 0
气相体积分数分布

图 6　相同空穴参数下的空穴分布($K=1.33$)

4　喷孔几何尺寸对空穴流动影响分析

同样针对 VCO 类型喷嘴,在针阀升程 $h=0.3\ \text{mm}$ 条件下,分别改变喷孔入口圆角半径和

445

喷孔倾角进行模拟计算,分析喷孔内部空穴流动现象。

图 7 为不同入口圆角半径下的压力和空穴分布的计算结果。可以看出,喷孔入口圆角半径的不同对压力和空穴分布有很大影响。随着喷孔入口圆角半径的减小,喷孔入口拐角处压力降明显减小,喷孔内部空穴延伸长度增加,乃至延伸出喷孔出口,形成"超空穴"现象,使得喷孔出口液流非常紊乱。这显然会加速油束的破裂及喷雾油滴的细化,最终使喷雾锥角增加、喷雾液滴尺寸减小,取得相对良好的雾化特性。从该角度分析得出:喷嘴设计时喷孔入口圆角半径应尽量减小,形成锐边过渡。这显然与传统上所要求的喷孔入口圆边过渡的要求不同。

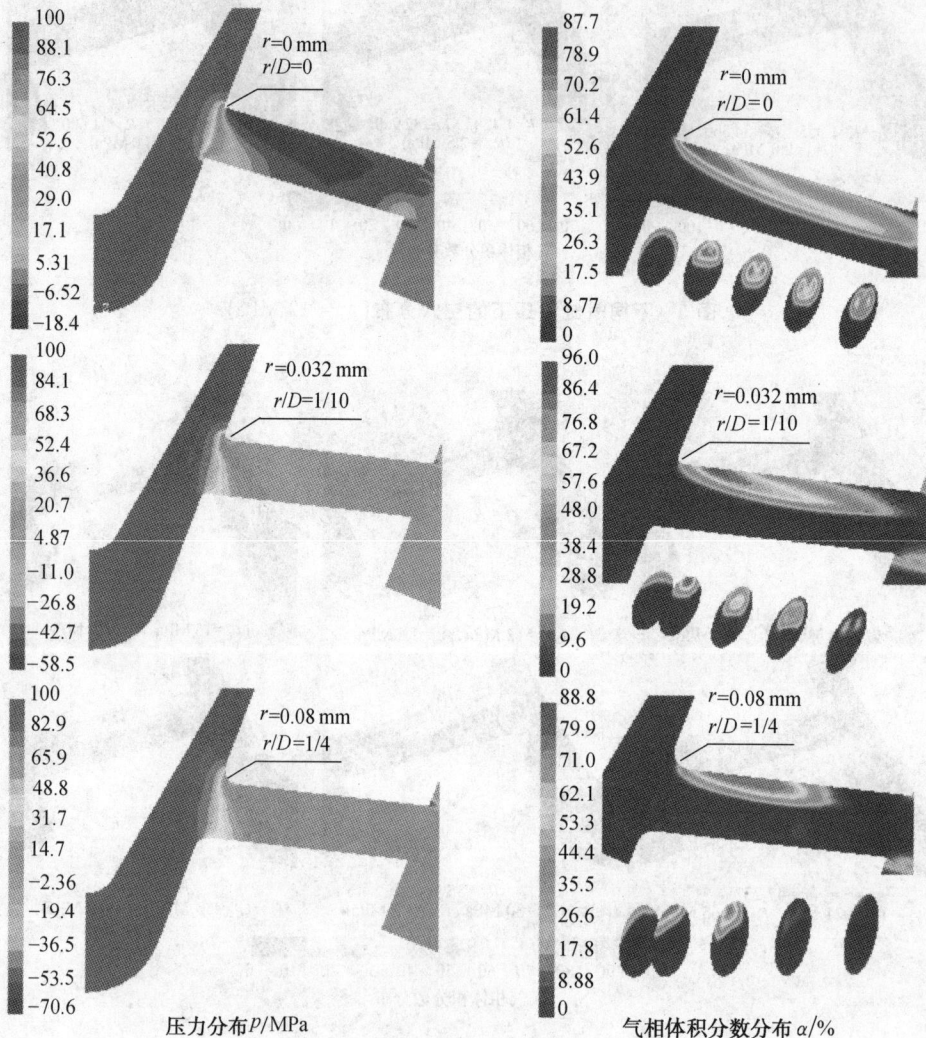

压力分布P/MPa   气相体积分数分布α/%

图7 不同喷孔入口圆角半径下的压力、空穴分布

图 8 为不同喷孔倾角(84°,78°,74°)下的压力和空穴分布情况。由压力分布看出,大倾斜角喷嘴与小倾斜角喷嘴相比,喷孔入口拐角处压力降会稍有减小,即喷孔倾角为 74°时,最低压力比 84°和78°时的都要小,空穴情况更为严重,这从空穴分布图中也得到了验证。只是在如此高的喷油压力下,下游气缸压力如此之低,使喷孔内部流速非常高,又由于喷孔入口锐边

过渡(锐边过渡更易发生空穴),喷孔内部空穴现象会非常严重,所以差别均不是很大。

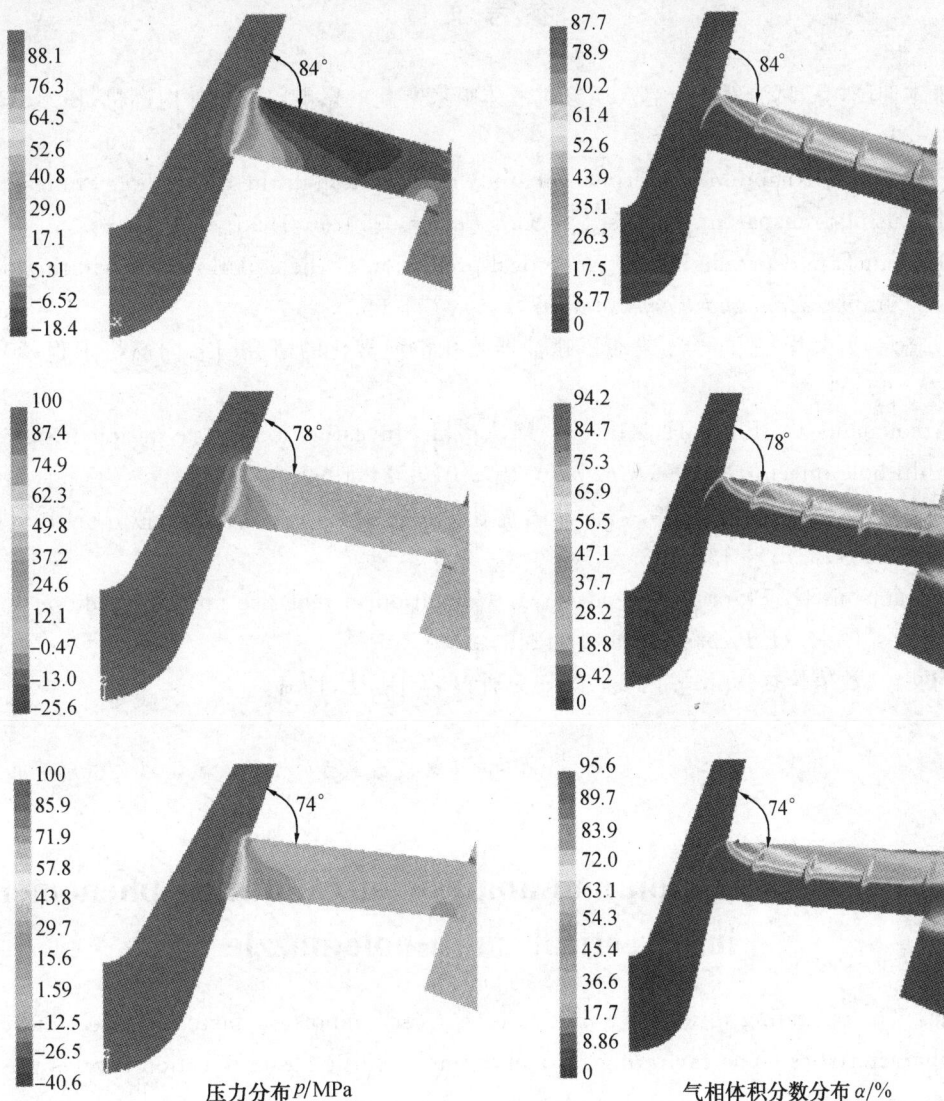

压力分布p/MPa   气相体积分数分布α/%

图 8 不同喷孔倾角下的压力、空穴分布

5 结 论

空穴现象的存在对供油系统中机械零件有负面影响,而对喷雾分裂过程会起到积极作用。采用混合多相流空穴模型进行多孔垂直喷嘴的三维数值模拟,清楚揭示了喷嘴内部流动的三维特征和空穴分布位置及形状,由此可知喷嘴喷孔内部空穴层分布情况很大程度上依赖于上游喷油压力和下游气缸背压以及喷嘴头部几何形状。喷孔入口圆角半径对喷孔内部空穴分布的影响最为显著,入口锐边过渡使喷孔空穴层基本都可延伸至喷孔出口,形成"超空穴"现象,加速液流紊乱,提高雾化质量。

参 考 文 献

［1］玉木伸茂,等.喷油嘴喷孔空穴现象对液体喷束雾化的影响[J].国外内燃机,1998,338(8)：23-29.

［2］Chaves H，Knapp M. Experimental study of cavitation in the nozzle hole of diesel injection using transparent nozzles[J]. *SAE Paper*，950290,1995.

［3］Schmidt D P，Corradini M L. Analytical prediction of the exit flow of cavitating orifices [J]. *Atomization and Sprays*，1997,7(6)：54-65.

［4］成晓北,黄荣华,王志,等.柴油机喷油器喷孔空泡雾化的研究[J].内燃机工程,2002,23(2)：60-64.

［5］Arcoumanis C，Flora H，Gavises M，et al. Investigation of cavitation in a vertical multi-hole injector[J]. *SAE Paper*，1999-01-0524,1999.

［6］何志霞,李德桃,胡林峰,等.喷油器喷嘴孔内部空穴两相流动数值模拟分析[J].内燃机学报,2004,22(5)：433-438.

［7］Arcoumanis C，Flora H，Gavaises M. Cavitation in real-size multi-hole diesel injector nozzles[J]. *SAE Paper*，2000-01-1249，2000.

［8］陶文铨.数值传热学[M].西安：西安交通大学出版社,1988.

（本文原载于《农业机械学报》2006年第2期）

Numerical simulation and analysis of cavitating phenomena in a vertical multi-hole nozzle

Abstract：The cavitating flow in a nozzle hole is a very important factor to affect diesel fuel spray characteristics. The cavitating flow phenomena in the diesel injection nozzles was analyzed. Based on this,using mixture multi-phase cavitating flow model，3-D numerical simulation was done to the fully developed cavitating flow in nozzle holes of a vertical multi-hole injector. The effects of injection pressure, back pressure, turning angle of the orifice and hole inclination angle on the cavitation phenomena in nozzle holes were analyzed. The results validated that the lower cavitation parameter was, the longer extending length of cavitation film could be, exit flow of holes with sharp-edged entrance shape would be more turbulent，and then accelerate the break up of spray accordingly.

高压共轨式柴油机多次喷射的数值模拟

梁凤标,李德桃,刘久斌,潘剑锋

[摘要] 构建了高压共轨喷射系统的计算模型,并在试验台架上进行了试验验证。对喷油器的电磁、液力和机械特性进行了模拟计算,分析了多次喷射的机理,确定了多次喷射的喷油定时和喷油脉冲间隔。

高压共轨燃油喷射系统在实现高压喷射的同时可独立控制喷油量、喷油正时与喷油规律,而不受柴油机转速与负荷的限制。该系统采用 120～180 MPa 的喷射压力,喷射过程的开始和结束一般由高速强力电磁阀控制。喷油器的性能直接决定着整个喷射系统的性能,因此,提高电磁阀的响应特性,精确控制多次喷射正时和喷油间隔,是系统开发的重点和难点。

本文对强力电磁阀的性能进行深入研究,提出在电磁阀通电结束时刻增加一反向电压,以提高电磁阀的响应速度;对喷油系统进行模拟计算,分析多次喷射的实现机理,以确定多次喷射的时间间隔和喷油正时。

1 高压共轨系统和喷油系统计算模型

图 1 是根据日本电装公司 ECD-U2 系统开发的高压共轨喷射系统。该系统由高压泵、低压输油泵、滤清器、共轨管、电磁阀控制的喷油器、电控单元(ECU)和各种传感器组成。系统电控单元根据柴油机工作状况和其他信息(如油温、压力等),依据给定的油压脉冲图谱,通过高压输油泵上的压力控制阀(PCV)来调整油泵供油量,以改变共轨管中的油压,因此油压与柴油机转速无关。该系统通过高速电磁阀可实现对喷射定时、喷射量及喷射速率的柔性控制。

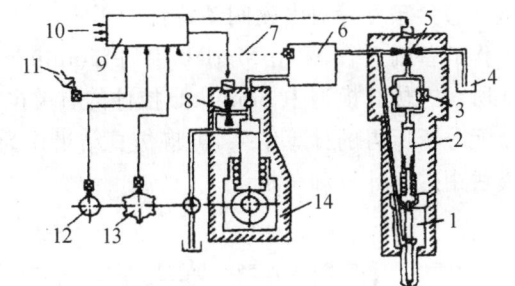

1—喷油器 2—液压柱塞 3—单向孔板 4—回油
5—三通阀 6—共轨 7—燃油压力传感器
8—泵控制阀 9—ECU 10—辅助信息
11—发动机负荷信息 12—气缸检测器 13—发动机转速和凸轮轴转角传感器 14—高压供油泵
图 1 高压共轨系统示意图

本文采用 Hydsim 软件构建了喷油系统计算模型,如图 2 所示。模型由模拟喷油系统实体单元的各种液力、机械元件组合而成,各个元件由输入、输出函数,以及内部的液力、机械计算方程函数组合成为计算模块,通过液力或机械连接组合成计算模型。计算元件的各种输入参数需根据喷油系统结构的实际参数和运行条件进行计算,而其内部函数的计算方程主要有[1]:

流体连续性方程

$$\oint \rho u \, \mathrm{d}A = 0$$

流体可压缩性方程

$$\mathrm{d}p = -\frac{E}{V}\mathrm{d}V$$

运动方程

$$m\frac{\mathrm{d}^2 x}{\mathrm{d}t^2} = F_0 + \sum pA_\mathrm{p} - Kx - C_\mathrm{r}\frac{\mathrm{d}x}{\mathrm{d}t}$$

式中 ρ 为燃油密度；u 为燃油流速；A 为流通面积；p 为燃油压力；V 为控制体容积；E 为燃油弹性模量；m 为活塞等运动件质量；x 为活塞等位移；F_0 为摩擦力；A_p 为受力面积；K 为弹簧刚度；C_r 为阻力系数。

喷嘴的流量方程

$$Q = \sqrt{\frac{\dfrac{2}{\rho_\mathrm{av}}(p_\mathrm{in} - p_\mathrm{out})}{\dfrac{\xi_\mathrm{seat}}{A_\mathrm{seat}^2} + \dfrac{\xi_\mathrm{k\text{-}inlt}}{A_\mathrm{holes}^2} + \dfrac{\xi_\mathrm{holes}}{A_\mathrm{holes}^2}}}$$

式中 ρ_av 为流体的平均密度；A_seat 为针阀座的面积；A_holes 为喷孔横截面面积；p_in 为针阀进口压力；p_out 为针阀出口压力；ξ_seat，$\xi_\mathrm{k\text{-}inlt}$，$\xi_\mathrm{holes}$ 为流量系数。

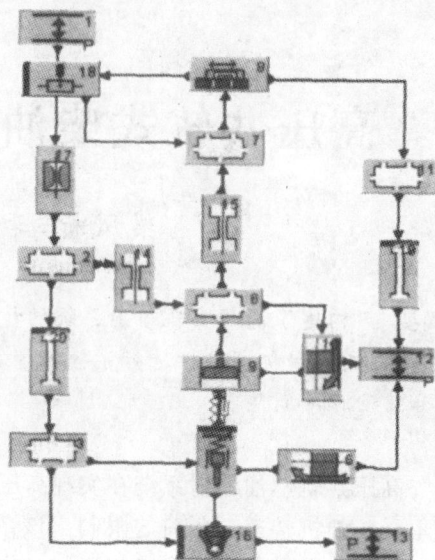

1,12,13—压力边界　2,3,6,7,11—容积
4—针阀体　5,19—泄漏槽　8—电磁执行器
9—活塞组　10,14,15,20—刚性管
16—喷嘴　17—节流器

图 2　共轨系统的计算模型

Hydsim 计算软件对电磁阀模块的计算进行了较大简化，没有考虑电磁材料的磁滞曲线、漏磁和边界条件等对电磁阀动特性的影响，这使模型的计算产生较大的偏差。本文在以前所做工作的基础上[2]，采用 Matlab 中的 Simulink 仿真程序对电磁阀的动特性进行动态仿真，如图 3 所示。仿真模型中 Simulink 控件的有关电磁阀特性的传递函数和数据都采用文献[2]中有限元分析所得的计算结果，再将仿真结果作为喷油系统模型中电磁阀的输入参数，代入电磁阀模块中进行计算和分析。

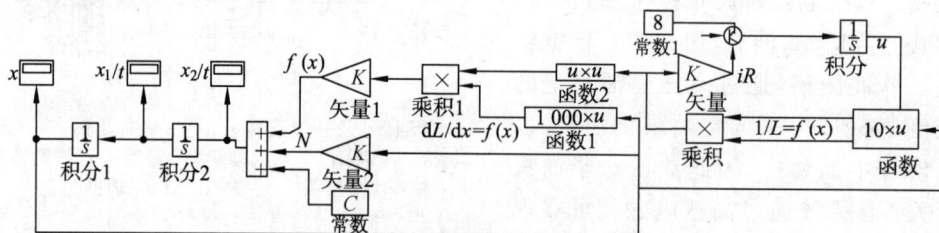

图 3　电磁阀的计算模型

在电磁场计算中，磁场和电场的偏微分方程为

$$\nabla^2 B - \mu\varepsilon\frac{\partial^2 B}{\partial t^2} = -\mu I$$

$$\nabla^2 \phi - \mu\varepsilon\frac{\partial^2 \phi}{\partial t^2} = -\frac{\rho_I}{\varepsilon}$$

式中 B 为矢量电势；ϕ 为标量电势；I 为传导电流密度矢量；ρ_I 为电荷体密度；μ，ε 为介质的磁导率和介电常数。

边界条件为诺依曼边界条件

$$\frac{\partial \phi}{\partial \boldsymbol{n}}\bigg|_{\Gamma} + f(\Gamma)\phi\big|_{\Gamma} = h(\Gamma) \tag{7}$$

式中 Γ 为诺依曼边界;$\boldsymbol{n}$ 为边界 Γ 的外法线矢量;$f(\Gamma)$,$h(\Gamma)$ 为一般函数,可为常数或零。

衔铁位移方程式为

$$F_x = m_{xt}\frac{\mathrm{d}^2 x_{xt}}{\mathrm{d}t^2} + F_f(x_{xt}) + F_f\left(\frac{\mathrm{d}x_{xt}}{\mathrm{d}t}\right)$$

式中 m_{xt} 为电磁铁衔铁所带动的运动部分质量;x_{xt} 为电磁铁衔铁的位移;$F_f(x_{xt})$,$F_f\left(\dfrac{\mathrm{d}x_{xt}}{\mathrm{d}t}\right)$ 为与衔铁运动部分的位移和速度有关的反力。

2 计算模型的试验验证

为验证所建模型的准确性,在共轨试验台架上对装有该高压共轨喷油系统的直喷式柴油机进行了试验研究。表 1 为不同共轨压力和喷油脉冲条件下的喷油量试验值,为多次测量的平均值;图 4 为计算值与表 1 中所示试验值的差量(即变化率)。

表 1　不同共轨压力和喷油脉冲条件下喷油量

试验号	共轨压力/MPa	预喷油量/mg	主喷油量/mg
1	30	0.18	
2	30		12.2
3	30	0.25	15.6
4	30	0.30	24.8
5	100	2.50	
6	100		67.9
7	100	2.80	78.5
8	100	3.20	82.5
9	140	3.46	
10	140		80.5
11	140	3.52	98.4
12	140	3.60	105.6

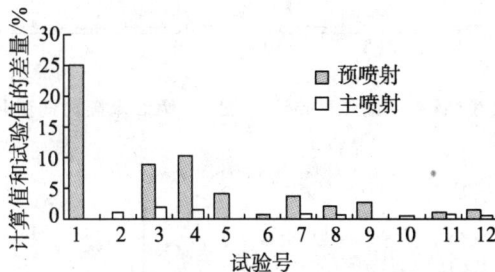

图 4　计算值与试验值的差量示意图

从图 4 可以看出,在不同的共轨压力和喷油量条件下,预喷油量差量范围为 25% 到 1.2%,主喷油量差量范围为 1.5% 到 0.3%。共轨压力 30 MPa 时试验值和计算值的差量达到了 25％,一方面是由于 30 MPa 接近于喷油系统的最小开启压力,共轨系统工作不稳定;另一方面也是由于针阀、电磁阀与阀座之间的碰撞使得低压下预喷油量难以精确控制。在较高压力(80 MPa 以上)和喷油量高于 15 mg(主要是在主喷中)时,计算值和试验值较为吻合,差量在 0.5％左右,证明了模型是准确的,可以进行参数分析。要进一步减小其差量是相当困难的,因为在模型的建立过程中,对一些参数和边界条件进行了简化和理想化,经过液力元件和机械元件的传递,差量往往会增大。本文所建模型在较高喷射压力和喷油量条件下的计算值较为准确。

3　计算分析

Imarisio 和 Bianchi 等指出高压共轨式柴油机喷油系统中新型电磁阀必须具有高速响应性能和精确的多次喷射控制能力[3,4]。由于电磁、机械元件的迟滞,电磁阀驱动电压的开始、结束时刻与喷油器的开始、结束时刻并不一致。为提高电磁阀的响应速度,电磁线圈中的电流脉冲如

图 5　驱动电流随时间的变化值

图 5 所示,采用较高的 8 A 驱动电流,以提高电磁阀的开启响应速度[2],并在结束时刻增加一反向电压,以减小阀的关闭时刻[5],从而提高电磁阀响应速度,实现多次喷射和小油量喷射。

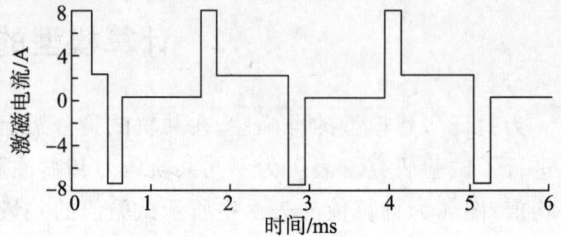

在此驱动脉冲作用和 140 MPa 共轨压力下,喷油系统的喷油规律如图 6～8 所示,2 次喷射时间间隔分别为 1.8 ms,1.4 ms 和 1.0 ms。从图 6 和图 7 可看出,当喷油间隔在大于1.4 ms 而小于 2 次喷射之间的时间间隔要求(1.8 ms)时,2 次主喷射的喷油规律基本相同。两者之间存在着微小的差别,这是由于针阀和柱塞的运动使控制室的压力产生波动,从而导致两次主喷射的压力条件不同。由图 8 可以看出,当间隔小于 1.0 ms 时,2 次主喷射连在一起,系统不能稳定工作。

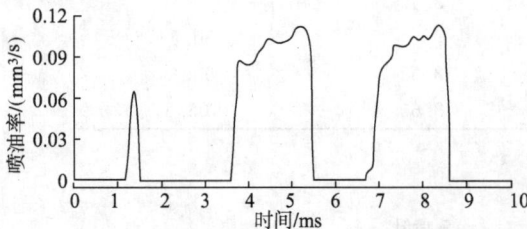

图 6　喷油率随时间变化的曲线(脉冲间隔 1.8 ms)

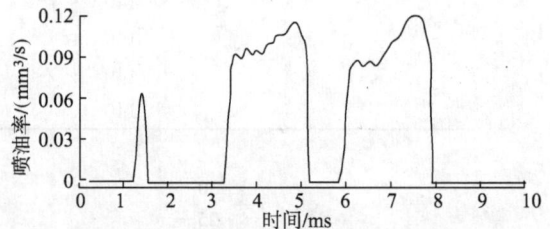

图 7　喷油率随时间变化的曲线(脉冲间隔 1.4 ms)

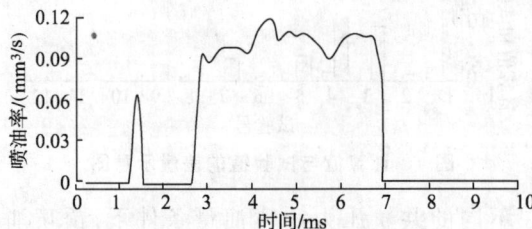

图 8　喷油率随时间变化的曲线(脉冲间隔 1.0 ms)

当 2 次主喷射之间间隔在 1.4 ms 以上时,喷油器针阀有足够的时间回位并消除压力波动,喷油规律基本相同;而如图 9 所示,当间隔小于 1.0 ms 时,第 2 次主喷射开启时刻针阀并没有完全落座,从而使前后主喷射的初始压力不同,喷油率也不相同。

脉冲间隔为 1.8 ms 和 1.0 ms 时,控制室压力变化曲线的比较如图 10 所示。脉冲间隔为 1.8 ms 时,2 次主喷射的控制室压力几乎相同;而脉冲间隔为 1.0 ms 时,第 2 次主喷射的控制室压力并没有恢复到初始水平,使 2 次主喷射连接在一起。因此 2 次喷射的时间间隔不能小于 1.0 ms。在间隔大于 1.4 ms 时可最大限度地消除所有的不稳定现象,从而实现可重复的多次喷射。

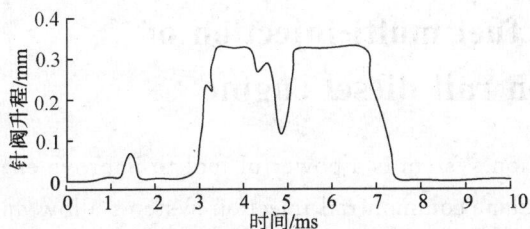

图 9　针阀升程随时间变化曲线(脉冲间隔 1.0 ms)　　图 10　控制室压力曲线(共轨压力 140 MPa)

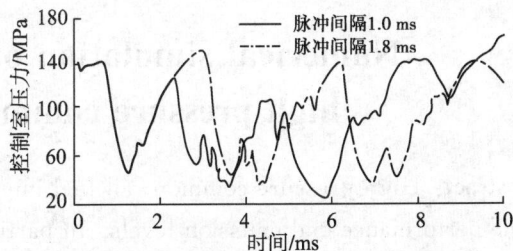

2 次喷射之间的脉冲间隔还与共轨压力、发动机负荷等工况有关,同样可以用本文所建模型进行模拟计算,以得到高压共轨系统的预喷、主喷和后喷的喷油图谱。

4　结　　论

(1) 建立了高压共轨燃油喷射系统的计算模型,模型的计算值与试验值的差量在 5% 左右。

(2) 对共轨压力 140 MPa 下的喷油规律进行了详细分析,研究表明:在电磁阀通电结束时刻对电磁线圈加上反向电压提高了电磁阀的响应速度,从而使 2 次喷射的时间间隔达到 1.4 ms左右,喷油系统可实现反复的多次喷射;喷射间隔的极限值不能小于 1.0 ms。

(3) 应用所建模型可对不同工况下的喷油特性进行模拟计算,从而为制定喷油控制策略提供依据。

参 考 文 献

[1] 王钧效,陆家祥,张锦杨,等. 柴油机喷油过程模拟计算中的几个经验公式研究[J]. 车用发动机,2001(5):6-11.

[2] 梁凤标,李德桃,胡林峰,等. 柴油机高速强力电磁阀的数值模拟[J]. 农业机械学报,2005,36(2):8-11.

[3] Nehmer D A, Reltz R D. Measurement of the effect of injection rate and split injections on diesel engine soot and NO$_x$ emissions[J]. *SAE Paper*, 940668, 1994.

[4] Bianchi G M, Pelloni P, Filicori F, et al Optimization of the solenoid valve behavior in common-rail injection systems [J]. *SAE Paper*, 2000-01-2042,2000.

［5］Bianchi G M，Falfari S，Pelloni P，et al. Development of a dynamicmodel for studying the 1st generation of common rail injectors for HSDI diesel engines［C］//*Capri*，*ICE*，2001.

［6］何志霞，袁建平，李德桃. 垂直多孔喷嘴内部流动空穴现象数值模拟分析［J］. 农业机械学报，2006，37(2)：4-8.

（本文原载于《农业机械学报》2006 年第 9 期）

Numerical simulation of fuel multi-injection of high pressure common rail diesel engine

Abstract：High pressure common rail fuel injection system is a powerful tool to improve engine performance and emission levels. In particular，common rail injection systems allow an almost completely flexible fuel injection event in DI diesel engines by pemitting a free mapping of the start of injection，injection pressure and injection rate. Precise control of the fuel injection quantity，injection timing and injection period between the pilot injection，main injection and late injection is the key technology in the design of the injector. In this research，numerical models of the common rail injection system were developed. Validation compared with experiments has been performed. The model has been used to provide insight into the operation conditions of the injector and highlight the application to injection system design. The models can be used to simulate the electro-fluid-mechanical behavior of the injector.

IV. 微型动力机电系统的研究

Investigation on Power MEMS

微动力机电系统和微发动机的研究进展[①]

李德桃,邓　军,潘剑锋,杨文明,薛　宏

[摘要]　微动力机电系统和微发动机是动力装置发展的第四个里程碑。本文先阐述了该系统的重要意义,进而描述了它的发展概况,指出了该系统面临的挑战和解决的基本途径。微动力机电系统的应用前景是广阔的,对社会和经济的影响是巨大的。

近年来,在国外,尤其是美国,相继开展了微动力机电系统(Power MEMS)和微发动机(Micro-engine)的研究工作[1−9]。其共同特征是利用碳氢燃料,在一个微型的燃烧器中燃烧放热。从动力机械发展的历史进程看,每当能源装置的能量密度产生一个飞跃,都会给社会和经济带来深远的变革。18世纪的蒸汽发动机,以0.005 W/g的能量密度为标志,引发了当时的工业革命。从19世纪到20世纪中叶,内燃机的发展使能量密度达到了0.05～1.0 W/g,从而使整个交通运输发生了巨变。20世纪研发的航空航天发动机使能量密度进一步上升到10 W/g。喷气式飞机大大地缩短了整个世界的距离。微动力装置的能量密度将冲破100 W/g的大关。可以说,它是动力机械发展的第四个里程碑。不难想象,它给现代社会带来的影响将是重大而深远的。

我们知道,氢气和烃类燃料的能量密度大约是当前最好的锂化学电池的10～100倍(图1),因此,使用电池的电子设备,如蜂窝电话、照相机等,如果用微发动机驱动,将可以极大地提高使用效率。而且采用碳氢燃料的微动力装置还具有成本低廉、电压稳定的优点。微发动机将广泛应用于工业、农业、环境保护、医疗卫生等各行各业的电子器件中,还可作为微型汽车、微型飞机、微型泵等微型机械的动力。这种微型、携带式的动力装置的研制成功,将不仅对微机电系统产业,而且对以现代微电子、信息和生物技术为支柱的产业产生巨大的影响。由此可见,开展微动力机电系统和微发动机的研究具有重大的意义。

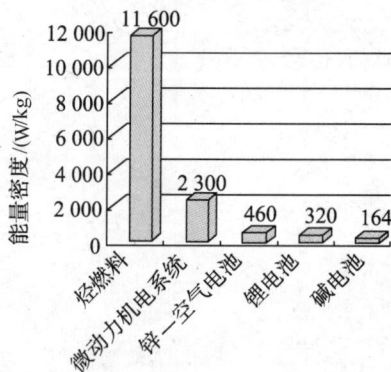

图1　微发动机同电池的能量密度对比图

---

①　据我们所知,本文在国内首次报道了微动力机电系统的发展和已取得的成果(含作者科研团队的成果),随后,国家自然科学基金委首次资助李德桃教授进行这方面的研究,开微动力系统研究之先河。8年来,通过国际合作,先后获得了三项国家自然科学基金的资助,并在国内首创了微热光电系统原理性样机,2004年,系统整体转换效率已达到0.81%,而2003年MIT的微结构热光电系统的整体效率也仅为0.08%。

——编者注

1　微发动机发展概况

Epstein 和 Groshenry 等人首先设计了一种每小时消耗燃料 7 g，产生 10～50 W 电功率的微涡轮机[1-3]。该装置如图 2 所示，主要包括压气机、燃烧室、涡轮机和起动电机/发电机等部分，总体积为 1 cm³，燃烧室容积为 0.07 cm³。他们完成了该装置的可行性研究、初步设计和性能估计。

稍后，Waitz 等人提出了相应的燃烧室工作原理和燃烧策略[4]。Mehra 等人首次采用微加工技术，设计制造了一个容积小于 0.07 cm³ 的硅基燃烧室（图 3），初步实现了预混氢-空气的稳定燃烧。经测试，出口燃气温度达到了 1 800 K，燃烧室的效率达 40%～60%，工作了 15 小时。这表明硅基燃烧室在这种环境下有着令人满意的性能，从而使采用微加工技术制造硅基微涡轮机迈出了重要的一步。

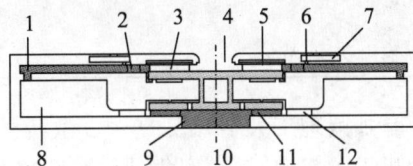

1—火焰稳定器　2—扩散叶片　3—转子叶片　4—进气口
5—起动器　6—燃料喷孔　7—燃料汇流腔　8—燃烧室
9—排气口　10—转子中心线　11—涡流转子叶片
12—涡轮导向叶片

图 2　微涡轮机示意图

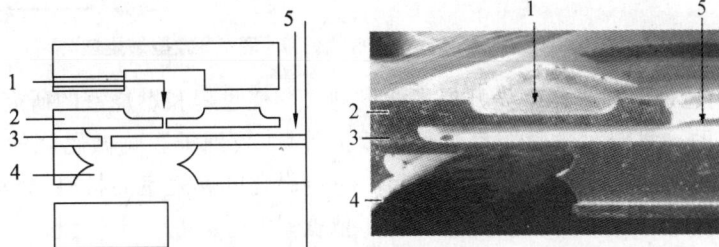

1—氢　2—燃料分喷板　3—混合气进口　4—燃烧室　5—空气

图 3　微燃烧室(1/2)结构示意图和横截面照片

Mehra 等人还采用深度活性离子蚀刻(Deep Reactive Ion Etching)和多层融合技术成功制造出较完整的微发动机，并实现了非预混氢-空气的燃烧[6]。该发动机可视为高能量密度（约 2 000 MW/m³）微发动机的最初示范。

图 4 和图 5 分别为巨型和微型燃气轮机的结构示意图，其主要参数的对比在表 1 中列出。

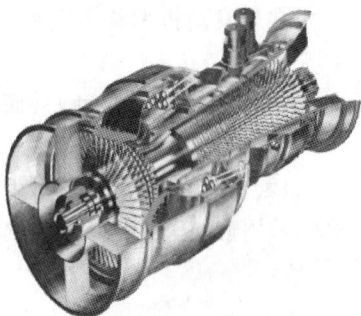

图 4　巨型燃气轮机

图 5　微型燃气轮机

表 1　巨型和微型燃气轮机参数对比

参　数　值	机　　型	
	巨　　型	微　　型
进口直径/mm	2×10^3	2
空气流量/(g/s)	5×10^5	0.25
重　　量/g	4×10^8	1
输出功率/W	1.5×10^8	50

图 6 是加州理工大学开发的一台微转子发动机。它能发出 $10\sim100$ MW 的功率。由于转子发动机无阀门，较易微型化，因此，它也是一种有发展前途的微动力装置。

图 7 是微发动机所用的微燃烧室。这种结构设计热损失很小。

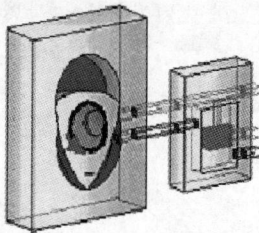

图 6　微转子发动机

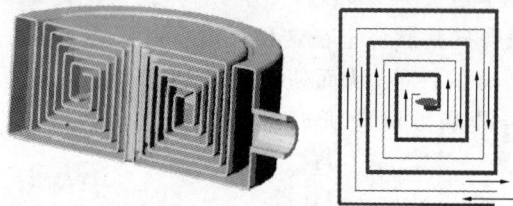

图 7　微型燃烧室

在国内，我们通过国际合作率先开展了微发动机燃烧过程和燃烧室的研究，并得到国家自然科学基金的资助。我们对微燃烧室进行了模拟计算，找到了影响微燃烧的 5 个主要因素：燃烧室的结构尺寸、燃料的性质、混合气质量流率、空燃比和燃烧室的材料，分别分析了这些因素对微燃烧的影响，从而为设计微燃烧室提供了依据[10,11]。

2　面临的挑战和解决的途径

微发动机的研究和开发尚处于起步阶段，尤其是对微燃烧的机理以及燃烧稳定性的认识还很不足。与传统发动机燃烧过程相比，微燃烧至少面临以下三方面的挑战：

(1) 微发动机中的混合和燃烧的驻留时间是常规发动机的 $1/100\sim1/10$，大大限制了燃料同空气混合和化学反应的时间，因此不易实现完全燃烧。

(2) 由于微发动机的面积与体积之比约为 500 m^{-1}，而常规发动机燃烧室则为 $3\sim5$ m^{-1}，因此微发动机的热损失很大。

(3) 点火和燃烧的淬熄(Flame Quenching)特征尺寸已经接近燃烧室尺寸，从而使燃烧不稳定性急剧增加。

上述问题，通过选择火焰界限宽、反应时间短的燃料(如氢)并加催化剂，采用耐高温材料(如硅)和特殊的结构设计等办法可以得到解决。微发动机采用硅材料，比金属基发动机能耐更高的温度，而且对减少排气中的 NO_x 有利。

3 结束语

微发动机的发展虽然初露端倪,但它作为高新技术的动力装置已经显示出巨大的发展潜力和广阔的应用前景。

通过采用硅基材料和相应的微加工技术,采用适当的燃烧策略,如使用氢燃料、预混合稀薄燃烧,初步实现了微小体积内的稳定燃烧。今后的工作主要集中在进一步对微小体积内的流体流动、燃料的喷射、混合、点火、催化反应以及燃烧室材料的氧化进行更深入的研究,以改进微发动机的性能,提高微发动机的寿命。另一方面的研究集中在碳氢燃料的燃烧,以使微发动机能够使用更容易得到的碳氢燃料。

参 考 文 献

[1] Epstein A H, Senturia S D, et al. Power MEMS and microengines[C]// *IEEE Transducers '97 Conference*, Chicago, IL, 1997: 753–756.

[2] Epstein A H, Senturia S D, et al. Micro-heat engines, gas turbines, and rocket engines—the MIT microengine project[C]// *28th AIAA Fluid Dynamics Conference*, Snowmass, CO, 1997: 1–12.

[3] Groshenry C. Preliminary study of A micro-gas turbine engine[D]. S. M. Thesis, Cambridge: Mass Inst Technol, 1995.

[4] Waitz I A, Gautam G, Tzeng Y S. Combustors for a micro gas turbine engines, (Invited paper) international symposium on Micro-Electro-Mechanical Systems(MEMS): *ASME* 1996 *Intemational Engineering Congress and Exposition*, 17–22 November, Atanta, Georgia, 1996[C]// *ASME Journal of Fluids Engineering*, 1998, 120: 109—117.

[5] Mehra A, Waitz I A. Development of a hydrogen combustor for a microfabricated gas turbine engine[C]// *Solid-state Sensor Actuator Workshop*, Hilton Head, SC, 1998.

[6] Mehra A, Arturo A. Microfabrication of high temperature silicon devices using wafer bonding and deep reactive ion etching[J]. *IEEE Journal of Microelectromechanical Systems*, 1999, 8(2): 152–160.

[7] Mehra A, Waitz I A and Schmidt M A. Combustion tests in the static structure of a 6-wafer micro gas turbine engine[C]// 1999 *Solid State Sensor and Actuator Workshop*, 1999: 1–4.

[8] Mehra A, Zhang X, Ayon A, Waitz I, Schmidt M, Spadaccini C. A 6-wafer combustion system for a silicon micro gas turbine engine[J]. *Journal of Microelectromechanical Systems*, 2000, 9(4): 517–527.

[9] Mehra A, Zhang X, Ayon A, Waitz I, Schmidt M. A through-wafer electrical interconnect for multi-level MEMS devices[J]. *Journal of Vacuum Science & Technology B*, 2000, 1(5): 2583–2589.

[10] 李德桃, 等. 微发动机的初步研究[C]// 中国汽车工程学会发动机分会, 中国内燃机学会煤气机汽油机分会年会学术论文, 2001.

[11] 李德桃，等. 微型发动机燃烧室的模拟研究.（待发表）

（本文原载于《世界科技研究与发展》2002 年第 1 期）

Development of research on power MEMS and micro-engine

Abstract：Power MEMS and micro-engine are the forth milestone of the development of power devices. In this paper，the significance of developing this kind of system is delineated. Then the advancement in this field is described. Some challenges faced when developing this system and promising approaches to these challenges are put forward. Power MEMS will have widely future application will have significant influence to society and economy.

微型发动机燃烧室的模拟研究

李德桃,邓 军,潘剑锋,杨文明,薛 宏

[摘要] 简要介绍了微型发动机的研究意义和应用前景。采用模拟软件对微型发动机燃烧室中的燃烧现象进行了模拟,并对影响燃烧的主要因素进行了分析,为微型燃烧室的设计提供了依据。

近年发展起来的微机电系统(MEMS)和微加工技术,为微型发动机的研究和开发创造了条件。所谓微型发动机是指以微机电系统和微加工技术的发展为基础,体积从不到 1 cm³ 到几 cm³,能产生大约 10~100 W 功率的热力发动机。

众所周知,氢气和烃类燃料的能量密度大约是当前最好的锂化学电池的 10~100 倍,因此,使用电池的电子设备,如果用微型发动机驱动,将可以极大地提高使用效率。而且采用碳氢燃料的微型动力装置还具有成本低廉、电压稳定的优点。因而微型发动机将广泛应用于工业、农业、环境保护和医疗卫生等各行各业的电子器件中,还可作为微型汽车、微型飞机和微型泵等微型机械的动力。这种微型、携带式的动力装置一旦研制成功,将不仅对微机电系统产业,而且对以现代微电子、信息和生物技术为支柱的产业产生巨大的影响。

在过去的几年中,国外,尤其是美国的一些大学和研究所[1-4],如 MIT、加大伯克莱分校、密执安大学等相继开展了这方面的研究。美国的国家科学基金(NSF)等机构均拨款资助微型发动机的研究。然而,据我们所知,国内在这一领域还是空白。因此,开展对微型发动机的研究具有重大的意义。

燃烧室是微型发动机最重要的部分。由于常规燃烧室和微燃烧室的几何参数、工作条件有极大的区别,在建造常规发动机中积累的知识和经验不能直接用于微型发动机燃烧室[5]。首先,燃料和空气在燃烧室中的驻留时间减少,会影响燃料混合和化学反应所能占用的时间。其次,随着尺寸的减小,微燃烧室的面积-容积比迅速增加,由此产生的高传热损失可能会降低微燃烧室的效率,并且微小尺寸产生的火焰淬熄可能会影响燃料的燃烧极限。因此,了解影响微燃烧现象的因素,研究合适的微燃烧室燃烧方案,在有限的驻留时间内实现完全燃烧非常必要。针对这些问题,我们先通过模拟计算,对微型发动机中的燃烧进行了分析研究,探讨了影响燃烧的主要因素,为燃烧室的设计打下基础。

1 微型发动机燃烧的模拟

为了描述微燃烧中涉及的基本热现象,使用 CFD 软件包 FLUENT 模拟在微燃烧室中的燃烧。

所研究的微发动机主要由压缩器、燃烧室、涡轮和启动电动机/发电机组成,如图 1 所示。

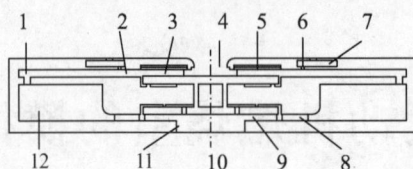

1—火焰稳定器　2—扩散叶片　3—转子叶片　4—进气口　5—起动器　6—燃料喷孔
7—燃料汇流腔　8—涡轮导向叶片　9—涡流转子叶片　10—转子中心线　11—排气口
12—燃烧室

图1　微燃烧室的结构

2　影响燃烧的主要因素分析

有5个主要因素影响燃烧现象：① 燃烧室的形状和尺寸；② 燃料种类；③ 质量流率；
④ 燃料空气比；⑤ 壁面材料。

2.1　燃烧室形状和尺寸的影响

为了比较燃烧室形状和尺寸的影响,我们用三维模型模拟了3种不同的燃烧室。图2～4
是3种燃烧室的网格图,其尺寸详细说明在表1中分别列出。

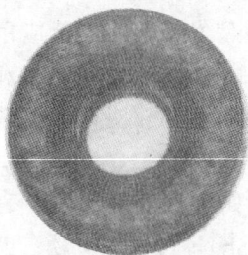

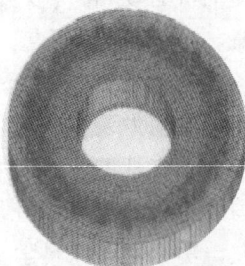

图2　A型燃烧室　　　　　图3　B型燃烧室　　　　　图4　C型燃烧室

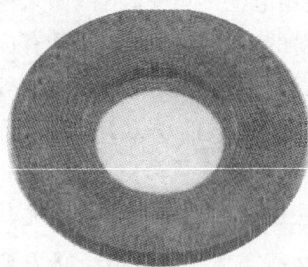

表1　3种不同尺寸微燃烧室的详细说明

	燃烧室内径 d/mm	燃烧室外径 d/mm	燃烧室高度 h/mm	容积 V/mm³	面积 A/mm²
A 型	5.0	12.0	1.4	130	262
B 型	5.0	12.0	3.0	280	347
C 型	9.6	18.4	1.0	193	475

表2为当工作压力为 0.4 MPa,质量流量为 $0.15\ g\cdot s^{-1}$,氢气/空气比为 0.4 时模拟的出
口平均温度和最高温度,结果表明:A 型的出口平均温度最高,因而其效率也最高;B 型和 C 型
几乎相等。这是由燃烧室的表面和容积决定的。由于 A 型的表面积最小,通过表面的热损失
也最小。同时,随着燃烧室容积的减小,混合物的驻留时间也减少,使得热损失减少。因此,A
型的效率高于另外两型。需要指出的是:混合物必须在有效驻留时间内完全燃烧。也就是说,
燃烧室的容积必须足够大以保证完全燃烧。尽管 B 型容积大于 C 型,但 B 型的表面积比 C 型
小,因此这两种燃烧室效率几乎相同。

表 2　出口平均温度和燃烧室内最高温度

燃烧室类型	出口平均温度 $T_{o,av}$/K	最高温度 T_{max}/K
A 型	1 453.1	1 660.3
B 型	1 383.9	1 615.1
C 型	1 384.8	1 506.3

从以上的模拟结果可知,当质量流量相同,并且有效驻留时间足够长时,为了获得高效率,必须在保证能提供足够驻留时间的容积条件下,尽力减小燃烧室的表面积和容积之比以降低热损失。

2.2　不同燃料的影响

所选的燃料——氢气和一种典型的烃类进行了比较,结果如表 3 所示。从比较结果可以发现,氢气的热值高、蒸发和扩散速度快、反应时间短、燃烧范围宽广、所需点火能量低和向环境辐射热量少。最重要的是,烃类燃料的燃烧通常需要一个较浓的初始燃烧区域,紧接着一个较稀的区域,而氢气宽广的燃烧范围免除了这一要求。氢气的以上特性使得我们可以设计更小的燃烧室产生相同的功率输出,因此选择氢气作为燃料。

表 3　所选微型发动机燃料的比较

燃料特性	氢气/空气混和物	烃类/空气混和物
燃料组成	H_2	$CH_{1.8}$
燃料热值 e/(kJ·g^{-1})	120	42.8
扩散速度 v_d/(cm·s^{-1})	2	0.2
燃烧极限 ϕ/%	4～75	0.6～4.0
蒸发速度 v_v/(cm·min^{-1})	2.5～5.0	0.05～0.50
最小点火能量 E_f/mJ	0.02	0.25
自点火温度 T_f/K	858	500
0.5 MPa 下特征反应时间 τ/ms	0.001	0.01
火焰传播速度 v_t/(cm·s^{-1})	300	20
绝热火焰温度 T_{mas}/K	2 318	2 200
辐射热能比例 K_{th}/%	17～25	30～42

2.3　混合气质量流量的影响

根据模拟结果和相关分析,混合物在微燃烧室中的有效驻留时间影响效率。因此,在完全燃烧的前提下,可以提高混合气的质量流量来减少有效驻留时间,从而减少通过燃烧室壁面的热损失。实际上,这和在固定质量流量时减小燃烧室的容积效果相同,见图 5。

2.4　氢气/空气比 R 的影响

氢气/空气比 R 是影响出口平均温度的一个重要因素。从图 6 和 7 可见,当氢气/空气比增加

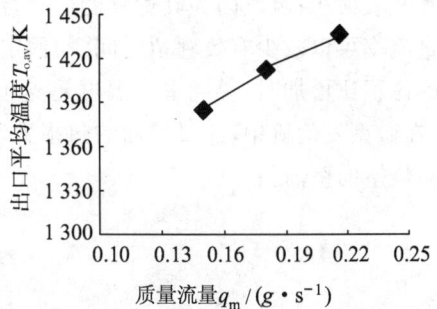

图 5　出口平均温度随质量流量的变化

0.1,平均温度增加超过 100 K,而燃烧室内最高温度变化更加明显。当 R 达到 0.7 时,燃烧室内最高温度超过 2 000 K。因此,在不同的条件下,必须选择不同的比值以使材料能承受产生的高温。

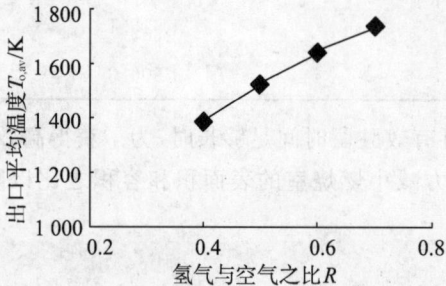

图 6　出口平均温度随氢气/空气比值的变化　　图 7　燃烧室最高温度随氢气/空气比值的变化

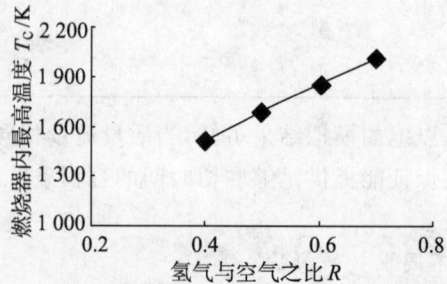

2.5　壁面材料的影响

在设计常规发动机时,通过燃烧室壁面传热产生的能量损失通常忽略掉。但对于微燃烧室来说,其面积-体积比与水力直径的倒数成正比,与传统发动机相比增大了约 100 倍,因而传热损失可能造成严重的问题。分别对对流和辐射对燃烧室温度分布的影响进行了模拟计算。燃烧室内最高温度和出口平均温度在表 4 中列出。结果表明:通过微燃烧室壁面的热损失很大,降低了燃烧室内的温度,特别是出口平均温度。因此需要选择有低传热系数和低热辐射的材料作为燃烧室的壁面。

表 4　考虑传热与不考虑传热时的温度

	最高温度 T_{max}/K	平均温度 T_{av}/K
考虑传热	1 506.3	1 384.8
没有辐射	1 587.1	1 544.0
绝　　热	1 616.2	1 591.6

3　结　　论

为了提高微燃烧室的效率,需要在保证足够的有效驻留时间的条件下尽量减小燃烧室的容积和表面积;氢气因为其优良的燃烧特性,比烃类更适合作为微型发动机的燃料;通过提高质量流率可以减少有效驻留时间,与固定质量流率时减小燃烧室的容积效果相同;当氢气与空气的体积比增加时,燃烧室内温度迅速上升,需要选择合适的混合比使材料能承受产生的高温;在微型发动机中,通过壁面的热损失非常巨大,需要选择有低传热系数和低辐射的材料作为燃烧室的壁面。

参 考 文 献

[1] Mehra A, Zhang X, Ayon A, et al. A through-wafer electrical interconnect for

multi-1evel MEMS devices[J]. *Journal of Vacnum Science & technology B*,2000,1 (5):2583-2589.

[2] Epstein A H,Senturia S D,Midanni Al,et al. Micro-heat engines,gas turbines and roc-ket engines—the MIT micro-engine project:AIAA Paper 97-1 773[C]//*28th AIAA Fluid Dynamics Conference*, Snowmass, CO,1997:1-12.

[3] Mehra A,Zhang X,Ayon A,et al. A 6-wafer combustion system for a silicon micro-gas turbine engine[J]. *Journal of Micro-electromechanical Systems*,2000,9(4): 517-527.

[4] Mehra A,Waitz I,Schmidt M.Combustion tests in the static structure of a 6-wafer mi-cro-gas turbine engine[C]//*1999 Solid State Sensor and Actuator Workshop*,1999.

[5] 李德桃,杨文明,薛宏,等. 微发动机的初步研究[C]//中国汽车工程学会发动机分会,中国内燃机学会煤气机汽油机分会年会学术论文,贵阳,2001.

(本文原载于《机械工程学报》2002 年第 10 期)

Simulation research on combustion chamber of micro-engines

Abstract:The research and application of micro-engines are introduced. The combustion in micro-combustor is simulated with simulating software. The influence of combustion chamber's structure, fuel, mass flow rate, H_2/air ratio, material of wall on combustion is discussed analytically. It provides foundation to the design of micro-combustion chambers.

Microscale combustion research for application to micro-thermophotovoltaic systems

Yang W M, Chou S K, Shu C, Xue H, Li Z W, Li D T, Pan J F

Abstract

A novel power MEMS concept, a micro-thermophotovoltaic(TPV)system,is first described in this work,which would use hydrogen as fuel and would be capable of delivering $3\sim10$ W electrical power in a package less than 1 cubic centimeter in volume. A micro-combustor is one of the most important components of a micro-TPV system. A high and uniform temperature distribution along the wall of the micro-combustor is required to get a high electrical power output. However, sustaining combustion in a MEMS size combustor will be largely affected by the increased heat losses due to the high surface to volume ratio, which tends to suppress ignition and quench the reaction. In order to test the feasibility of combustion in micro-devices and determine the relevant factors affecting micro-combustion, numerical and experimental work was performed. The results indicated that a high and uniform temperature could be achieved along the wall of the flame tube.

1 Introduction

With the demand for smaller scale and higher energy density power sources, traditional batteries cannot satisfy the need, which urges the development of micro-power devices or power MEMS. These systems use hydrogen or hydrocarbon as fuel and are characterized by thermal, electrical and mechanical power densities of $1\sim20$ W in sub-centimeter size packages[1−5]. It is well known that the use of combustion processes for electrical power generation provides enormous advantages over batteries in terms of energy storage per unit mass and in terms of power generation per unit volume. Furthermore, the advantages of hydrocarbon fuels include low cost, improved voltage stability, no memory effect and instant recharge.

Nomenclature

A	pre-exponential factor for reaction rate
a	constant
b	constant
C_j	concentration of jth gas phase species
Da_h	Damkohler number
E_a	activation energy
$\dot{m}$	mass flow rate
N_r	number of gas phase reactions
n_g	total number of gas phase species
n_{ij}	reaction rate coefficient of reaction i and species j in reaction power matrix
P	pressure
R	gas constant
R_k	net difference between r_{kf} and r_{kb} of kth reaction
r_i	rate of reaction i
r_{kb}	backward reaction rate
r_{kf}	forward reaction rate
T	temperature
V	volume
$\dot{w}_j$	jth species generation rate

Greeks

α_{ij}	third body enhancement factor of jth species in ith reaction
β	temperature exponent
v_{jk}	stoichiometric coefficient of species j in reaction k
$\tau_{residence}$	residence time
$\tau_{reaction}$	reaction time

The micro-gas turbine engines, microrockets and micro-rotary internal combustion engines are typical micro power devices being developed by MIT and Berkeley[1-4]. These microengines employ scaled down versions of existing macroscale devices, in particular internal combustion engines. However, at microscale, these devices experience more difficulties with heat loss, friction, sealing, fabrication, assembly etc. than their macroscale counterparts.

The microscale spiral counter flow heat recirculating combustor is another concept of power MEMS developed by Sitzki et al. [5]. It does not involve any

467

moving parts. Electrical power is generated by thermoelectric elements embedded in the walls between cold reactants and hot products, but the three dimensional structures of a counter flow heat exchanger and combustor make the fabrication complex. In addition, the current thermoelectric technology only has a maximum energy conversion effciency of 4%.

2 Development of micro-thermophotovoltaic system

Different from the previous studies just mentioned, the ultimate goal of this research is to develop a novel micro-power system, a micro-thermophotovoltaic (TPV) system, which uses photovoltaic cells to convert heat radiation, e. g. from the combustion of fossil fuels, into electricity. The concept of TPV energy conversion was first proposed in the 1960s[6,7]. It is only in recent years that technological improvements in the field of low band gap photovoltaic cells and high temperature materials have evoked a renewed interest in TPV generation of electricity[8,9]. The micro-TPV system we are developing originates from this concept.

The micro-TPV system consists of three main parts: a heat source, a micro-flame tube combustor (the wall of the microcombustor would be made of selective emitting materials, such as $Er_3Al_5O_{12}$ and Co doped MgO, etc.) and a photovoltaic array made of low band gap materials, such as GaSb (0.72 eV),GaInAsSb (0.5 eV) and so on. Fig. 1 shows the basic design of the micro TPV system. The volume of the microcombustor is 71 mm³. Hydrogen and air are mixed in a micromixer and then enter the microcombustor and combust. When the wall, i. e. the selective emitter, is heated to a sufficiently high temperature, it emits photons, most of them having an energy greater than the band gap energy of the photovoltaic materials due to the selective function of the emitter. Therefore, when they impinge on the photovoltaic array, they evoke free electrons and produce electrical power output under the action of a P-N junction. Because the system

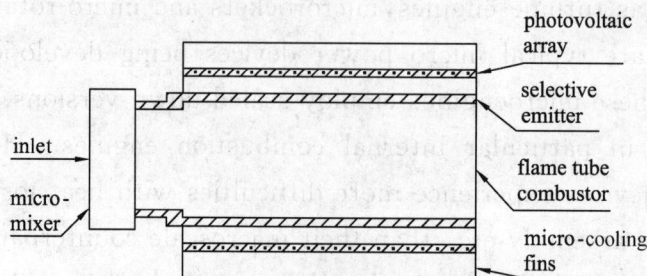

Fig. 1 Basic design of micro-TPV system

does not involve any moving parts, its fabrication and assembly are relatively easy, and its operation is expected to be more reliable. At the same time, it possesses relatively high efficiency of energy conversion[9]. As a result, it can be more commonly used in commercial electronics and microdevices, in which convenient and inexpensive production, reliable operation and low maintenance cost are the key factors of success.

As one of the most important components of the micro-TPV system, the microcombustor must be developed first. The most challenging issue in microcombustor design is maintaining an optimal balance between sustaining combustion and maximizing heat output. According to the cube-square law, as the size of the combustor is reduced by a factor of 100, the surface and volume will decrease by 4 and 6 orders of magnitude, respectively, and thereby, the surface to volume ratio will increase by a factor of 100. So, sustaining combustion in a MEMS size combustor will be largely affected by the increased heat losses due to the high surface to volume ratio, which tends to suppress ignition and quench the reaction. On the other hand, the power output of a micro-TPV system depends on the temperature and the size of surface. High surface to volume ratio is very favorable to the output of power density per unit volume. This is the most attractive feature of a micro-TPV system. In order to test the feasibility of combustion in microdevices and determine the relevant factors affecting microcombustion, numerical and experimental work on a micro-flame tube combustor was performed.

3   H₂/air combustion mechanism

In microcombustor design, the choice of fuel plays a key role. Compared to a conventional combustor, a microcombustor is more highly constrained by inadequate residence time for complete combustion and high rates of heat transfer from the combustor. This fundamental time constraint can be quantified in terms of a homogeneous Damkohler number Da_h, the ratio of gas residence time to the characteristic chemical reaction time[3]. To ensure complete combustion, Da_h must be greater than unity.

$$Da_h = \frac{\tau_{residence}}{\tau_{reaction}} \qquad (1)$$

with

$$\tau_{residence} \approx \frac{VP}{mRT} \qquad (2)$$

$$\tau_{\text{reaction}} \approx \frac{[\text{fuel}]_0}{A[\text{fuel}]^a[O_2]^b e^{-E_a/RT_0}} \tag{3}$$

In this work, hydrogen was chosen as the fuel because of its high heating value, fast diffusion velocity and short reaction time[3].

Gas phase kinetics of hydrogen oxidation reactions has been widely studied, and the typical mechanism (shown in Tab. 1), consisting of 19 reversible reactions and nine species, is quoted in our study[10,11]. Five reactions in this mechanism involve third body collisions, and their reaction rates are expressed as:

$$r_i = k_i \prod_{j=1}^{n_g} C_j^{n_{ij}} \sum \alpha_{ij} C_j \tag{4}$$

The other gas phase reactions are written as:

$$r_i = k_i \prod_{j=1}^{n_g} C_j^{n_{ij}} \tag{5}$$

Tab. 1　Gas-phase reaction mechanism of hydrogen oxidation[10,11]

Reactions	A	β	$E_a/(\text{cal/mol})$
$O_2 + H = OH + O$	5.13E+16	−0.816	16 507.0
$H_2 + O = OH + H$	1.18E+10	1.00	8 842.7
$H_2 + OH = H_2O + H$	1.17E+09	1.30	3 626.0
$OH + OH = H_2O + O$	6.00E+08	1.30	0.0
$H_2 + O_2 = OH + OH$	1.70E+13	0.00	47 780.0
$H + OH + M = H_2O + M^a$	7.50E+23	−2.60	0.0
$O_2 + M = O + O + M$	1.85E+11	0.50	95 560.0
$H_2 + M = H + H + M^b$	2.23E+12	0.5	92 600.0
$H + O_2 + M = HO_2 + M^c$	2.10E+18	−1.00	0.0
$H + O_2 + O_2 = HO_2 + O_2$	6.70E+19	−1.42	0.0
$H + O_2 + N_2 = HO_2 + N_2$	6.70E+19	−1.42	0.0
$HO_2 + H = H_2 + O_2$	2.50E+14	0.00	1 900.0
$HO_2 + H = OH + OH$	2.50E+13	0.00	700.0
$HO_2 + O = OH + O_2$	4.80E+13	0.00	1 000.0
$HO_2 + OH = H_2O + O_2$	5.00E+13	0.00	1 000.0
$HO_2 + HO_2 = H_2O_2 + O_2$	2.00E+12	0.00	0.0
$H_2O_2 + M = OH + OH + M$	1.30E+17	0.00	45 500.0
$H_2O_2 + H = H_2 + HO_2$	1.60E+12	0.00	3 800.0
$H_2O_2 + OH = H_2O + HO_2$	1.00E+13	0.00	1 800.0

Rate constants are given in the form of $k = AT^\beta \exp(-E_a/T)$.

ᵃ Enhancement factors: $H_2O = 20.0$.

ᵇ Enhancement factors: $H_2O = 6.0, H = 2.0, H_2 = 3.0$.

ᶜ Enhancement factors: $H_2O = 21.0, H_2 = 3.3, O_2 = 0.0, N_2 = 0.0$.

α_{ij} is unity except as specified in Tab. 1. Given the rate constant data in Tab. 1, the species generation rate can be written as:

$$\dot{w}_j = \sum_{k=1}^{N_r} v_{jk} R_k \tag{6}$$

with R_k being expressed as:

$$R_k = r_{kf} - r_{kb} \tag{7}$$

4 Experimental set-up

To investigate the stability of the flame and validate the simulation results, at the same time, for simplicity of fabrication, three kinds of different stainless steel flame tube combustors (one is straight tube of 2.2 mm diameter, the other two are flame tubes with a sudden expansion step, having the diameter of 3 mm) were fabricated and tested. The construction and specifications of the three flame tube combustors are given in Fig. 2. The mass flow rates of hydrogen and air were controlled accurately by two sets of electronic mass flow controllers, through which the H_2/air ratio can also be adjusted. The distributions of temperature, both on the exit plane and along the wall of the flame tube, were measured by 0.203 mm diameter type K thermocouple.

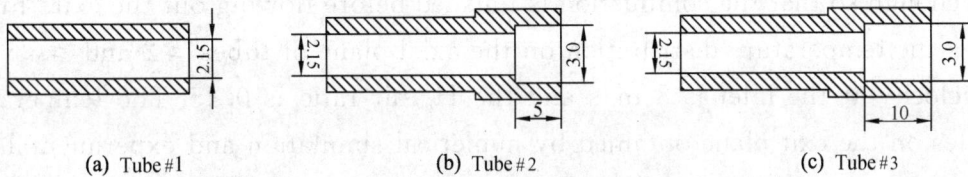

(a) Tube #1   (b) Tube #2   (c) Tube #3

Fig. 2 **Configuration and specifications of three combustors** (unit: $m \times 10^{-3}$)

5 Results and discussion

According to the simulation results and experimental data of the three different flame tube combustors, when the flow speed at the inlet drops to 1.3 m/s, because of the great heat loss, the flame extinguishes after a short term combustion, which indicates that combustion cannot be sustained in such small flame tubes when the flow rate is too low.

When the flow speed exceeds 8 m/s, combustion does not take place inside the straight tube under any conditions, but it may take place outside of the tube. Therefore, stable combustion can only be obtained in the straight tube with flow speeds at the inlet varying from 1.5 to 8 m/s, but the position of the flame core is different for each speed. Furthermore, with increasing flow rate, a higher H_2/air ratio is required to get stable combustion in the flame tube, or the flame will be blown out the exit[12]. At 2 m/s, the peak temperatures on the wall occur at about 5,14 and 25 mm below the exit plane when the H_2/air ratios are 0.45,0.5 and 0.55, respectively. This indicates that ignition occurs earlier, and increasing H_2/air ratios can increase the combustion rate.

From the above study, we know that there are some disadvantages with the straight flame tube combustor: First, it is diffcult to control the position of the flame, which keeps changing with the variation of working condition. Second, combustion cannot be sustained in a straight tube when the H_2/air ratio is not high enough, especially for high flow rates. So, we designed another kind of flame tube with a sudden expansion step, which is used to facilitate recirculation along the wall, thereby enhancing combustion completeness around the rim of the flow tube and ensuring stable combustion.

Comparing to straight tube #1, tubes #2 and #3 with a sudden step can work steadily under a much wider flow rate and wider H_2/air ratio. Stable combustion has been obtained with flow speeds at the inlet varying from 1.5 to 20 m/s and H_2/air ratios varying from 0.45~1.0. At the same time, the sudden step is very useful in ensuring that the flame occurs in the tube downstream of the step.

It should be mentioned that the tube downstream of the sudden step must be long enough so that the combustion is finished before flowing out the exit. Fig. 3 shows the temperature distribution on the axial plane of tubes #2 and #3 when the velocity at the inlet is 8 m/s and the H_2/air ratio is 0.45. The temperature profiles on the exit plane obtained by numerical simulation and experimental testing are given in Fig. 4. From Fig. 3(a), we can see clearly that part of the fuel has not been combusted before flowing out the exit, and thus, the temperature in the centerline of the exit plane is very low (see Fig. 4(a)). In contrast, the result in Figs. 3(b) and 4(b) is much better. Furthermore, we can further improve the completeness of combustion by increasing the H_2/air ratio (see Fig. 5[12]).

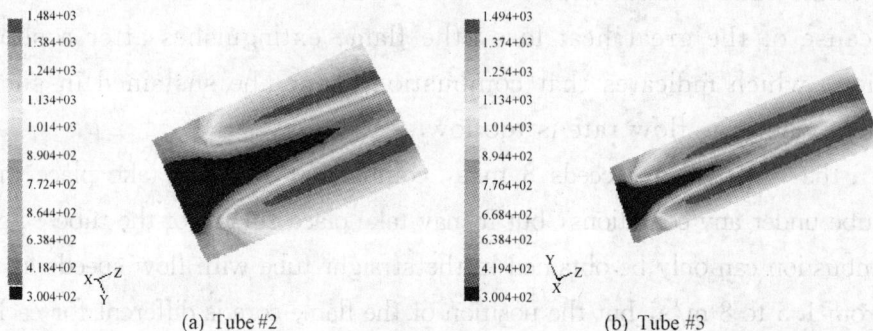

(a) Tube #2   (b) Tube #3

Fig. 3 Distribution of temperature on axial plane (Velocity: 8 m/s, H_2/air ratio: 0.45)

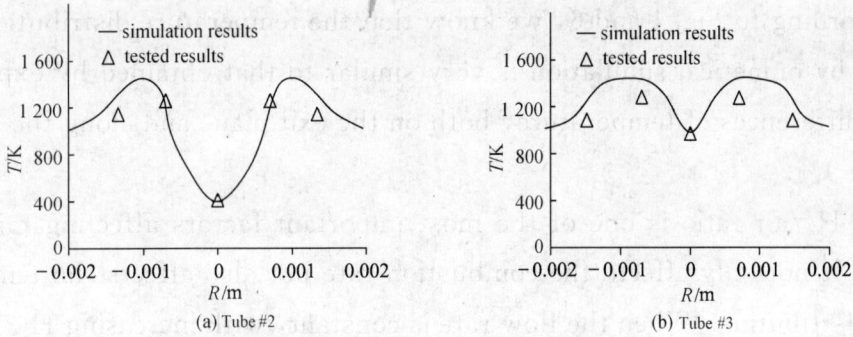

Fig. 4   Temperature profile on exit plane

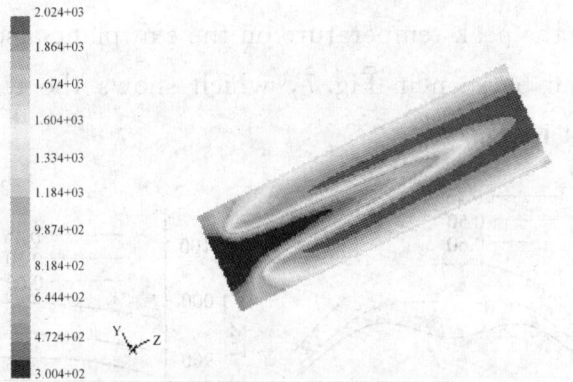

Fig. 5   Distribution of temperature on axial plane（tube ♯3）
（velocity：8 m/s，H_2/air ratio：0.75）

From Fig. 3 and 5, we also observe another interesting phenomenon. The combustion takes place near the wall rather than the centerline of the flame tube at the beginning. Thus, the fuel/air mixture around the centerline is heated and accelerated by the surrounding combustion products and flows quickly to near the end of flame tube and combusts there, which is favorable for maintaining a uniform temperature along the wall. The distributions of temperature on the wall of the flame tube are given in Fig. 6. Both the simulation and experimental results indicate that the maximum difference of temperature along the wall is less than 5%. This feature of the micro-flame tube combustor is very important to the design of micro-TPV system.

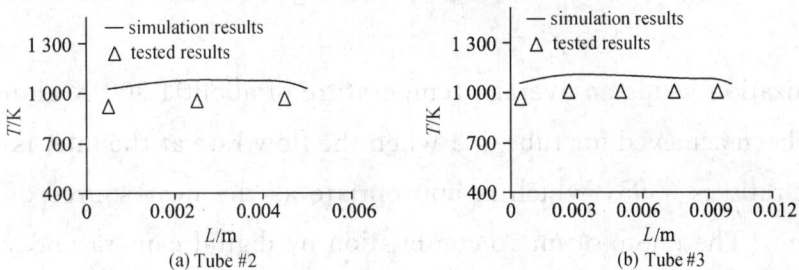

Fig. 6   Distribution of temperature along the wall（H_2/air ratio：0.45；velocity at inlet：8 m/s）

According to Fig. 4 and 6, we know that the temperature distribution profile obtained by numerical simulation is very similar to that obtained by experiment, and the differences of temperature, both on the exit plane and along the wall, are less than 9%.

The H_2/air ratio is one of the most important factors affecting micro-combustion. It not only affects the combustion rate but also affects the temperature and its distribution. When the flow rate is constant, with increasing H_2/air ratio, both the temperatures on the exit plane and along the wall increase drastically, and the position of the peak temperature on the exit plane also moves towards the centerline. This can be seen in Fig. 7, which shows the test results when the velocity at the inlet is 5 m/s.

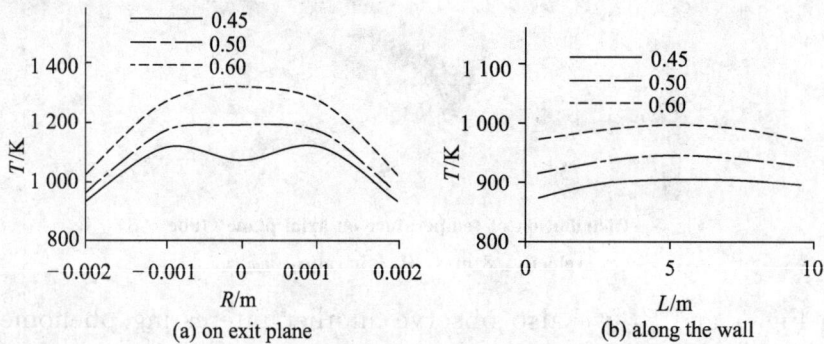

(a) on exit plane   (b) along the wall

Fig. 7   Distribution of temperature（velocity at inlet: 5 m/s）

Flow rate is another important factor affecting the temperatures, both on the exit plane and along the wall[12]. Fig. 8 shows the variation of mean wall temperature with velocity. With increasing velocity at the inlet, the total energy released by combustion increases, and thereby, the mean wall temperature increases. However, as the flow rate increases further, the effective residence time of the fuel decreases, worsening the completeness of combustion, lowering the efficiency and the temperature on the wall, even quenching the flame. This situation should be avoided.

Optimization yields an average temperature of about 1 300 K along the wall, which has been achieved for tube #3 when the flow rate at the inlet is 12 m/s and the H_2/air ratio is 0.95, which is appropriate as the heat source of the micro-TPV system. The photo of micro-combustion by digital camera under these conditions is shown in Fig. 9.

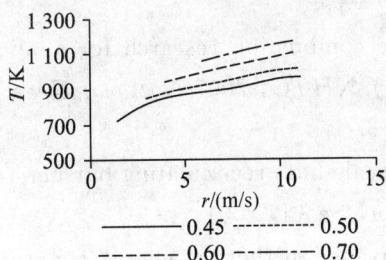

Fig. 8　Variation of mean wall temperature with velocity

Fig. 9　The picture of micro-combustion

6　Conclusions

High surface to volume ratio is very favorable to the output of power density per unit volume, though it tends to suppress ignition and quench the reaction in microdevices. The above studies indicate that stable combustion can be achieved in a small tube under a sudden step with a wider flow rate and wider H_2/air ratio than in a straight tube. Furthermore, the sudden step is very useful in controlling the position of the flame. Combustion takes place near the wall rather than the centerline of the flame tube at the beginning. Thus, the part of the fuel/air mixture around the centerline is heated and accelerated by the surrounding combustion products and flows quickly to near the end of flame tube and combusts there, which is favorable to keep the uniform temperature along the wall. The H_2/air ratio not only affects the temperatures, both on the exit plane and along the wall, but also affects the position of the flame core. Flow rate is another important factor affecting microcombustion in a flame tube. An average temperature of about 1 300 K along the wall has been obtained with a flow speed of 12 m/s at the inlet and the H_2/air ratio of 0. 95, which is appropriate as the heat source of the micro-TPV system.

References

[1] Epstein A H, Senturia S D, et al. Micro-heat engines, gas turbines, and rocket engines—the MIT Microengine project: AIAA Paper 97-1773[J]. *American Institute of Aeronautics and Astronautics*, 1997: 1–12.

[2] Epstein A H, Senturia S D, et al. Power MEMS and microengines[C]// *IEEE*

Transducers '97 Conference, Chicago, IL, 1997: 753-756.

[3] Waitz I A, Gauba G, et al. Combustors for micro-gas turbine engines[J]. *ASME J Fluids Eng*, 1998, 120:109-117.

[4] Kelvin Fu, Knobloch A J, et al. Microscale combustion research for applications to MEMS Rotary IC Engine[C]// *Proceedings of NHTC*, 2001 *National Heat Transfer Conference*. Anaheim, CA:[s. n.], 2001: 1-6.

[5] Sitzki L, Borer K, et al. Combustion in microscale heat-recirculating bur-ners[C]// *The Third Asia-Pacific Conference on Combustion*, Seoul, 2001: 1-4.

[6] White D C, Wedlock B D, Blair J. Recent advance in thermal energy conversion[C]// *15th Annual Power Sources Conference*, Atlantic City, 1961: 125-132.

[7] Guazzoni G, Kittl E, Shapiro S. Rare earth radiators for thermophotovoltaic energy conversion [C]// *International Electron Devices Meeting*, Washington, DC, 1968: 130-132.

[8] Ferguson L G, Fraas L M. Thereoretical study of Gasb PV cell efficiency as a function of temperature[J]. *Solar Energy Mater Solar Cells*, 1995, 39: 11-18.

[9] White D C, Hottel H C. Important factors in determining the efficiency of TPV systems[C]// *First NREL Conf Thermophotovoltaic Generation of Electricity*, Copper Mountain, CO, 1995: 425-456.

[10] Markatou P, Pfefferle L D, Smooke M D. The influence of surface chemistry on the development of minor species profiles in the premixed boundary layer combustion of H_2/air mizture[J]. *Combust Sci Tech*, 1991, 79:237-268.

[11] Warnatz J, Allendorf M D, et al. A model of elementary chemistry and fiuid mechanics in the combustion of hydrogen on platinum surfaces[J]. *Combust Flame*, 1994, 96:393-406.

[12] Yang Wenming. Simulation of micro-combustion in micro-combustor and relevant experiments[R]. Annual report, National University of Singapore, 2001: 1-18.

(From: *Energy Conversion and Management*, 2003)

应用于微热光电系统的微尺度燃烧研究

[摘要] 本文描述了一种新颖的 MEMS 动力源概念,即微热光电(TPV)系统。该系统将使用氢气作为燃料,每立方厘米体积能够发出 3~10 W 的电力。微燃烧室是该系统中最重要的元件之一,为了获得较高的电能输出,燃烧室壁面的温度分布要求高而且均匀。不过,由于微燃烧室面容比大,热损失显著增加,这会导致着火困难并使火焰熄灭。为了测试微燃烧室内燃烧的可行性和确定影响燃烧的有关因素,我们进行了实验和数值模拟,结果表明燃烧室壁面能够得到我们要求的高温,且温度分布均匀。

微型发动机的燃烧模型和数值模拟

李德桃,潘剑锋,邓　军,杨文明,薛　宏

[摘要]　提出了微型发动机燃烧过程的数学模型,并对采用氢气为燃料的微型涡轮机中环形燃烧室内的燃烧,建立了阿累尼乌斯有限反应速率模型,在考虑对流和辐射传热损失的条件下,采用 CFD 软件包 FLUENT 进行了数值模拟。根据所得到的模拟结果分析了影响燃烧的主要因素:结构、燃料、质量流率、H_2 与空气比和壁面材料,提出了克服尺寸减小对燃烧带来的困难的途径。

近年来,在国外,尤其是美国,相继开展了微动力机电系统(Power MEMS)和微型发动机(Micro-engine)的研究工作[1-3]。其共同特征是利用碳氢燃料,在一个微型的燃烧器中燃烧放热。使用碳氢燃料的微型发动机即使在热效率很低的情况下也具有比现有的电池高出数倍的能量密度。从动力机械发展的历史进程看,每当能源装置的能量密度产生一个飞跃,都会给社会和经济带来深远的变革。18 世纪的蒸汽发动机,以 0.005 W/g 的能量密度为标志,引发了当时的工业革命。从 19 世纪到 20 世纪中叶,内燃机的发展使能量密度达到了 0.05～1.0 W/g,从而使整个交通运输发生了巨变。20 世纪发明的航空航天发动机使能量密度进一步上升到 10 W/g。喷气式飞机大大地缩短了整个世界的距离。微动力装置的能量密度将冲破 100 W/g 的大关。可以说,它是动力机械发展的第四个里程碑。给现代社会带来的影响将是重大而深远的。

微型发动机的研究尚处于起步阶段,微型发动机热力循环的选择、燃烧系统的研究尚处于探索之中。当前微型发动机的几个主要发展方向有微型涡轮机、三角转子发动机和采用新材料直接将热能转化为电能的发动机。本文对微型发动机中的燃烧进行了模拟计算。图 1 为 MIT 研究开发的微型涡轮机的结构示意图。该发动机主要由压缩器、燃烧室、涡轮和启动电动机/发电机组成。由于以光刻技术为基础的微加工方法更适合于二维或准二维结构的几何形状,同时从减少传热损失的考虑出发,本文选择了环形燃烧室(图 2)作为模拟计算的对象。

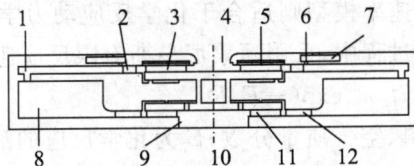

1—火焰稳定器　2—扩散叶片　3—转子叶片　4—进气口
5—启动器　6—燃料喷孔　7—燃料汇流腔
8—燃烧室　9—排气口　10—转子中心线
11—涡轮转子叶片　12—涡轮导向叶片

图 1　微燃烧室的结构

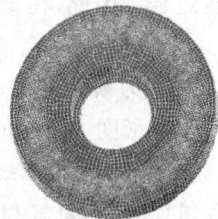

图 2　燃烧室形状

1 燃烧模拟

1.1 三维流动的控制方程

本文选择了"炸面包圈"式的环形燃烧室作为模拟计算的对象。选择氢气作为燃料,理想配比下氢气与空气燃烧的化学反应方程式为

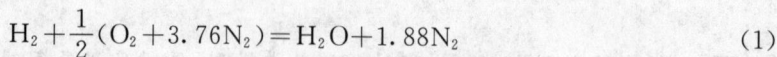

$$H_2 + \frac{1}{2}(O_2 + 3.76N_2) = H_2O + 1.88N_2 \tag{1}$$

在数值模拟中,流动和燃烧反应由质量、动量、能量和化学组分的守恒方程描述,并与燃烧反应模型联立求解。选择粘性层流流动模型,以张量表示的支配方程的一般形式为

$$\frac{\partial}{\partial t}(\rho\phi) + \frac{\partial}{\partial x_i}(\rho u_i\phi) = \frac{\partial}{\partial x_j}\left(\Gamma_\phi \frac{\partial \rho u_i\phi}{\partial x_j}\right) + S_\phi \tag{2}$$

式中 ρ 为燃气混合物的密度;u_i 为 i 方向的速度分量;t 为时间间隔;ϕ 可表示为下列各个变量:(1) 3 个方向的速度分量,(2) 化学元素 m_i 的质量分率,(3) 焓 h。各个变量的扩散系数 Γ_ϕ 和源项 S_ϕ 的表达式在表 1 中列出,其中 Sc 为 Schmit 数;Hc 为燃烧热;Pr 普朗特数。

表 1 控制方程各变量的扩散系数和源项

ϕ	Γ_ϕ	源项 S_ϕ
连续性 1	0	0
动量 u_i	μ	$-\dfrac{\partial p}{\partial x_i} + \dfrac{\partial}{\partial x_j}\left(\mu\dfrac{\partial u_j}{\partial x_i}\right) + (\rho - \rho_{\text{ref}})g_i$
组分 m_{fu}	μ/Sc	R_{fu}
组分 m_{ox}	μ/Sc	$r_{\text{ox}} \cdot R_{\text{fu}}$
焓 h	μ/Pr	$R_{\text{fu}} \cdot H_c$

1.2 燃烧模型

燃烧模型考虑氢气和空气的化学反应,化学反应中的生成物和产生的能量分别在元素和能量守恒方程式中的源项中给出,从而使支配方程式封闭。在数值燃烧模拟中,比较成熟和被广泛采用的燃烧模型包括涡破碎模型(Eddy-break-up)[4]和阿累尼乌斯有限反应速率模型(Arrhenius Finite Reaction Rate)[5]。涡破碎模型适合于反应物的紊流扩散和混合起支配作用的化学反应过程,而阿累尼乌斯有限反应速率模型则适合于化学反应动力学为支配因素的化学反应过程。因此,在微型发动机的燃烧过程中,采用阿累尼乌斯有限反应速率模型,即

$$R_{\text{fu}} = -A\rho^{\alpha+\beta} m_{\text{fu}}^\alpha m_{\text{ox}}^\beta \exp(-E/RT) \tag{3}$$

式中 A,α 和 β 为模型常数;m_{fu} 和 m_{ox} 为燃料、空气质量分数;E 为化学反应的活化能;ρ,T 和 R 分别为气体密度、温度和气体常数。

1.3 边界条件及计算方法

在数值计算中,氢气和空气的进口速度由质量流率给出,各个变量的出口条件由背压(大气压)及质量守恒决定,壁面采用无滑移速度,考虑对流和辐射传热损失。采用 CFD 软件包 FLUENT 进行模拟,采用非结构式网格,在计算过程中收敛条件规定各个支配方程迭代计算的余量至少小于 10^{-4}。

2 模拟结果

影响燃烧的因素主要有：燃烧室的形状和尺寸；燃料种类；质量流率；燃料空气比；壁面材料。在模拟中分别选择了不同的燃烧室尺寸、质量流率、燃空比和壁面材料以分析这些因素对燃烧的影响情况。表2为3种微燃烧室的尺寸，表3为3种燃烧室的出口平均温度和最高温度的模拟计算结果。

表2 3种不同尺寸微燃烧室的详细说明

参　　　数	A 型	B 型	C 型
燃烧室内径 d_i/mm	5	5	9.6
燃烧室外径 d_e/mm	12	12	18.4
燃烧室高度 h/mm	1.4	3	1
容积 V/mm³	130	280	193
表面积 A/mm²	262	347	475

表3 出口平均温度和燃烧室内最高温度

燃烧室类型	出口平均温度 T_o/K	最高温度 T_{max}/K
A	1 453.1	1 660.3
B	1 383.9	1 615.1
C	1 384.8	1 506.3

图3为燃烧室出口平均温度在不同燃空比下的模拟结果。

表4为传热和辐射损失对燃烧室最高温度和平均温度影响的模拟结果。

表4 考虑传热与不考虑传热时燃烧室温度的模拟结果

条件	最高温度/K	平均温度/K
考虑传热	1 506.3	1 384.8
没有辐射	1 587.1	1 544.0
绝　热	1 616.2	1 591.6

图4为模拟得到的 A 型燃烧室通过中心横截面上的温度场分布。

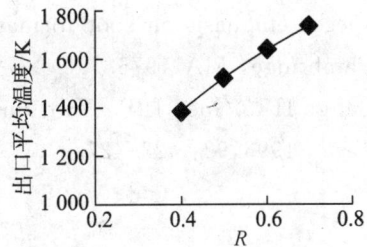

图3 出口平均温度随 H_2/Air 比值 R 的变化

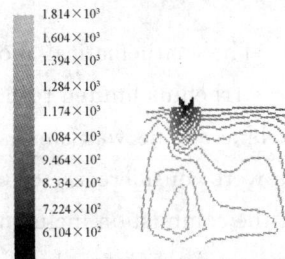

图4 燃烧室通过中心横截面上的温度场

3 结 论

从模拟结果可见,随着燃烧室尺寸的减小,燃料和空气在燃烧室中的驻留时间减少,影响了燃料混合和化学反应所能占用的时间。另外,随着尺寸的减小,微燃烧室的面积与容积比迅速增加,由此产生了高传热损失,降低了微燃烧室的效率,并且微小尺寸产生的火焰淬熄可能会影响燃料的燃烧极限。为了提高微燃烧室的效率,需要在保证足够的有效驻留时间的条件下尽量减小燃烧室的容积和表面积;通过提高质量流率可以减少有效驻留时间,和固定质量流率时减小燃烧室的容积效果相同;当氢气/空气的混合比增加时,燃烧室内温度迅速上升,需要选择合适的混合比使材料能承受高温;在微型发动机中,通过壁面的热损失非常大,需要选择有低传热系数和低辐射的材料作为燃烧室的壁面。

参 考 文 献

[1] Waltz I A, Gauba G, Tzeng Y S. Combuster for micro-gas turbine engines[J]. *Journal of Fluids Engineering*, ASME, 1998,120:109–117.

[2] Tu K, Knoblch. Micro-scale combustion research for application to MEMS rotary IC engine [C]// *Proceeding of NHTC*, 2001 *National Heat Transfer Conference*. Anaheim, CA: [s. n.], 2001:1–6.

[3] Sitizki L, Borer K, Schuster E, et al. Combustion in micro-scale heat recalculating burners [C]// *The Third Asia-Pasfic Conference on Combustion* 2001, Seoul,Korea, 2001.

[4] Magnussen B F,Hjertager B H. On mathematical models of turbulent combustion with special emphasis on soot formation and combustion [C]//16*th Symp on Combustion*, Cambridge, MA,1976.

[5] Lange H C,Goey L P. Two dimensional methane/air flame [J]. *Combustion Sci and Tech*, 1993,92: 423–427.

(本文原载于《燃烧科学与技术》2003 年第 2 期)

Combustion Model and Numerical Simulation of Micro-Engine

Abstract: The mathematical model of combustion process in micro-engine was proposed. Based on Arrhenius limited reaction rate model, the combustion of hydrogen in combustor of micro-turbine engine was simulated with FLUENT, in which the convection and radiation heat loss were considered. According to the results of simulation, the main factors which influenced the combustion including structure, fuel, mass flow rate, H_2/air ratio and material of combustor were analyzed. Some methods to overcome the challenge bring from the deflation of scale are presented.

开发微型发动机燃烧器遇到的问题和解决途径

张孝友,卫　尧,李德桃,唐维新,张海燕,邓　军,潘剑锋

[摘要]　论述了当前微型发动机燃烧器研究所遇到的主要问题及解决的途径,介绍了国外正在研究的三种微型发动机燃烧器。这三种微型燃烧器反映了微型发动机开发和应用的良好前景。在此基础上研制了一种新型的微型 TPV 燃烧系统,该系统具有热量利用效率高、制造成本低等优点。

当前最好的锂化学电池的能量密度只有 0.50 MJ/kg,而碳氢燃料的能量密度可以达到 45 MJ/kg 左右。因此,即使热能转化为电能的效率只有 10%,碳氢燃料所能提供的能量密度也是电池的 10 倍左右[1]。考虑到碳氢燃料在能量密度方面的优势,人们提出了微型发动机的概念[2]。所谓的微型发动机就是指以近几年发展起来的微机电系统(MEMS)和微加工技术为基础,体积在 1 cm³ 量级上,能产生大约 10~100 W 功率的热力发动机。由于微型发动机燃烧器的燃烧空间极小,燃烧器内的最高温度可达 1 600 ℃左右,常规发动机的燃烧器材料很难适应这么高的温度。其次,在微型燃烧器内燃料燃烧的驻留时间也急剧缩短,很难保证燃料能完全燃烧。此外,随着燃烧空间的缩小,燃烧器的面积-容积比增大,从而使热量损失比较严重。所以,在微体积内如何使燃料保持稳定的燃烧,采用何种措施提高微型燃烧器内热量利用效率等,给微型发动机燃烧器的研究提出了挑战。

1　微型发动机燃烧器的开发研究

1.1　材料与加工

常规发动机燃烧器所用的材料一般为铸铁或铝合金,部分零件则使用钛、镍钴合金等,这些材料对温度和应力都有严格的要求与限制,以至于在常规发动机中都有较强的冷却系统来降低燃烧器壁面的温度。由于微型发动机的体积很小,一般不宜再设计冷却系统,造成燃烧器的壁面温度急剧升高,所以制造微型燃烧器的材料必须具有较强的耐高温性能。近几年研制的耐火陶瓷(如氮化硅(Si_3N_4)和碳化硅(SiC))在微尺寸范围内具有承受的应力高、可适应温度范围广、机械性能以及抗几何变形能力强等优点。这正是微型发动机燃烧器的壁面材料所应具有的性能。

常规发动机零部件的加工精度一般在毫米级上,而微型发动机零部件的尺寸都在 10~100 μm 级上。在如此小的基准尺寸上要保持较高的加工精度,其难度是很大。常规的机械加工方法不能用来制造微型发动机的零部件,只有使用超强等离子体的化学蒸发处理技术、深度活性离子蚀刻技术、电子放电技术以及电化学加工技术等才能达到这样的加工精度。

1.2　驻留时间

在微型发动机燃烧器中最重要、最具有技术挑战性的一点就是如何提高和分配燃烧的驻

留时间。驻留时间一般包括燃料的混合时间和化学反应所用的时间。其中，化学反应所用的时间只有几百 μs 甚至更少，大部分时间都用于燃料的混合。如果将常规燃烧器的容积减小 500 倍，并保持相同的单位面积质量流率，那么在微型燃烧器内的驻留时间就为 $0.05\sim 0.1$ ms，这和碳氢燃料的化学反应特征时间（$0.01\sim0.1$ ms）处于同一数量级上。显然，在这么短的驻留时间内燃料的混合和燃烧都是不充分的。

由 Kerrebrock 给出的燃烧器驻留时间的简化公式[3]：

$$\tau_{res} \propto \frac{L \cdot (A_b/A_2) \cdot \pi_c^{1/\gamma}}{\dot{m}/A_2}$$

式中 L 为长度；A_b 为燃烧器横截面积；A_2 为压缩器流通面积；π_c 为压缩比；γ 为多变指数；$\dot{m}$ 为空气流通速率；τ_{res} 为驻留时间。

假设两者的单位面积质量流率 $\dot{m}/A_2$ 相同时，受发动机的总体尺寸的约束，压缩器的流通面积 A_2 和燃烧器的特征长度 L 以及多变指数 γ，基本上都保持不变。因此，只有增大燃烧器相对于发动机的尺寸 A_b，才能提高燃烧的驻留时间，从而保证燃料在燃烧器内充分燃烧。

另一方面，也可以从减少燃料在燃烧器内的混合时间以及燃料燃烧的化学反应时间的角度来缩短燃料燃烧的驻留时间，从而使燃料在微型燃烧器内充分燃烧。例如：采取稀燃技术、提高燃烧器内的工作压力和混合气温度、以及在燃料和空气进入燃烧室前就将两者进行充分的混合等方法，都可以减少燃料的混合时间。在减少燃料的燃烧时间方面，可以采用催化燃烧或者使用燃烧时间更短的燃料，如氢气、氢水混合气等。

1.3　大的面积-体积比造成的传热损失

由于微型发动机体积的缩小，使得面积-容积比急剧增大，进而造成较大的热量损失。传热损失不仅降低微型发动机燃烧器的效率而且还影响到燃料燃烧的稳定性。减少传热损失应从以下几个方面来考虑：（1）降低燃烧混合气与燃烧器壁面之间的温差，可以减少由于温差传热造成热量损失。为此应尽量使用不需冷却的耐火陶瓷，提高燃烧器壁面的温度来降低两者之间的温差。（2）应用催化燃烧技术可以降低燃料的着火温度，进而可以降低整个燃烧器内的温度，减少传热损失。（3）由于传热速率与气体的流速以及传热距离成正比，而与气体的运动粘度成反比，因此降低进入燃烧器的混合气的流速、合理设置燃烧火焰中心的位置以及使用适当的混合气浓度都可以降低传热速率，从而减少传热损失。

2　三种微型燃烧器燃烧设计方案

2.1　微型燃气轮发动机燃烧器

图 1 是微型燃气轮发动机燃烧器的设计方案，包括径流式压缩器、燃烧器、径流式涡轮和启动器/电机等部件[3]。整体尺寸大约为直径 1 cm，高为 3 mm。该微型发动机是由六个硅基薄片逐层叠加熔合粘接而成，并根据发动机的整体结构来分别设计每个薄片的具体外形和加工尺寸。空气经压缩机压缩后与燃料进行混合，通过火焰稳定器到达燃烧器内进行燃烧，混合气体经燃烧膨胀推动涡轮机的叶片高速旋转。这样就将燃料的化学能直接转化为机械能，并通过涡轮机轴对外做功。Epstein 等人以氢气为燃料所作的初步分析表明，当速度达到 500 m/s，压缩比为4.5：1，涡轮进气温度为 1 600 K，进气口的面积为 1 mm² 时，该微型燃气轮机能产生 $10\sim20$ W 的功率。

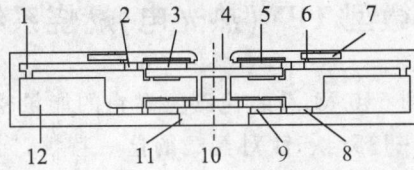

1—火焰稳定器　2—扩散叶片　3—转子叶片　4—进气口　5—起动器　6—燃料喷孔　7—燃料汇流腔　8—涡轮导向叶片　9—涡轮转子叶片　10—转子中心线　11—排气口　12—燃烧室

图 1　微型燃气轮机燃烧器示意图

2.2　微型三角转子发动机燃烧器

图 2 为该微型三角转子发动机燃烧器设计示意图[4]。转子中心与燃烧器中心有一定的偏心距。在转子旋转的过程中,转子的三个顶点将燃烧器内腔分隔成三个封闭、独立的空间。每一空间都依次经历吸气、压缩、点火做功以及排气等四个过程,从而完成一个工作循环,即转子每转一圈在外旋轮形燃烧器中进行三个这样的工作循环。当转子的扫气半径、转子的宽度以及两轴之间的偏心距确定后,发动机的其他参数(如排量、压缩比、转子最大速率等)也就确定了。初步设计的微转子发动机的扫气容积为 $77.5\ mm^3$,微型燃烧器的特征长度为 $0.45\ mm$,输出的功率可以达到 $13.9\ W$。

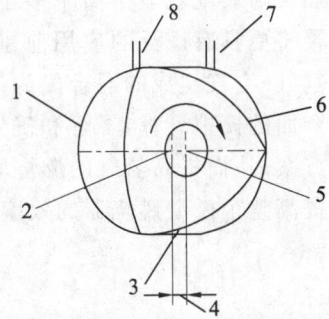

1—外旋轮形燃烧室　2—曲柄中心
3—点火起始处　4—偏心距　5—转子中心
6—转子　7—进气口　8—排气口

图 2　微型转子发动机燃烧器的示意图

2.3　微型"瑞士面包卷"燃烧器

图 3 为微型"瑞士面包卷"燃烧器的工作原理示意图[1]。这种经电化学方法加工所得的"瑞士面包卷"燃烧器是一种没有运动部件和不需要高精度加工与装配的能源装置,可以避免在摩擦、泄露、制造以及装配等方面所遇到的许多困难。此结构最显著的优点是具有充分对流换热面积,燃烧释放的热量在排出过程中经面包卷的壁面与反应物进行充分的热量交换,从而增大了反应物的总能量(化学能+热能),同时也降低了微型燃烧器由于尺寸急剧缩小而带来的热量损失。在面包卷燃烧器壁内镶有热电材料,可以直接将热能转化成电能。

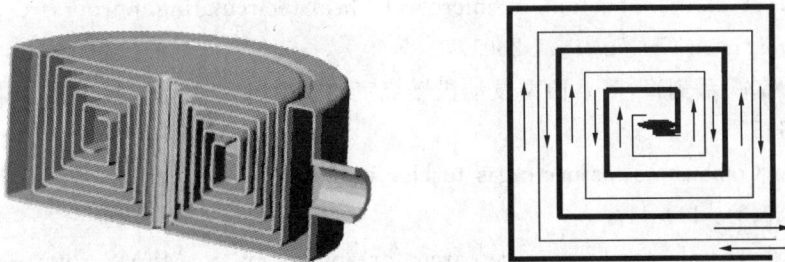

图 3　"瑞士面包卷"燃烧器的示意图

3 微型 TPV(热光电)燃烧系统

氢气的能量密度可以达到 550 MJ/kg,比碳氢燃料的能量密度要高 10 倍,但氢气的驻留时间却是碳氢燃料的 $1/50\sim1/5$[5]。针对氢气的这一特性,同时考虑到国内的机械加工水平,提出了新型的微型 TPV(热光电)燃烧系统,将燃料燃烧产生的热量转化为燃烧器的热辐射,并激活 TPV 材料产生电能。该系统最大的优势:(1) 将大多数燃烧器中由于大的面积-容积比造成的传热损失这一不利因素转化为有利因素,可以进一步利用发动机燃烧产生的热量损失来加热微型发动机燃烧器的壁面,增大微型发动机燃烧器的热辐射,提高微型发动机燃烧器利用效率;(2) 该系统不包括任何运动部件,相对于上述三种微型发动机而言其加工与装配就显得简单多了,从而可以降低整个装置的制造成本。因此,这种微型动力装置将更具有广泛的应用前景以及较高的商业开发价值。为此着重在简易的微型管道内以氢气-空气组成的混合气为燃料进行了燃烧试验,并在此基础上对微体积内的燃烧进行更全面的模拟计算,了解和确定影响微体积内氢气燃烧的诸多因素,开发新的微燃烧器燃烧方案,从而提高氢气的燃烧驻留时间,使得氢气在有限的驻留时间内实现完全燃烧,进一步提高整个微型燃烧器的效率。

4 结 论

(1) 微型发动机燃烧器的体积很小,燃料燃烧的温度很高,一般的金属材料都不能适应其内部的工作环境,只有选用耐高温的碳化硅材料。

(2) 微型燃烧器内燃料燃烧的完全程度与混合气体在燃烧器内的驻留时间成正向比例关系,应尽量增大燃料在其内部的驻留时间。

(3) 在其他方案中普遍存在的热量损失问题,随着体积的急剧减小,越来越成为微型燃烧器研究的瓶颈,而本文中所探讨的微型 TPV 系统就可以将这一不利因素转化为有利因素。

(4) 本文介绍了三种国外最新研究的微型发动机燃烧器的设计思路和方案。这三种方案为目前微型燃烧器的研究提供了经验。

参 考 文 献

[1] Lars S, et al. Combustion in microscale heat-recirculating burners[C]// *The 3rd Asia-Pacific on Combustion*, 2001.

[2] 李德桃,邓军,潘剑锋,等. 微型发动机燃烧室的模拟研究[J]. 机械工程学报,2002,23(10):56-59.

[3] Waitz A. Combustors for micro-gas turbine engines[J]. *Journal of Fluids Engineering*, 1998,120(3):109-117.

[4] Fu K. Microscale combustion research for application to MEMS rotary engine[C]// *2001 National Heat Transfer Conference*, Anahcim, CA, 2001.

[5] Spadaccini. Power density silicon combustion systems for micro-gas turbine engines [R]. ASME International Gas Turbine Institute, TURBO EXPO, 2002.

（本文原载于《内燃机工程》2003 年第 4 期）

Problems in the research of micro-engine combustors and the approaches to solving them

Abstract: Problems in the research of micro-engine combustors and the approaches to solving them are reported in this paper, and we introduce three types of micro-engine combustors being devloped. These combustors reflect a good prospects in the research and application of micro-engines. Base on this, a new micro-PTV system is described. This system has high efficiency of burner and low cost of fabrication.

微型火焰管中燃烧的研究

邓　军,李德桃,潘剑锋,唐维新,杨文明,薛　宏

[摘要]　提出了一种新型的动力系统观念,即微型热光电 TPV 系统。微型燃烧室是微型 TPV 系统中最重要的部分之一。为了获得较高的能量转换效率,需要使燃烧器壁面四周处于较高且分布均匀的温度状态。尺寸效应对微型燃烧室中的持续燃烧带来了很大的影响。为了分析微型燃烧器中燃烧的可行性和有关影响因素,在不同工况下进行实验。结果表明,在一定的流量和混合比范围内,可以在微型火焰管内维持稳定的燃烧,高温能够在燃烧室四周均匀分布。

近年来,国外出现了几种不同概念的微动力机电系统(Power MEMS System)。如微型燃气轮机、微型三角转子发动机、微尺度螺旋形热循环燃烧室等[1-3]。然而,由于尺寸微型化,这些装置仍存在很多问题,如热损失、摩擦、密封、加工、装配等等。

1　微型 TPV 系统的发展

在微型热光电 TPV(Thermo Photo-Voltaic)系统中先燃烧化石燃料,把储存在其中的化学能转换成热能,然后通过光电池,转换成电能。这种 TPV 能量转换的设想是在 19 世纪 60 年代第一次提出的。最近这几年,由于低频带间隙的光电池和耐高温材料这两个领域的技术上的重大进步,又引起人们研究 TPV 系统发电的兴趣。我们发展的这种微型 TPV 系统的设计思路,就是来自这个设想。

该 TPV 系统由三个主要的部份组成:微混合器、微型的火焰管燃烧室、光电阵列。微型 TPV 系统的基本原理如图 1 所示。微型火焰管的管壁是由耐热材料 SiC 和选择性辐射材料,如 $Er_3Al_5O_{12}$ 和 Co-doped MgO 等组成。氢气和氧气是在一个微型混合器中混合的,然后进入微型的火焰管中燃烧。光电阵列是由低频带间隙材料制成的,其材料如 GaSb(0.72 eV),Ga-InAsSb(0.5 eV)等等。当火焰管的壁面,也就是说,辐射材料被加热到一定高温时,它会放出光子,而大多数光子的能量比低频带间隙的光电材料的能量还要大。所以,当它们撞击光电阵时,会释放出自由电子,在光电管的作用下,产生电能输出。因为整个系统没有运动部件,制造

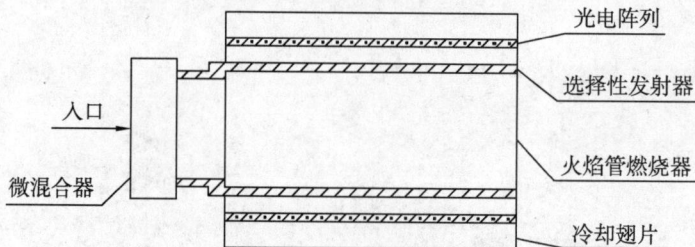

入口

微混合器

光电阵列

选择性发射器

火焰管燃烧器

冷却翅片

图 1　微型 TPV 系统的基本设计

和装配较容易,操作更可靠。同时,它具有较高的能量转换率。这些因素都是微动力装置研发成功并推广使用的关键[4]。

2 微型燃烧器的原理和基本设计

微型燃烧器是 TPV 系统的一个重要组成部分。必须根据微尺度燃烧的特点选择设计合适的燃烧室形状。在微燃烧室设计中,最具挑战性问题是维持持续燃烧和最大热效率之间的最佳平衡。根据平方与立方的法则,当燃烧器尺寸减少一半,表面积和体积将分别减少 3/4 和 7/8,从而,面容比将增加 1 倍。大面容比导致了大的热损失,从而容易产生火焰淬熄或燃烧不能持续。所以,在一个 MEMS 级的燃烧室中的持续燃烧将受到很大的影响,这是由于逐渐增加的热损失之故。另外,微系统的动力输出将取决于温度和表面积的大小。大面容比有利于提高每单位体积的动力密度输出。这就是微型 TPV 系统的最吸引人的地方。为了测试微装置中燃烧的可行性和测定影响微燃烧的相关因素,作者对微火焰管中的燃烧进行了实验。

在微燃烧室设计中,燃料的选择起了一个关键的作用。对比于传统的燃烧室,微燃烧室在两个方面有限制,即对不充足的反应时间和燃烧器的高的热转换效率有限制。为了保证完全燃烧,燃烧器中的 Damkohler 数 Da_h 即特征化学反应时间 $\tau_{reaction}$ 与气体停留时间 $\tau_{residence}$ 的比率必须大于 1。

$$Da_h = \frac{\tau_{residence}}{\tau_{reaction}} \tag{1}$$

其中

$$\tau_{residence} \approx \frac{VP}{mRT} \tag{2}$$

$$\tau_{reaction} = \frac{[fuel]_0}{A[fuel]^a[O_2]^b e^{-Ea/RT_0}} \tag{3}$$

在反应中选氢气作燃料。这是因为相对于碳氢燃料,氢气具有更宽广的可燃范围,较高的热值,较快的扩散速度和较短的反应时间[5]。

氢气的氧化反应的气相动力学已被广泛研究,我们用 CFD 软件对燃烧器中的燃烧进行了数值模拟,以了解燃烧室中温度的分布状况,并根据模拟结果对燃烧室的尺寸进行初步选择。

3 实验装置

为了研究各种流动参数和结构参数对燃烧的影响,及证实模拟计算的结果,我们制造了 SiC 火焰管燃烧室和相应的实验装置,如图 2 所示。预混合的氢气和氧气通过一个喷口喷入圆柱形的火焰管。火焰管一个是内径为 3.0 mm,长度为 18 mm 的 SiC 直管,内部没有或有不同长度的后退台阶以进行对比。氢气/氧气的质量流量由两组电

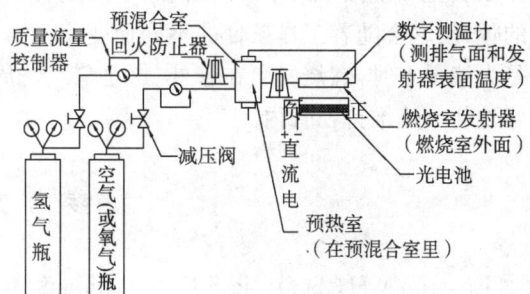

图 2　实验装置示意图

子质量流控制器精确测量。通过该装置，氢气/氧气的流量及其混合比也可以被调整。实验中采用了直径分别为 0.6，0.9，1.5 mm 的三种不同喷口。在壁面外侧等间距地布置了 5 个 K 型热电偶，火焰管出口温度由另一个可改变位置的 K 型热电偶测定。

4　结论与讨论

图 3 和图 4 分别为不同流量和不同氢气/氧气混合比下得到的部分实验测量结果。根据火焰管燃烧室的模拟计算的结果和实验数据我们可以作出以下结论：在一定的流量和混合比范围内，混合气可以在管内燃烧。火焰管中的燃烧情况受混合气流量和氢气与氧气的混合比影响很大，对于不同的喷口直径有相似的变化规律。

流量是影响出口面和管壁上温度的一个重要因素。要使壁面维持在较高的温度，需要保证一定的流量。当流量太小时，由于尺寸效应产生的巨大的热损失，在微火焰管中燃烧不能持续。喷口直径为 0.9 mm 时，混合气流量小于 3.33 mL/s，无论混合比多大，在火焰管内都不能形成火焰，燃烧只能在出口外进行。随着进口流速的增加，由燃烧产生的能量也逐渐增加，因此，平均管壁温度也增加。然而，当流量进一步增加，燃料的有效反应时间减少，使燃烧进一步变差，从而降低了效率和管壁的温度，甚至使火焰熄灭，这种情况必须避免。

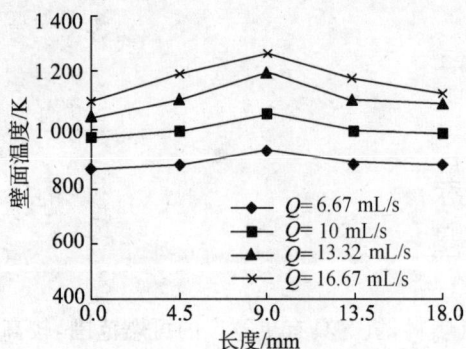

图 3　不同流量下的管壁温度分布
（H_2/O_2 混合比为 3∶1，喷口直径 0.9 mm）

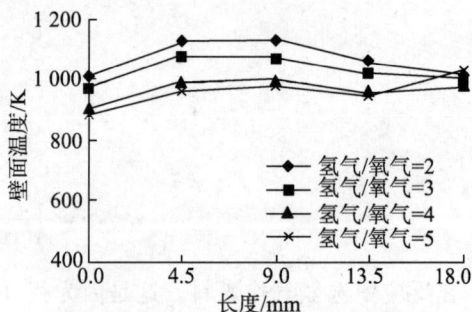

图 4　不同混合比下的管壁温度分布
（总流量 10 mL/s，喷口直径 0.9 mm）

氢气与氧气的混合比是另一个重要的影响因素。当混合气流量一定，氢气与氧气的比接近于 2 时得到的壁面温度最高。当混合比增大时，温度最高点逐渐向出口移动，直至在管外燃烧。

从实验和模拟结果来看，我们还了解到直火焰管燃烧室有些不利因素。第一，很难控制火焰的位置，它会随着工作条件的不同而变化。第二，当氢气与氧气比离理论当量比较远时，尤其是大流量上时，燃烧发生在管外，这是氢气与周围空气燃烧。因此需要改变火焰管的设计以更好地使火焰维持在管内。

参　考　文　献

[1] Epstein A H, Senturia S D, Al Midani O, et al. Micro-heat engines, gas turbines, and rocket engines—the MIT micro-engine project[J]. *American Institute of Aeronautics*

and Astronautics，1997：1-12.

[2] Waitz I A，Gauba G，Tzeng Y S. Combustors for micro-gas turbine engines[J]. *ASME Journal of Fluids Engineering*，1998，120(3)：109-117.

[3] Kelvin Fu，Knobloch A J，Cooley B，et al. Micro-scale combustion research for applications to MEMS rotary IC engine[C]// *Proceedings of NHTC*，2001 *National Heat Transfer Conference*. Anaheim，CA：[s. n.]，2001：1-6.

[4] Yang W M，Chou S K，Li D T，et al. Micro-scale combustion research for application to micro-thermophotovoltaic systems[J]. *Energy Conversion and Management*，2003，44(16)：2625-2634.

[5] 李德桃，邓军，潘剑锋，等. 微型发动机燃烧室的模拟研究[J]. 机械工程学报，2002，38(10)：59-61.

（本文原载于《工程热物理学报》2004 年第 3 期）

The investigation of combustion in micro-flame tube combustors and the approaches to solving them

Abstract：A novel concept of micro-power system，thermo photovoltaic system，is proposed in this paper. Micro-combustor is the key component of this system. In order to obtain relatively high efficiency，uniform temperature distribution along the wall of combustor is needed. Scale effects have significant influence on the persistent combustion in micro-combustor. Various experiments have been done to analyse the feasibility of combustion in micro-combustor. The results indicate that steady combustion can be kept in micro-flame tube and high temperature can be kept along the wall.

微热光电系统燃烧的若干
影响因素的试验研究

潘剑锋,李德桃,邓　军,张孝友,卫　尧,唐维新

[摘要]　微热光电(MTPV)系统是一种新型微动力机电系统。通过试验研究,分析阐述了不同氢氧混合气流量、氢氧混合比等因素对 MTPV 燃烧器内的燃烧状况、管壁以及出口端面的温度分布的影响。试验证明,氢氧混合气在微型燃烧器内易实现稳定燃烧。当流量为 2.167×10^{-5} m^3/s,混合比为 $2:1$ 时,微型燃烧器的壁面温度可以达到 1 300 K 左右。

对近年提出的微动力机电系统(Micro-MEMS)的研究取得了迅速进展[1-3],微热光电系统(MTPV)是其中的一种[4]。由于它没有运动件,制造和装配较容易,操作更可靠,并具有较高的能量转换率,所以应用前景广阔。在已有工作的基础上,我们对影响该系统燃烧的若干因素进行了试验研究,报道部分研究结果。

1　微型 TPV 系统的工作原理

该微 TPV 系统由三个主要的部分组成：微混合器、微火焰管燃烧室和光电池。其基本设计如图 1 所示。

微火焰管燃烧室的管壁由辐射性材料组成,光电池则由低频带隙材料制成。氢气和氧气在一个微型混合器中混合后,进入微火焰管燃烧室中燃烧。当壁面,也就是说,选择性辐射材料被加热到一定高温时,就会放出光子。当它们撞击光电池时,会激发出自由电子,在二级管的作用下输出电能。

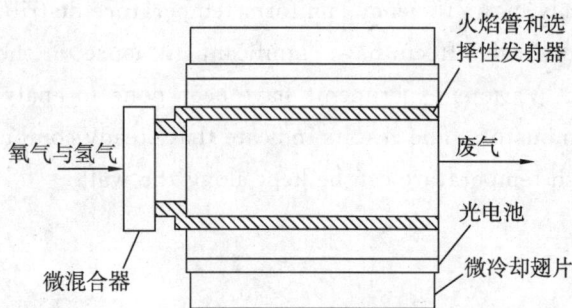

图 1　微 TPV 动力系统

2　试验方法

微燃烧管的材料选用具有较强的耐高温性能的陶瓷 SiC,其基本尺寸：长为 18 mm、内径 3 mm、外径 5 mm,混合气入口处孔径为 0.6 mm。选用 MKS 质量流量控制器测试并控制氢气和氧气的流量,改变混合气的混合比。管壁和出口端面的温度分布测量选用 0.18 mm 的 K 型热电偶。氢、氧入口压力设定为 0.12 MPa,环境温度为 293 K。针对不同的混合气入口流速,分别调整氢氧混合比,观测随着混合比的变化,管内的燃烧状况、管壁和出口端面的温度分

布,由此可得到一组合理的微燃烧管的结构设计和运行参数。

3　试验结果和讨论

当混合气的流量在 $3.33\times10^{-6}\sim6.67\times10^{-6}$ m^3/s 时,管壁和出口端面的温度分布示于图 2 和图 3。

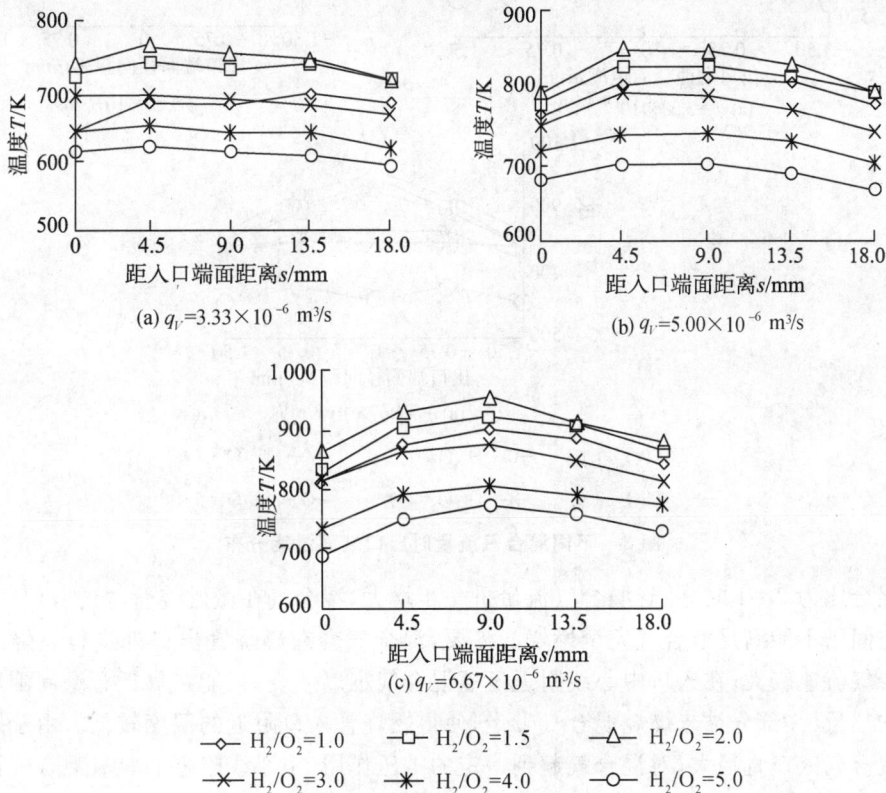

(a) $q_V=3.33\times10^{-6}$ m^3/s

(b) $q_V=5.00\times10^{-6}$ m^3/s

(c) $q_V=6.67\times10^{-6}$ m^3/s

◇— H$_2$/O$_2$=1.0　　□— H$_2$/O$_2$=1.5　　△— H$_2$/O$_2$=2.0
✳— H$_2$/O$_2$=3.0　　✳— H$_2$/O$_2$=4.0　　○— H$_2$/O$_2$=5.0

图 2　不同混合气流量时的管壁温度分布

由于此时氢氧混合气流量很小,其入口处的流速也较小。当混合气的混合比为 2:1,由图 2 可看出最高温度分别位于距入口端 4.5 mm,4.7 mm,9.0 mm 处,这说明混合气在微燃烧管入口处附近就完全燃烧了。相对于其他混合比的燃烧状况,此时管壁和出口端面的平均温度较高,而且出口端面的温度分布是沿半径方向逐渐降低的(见图 3)。

当氢氧混合比小于 2:1 时,由于管内放热量的降低,管壁平均温度逐渐降低,但管壁和出口端面的温度变化规律相似。

当氢氧混合比大于 2:1 时,由于氢气不能完全在微燃烧管内燃烧,燃烧放热量一部分被氢气带走,管壁的温度会逐渐降低。另外随着混合气流量增大,同一混合比状态下的火焰中心位置也有向燃烧管出口处移动的趋势,且此趋势与混合比为 2:1 的着火中心运动趋势一致。但值得注意的是,在混合比大于 4:1 时,管壁温度分布的差异在降低(见图 2),这是因为此时氢气量很大,在管内不能完全燃烧的氢气吸收热量后,在出口端面附近与外界空气中的氧气进行反应,使出口端中心处的温度逐渐升高,并且沿半径方向温度梯度减小(见图 3)。

当混合气的流量大于 8.33×10^{-6} m³/s 时，其管壁和出口温度分布如图 4、图 5 所示。

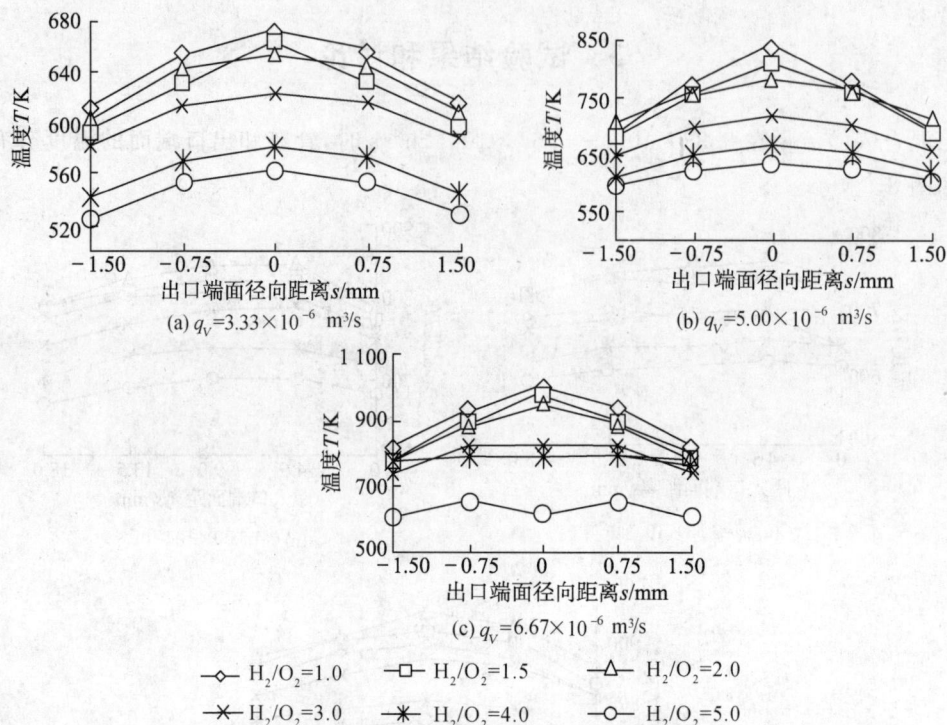

(a) $q_V = 3.33 \times 10^{-6}$ m³/s

(b) $q_V = 5.00 \times 10^{-6}$ m³/s

(c) $q_V = 6.67 \times 10^{-6}$ m³/s

—◇— H₂/O₂=1.0　　—□— H₂/O₂=1.5　　—△— H₂/O₂=2.0

—×— H₂/O₂=3.0　　—✳— H₂/O₂=4.0　　—○— H₂/O₂=5.0

图 3　不同混合气流量时出口端面温度分布

当混合比为 2∶1 时，随着混合气流量进一步增大，混合气在微燃烧管内的驻留时间减小，管内的空间已不能满足混合气完全燃烧的需要，混合气会在燃烧管出口外进行燃烧。由于入口处混合气流速较大，在入口中心线附近氢氧混合气难以燃烧，只能在靠近管壁周围进行。燃烧产生的热量，一部分被未燃烧混合气带走，使得燃烧管入口附近的温度较低。由于燃烧管的内径比混合气入口直径大，对混合气起到一定的扩压作用，中心线附近的氢氧混合气被周围的燃烧产物加热和加速，然后快速流向火焰管的尾部燃烧。这样的燃烧分布，有利于使管壁维持均匀的温度。

当流量不太大时，混合气能在微燃烧管内完全燃烧，如图 4a 所示。当 $q_V = 8.33 \times 10^{-6}$ m³/s 时，出口端面的温度分布还是中心点最高，沿半径方向逐渐降低，但是当流量达到 $q_V = 1.00 \times 10^{-5}$ m³/s(见图 4b)时，由于微燃烧管内混合气的流速很大，混合气不能在燃烧管内完全燃烧，使得出口端面的温度分布呈现出中心温度最低，沿半径方向温度逐渐升高的趋势。

当混合比小于 2∶1 时，随着混合气流速的增大，火焰核心的位置继续向微燃烧管的出口端面转移，燃烧管入口附近的管壁温度较管壁出口端面低很多(见图 4)。随着混合比的进一步降低，使燃烧管相同位置的管壁温度更低，出口端面的中心温度最高，并沿半径方向温度降低的梯度越大(见图 5)。

当混合比大于 2∶1 时，在微燃烧管的入口处，随着氢气量和流速的增大，火焰位置后移，使得入口附近管壁的温度降低，而且，随着混合比的增大，氧气的减少，在管内燃烧的氢气量也会减少，使得在微燃烧管壁面相同位置上的温度逐渐降低(见图 4)。相对于微燃烧管内有限的燃烧空间，混合气已不能在其内部完全燃烧，剩余的氢气在出口处与外界空气混合，进行二次燃烧，使燃烧管出口端面的温度分布呈中心温度低,沿半径方向温度逐渐升高的趋势(见图5)。

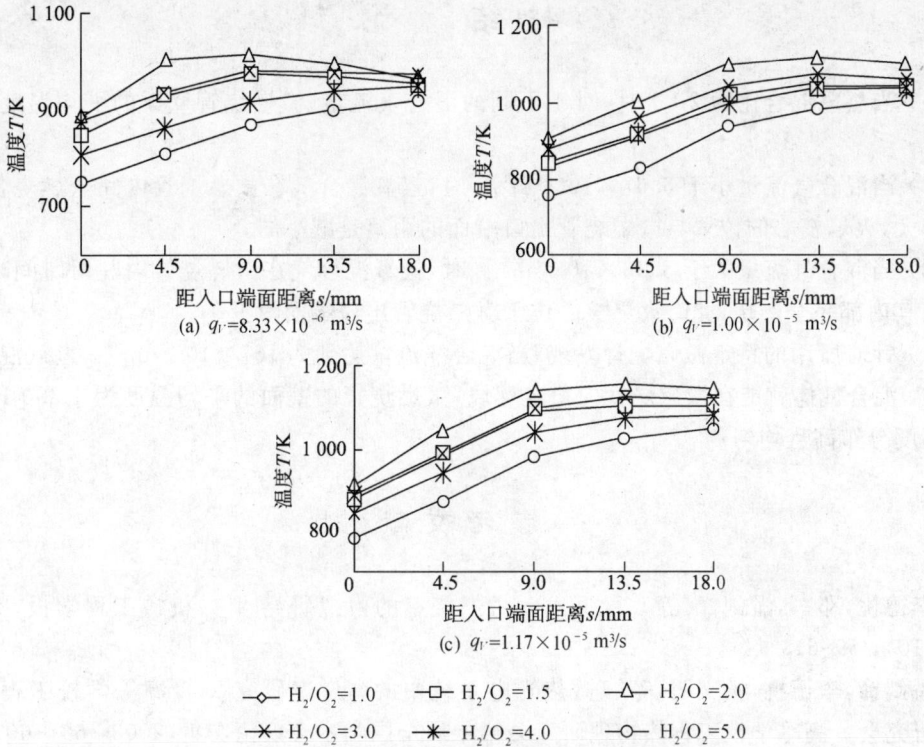

图 4　不同混合气流量时的管壁温度分布

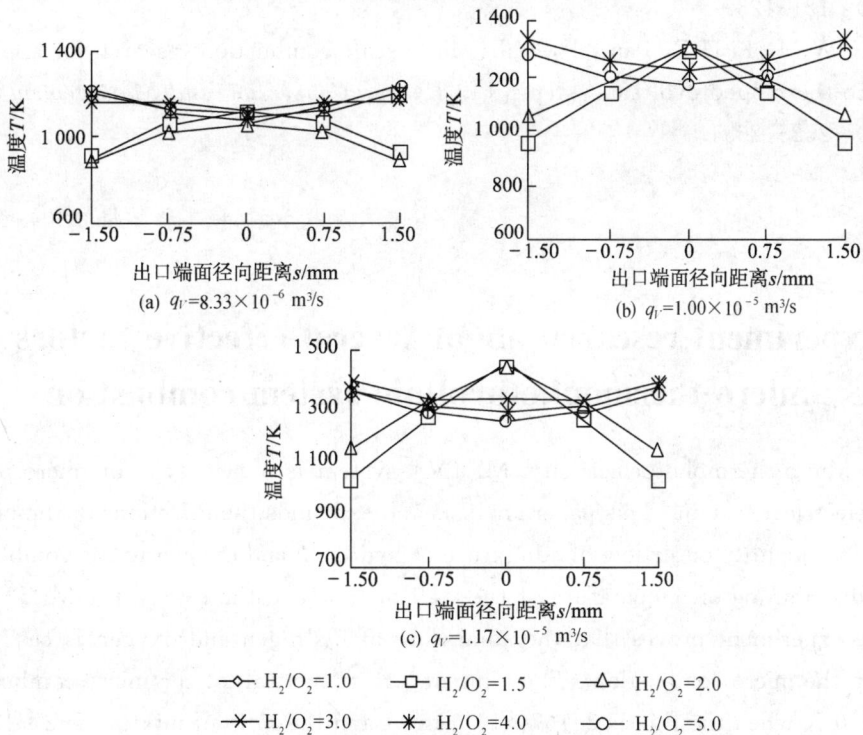

图 5　不同混合气流量时出口端面温度分布

4 结　论

　　(1) 当氢氧混合比为 2∶1 时，对于不同的混合气流量，微燃烧管壁面的平均温度都是最高的。

　　(2) 当混合气流量小于 6.67×10^{-6} m³/s 时，氢氧混合气在微燃烧管内完全燃烧，随着流量的增大，火焰核心的位置与微燃烧管出口端面的距离是减小的。

　　(3) 当混合气流量大于 8.33×10^{-6} m³/s 时，氢氧混合气在微燃烧管内驻留时间减少，已不能在其内部完全燃烧，此时火焰核心位于微燃烧管出口端面附近。

　　(4) 针对所用的特定微燃烧管发现，当混合气流量达到 2.167×10^{-5} m³/s、氢氧混合比为 2∶1 时，混合气体就能在微燃烧管内稳定燃烧，微燃烧管的壁面的平均温度为 1 300 K 左右，壁面温度分布高且均匀。

参 考 文 献

[1] 李德桃，邓军，潘剑锋，等. 微型发动机燃烧室的模拟研究[J]. 机械工程学报，2002，38(10)：59-61.

[2] 潘剑锋，李德桃，杨文明，等. 微热光电系统中的微燃烧研究[C]//高等学校工程热物理研究会. 高等学校工程热物理研究会第十届全国学术会议论文集，2003：601-607.

[3] 李德桃，潘剑锋，邓军，等. 微型发动机燃烧模型和数值模拟[J]. 燃烧科学与技术，2003，9(2)：121-123.

[4] Yang W M，Li D T，Pan J F，et al. Micro-scale combustion research for application to micro-thermophotovoltaic systems[J]. *Energy Conversion and Management*，2003，44(16)：2625-2634.

(本文原载于《机械工程学报》2004 年第 12 期)

Experiment research about several effective factors of micro-thermophotovoltaic system combustion

Abstract：Micro-thermphotoelectricity(MPTV) system is a new type of micro-power mechanical-electrical system. This paper analysis and expounds the infections of the factors that the different quantity of airflow of admixture of hydrogen and oxygen to the combustion status, the distributing of temperature of the wall of tube and the exit in the MPTV combustion. The experiment proves that the admixture of hydrogen and oxygen is easily burning steadily in the micro-combustion. The temperature of the wall of the micro-combustion may attain 1 300 K when the flux is 2.167×10^{-5} m³/s and the ratio of mixture is 2∶1.

微热光电系统燃烧器的研究

潘剑锋,李德桃,杨文明,薛　宏

[摘要]　介绍了一种新颖的微动力机电系统,即微热光电系统。该系统可以取代电池作为各种微型机械的动力,广泛应用于民用与军工部门。微燃烧器是微热光电系统的一个重要元件,其外表面必须产生一个高而均匀的温度分布,以尽可能高的输出辐射能。结合三维数值模拟与实验对微燃烧室进行了研究。研究结果表明,带有台阶的微火焰管燃烧室非常适合于微热光电系统的应用。

随着微机电系统的发展,对高功率密度的微动力机电系统的需求日益迫切[1]。美国和欧洲一些国家正在研制微型燃气轮机和微型转子发动机[2,3],这些微动力装置在摩擦、密封、生产和装配等方面遇到了传统动力装置所未遇到的困难。而微尺度热电转换装置因无运动部件而不存在上述困难[4],但其热交换器和燃烧室三维结构的加工很困难,此外,现有热电技术的能量转换率不高。

本文介绍了笔者研发的一种新颖的微动力装置——微热光电系统(Micro-TPV),该系统利用光电池将高温微燃烧室的辐射能直接转换为电能。由于它既无任何运动部件,又可以获得较高的能量转换率,故与其他微动力装置相比有明显的优越性,应用前景广阔。

1　原　　理

图 1 为 Micro-TPV 系统原理示意图。该系统由微混合器、微火焰管燃烧室和光电池 3 个部分组成。燃料(氢气或烃类)和氧化剂(氧气或空气)在微混合器内混合后,进入微火焰管燃烧。燃烧释放出来的热会使火焰管温度升高,从而辐射出大量光子,当这些光子碰到光电池时,就会激发出自由电子,并输出电能。

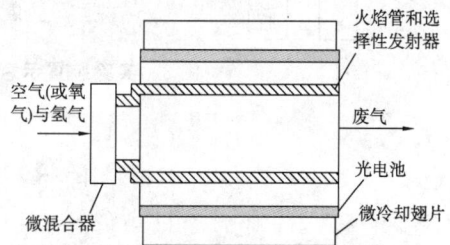

图 1　Micro-TPV 系统原理图

2　设　　计

微燃烧器是 Micro-TPV 系统的核心部分。在设计微燃烧器时,关键是维持持续燃烧和输出最大热效率之间的最佳平衡。当燃烧器尺寸缩小 1 个数量级时,表面积和体积将分别缩小 2 个与 3 个数量级,从而面容比将增加 1 个数量级。这会显著增大通过壁面的热损失,而大的热量损失会抑制混合气的点火以及使反应发生淬熄[5]。所以,在一个 MEMS 级的燃烧室中的稳定燃烧将受到很大的影响。另一方面,微热光电系统的动力输出取决于燃烧室壁面温度和表面积的大小,大面容比有利于提高单位体积的功率密度输出。为了测试在微型装置中燃烧的可行性并确定影响微燃烧的相关因素,设计了 3 种不同的微火焰管燃烧室,并进行了三维模拟计算。

为了研究火焰管中燃烧稳定性和验证模拟结果的准确性,采用不锈钢设计制造了火焰管,一种为直管,另外两种为内径带突扩部分的火焰管(见图 2)。

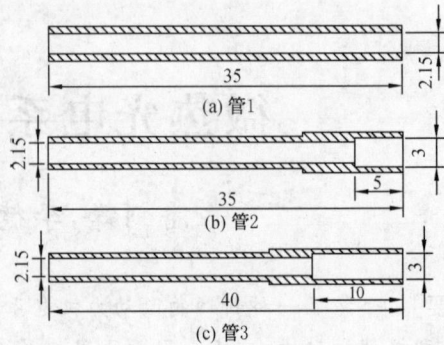

图 2　火焰管的结构

3　实　验

图 3 为实验装置示意图。氢气和氧气(或空气)经过减压阀进入质量流量控制器,再进入预混合器中进行混合。混合后的气体经管路导入火焰管,在距火焰管外壁面 20 mm 处布置了光电池,产生的电压和电流可以用测量电路测量。氢气和空气(或氧气)的质量流量由两组电子控制流量计精确控制。通过该装置,氢气/空气(或氧气)的配比也可以很方便地进行调整。在火焰管的排气口和管壁上的温度分布状况由直径为 0.203 mm 的 K 型热电偶测定。图 4 为火焰管燃烧的照片。

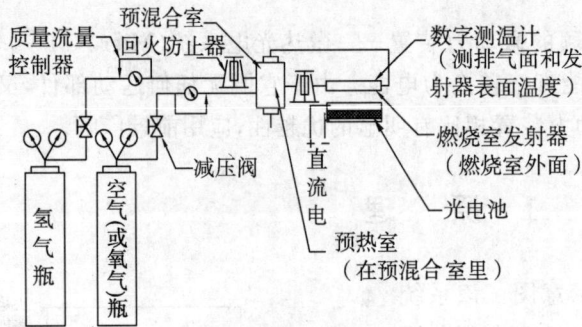

图 3　实验装置示意图

图 4　微燃烧室照片

4　结果及讨论

首先采用空气作为氧化剂,火焰管材料为不锈钢。根据 3 种不同火焰管的模拟结果和实验数据可以作出以下结论:当进口的流速低于 1.3 m/s 时,由于产生的热量小于通过管壁损失的热量,燃烧很快熄灭,不能持久。当流速超过 8 m/s 时,在任何条件下,燃烧不会在直管内发生,但它可以在管外发生。只有当进口流速在 1.5～8 m/s 之间时,稳定的燃烧才能在直管内发生。火焰核心的位置相对于不同的进口流速是变化的,随着流量逐渐增加,需要更大的氢气/空气的配比,才能在火焰管中获得稳定燃烧。否则,火焰将会被吹灭或吹出排气口。

通过分析,我们了解到直火焰管燃烧室有些不利因素,所以,设计了另外一种截面突变火焰管。它可以用来在台阶处产生涡流,减缓流动速度,增强燃料和空气的混合。由此,加强管壁边缘燃烧的完善性以确保稳定地燃烧。

相对于直管 1,截面突变火焰管 2 和 3 可以在更广的流量范围和更广的氢气/空气比率范围内稳定地工作。当进口的流速在 1.5～20 m/s,氢气/空气的比率在 0.45～1.0 时,能够在火焰管内获得稳定的燃烧。同时,截面突变火焰管在保证火焰发生在截面突变的下游燃烧室中起着非常重要的作用。必须注意的是,截面突变火焰管的下游管段必须足够长,使燃烧在气流流出出口前完成。图 5 为模拟计算得到的当进口流速达 8 m/s 及氢气/空气比率为 0.45 时,管 2 和管 3 的轴向剖面上的温度分布。很明显,在管 2 中的燃烧因为燃烧室太短,没有完成燃烧,这会降低燃烧效率。由数值模拟和实验得到的出口面上的温度曲线见图 6。

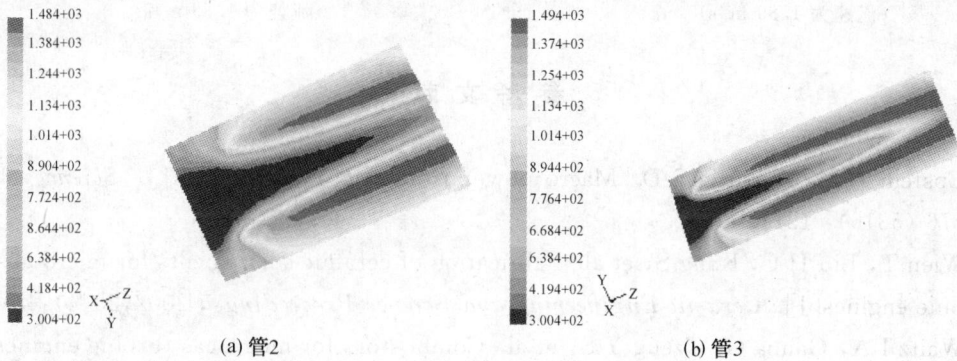

(a) 管2　　　　　　　　　　　　　　　(b) 管3

图 5　　中心轴平面的温度分布(单位: K)

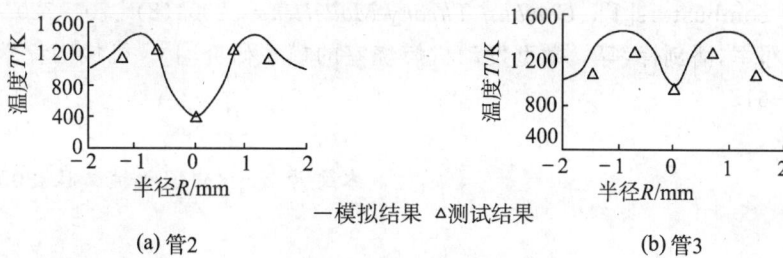

一模拟结果　△测试结果

(a) 管2　　　　　　　　　　　　　　　(b) 管3

图 6　　出口温度分布

通过对样机的初步测试,当氢气流率为 0.07 g/s,氢气与氧气的混合比 ε 为 0.9 时,Micro-TPV 系统能在一个容积为 0.113 cm³ 的微燃烧室内产生 1.02 W 的电力输出,相应地,开路电压和短路电流分别为 2.28 V 和 0.59 A。如果在 Micro-TPV 系统中采用热激发材料——掺 Co/Ni 的 MgO 的发射极,在容积为 0.155 cm³ 的微燃烧室内输出电功率能增至 5.5 W。

笔者还采用氧气作为氧化剂进行了另一次测试,同时实验时火焰管材料选用辐射性能好的 SiC。SiC 火焰管基本尺寸如下:长为 18 mm,内径为 3 mm,外径为 5 mm,混合气入口处孔径为 0.9 mm,混合气经小喷嘴进入直管火焰管。在流速小于 0.47 m/s 时不能在火焰管中维持火焰,流量增大时温度逐渐升高。氢气与氧气的混合比 ε 对温度也有非常大的影响。当混合气流量一定,氢气与氧气的物质的量之比接近于当量混合比 2 时得到的壁面温度最高,产生的光电池电压也最高。当混合比增大时,温度最高点逐渐向出口移动,当混合比超过 5 以后,燃烧完全发生在管外。图 7 和图 8 分别为不同混合比 ε 下的管壁温度和单片光电池产生的输出电压。

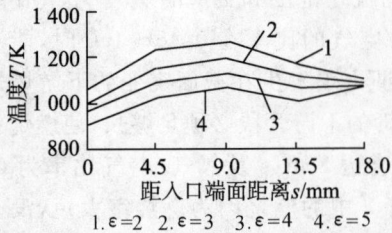

图7　在不同混合比 ε 下的直管管壁温度分布
（流速为 1.89 m/s）

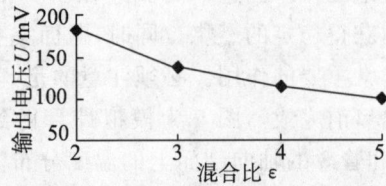

图8　在不同混合比 ε 下的单片光电池获得的输出电压
（流速为 1.89 m/s）

参 考 文 献

［1］Epstein A E，Senturia S D. Macro power from micro-machinery[J]. *Science*，1997，276（5316）：1211.

［2］Wien T，Liu H C，Kang S，et al. Fabrication of ceramic components for micro-gas turbine engines[J]. *Ceramic Engineering and Science Proceedings*，2002，23（4）：43-50.

［3］Waitz I A，Gauba G，Tzeng Y S，et al. Combustors for micro-gas turbine engines[J]. *ASME Journal of Fluids Engineering*，1998，120（1）：109-117.

［4］Chen M，Buckmaster J. Modelling of combustion and heat transfer in 'Swiss roll' micro-scale combustors[J]. *Combust Theory Modelling*，2004（8）：701-720.

［5］李德桃,邓军,潘剑锋,等. 微型发动机燃烧室的模拟研究[J]. 机械工程学报,2002,38（10）:59-61.

（本文原载于《中国机械工程》2005 年第 6 期）

Study on micro-combustor for micro-thermophotovoltaic system

Abstract：A novel power MEMS device，micro-thermophotovoltaic（micro-TPV）system is introduced in this work. It can replace traditional batteries as the power source of all kinds of micro-devices for civilian and military application. As one of the most important components of micro-TPV system，micro-combustor should be able to produce a high and uniform temperature distribution along the outer surface to ensure a high radiation power density. To testify the feasibility of micro-combustion，a lot of 3-D numerical simulation and experimental tests were carried out. The results indicate a micro-flame tube combustor with a backward facing step is one of the most effective combustors for micro-TPV application.

微热光电系统原型的设计制造和测试

潘剑锋,杨文明,李德桃,黄　俊

[摘要]　本文首次描述了一种新颖的微热光电(微 TPV)系统原型的设计、加工过程和测试结果。该系统由一个 SiC 发射极、一个简单的 9 层绝缘过滤器和一个 GaSb 光电池环路组成。当 H_2 流率为 3.443 9 g/h,H_2/O_2 比为 1.8 时,微 TPV 系统能在一个容积为 0.091 cm^3 的微型燃烧室内产生 0.562 W 的电力输出,相应的开路电压和短路电流分别为 1.14 V 和 0.65 A。该项工作使得我们在不久的将来用微 TPV 系统代替电池作为微机械装置的动力源成为可能。

随着机械和机电工程装置的小型化趋势日益增大,各种微装置如微型泵、微型引擎、微型机器人、微型航天器和微型飞机正在开发中。然而,这些装置的小型化受到了可供使用的动力系统(电池)的限制。它们在质量和容积上占据了整个装置的大部分。典型的轻便机械装置也面临着充电时间长和需要频繁充电的问题。减小系统的重量和延长运行时间的需求促进了一种新型 MEMS 装置、微动力发生器的开发[1]。

众所周知,烃燃料的能量密度比最好的电池要高 10~100 倍。因此,利用化学燃料的高能密度来产生动力成为一项有吸引力的取代电池的技术。微燃气轮机、微型三角转子发动机(Wankel 型)、微热电系统[2]和微燃料电池[3]是最近正在开发的典型的微动力发生器。微TPV 系统[4]是我们正在研发的另一种微动力发生器。该系统利用光电池把烃燃料燃烧释放出的热辐射能转化为电能。

Rumyantser 等人[5]也开发了一种便携式 TPV 发生器。然而,它的总体尺寸是 100 cm^3 数量级甚至更大。这里所描述的微 TPV 系统是立方厘米数量级的。微燃烧室的大面容比使得微 TPV 系统的研究在单位容积的能量密度输出方面特别有吸引力。本文详细地描述了微TPV 系统原型的设计、加工过程和测试结果。

1　微 TPV 系统的设计和制造

我们已制成一个微 TPV 动力发生器的样机并进行了测试。该系统主要由以下部件组成:(1) 一个圆柱形 SiC 发射器(即微燃烧室);(2) 一个简单的 9 层绝缘过滤器;(3) 一个 GaSb 光电池回路。

图 1 为微 TPV 系统的示意图。微 TPV 系统的一张照片如图 2 所示。该系统不包括任何运动部件。它的制造和装配相对简单。因此,它能广泛地应用于商业电子和微装置。

在微尺度上,燃烧器内燃料能否充分燃烧受驻留时间的影响很大。因此,我们选择热值高、反应时间短的氢气作为燃料。H_2/O_2 混合物在微圆柱形 SiC 燃烧器内燃烧。当燃烧室壁,即辐射器被加热到一定温度时,它散发出大量光子。过滤器能将大部分能量低于 PV

电池带隙的光子反射回辐射器,尽可能多地传送能量高于 PV 电池带隙的光子。当这些有足够高能量的光子撞击 GaSb 光电池时,它们能激发出自由电子并在 P-N 结作用下产生电流。

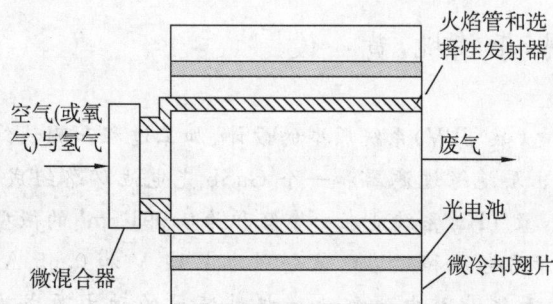

图 1　微TPV系统示意图

图 2　微TPV系统原型

1.1　微圆柱形 SiC 辐射器

辐射器是微 TPV 系统最关键的部分。有两种不同的辐射器:宽带辐射器和选择性辐射器。黑体是一种典型的宽带辐射材料,它的辐射能力近似于石墨或煤烟覆盖的表面。然而,实际应用中重要的宽带辐射材料是碳化硅 SiC,发射率 $\varepsilon_{sic} \approx 0.9$。选择性辐射器在 PV 电池有效的光谱范围内显示出很高的辐射率,在其他范围内则比较低。最近 10 年,已经开发出很多制造选择性辐射器的方法。一种熟悉的方法是利用稀土材料的氧化物,如氧化铒(Er_2O_3)和氧化镱(Yb_2O_3)。另一种方法是对辐射器表面进行微加工[6]。最近,一种新的热激发材料——掺 Co/Ni 的 MgO 的辐射器在华盛顿大学研制成功[7],这种辐射器展现了更高的光谱发射效率。为了使加工和装配简单,这里我们选择 SiC 作为微 TPV 样机的辐射器。

设计微燃烧室时遇到的主要挑战在于维持持续稳定燃烧和获得最大热量辐射输出之间的最佳平衡。大的面容比对于单位体积内能量输出密度是非常有益的。然而,持续稳定的燃烧受热量损失增大的影响很大,甚至导致着火困难和淬熄。为了证实微燃烧的可行性并优化微燃烧室的设计,我们进行了一系列数值模拟和实验工作。结果显示,一个带有突扩台阶的微圆柱形燃烧室是应用于微 TPV 系统的最简单而有效的结构之一[8]。突扩台阶有助于使沿壁面产生回流从而加强管内气流边缘的混合过程。因此,能在微燃烧室壁面获得高且稳定的温度分布。在一个容积为 0.091 cm³,直径为 2.4 mm,长为 20 mm,壁面壁厚为 0.8 mm 的微圆柱形 SiC 燃烧室内,当 H_2 流率为 3.443 9 g/h,H_2/O_2 为 1.8 时,在壁面上能获得 1 275 K 的平均温度。图 3 显示了在上述条件下燃烧的照片,由数码相机拍得。

图 3　微燃烧室照片

（H_2 流率为 3.443 9 g/h,

H_2/O_2 为 1.8）

1.2　一个 GaSb PV 电池回路

微 TPV 系统的另一个关键部分是一个低能带隙的 PV 电池回路。和太阳能的光电转化相比,以 1 000~1 600 K 热源散发出的光子分布在较低能量和较长波长范围内。这使得为了达到最高效率和最大能量密度,将低能带隙半导体应用于 TPV 能量转换是必要的。尽管 TPV 能量转换的概念在 20 世纪 60 年代就被第一次提出[9],但直到最近几年在低能带隙光电池和高温材料领域内的技术进步才使人们重新产生了用 TPV 发电的兴趣。GaSb,GaInAs 和 InGaAsSb 是最近正在开发的应用于 TPV 的典型的低能带隙 PV 电池。为了和下一部分将要介绍的过滤器相对应,我们设计并制造了一个 GaSb PV 电池回路用于微 TPV 系统。GaSb PV 电池可以对波长为 $1.8~\mu m$ 以下的光子产生反应。形成 P-N 结的过程是采用在 GaSb 的 N 基层进行 Zn 扩散的技术,因而,成功地避免了采用昂贵的半导体层的外延生长技术。Zn 扩散过程是在一个被称为“pseudo-closed”的盒内进行的,扩散源为 Zn 和锑的混合物。

图 4 显示了 GaSb PV 电池大致的制造过程。首先在掺有 Te 的 N 型 GaSb 晶片表面覆盖了氮化硅防扩散绝缘层,再用标准影印石版术在绝缘体上开孔,然后在开孔的表面上进行锌扩散形成 P-N 结,扩散形成了一个掺 Zn 的 P 型 GaSb 发射层。

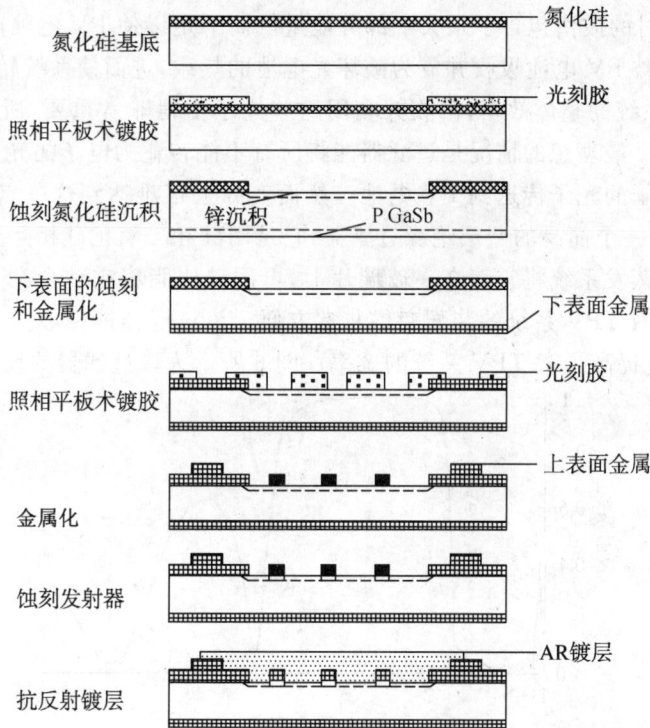

图 4　GaSb 光电池大致的制作过程

晶片的前端模板由光敏电阻保护,而晶片背面的 P-N 结通过非选择性蚀刻技术除去。然后晶片的背部就被镀上金属作为正电极,再应用标准的照相平板印刷术和金属汽化的方法来给晶片前端镀上金属作为负电极。最后,对发射层进行蚀刻和镀上非反射(AR)层使光电流最大。GaSb 电池的量子效率如图 5 所示。

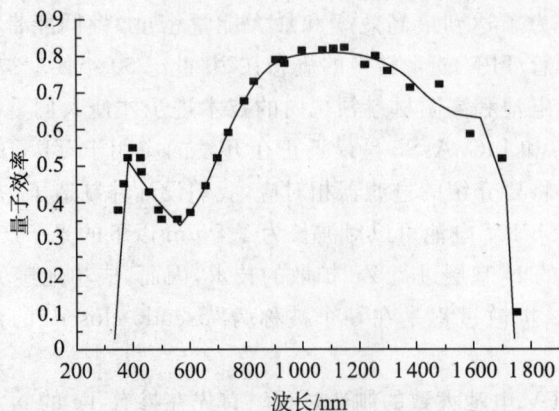

图 5　GaSb 光电池的量子效率

1.3　一个 9 层绝缘过滤器

微 TPV 系统的第三个关键部分是一个简单的 9 层绝缘过滤器。SiC 辐射器在运行温度为 1 000～1 600 K 时的波谱包含了很大一部分能量低而不足以使 PV 电池产生自由电子的光子。这部分能量会被 PV 电池吸收并成为破坏光电池的热载，进而使系统转换效率急剧下降。为了提高微 TPV 系统的整体效率，回收并利用这些光子变得非常重要，所以应在微 TPV 系统采用一个过滤器。最理想的情况是过滤器能将所有不能转化为电子的光子反射回辐射器并将所有可激发出电子的光子传送到 PV 电池。然而实际上很难达到这一点。这里，我们在微 TPV 系统中采用了一个简易的 9 层绝缘过滤器，它是用硅和二氧化硅相互叠加制成的。利用一个常规的电子束蒸发系统将它镀在一玻璃片上，再用硅树脂将它粘合在 GaSb PV 电池上，这种设计方法使得微 TPV 系统的装配变得非常方便。该过滤器能回收分布在 1.8～3.5 μm 波段上的能量，因此提高了微 TPV 系统的效率。图 6 所示为该过滤器的反射率。

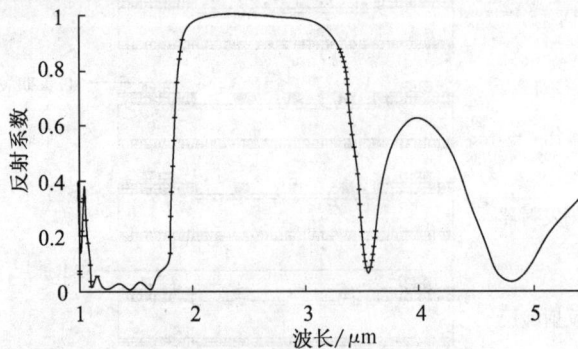

图 6　9 层绝缘过滤层的反射

2　结果和讨论

带过滤器的 GaSb PV 电池回路首先用太阳能辐射模拟器进行了测量。通过调整模拟器辐射光的强度能使 PV 电池得到不同的电流密度。

图 7 所示为 GaSb PV 电池回路的 I-U 曲线，结果显示 GaSb PV 回路具有非常好的光电转换性能。在光照强度相当于 1 573 K 的黑体辐射时，该回路可产生 6.36 W 的电能输出。

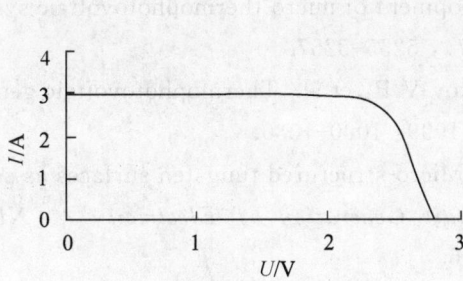

图 7　带过滤器的GaSb光电池回路的电流-电压曲线　　图 8　不同流率和氢氧混合比下的最大输出功率

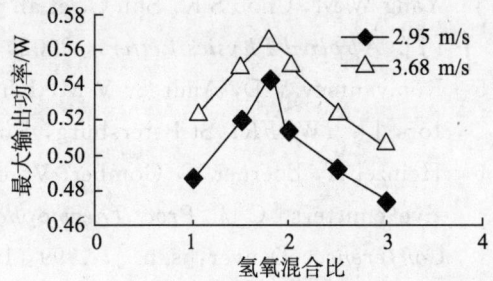

然后测量了各种流率和 H_2/O_2 比率下，微 TPV 系统样机的功率输出。图 8 显示了在不同流率和 H_2/O_2 下的最大电功率输出。随着流率和 H_2/O_2 的增大，最大电功率输出急剧上升，这是由于有更多的燃料参与燃烧。当氢气流率为 3.443 9 g/h，H_2/O_2 为 1.8 时，微 TPV 系统达到 0.562 W 的电功率输出。PV 电池的温度能通过散热片在强迫空气对流冷却下保持在 60 ℃ 左右。相应的开路电压和短路电流为 1.14 V 和 0.65 A。应该注意的是，就像世界上其他正在研究的微动力装置一样，微 TPV 系统的研究才刚刚起步，所以效率还比较低。然而，和其他微动力系统样机相比，它还是好的，到目前为止，该系统已稳定工作 100 小时以上。

为了进一步增大输出电功率密度和提高微 TPV 系统的效率，必须在未来的设计中采用选择性辐射器来代替 SiC 宽频辐射器。如果在微 TPV 系统中采用热激发材料——掺 Co/Ni 的 MgO 的辐射器，在相同的燃烧室效率下，通过以前的论文中所描述的数学模型计算，能在容积为 0.155 cm^3 的微燃烧室内达到 5.5 W 的电功率输出。

3　结　论

本文首次描述了一种新颖的微热光电（微 TPV）系统样机的设计、制造和测试结果。该系统由一个 SiC 辐射器、一个简单的 9 层绝缘过滤器和一个 GaSb 光电池阵组成。当 H_2 流率为 3.443 9 g/h，H_2/O_2 为 1.8 时，微 TPV 系统能在一个容积为 0.091 cm^3 的微型燃烧室内产生 0.562 W 的电力输出，相应的开路电压和短路电流分别为 1.14 V 和 0.65 A。如果在微 TPV 系统中采用热激发材料——掺 Co/Ni 的 MgO 的辐射器，在容积为 0.155 cm^3 的微燃烧室内输出电功率能增至 5.5 W。这项工作使得我们有可能在不久的将来用微 TPV 系统来代替电池作为微机械装置的动力源。

参 考 文 献

[1] Epstein A H，Senturia S D. Macro power from micro-machinery[J]. *Science*，1997，276(5316)：1211.

[2] 李德桃，邓军，潘剑锋，等. 微动力机电系统和微发动机的研究进展[J]. 世界科技研究与发展，2002，24(1)：24-27.

[3] Lee S J，Chang C A，Cha S W，et al. Design and fabrication of a micro-fuel cell array with "Flip-Flop" interconnection[J]. *Journal of Power Sources*，2002，112(2)：410-418.

[4] Yang W M，Chou S K，Shu C，et al. Development of micro-thermophotovoltaic system [J]. *Applied Physics Letters*，2002，81(27)：5255-5257.

[5] Rumyantsev V D，Andreev V M，Khvostikov V P，et al. Thermophotovoltaic genera-tors[J]. *IWRFRI*，St Petersburg，Russia，1999：1000-1030.

[6] Heinzel A，Boerner V，Gombert V，et al. Micro-structured tungsten surfaces as selec-tive emitters [C]// *Proc Thermophotovoltaic Generation of Electricity* 4th *NREL Conference*. Denver：[s. n.]，1999：191-196.

[7] Ferguson L G，Dogan F. A highly efficient NiO-Doped MgO matched emitter for thermopho-tovoltaic energy conversion[J]. *Materials Science and Engineering*，2001，83(B)：35-41.

[8] Yang W M，Chou S K，Shu C，et al. Micro-scale combustion research for application to micro-thermophotovoltaic systems[J]. *Energy Conversion and Management*，2003，44 (16)：2625-2634.

[9] White D C，Wedlock B D，Blair J. Recent advance in thermal energy conversion [C]// 15th *Annual Power Sources Conference*，Atlantic，NJ，1961：125-132.

（本文原载于《工程热物理学报》2005 年第 5 期）

Design，fabrication and testing of a prototype micro-thermophotovoltaic system

Abstract：The design，fabrication and tesing of a novel prototype micro-thermophotovoltaic (micro-TPV) system is first described in this paper. The system is made of a SiC emitter，a simple 9 layer dielectric filter and a GaSb Photovoltaic cell array. When the flow rate of hy-drogen is 3. 443 9 g/h and the H_2/O_2 is 1. 8，the micro-TPV system is able to deliver an elec-trical power output of 0. 562 W in a micro-combustor of 0. 091 cm^3 in volume. The open-circuit electrical voltage and short-circuit current are 1. 14 V and 0. 65 A respectively. It makes us possible to replace batteries with micro-TPV systems as the power of micro-me-chanical devices in near future.

微热光电系统中柱型燃烧室的试验

黄　俊,潘剑锋,唐维新,杨文明,薛　宏,李德桃

[摘要]　微燃烧室是微热光电(MTPV)系统的一个重要组成部分。对一种薄壁柱型微燃烧室进行了试验,得到了氢氧混合比 ε 和喷嘴孔径 Φ 对微燃烧室外壁面和出口端面温度分布影响的规律。在不同流量、不同喷嘴孔径下,当 ε 为 1.8∶1 时,获得的壁面温度平均值最高。当 Φ 为 0.7 mm,ε 为 1.8 时,在微燃烧室获得高且分布均匀的壁面温度,适合于 MTPV 系统应用。

随着微制造技术的进步,微动力机电系统(MEMS)的发展正在不断加速[1-3]。微热光电(MTPV)系统也是其中的一种[4]。它没有运动部件,且不需要高加工精度与装配精度,具有较高的能量转换效率。在 MTPV 系统中,微燃烧室是最为重要的组成部分之一,在微燃烧室内实现稳定燃烧,获得高且分布均匀的壁面温度是 MTPV 系统的一个关键问题。文献[5]中虽然研究了 MTPV 系统燃烧的若干影响因素,但对燃烧室壁厚、喷嘴孔径 Φ 等重要结构因素未作研究。本文则在更宽的混合比 ε 范围内重点研究壁厚、喷嘴孔径 Φ 等重要因素对微尺度燃烧的影响。

1　MTPV 系统的工作原理

MTPV 系统的基本原理是使用光电池把燃料燃烧所产生的热能转换成电能。该 MTPV 系统主要由微燃烧室、光电池等部件组成。它的基本原理如图 1 所示。

氢气和氧气在一个微型混合器中混合后,进入微燃烧室中燃烧。微燃烧室的壁面由辐射性材料组成,壁面加热到一定高温时放出光子。光子撞击光电池激发出来的自由电子在二极管的作用下输出电能。光电池则由低频带隙材料制成,它的表面有一过滤层,该过滤层使辐射性材料发出的有效辐射被光电池吸收,而无效辐射则被反射回辐射性材料,从而减少能量损失。

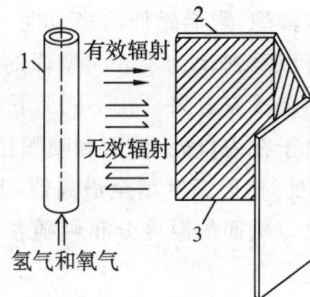

1—微燃烧室　2—光电池　3—过滤层
图 1　MTPV 系统原理图

2　微燃烧室的试验

在本试验中,选用具有耐高温性能的 SiC 陶瓷作为微燃烧室的材料,微燃烧室的结构为薄壁柱型,其尺寸如图 2 所示。

在微燃烧室入口处采用不同孔径 Φ 的喷嘴,并改变氢氧混合气的体积混合比 ε 进行氢氧燃烧试验,观测微燃烧室内

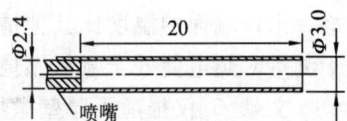

图 2　薄壁柱型微燃烧室结构图

燃烧状况的变化,采用 K 型铠装热电偶测量壁面和出口端面温度分布,由此为薄壁柱型微燃烧室的结构设计提供依据。氢气、氧气入口压力设定为 0.12 MPa,环境温度为 293 K。试验中通过 MKS 质量流量控制器来实现氢气、氧气的流量控制和氢氧混合比 ε 的调节。

3   结果与讨论

通过试验发现,氢氧混合气在薄壁柱型微燃烧室内易实现稳定燃烧。氢氧体积混合比 ε 以及喷嘴孔径 Φ 对微燃烧室内的燃烧有重要影响。

3.1   氢氧混合比 ε 对微燃烧的影响

当氢氧混合气体的入口流量 q 为 500 mL/min,采用的喷嘴孔径 Φ 为 0.7 mm 时,在不同混合比 ε 下壁面和出口端面温度分布如图 3 和图 4 所示。

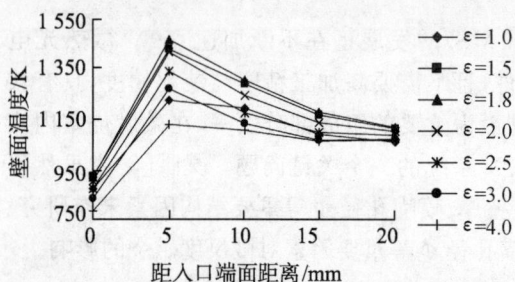

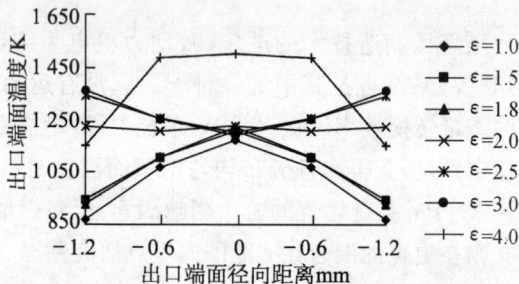

图 3   混合气流量为 500 mL/min 时壁面温度分布图   **图 4   混合气流量为 500 mL/min 时出口端面温度分布图**

在微燃烧室内,随着尺寸缩小,面容比增大,使得壁面散热损失急剧增大,会导致点火困难和反应淬熄,影响燃烧的稳定性。

可以看到,ε 为 1.8 时壁面的温度最高,这说明微燃烧室内需要在富氧条件下才能达到最好的燃烧反应效果。在不同 ε 下,壁面温度的分布趋势相同:燃烧室入口处壁面温度值最低,这是由于在入口处,壁面和喷嘴接触,喷嘴的热传导带走了一部分热量所造成的;在距入口约 5 mm 处,壁面温度增至最高值,此后温度值沿壁面逐渐降低。

出口端面的温度分布则随着 ε 的变化而不同,在 ε 为 1.8 时,出口端面温度呈中间高、两边低的分布。

当 ε 小于 1.8 时,由于燃烧室内氢气量较小,所能放出的热量偏低,所以壁面平均温度有所降低,但变化规律相似。出口端面温度数值也降低,分布趋势均是中间高、两边低。

当 ε 大于 2 时,随着氢气所占的比例升高,部分氢气未在燃烧室内燃烧就直接排出燃烧室外,同时还带走了一部分气体燃烧时放出的热量,进而降低了壁面的温度分布。对于出口端面的温度分布则正好相反,由于氢气比例升高,燃烧室内氢气不能完全反应,部分未燃氢气在燃烧室出口处与外界空气中的氧气发生燃烧反应。试验中,能在出口处观察到氢、氧燃烧的火焰。出口端面的温度比其他情况下要高很多,而且会出现两边高、中间低的温度分布。这是由于测量时热电偶处于氢氧燃烧的火焰中,而火焰的外焰温度比火焰核心温度高的缘故。

文献[5]仅报道了厚壁微燃烧室的试验情况。这里把它和薄壁微燃烧室在相同流量、相同混合比等情况下的温度分布数据对比发现,薄壁微燃烧室所得的壁面平均温度值比厚壁微燃烧室要高出 150 K 左右,这说明壁厚对微燃烧室的影响非常大,采用薄壁燃烧室更有利于

MTPV 系统。但由于实际加工困难等因素影响,未能采用壁厚更小的燃烧室进行试验。

在文献[5]中得到当 ε 为 2 时,壁面温度值最高的结论。本文则对薄壁柱型微燃烧室在更宽的 ε 范围内进行试验,发现当 ε 为 1.8 时,燃烧室壁面温度值比 ε 为 2 时更高,从而得到流量对微尺度燃烧更为精确的影响规律。

3.2 喷嘴孔径 Φ 对微燃烧的影响

对薄壁柱型微燃烧室分别采用孔径 Φ 为 0.5,0.7,0.9 mm 的喷嘴进行试验,研究喷嘴孔径 Φ 对微尺度燃烧的影响。图 5 和图 6 是在燃烧状况较好、混合气流量为 600 mL/min 时壁面和出口的温度分布。

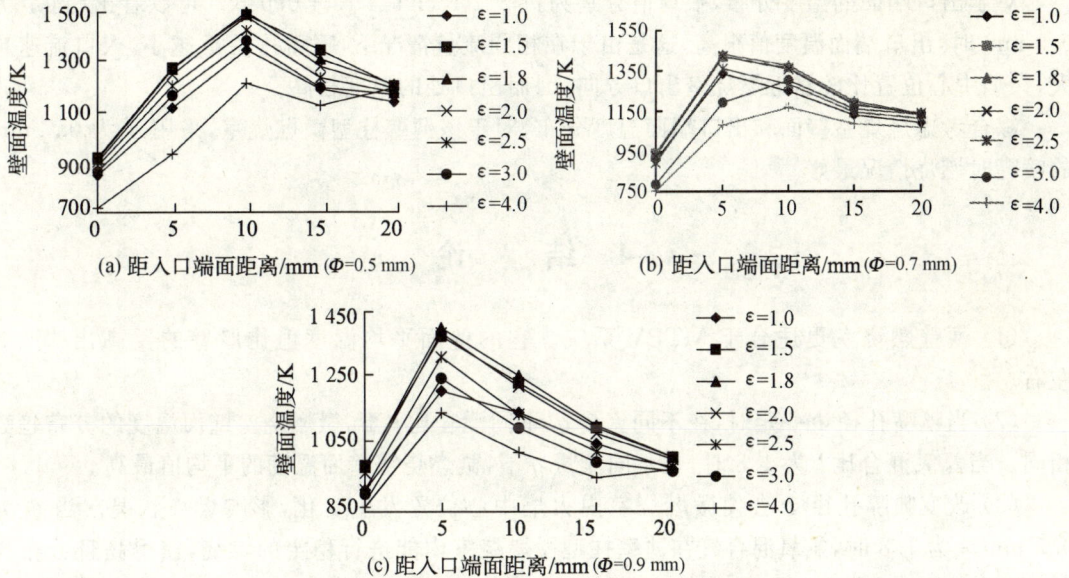

图 5　不同喷嘴孔径时壁面温度分布图

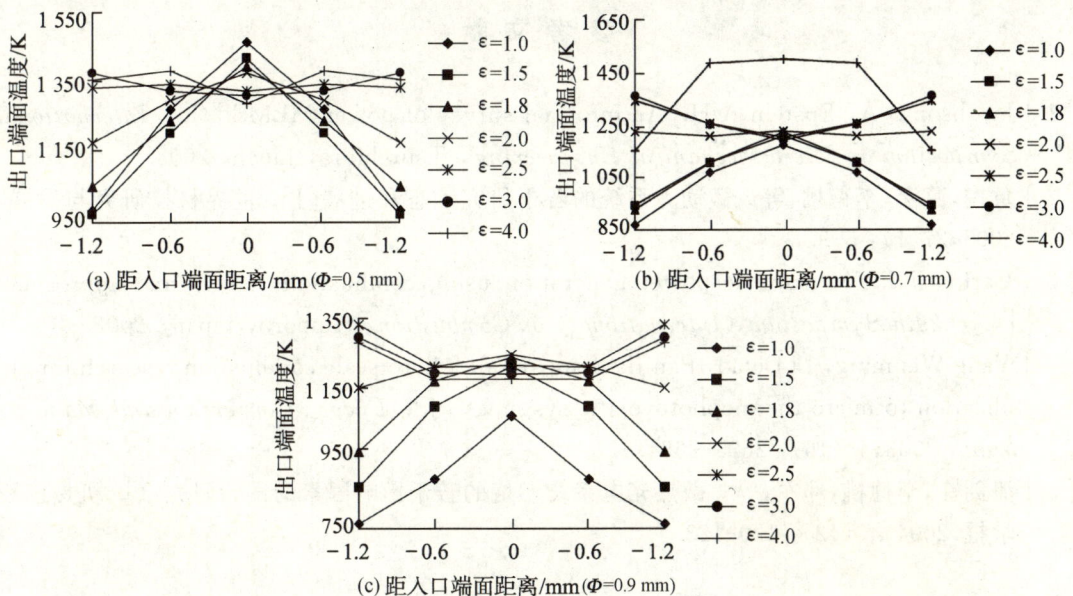

图 6　不同喷嘴孔径时出口端面温度分布图

当 q 为 600 mL/min, ε 为 1.8 时,3 种喷嘴条件下得壁面温度平均值分别为:1 195 K, 1 216 K,1 182 K。从数值大小来看,3 种情况温度值差别不大,但是从温度分布图可看出,燃烧室壁面的最高温度点发生了变化。当 Φ 为 0.5 mm 时,在相同流量下,入口速度较大,使火焰中心位置向燃烧室出口方向移动,因而温度最高点也随之移动,处于离燃烧室入口约 10 mm 处。而当 Φ 为 0.7 mm 和 0.9 mm 时,最高温度点均出现在距燃烧室入口约 5 mm 处,且当 Φ 为 0.9 mm 时,由于火焰中心位置进一步靠近入口处,使得壁面各点温度差值变大,均匀性差。这说明改变喷嘴孔径 Φ 会使柱型微燃烧室内火焰中心位置发生变化,影响燃烧状况。

对于出口端面的温度分布,平均值分别为:1 212 K,1 144 K,1 090 K。可以看出,当 Φ 为 0.5 mm 时,出口端面温度值很高,这是由于在相同流量情况下,喷嘴孔径 Φ 越小,入口流速越大,火焰中心位置比较靠近燃烧室出口方向,因而出口处的温度较高。

综合考虑燃烧室壁面和出口端面温度分布,对于该薄壁柱型微燃烧室,采用 Φ 为 0.7 mm 的喷嘴时燃烧情况最好。

4 结　论

(1)薄壁燃烧室更适合于 MTPV 系统。它的壁面平均温度值比厚壁的要高出 150 K 左右。

(2)当喷嘴孔径 Φ 一定时,在不同流量 q,不同混合比 ε 下,微燃烧室壁面温度的分布趋势相同。当氢氧混合比 ε 为 1.8 时,在不同流量 q 下,微燃烧室壁面温度的平均值最高。

(3)改变喷嘴孔径 Φ 会使微燃烧室内火焰中心位置发生变化,影响燃烧状况。当 Φ 为 0.7 mm, ε 为 1.8 时,氢氧混合气在薄壁柱型微燃烧室内能进行稳定的燃烧,获得高且分布均匀的壁面温度,适合于 MTPV 系统应用。

参 考 文 献

[1] Jacobson S A, Epstein A H. An informal survey of power MEMS [C]// *International Symposium on Micro-Mechanical Engineering*, Tsurchiura, Japan, 2003.

[2] 黄俊,薛宏,李德桃,等. 微动力系统的若干研究动态和进展[J]. 世界科技研究与发展, 2005,27(1):5-9.

[3] Carlos F P. Micro-scale power generation using combustion:issues and approaches [C]// *29th Symposium (International) on Combustion*, Sapporo, Japan, *2002*.

[4] Yang Wenming, Li Detao, Pan Jianfeng, et al. Micro-scale combustion research for application to micro-thermophotovolatic systemes [J]. *Energy Conversion and Management*, 2003,44(16):2625-2634.

[5] 潘剑锋,李德桃,邓军,等. 微热光电系统燃烧的若干影响因素的试验研究[J]. 机械工程学报,2004,40(12):120-123.

(本文原载于《农业机械学报》2006 年第 3 期)

Experimental research on cylindrical combustor of micro-thermophotovoltaic system

Abstract: Micro-combustor is one of the most important components of micro-thermophotovoltaic (MTPV) system. The effect of the hydrogen/oxygen mixture ratio ε and the diameter of nozzle Φ on the temperature distribution both along the wall and on the exit plane of micro-combustor was obtained through the experiment of a thin cylindrical micro-combustors. The average temperature along the wall of micro-combustor was the highest when ε was 1.8. When Φ was 0.7 mm, ε was 1.8, a high and uniform temperature distribution was achieved along the wall of micro-combustor, which was for application to MTPV system.

Effects of major parameters on micro-combustion for thermophotovoltaic energy conversion

Pan J F , Huang J , Li D T , Yang W M , Tang W X , Xue H

Abstract

For a micro-thermophotovoltaic (TPV) energy conversion device, high surface to volume ratio in the micro-combustor provides a great potential to achieve high surface radiation power output per unit energy input. This work investigated experimentally the effects of three major parameters on micro-combustion, namely hydrogen to oxygen mixing ratio, nozzle to combustor diameter ratio, and wall thickness to combustor diameter ratio. The results show that the high average wall temperature can be achieved at slightly oxygen rich mixing ratios. Nozzle to combustor diameter ratio affects both the magnitude and uniformity of wall temperature distribution. The newly designed thin wall combustor which yields a reduction of axial heat conduction loss is able to increase wall temperature more than 150 K. Optimized design of these parameters will have significant impact on the enhancement of radiation heat output in micro-TPV energy conversion.

1   Introduction

The micro-thermophotovoltaic(TPV) system is a direct energy conversion device proposed recently[1,2]. The system uses PV cells to convert thermal radiation energy from combustion of fossil fuel into electricity. The micro-TPV system mainly consists of a fuel source, a combustor with its wall served as emitter, and a PV array. The amount of radiation energy convertible by the PV cell grows with the emitter temperature and with decreasing bandgap energy of the PV cell material. The typical PV cell such as gallium antimonide(GaSb) has a low energy bandgap of 0. 72 eV or wavelength of 1. 8 μm . In the case of a blackbody radiation spectrum, 9. 6% of photons have sufficient energy to clear the bandgap hurdle at an emitter temperature of 1 500 K, and it increases to 21. 8% at 2 000 K. Therefore, while researches are focused on development of spectrally selective and matched emitter for higher energy conversion efficiency[3,4], it is equally important to enhance combustion temperature and promote combustion efficiency of micro-TPV sysyem.

As micro-TPV relies on the radiation of combustor surface(emitter), the high surface to volume ratio in a micro-combustor provide an ample opportunity for the system to achieve higher radiation power output per unit energy input than that in macroscale. Since the system has no moving parts and does not require high precision fabrication,it does not suffer the penalties due to increasing friction loss and/or excessive heat transfer from hot to cold end that are usually encountered when a macroscale engine is scaled down. The most challenging issue here is to keep an optimal balance between sustainable combustion at micro-scale and maximizing radiation heat output per unit volume. Excessive heat loss through the wall of combustor may suppress ignition and quench the reaction. Furthermore, with decrease in combustor size, there is less residence time for mixing and combustion process. For combustion at micro-scale, Waitz et al. [5] studied a micro-combustor for micro-gas turbine engine. Fu et al. [6] investigated quench diameter in micro-scale combustion. They concluded that the barrier for quenching distance can be effectively removed by reducing heat loss or preheating the unburnt mixture. For TPV application, a micro-combustor with a backward facing step was developed by Yang et al. [7],who showed that the backward facing step is capable of enhancing mixing process and prolonging residence time. Li et al . [8] analyzed the combustion at micro-scale using scale analysis and numerical simulation to relate flame temperature explicitly to external wall temperature of the combustor. An excellent review on recent developments on the design of combustor for TPV system can be found in Colangelo et al. [9].

In this work, we investigated experimentally several important parameters in micro-scale combustion. They include stoichiometric mixing ratio of hydrogen to oxygen mole fraction, ratio of injection nozzle to combustor diameter, and ratio of wall thickness to combustor diameter. The objective of the study is to quantify the effects of these important parameters on micro-combustion process and combustor design. Their performance is evaluated under various operation conditions on a criterion to achieve high and relatively uniform surface temperature along the wall of micro-combustor.

2　Micro-combustor and experimental set-up

Micro-combustors made of silicon carbon (SiC) with different geometric dimensions were designed and are shown in Fig. 1. Yang et al. [7] suggested that a backward facing step is able to generate recirculation of combustion mixture near the wall, hence to promote the combustion around the rim of the cylindrical combustor. In the present design, the backward facing step is replaced by a nozzle shown in Fig. 1a, which simplified the structure of the micro-combustor. Different nozzle diameter can be easily designed and manufactured. Three different nozzle diameters, namely 0.5 mm, 0.7 mm and 0.9 mm were designed for the test. The ratio of nozzle to combustor diameter is equivalent to the step to channel height in the mico-combustor designed with a backward facing step[7]. The combustor in Fig. 1b is an early version of the combustor design. It has very similar structure but a wall thickness of 1 mm. We intended to use it to investigate the effect of thermal conduction in the wall of the combustor in both radical and axial direction due to the change of wall thickness.

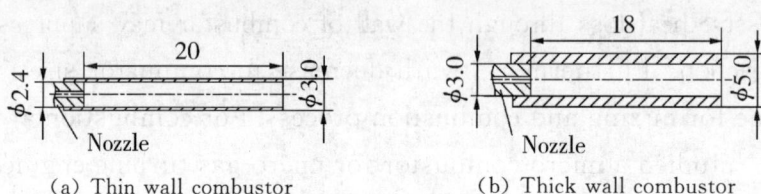

(a) Thin wall combustor   (b) Thick wall combustor

Fig. 1   **Dimensions of micro-combustors (unit: mm)**

The schematic of experimental set-up is shown in Fig. 2. Pure hydrogen(> 99.99%) and oxygen (>99.9%) were supplied from two pressurized bottles for combustion. They were premixed in a plenum at prescribed flow rate controlled by separate volumetric flow controllers (Type 1 640 A, MKS). Five K-type thermocouples of 0.18 mm in diameter were employed to measure surface temperature along the combustor wall. Another K-type thermocouple was installed on a linear positioner so it could transverse at the exit plane of combustor to measure temperature distribution. The measurement uncertainty was estimated at 5% for the thermocouples and 3% for the volumetric flow rate.

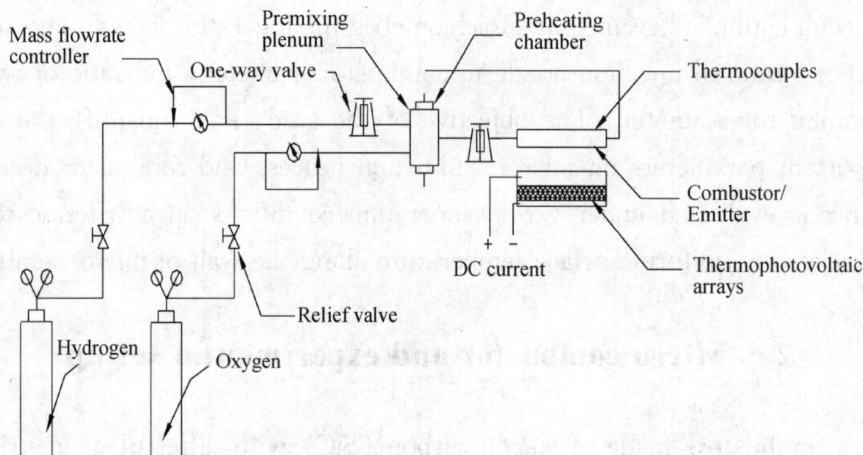

Fig. 2   **Schematic of experimental set-up**

3   Results and discussion

The experiments were conducted systematically to investigate the effects of hydrogen to oxygen mixing ratio, nozzle to combustor diameter ratio, and wall thickness to combustor diameter ratio on temperature distributions along the wall of the combustor and at exit plane. For micro-TPV application, combustion efficiency is defined as the desired output from combustion, which is the net radiant energy emitted from outer surface of combustor, over the requested input determined from the heating value of fuel. Although the overall efficiency of the system is also related to spectral distribution of emitter material and PV cell efficiencies[2], a uniform and high surface temperature of combustor is critical for radiant energy output. The exit temperature is strongly correlated to the wall temperature as an indication of completeness of combustion. On the other hand, the exit temperature also represents amount of heat exhausted from combustor if flow rate remains as constant.

3.1   Effect of hydrogen to oxygen mixing ratio

The effect of hydrogen to oxygen ratio ε was studied on a micro-combustor shown in Fig. 1a with the nozzle diameter of 0.7 mm. The combustions under several of hydrogen to oxygen ratios were tested at two different volumetric flow rates namely 500 mL/min and 600 mL/min. The measured surface temperatures along the combustor wall are shown in Fig. 3. The temperature distributions at exit plane are plotted in Fig. 4. From Fig. 3, it is found that the highest surface temperature occurs at hydrogen to oxygen ratio $\varepsilon = 1.8 : 1$ for both flow rates. Mixing ratio $\varepsilon = 1.5 : 1$ is the second highest. This suggests that combustion in micro-combustor is most favorable at slightly oxygen rich circumstance.

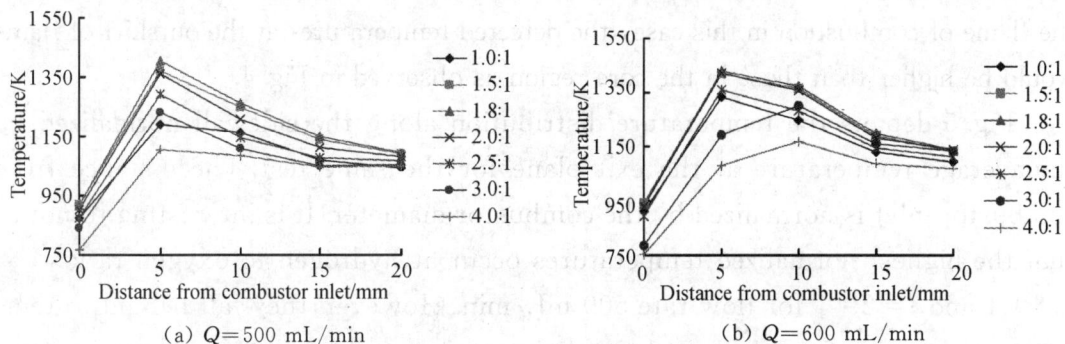

(a) $Q = 500$ mL/min   (b) $Q = 600$ mL/min

Fig. 3   **Temperature profiles along sidewall for different flow rates**

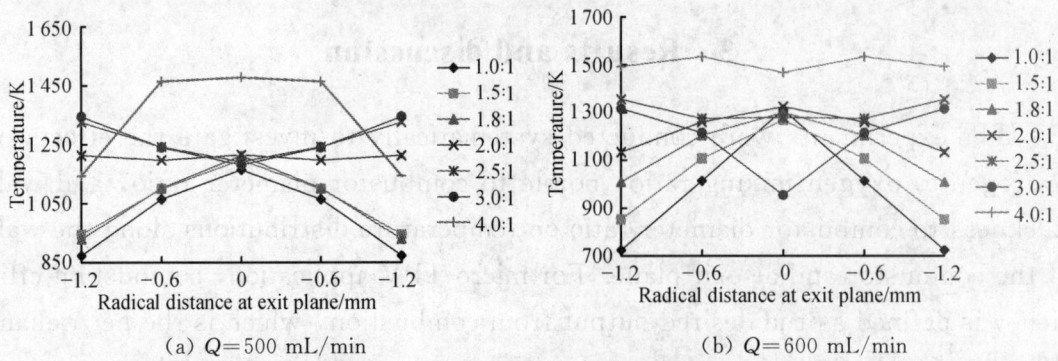

(a) $Q=500$ mL/min (b) $Q=600$ mL/min

Fig. 4 Temperature distribution at exit plane for different flow rates

The distribution of temperature along the wall is consistent at both flow rates and different hydrogen to oxygen ratios in Fig. 3. The temperature at inlet plane of nozzle is the lowest due to heat conduction to nozzle. At the region about 5 mm away from the inlet plane, that is where the center of combustion flame possibly located, surface temperature reaches the highest value, and decrease gradually afterward. On the other hand, in Fig. 4, the temperature distribution at exit plane is largely different with variation of hydrogen to oxygen ratios. At mixing ratio of 1. 8 ∶ 1, the temperature distribution is close to parabolic; temperature has highest value at the center and is lower near the wall. When the ratio gets smaller, the heat release during combustion is low due to lower fuel fraction. Therefore, the flame temperature and surface temperature are reduced although the temperature profile at the exit appears similar. If the ratio gets higher than 2 ∶ 1, combustion becomes incomplete due to excessive hydrogen in the combustor. The excessive hydrogen absorbs heat in the combustor and is directly exhausted. Once the unburnt hydrogen approaches to the exit of combustor, the highly heated hydrogen automatically react with oxygen from ambient. This phenomenon was observed during the experiment. Combustion flame at exit was clearly visible. Hence, higher exit temperatures were recorded. In addition, since thermocouple transverses in the flame of combustion in this case, the detected temperatures at the outskirt of flame would be higher than those in the core region as observed in Fig. 4.

Fig. 5 depicts the temperature distribution along the sidewall normalized by the average temperature at the exit plane for the same test. The distance from combustor inlet is normalized by the combustor diameter. It is interesting to notice that the highest normalized temperatures occur at hydrogen to oxygen ratios $\varepsilon=$ 1. 8 ∶ 1 and $\varepsilon=2$ ∶ 1 for flow rate 500 mL/min. However, they fall behind to temperatures at $\varepsilon=1.$ 5 ∶ 1 and $\varepsilon=1.$ 0 ∶ 1 due to increased exit temperature at flow rate of 600 mL/min. As we mentioned previously, normalized temperatures reflect the fraction of radiation heat output to exhaust. The results indicate the efficiency of combustion is sensitive to the change of flow rate.

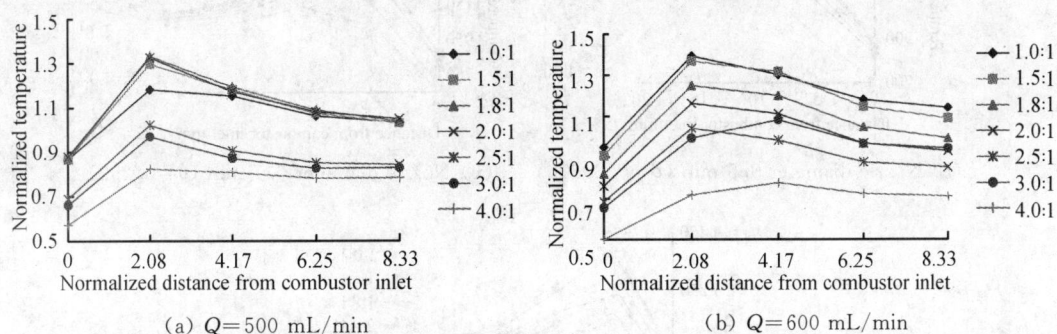

(a) $Q=500$ mL/min (b) $Q=600$ mL/min

Fig. 5 Normalized sidewall temperature profiles for different flow rates

3.2 Effect of nozzle to combustor diameter ratio

For the thin wall micro-combustor in Fig. 1a, three different nozzle diameters, that is, 0.5 mm, 0.7 mm and 0.9 mm were designed, which corresponds to the nozzle to combustor diameter ratio $\delta=0.208$, 0.292, and 0.375 respectively. Temperature distributions along the sidewall and at exit plane are shown in Figs. 6 and 7 at the flow rate of 600 mL/min.

In Fig. 6, the highest surface temperature occurs at hydrogen to oxygen ratio $\varepsilon=1.8:1$. This indicates that the favorable combustion mixing ratio at slightly oxygen rich condition is consistent for different nozzle sizes as well. The average temperatures at wall surface for the three nozzle-to-combustor diameter ratios are 1 195 K, 1 216 K and 1 182 K respectively. The difference among these values is insignificant, but the locations where the highest temperatures appear are different. For the case $\delta=0.208$, the highest temperature occurs at 10 mm downstream from the inlet of nozzle due to higher initial velocity at the nozzle, while for the cases of $\delta=0.292$ and $\delta=0.375$, the temperatures are peaked at about 5 mm downstream. Comparing Fig. 6b and c, it is obvious that the center of combustion flame is closer to the nozzle for the case of $\delta=0.375$, which leads to a distinguish temperature gradient along the sidewall. It should be noted that the location of combustion flame depends on the nozzle to combustor diameter ratio and volumetric flow rate of the mixture in a cylindrical combustor. To ensure complete combustion in the combustor and uniform distribution of temperature, it is desirable for the combustion flame to be located shortly after the injection of mixture from the nozzle like the case we observed in Fig. 6b. At the exit plane, the temperatures are higher for the case $\delta=0.208$ in Fig. 7. This suggests that larger portion of heat is exhausted.

515

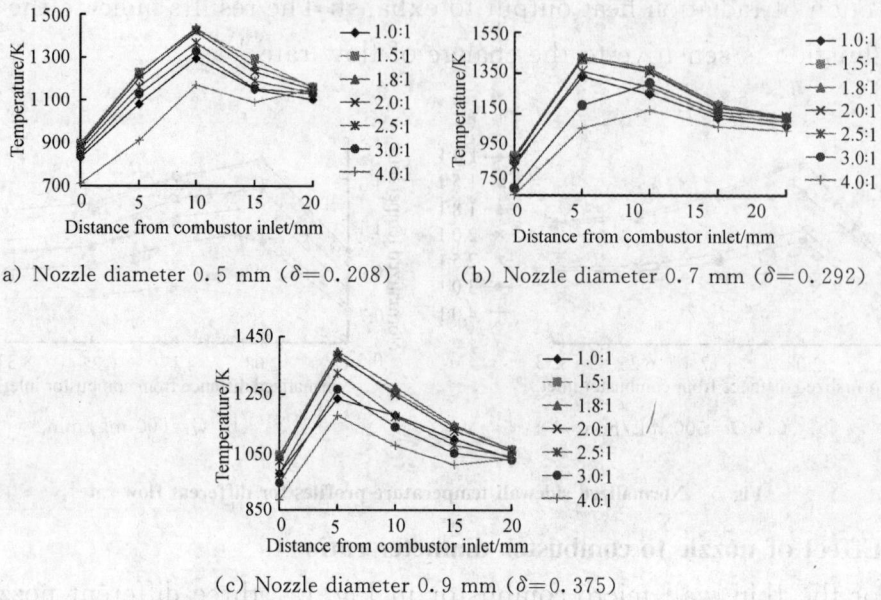

(a) Nozzle diameter 0.5 mm ($\delta=0.208$) (b) Nozzle diameter 0.7 mm ($\delta=0.292$)

(c) Nozzle diameter 0.9 mm ($\delta=0.375$)

Fig. 6   Temperature profiles along sidewall for different nozzle diameters

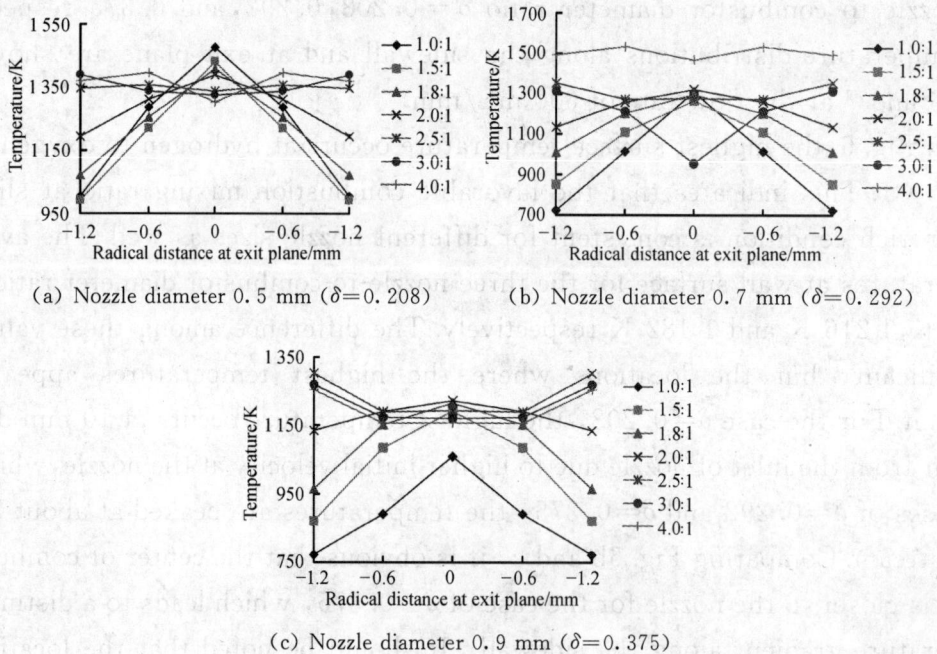

(a) Nozzle diameter 0.5 mm ($\delta=0.208$) (b) Nozzle diameter 0.7 mm ($\delta=0.292$)

(c) Nozzle diameter 0.9 mm ($\delta=0.375$)

Fig. 7   Temperature distribution at exit plane for different nozzle diameters

Fig. 8 depicts the temperature distribution along the sidewall normalized by the averaged temperature at the exit plane for the same test. It is further confirmed that normalized temperatures are higher for oxygen rich mixing ratios, typically $\varepsilon = 1.8:1$ and $\varepsilon=1.5:1$. As for the mixing ratio $\varepsilon=1.0:1$, the absolute tempera-

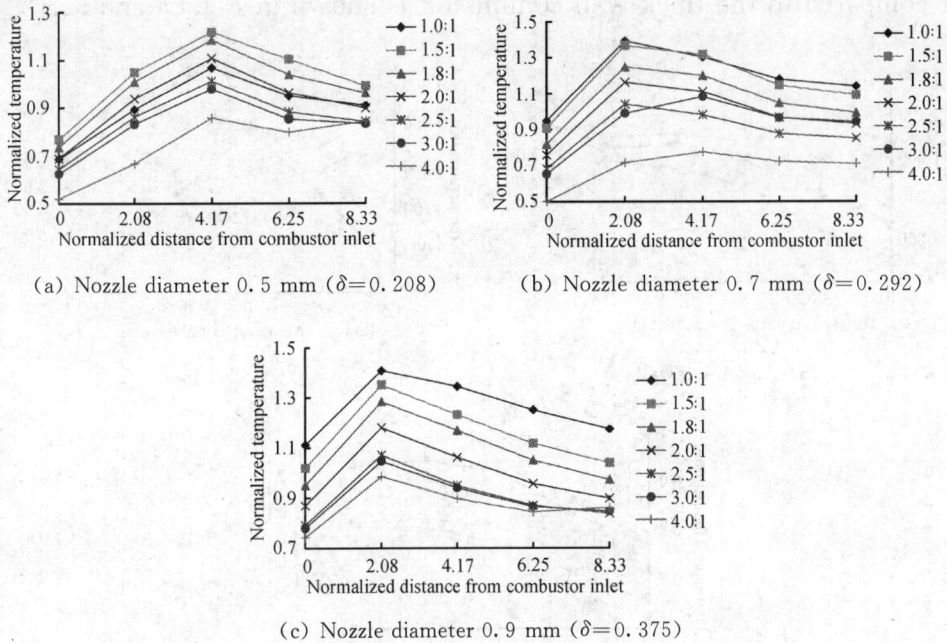

(a) Nozzle diameter 0.5 mm ($\delta=0.208$)

(b) Nozzle diameter 0.7 mm ($\delta=0.292$)

(c) Nozzle diameter 0.9 mm ($\delta=0.375$)

Fig. 8　Normalized sidewall temperature profiles for different nozzle diameters

ture are relatively low for radiation output although it has the lowest exit temperature at the cases of $\delta=0.292$ and $\delta=0.375$.

3.3　Effect of wall thickness to combustor diameter ratio

Two micro-combustors were selected to study the effect of wall thickness to combustor diameter ratio at flow rate of 600 mL/min. One is shown in Fig. 1a with nozzle to combustor diameter ratio of $\delta=0.208$ and the other in Fig. 1b with $\delta=0.20$. However, the former featuring a thin wall structure has wall thickness of 0.3 mm and diameter of 2.4 mm while the later has a relative thick wall of 1 mm and diameter of 3.0 mm . They corresponds to wall thickness to combustor diameter ratio of $\psi=0.125$ and $\psi=0.333$ respectively.

Temperature distributions along the side wall and exit plane were plotted in Fig. 6a, 7a and 8a for the thin wall combustor, and in Fig. 9 for the thick wall combustor. The comparison shows that the average wall temperature for the thin wall combustor is about 150 K higher than that of the thick wall combustor for tests with the same hydrogen to oxygen ratio. Obviously the thin wall structure has reduced heat loss significantly. This implies that the heat conduction in axial direction plays an important role in micro-combustor. The average temperature at exit plane for the thin wall combustor is only about 50 K higher, which results in a non-dimensional temperature ratio for the thin wall combustor to be much higher compared to the thick wall combustor as shown in Fig. 8a and 9c.

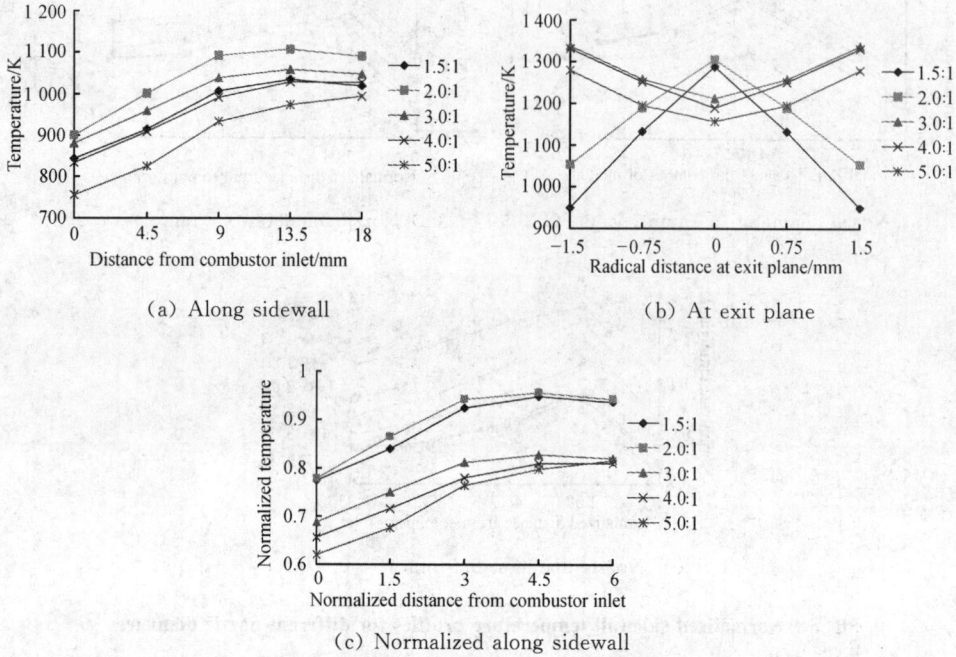

(a) Along sidewall

(b) At exit plane

(c) Normalized along sidewall

Fig. 9　Temperature profiles for thick wall combustor ($\phi = 0.333$) at flow rate of 600 mL/min

To assess the axial heat conduction loss, we consider an energy balance in the wall of a combustor in cylindrical coordinates:

$$\frac{\partial}{\partial x}\left(k_w \frac{\partial T_w}{\partial x}\right) + \frac{1}{r}\frac{\partial}{\partial r}\left(k_w \frac{\partial T_w}{\partial r}r\right) = 0 \tag{1}$$

where T_w is the temperature of wall, k_w is thermal conductivity, r and x are coordinates in radial and axial directions respectively. Following the procedure of dimensional analysis[10], we choose wall thickness b as radial length scale, and combustor diameter d_0 as axial length scale as it confines the flame size in axial direction. By non-dimensioning Eq. (1), it is not difficult to find out that the ratio of axial to radial heat conduction rate in the wall is proportional to non-dimensional parameter $(b/d_0)^2$. In the present case, $(b/d_0)^2$ is 0.015 6 for the thin wall combustor, and 0.111 1 for the thick wall combustor, which is over 7.1 times higher. The non-dimensional parameter $(b/d_0)^2$ also implies the weight of axial heat conduction is only 1.5% of total heat conduction for the thin wall combustor, which is negligible. But the weight increases to 11.1% for the thick wall combustor. This explains qualitatively why the average temperature increases for the thin wall combustor. For a typical micro-combustor with a flame temperature of 1 500 K[8], a 10% net reduction of axial heat loss predicts an increase of wall temperature for about 150 K, which is consistent with our experimentalresult. However, the thin wall structure is restricted by manufacturing and strength of material limits.

4　Conclusions

For micro-combustors ranged from 1 to a few millimeters in characteristic dimension, continuum mechanics is still valid for momentum and energy transport process. However, changes in dominating forces governing the mechanism of momentum and energy transport have become evident to impact characteristics of the phenomenon. This work has been dedicated to investigate the effects of major parameters on micro-combustion. The findings are summarized as follows:

(1) Slightly oxygen rich mixing rate ε at 1. 8 : 1 results in the highest wall surface temperature, which is consistent throughout different flow rates and nozzle to combustor diameter ratios.

(2) Changing volumetric flow rate and nozzle to combustor diameter ratio for a given flow rate leads to relocation of combustion flame in a cylindrical combustor, and hence affects the distribution of wall temperature. Considering both wall temperature magnitude and uniformity, an optimal ratio exists. For example, the best performance appears at ratio of 0. 292 with flow rate 600 mL/min.

(3) It is shown that the axial heat conduction is sensitive to the wall thickness through a non-dimensional parameter $(b/d_0)^2$. The thin wall combustor has resulted in an increase of surface temperature about 150 K, which is significant for improving photovoltaic energy conversion efficiency.

The present work is useful for understanding transport phenomena and optimal design of combustion in development of the micro-TPV system.

Acknowledgments

The authors wish to acknowledge the research grant from Natural Science Foundation of China 50376020 and Innovative Research Grant of Jiangsu Province for Graduates xm04-29.

References

[1] Yang W M, Chou S K, Shu C, Li Z W, Xue H. Development of micro-thermophoto-

voltaic system[J]. *Applied Physics Letters*，2002,81：5255-5257.

[2] Xue H，Yang W M，Chou S K，Shu C，Li Z W. Micro-thermophotovoltaics power system for portable MEMS devices[J]. *Micro-scale Thermophysical Engineering*，2005，9：85-98.

[3] Ferguson L G，Dogan F. Spectrally selective，matched emitters for thermophotovoltaic energy converstion processed by tape casting[J]. *Journal of Material Science*，2001，36：137-146.

[4] Licciulli A，Diso D，Torsello G，Tundo S，Maffezzoli A，Lomascolo M，Mazzer M. The challenge of high performance selective emitters for thermophotovoltaic applications[J]. *Semiconductor Science and Technology*，2007,18：174-183.

[5] Waitz I A，Gauba G，Tzeng Y S. Combustors for micro-gas turbine engines[J]. *Journal of Fluids Engineering*，ASME,1998,120：109-117.

[6] Fu K，Knobloch A J，et al. Micro-scale combustion research for applications to MEMS Rotary IC Engine[C]// *Proceedings of NHTC*，2001 *National Heat Transfer Conference*. Anaheim，CA：[s. n.]，2001：1-6.

[7] Yang W M，Chou S K. Shu C，Li Z W，Xue H. Combustion in micro-cylindrical combustors with and without a backward facing step[J]. *Applied Thermal Engineering*，2002,22：1777-1787.

[8] Li Z W，Chou S K，Shu C，Yang W M，Xue H. Predicting the temperature of a premixed flame in a micro-combustor[J]. *Journal of Applied Physics*，2004,96：3524-3530.

[9] Colangelo G，Risi de A，Latorgia D. New approaches to the design of the combustion system for thermophotovoltaic applications[J]. *Semiconductor Science and Technology*，2003,18：s262-s269.

[10] Incropera F P，Dewitt D P. *Heat and Mass Transfer*[M]. 5th ed.. John Wiley & Sons Inc,2002.

(From：*Applied Thermal Engineering*，2007)

在微热光电系统的能量转换中若干主要参数对微燃烧的影响

[摘要]　在微热光电动力系统中,燃烧室的面容比很大,这增强了从壁面辐射的能量,从而增加整个系统的能量密度。通过实验研究了三个主要参数对微燃烧的影响,即氢氧混合比、喷嘴直径与燃烧室直径的比率、壁厚与燃烧室直径的比率。结果表明,在富氧情况下能获得较高的壁面平均温度。喷嘴直径与燃烧室直径的比率对壁温的数值和分布均有影响。新设计的薄壁燃烧室由于减少了轴向导热损失,能使壁温提高150 K以上。三个参数的优化设计对加强微热光电系统的辐射能量输出有重要的意义。

V. 其他相关研究

Other Investigation

旋进型旋涡流量计内流体运动规律的研究 *①

李德桃

[摘要] 旋进型旋涡流量计是一种新型流量计,被称为 70 年代的理想的流量计之一。但其内部机理尚待弄清楚。

本文利用高速摄影研究了流体在此流量计中的运动规律,建立了一个较完整、较精确、较具体的物理模型。

本文针对此物理模型进行了理论分析,获得了此流量计的两个基本公式。其中一个公式揭露了旋涡强度同各几何参数和运动参数间的关系;另一个公式表达了涡核的进动规律。

1 任务的提出

旋进型旋涡流量计(简称旋涡计)是近年来发展起来的一种新型流量计(见图 1)。它具有精度高、量程宽、无可动部件、维护简便、脉冲输出等一系列优点。与卡门涡列型旋涡流量计同被称为 70 年代的理想流量计[1]。这种流量计可以在较宽的雷诺数范围内输出同流量成线性关系的脉冲频率,这就便于同数字仪表或电子计算机配套,从而为流量测量的自动化和计算机化提供了一种有效的方法[2]。

图 1 φ50,φ80,φ100,φ150 四种口径的旋涡计

* 参加试验研究的还有鲍友良、石明华、严明和袁家麟等同志。高速摄影过程中曾得到下列同志的帮助:镇江农机学院吴守一教授;上海煤矿机械研究所高济宁、赵寿福;上海内燃机研究所张开慧;上海电影制片厂戴胜潮、戈永良;上海实验生物研究所谢世栋;上海照相机三厂黄子英。上海交大江可宗教授和镇江农机学院潘君拯教授对理论分析提出了宝贵意见。该仪器由我国常州热工仪表厂生产。

① 1975 年,上海工业自动化仪表研究所邀李德桃主持该项研究,本文奠定了该类流量计的理论基础,曾在全国仪器仪表学会成立大会上宣读,上海工业大学曾邀李教授作专题宣讲。

——编者注

但是,关于流体在此流量计内的运动规律还不太清楚。Chanaud R C 虽对所谓旋涡梢进行过分析,可是他采用的是普通摄影法[3],他提出的物理模型不完整、不具体。Киясбейлеи А Ш 和 Перельштей М Е 根据流体动力学对这类流量计的工作原理进行过分析[4],但除了两张用阴影法摄得的照片外,没有提供更多的实验依据和说明。佐乌聪夫等人所提出的物理模型[5],也不能对该流量计内发生的基本现象进行合理的、圆满的解释,因为他们所依据的物理模型的缺陷造成了一些理论上的错误。

在这种情况下,为了弄清流体在本流量计中的基本运动规律,为了给改进和完善这种流量计提供充分的实验和理论依据,我们利用高速摄影研究了流体在流量计内的运动状态,获得了一个较精确的、较完整的物理模型;并在此基础上进行了理论分析。理论分析的结果同各种实验研究的结果是一致的。

2 研究方法和设备

为了对流体在旋涡流量计内的全过程的每一步进行深入研究,首先要获得一个正确的物理模型。实践证明,高速摄影不但能提供一个正确的物理模型,显示出时—空过程中的每一瞬间,满足研究者对流体运动作"亲眼"观察的要求,同时还能对某些运动参数进行精确测量。

流动过程摄影的两种基本设备是,一个可以观察流量计内部的透明壳体和一架高速摄影机。我们所采用的透明壳体,内部几何形状和尺寸与产品 φ50 旋涡计相同,外部为方形,材料是用有机玻璃制成的。所用高速摄影机有两种:一是苏制 CKC-1M 型,其最高摄影频率为 4 000 张/秒,我们用的摄影频率为 1 500～3 000 张/秒;二是德制 PZ-35 型,最高摄影频率为 40 000 张/秒,胶片能放映的最高频率为 2 000 张/秒,我们采用的摄影频率为 1 000 张/秒。我们之所以没有采用更高的摄影频率,是因为本流量计的频率输出,一般在 320 周/秒以下。这就是说,涡核每一个进动过程,可以有几张至十几张照片来显示。我们将胶片常速放映时,把过程细节形成一段慢动作电影,可以清楚地看出旋涡的形成、运动和进动过程。这也说明,我们采用的摄影频率是适当的。

我们用的是黑白胶卷,16 mm 为 19DIN(定),35 mm 为 24DIN(定)。

在摄影技术方面,显示流体运动形态的方法有很多种[6]。开始时,我们试用过纹影法。由于我们所用透明壳体的内腔直径小(喉部直径为 35 mm),光线通过它时散射很厉害;加上所通过的流体的密度梯度也较小,故未能清晰地显示出流体的流动形态。后来,对于气流,我们采用过两种方法:一是在流量计的进口上游加入烟,或在上游流道中直接设置一个微型烟发生器。由于烟进入流量计后很快就与被测流体混合成同一种颜色的气体,且透明壳体内腔很快被烟污染,故亦未能显示出流体的流动形态。此法基本上没有成功。另一个方法是在导流架(螺旋叶片)的后锥体上或流道内,粘一根或数根细丝线,借它们来显示流体的运动形态。实践表明,这是显示气流在本流量计内的运动形态的有效方法之一。对于水流,在流量计上游加入的有色液体,通过螺旋叶片后,同样很快与被测水流混合,使流量计内成为同一色液流,不能显示出流体的运动形态。后来,我们利用一种专门装置,把同水流压力相同(或稍高)的气体,从螺旋叶片的上游或其后锥体的中心孔注入,因而形成许多直径约 3 mm 以下的小气泡,借气泡来显示旋涡的形成和运动规律。实践证明,此法是成功的。它不仅能为我们提供一个清晰的物理模型,而且能对有关参数提供数量上的依据。

3 高速照片的分析和物理模型的建立

我们摄得的数十卷计数十万张照片,从不同侧面提供了一个较完整的物理模型。现按不同拍摄条件和不同工况,分别挑选若干张,从不同侧面来说明主要现象和规律。

3.1 旋涡的形成

从图2至图5用不同显示方法所获得的照片,都可以看出流体流过螺旋叶片后按螺旋线运动。这一点,放电影时可以看得更清楚。形象地说,从螺旋叶片流出的各股流体,如同若干股绳拧在一起。事实上,我们曾在叶片间的各流道内,粘一根细丝线。细丝线明显地按螺旋线运动(图3),并拧成一股。这种运动,可以看作前进运动和平面涡的叠加。平面涡的涡核由图2、图4和图5的气泡所显示。

拍摄速度:1 000张/秒
水流量:0.956 L/s
α(螺旋角)=60°,N(叶片数)=6片
β(后锥角)=30°,θ(扩大角)=60°

图2 用气泡一细丝线共同显示液流的旋涡形成

拍摄速度:2 000张/秒
气流量:2.85 L/s
α=60°,N=5片
β=30°,θ=60°

图3 用细丝线显示气流的旋涡形成

3.2 旋涡的发展

在流量计的收缩段和喉部,螺旋的直径随着管内径渐缩而渐缩,涡核的直径也沿着流动方向逐渐缩小。这是我们通过高速摄影新发现的一个现象。此现象通过理论分析得到了说明。

旋涡一进入扩大段,流速就急剧下降;同时,对于旋涡中心区的压力比周围的压力低,于是产生一种如图4和图7所示的回流(此回流只有在银幕上方可以清楚看到,它由扩大段下游涡核散开来的单个气泡的回流所显示);并形成一流速为零的圆锥面。

3.3 涡核的进动

照片和电影都清楚地表明,象刚体一样旋转的涡核,在扩大段产生一种类似于陀螺的进动。由于:

(1)进动锥的顶点(固定点)同陀螺的重心不重合;

(2)进动角θ=常数;

(3)从高速照片的时间信号算得,当流量计输出的频率信号稳定时,陀螺的自动角速度和进动角速度都为常数。

所以,我们确认涡核的进动是类似于重力陀螺受迫稳定进动。

从图 2 至图 5 都可看出,涡核是贴近扩大段壁面进动的。因此,随着扩大角减小,进动锥要拉长,即固定点向上游移动。

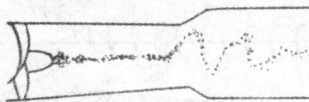

拍摄速度:1 000 张/秒
水流量:3.45 L/s
$\alpha=60°,N=6$ 片
$\beta=30°,\theta=60°$

图 4　用大量气泡显示涡核在扩大段散开后的流态

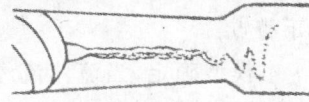

拍摄速度:1 000 张/秒
气流量:1.662 L/s
$\alpha=45°,N=5$ 片
$\beta=30°,\theta=60°$

图 5　用少量气泡显示液流的旋涡形成

3.4　旋涡的消失

在扩大段后部(下游),涡核自行散开。充满在扩大段及其下游空间的流体,本身也作一种同涡核转向相同的螺旋线运动(图 4 和图 5)。在仪表的下游装与不装螺旋叶片,对后续仪表产生明显的影响,也感受到这种螺旋运动的存在。

图 2 至图 5 都是显示流量计正常工况时的流态。

根据高速照片(图 6)所显示的情况,和对示波器上波形的观察,我们发现,常常在下列两种情况下,流量计内出现异常流态:

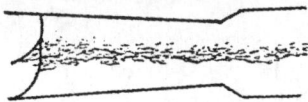

(a) $\alpha=15°,\beta=0°$

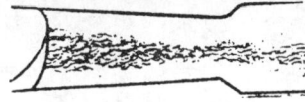

(b) $\alpha=30°,\beta=0°$

拍摄速度:1 000 张/秒;水流量 3.01 L/s

图 6　当叶片螺旋角 α 太小时流量计内出现异常流态

(1)当螺旋角 α 太小时,如图 6a 和图 6b,由于流体经螺旋叶片后获得的圆周分速太小,流体几乎不作螺旋线运动,不形成规则的旋涡,流线接近于直线;这时,示波器上也不出现有旋涡进动时那样的方波。我们通过多次观察和记录认识到,螺旋角 α 是影响流态的一个最主要的参数。

(2)当后锥角 $\beta=0°$ 时,由于流体流经的叶片之间的通道截面始终保持不变。这样,流体不仅不能获得加速,而且由于摩擦效应而作减速运动;另一方面,同一叶片两边的流体在离开叶片时存在较大的速度差。这两种情况都容易引起严重的边界层分离现象[7],使流体破裂出许多不规则的小旋涡,这就导致仪表常数的非线性急剧增加。关于边界层分离的问题,需要再作专题研究。

通过上述分析,我们获得一个较完整、较精确、较具体的物理模型(图7)。

此模型与国外所提出的模型相比,具有以下特点:

(1)本模型确定,当流体流经螺旋叶片后,都作螺旋线运动;而不像文献[5]的模型那样,尚有接近于直线的流动。流体的螺旋线运动,可以看作前进运动和平面涡的叠加。

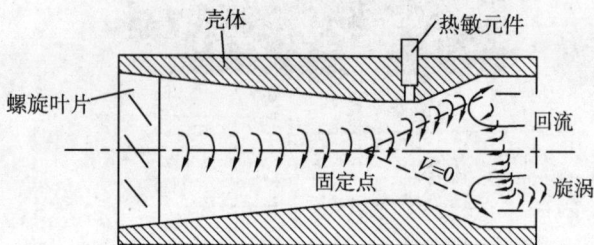

图 7　流体在流量计内的运动状态

(2)从流量计的收缩段至喉部,涡核直径沿流动方向逐渐缩小。在喉部,涡核直径达最小值。

(3)有关文献提出的模型,几乎都认为涡核在扩大段作正规进动[3-5],即涡核的自转轴恒通过涡核的重心。按照我们的模型,涡核的进动类似于重力陀螺受迫稳定进动。

(4)按照佐乌聪夫的模型[5],固定点是在因回流而形成的一个速度为零的圆缘面上,此圆缘的曲率半径较大;而按照我们的模型,固定点是在一个因回流而形成的速度为零的圆锥面上,且就是圆锥体的顶点。

4　理论分析及其结果

4.1　旋涡速度

我们把图7的流态抽象成图8所示的旋涡及其进动的模型。

如前所述,我们把流体流经螺旋叶片后的运动,看作平面涡和前进运动的叠加。平面涡由一个像刚体一样旋转的涡核和流速向外逐渐降低的环流所组成(图8)。

今把流体当作理想流体处理,我们可根据图9导出流体微团绕管轴线(旋涡中心线)旋转角速度的公式[8]:

$$\omega=\frac{1}{2}\left(\frac{\partial V_s}{\partial r}+\frac{V_s}{r}-\frac{1}{r}\frac{\partial V_r}{\partial \theta}\right) \tag{1}$$

当 $V_r=0$ 时,有

$$\omega=\frac{1}{2}\left(\frac{\mathrm{d}V_s}{\mathrm{d}r}+\frac{V_s}{r}\right) \tag{2}$$

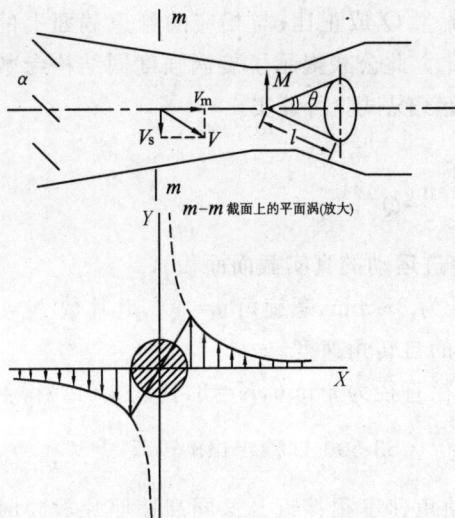

图 8 旋涡及其进动的模型

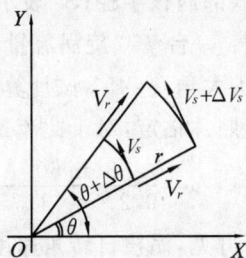

图 9 在直角座标系上导出ω的公式

因此,对于速度 V_s,得到方程 $\dfrac{\mathrm{d}V_s}{\mathrm{d}r}+\dfrac{V_s}{r}-2C=0$,积分得

$$V_s=Cr+\frac{C_1}{r} \tag{3}$$

在旋涡轴上,速度为0,所以接近于旋涡轴的区域(即涡核区)$C_1=0$,此区域的量 C 以 C_0 表之,于是有

$$V_s=C_0r=\omega r \tag{4}$$

对于涡核以外的区域,$\omega=C=0$,故有

$$V_s=\frac{C_1}{r} \tag{5}$$

这里,假设在涡核与环流边界上的流动,同时满足式(4)和式(5),则有

$$C_1=r_m^2\omega \tag{6}$$

式中 r_m 表示涡核半径。

若以 V_m 表示在此边界上的流体沿管轴方向的前进速度,则从式(4)有

$$\omega=\frac{V_m\tan\alpha}{r_m} \tag{7}$$

若令流体沿管轴方向的平均速度为 V_{cm},则得

$$V_m=KV_{cm} \tag{8}$$

式中 K 为比例系数,其值大于1。

将式(8)代入式(7),得

$$\omega=\frac{KV_{cm}\tan\alpha}{r_m} \tag{9}$$

由于涡核本身不作前进运动,故流体前进运动的真实截面面积为

$$S=\pi(R_m^2-r_m^2) \tag{10}$$

式中 R_m 代表截面 $m-m$ 上的管道半径。由此,从式(9)可得

$$\omega=\frac{KV_{cm}S\tan\alpha}{r_mS}=\frac{K\tan\alpha}{r_mS}Q \tag{11}$$

式(11)表明,旋涡强度与螺旋角 α 和容积流量 Q 成正比;与相应的管道截面上的涡核半径 r_m 和流体前进运动的真实截面面积 S 成反比。此公式揭示了旋涡强度同结构参数和运行参数之间的关系;并完全为变结构参数试验和变工况试验所证实。

对于喉部而言,式(11)可写成:

$$\omega = \frac{K\tan\alpha}{r_{m0}S_0}Q \tag{12}$$

式中 r_{m0} 表示喉部涡核半径;S_0 表示喉部流体前进运动的真实截面面积。

计算举例:一台 $\phi50$ 旋涡流量计,喉部直径为 35 mm,螺旋角 $\alpha=60°$,叶片数 $N=6$ 片,后锥角 $\beta=30°$,扩大角 $\theta=45°$,试计算涡核进动时的自转角速度。

根据高速照片估定,对于此流量计,喉部涡核直径为 4 mm,$K=8$,代入式(12)得:

$$\omega = \frac{8\tan60°}{0.2\times\pi\times1.75^2}\times6.71\times10^3 = 53\,500 \text{ 1/秒} = 148 \text{ 转/秒}$$

由此计算可见,涡核自转角速度与涡核进动角速度很接近。这同高速照片放映时所看到的情况相符合;也与 Chanaud R C 的估计相符合[3]。

4.2 进动频率

如前所述,根据高速摄影所建立的物理模型,使我们有理由确认涡核的进动类型于重力陀螺受迫稳定进动。据此,我们便有涡核进动时的运动方程[9]

$$M = \dot{\psi}[I\omega + (I-I_1)\dot{\psi}\cos\theta]\sin\theta \tag{13}$$

式中 M 为回转力矩;$\dot{\psi}$ 为进动角速度;I 为涡核进动时绕自转轴的转动惯量。

$$I = \frac{\rho}{2}(\pi r_{m0}^2 l)r_{m0}^2$$

其中 ρ 为流体密度;l 为涡核进动部分的长度。

I_1 为涡核绕赤道轴的转动惯量:

$$I_1 = \frac{\rho}{12}(\pi r_{m0}^2 l)l^2$$

关于回转力矩,只有文献[4]明确提出由升力 p_n 所引起。我们认为:在图 10 所示的情况下,根据 magnus 效应,升力 p_n 是存在的。此外,根据我们所作改变扩大角 θ 的试验(图 11),可以看出,在同一流量下,随着 θ 角增加,进动频率也增加,这就说明作用在涡核上的力,除升力 p_n 外,还有质量力。不过,由图 11 可知,在同一流量下,θ 从 15° 增至 60°,进动频率 f 增加

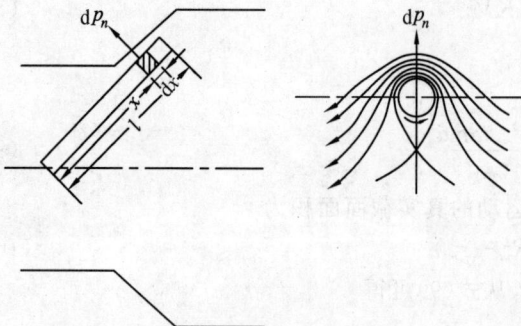

图 10　回转力矩的产生

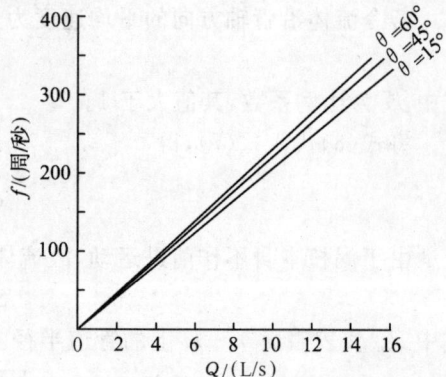

图 11　扩大角 θ 对进动频率的影响

10%左右。所以我们设作用在涡核上的总外力为 $k_0 p_n$（k_0 为考虑质量力而引进的一个常数）。根据图 11 的试验结果，$k_0 = 1.1 \sim 1.2$。

由图 10 得回转力矩

$$M = \int_0^l x \mathrm{d}(k_0 p_n) \tag{14}$$

令

$$\mathrm{d}p_n = \rho \Gamma V \mathrm{d}x$$

式中速度环量

$$\Gamma = 2\pi C_1 = 2\pi r_{m0}^2 \omega$$

充满扩大段的流体与涡核的相对速度

$$V = x \dot{\psi} \sin \theta$$

于是有

$$M = \int_0^l x k_0 \rho (2\pi r_{m0}^2 \omega)(x \dot{\psi} \sin \theta) \mathrm{d}x$$

积分并化简得

$$M = \frac{2\pi}{3} k_0 \rho r_{m0}^2 l^3 \omega \dot{\psi} \sin \theta \tag{15}$$

将式(15)代入式(13)，化简得

$$\dot{\psi} = \frac{(8k_0 l^2 - 6r_{m0}^2)}{(6r_{m0}^2 - l^2)\cos\theta} \omega \tag{16}$$

令 $l/r_{m0} = \lambda$，并将式(12)代入式(16)，化简得

$$\psi = \frac{K(8k_0\lambda^2 - 6)\tan\alpha}{(6 - \lambda^2) r_{m0} S_0 \cos\theta} Q \tag{17}$$

由于进动频率 $f = \dfrac{\psi}{2\pi}$，故有

$$f = \frac{K(4k_0\lambda^2 - 3)\tan\alpha}{(6 - \lambda^2)\pi r_{m0} S_0 \cos\theta} Q \tag{18}$$

对一台具体的流量计来说，上式右边除流量 Q 外都可看成常数，故可令

$$\frac{K(4k_0\lambda^2 - 3)\tan\alpha}{\pi(6 - \lambda^2) r_{m0} S_0 \cos\theta} = \xi \tag{19}$$

我们把 ξ 称之为仪表常数，于是得

$$f = \xi Q \tag{20}$$

式(18)或式(20)即流体在旋进型旋涡流量计内进动规律的数学表达式。它们是本流量计的另一基本公式。该公式指明了进动频率和流量之间的线性关系；揭示了各结构参数和运行参数对进动频率的影响；还表明对一台具体的流量计而言（螺旋角 α、扩大角 θ 等结构参数一定），输出频率只与容积流量有关，而与流体的粘度和密度无关。这两个公式表达的内容，完全为高速照片、校准曲线和变结构参数试验所证实。

5 结 语

(1) 我们利用高速摄影建立了流体在旋进型旋涡流量计内运动的物理模型。同国外提出的同类模型相比，本模型较精确、完整地描述了旋涡的形成、发展和进动的全过程；并纠正了国

外模型中的若干错误。

（2）在所建立的物理模型的基础上，我们对流体在本流量计内的运动规律进行了理论分析。理论分析的结果，不仅说明了物理现象之间的联系，而且对改善本流量计指明了方向。理论分析的结果不仅同高速照片相吻合，而且同校准曲线、变结构参数试验很好地相吻合。

（3）我们已完成的试验和理论研究工作，弄清了本流量计的一些主要内部机理和基本进动规律。还有一些问题尚不十分清楚，如螺旋叶片出口处的边界层分离、流量计内部的流场等，这些问题有待今后作进一步的试验和理论研究。

参 考 文 献

［1］上海工业自动化仪表研究所. 流量测量技术与仪表译文集［G］. 1972(4)：10.

［2］李德桃. 测量内燃机空气流量的新仪器［J］. 内燃机，1977(17)：64.

［3］Chanaud C R. Observations of the Oscillatory motion in certain swirl flows［J］. *J Fluids Mech*，1965(1).

［4］Киясбейли А Ш，Перелштей М Е. Вихревые счетчики-расходомеры［J］. *Машиностроение*，1974：30-35.

［5］佐鳥聡夫. スロールメータとるの応用［J］. 計測技術，1973.

［6］種子田定俊. 流体運動の Visualization［J］. 日本物理学会志，1968，23(6).

［7］Prantl L. 流体力学概论［M］. 郭永怀，译. 北京：科学出版社，1974.

［8］Фабрикант Н Я. *Аэродинамика*［M］. M Наука，1964：282.

［9］铁摩辛柯 S，Yang D H. 高等动力学［M］. 北京：科学出版社，1962.

（本文原载于《农业机械学报》1979 年第 2 期）

An investigation on the fluid motion in precession-type swirl-flowmeter

Abstract：Swirl-meter is a new flowmeter. It is called one of the ideal flowmeters in the last years.

In this article，the fluid motion in a flowmeter of this type was studied by highspeed photography. A more complete，more exact and more concrete physical model of fluid flow is proposed.

Two basic formulas are derived through theoretical analysis of the model. The relation between vortical intensity and geometric as well as moving parameters is expressed by one of the formulas while the law of precession of vortical nucleus is shown by the other one.

汽油机高压缩比快速稀燃系统及其爆震控制

黎　苏,黎晓鹰,韩显华,张泰民,李德桃

[摘要]　本文介绍了一种汽油机用高压缩比快速稀燃系统及其爆震控制点火系统。通过试验研究,对该系统的稀燃特性、爆震特性以及经济、动力性能等进行了分析。

汽油机采用高压缩比快速稀燃系统是改善其燃油经济性、降低有害排放的重要途径。由循环的热力学分析可知,提高压缩比和使混合气稀薄化可有效地提高汽油机的热效率[1]。同时,提高压缩比可缩短滞燃期、改善初期燃烧稳定性并扩大稀燃范围。但是,提高压缩比受到爆震的限制,特别是在低速大负荷区,存在扭矩下降的问题;而且燃烧稀混合气又受到着火与燃烧稳定性以及火焰传播速度下降的限制。本文仅介绍针对 92 型汽油机开发的高压缩比稀混合气燃烧系统、爆震控制点火系统以及所进行的试验研究工作。

1　高压缩比快速稀燃系统

为了解决高压缩比和稀燃带来的问题,我们研制了缩短火焰传播距离、增加挤气面积和减小挤气间隙的紧凑型燃烧室[2],并通过改进进气道形状,获得适当的紊流强度的方法,实现了高压缩比快速稀燃化。紧凑型燃烧室的结构如图 1 所示,燃烧室中绝大部分混合气集中在排气门下方的主燃室内,而排气门处温度很高可促进燃烧,温度较低的进气门则处于末端混合气区有利于冷却。紧凑型燃烧室与 492 型汽油机原浴盆形燃烧室相比,火焰传播距离缩短了 10.7%,挤气面积由 19% 增至 53%。由于挤气面积增大,所以挤气间

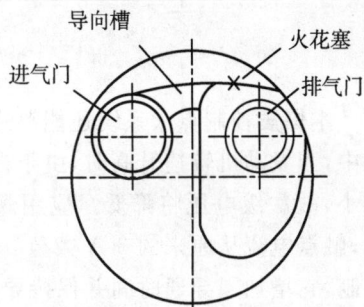

图 1　紧凑型燃烧室结构简图

隙成为重要的设计指标。本燃烧系统的挤气间隙为 1.15 mm(原浴盆形为 1.5 mm)。为了实现快速稳定稀燃,还必须使充气运动获得适当的涡流强度,充气所需的适当涡流强度是采用小型螺旋气道来实现的。有关紧凑型燃烧室燃烧性能的研究结果详见文献[3]。

快速稀燃系统采用了进气管电子燃油喷射系统,它具有燃油计量精确、稳定性高、雾化质量好等特点。为了减小进气阻力,提高充气效率,系统去掉了原化油器的大、小喉管,只保留节气门,以便对进气量加以控制。

2　爆震控制点火系统

高压缩比汽油机在全负荷下的爆震除受到燃烧系统和空燃比影响外,还受到点火特性的影响。选择汽油机压缩比和点火控制的有效方法是:提高压缩比,使其在小负荷下燃油经济性

显著改善,而在大负荷下只产生轻微爆震。这样的点火控制系统能够通过图2中的爆震控制器来检测爆震信号,经过处理后来控制点火执行机构推迟点火提前角以消除爆震[4]。由此可允许汽油机采用更高的压缩比,提高其使用经济性和动力性。

图2所示的爆震控制点火系统主要由爆震传感器和爆震控制器组成。爆震传感器的作用是检测爆震时产生的特殊频率的振动信号,将其转换成电信号后分别输入爆震控制器的检波电路和电压保持电路,两者的输出通过比较器进行比较,发生爆震时比较器输出为1(否则输出为0),通过判别电路使延时电路延时推迟点火角。本系统采用电压保持电路和比较器,对上一次爆震和本次爆震的强度进行比较,以判断爆震的发展趋势。推迟点火后若爆震消失,则延时电路逐渐恢复初始状态,点火提前角也恢复到原先的设定值。再发生爆震时再一次推迟点火,从而控制点火角在爆震界限附近波动。它利用判断输入信号频率和幅值的方法判断是否产生爆震。由于具有幅值判别功能,故可以选择允许的爆震强度。判别电路还被用来消除点火脉冲干扰及其他用电设备产生的随机干扰。

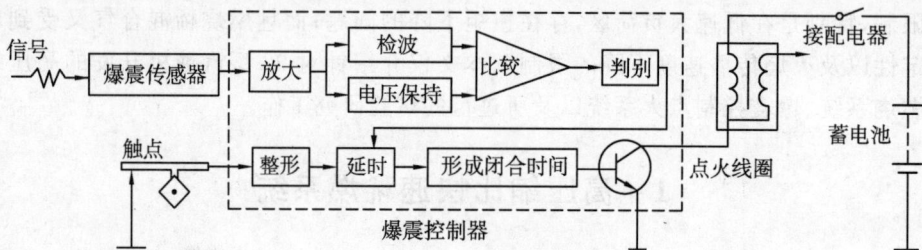

图 2　爆震控制点火系统

本爆震控制点火系统还能够克服传统机械式点火系统存在的一些缺陷。在机械式点火系统中,当发动机转速升高时,由于白金触点的相对闭合时间减小,使得点火线圈初级断电电流减小,故次级电压将降低。应用爆震控制点火系统后,白金触点只被用来产生控制点火的信号,触点电流从原来的 3 A 减至 0.1 A,触点不会烧蚀;点火线圈的通电则靠功率开关晶体管控制,它增加了导通时间并保持导通、截止周期恒定,故提高了次级电压;同时它允许通过大电流,可以用高能点火线圈,更有利于稀燃。此外,晶体管开关速度快,次级电压上升速度高,对火花塞积炭不敏感。

3　试验结果及分析

试验是在化油器式 92 单缸机基础上,采用高压缩比紧凑型燃烧室缸盖(压缩比为 10.26)、爆震控制点火系统和电子燃油喷射系统后进行的,试验使用 70 号汽油。为了便于说明,以下称文献[2]中不带爆震控制点火系统和电子燃油喷射系统的化油器式紧凑型快速燃烧室机为原机。

3.1　喷油时刻的选择和稀燃特性分析

首先进行了喷油时刻调整特性试验来选择最佳喷油时刻。图 3 所示是当发动机在 2 000 r/min、节气门全开、空燃比 $A/F=18.8$ 时的喷油时刻调整特性。由图可见,喷油时刻对发动机的动力性和经济性均有较大影响,当在排气上止点前 60° CA 时开始喷油,发动机的动力性和经济性最好；而在膨胀下止点前 90° CA 开始喷油时，发动机的动力性和经济性最差。以下试验均按最佳喷油时刻进行。

为了分析快速稀燃系统的稀燃性能，并确定一合适的混合气浓度，作为性能试验时控制混合比的依据，做了各种工况下的燃料调整特性试验。图 4 所示的是节气门全开、转速为 2 000 r/min 时的燃料调整特性。可以看出，快速稀燃系统的稀燃界限大大扩展，在 $A/F = 22.8$ 时发动机仍能稳定运转。快速稀燃系统的燃油消耗率在较大的空燃比范围内均保持较低水平，A/F 从 15.4 到 20 之间其燃油消耗率均低于 300 g/(kW·h)；最低油耗点在 $A/F = 18.1$，较原机 ($A/F = 17.6$) 增大。

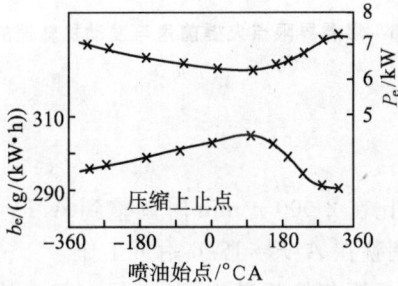

图 3 喷油时刻调整特性

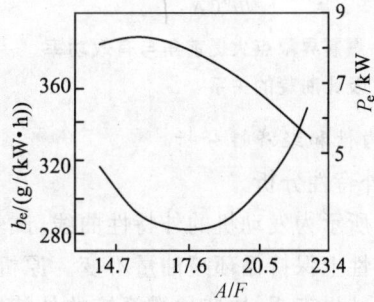

图 4 燃料调整特性

快速稀燃系统稀燃界限扩大的主要原因是由于采用了高压缩比紧凑型燃烧室结构，缩短了火焰传播距离，获得了较强的压缩涡流，并组织了适当强度的进气涡流，从而大大提高了火焰传播速度。同时，由于采用了汽油直接喷射，改善了混合气形成，进一步提高了稀燃性能。

节气门全开，各种转速下的燃料调整特性试验结果表明：$A/F = 18 \sim 19.3$ 时快速稀燃系统的燃油消耗率较低，且其最大功率与原机使用 $A/F = 17.6$ 时基本相同。因此，在以下的性能试验中以 $A/F = 18 \sim 19.3$ 作为所控制的混合比。

3.2 爆震特性

应用爆震控制点火系统后，我们可以容易地得到快速稀燃系统在各种混合气浓度下，出现爆震时的点火提前角（以下称爆震界限点火提前角）。在不同空燃比下，爆震界限点火提前角 θ_j 与有效功率 P_e 及比油耗 b_e 之间的试验关系见图 5。由图 5 可见，快速稀燃系统因压缩比较高，所以在其达到爆震界限点火提前角时发动机尚未达到最大功率（特别是在空燃比较小的情况下）。随着空燃比增大，爆震界限点火提前角明显增大。

快速稀燃系统的爆震界限点火提前角 θ_j 与发动机转速 n 的关系如图 6 所示。图中还给出了传统分电器点火提前角随转速的变化情况。从图中可以看出，快速稀燃系统的爆震界限点火提前角随发动机转速升高而增大，但其增加的速度比分电器离心调整机构所提供的点火提前角增加速度要小。这表明传统分电器的离心点火调整机构和快速稀燃系统不能匹配，即随着转速升高其点火提前角增加太快。本试验应用爆震控制点火系统有效地防止了发动机在全负荷高速时发生爆震，同时在部分负荷时可以适当加大点火提前角，进一步提高燃油经济性。由此满足了快速稀燃系统对点火的要求，提高了发动机运转的稳定性。

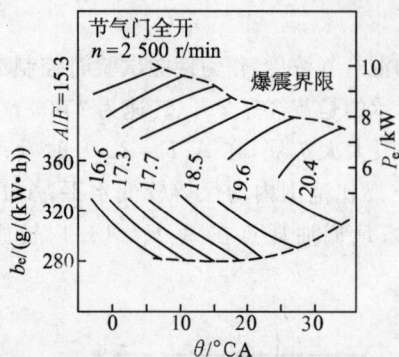

图 5　爆震界限点火提前角与有效功率
　　　及比油耗的关系

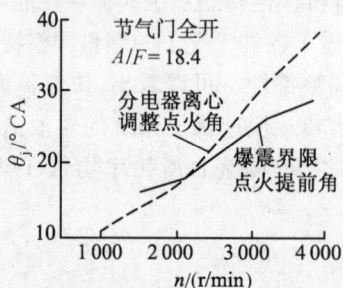

图 6　爆震界限点火提前角与发动机转速的关系

3.3　动力性和经济性分析

3.3.1　外特性分析

图 7 所示为发动机的外特性曲线,试验时空燃比在 2 000 r/min 时调整到 $A/F=18.8$,在整个外特性上保持循环喷油量不变。原机化油器调整在 $A/F=17.6$ 附近工作。

试验结果表明,快速稀燃系统的外特性功率和扭矩与原机基本相同,但在整个外特性上,燃油消耗率均有所降低,平均约下降 3.4%,且比油耗曲线随转速变化平坦,在 1 600~3 000 r/min 范围内比油耗均低于 300 g/(kW·h)。

表 1 和图 8 分别示出了本快速稀燃系统与原机外特性上的点火提前角 θ 和燃烧放热率 $dx/d\varphi$。从中可见:快速稀燃系统的燃烧放热速率与原机基本相同。由于应用了爆震控制点火系统和使混合气进一步稀化,因此快速稀燃系统的点火提前角有所增大,从而提高了燃烧的及时性,确保其经济性指标的改善。但由于使用 70 号汽油,快速稀燃系统的点火仍较晚,放热尚不够及时;如使用高辛烷值汽油并加大点火能量,其经济性将继续获得改善。

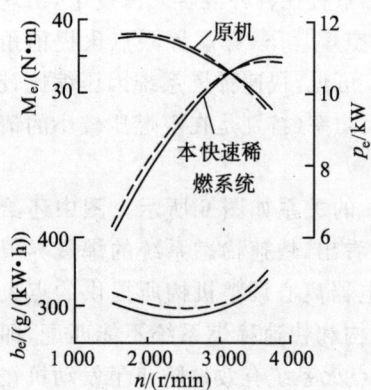

图 7　外特性曲线

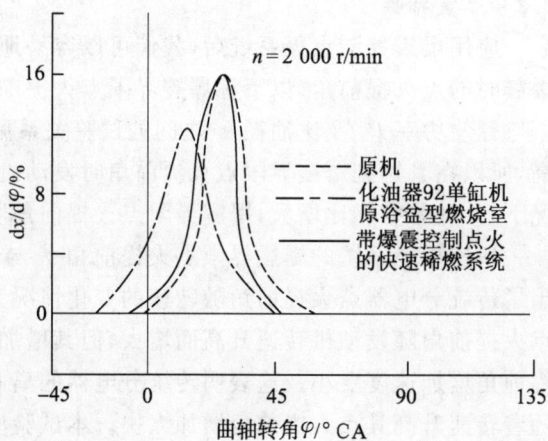

图 8　燃烧放热速率

表1 外特性上的点火提前角

转速 $n/(\text{r/min})$	点火提前角 $\theta/^{\circ}\,\text{CA}$	
	本快速稀燃系统	原快速稀燃系统
1 500	16.5	15
2 000	17	16
2 500	20	18
3 000	24	21.5
3 500	27	25.5
3 800	28	27

实测的快速稀燃系统外特性上连续 300 个循环的最高燃烧压力变动率如图 9 所示。在外特性上,最高燃烧压力循环变动率均低于 10%,随着转速的提高,循环变动率下降很快。

3.3.2 负荷特性分析

图 10 所示为 2 000 r/min 时的负荷特性。由图可见,本快速稀燃系统在各种负荷下的燃油经济性均较原机有所改善,功率在 2 kW 到 7.3 kW 之间,燃油消耗率平均降低约 5%;比油耗曲线随负荷变化平坦,特别在中小负荷燃油经济性获得较大改善。本快速稀燃系统的燃油经济性获得改善首先是由于使用爆震控制点火系统,改善了点火特性,提高了燃烧及时性。再者,采用了汽油喷射系统,燃油雾化状态得到改善,循环供油量的变动率降低,使混合气进一步稀薄化。

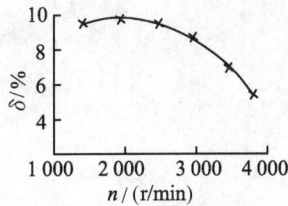

图 9 外特性上的循环变动率

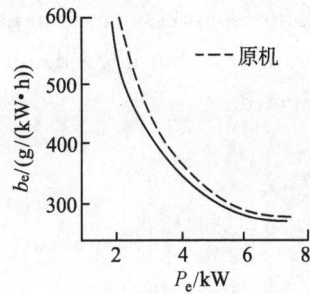

图 10 负荷特性

3.4 怠速排放

本快速稀燃系统在怠速工况下实测的排放指标如下:CO 排放量 0.3%,HC 排放量 290×10^{-6},远低于 GB3842—83 规定的 5%(CO)和 $2\,500 \times 10^{-6}$(HC)的怠速排放指标。

4 结 论

(1)采用高压缩比的紧凑型燃烧室结构,并组织适当强度的进气涡流和压缩挤流是实现快速稀燃技术的关键。本稀燃系统不同于以往的分层燃烧稀燃系统,它基本上不考虑混合气的分层,从而避免了燃烧分层在结构上的复杂性。试验证明,本稀燃系统在全部转速和负荷范围内均能达到稳定稀燃。

（2）快速稀燃系统应用汽油直接喷射可进一步提高稀燃系统的稀燃性能，但喷油时刻对发动机性能有较大影响。

（3）快速稀燃系统加上爆震控制点火系统，可以适当增大部分负荷的点火提前角，提高燃油经济性；有效地防止大负荷时发生爆震，提高了发动机运转的平稳性。如在此基础上实现电控高能点火和电控汽油喷射，必将进一步改善燃油经济性，降低有害排放，提高稀燃稳定性。

参 考 文 献

［1］Lean Combustion. A review[J]. *Automotive Engineering*，1984(2)；1984(3).

［2］黎志勤，等.汽油机用紧凑型快速燃烧室[J].内燃机工程 1985(3).

［3］黎志勤，等.492 汽油机燃烧系统改进的试验研究[J].北京汽车，1987(3)；1989(4)；1989(5).

［4］黎苏，等.高压缩比汽油机爆震的控制[J].汽车技术，1990(8).

（本文原载于《内燃机学报》1995 年第 4 期）

Rapid lean combustion system with high compression ratio for gasoline engines and detonation control

Abstract：The system consisted of a rapid lean combustion chamber with high compression ratio and a detonation suppressor used in gasoline engine is presented in this paper. The system performances of lean burning, detonation, fuel economy and power output etc. are tested and investigated.

关于建立和完善我国汽车排放法规
若干问题的探讨和建议

史绍熙,李德桃,郑　杰,吕兆华,龚金科

[摘要]　根据作者在国内外从事汽车排放研究工作的体会,论述了我国建立汽车排放法规的重要性和紧迫性;阐明了对汽车排放标准和法规及其制订和完善过程的意见;介绍了发达国家该法规的基本情况。在此基础上,为尽快制定和完善我国汽车排放法规提出了若干具体建议。

汽车排放对生态环境、人类健康和经济发展的严重影响,已经受到世界各国政府和人民的普遍关注。发达国家为解决此问题经过了多年努力,制定了行之有效的法规,并已取得了显著效果。然而大多数发展中国家,由于种种原因,这一问题仍没有得到应有的重视和解决。

根据美国的研究报道,大气中的污染物约 50% 来自汽车排放[1]。最新的研究表明,汽车的排放物质达 140 种之多[2],如碳氢化合物(HC)、一氧化碳(CO)、二氧化碳(CO_2)、氮氧化物(NO_x)、醚、甲醛(HCHO)等。其中有些是易挥发的有机物(VOC),这些物质与氮氧化物(NO_x)发生光化学反应形成地面臭氧层(Ozone),造成对人及其他动植物的危害。柴油机排放微粒中的多环芳烃(PAHs)如苯并 a 芘(Benzo(a)pyrene)是致癌物质,直接威胁人的生命,从而更加引起了人们对汽车排放的重视。

我国在解决汽车排放方面做了一些工作,但研究的深度和广度,尤其是对建立适合我国国情的排放法规的研究,远远不能适应蓬勃的经济和社会发展的需要。我国至今尚无一部比较完善的汽车排放法规,而另一方面,我国车辆的增加速度和大气污染程度却是惊人的。这就要求我们对于解决这一关系千秋万代问题的紧迫性有充分的认识。可以说,汽车排放立法和为此进行科学技术研究已到了刻不容缓的地步,否则,后果的严重性是难以估量的。

本文作者,有的在欧洲、日本和美国进行国际合作研究时涉及汽车排放问题,有的在美国对汽车排放进行了多年的研究,都非常关心我国汽车排放法规的研究和制定工作。现将我们共同讨论和研究的部分结果写成此文,期望能为尽快建立有中国特色的汽车排放法规起到一定的促进作用。

1　对汽车排放法规的基本认识

所谓"车辆排放",从广义上说,包括汽车、摩托车、工程车辆、拖拉机、机车、割草机等动力装置的排放物。当前车辆排放主要针对汽车而言,而对重型卡车、轻型车、轿车等不同种类的车辆,有其不同的排放标准和法规。

排放不仅指排气管尾气的排放,而且也包括各种动力装置的加油孔、机油滤清器、油箱等所挥发排放出来的物质,即所谓蒸发排放。总之,车辆排放是指使用车辆时,排放到大气中的所有有害物质。根据现阶段科技发展的水平,我们还只能对其中的某些物质,如 CO,HC,NO_x 和微粒等研究和制定检测标准和法规。有些物质,如甲醛等即将制定标准,还有些物质尚待我们去分析、研究和认识,还谈不上为它们制定检测标准和法规。

使用车辆已成为当代人类生活、生产、交流和社会发展的不可少的一部分,车辆排放问题也就成了人们不能回避的问题。解决得好,可获巨大的效益,解决得不好,则需付出昂贵的代价,早解决要比晚解决好。以美国洛杉矶地区为例,在控制空气污染方面花费 27.9 亿美元,但可获得节约医疗费 94 亿美元的效益。当然,车辆排放立法,不仅是净化空气所必需,也是发展生产(尤其是车辆生产)、繁荣经济和改善交通所必需。据估计,我国汽车的单车排放量为发达国家的数十倍,有些排放物甚至可达数百倍。作为我国支柱产业之一的汽车工业现在正面临"入关"后的国际竞争局面,如果我们不加紧研究和制定汽车排放法规,并与国际相应的法规"接轨",则我们将无法参与这一竞争。另一方面,由于我国无自己的排放法规,这不仅不能对进口车辆的排放加以检测和控制,而且进口车辆的排放控制装置也随之被出口国取消了。总之,车辆排放无法可依的局面再也不能继续下去了。

车辆排放标准,按照发达国家的惯例,是指车辆在实际行驶和操作条件下的排放标准。制定该标准,需要在实验室模拟不同的道路、车辆种类、燃料类别、车速变化历程等具体条件,进行排放物质的测量和分析。故这种测量属于动态的多参数测量,涉及一些高新技术。它既不同于企业为产品出厂而进行的静态排放的测量,也不同于研究发动机排放特性和原理而进行的测量。而且,各国根据本国国情所建立的排放检测车辆行驶模拟模式也是不同的(图1和图2)[3]。

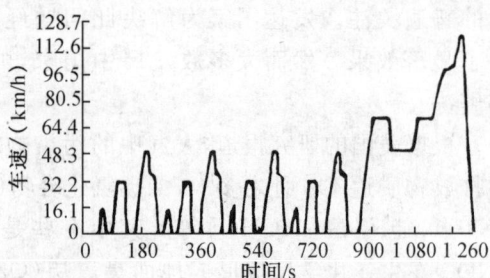

图 1　欧洲车辆行驶模拟模式

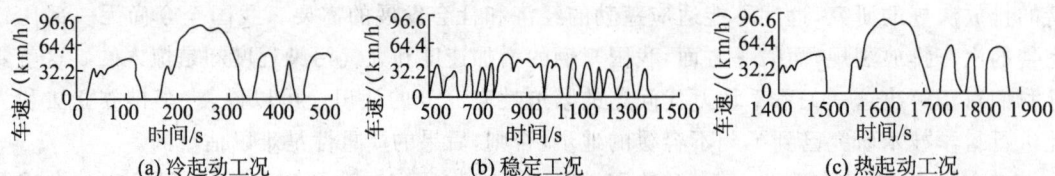

(a) 冷起动工况　　(b) 稳定工况　　(c) 热起动工况

图 2　美国车辆行驶模拟模式(FTP75)

车辆排放法规不同于一般法规,它有很强的科学性,必须建立在科学实验的基础上。在制定和不断完善的过程中,要经常进行测试;要设计和研制相应的测试装置;要解决测试中的各种科学技术问题;要研究各种条件下的排放模式;要建立数据库;还要结合国家有关政策和地区条件进行综合分析。在此基础上,制定排放标准,并逐步形成法规。立法后,仍要根据油料、道路和车辆等诸多因素的变化,以及执行过程中出现的问题不断地补充、修正和完善。因此可以说,车辆排放法规是自然科学技术、社会科学和国家政策相结合的研究成果,它反映一个国家和地区的科学技术水平、发展程度和人民的生活质量,是一项大的系统工程。

2 发达国家的汽车排放法规现状和发展

西方发达国家在工业革命完成后,生态环境受到严重破坏,各国相继投入巨大的人力和物力控制环境环染,保护生态平衡,制定了国家的环境保护法,其中包括车辆的排放法规。

这里主要以美国为例,来说明发达国家车辆排放法规的制定情况。因为,相对来说,美国的车辆排放法规是比较完整和系统的,它不仅制定了联邦的车辆排放法,而且各州政府还根据当地情况制定适合本州的车辆排放法规。最著名的是加州大气保护局(CARB)制定的车辆排放法规。它已被美国联邦政府和许多国家制定相应法规时参照和引用。

早在 1970 年,美国众议院修改了"净化空气法案"(CAA),提出要保护和提高国家大气的品质。其中要求车辆排放的一氧化碳(CO)和碳氢化合物(HC)在1970 年标准的基础上减少90%,而氮氧化物(NO_x)的排放也要有相应的减少。1990 年众议院再度修正"净化空气法案",并形成了正式的"净化空气修正案"(CAAA),同年 11 月经总统签署,正式成为国家法律。图 3 表示了根据联邦车辆排放法要求,以 1979—1980年为基准,到 2000 年左右,3 种排放物逐年减少量的对比[4]。

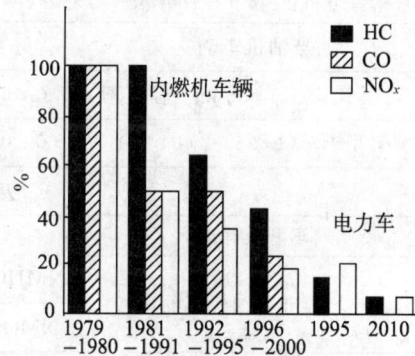

图 3　美国联邦车辆排放法规要求

在联邦政府推动下,各州也根据本地区大气污染情况,相应制定和准备制定本州法案。加州有几个地区地面臭氧水平较高,特制定了"加州净化空气法案"(CCAA),这个法案要求加州大气保护局采取一切必要的措施控制车辆排放,到2000 年把 HC 在现有水平上减少55%,把 NO_x 减少15%,其他排放物也要尽可能减少到最低水平。根据这一要求,加州大气保护局开展了一系列控制和减少车辆排放的研究项目。在此基础上形成了加州一整套排放法规。这套法规比联邦排放法规既严格又在某些方面具有灵活性。因而,目前,美国许多州都准备采用加州的排放法规,德克萨斯州(Texas)就是其中之一[5]。

表 1 中列出了美国联邦和加州有关轻型车辆的排放法规的标准(1991—2003 年)[6]。从1994 年开始实施,实施的前 3 年为过渡阶段。第 1 年(1994 年),出厂的新车的 40% 必须达到此标准;第 2 年(1995 年),80% 的新车必须达到此标准;第 3 年(1996 年)之后,100% 的新车达到此标准。而在第 2,3 年不能达到此标准的车辆,联邦政府也另有标准,采用表 1 中的"标准零"。同时,联邦政府还对使用中的车辆也制定了明确的排放标准。

在表 1 所列加州标准中,还列有 1990 年加州大气保护局首先提出的低排放、超低排放及零排放等标准。以轻型车辆为例,低排放车辆法案创造性地提出了车辆排放的 4 个新的分类:

(1) 过渡低排放车辆(TLEVs);

(2) 低排放车辆(LEVs);

(3) 超低排放车辆(ULEVs);

(4) 零排放车辆(ZEVs)。

在标准中,引人注目的变化有:(1) 非甲烷有机物(NMOG)或非甲烷碳氢化合物(NMHC)

表1 美国车辆排放标准

美国联邦车辆排放标准

车型及标准	A. 5年或8.045万公里				B. 10年或16.09万公里			
轻型(≤1 698 kg)非柴油机车	THCs	CO	NO_x					
标准零	0.255	2.11	0.62					
标准Ⅰ	NMHC	CO	NO_x	PART.	NMHC	CO	NO_x	PART.
轻型卡车(≤1 698 kg)、轻型车	0.16	2.11	0.25		0.19	2.61	0.37	
轻型卡车(1 698~2 604 kg)	0.20	2.73	0.44		0.25*	3.42*	0.60*	
柴油机车辆								
轻型卡车(≤1 698 kg)、轻型车	0.16	2.11	0.62	0.05	0.19	2.61	0.78	0.062
轻型卡车(1 698~2 604 kg)	0.20	2.73		0.05	0.25	3.11	0.60	0.062

美国加州车辆排放标准

车型及标准	A. 8.045万公里				B. 16.09万公里			
轻型非柴油机车辆	NMHC/ NMOG	CO	NO_x	HCHO	NMHC/ NMOG	CO	NO_x	HCHO
轻型卡车(≤1 698 kg)、轻型车	0.16	2.11	0.25	9.32	0.19	2.61	0.37	
过渡低排放车辆(≤1 698 kg)	0.075/ 0.078	2.11	0.25	9.32	0.094/ 0.097	2.61	0.37	11.19
低排放车辆(≤1 698 kg)	0.045/ 0.047	2.11	0.12	9.32	0.05/ 0.056	2.61	0.187	11.19
超低排放车辆(≤1 698 kg)	0.024/ 0.025	1.06	0.12	4.97	0.033/ 0.034	1.31	0.187	6.84
零排放车辆	0	0	0	0				
	NMHC	CO	NO_x	HCHO	NMHC	CO	NO_x	HCHO
轻型卡车(1 698~2 604 kg)	0.20	2.73	0.62	11.19	0.25	3.42	0.56	11.19

	A. 8.045万公里				B. 19.30万公里			
	NMOG	CO	NO_x	HCHO	NMOG	CO	NO_x	HCHO
过渡低排放车辆(1 698~2 604 kg)	0.099	2.73	0.44	11.19	0.12	3.42	0.56	14.29
低排放车辆(1 698~2 604 kg)	0.062	2.73	0.25	11.19	0.08	3.42	0.31	14.29
超低排放车辆(1 698~2 604 kg)	0.031	1.37	0.25	5.59	0.044	1.74	0.31	8.08

NMHC：非甲烷碳氢化合物(g/km)　　NO_x：氮氧化物(g/km)
NMOG：非甲烷有机化合物(g/km)　　PART：微粒(g/km)
CO：一氧化碳(g/km)　　HCHO：甲醛(mg/km)
* 为19.308万公里数据。

的标准代替了传统的碳氢化合物总量(THC)的标准。这是由于当前研究表明,在车辆排放的碳氢化合物中,甲烷并不对地面臭氧层的形成起重大作用,因而在碳氢化合物中暂不把甲烷作为排放标准。(2)首次提出了甲醛(HCHO)的排放标准。甲醛与车辆排放的其他物质相比,含量较小,但由于甲醛是重要的致癌物,所以加州的标准首先把它引入了排放法规。此外,在4种车辆排放分类中,提出的低排放车辆(LEVs)标准将于1997年开始实施,当年加州25%的新车必须达到这一标准,以后逐年增加。超低排放车辆(ULEVs)标准也将于1997年开始实施,当年加州2%的新车必须达到这一标准,以后逐年增加。分类中的零排放车辆(ZEVs)是指这些车辆的4种排放的含量为零。就目前的科学技术水平而言,只有电动车辆和太阳能车辆能达到此水平。但加州规定,到1998年底,零排放车辆至少应占当年总销售车辆的2%,到2003年,至少应占10%。

加州的这套排放法规的实施预期将使该州车辆排放的水平低于联邦法规控制的车辆排放的水平。

表2为德克萨斯州休斯敦地区车辆排放物分别按联邦标准和加州标准模拟计算的结果[5]。由表中可见,用加州低排放标准计算出来的轻型车辆每天的排放物要低于用联邦标准计算出来的结果。

表 2　美国休斯敦地区排放模拟计算

A. 用美国联邦标准计算休斯敦地区轻型车辆(3 850.5 kg)每天3种关键排放物的总重量				
	1990 年	2000 年	2005 年	2010 年	2015 年
NMOG(吨/天)	224.25	82.56	67.96	66.09	70.70
NO$_x$(吨/天)	212.95	195.86	178.58	185.02	201.10
CO(吨/天)	2 421.53	1 145.91	980.77	995.78	1 077.84
B. 用加州低排放车辆标准计算(最大值)休斯敦地区轻型车辆每天3种排放物的总重量				
	1990 年	2000 年	2005 年	2010 年	2015 年
NMOG(吨/天)	224.25	80.54	61.62	55.59	59.92
NO$_x$(吨/天)	212.95	191.48	163.07	157.76	164.90
CO(吨/天)	2 421.53	1 117.58	870.79	806.64	832.15

美国联邦政府和州政府已在实践中严格实施了车辆排放法规,并已取得了净化空气的显著成绩。但随着人们对生态环境和健康水平要求的提高,政府拟制定更严格的排放标准和法规,车辆制造业和燃料部门也在研究进一步的达标技术,如采用天然气作为一种代用燃料就是一例。

当前,美国正加强车辆代用燃料,如天然气、甲醇等的研究[4,7],表3中为一项天然气(CNG)及汽油对Voyager微型客车做排放对比试验的结果。试验不仅对常温(24 ℃)下的排放物进行了对比测试,而且还对低温运行(−5 ℃和−20 ℃)的状况下的排放物进行了对比测试。结果表明,在几种温度试验条件下,以天然气为燃料的排放水平大大低于以汽油为燃料的排放物水平。

表 3　天然气与汽油的排放物对比

		排放物/(g/km)				
		CO	CO$_2$	NO$_x$	THC	NMHC
24 ℃	天然气	0.156	229.3	0.025	0.068	0.006
	汽　油	1.21	302.7	0.14	0.174	0.14
	减少百分比	87%	24%	83%	61%	96%
−5 ℃	天然气	0.224	257.9	0.04	0.099	0.006
	汽　油	3.06	333.1	0.21	0.447	0.39
	减少百分比	93%	23%	79%	78%	98%
−20 ℃	天然气	0.21	289.6	0.06	0.08	0.006
	汽　油	5.87	356.1	0.39	0.92	0.83
	减少百分比	96%	19%	84%	91%	99%
超低排放标准		1.74		0.31		0.044*

* 为 NMOG(非甲烷有机物)

多项研究结果都已表明,天然气是一种比较有前途的代用燃料,它不仅可以减少对大气环境的污染,而且经济性也优于汽油。目前,美国联邦政府和州政府正制定和完善相应的代用燃料的排放法规,并着手为这些有前途的代用燃料提供配套的设施,诸如建立加气站,研究代用燃料的排放控制装置,修订、完善代用燃料排放的试验方式,等等。

此外,为更深入的认识车辆排放,以及寻求低排放的燃油,美国 3 家大汽车公司(通用、福特和克莱斯勒)和 14 家主要的石油公司(包括美孚等)于 1989 年 10 月发起了一个称为"车辆燃油及大气质量改善研究项目(AQIRP)"。该项目采用了更先进的试验技术和设备,对使用不同成份的燃油所形成的车辆排放物进行了更科学的分析研究,以便得到更新的控制技术,达到更严格的排放标准。

3　建立有中国特色的汽车排放法规的几点建议

3.1　建立国家级车辆排放实验室

如前所述,制定车辆排放法规是一个非常复杂的过程。随着城市、能源、交通事业的发展和环境污染状况的变化,该法规还要不断补充、修正和完善,所有这些工作,都必须建立在科学实验的基础上,并有其独特的目的、内容和分析方法。另一方面,排放法规涉及到不少的产业、部门和学科领域,还要研究和参照国外排放法规的变化和发展,因此必须设有专门的重点实验室来承担这一任务。只有这样,才能保证我国车辆排放法规加速制定,并随着实践的发展不断地完善。

以美国为例,制定车辆排放标准和法规的数据、信息都来源于这样的实验室。这些实验室都有非常先进的测试设备,训练有素的工作人员和各具特色的实验程序。美国联邦车辆排放法规和各州的有关法规,就是依据这些实验室不断提供的数据来制定、补充、修正和完善的,国家和地区用这些数据和有关信息作为制定环保政策和进行投资的重要依据,并对社会的发展进行某种预测。

3.2　建立车辆排放法规基金制

车辆排放法规有利于改善生态环境、人类健康和发展生产,但一部排放法规的制定和完善,需要一定数量的研究与活动经费。建立基金制来支持排放立法是一种行之有效的经验和途径。美国仅为制定天然气汽车排放标准,国家能源部、德州交通局和有关公司,向德州代用燃料实验室提供了约 200 万美元的实验和研究费。这些经费对加速制定有关标准和法规起了保证作用。根据我国国情,由国家、环保部门、汽车厂、石油部门等为车辆排放实验室联合提供筹建和运行经费,看来是势在必行的好措施。

3.3　各地区可以因地制宜制定自己的排放法规

我国幅员辽阔,各地区的自然环境、车辆种类及保有量、人口密度、大气污染程度等均有所不同,因此,各地在国家总的排放法规的指导和约束下,可以针对本地区的具体情况制定自己的排放法规。如前所述,加州排放法规,它比美国联邦的相应的法规更严格但又灵活。德州天然气汽车很有发展前途,因此,该州的有关实验室正在为制定天然气车辆的排放法规提供实验数据。国家车辆排放法规与各地区的相应法规互相补充,相互配套,形成一个较完整的车辆排放法规体系,这就更有利于促进我国大气环境的改善,生产的发展和人民健康水平的进一步提高。

3.4　建立车辆排放法规有效的执行和监督体制

我国各地环保部门应有分支机构专管车辆排放法规的执行,要严格执法,他们也要熟悉和研究车辆排放的有关问题。该机构要对人民负责,定期向相应级别的人民代表大会及其常务委员会汇报本地区车辆排放法规的执行情况和大气质量的状况,并定期予以公布。

4　结　束　语

我国面临车辆排放大气污染问题的严重挑战,首先要进一步理解这一问题的重要性和紧迫性,要吸取发达国家解决这方面问题的经验和教训,要结合国情作深入细致的调查研究,要尽快建立国家和省级实验室,为加速制定排放标准和法规提供科学实验的数据和信息。我们期望在各方面的协同努力下,尽早制定出一部有中国社会主义特色的车辆排放法规,并使其在实践中逐步发展和完善,为改善人民的生活和工作环境,提高生产水平,为改革开放,造福子孙而奋斗。

在本文的撰写过程中,美国德克萨斯大学发动机基础燃烧和汽车研究中心主任 Matthews R D 教授为我们提供了帮助和资料,在此表示衷心感谢!

参 考 文 献

[1] Garry Mauro. Why alternative fuels? Quality of life and economic development[J]. *Enviro Nomics*, 1994, 3(2).

[2] Byrns B R, Benson J D. Description of auto/oil air quality improvement research

program[J]. *SAE paper*, 912320,1991.

[3] Charlton S J, et al. An investigation of the emission characteristics of the passenger car IDI diesel engine, proc. of institution of mechanical engineers, combustion in engines [J]. *C IMech E*, 1992,C448/025.

[4] Roberta J Nichols. The challenges of change in the auto industry: why alternative fuels? alternate fuels, engine performance and emission[J]. *ICE*,1993,20.

[5] Pechan E H. Adapting the california low emission vehicle program in Texas—an evaluation[R]. *Report of Association*, Inc, 1993.

[6] Federal Register Part Ⅳ, Environmental Protection Agency, USA, 1992.

[7] Sinck J C A, Lynm R Helpard. *Living with a Natural Gas Vehicle* [M]. World Publishing Company, Inc, 1994.

（本文原载于《内燃机学报》1996 年第 2 期）

Study on establishment and improvement of emission regulations for motor vehicles in China

Abstract: The importance and urgency of establishing the emission regulations for motor vehicles in China are expounded on the basis of authors' experience, both at home and abroad. Essential understandings, formulations and improvements of the emission regulations and standards are presented. After introducing and discussing the main emission regulations of developed countries, this paper puts forward some specific proposals in order to facilitate the establishment of China's own motor vehicle emission regulations.

有机硝酸酯类柴油十六烷值改进剂的研究

董　刚，李德桃，吴志新，陈嘉生

[摘要]　研制了一种成分为脂环族硝酸酯的柴油十六烷值改进剂，并测定出这种改进剂的一系列理化性质。分析了有机硝酸酯缩短柴油着火滞燃期的化学动力学机理。实机研究了这种改进剂对柴油十六烷值和燃烧特性的改善效果，以及对柴油机燃油消耗量和烟度的影响。研究结果表明，脂环族硝酸酯作为柴油十六烷值改进剂具有很高的使用价值。

商品柴油中十六烷值较低的催化裂化柴油的调合比例不断增加以及原油质量的不稳定，使柴油的十六烷值具有降低的趋势。低十六烷值的柴油在使用中会带来着火滞燃期长，柴油机工作粗暴，排放增加，以及冷起动困难等问题。在柴油中加入十六烷值改进剂是提高柴油十六烷值，从而解决以上问题的经济、有效和简便的方法。因此，柴油十六烷值改进剂的研制是一个具有重要意义的课题。

柴油十六烷值改进剂以有机硝酸酯和有机过氧化物为主，其中过氧化物具有爆炸性，受到一定的使用限制，因而有机硝酸酯类十六烷值改进剂得到了迅速发展并已商品化。例如美国Ethyl公司生产的 DⅡ-2 有机硝酸酯类系列十六烷值改进剂得到了广泛应用；虽然国内研究起步较晚，但齐鲁石化公司研制的硝酸异辛酯类十六烷值改进剂也取得了一定效果[1]，填补了我国在这方面的空白。此外，国内外对柴油机十六烷值改进剂作用机理的研究也不充分，因而有必要进一步探讨。

本文提出了一种国内尚未研究的以脂环族硝酸酯为组分的柴油十六烷值改进剂，并给出了该剂理化性质的一些测试值。对有机硝酸酯的作用机理进行了化学动力学分析，实机测试了使用该剂改进柴油十六烷值及降低柴油机油耗率和烟度的效果，结果表明，这种十六烷值改进剂具有显著的使用效果。

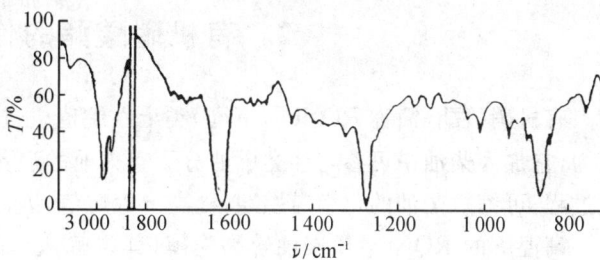

T-透过率　　$\bar{\nu}$-波数

图1　c-RONO$_2$ 的红外光谱图

1　脂环族硝酸酯的研制

采用硝硫混酸硝化的合成工艺对脂环族硝酸酯（简称 c-RONO$_2$，下同）进行了合成试验。图1给出了的合成的 c-RONO$_2$ 的红外光谱图，通过与文献[2]的图谱比较可确定该物质具有环状分子结构。文献[3]表明，这种结构具有较好的稳定性，不易水解和皂化，尤其适于船用柴油机的使用，因而具有独特的优点。

为初步制定该产品的质量标准,对 c-RONO$_2$ 的一系列理化性质进行了测试,其结果见表 1。

由表 1 可知,c-RONO$_2$ 作为十六烷值改进剂与柴油相比,具有粘度低、凝点低、饱和蒸汽压高、闪点接近等优点,因而对柴油雾化和燃烧前的物理准备过程较为有利。此外,c-RONO$_2$ 制备简单、可靠、成本低,与柴油具有良好的互溶性,对柴油机气缸无腐蚀作用,具有良好的理化特性。

表 1　c-RONO$_2$ 的理化性质测试结果

项　目	指　标	测试方法
外观	淡黄色透明液体	目测
纯度/%	≥98	气相色谱
密度/(g/cm^3)	1.102 4(20 ℃)	比重瓶法[4]
运动粘度/(mm^2/s)	1.673 0(20 ℃)	乌氏粘度计法[4]
沸点/℃	86~88(3.19kPa)	减压蒸馏法
凝点/℃	<−20	毛细管法
折射率	1.453 7(24 ℃)	阿贝折射仪
饱和蒸汽压/kPa	2.51(70 ℃)	
	4.31(80 ℃)	静态法[5]
	6.29(90 ℃)	
汽化热/(kJ/kg)	271.0	静态法[5]
闪点/℃	70(闭合)	GB261—83 闭口闪点测试法

2　有机硝酸酯的作用机理

有机硝酸酯(简称 RONO$_2$,下同)类十六烷值改进剂的作用机理目前尚未完全明确,一般认为,它加入柴油中可参与烃类燃油分子着火前的退化分支反应,有效地改变了链式反应的引发过程,可缩短柴油的化学滞燃期,提高十六烷值,因而发动机起动容易、工作柔和。

高温下的 RONO$_2$ 比柴油容易裂解,其裂解速率的决定步骤是 RONO$_2$ 中 O—N 键均裂的可逆反应(0′),即

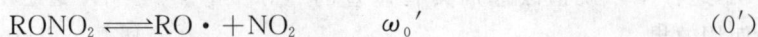

$$RONO_2 \rightleftharpoons RO· + NO_2 \qquad \omega_0' \qquad (0')$$
$$NO_2 + RH \longrightarrow HNO_2 + R· \qquad (0'')$$

式中 RO· 为烷氧自由基;RH 为烃类燃油分子;R· 为烷基自由基;ω_0' 为反应(0′)的裂解速率。

反应(0′)的活化能比柴油裂解的活化能低,反应容易引发,生成的 NO$_2$ 以极快的速率从 RH 中夺取 H 原子而生成 R·,从而导致柴油着火前发生退化分支反应。令柴油本身的裂解反应为

$$RH + O_2 \longrightarrow R· + HOO· \qquad \omega_0 \qquad (0)$$

式中 HOO· 为过氧化氢自由基;ω_0 为反应(0)的裂解速率。

由于其裂解速率 ω_0 要比式 $(0')$ 中的 $RONO_2$ 的柴油裂解速率 ω_0' 小,即

$$\omega_0' > \omega_0 \tag{1}$$

又根据谢苗诺夫的链反应理论,氧化反应初期的退化分支反应诱导期(即化学滞燃期)τ 可表达为[6]

$$\tau = \int_0^\eta \frac{d\eta}{\sqrt{\eta - (1/2 - \varpi)(1 - e^{-2\eta})}} \tag{2}$$

式中 η 为与退化分支反应浓度有关的无量纲参数;ϖ 为与裂解引发速率有关的无量纲参数。其中

$$\varpi = \frac{\omega_0 k_1}{k_1 k_2 [RH]} \tag{3}$$

式中 k_1 为链增长反应速率常数;k_2 为退化分支反应速率常数;$[RH]$ 为燃油浓度。

因此,将式(1)代入式(3)则有

$$\varpi_0' > \varpi_0 \tag{4}$$

式中 ϖ_0' 为加入 $RONO_2$ 的柴油中与裂解引发速率有关的无量纲参数;ϖ_0 为未加入 $RONO_2$ 的柴油中与裂解引发速率有关的无量纲参数。

将式(4)代入式(2)则有

$$\tau' > \tau \tag{5}$$

式中 τ' 为加入 $RONO_2$ 的柴油反应初期的退化分支反应诱导期。

以上分析表明:柴油中加入一定量的 $RONO_2$,可缩短柴油在焰前氧化反应中的化学滞燃期。

3 脂环族硝酸酯类十六烷值改进剂在柴油机上的应用研究

3.1 柴油着火滞燃期和燃烧特性

试验用机为湖南动力机厂 6105Q-1C 型直喷式柴油机,其技术规格见表 2。

表 2 试验用机的技术规格

项目	指标	项目	指标
型号	6105Q-1C		
型式	四冲程,水冷,直列	标定功率/kW	103
燃烧室型式	直喷式	标定转率/(r/min)	2 800
缸径/mm	105	活塞平均速度/(m/s)	11.2
行程/mm	120	压缩比	17
缸数	6	喷油泵型式	A 型

采用 EAS-900 型发动机分析系统,分别测试了在标定工况($n = 2\ 800$ r/min,$P_e = 130$ kW)和最大扭矩工况($n = 1\ 800$ r/min,$P_e = 73$ kW)下,0 号柴油(称基础油)与添加 0.1%(质量百分比,下同)$c\text{-}RONO_2$ 十六烷值改进剂的 0 号柴油(称加剂油)的示功图对比曲线。图 2 给出了试验测试系统的装置图。图 3 和图 4 分别给出了两种工况下的示功图对比曲线及压力升高率对比曲线。

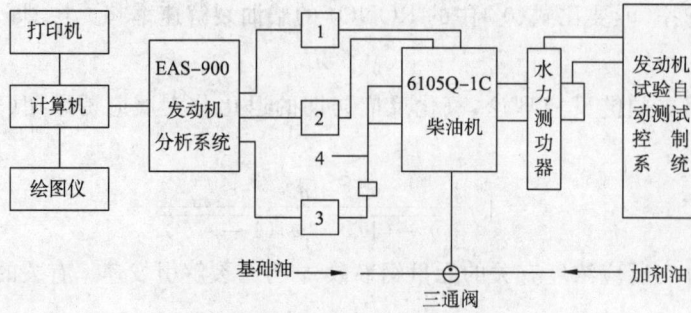

1—针阀升程仪　2—稳零电荷放大器　3—曲轴转角发生器　4—TDC 和角度信号光电盘

图 2　试验测试系统装置图

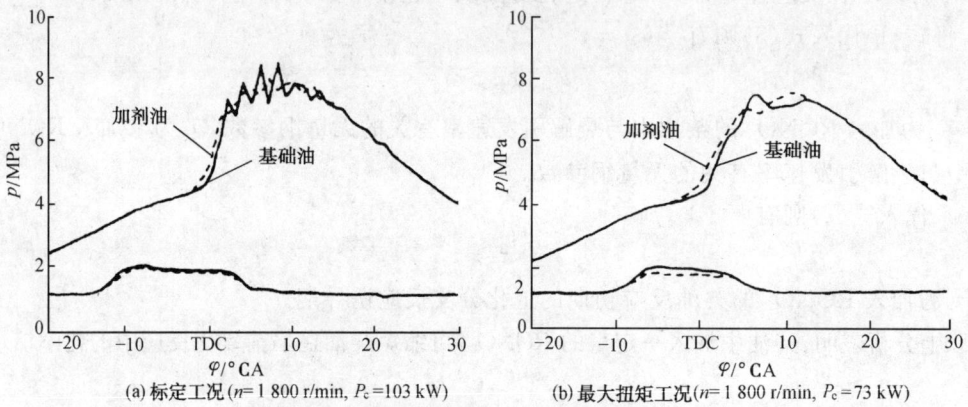

(a) 标定工况(n=1 800 r/min, P_e=103 kW)　　(b) 最大扭矩工况(n=1 800 r/min, P_e=73 kW)

图 3　不同工况下的示功图对比曲线

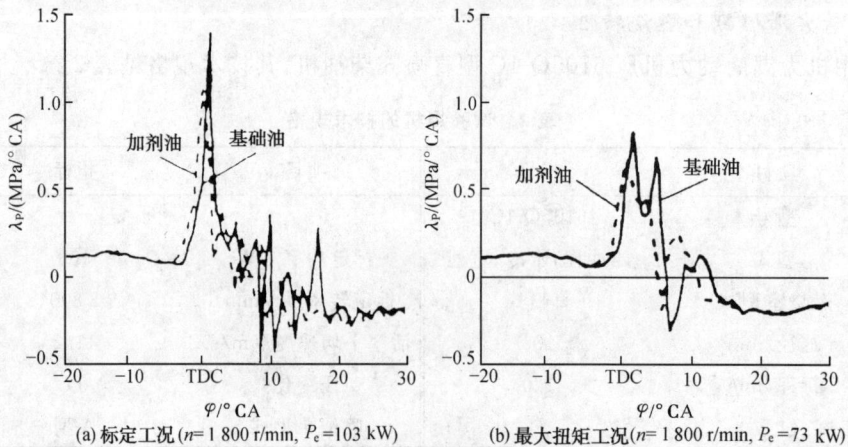

(a) 标定工况(n=1 800 r/min, P_e=103 kW)　　(b) 最大扭矩工况(n=1 800 r/min, P_e=73 kW)

图 4　不同工况下的压力升高率 λ_P 对比曲线

由此可见,加剂油的添加量为 0.1% 时,与基础油相比,其着火滞燃期在上述两种工况下均有明显缩短,最高燃烧压力降低,最大压力升高率以及燃烧振动幅度均有所下降。可以认为,c-RONO_2 加入柴油中提高了柴油十六烷值,柴油机起动容易,工作柔和,噪声减小,爆燃现象缓和。

3.2 柴油机油耗率和烟度

将 c-RONO$_2$ 分别以 0.06％和 0.1％加到柴油中,在 S195 涡流室式柴油机上进行负荷特性对比试验,以研究 c-RONO$_2$ 对柴油机油耗率和烟度的改善效果。图 5 给出了这两种添加量下,S195 柴油机在标定转速($n=2\,000$ r/min)下的负荷特性对比曲线。

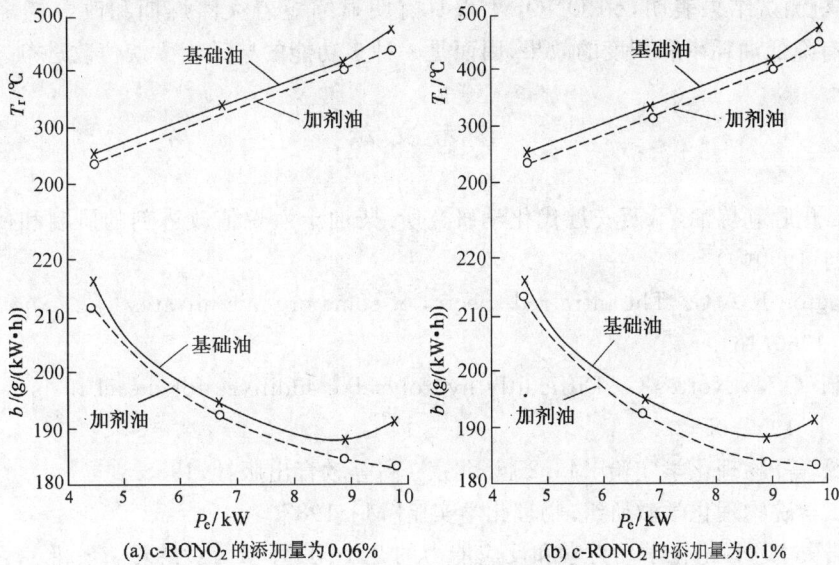

(a) c-RONO$_2$ 的添加量为 0.06%　　　　(b) c-RONO$_2$ 的添加量为 0.1%

图 5　标定转速($n=2\,000$ r/min)的负荷特性对比曲线

表 3 给出了柴油中加入 c-RONO$_2$ 的节油率和消烟率。

表 3　加入 c-RONO$_2$ 的柴油节油率和消烟率

标定转速 $n/(\text{r/min})$	有效功率 P_e/kW	添加量 0.06％		添加量 0.1％	
		节油率/%	消烟率/%	节油率/%	消烟率/%
	9.72	4.35	62.5	3.93	64.3
	8.83	1.13	61.5	1.25	51.5
2 000	6.66	0.81	—	1.30	—
	4.42	1.92	—	1.08	—
平均值		2.05	—	1.89	—

由此可见,使用添加有 c-RONO$_2$ 的柴油在标定转速下,其负荷特性中各个工况均可降低油耗率,低负荷时油耗率降低不明显,高负荷时油耗率降低较为明显。两种浓度下的平均节油率分别为 2.05％和 1.89％,最大节油率可达到 4.35％。两种浓度下的加剂油均可使柴油机排温降低,烟度降低 50％以上。

因此可认为,c-RONO$_2$ 除可缩短柴油着火滞燃期(即提高柴油十六烷值)和改善柴油机燃烧特性外还具有降低油耗率和烟度的多种效果。因此,c-RONO$_2$ 十六烷值改进剂的研究对国民经济的发展具有重要意义。

4　结　　论

(1) 提出一种具有环状分子结构的有机硝酸酯(c-RONO$_2$)作为柴油的十六烷值改进剂,

并取得了这种改进剂的一系列理化性质测试值。测试结果表明,这种改进剂具有许多优越和独特的性质,宜于推广使用。

（2）对有机硝酸酯（RONO$_2$）类柴油十六烷值改进剂作用机理的探讨表明,RONO$_2$的热裂解加速了柴油氧化反应初期退化分支的引发反应,缩短了氧化反应的诱导期。

（3）实机测试结果表明,c-RONO$_2$不仅具有明显缩短着火滞燃期和改善燃烧特性的效果,而且具有降低油耗率和烟度的效果,因而是一种多功能的柴油十六烷值改进剂。

参 考 文 献

［1］齐鲁石化胜利炼油厂,西安近代化学研究所.柴油十六烷值改进剂的研制和使用［J］.石油炼制,1990(9).

［2］Carrington R A G. The infra-red spectra of some organic nitrates［J］. *Spectrochimica Acta*, 1960(16).

［3］Schickh O V, Nottes G. Difficultly hydrolysable addifives for diesel fuels US,2, 905, 540［P］: 1959-09-22.

［4］罗澄源,等.物理化学实验［M］.2版.北京:高等教育出版社,1985.

［5］华东工学院物理化学教研组.物理化学实验［M］.1987.

［6］谢苗诺夫 H H.论化学动力学和反应能力的几个问题［M］.黄继雅,译.北京:科学出版社,1962.

（本文原载于《内燃机工程》1996 年第 4 期）

A study on organic nitrate as a cetane number improver for diesel fuel

Abstract：A diesel fuel cetane number improver, which constituent is a naphthenic nitrate, is developed, and physico-chemical properties of this improver is determined. The chemical kinetic mechanism of organic nitrate for shortening ignition delay is analyzed. The effects of this improver on diesel fuel cetane number and combustion charateristics as well as on fuel consumption and smoke are also inverstigated in the engine. The results show that naphthenic nitrate is a valuable additive in practical use.

着火促进剂作用机理的数值研究

董　刚,夏兴兰,李德桃,杨文明

[摘要]　在 Shell 着火模型的基础上,分析了着火促进剂的作用机理,将其纳入 Shell 模型中进行了数值计算。计算结果表明,着火促进剂可缩短燃油着火前的冷焰诱导期,其效力受促进剂热解动力学参数的控制,并随其添加浓度的增加而增加。

着火促进剂是用于提高柴油十六烷值,改善其着火品质的一类重要的柴油添加剂。近期研究还表明,着火促进剂对降低柴油机 NO_x 等排放具有明显作用[1]。由于目前世界范围内的柴油十六烷值有下降的趋势,且各国排放法规也日趋严格,因此这类添加剂的使用和研究正日益受到人们的重视。

建立合理的数学物理模型,从理论上对着火促进剂的作用机理进行研究是尚待深入的一个内容。可以相信,随着这方面研究工作的开展,对澄清这类添加剂改善燃油着火品质的作用机理具有重大意义,同时也为研制新型高效的着火促进剂提供了理论基础和实施依据。

本文在燃油氧化多步化学动力学模型(Shell 模型)的基础上,对几种不同着火促进剂的作用机理进行了数值计算和分析,并对其结果进行了探讨。

1　Shell 着火模型的建立

柴油在发动机燃烧室高温高压条件下,经过复杂的物理和化学过程,以热自燃的形式开始着火燃烧。在着火滞燃期的化学过程中,柴油(碳氢化合物)与氧气以退化分支的链式方式进行反应,反应放出的热量进一步加速链式反应速率,最终使柴油着火。为研究这一过程,英国 Shell 石油公司的 Halstead 等人于 1977 年提出了著名的 Shell 着火模型[2]。

Shell 模型建立了一个需要化学动力学反应步骤和传热过程参数最少,但却能全面反映高压下燃油两阶段着火基本特征和负温度系数现象的通用数学模型。在这一模型中,燃油分子以链式过程被氧化,并通过自由基 R 来传播,同时伴有热量的生成;链式化学反应的加速则是由于退化分支剂 B 通过退化分解反应的形成所导致的;包括 R 的线性和二次终止反应描述了第 1 阶段(冷焰)的着火;中间产物 Q 的形成描述了第 2 阶段热自燃的形成,它提供了冷焰之后更高温度下分支剂 B 的第 2 个源泉;化学放热则依据于能量守恒关系,其中散热项应用了谢苗诺夫传热模型。Shell 模型在形式上包含 8 个化学反应和 5 种物质(燃油、氧气、R、B 和Q),每个反应的速率常数为 Arrhenius 形式,整个多步反应体系通过与快速压缩机比较设置了 26 个参数。Shell 模型目前已成功地应用于模拟汽油机爆震和柴油机着火,表明了该模型具有描述燃油一般氧化特性的能力[3]。

图 1 给出了本文建立的 Shell 模型,其压缩末端(在 12 ms 处)气体温度为 700 K 时,均相工质温度和组分 R,B 和 Q 的变化历程。由此可见,活性自由基 R 和退化分支剂 B 的浓度在

迅速增加到一个最大值后又迅速下降,表明燃油氧化经历了所谓的冷焰阶段(τ_1)(以 B 的最大浓度为冷焰的形成标志),随后再进入热焰阶段(τ_2),并最终在组分浓度和工质温度迅速上升时达到着火,根据文献[2],以温升率大于 10^5 K/s 作为着火的标志。图 1 还清晰地显示了这一典型条件下温度的两阶段变化。

图 2 为本文建立的模型与原始 Shell 模型在不同压缩末端温度下,总诱导期(τ)、冷焰诱导期(τ_1)和热焰诱导期(τ_2)变化对比关系。尽管本文对 Shell 模型中的散热系数项进行了简化,但除在低温时两者有些差别外,本模型与 Shell 模型仍具有很好的一致性。图 2 还反映了燃油氧化的负温度系数现象,即燃油氧化速率随温度的增加反而减小,总诱导期延长。当温度进一步增加时,氧化速率则重新加快,总诱导期又重新减小。

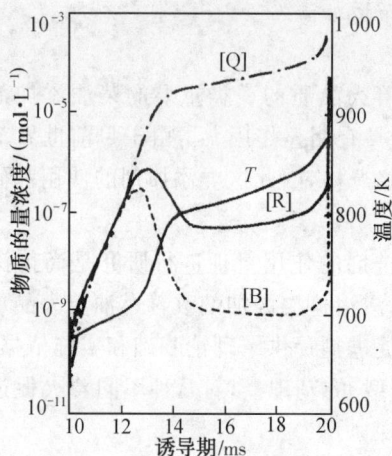

图 1 燃油组分 R,B 和 Q 以及工质温度 T 随时间的变化关系(压缩末端温度为 700 K)

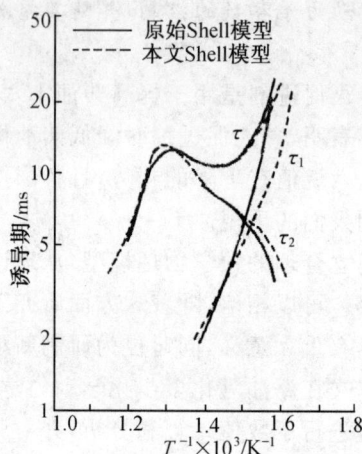

图 2 Shell 模型中诱导期的对比关系

由于本文建立的模型全面反映了 Shell 模型的功能,因此,将以这一模型为基础,对其进行改进以模拟零维条件下着火促进剂添加到燃油中,影响燃油氧化历程的作用机理。

2　着火促进剂作用机理的分析与数值计算

2.1　着火促进剂的作用机理

目前有效的着火促进剂主要以有机硝酸酯类物质和有机过氧化物为主。这两类物质的热解动力学研究表明[4-6],单分子离解是反应速率的决定性步骤。对硝酸酯,这一反应如下:

$$\text{RONO}_2 \longrightarrow \text{RO} + \text{NO}_2 \tag{1}$$

式中 RONO_2 为有机硝酸酯;RO 为烷氧自由基。

对过氧化物,这一反应如下:

$$\text{ROOR}' \longrightarrow \text{RO} + \text{R}'\text{O} \tag{2}$$

式中 ROOR' 为有机过氧化物;$\text{RO}, \text{R}'\text{O}$ 为不同基团的烷氧自由基。

上述反应分解得到的物质具有极高的反应活性,可参与到接下来的反应中,从而促进燃油焰前氧化的退化分支反应。

2.2 着火促进剂作用机理的数值研究

由于反应(1)和反应(2)分解的活性自由基与燃油氧化形成的自由基具有类似的性质,因此可纳入 Shell 模型的活性自由基 R 中。当含有少量着火促进剂的燃油随温度和压力升高时,着火促进剂的热解反应(1)和(2)成为燃油氧化的重要引发方式,通用形式如下:

$$I \xrightarrow{k_1} 2R \qquad (3)$$

式中 I 为着火促进剂;R 为活性自由基(与 Shell 模型中的 R 相同)。

将反应(3)纳入本文建立的模型中,k_1 为促进剂热解速率常数,有以下 Arrhenius 形式:

$$k_1 = A \cdot e^{-E/(RT)} \qquad (4)$$

式中 A 为指前因子,s^{-1};E/R 为活化温度,K;T 为工质的瞬时温度,K。

对这一改进的模型所组成的微分方程求数值解,可得到含促进剂的燃油组分的氧化历程及工质温度的变化关系。本文考虑了 3 种不同的着火促进剂,并分别在不同浓度下进行了数值计算,这 3 种促进剂及其热解速率常数见表 1。

表 1　不同着火促进剂及其热解速率常数

着火促进剂	A/s^{-1}	$E/(kJ \cdot mol^{-1})$	来源
环己基硝酸酯(CHN)	$10^{17.4}$	180	[4]
2-乙基己基硝酸酯(ION)	$10^{15.4}$	170	[5]
二叔丁基过氧化物(BOOB)	4×10^{15}	156.6	[6]

3　数值结果分析

3.1 不同促进剂对着火化学过程的影响

图 3 给出了不同促进剂(添加量为 0.5%,体积比)对燃油氧化过程中的自由基 R、退化分支剂 B、中间产物 Q 以及工质温度的影响情况。该图表明,3 种着火促进剂加速燃油氧化反应(缩短着火诱导期)的效力顺序为 BOOB>CHN>ION。图 3a、图 3b 和图 3c 表明,3 种促进剂均可使 R,B 和 Q 的浓度提前增加,浓度也有所变大。由于退化分支反应是燃油氧化过程的最基本的特性,因此退化分支剂 B 的变化表征了燃油氧化特征的变化。图 3b 表明,随着促进剂效力的提高,B 的浓度在冷焰诱导期内不断增加,其最大浓度值不断提前和变大,在 B 达到最大浓度后的变化趋于一致,到着火前(B 重新变大的时刻)的这段时期(热焰诱导期),与不加促进剂的纯燃油的这段时期大致相同。由此可以得到,着火促进剂的作用主要是加快了冷焰诱导期内的反应。表 1 指出,BOOB 的热解活化能最小,其热解反应最容易发生,因而表现出最大效力;CHN 的热解活化能虽然要比 ION 的大一些,但其指前因子是 ION 的 100 倍,因而CHN 表现出比 ION 更大的效力。从数值计算的结果看,指前因子 A 和活化能 E 的大小对着火促进剂具有综合性的影响,活化能越低或指前因子越大,促进剂的效力就越强。图 3d 还表明,燃油中添加着火促进剂后,均相工质的温度上升得更快,与纯燃油氧化的工质温度相比,其两阶段温升不很明显,反应温度有向单阶段温升变化的趋势。

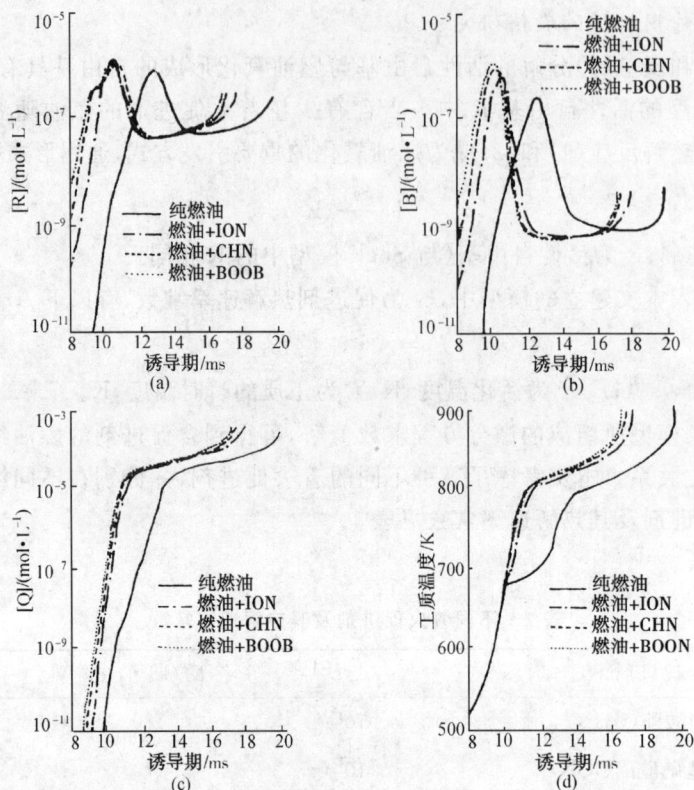

图 3　不同的着火促进剂对燃油氧化中的 R,B,Q 和工质温度变化的影响

3.2　着火促进剂浓度变化对着火化学过程的影响

图 4 给出了不同着火促进剂在不同添加浓度下对着火诱导期的影响。可以看出,随着其添加浓度的增加,3 种着火促进剂均使燃油着火诱导期不断缩短,但缩短的程度逐渐平缓。此外,不同的促进剂,其效力的不同也可在诱导期缩短的百分数上反映出来。

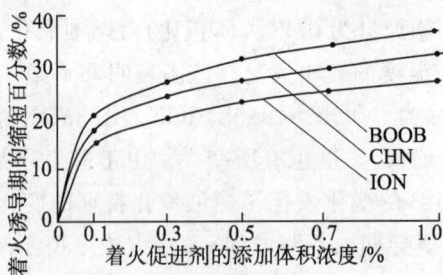

图 4　着火促进剂的添加量对燃油着火
诱导期的缩短百分数的影响

4　结　论

（1）在正确分析着火促进剂热分解机理的基础上,对 Shell 着火模型进行了建立和改进,使之能有效地模拟在着火促进剂作用下燃油的氧化行为。同时也表明 Shell 着火模型具有通

用性和易改进性。

（2）计算结果表明，着火促进剂可缩短燃油的诱导期，其中主要是冷焰诱导期，这一作用是通过着火促进剂首先发生热分解而释放自由基所导致的。

（3）不同的着火促进剂具有不同的效力，主要受其热解动力学参数的控制；着火促进剂的效力还随着其添加浓度的增加而有所增加。

参 考 文 献

[1] Ullman T L，Spreen K B，Mason R L. Effects of cetane number，cetane improver，aromatics，and oxygenates on 1994 heavy-duty diesel engine emissions[J]. *SAE Paper*，941020，1994.

[2] Halstead M P，Kirsch L J，Quinn C P. The autoignition of hydrocarbon fuels at high temperature and pressure-fitting of a mathematical mode[J]. *Combust Flame*，1977（30）：45-60.

[3] Kong S C，Reitz R D. Multidimensional modeling of diesel ignition and combustion using a multistep kinetics model[J]. *J Eng Gas Turb Power*，*Trans ASME*，1993，115：781-789.

[4] Hiskey M A，Brower K R，Oxley J C. Thermal decomposition of nitrate esters[J]. *J Phys Chem*，1991，95：3955-3960.

[5] Pritchard H O. Thermal decomposition of isooctyl nitrate[J]. *Combust Flame*，1989（75）：415-416.

[6] Kirsch L J，Rosenfeld J L J，Summers R. Studies of fuel injection into a rapid compression machine[J]. *Combust Flame*，1981(43)：11-21.

（本文原载于《燃烧科学与技术》1998 年第 4 期）

Numerical studies of the mechanisms for ignition improvers

Abstract：The mechanisms of ignition improvers are analysized，and the numerical calculation is made by adding the mechanisms to ignition model——the Shell model. The results show that all three improvers can shorten cool flame periods before ignition，their effectiveness is controlled by thermal decomposition kinetical parameters of improvers，and enhanced with increasing of their adding concentrations.

柴油喷雾着火过程的化学动力学模拟

董　刚，吴志新，夏兴兰，李德桃

[摘要]　在直喷式柴油机准维燃烧模型的基础上，建立了能描述柴油机着火前燃油的氧化历程及着火滞燃期的多步着火化学动力学子模型，它不仅可以很好地模拟实际柴油机的燃油氧化过程，而且可以较好地反映着火滞燃期随各种因素的变化关系，比以往的描述着火滞燃期的经验模型更具有合理性和先进性。整个燃烧模型的计算结果及其与试验结果的对比还表明，本模型能够合理地预测柴油机的燃烧过程。

烃类燃料的氧化动力学模拟是研究着火和燃烧现象的有力工具。从 80 年代开始，这一研究不断受到国外的重视，并得到了一定的应用。然而，由于实际使用的烃类燃料（如柴油）氧化机理十分复杂，为减轻模拟计算的工作量，这一研究主要局限于零维氧化过程的模拟[1]，而对于柴油机这样较为复杂的物理环境，其模拟研究尚不够深入，因而研究柴油机中燃料氧化过程对从本质上了解着火和燃烧过程无疑具有重大的实际意义。

着火滞燃期是柴油机燃烧过程的重要时期，它对其后的燃烧过程以及有害物质的排放均有显著影响。在目前的柴油机燃烧模型中，滞燃期的计算大多采用简单的经验拟合式进行[2]，柴油在这段时期内发生的氧化动力学反应却很少能被表征，因而探讨柴油着火前的氧化历程，并精确预测其着火滞燃期是亟待深入研究的内容。另一方面，柴油机准维燃烧模拟的优点在于：它既可考虑实际燃料的通用氧化动力学模型，从而反映燃料的氧化特性，又可相对容易地考虑空间的不同以及柴油雾化、蒸发、与空气混合、传热传质等物理现象，因而模型结构相对简单，计算工作量小，对计算机硬件要求不高。基于以上考虑，作者建立了一个能够模拟柴油着火过程的直喷式柴油机准维燃烧模型，并应用该模型进行了模拟计算，将计算结果与实测结果进行了比较。

1　模型的建立

1.1　直喷式柴油机准维燃烧模型的描述

本模型假设燃料以一定的压力和喷油规律喷入燃烧室，按时间发展分为若干个小区，这些小区在今后的过程中虽有空间位置的超越，但无相互混合，为研究方便，喷雾在径向不再区分，同时空气单独作为一个区处理，见图 1。假定每个小区有各自的油束破碎、油滴蒸发、空气卷吸、着火燃烧以及放热传热等过程，每个小区内温度均匀一致，卷吸的空气和已蒸发的燃油蒸汽瞬间达到均匀混合，并同时开始进行氧化

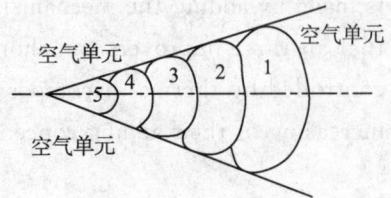

图 1　小区的划分

动力学反应，当其中一个小区着火后，认为火焰传播速度很快，因而导致其他小区也开始着火燃烧。着火后的小区温度迅速上升，体积急剧膨胀，小区之间不断发生能量传递，使整个燃烧

室内的温度和压力升高。

在上述过程中,燃料的油束破碎、空气卷吸以及整个热力学过程是以广安模型[3]为基础加以考虑的;油滴的蒸发采用了作者建立的油滴蒸发模型[4]来加以描述;燃料着火前的氧化过程则采用了 Shell 多步氧化反应动力学模型[5]进行描述;燃料的放热过程则通过下面的单步燃烧反应来加以描述:

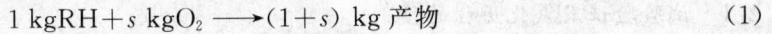

$$1 \text{ kgRH} + s \text{ kgO}_2 \longrightarrow (1+s) \text{ kg 产物} \tag{1}$$

其燃烧反应速率 W 可表达如下:

$$W = A\rho^2 Y_{RH}^a Y_{O_2}^b \exp{-E/(RT)} \tag{2}$$

式中 RH 为燃料;O_2 为氧气;A 为指前因子;ρ 为密度;Y 为质量分数;E/R 为活化温度;T 为工质的瞬时温度;a,b 为指数。式(2)中的常数 $A,E/R,a$ 和 b 则依据文献[6]加以选取。

1.2 氧化反应动力学子模型的建立

为模拟柴油的氧化过程和着火滞燃期,采用了 Halstead M P 的 Shell 模型[5]。该模型建立了一个化学动力学反应步骤和传热过程参数最少,但却能全面反映高压下燃料两阶段着火基本特性和负温度系数现象的通用零维数学模型。模型在形式上包含 5 种物质(燃料 RH、氧气 O_2、活性物质 R、退化分支产物 B 和中间产物 Q)。根据质量作用定律和能量守恒方程,Shell 模型可表达成如下的动力学方程形式:

$$d[R]/dt = 2(k_q[RH][O_2] + k_B[B] - k_t[R]^2) - 2f_3k_p[R] \tag{3}$$

$$d[B]/dt = f_1k_p[R] - f_2k_p[R][Q] - k_B[B] \tag{4}$$

$$d[Q]/dt = f_4k_p[R] - f_2k_p[R][Q] \tag{5}$$

$$d[O_2]/dt = -pk_p[R] \tag{6}$$

$$\frac{dT}{dt} = \frac{1}{C_v n_{tot}}\left(Q_K - Q_W - \frac{n_{tot}RT}{V}\frac{dV}{dt}\right) \tag{7}$$

式中 k_q, k_B, k_t 和 k_p 以及 f_1, f_2, f_3 和 f_4 分别代表反应的表现速率常数,这些常数以 Arrhenius 形式给出[5];$[\;]$表示物质浓度;T 为工质温度;C_v 为工质等容比热;n_{tot} 为工质总摩尔数;V 为工质体积;R 为通用气体常数;Q_K 为化学反应放热项;Q_W 为散热项。

式(3)中的[RH]可通过下式求得:

$$[RH] = ([O_2] - [O_2]_{t=0})/p + [RH]_{t=0} \tag{8}$$

式中 p 为每摩尔燃料需消耗的 O_2 量。

将式(8)代入式(3),则联立式(3)~式(7)组成的 5 个方程式可求解出物质 R,B,Q,O_2 的物质的量浓度以及温度 T 随时间的变化关系。为解决上述常微分方程组的"Stiff"特性,本文使用了依据于 Radau 求积公式的 Ehle II A 类方法[7],因而具有 S-稳定性和 Stiff 精确性,获得了合理的数值解。图 2 为本文建立的模型与原始 Shell 零维模型在不同压缩末端温度 T_c 下,总着火滞燃期(τ)、冷焰着火滞燃期(τ_1,以 R 的最大生成浓度为其结束点)和热焰着火滞燃期(τ_2,以着火点为其结束点)的变化对比关系,尽管本文对 Shell 模型中的散热项 Q_W 进行了简化,但除在低温时两者有些差别外,本模型与 Shell 模型仍具有很好的一致性。

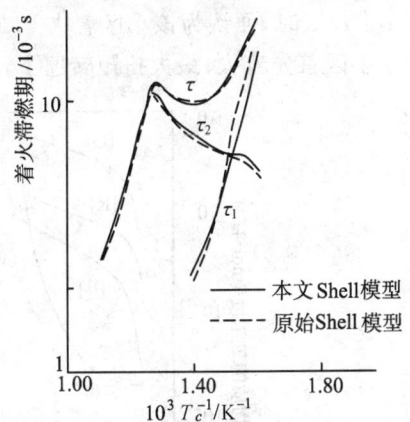

图 2 不同压缩末端温度下,Shell 模型着火诱导期变化关系的对比

将该模型以子模型的方式移植到准维模型中,不仅可代替以往经验的着火滞燃期子模型,而且可模拟在这段时期内的燃料氧化历程。

2　模拟计算结果与分析

2.1　试验验证及氧化历程的预测

为验证上述准维模型的作用和精确性,对一台 6105Q-1C 型直喷式柴油机的燃烧过程进行了实测和分析,试验装置及设备已在文献[8]中描述。

选取该发动机标定转速(2 800 r/min)和最大扭矩转速(1 800 r/min)作为模拟计算的对比工况,其试验结果和计算结果的对比见图 3。由此可见,在这两种工况下,模拟计算结果与试验结果符合较好。

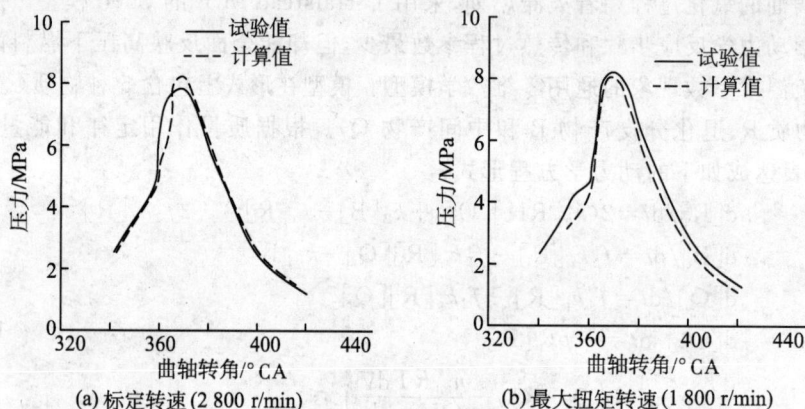

(a) 标定转速(2 800 r/min)　　　　(b) 最大扭矩转速(1 800 r/min)

图 3　柴油机气缸压力的对比结果

在模型中,标定转速(2 800 r/min)下的喷雾按时序划分为 52 个小区,最大扭矩转速(1 800 r/min)下的喷雾按时序划分为 40 个小区(见图 1),其中当某个小区内的温升率达到 10^5 K/s 时,便认为该小区着火[5],从而引发柴油机着火燃烧。计算结果表明,标定转速下,第 8 小区首先着火;最大扭矩转速下,第 4 小区首先着火,图 4a 和 4b 分别给出了这两个工况下

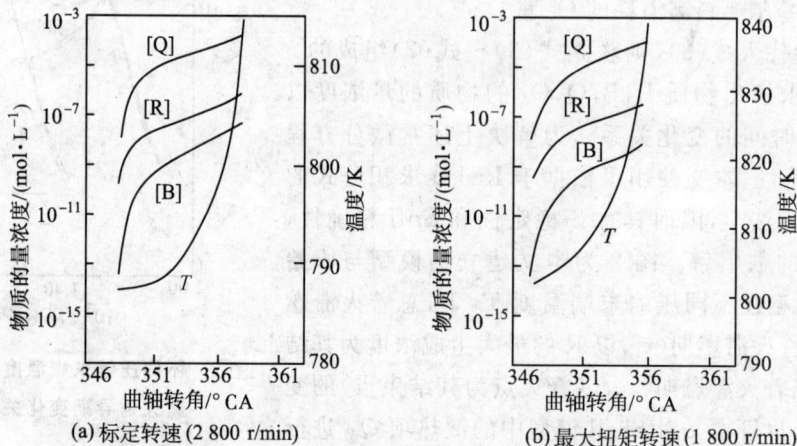

(a) 标定转速(2 800 r/min)　　　　(b) 最大扭矩转速(1 800 r/min)

图 4　喷雾首先着火的小区中物质 R,B,Q 浓度和温度 T 的变化历程

首先着火的小区内物质 R,B,Q 的浓度以及温度 T 随曲轴转角的变化关系。

由图 4 可知,在两种转速下的着火滞燃期内,燃料氧化过程中产生的活性物质 R、退化分支产物 B 以及中间产物 Q 随反应的进行呈先快后慢的趋势积累,这表明在着火滞燃期内,燃料以化学动力学反应的方式不断进行,反应放热同时还导致了小区内温度的不断升高。

2.2 变参数条件下着火滞燃期的计算结果

在与柴油机着火滞燃期有关的诸运转因素中,进气温度、进气压力、喷油始点温度及喷油提前角具有重要和本质性的影响。本文计算了在上述运转因素变化时,着火滞燃期的变化结果,如图 5~图 8 所示。

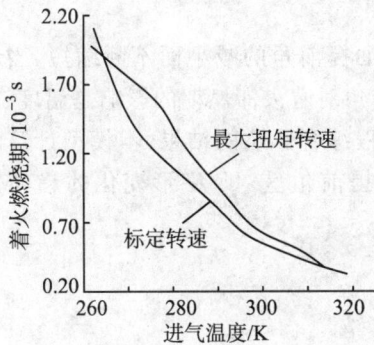

图 5　着火滞燃期随进气温度变化的计算结果

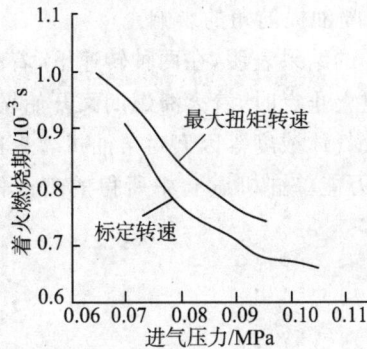

图 6　着火滞燃期随进气压力变化的计算结果

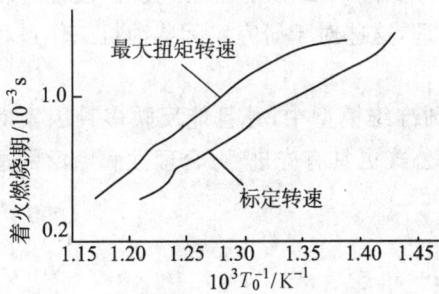

图 7　着火滞燃期的对数随喷油始点温度 T_0 的倒数变化的计算结果

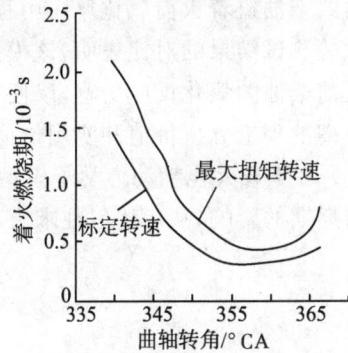

图 8　着火滞燃期随喷油提前角变化的计算结果

(1) 进气温度的影响。

图 5 的结果表明,在两种转速下,随着进气温度的增加,工质温度增加,从而加速了着火前燃油的蒸发过程和化学动力学反应速率,因而着火滞燃期均明显缩短,其中在标定转速下,着火滞燃期受较低进气温度的影响比受较高进气温度的影响要大。这说明,进气温度对着火滞燃期有重要影响,作者计算得出的结论与文献[9]在万能单缸机试验中得出的规律是一致的。

(2) 进气压力的影响。

图 6 的结果表明,在两种转速下,随着进气压力的增加,着火滞燃期也都有所缩短,但其缩短幅度与受进气温度的影响相比要小一些。这一结果说明,进气压力也是影响着火滞燃期的因素之一,压力增加将大大增加反应物分子之间有效碰撞的几率,因此使得反应速率加快,着

火滞燃期缩短。该计算结果得出的规律与文献[9]中的试验所得规律也是一致的。

（3）喷油始点温度的影响。

图 7 给出了两种转速下的 $\lg \tau_i$ 与 $1/T$ 的变化关系，结果表明，在这两种转速下，$\lg \tau_i$ 与 $1/T$ 基本上呈线性关系变化：

$$\lg \tau_i \propto 1/T \tag{9}$$

即有

$$\tau_i \propto e^{1/T} \tag{10}$$

这一关系与文献[2]中提出的着火滞燃期经验公式是一致的。由此可知，在作者建立的模型中，其着火滞燃期的变化较好地符合了前人经验公式的一般形式，这表明，本模型中燃油多步着火化学动力学子模型是合理的。

（4）喷油提前角的影响。

图 8 的结果表明，在两种转速下，着火滞燃期随喷油提前角的减小而不断缩短，当喷油提前角超过上止点时，着火滞燃期又开始增加。喷油提前角对着火滞燃期的影响是温度、压力以及混合气气体浓度等因素对柴油机着火前物理化学过程综合影响的结果，本模型计算与文献[2]中在万能单缸机试验中所得的着火滞燃期随喷油提前角呈"U"字形变化的趋势是相吻合的。

3　结　　论

（1）将描述柴油机燃烧过程的广安模型与多步着火化学动力学子模型结合起来，对实际的直喷式柴油机着火前氧化反应历程以及整个燃烧过程进行了数值模拟研究。模拟计算的结果与试验测试结果的对比表明，该模型除能够精确地反映实际柴油机的燃烧过程外，还能够预测着火前柴油的氧化反应历程，以便于分析燃烧氧化机理，这些对于研究实际柴油机工作过程具有较高的理论分析价值和实用价值。

（2）多步着火化学动力学子模型能够很好地移植到准维模型中，并且能反映多种因素对着火滞燃期的影响，从而比以往求解着火滞燃期的经验公式更具有先进性、合理性和广泛的适用性。

参 考 文 献

[1] Someya T. *Advanced combustion science*[M]. Tokyo：Springer-Verlag，1993：137-161.

[2] 何学良，李疏松. 内燃机燃烧学[M]. 北京：机械工业出版社，1990：246-270.

[3] Hiroyasu H，Kadota T. Models for combustion and formation of nitric oxide, soot in D I diesel engines[J]. *SAE Paper*，760129.

[4] 夏兴兰，李德桃，董刚，等. 一个简化的柴油机油滴燃烧模型[J]. 小型内燃机，1996，25(3)：6-10.

[5] Halstead M P，Kirschl J，Quinn C P. The autoignition of hydrocarbon fuels at high temperatures and pressures-fitting of a mathematical model [J]. *Combust Flame*，1977(30)：45-60.

[6] Meguerdichian M，Watson N. Prediction of mixture formation and heat release in diesel

engines[J]. *SAE Paper*, 780225.

[7] Prathero A, Robinson A. On the stability and accuracy of one-step methods for solving stiff systems of ordinary differential equations [J]. *Math Comp*, 1974, 28(125): 145-162.

[8] 董刚,李德桃,吴志新,等. 有机硝酸脂类十六烷值改进剂的研究[J]. 内燃机工程,1996, 17(4): 12-27.

[9] 何学良,顾德明,吴吉湘. 运转因素对柴油机滞燃期影响的试验研究[J]. 小型内燃机, 1984(3): 12-17.

(本文原载于《内燃机学报》1999 年第 3 期)

The chemical kinetical modeling of diesel fuel ignition process

Abstract: Based on the quasi-dimensional combustion model of D. I. diesel engines, a multi-step ignition chemical kinetical model was proposed, which could not only describe the fuel oxidation before ignition, but also reflect the changes of the ignition delay time along with the various factors. This model is more reasonable and advanced than the experienced models which only describe the ignition delay time. Also, the whole combustion model predicted the combustion process of a D. I. diesel engine reasonably compared to the test results.

利用实测放热规律研究十六烷值改进剂对柴油机着火特性和燃烧过程的影响

董　刚,李德桃,陈义良

[摘要]　通过基础油和加剂油实测放热规律的对比,就自行研制的 c-RONO₂ 柴油十六烷值改进剂对柴油机着火和燃烧过程的影响进行了深入的分析。结果表明,十六烷值改进剂能加速柴油的氧化过程,缩短着火滞燃期,使预混合燃烧期和扩散燃烧期的放热百分比得到合理分配,从而改善了柴油机的着火和燃烧特性。

柴油十六烷值改进剂是用于提高柴油十六烷值、改善其着火品质的一类重要柴油添加剂。近年来,由于世界各国对柴油机排放法规的日趋严格,对柴油品质也提出了更高的要求,因此,柴油十六烷值改进剂的研究也日益受到重视。研究表明[1,2],柴油十六烷值改进剂添加到柴油中,在柴油机工作过程中,通过热分解反应及随后的二次分解反应,可以加速柴油的氧化过程,缩短着火滞燃期,从而对后继燃烧过程也有显著影响。燃烧放热规律是分析和评估发动机燃烧过程的重要手段,国内曾报道利用这种方法对节油消烟型柴油添加剂进行过研究[3],但尚未有对十六烷值改进剂进行研究的报道。因此,为进一步探讨柴油十六烷值改进剂对柴油机着火和燃烧过程的影响,作者将自行研制的有机硝酸酯类十六烷值改进剂 c-RONO₂ 加入柴油(文中简称加剂油)中,对柴油机燃烧放热规律的影响进行了计算和分析。

1　燃烧放热规律的计算

对 0 号柴油(称基础油)和加剂油,在柴油机标定工况 2 800 r/min 和最大扭矩工况 1 800 r/min下,分别进行了示功图测录,并根据实测示功图,用自行编制的燃烧放热规律计算程序进行了放热规律的对比计算。

1.1　示功图的测录

试验机为湖南动力机厂生产的 6105Q-1C 型直喷式柴油机,示功图测试仪器为上海内燃机研究所开发的 EAS-900 发动机测试分析系统[4]。十六烷值改进剂分别以三种不同的添加浓度加到 0 号柴油中,三种添加浓度分别为 0.1%,0.3% 和 0.5%(质量百分比)。试验先从基础油开始进行,再按浓度由低到高的次序在同一工况下测录加剂油的示功图,测试完成后,改变工况重复上述过程。每次换油后,柴油机继续工作约 3 min,使留在油管、油泵和喷油嘴内的上次测试的燃油完全消耗后再继续测试。图 1 分别给出了这两种工况下基础油和加剂油示功图的对比结果,其中标定工况下两种油的喷油提前角均为 346°CA,后一工况下的喷油提前角均为 349°CA。为便于和下面结果进行比较,图 1 给出了 p-φ 曲线的高峰部分(350~400°CA)。

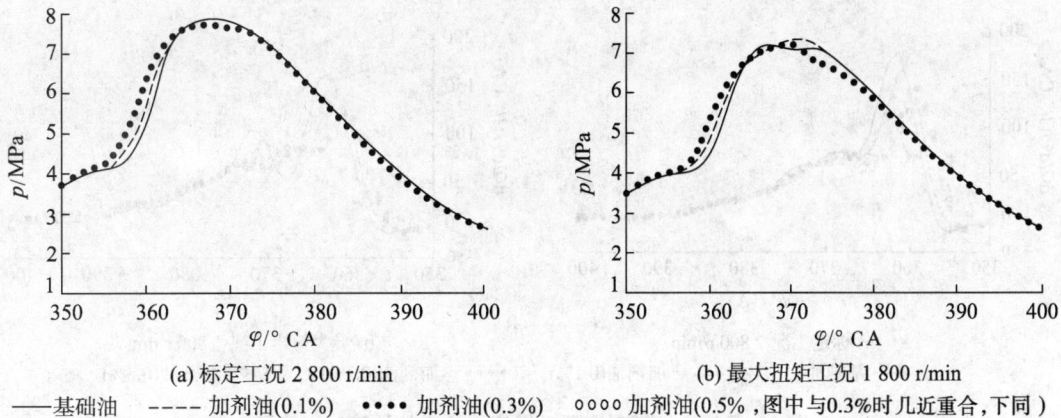

(a) 标定工况 2 800 r/min   (b) 最大扭矩工况 1 800 r/min

——基础油   ---- 加剂油(0.1%)   ●●●● 加剂油(0.3%)   ○○○○ 加剂油(0.5%，图中与0.3%时几近重合，下同)

图 1　基础油和加剂油实测示功图曲线

1.2　燃烧放热规律的计算

在直喷式柴油机的放热规律计算中作了如下假定：

（1）燃烧过程中气缸内的工质是均匀的，即在每一瞬间，工质成分、压力和温度处处相同。

（2）气缸内工质为理想气体，满足气体状态方程式，工质的比热仅与气体温度和成分有关。

（3）燃气由理论混合比完全燃烧的燃烧产物和剩余空气组成，不考虑燃油的高温裂解。

（4）燃烧室周围零件温度视为均匀的常数，并等于 T_w（壁面平均温度），T_w 根据经验公式选取

$$T_w = 373 + n p_{me} \tag{1}$$

式中 p_{me} 为平均有效压力，kPa；n 为经验值，对不同的燃烧室周围零件取值不同[5]。

（5）忽略工质泄漏。

根据热力学第一定律，参照文献[5]的燃烧放热率的计算公式和瞬时累积放热百分比表达式编制了放热规律程序并进行了计算，其中燃烧放热率计算公式中壁面传热项 $dQ/d\varphi$ 中的瞬时传热系数 α_g 按 Sitkei G 的经验公式选取[6]

$$\alpha_g = 0.205(1+b)\frac{p^{0.7}C_m^{0.7}}{T^{0.2}de^{0.3}} \tag{2}$$

式中 C_m 为活塞平均速度，m/s；p 为气缸瞬时压力，MPa；T 为气缸瞬时温度，K；de 为活塞至气缸顶端的距离，m；b 为经验常数。

在计算中，计算始点选在上止点前 20° CA，计算终点选在上止点后 60° CA，计算步长为 1° CA。

2　计算结果与分析

2.1　计算结果

图 2 和图 3 分别为利用图 1 测试的示功图计算的标定工况（2 800 r/min）和最大扭矩工况（1 800 r/min）下基础油和加剂油的瞬时燃烧放热率 $dQ/d\varphi$ 和燃烧温度 T 曲线。用表 1 对上述计算结果进行了总结。

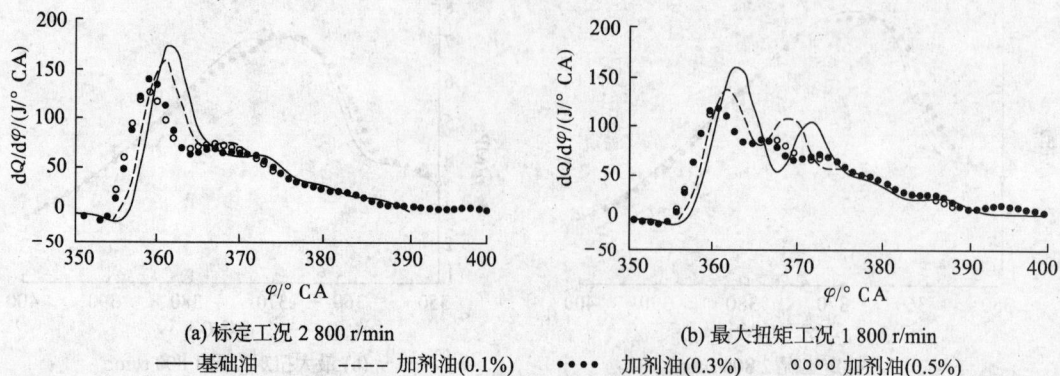

(a) 标定工况 2 800 r/min   (b) 最大扭矩工况 1 800 r/min

—— 基础油   ---- 加剂油(0.1%)   ●●●● 加剂油(0.3%)   ○○○○ 加剂油(0.5%)

图 2   基础油和加剂油瞬时燃烧放热率的计算对比结果

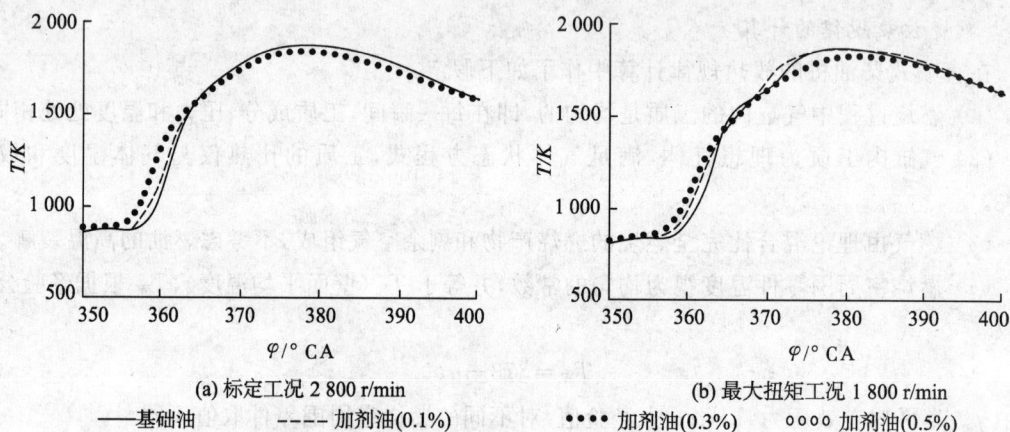

(a) 标定工况 2 800 r/min   (b) 最大扭矩工况 1 800 r/min

—— 基础油   ---- 加剂油(0.1%)   ●●●● 加剂油(0.3%)   ○○○○ 加剂油(0.5%)

图 3   基础油和加剂油燃烧温度的计算对比结果

表 1   基础油和加剂油燃烧放热规律计算结果对比

燃烧特性	标定工况				最大扭矩工况			
	基础油	加剂油			基础油	加剂油		
		0.1%	0.3%	0.5%		0.1%	0.3%	0.5%
最高瞬时放热率/(J/° CA)	174.24	159.25	140.70	127.72	161.87	138.28	120.17	116.94
曲轴转角/° CA	362	361	359	359	363	362	361	361
最高燃烧温度/K	1 881	1 877	1 843	1 845	1 861	1 863	1 812	1 828
曲轴转角/° CA	378	378	377	378	380	380	381	381

由图 2 可以看出,整个燃烧过程分为预混合燃烧阶段和扩散燃烧阶段,预混合燃烧放热峰(左起第一峰)要比扩散燃烧放热峰(左起第二峰)高,这一特点在最大扭矩工况时表现尤为明显。以两峰相交点作为预混合燃烧和扩散燃烧的分界点,可以得到不同燃烧方式所放热量的比例,见图 4。

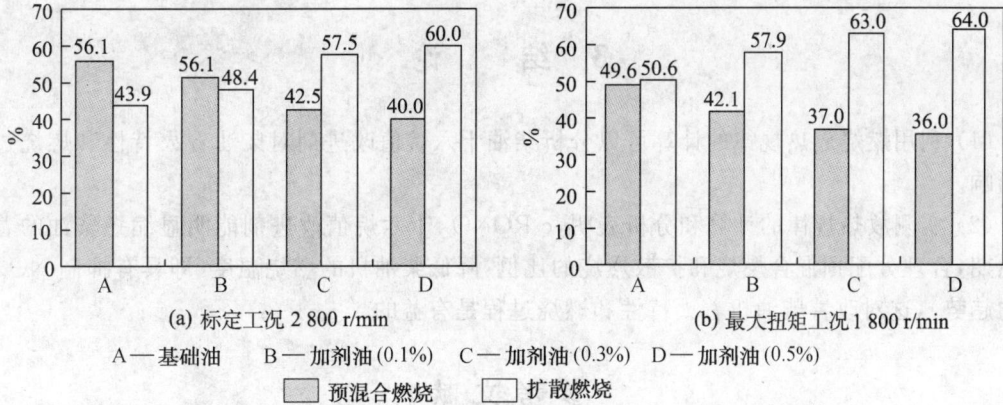

(a) 标定工况 2 800 r/min　　　　　(b) 最大扭矩工况 1 800 r/min

A—基础油　　B—加剂油(0.1%)　　C—加剂油(0.3%)　　D—加剂油(0.5%)

预混合燃烧　　扩散燃烧

图 4　预混合燃烧和扩散燃烧的放热百分量

2.2　燃烧放热率的对比分析

图 2a 表明,在标定工况下,加剂油要比基础油提前开始放热,且随着添加浓度的增加,提前开始放热的趋势更加明显,这说明十六烷值改进剂的加入可促使柴油提前开始燃烧。由表 1 还可看出,随着加剂油中改进剂浓度的增加,最高瞬时燃烧放热率的峰值(即第一峰值)不断减小,其所处的曲轴转角也不断变小,在大致相同的预混合燃烧期(因为其所处的曲轴转角的变化趋势与着火点的变化趋势一致)内,预混合燃烧放热所占的比例也不断减少,扩散燃烧放热所占的比例则不断增加,图 4a 的结果表明,随着加入的十六烷值改进剂的浓度的增加,预混合燃烧的比例从基础油的 56.1% 下降至含 0.5% 加剂油的 40.0%,而扩散燃烧的比例则从 43.9% 上升至 60.0%。图 2a 还表明,随着改进剂浓度的增加,加剂油的扩散燃烧放热率峰(即第二峰)逐渐向上止点附近靠拢,而峰以后的放热率与基础油没有明显差别,这说明扩散燃烧前期,燃烧放热率有所提高,使得整个燃烧放热过程更加集中于上止点附近,因而达到了合理组织燃烧、提高柴油机热效率的目的。

图 2b、图 4b 和表 1 指出,在最大扭矩工况下,加剂油具有与在标定工况下相同的影响燃烧放热的趋势。但由于此时喷油提前角减小,喷油落后,再加之转速下降减缓了空气运动,因此喷入气缸的燃油预混合不如标定工况下的充分,这就导致在扩散燃烧阶段有较为明显的放热峰的现象。在这种情况下,扩散燃烧放热峰向上止点附近靠拢更加明显,两峰之间的差别逐渐缩小,整个燃烧放热过程也更为集中。

2.3　燃烧温度的对比分析

由图 3 的燃烧温度曲线同样可以看出,在两种工况下,随着改进剂浓度的增加,加剂油的燃烧温度不断提前上升,这说明加剂油着火不断提前,着火滞燃期不断缩短。着火后不久,改进剂浓度越高,则燃烧温度整体上越呈下降趋势。由表 1 看出,在标定工况下,最高燃烧温度的下降随改进剂浓度的增加呈单调变化,浓度大于 0.3% 时,温度下降接近 40 K;在最大扭矩工况下,加剂油的最高燃烧温度也有十分明显的下降,其中浓度为 0.3% 时,温度下降接近 40 K;在最大扭矩工况下,加剂油的最高燃烧温度也有十分明显的下降,其中浓度为 0.3% 时,最高燃烧温度下降可接近 50 K。其原因是由于着火滞燃期缩短,着火前累积的燃油蒸汽量减少,当达到自燃条件时,预混合燃烧的比例减小,放热量减少,燃烧温度降低,因温度的滞后效应,可使得在整个燃烧过程中,加剂油均保持较低的燃烧温度。

3 结 论

(1) 利用燃烧放热规律的计算可以分析柴油十六烷值改进剂对柴油着火特性和燃烧过程的影响。

(2) 实测放热规律的计算和分析表明，c-RONO$_2$ 十六烷值改进剂能明显缩短柴油的着火滞燃期，合理分配预混合燃烧和扩散燃烧的比例，降低柴油机的燃烧温度（即具有抑制 NO$_x$ 生成的趋势），这对改善柴油机着火特性和燃烧过程是有益的。

参 考 文 献

[1] Clothier PQE, Moise A and Pritchard H O. Effect of free-radical release on diesel ignition delay under simulated cold-starting conditions[J]. *Combustion and Flame*, 1990, 81: 242.

[2] Li T M and Simmons R F. The action of ignition improvers in diesel fuel[C]// *The Combustion Insititute*, 21th Symposium (Int.) on Combustion, Pittisburgh, 1986, 455-462.

[3] 冯明志，肖福明，金晶，等. 利用放热规律计算研究柴油添加剂对燃烧过程的影响[J]. 内燃机学报，1996,14(1): 19-24.

[4] 董刚，李德桃，吴志新，等. 有机硝酸酯类柴油十六烷值改进剂的研究[J]. 内燃机工程，1996,17(4): 21-27.

[5] 林杰伦. 内燃机工作过程数值计算[M]. 西安：西安交通大学出版社，1986.

[6] 蒋德明. 内燃机原理[M]. 2 版. 北京：机械工业出版社，1988.

（本文原载于《内燃机工程》2000 年第 2 期）

Study on the effects of cetane number improver on ignition characteristics and combustion processes of diesel engine by measured heat release rates

Abstract: Effects of self-developed c-RONO$_2$ diesel fuel cetane number improver on ignition and combustion process of engine have been analyzed by comparing the measured heat release rates between the base fuel and additive-fuel blend. The results show that the cetane number imporver can accelerate the oxidation of diesel fuel, shorten the ignition delay time, and properly distribute the percentage of heat release between premixing and diffusion combustion periods, as a result, the ignition and combustion characteristics of diesel fuel have been improved.

6110 型柴油机机体组件的有限元分析

曹茉莉,卜安珍,李德桃,姜树李,王　辉

[摘要]　利用有限元分析技术对 6110 型柴油机的机体、缸盖、曲轴、主轴承盖、缸套、飞轮壳等柴油机主要零部件的组合部件进行了结构强度和刚度分析,获得了结构改进的依据。利用专业的三维造型软件 Pro/Engineer 建立了较为精确的机体、缸盖、曲轴、主轴承盖、缸套、飞轮壳等零件的三维模型;用试验方法获得边界条件,并用试验结果标定有限元计算模型,从而获得了较好的计算结果。

1　有限元模型

本文采用三维模型坐标,如图 1 所示。

在本文中,柴油机机体、气缸套、飞轮壳、曲轴等计算组件均采用 4 节点 4 面体单元,4 面体单元不如 6 面体单元的计算精度高,但易于自动生成网格;4 节点 4 面体不如 10 节点 4 面体计算精度高,但可大大减少节点数,从而节省计算时间*。生成网格时采用对关键部位加密网格的方法来提高计算精度。考虑到计算规模的问题,组合部件之间的联接关系采用了接触单元与节点耦合相结合的处理方法。由于计算中气缸盖不是主要分析对象,而是作为机体组件的支承边界,考虑气缸盖主要是由复杂的板壳构成,为减少节点数,采用三维 4 节点板壳单元。缸盖螺栓、主轴承盖螺栓采用实心圆柱形梁单元模拟,缸盖中螺栓孔、推杆孔等结构采用空心圆柱结构的梁单元摸拟。由于本次计算模型太大,采用了子结构技术进行计算分析,将整个模型分割成几部分,分别组合成超级单元。

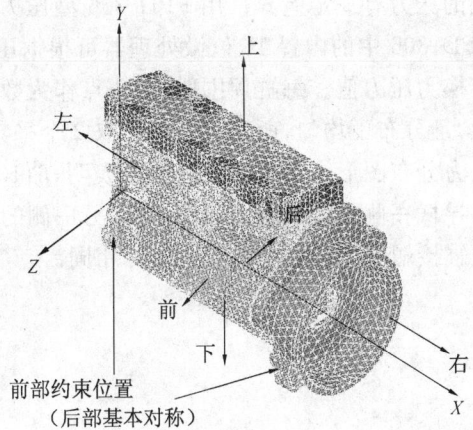

图 1　有限元模型

为了便于描述,称机体正常安装方位的上方为上,下方为下,曲轴自由端一侧的机体为左,曲轴输出端一侧的机体为右,进气管所处的一侧为前,排气管所处的一侧为后。

---

*　推荐采用高阶单元。由于当时(1998—1999 年)计算机硬件条件的限制,为使计算能够进行下去,有限元模型方程组的自由度必须控制在一定数量之内,因而采用了低阶 4 面体单元。

2　计算工况

计算工况分为 2 类:非工作工况和工作工况。

非工作工况的计算主要用于模型的标定,该工况下柴油机不工作,机体组件只受缸盖螺栓、主轴承盖螺栓预紧力及配气机构弹簧力的作用,记为工况 1。

工作工况:选取第 3 缸爆发(第 3 缸 370°CA)和第 4 缸爆发(第 4 缸 370°CA)为计算工况,记为工况 2 和 3。为了解工作状态下机体组件的温度对其变形和应力的影响,在不考虑温度的情况下,再进行第 3、第 4 缸爆发的有限元分析,记为工况 2-1 和 3-1。

3　试　　验

3.1　机体顶平面与气缸垫之间的接触压力

测试接触压力的目的有 3 个:(1)了解该接触压力的分布情况;(2)标定模型;(3)为非工作状态的计算提供边界条件。用 FUJI 感压法进行测试。测试的缸盖螺栓的拧紧力矩为 19 kg·m。图 2 为试验后的感压纸,可以明显看出每缸周围的接触压力分为 2 个圆环区域,气缸垫上的 3 个水孔周围密封处也有明显的压力圆,且靠近气缸孔一侧的压力与远离气缸孔一侧的压力有一定差异。用 FPD-306 型压力读取仪与 FPD-305 型显像密度仪采集压力数据,FPD-306 中的内置 16 位微处理器可根据压力/密度曲线将 FPD-305 测得的显像密度值快速转换为压力值。每缸周围取 12 个点作为数据采集点,气缸垫上的 3 个水孔边缘上的压力比较大,且分布不均匀,所以每只水孔取 2 个采集点。图 3 为根据采集的数据绘制的第 4 缸周围的压力分布图形。多次测试的结果表明,6110 型柴油机机体顶平面的接触压力分布不均匀,气门导杆一侧的接触压力比较小,而另一侧的接触压力较大。每缸周围的接触压力不均匀,而且各缸接触压力分布的图形也不尽相同。

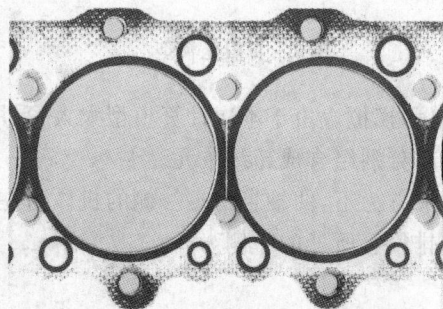

图 2　第 4 缸周围的接触压力(感压纸)　　　图 3　第 4 缸周围接触压力的分图形

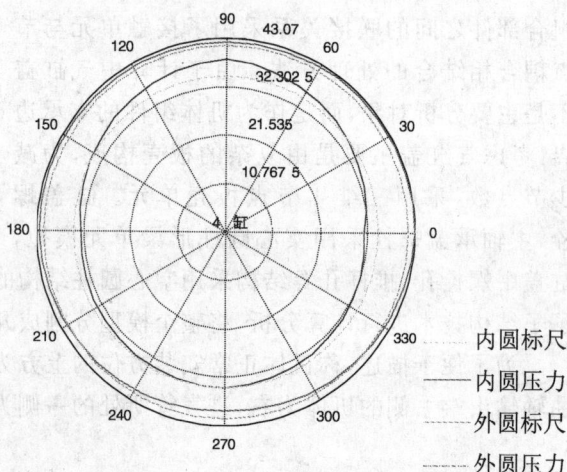

3.2　缸盖螺栓的拉力试验

该项试验的目的是了解以一定力矩拧紧缸盖螺栓时,在螺栓内部产生的拉力,为有限元分析提供较为准确的边界条件。在用扭力扳手拧紧螺栓的条件下,经过多次测试,发现以相同的拧紧力矩拧紧缸盖螺栓时,不同位置的螺栓预紧力的值各不相同。每次测试结果的最大相对偏差达 14.96%。不同条件下,同一螺栓编号位置的预紧力的值也各不相同,最大相对偏差达 15.88%,总体最大相对偏差为 18.125%。可见螺栓预紧力的值是随机变量,它与材料、螺栓、螺纹孔的加工精度、表面粗糙度以及两者的配合、装配条件等因素均有很大关系。因此用这种方法进行安装时,很难寻找到螺栓预紧力矩的大小规律。

3.3 气缸套内孔变形试验

该项试验的主要目的是了解装配前后气缸的变形情况,标定有限元模型的试验数据。分别在自由状态、装入气缸套未安装气缸盖(无缸盖螺栓预紧力)以及装入气缸套并安装气缸盖(有螺栓预紧力)条件下进行气缸套内孔测量。从多次测量的结果看,第4缸气缸套变形后的形状均较为一致,如图4所示。

位置1
位置2
位置3
位置4
位置5
位置6
位置7

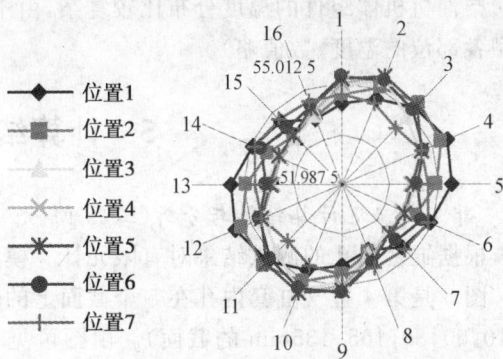

图 4　第 4 气缸套内孔的变形

4　边界条件

4.1　位移边界条件

工作状态下,发动机安装在车架上,由机体和飞轮壳上 4 个支承位置上的 16 个螺栓连接。在这 16 个固定螺栓的作用部位,约束左前部 4 个螺栓的 6 个自由度,释放右前部 4 个螺栓(在飞轮壳上)的 X 方向平移自由度,释放左后部 4 个螺栓的 Z 方向平移自由度,释放右后部 4 个螺栓的 X,Z 方向自由度。

4.2　力边界条件

力边界条件主要有螺栓预紧力、曲轴载荷、气体压力和运动部件惯性力、凸轮轴载荷。

4.2.1　缸盖螺栓预紧力(所有工况)

根据缸盖螺栓预紧力测试结果,在 19.0 kg·m 的拧紧力矩下,平均螺栓预紧力为 56.154 kN,分别作用在 26 个缸盖螺栓相应的模型上。

4.2.2　主轴承盖螺栓预紧力

主轴承盖螺栓的拉力未进行测试,根据螺栓与拧紧力矩的经验公式[2]进行估算。主轴承盖螺栓的拧紧力矩为 25 kg·m,螺栓为 M16 时,螺栓拉力为 64.65 kN,分别作用在 14 个主轴承盖螺栓相应的模型上,主轴承盖螺栓拉力也按集中力施加。

4.2.3　工况 2,3,2-1,3-1 的气体压力和运动部件惯性力

对于工作工况,根据示功图确定气体压力,分别作用在相应气缸内;根据曲轴连杆机构运动学计算惯性力,与气体压力合成后计算出活塞侧推力和连杆力,分别施加在相应气缸套内壁的适当位置和连杆轴颈上。由于曲轴不是这次计算的主要分析对象,故只将连杆切向力分解为 Z,Y 两个方向的分量,简单地平均分配在连杆轴颈表面的节点上(数值略)。

4.2.4 凸轮轴载荷

对于工作工况,根据计算工况的凸轮轴转角对应的配气机构零部件的相应位置(数值略)。

4.2.5 曲轴扭矩

这里 $M=132\,750\text{ N}\cdot\text{m}$。在曲轴输出端面上施加该扭矩。

4.3 温度边界条件

柴油机机体组件的温度分布比较复杂,由于人力、成本、时间的限制,只粗略地确定了计算模型各部位的温度(数值略)。

5 计算结果与分析

5.1 非工作工况计算结果与分析(工况1)

根据缸套内孔的测试结果对有限元计算模型进行标定。

图5是第4缸气缸套内孔在5个截面上的位移图(截面1—5分别代表气缸套内孔距顶平面 10,50,80,105,135mm 的截面)。由图可见,气缸套上部变形较大,下部变形较小,符合受压圆筒的变形规律。各气缸套在5个截面上的变形总体上有 Z 轴方向被拉长、X 轴方向被压缩的趋势,与测试结果的变形趋势较为一致,可见所建立的计算模型比较合理。

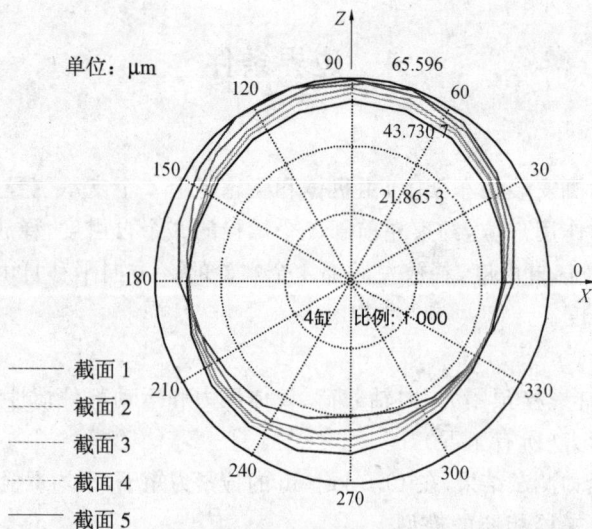

图5 非工作工况下第4缸5个截面上的径向位移

5.2 工作工况的计算结果与分析

由于篇幅的限制,这里只给出工况2的部分计算结果。图6为机体顶平面 Y 方向的位移图。可见3缸爆发时,3,4,5缸附近前部的 Y 方向变形比较大,这主要是受气体爆发压力的影响。

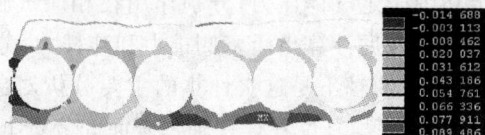

图6 机体顶平面 Y 方向位移图(工况2)

5.2.1 变形

（1）气缸套的变形。

缸套的下部位移明显大于上部位移，且各缸套在不同截面上位移的大小和方向均有不同。

1 缸位于 70° CA ABDC，处于自由排气与强迫排气阶段之交，缸内气体压力较高，气缸壁温度较高，此时活塞销的位置距机体顶平面 158.59 mm，活塞侧推力就作用在这个位置附近。相对于其受力位置，机体对其支承的位置较远，支承相对较弱。位移主要是朝着负 Z 方向，表明活塞的侧推力对气缸套变形的影响很大。

6 缸与 1 缸活塞所处的位置相同，但由于处于进气行程结束、压缩行程之初，气缸套的变形与 1 缸有着较大的差异。

3 缸处于爆发膨胀行程之初，缸内温度高、压力大，但由于活塞侧推力的作用位置与机体对其支承的位置很近，因而 3 缸的变形反而较小。

4 缸虽与 3 缸活塞位置相同，但由于 4 缸处于排气末尾、进气之初、气门叠开的扫气过程中，气缸内的压力与大气压力几乎相等，缸内温度也较低，气缸套的变形比 3 缸小许多，在所有的气缸套中，4 缸的变形是最小的。

在计算的曲轴转角下，2，5 缸的气缸套变形方向与 1，6 缸相反，从数据上看侧压力较小，但由于机体前端开有机油冷却器大窗口，刚度较后端差，故在较小的侧压力下仍有较大的变形，尤其是缸盖螺栓力传递线中断，在侧壁上造成弯曲变形，使气缸套支承刚度下降。2，5 缸活塞的位置最低，气体压力、活塞侧推力的作用位置均远离机体对缸套的支承，因此尽管 2 缸的气体压力很小，缸内温度较低，其变形却比较大。5 缸的气体压力比较大，缸内温度较高，活塞侧推力的作用位置又远离支承，故 5 缸的变形较大。

由此可见，活塞侧推力、缸内气体压力、气缸套的温度对缸套变形均有着明显的影响。另外，以上现象还表明机体对气缸套的支承刚度与支承位置与气缸套变形有着密切的关系。图 7 所示为第 3 气缸套的内孔在 5 个截面上的变形。

（2）机体的变形。

机体在 X 方向上以负向位移为主，只有左侧 2，3，4 缸裙部下缘的位移为 X 正向，其中 3，4 缸之间隔墙的下缘 X 向变形最大。机体在 1 缸附近有较小范围的负 Y 方向位移，2 缸附近的位移较小，4-6 缸的位移较大，且以正 Y 方向的位移为主。

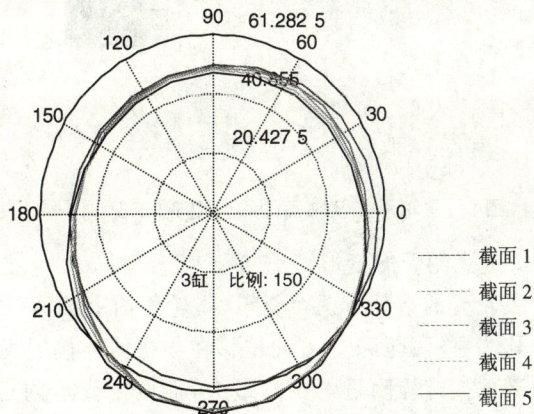

图 7　第 3 缸气缸套内孔 5 个截面的变形图（工况 2）

这表明，除了 3 缸爆发气体压力较大的原因外，支承对机体的位移有着很大的影响。1，2 缸附近的支承对机体刚度起到加强作用，机体另两个支承在飞轮壳上，飞轮壳与机体靠若干个 M10 的螺栓连接，因而对机体的支承较弱，导致整个模型变形较大的部位比较靠近飞轮壳。机体总体上以正 Z 向的位移为主，在 4 缸附近的裙部正 Z 向位移最大，1，6 缸附近上部的较小范围为负 Z 方向位移，这主要是由于活塞侧推力通过气缸套传递到机体上部、连杆切向力的 Z 向分力通过曲轴传递到主轴承座在机体上作用的结果。

图 8 为机体的综合变形图。由图不难看出，机体在 5 缸附近的综合位移最大，而 1 缸附近的综合位移最小。

（3）主轴承盖的变形。

主轴承盖的变形如图 9 所示。3,4 缸之间的主轴承盖在 X 正向的位移最大,2,3 缸之间的主轴承盖在 X 负向的位移最大,这是通过曲柄连杆机构传递到主轴承座和主轴承盖的连杆力使它们有沿 X 方向向两侧分开的趋势,而 3 缸爆发时,它对应的曲拐连杆力最大,在主轴承座和主轴承盖上产生的支反力也最大,导致它们的变形也最大。其他主

图 8 机体组件的综合变形（工况 2）

轴承座和主轴承盖的 X 方向位移均较小。主轴承座和主轴承盖的 Y 位移以正方向位移为主,最大 Y 方向正位移发生在第 4,5 缸之间的主轴承盖上,这主要是由发动机的工作状态和约束条件决定的。主轴承座和主轴承盖的 Z 方向位移以正向位移为主,这主要是由于连杆切向力的 Z 分量传递到主轴承座产生的结果。加强机体主轴承座的刚度,可以减小该处的位移。

图 9 主轴承盖及轴瓦的综合变形（工况 2）

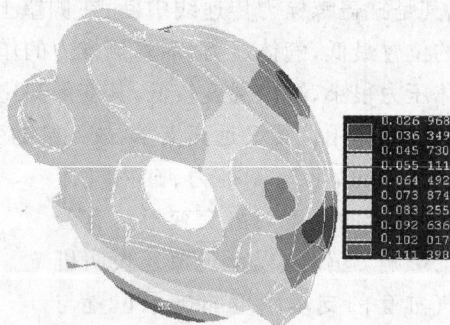

图 10 飞轮壳综合变形（工况 2）

（4）飞轮壳的变形。

飞轮壳的下部有两个支承,飞轮壳前上部 X 方向的位移以正向为主,后下部以负 X 方向的位移为主;飞轮壳 Y 方向位移在与机体连接的飞轮壳盖板附近,以 Y 正向位移为主,其中盖板前部联接螺栓附近的位移最大,而两个支承附近的位移很小;在飞轮壳前部的支承附近 Z 方向位移很小,后部支承附近的 Z 方向位移最大,该处刚好处在机体和飞轮壳的联接螺栓处,较大的位移易引起飞轮壳破坏;从整体上看,飞轮壳支承以下的部位综合位移较大,支承以上的部位综合位移较小(参见图 10)。

（5）曲轴的变形。

曲轴 X 方向的位移以正向为主,最大正 X 方向位移发生在 3,4 缸之间主轴颈两侧的平衡轴上,只有第 3 曲拐附近有负向位移,其余部位的 X 方向

图 11 曲轴综合位移图（工况 2）

位移介于两者之间；曲轴 Y 方向的位移以正向为主，4，5，6 缸曲轴的正 Y 向位移较大，1，2，3 缸曲轴的 Y 向位移较小；曲轴的 Z 方向的位移以正向为主，4 号主轴颈附近的正 Z 向位移较大，1，7 主轴颈附近的 Z 正向位移较小；图 11 为曲轴的综合位移图，可见第 4 缸曲轴附近的综合位移最大，1 缸附近的综合位移最小。

5.2.2 应力

（1）气缸套应力。

气缸套的拉应力出现在气缸套的前、后侧面上，可见活塞侧推力是工作状态下的气缸套应力产生的主要原因，气缸套上部（凸缘附近）的应力是气缸套受力与机体的支承共同作用的结果（图略）。

（2）飞轮壳应力。

在飞轮壳施加约束的两个支承附近和飞轮壳与机体联接的螺栓附近应力较大，只有很少的部位有压应力，其余部分的应力很小（图略）。

（3）机体应力。

整体上机体应力不大，机体顶部缸盖螺栓孔附近以压应力为主。与非工作工况相比，压应力的数值略有下降，这是由于气体压力将气缸盖向上抬起，部分地释放了缸盖螺栓预紧力引起压力下降的缘故。机体上螺栓力作用点附近的应力以拉应力为主，且在螺栓凸缘与各气缸之间，隔墙间的过渡处有应力集中现象。另外，机体上的支承部位附近也有应力集中现象，机体工作状态下的受力最终传递到支承部位，在支承部位附近产生较大的应力。主轴承座是刚度较差和应力集中的区域，尤其在主轴承盖止口处，有较明显的应力集中存在（图 12～图 16）。

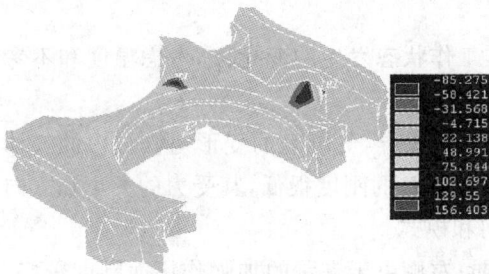

图 12　机体 1 段 3，4 缸隔墙应力（工况 2）

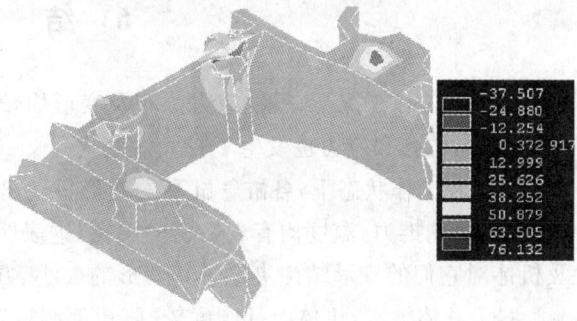

图 13　机体 2 段 3，4 缸隔墙附近的应力（工况 2）

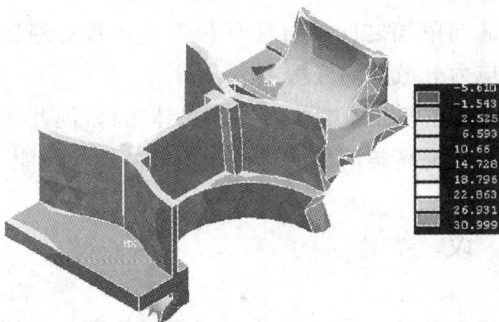

图 14　机体 3 段 3，4 缸隔墙附近的应力（工况 2）

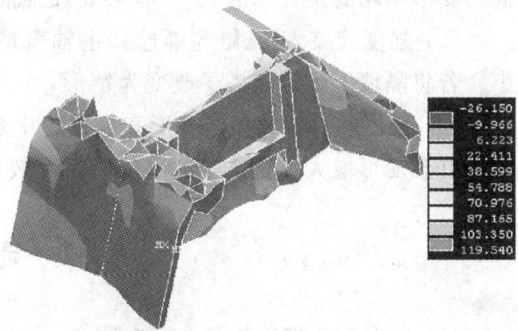

图 15　机体 4 段 3，4 缸隔墙附近的应力（工况 2）

（4）主轴承盖应力。

主轴承盖螺栓力施加点与主轴承盖结合面之间的部位有数值较小的压应力，与非工作工况相比，此处的压应力减小许多，这是由于在工作状态下，曲轴将主轴承盖向下拉，部分地释放了主轴承盖螺栓预紧力所致。各轴瓦上的应力分布也有差异，这是由于柴油机各缸工作状态不同，所处的曲轴转角也不同，通过曲轴传递的切向连杆力在主轴承座和主轴承盖上的作用位置不同，产生的应力分布亦不同所致（见图17）。

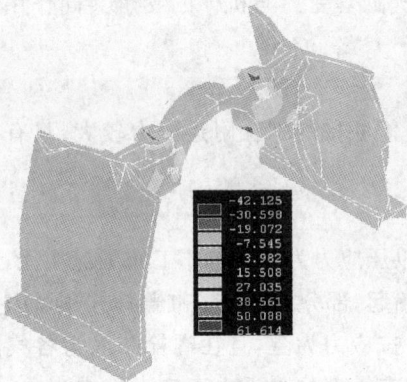

图 16　机体 5 段 3,4 缸隔墙附近的应力（工况 2）　　图 17　3,4 缸之间的主轴承盖应力（工况 2）

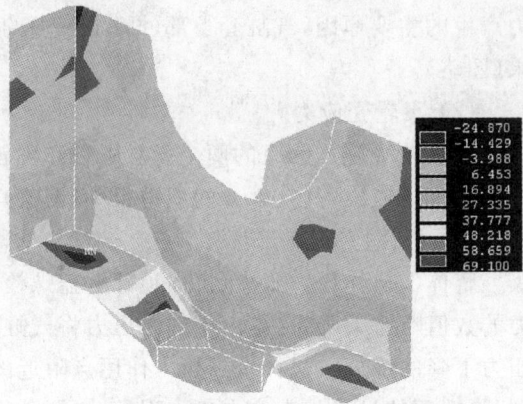

6 结　论

（1）在工作工况下，气缸应力分布及变形均较非工作状态发生的变化大，考虑温度和不考虑温度时，位移和应力也发生了较大的变化。

（2）在工作状态下，各缸气缸套的变形呈现出较大的差异，气缸套的变形与其缸内的气体压力、活塞侧推力、温度均有着密切的关系，这表明气缸本身的刚度很低，其受力的大小和方向及机体对它们的支承结构决定了其变形的大小、方向和模式。

（3）总体上看，机体上部刚度较好，裙部刚度较弱；左侧由于支承的加强作用而刚度较好，右侧由于支承太远而刚度较差；后部刚度较好，前部由于开有机油散热器孔，刚度较差，是造成缸套变形不均的主要原因之一，有必要加强前部的刚度。

（4）强度校核表明，缸盖螺栓和主轴盖螺栓凸缘与隔墙之间的过渡部位安全系数普遍较小。若将隔墙上的螺栓搭子改变为铲形，可将部分应力分散到隔墙板上。

（5）机体局部有较大的变形和应力集中现象，主要在气缸盖螺栓附近和机体裙部，因此导致气缸套变形增大和机体裙部出现"音叉"效应，不利于提高柴油机的排放和裙部的抗噪性能。

7 建　议

（1）所有缸盖螺栓预紧力同时降低，可望改善气缸套的变形。因此在保证可靠密封的前提下，可适当降低缸盖螺栓预紧力。

（2）适当加强机体对气缸套的支承，可减小气缸套的变形。

（3）螺栓凸缘的结构改为铲形，可望改善过渡部位的应力分布，增大安全系数。

（4）改变主轴承盖止口处加强筋的结构，可提高该处的强度和刚度，减小该处的变形和裙部的"音叉"效应，降低裙部的噪声。

（5）采用机体与气缸套止口高度按公差分组选配的方法，可望在不改变结构和加工精度的情况下减小气缸套的变形。

参 考 文 献

［1］曹茉莉.有限元计算与测试分析在直喷式柴油机结构设计中的应用［D］.镇江：江苏大学,2000.

［2］徐灏.机械设计手册(3)［M］.北京：机械工业出版社,1991.

（本文原载于《内燃机学报》2002 年第 5 期）

A FEA on the assembly parts of 6110 diesel engine

Abstract：FEA technology is used to perform the structural intensity and stiffness analysis on the assembly parts of cylinder block，cylinder head，crank shaft，main bearing covers，cylinder liners，and flywheel shell，etc. of the 6110 diesel engine. The results of FEA are based on the FE-models and boundary conditions applied on the FE model. The precise 3D models of the parts mentioned above are modeled with Pro/Engineer. The boundary conditions are obtained by test. Models used in the calculation are calibrated with the results of the test.

锅炉给水泵控制系统仿真研究

刘久斌,李德桃

[摘要] 现有给水泵系统都采用单变量系统进行控制,但该系统本质上是多变量系统,用单变量方法解决多变量问题给系统安全经济运行带来隐患。本文将锅炉变速给水泵控制系统视为多变量控制系统进行了讨论,用 Bristol-skinkey 方法进行了分析,应用 V 规范解耦环节结构解耦,利用总和时间常数法对等效对象进行了校正、整定,并进行了计算机仿真,给出了仿真曲线。结果表明,系统间能做到分别调节,相互作用小,无静态偏差。分析和整定方法适合于给水泵系统,控制品质好,符合工程实际,有实用价值。

在利用变速给水泵控制锅炉给水时,往往通过控制泵的转速和给水调节阀的开度来维持泵出口压力和给水流量,并保证泵运行在安全工作区。目前,变速给水泵的给水控制系统通常有两种类型,一种为一段控制系统,另一种为两段控制系统。两段控制系统有两个控制回路,一个回路通过控制给水调节阀改变给水流量以维持水位,一个回路通过控制给水泵转速以维持给水调节阀前后的压差间接维持泵出口压力。一段控制系统也有两个控制回路,一个回路通过控制给水泵转速改变给水流量以维持水位,另一个回路通过控制给水调节阀以维持泵出口压力。实际上,这两种控制系统均采用单变量控制,而泵出口压力和给水流量与阀门调节电流和转速调节电流之间本质上是多变量系统。采用单变量控制实施多变量系统的控制存在缺陷,原因是该方案不能消除系统间的相互影响,影响了控制质量,从多变量系统角度考虑这一问题,可消除系统间的相互影响,从而提高控制质量。本文试图利用 Bristol-skinkey 方法[1]分析给水泵控制系统并且用总和时间常数法对解耦后等效对象进行整定和仿真,这是一种新的尝试。

1 锅炉给水系统模型及分析

锅炉给水系统可用图 1 表示。

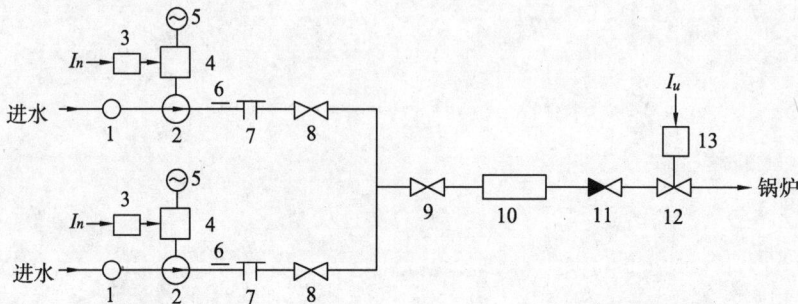

1—前置泵　2—调速给水泵　3,13—执行器　4—执行机构　5—电动机
6—压力测点　7—孔板　8,9—闸阀　10—高加　11—止回阀　12—给水调门

图 1　锅炉给水泵系统

DG-450-180 型调速给水泵系统现场试验的结果如下：

$G_{11}(S)=Q(S)/I_n(S)=0.04/(S+0.025)(S+0.0625)(S+1.0)$

$G_{12}(S)=Q(S)/I_u(S)=0.158/(S+0.05)(S+1.0)$

$G_{21}(S)=P(S)/I_n(S)=0.032/(S+0.025)(S+0.0625)(S+1.0)$

$G_{22}(S)=P(S)/I_u(S)=-0.068/(S+0.05)/(S+1.0)$

其中 Q 为给水流量；P 为泵出口压力；I_n 为转速调节电流；I_u 为阀门调节电流。

设

$$C(S)=\begin{bmatrix}Q(S)\\P(S)\end{bmatrix} \quad M(S)=\begin{bmatrix}I_n(s)\\I_u(S)\end{bmatrix} \quad G(S)=\begin{bmatrix}G_{11}(S) & G_{12}(S)\\G_{21}(S) & G_{22}(S)\end{bmatrix} \quad C(S)=G(S)M(S)$$

根据 Bristol-skinkey 方法可得

$$\Phi=\begin{bmatrix}25.6 & 3.16\\20.5 & -1.36\end{bmatrix}$$

$$\Psi=(\Phi^{-1})^{\mathrm{T}}=1/99.5\begin{bmatrix}-1.36 & -20.5\\-3.16 & 25.6\end{bmatrix}$$

$$\Lambda=\Phi*\Psi=\begin{bmatrix}0.35 & 0.65\\0.65 & 0.35\end{bmatrix}$$

可见 Q 的控制量应为 I_u，P 的控制量应为 I_n，组成 P 规范控制对象形式：

$$\begin{bmatrix}Q(S)\\P(S)\end{bmatrix}=\begin{bmatrix}P_{11}(S) & P_{12}(S)\\P_{21}(S) & P_{22}(S)\end{bmatrix}\begin{bmatrix}I_u(S)\\I_n(S)\end{bmatrix}=\begin{bmatrix}G_{12}(S) & G_{11}(S)\\G_{22}(S) & G_{21}(S)\end{bmatrix}\begin{bmatrix}I_u(S)\\I_n(S)\end{bmatrix}$$

2 整定方法

总和时间常数整定方法由德国学者提出[2]，适合于阶跃反应曲线为 S 型的自平衡能力对象，设被控对象阶跃反应曲线的传递函数为

$$G(s)=\frac{K_o(t_1S+1)(t_2S+1)\cdots(t_mS+1)}{(T_1S+1)(T_2S+1)\cdots(T_nS+1)}e^{-ts} \tag{1}$$

总和时间常数 T 定义为

$$T=T_1+T_2+\cdots+T_n-(t_1+t_2+\cdots+t_m)+t \tag{2}$$

再定义：

$$y_1(t)=\int_0^t[K_0-y(v)]\mathrm{d}v \tag{3}$$

有如下结论：

A_0 为阶跃反应曲线图 2 中的斜线面积：

$$A_0=y_1(\infty)=K_oT \tag{4}$$

针对有自平衡能力对象，实际整定时，若其传递函数未知，可试验求取其阶跃反应曲线，求得 K_0，A_0，进而得知 T；若其传递函数已知，可容易得知 K_0，T；经理论推导有表 1 所示结果。在 K_0，T 已知的情况下，可按表 1 整定调节参数。

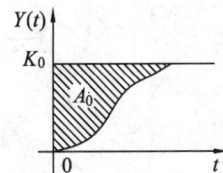

图 2　被控对象的阶跃
反应曲线

表 1　参数整定公式

类型	参数		
	K_p	T_i	T_d
P	$1/K_o$		
PI	$0.5/K_o$	$0.5T$	
PID	$1/K_o$	$0.66T$	$0.167T$

3　给水泵系统的控制和仿真[3]

应用 V 规范解耦环节结构解耦：

$$\begin{bmatrix} W_{11} & W_{12} \\ W_{21} & W_{22} \end{bmatrix} = \begin{bmatrix} P_{11} & P_{12} \\ P_{21} & P_{22} \end{bmatrix} \begin{bmatrix} 1 & V_{12} \\ V_{21} & 1 \end{bmatrix}$$

由解耦条件并忽略高次项得

$$V_{12} = -P_{12}/P_{11} = -8.1(20S+1)/(40S+1)(16S+1)$$

$$V_{21} = -P_{21}/P_{22} = 0.066\,4(56S+1)/(20S+1)$$

$$W_{11}(S) = (P_{11}+V_{21}P_{12})/(1-V_{21}V_{12})$$

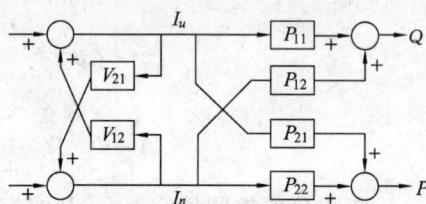

图 3　V 规范解耦环节结构解耦图

$$= 3.27(13.5S+1)(30.6S+1)/(S+1)(20S+1)(9.1S+1)(46.9S+1)$$

$$W_{22}(S) = (P_{22}+V_{12}P_{21})/(1-V_{21}V_{12}) = 4.9/(40S+1)(16S+1)(S+1)$$

至此多变量对象已解耦成以 $W(S)$ 对角线元素组成的两个单变量对象，可按两个单回路整定利用总和时间常数的整定法以 $W_{11}(S)$，$W_{22}(S)$ 为对象进行整定。

考虑 $W_{11}(S)$：其总和时间常数 $T = (1+20+9.1+46.9)-(13.5+30.6)=22.9$

$$K_0 = 3.27$$

所以调节器 T_1：$K_{p1}=0.31$，$T_{i1}=15.1$，$T_{d1}=3.8$

考虑 $W_{11}(S)$：其总和时间常数 $T = 40+16+1=57$

$$K_0 = 4.9$$

所以调节器 T_2：$K_{p2}=0.2$，$T_{i2}=37.6$，$T_{d2}=9.5$

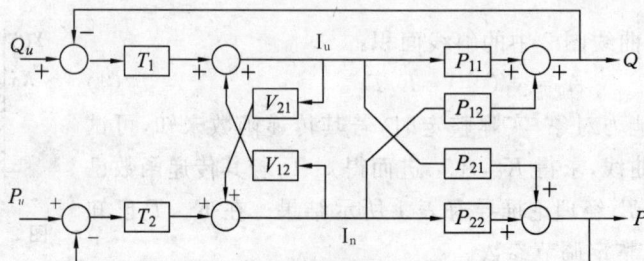

图 4　最终的控制系统框图

仿真结果如下：

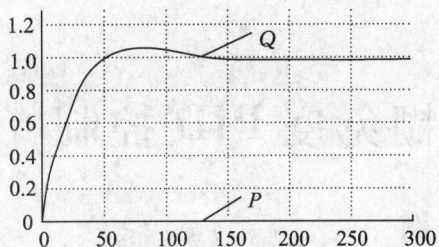

图 5　给水流量定值阶跃扰动下的反应曲线

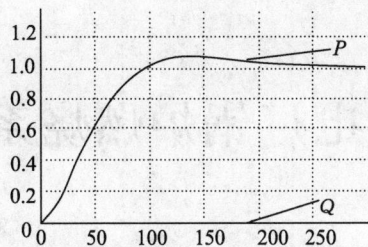

图 6　泵出口压力定值阶跃扰动下的反应曲线

4　结　　论

（1）锅炉给水泵控制系统作为多变量进行考虑符合实际情况，解耦后的系统能消除了系统间的相互影响，使控制更加合理。

（2）用 Bristol-skinkey 方法分析了组成锅炉给水泵控制对象，科学合理，引入总和时间常数法对解耦后的等效对象进行整定，方法简便易行。

（3）结果表明，最终组成的控制系统，工程上易于实现，控制质量好，能够满足工程要求。

参 考 文 献

［1］Bristol E H. What happens when you decouple? —a qualitative analysis ［C］. *Proc Second Advanced Control Conference* ,1998：111–119.

［2］Udo Kuln. Ein praxisnahe einstellregel fuer PID-regler：die T-summen-regel ［J］. *Automatisier Umgste Chmische Praxis* ,1995(5).

［3］丁轲轲,刘久斌,林青. 热工过程自动调节[M].北京：中国水利水电出版社,2000.

（本文原载于《动力工程》2004 年第 4 期）

Simulation on boiler feed-pump control system

Abstract：The feed pump system is controlled by single loop control theory , but it is truly multivariable system , so feed-pump control system in use exist problems about safety and efficiency. This paper discusses control system of boiler feed-pump with variable speed and analysis the controlled object by Bristol-skinkey , and the controlled object is decoupled by *V* standard structure , decoupled equivalent controlled objects are tuned and adjusted by sum time constant approach , the control system are simulated by computer , step response curves are given. Result shows that control system can be regulated respectively , interaction can be eliminated , steady state error is zero , it is better quality. The approaches adapt to feed-pump control system and can meet need of engineering , so those possess practical significance.

电厂锅炉燃烧系统的模糊免疫 PID 控制

刘久斌,李德桃

[摘要] 本文讨论了单元机组采用炉跟机负荷控制方式时的锅炉燃烧控制系统,针对风扇磨直吹制锅炉汽压调节对象惯性大、给煤机给煤量与控制电流呈非线性等特点,分析了对象的数学模型,将高阶惯性对象等效为带纯迟延的一阶惯性系统;利用借鉴生物系统的免疫机理而设计出的一种非线性控制器,再使用 Zadeh 的模糊逻辑 AND 操作并采用常用的 mom 反模糊化方法得到模糊控制器的输出 $f(u(k),\Delta u(k))$。通过计算机仿真,给出了仿真曲线。结果表明,汽压系统能做到稳定性高,鲁棒性强,调节及时,反应速度快,动态偏差小,无静态偏差。该分析和控制方法非常适合于燃烧控制系统,控制品质好,符合工程实际,有实用价值。

锅炉燃烧过程自动控制的任务在于使锅炉的燃烧工况与锅炉的蒸汽负荷要求相适应,同时保证锅炉燃烧过程能安全经济运行。因此,当锅炉的负荷改变时需要进行燃烧的调整。每台锅炉燃烧过程的具体控制任务及控制策略因燃料种类、制粉系统、燃烧设备以及锅炉的运行方式不同而异。燃料量控制就是使进入锅炉的燃料燃烧所产生的蒸汽量满足外部负荷要求。燃料量控制是锅炉控制中最基本也是最主要的一个系统。因为给煤量的多少既影响主汽压力,也影响送、引风量的控制,还影响到汽包中

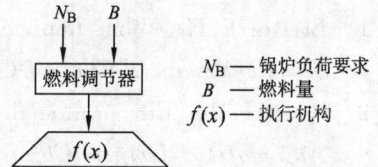

图 1 燃料量控制策略

N_B —— 锅炉负荷要求
B —— 燃料量
$f(x)$ —— 执行机构

蒸汽蒸发量及汽温等参数,所以燃料量控制对锅炉运行有重大影响。各种燃料控制可用图 1 简单表示。当单元机组采用机跟炉负荷控制方式时,锅炉调节机组负荷,汽机调节汽压,直接将电网的负荷要求 N_0 作为锅炉负荷要求信号;当单元机组采用炉跟机负荷控制方式时,汽机调节机组负荷,锅炉调节汽压,由于锅炉出口汽压是表征锅炉生产的蒸汽量与汽机耗汽量之间的平衡指标,所以取锅炉出口汽压作为锅炉负荷要求信号;当单元机组采用机炉协调负荷控制方式,负荷控制系统(主控系统)的锅炉主控信号作为锅炉负荷要求信号。本文主要讨论炉跟机负荷控制方式时汽压调节控制系统。

1 锅炉汽压系统模型及分析

某电厂锅炉汽压系统可用图 2 表示。

常规调节方案是:当发生扰动主汽压偏离给定值时,调节系统改变给煤机

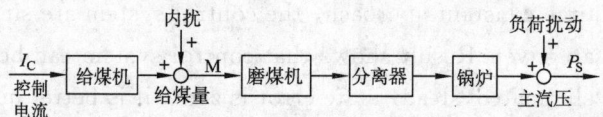

图 2 锅炉汽压系统

的控制电流 I_C 进而改变给煤量 M，于是磨煤机输出的煤粉量也相应发生变化，经分离器后，细度合格的煤粉由喷燃器送入炉膛燃烧，从而主汽压得到调节。可见，控制电流 I_C 改变时，主汽压发生变化要经过给煤机、磨煤机、分离器及锅炉的燃烧、热量/汽压转换等过程，整体看来调节对象属于高阶大惯性对象。另外，由于原煤的物理状况变化和煤种变化等原因，给煤机的控制电流与给煤量之间的关系呈非线性，甚至有时呈不确定关系。某电厂给煤机的控制电流与主汽压的数学模型如下：

$$W(s) = \frac{P_s(s)}{I_C(s)} = \frac{1.5}{(80s+1)^3} \quad \text{(MPa/mA)} \tag{1}$$

式中 P_s 为主汽压（过热器出口汽压，MPa）；I_C 为给煤机的控制电流（控制输出信号，mA）

$W(s)$ 可等效为 $G(s)$：

$$G(s) = \frac{K}{(T_c s+1)} e^{-Ts} \tag{2}$$

对式(1)求单位阶跃响应曲线，进而得

$$T_c = 289 \text{ s} \qquad\qquad T = 83 \text{ s}$$

所以，调节对象的等效传递函数为

$$G(s) = \frac{1.5}{(289s+1)} e^{-83s} \tag{3}$$

2　模糊免疫控制

免疫是生物体的一种特性生理反应。生物的免疫系统对于外来侵犯的抗原，可产生相应的抗体来抵御。抗原和抗体结合后，会产生一系列的反应，通过吞噬作用或产生特殊酶的作用而毁坏抗原。生物的免疫系统由淋巴细胞和抗体分子组成，淋巴细胞又由胸腺产生的 T 细胞（分别为辅助细胞 T_H 和抑制细胞 T_S）和骨髓产生的 B 细胞组成。当抗原侵入肌体并经周围细胞消化后，将信息传递给 T 细胞，即传递给辅助细胞 T_H 和抑制细胞 T_S，然后刺激 B 细胞。B 细胞产生抗体以消除抗原。当抗原较多时，肌体内的 T_H 细胞也较多，而 T_S 细胞却较少，从而会产生较多的 B 细胞。随着抗原的减少，肌体内的 T_S 细胞增多，它抑制 T_H 细胞的产生，从而 B 细胞也随着减少。经过一段时间间隔后，免疫反馈系统便趋于平衡。抑制机理和主反馈机理之间的相互协作，是通过免疫反馈机理对抗原的快速反应和稳定免疫系统完成的[2]。免疫 PID 控制器是借鉴上述生物系统的免疫机理而设计出的一种非线性控制器：假设第 k 代的抗原数量为 $\varepsilon(k)$，由抗原刺激的 T_H 细胞输出 $T_H(k)$，T_S 细胞对 B 细胞的影响为 $T_S(k)$，则 B 细胞接受的总刺激

$$S(k) = T_H(k) - T_S(k) \tag{4}$$

式中 $T_H(k) = k_1 \varepsilon(k)$；$T_S(k) = k_2 f(\Delta S(k)) \varepsilon(k)$。

若以抗原的数量 $\varepsilon(k)$ 作为偏差 $e(k)$，B 细胞接受的总刺激 $S(k)$ 作为控制输入 $u(k)$，则有如下的反馈控制规律：

$$u(k) = K(1 - \eta f(u(k), \Delta u(k))) e(k) = k_{p1} e(k) \tag{5}$$

式中 $k_{p1} = K(1 - \eta f(u(k), \Delta u(k)))$；$K = k_1$ 为控制反应速度；$\eta = k_2/k_1$ 为控制稳定效果；$f(u(k), \Delta u(k))$ 为一非线性函数。利用模糊控制器可逼近非线性函数 $f(u(k), \Delta u(k))$：每个输入变量被两个模糊集模糊化，分别是正（P）和负（N）；输出变量被三个模糊集模糊化，分别是

正（P）、零（Z）和负（N）；以上隶属度函数都定义在整个$(-\infty,+\infty)$区间。模糊控制器采用以下规则：

(1) if u is P and Δu is P then $f(u,\Delta u)$ is N

(2) if u is P and Δu is N then $f(u,\Delta u)$ is Z

(3) if u is N and Δu is P then $f(u,\Delta u)$ is Z

(4) if u is N and Δu is N then $f(u,\Delta u)$ is P

各规则中，使用 Zadeh 的模糊逻辑 AND 操作并采用常用的 mom 反模糊化方法得到模糊控制器的输出 $f(u(k),\Delta u(k))$。

常规增量式 PID 控制器离散形式为

$$u(k)=u(k-1)+\Delta u(k) \tag{6}$$

$$\Delta u(k)=k_p(e(k)-e(k-1))+k_i e(k)+k_d(e(k)-2e(k-1)+e(k-2))$$
$$=k_p((e(k)-e(k-1))+(k_i/k_p)e(k)+(k_d/k_p)(e(k)-2e(k-1)+e(k-2))) \tag{7}$$

式中 k_p,k_i,k_d 分别为比例、积分和微分系数。基于免疫反馈原理的控制器实际上就是一个非线性 P 控制器，其比例系数 $k_{p1}=K(1-\eta f(u(k),\Delta u(k)))$，$k_{p1}$ 随 PID 控制器输出 $u(k)$ 及其增量 $\Delta u(k)$ 变化而变化，K 为增益，则免疫 PID 控制器的输出为

$$u(k)=u(k-1)+k_{p1}((e(k)-e(k-1))+(k_i/k_p)e(k)+(k_d/k_p)(e(k)-2e(k-1)+e(k-2)))$$
$$=u(k-1)+k_{p1}((e(k)-e(k-1))+K_I e(k)+K_D(e(k)-2e(k-1)+e(k-2))) \tag{8}$$

式中 $K_I=k_i/k_p$；$K_D=k_d/k_p$。

3 汽压控制和仿真

汽压控制系统框图如图 3 所示：

模糊推理控制器以 u 和 Δu 作为输入，PID 参数作为输出。利用上述模糊规则在线对 PID 参数进行修改，PID 算法取式（8），便构成了模糊免疫控制系统。对上述系统进行编程仿真计算，取采样时间为 4 s，$K=0.1$，$\eta=0.8$，

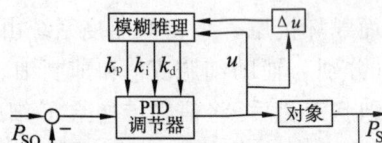

图 3 汽压控制系统方框图

输入信号 P_{SO} 为单位阶跃信号，$K_I=0.55$，$K_D=2.4$ 仿真结果为：模糊免疫控制 k_{p1} 的变化如图 4 所示，模糊免疫控制汽压 P_S 的阶跃响应图 5 所示。另外 $u,\Delta u$ 和 $f(u(k),\Delta u(k))$ 的隶属函数曲线由于篇幅关系略。

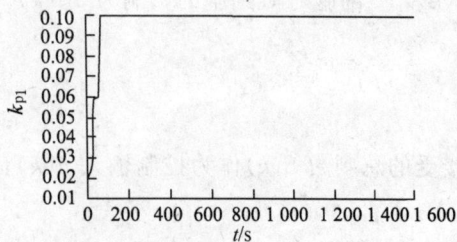

图 4 模糊免疫控制 k_{p1} 的变化

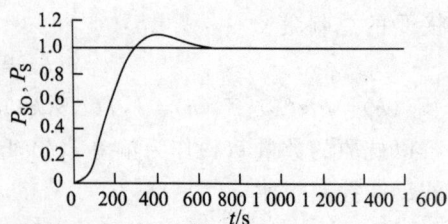

图 5 模糊免疫控制汽压 P_S 的阶跃响应

为验证控制系统的鲁棒性，在上述同样情况时在时间 800 s 瞬间加一幅值为 1 的干扰，结果响应曲线很快恢复平稳，其响应曲线如图 6，可见鲁棒性强。

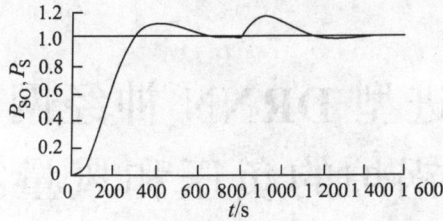

图6　扰动时模糊免疫控制汽压 P_S 的阶跃响应

4　结　论

(1)风扇磨直吹制锅炉汽压调节对象惯性大、给煤机给煤量与控制电流呈非线性,可以将其视为非线性问题,模糊免疫控制是一种可选的理想设计方法。

(2)模糊免疫控制方法科学合理,整定和仿真简便易行。

(3)最终组成的控制系统,工程上易于实现,控制质量好,鲁棒性强,汽压系统能做到稳定性高,调节较及时,反应速度较快,动态偏差小,无静态偏差,能够满足工程要求,有实用价值。

参 考 文 献

［1］Jerne N K. The immune system[J]. *Scientific American*,1993,229(1)：52-62.

［2］De Castro L N,Zuben F J V. Artificial immune system：part 1——basic theory and application［R］. Technical Recort,Tr-Dca 01/99,1999.

［3］丁轲轲,刘久斌,林青. 热工过程自动调节[M]. 北京：中国水利水电出版社,2000.

［4］李士勇. 模糊控制、神经控制和智能控制[M]. 哈尔滨：哈尔滨工业大学出版社,1998.

(本文原载于《动力工程》2005 年第 5 期)

Fuzzy-immune control on boiler combustion system

Abstract：The boiler combustion control system is discussed in case of BF load control mode. Combustion object mathematical model of direct-fired boiler with fan mill is analyzed. A nonlinear controller designed by use of fuzzy-immune principle is in use. $f(u(k),\Delta u(k))$ is acquired by fuzzy logic AND operation of Zadeh and mom defuzzy method. The combustion control system are simulated by computer, pressure step response curves are given. Result shows that control system has strong stability and robustness,it can be regulated in time, dynamic error is small，steady state error is zero,it is better quality. The approach adapts to combustion control system and can meet need of engineering,so it possesses practical significance.

利用改进型 DRNN 神经网络控制
锅炉的负压和风量

刘久斌,李德桃

[摘要] 火电厂锅炉负压和送风系统是具有多变量、非线性及时变参数的受控对象,利用改进型 DRNN 神经网络来辨识系统模型,进而对 PID 控制器参数进行整定,实现多变量解耦控制。对负压和送风控制系统进行了设计和仿真研究,通过计算机仿真给出了仿真曲线。结果表明,系统达到了解耦目的,系统能做到稳定性高,鲁棒性强,调节及时,反应速度快,动态偏差小,无静态偏差。该分析和控制方法适合于负压和送风控制系统,符合工程实际,控制品质好,有实用价值。

负荷变化影响燃料量控制,同时影响送、引风量的控制。从经济性角度看,当锅炉的负荷变化时,需要进行燃烧调整,使锅炉的燃烧工况与锅炉的蒸汽负荷要求相适应,燃料量变化时,应及时改变进入炉膛的空气量,以保证燃料的完全燃烧和排烟损失最小,送风量控制应适应锅炉燃烧过程的经济性要求。从安全性角度看,引风量控制应使引风量与送风量相适应,并保持炉膛压力在要求的范围内。炉膛压力反映引风量与送风量相适应的程度,在送风量一定时,炉膛压力高(炉膛负压低)说明引风不足,炉膛压力低(炉膛负压高)说明引风过大,炉膛压力关系到锅炉运行的安全,负压控制应适应锅炉燃烧过程的安全性要求。目前,电厂中送风量控制系统有以下几种类型:燃烧率指令——风量系统;燃料量——风量系统;蒸汽量——风量系统;热量——风量系统;氧量——风燃比系统;带氧量校正的送风控制子系统。引风量控制系统以炉膛压力为反馈信号控制引风,并引入送风调节器输出或送风档板开度为前馈信号。可见,负压和风量控制系统均分别考虑控制方案,而负压和风量系统本质上是多变量系统,理论和实践表明:采用单回路控制方案实施多变量系统的控制存在缺陷,原因是该方案不能消除送、引风量系统间的相互影响,影响了控制质量。从多变量系统频域法角度考虑这一问题,通过解耦可消除送、引风量系统间的相互影响,从而提高控制质量,但解耦方法及其实现较为复杂。本文利用改进型 DRNN 神经网络的特点,对负压和风量控制进行探讨。

1 改进型 DRNN 神经网络、Jacobian 信息辨识及控制原理

1.1 DRNN 神经网络算法[1]

DRNN 神经网络有输入层、隐层(回归层)和输出层三层。在 DRNN 神经网络中,$I = [I_1, I_2, \cdots, I_n]$ 为网络输入量,$I_i(k)$ 为输入层第 i 个神经元的输入,$X_j(k)$ 为网络回归层第 j 个神经元的输出,$S_j(k)$ 为第 j 个回归神经元输入的总和,$f(\cdot)$ 为双 S 函数,$O(k)$ 为 DRNN 网络的输出。

$$O(k) = \sum_j W_j^o X_j(k) \tag{1}$$

$$X_j(k) = f(S_j(k)) \tag{2}$$

$$S_j(k) = W_j^D X_j(k-1) + \sum_i W_{ij}^I I_i(k) \tag{3}$$

式中 W^I, W^D 和 W^O 分别为输入层、回归层和输出层的权值向量。

1.2 Jacobian 辨识

设 $u(k)$ 和 $y(k)$ 分别为被控对象的输入和输出,将 $u(k)$ 和 $y(k)$ 作为 DRNN 的输入,$O(k)$ 作为 DRNN 的输出,辨识误差 $em(k)$ 和指标 $Em(k)$ 分别为

$$em(k) = y(k) - O(k)$$

$$Em(k) = \frac{1}{2} em(k)^2$$

学习算法采用梯度下降法

$$W_j^O(k) = W_j^O(k-1) + \eta_o \Delta W_j^O(k) + \alpha(W_j^O(k-1) - W_j^O(k-2)) \tag{4}$$

$$W_{ij}^I(k) = W_{ij}^I(k-1) + \eta_i \Delta W_{ij}^I(k) + \alpha(W_{ij}^I(k-1) - W_{ij}^I(k-2)) \tag{5}$$

$$W_j^D(k) = W_j^D(k-1) + \eta_d \Delta W_j^D(k) + \alpha(W_j^D(k-1) - W_j^D(k-2)) \tag{6}$$

式中 η_i, η_d, η_o 分别为输入层、回归层和输出层的学习步长;α 为惯性系数。

Jacobian 信息为 $\qquad \partial y / \partial u \approx \partial O / \partial u = \sum_j W_j^O f'(S_j) W_{ij}^I \tag{7}$

当学习步长 η 加大时,可使收敛速度加快,但易产生振荡和不稳;反之,当 η 减小时,可维持算法的稳定但可能收敛缓慢。当 $\partial Em/\partial w = 0$ 时,权值的梯度调整就卡在了局部极值点,这时可以采用加入动量项加以改善。α 为动量因子,加入动量项后表明权值修正不仅取决于梯度,还取决于上一步权值的变化增量。为进一步优化算法,对式(4),(5)及(6)改进为式(4a),(5a)及(6a),成为带动量项的 PID 梯度优化算法。它根据 PID 控制原理,对具有纯积分性质的普通梯度法再引入比例及微分控制作用。为保证算法渐近稳定,取 $0<\alpha<1$。理论证明 PID 梯度优化算法比普通梯度算法具有更快的跟踪性和收敛性,且具有更大的可能性跳出局部极值的陷阱。

$$W_j^O(k) = W_j^O(k-1) + \eta_{op}(\Delta W_j^O(k) - \Delta W_j^O(k-1)) + \eta_{oi}\Delta W_j^O(k) + \eta_{od}(\Delta W_j^O(k) -$$
$$2\Delta W_j^O(k-1) + \Delta W_j^O(k-2)) + \alpha(W_j^O(k-1) - W_j^O(k-2)) \tag{4a}$$

$$W_{ij}^I(k) = W_{ij}^I(k-1) + \eta_{ip}(\Delta W_{ij}^I(k) - \Delta W_{ij}^I(k-1)) + \eta_{ii}\Delta W_{ij}^I(k) + \eta_{id}(\Delta W_{ij}^I(k) -$$
$$2\Delta W_{ij}^I(k-1) + \Delta W_{ij}^I(k-2)) + \alpha(W_{ij}^I(k-1) - W_{ij}^I(k-2)) \tag{5a}$$

$$W_j^D(k) = W_j^D(k-1) + \eta_{dp}(\Delta W_j^D(k) - \Delta W_j^D(k-1)) + \eta_{di}\Delta W_j^D(k) + \eta_{dd}(\Delta W_j^D(k) -$$
$$2\Delta W_j^D(k-1) + \Delta W_j^D(k-2)) + \alpha(W_j^D(k-1) - W_j^D(k-2)) \tag{6a}$$

1.3 控制原理

图 1 为多变量控制原理图,图 1 中 DRNN1 和 DRNN2 为神经网络,r_1 和 r_2 为系统定值输入,u_1 和 u_2 为控制器 PID1 和 PID2 的输出,y_1 和 y_2 为系统输出,其参数为 k_{p1}, k_{i1}, k_{d1} 和 k_{p2}, k_{i2}, k_{d2},由 DRNN 整定,其中 u_1 (u_2 类同)的控制算法如下:

$$u_1(k) = k_{p1}(k)x_1(k) + k_{i1}(k)x_2(k) + k_{d1}(k)x_3(k) \tag{8}$$

$$k_{p1}(k) = k_{p1}(k-1) + \eta_p error_1(k)\frac{\partial y_1}{\partial u_1}x_1(k) \tag{9}$$

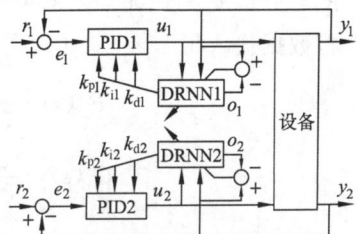

图 1 多变量控制原理图

$$k_{i1}(k) = k_{i1}(k-1) + \eta_i error_1(k)\frac{\partial y_1}{\partial u_1}x_2(k) \tag{10}$$

$$k_{d1}(k) = k_{d1}(k-1) + \eta_d error_1(k)\frac{\partial y_1}{\partial u_1}x_3(k) \tag{11}$$

$$x_1(k) = error_1(k) = r_1(k) - y_1(k) \tag{12}$$

$$x_2(k) = \sum_i^k error_1(i) \times T \tag{13}$$

$$x_3(k) = (error_1(k) - error_1(k-1))/T \tag{14}$$

T 为采样时间,$\partial y/\partial u$ 由式(7)计算。

2　负压和送风对象的数学模型[2-4]

锅炉负压和送风对象可用图 2 简单表示。

图 2 中,Q_s 为送风流量;Q_g 为引风流量;U_s 为送风开度;U_g 为引风开度;P_f 为炉膛压力。

负压和送风对象的数学模型可用图 3 表示。

R_s 为送风通道流动阻力系数,即 U_s 阶跃扰动时,P_f 与 Q_s 的静态值之比;

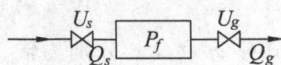

图 2　负压和送风对象

R_g 为引风通道流动阻力系数,即 U_g 阶跃扰动时,P_f 与 Q_g 的静态值之比;

K_s 为送风调节机构档板特性,即 U_s 阶跃扰动且炉膛压力不变时,Q_s 的静态放大倍数;

K_g 为引风调节机构档板特性,即 U_g 阶跃扰动且炉膛压力不变时,Q_g 的静态放大倍数;

C 为炉膛容积系数。

图 3 可等效图 4,可见负压和送风对象是一个多变量对象。

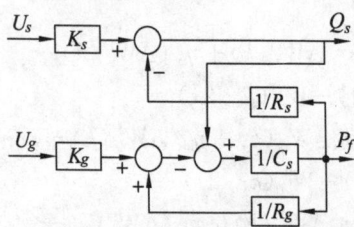

图 3　负压和送风对象方框图

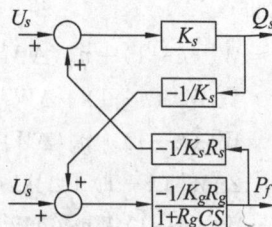

图 4　负压和送风对象等效方框图

现场试验数据如下:

$R_s = 3.1, R_g = 16, C = 2.3, K_s = 2.5, K_g = 2.1$

将数据代入得

$$P_f(s) = \frac{40.3}{37S + 6.24}U_s(s) - \frac{33.6}{37S + 6.24}U_g(s) \tag{15}$$

$$Q_s(s) = \frac{2.5 + 92.5S}{37S + 6.24}U_s(s) + \frac{10.92}{37S + 6.24}U_g(s) \tag{16}$$

3 仿真计算

针对表达式(15)和(16)编程,将它们作为图 1 中的对象进行仿真计算,程序框图如图 5。

仿真时分别考虑炉膛压力定值变化和风量定值变化两种情况,仿真结果为:炉膛压力定值变化时 P_f 和 Q_s 的响应曲线如图 6,整定出的调节器参数随时间变化时曲线如图 7 和图 8。风量定值变化时 P_f 和 Q_s 的响应曲线如图 9,整定出的调节器参数随时间变化曲线如图 10 和图 11。

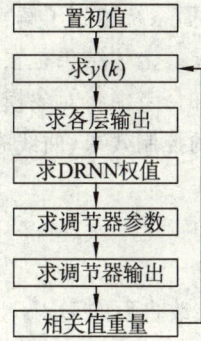

图 5 P_f 和 Q_s 仿真程序框图

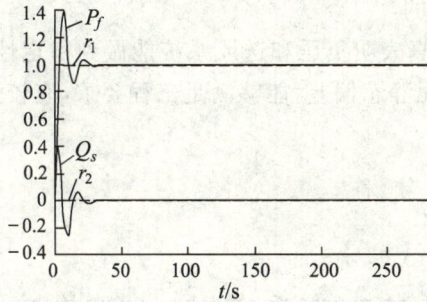

图 6 炉膛压力定值变化时 P_f，Qs 的阶跃响应曲线

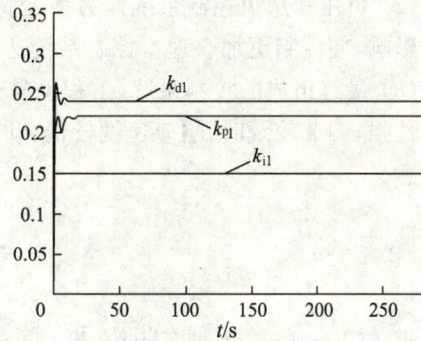

图 7 炉膛压力定值变化时 k_{p1}，k_{i1}，k_{d1} 的变化曲线

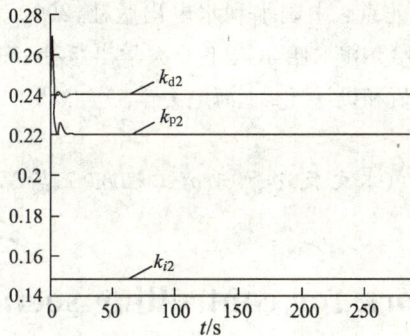

图 8 炉膛压力定值变化时 k_{p2}，k_{i2}，k_{d2} 的变化曲线

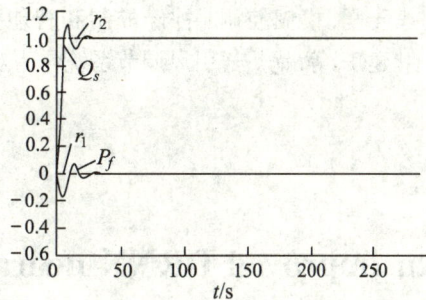

图 9 风量定值变化时 P_f，Qs 的阶跃响应曲线

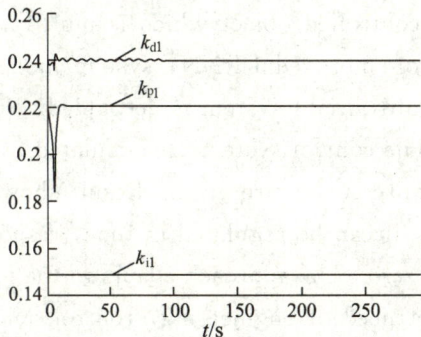

图 10 风量定值变化时 k_{p1}，k_{i1}，k_{d1} 的变化曲线

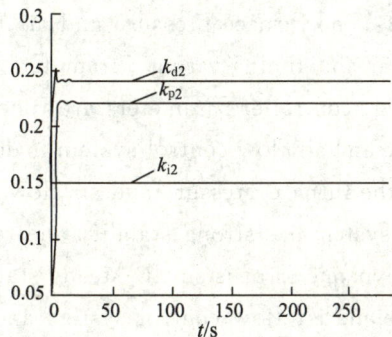

图 11 风量定值变化时 k_{p2}，k_{i2}，k_{d2} 的变化曲线

可见,系统达到了解耦目的,系统能做到稳定性高,调节及时,响应速度快,动态偏差小,无静态偏差。上述结果是根据机组正常运行工况对象模型取得的,当机组状态变化时动态特性变化,这时,考虑将对象模型参数变化±15%,将模型代入控制系统进行仿真,结果表明仍能维持很好的控制效果(曲线略),说明控制系统鲁棒性强。

4 结 论

(1) 改进后的 DRNN 神经网格控制方法有更快的跟踪性和收敛性,且具有更大的可能性跳出局部极值的陷阱,科学合理,用于对非线性多变量对象解耦控制是一种可选的理想设计方法,整定和仿真简便易行。

(2) 负压和送风控制系统作为多变量进行考虑符合实际情况,通过解耦能消除系统间的相互影响,使控制更加合理。控制系统达到了解耦目的。

(3) 最终组成的控制系统,工程上易于实现。仿真表明负压和送风系统能做到稳定性高,鲁棒性强,调节较及时,响应速度较快,动态偏差小,无静态偏差,能够满足工程要求,有较强的实际意义。

参 考 文 献

[1] Ku Chao Lee, Lee Kwang Y. Diagonal recurrent neural net-works dynamics systems control [J]. *IEEE Transactions on Neural Networks*, 1995, 6(1).

[2] 丁轲轲,刘久斌,林青. 热工过程自动调节[M]. 北京:中国水利水电出版社,2000.

[3] 李士勇. 模糊控制:神经控制和智能控制[M]. 哈尔滨:哈尔滨工业大学出版社,1998.

[4] 叶建华,等. 锅炉送引风调节系统作为双变量的分析和设计[J]. 上海电力学院学报,1985(1).

(本文原载于《动力工程》2005 年第 6 期)

An improved DRNN neural-network for controlling suction pressure and air flow of boiler

Abstract:The furnace pressure and air flow is a controlled object which is multivariable, Nonlinear and timing-varing parameters. By use of improved DRNN, system models are identified, controller parameters are tuned, and multivairiabl system is decoupled. Furnace pressure and air flow control system is designed. The control system are simulated by computer, the furnace pressure and air flow step response curves are given. Result shows that control system has strong stability and robustness, it can be regulated in time, response is rapid, dynamic error is small, steady state error is zero. The approach adapts to the furnace pressure and air flow control system and can meet need of engineering, control quality is high, so it possesses practical significance.

附 录

Appendixes

附 录 A

未选入的学术论文题录

1. 李德桃. 对农用排灌内燃机的要求. 中国农业机械,1962(1):20-21.

2. 李德桃. 105 系列柴油机分析. 中国农业机械,1962(5):19-20.

3. 李德桃. 内燃机的使用寿命. 中国农业机械,1966(4):38-39.

4. 李德桃. 测量内燃机空气流量的新仪器. 内燃机,1977(17):64-72.

5. 李德桃. 利用二维液流动模型研究吊钟型涡流室内的空气运动. 内燃机工程,1983,4(1):12-16.

6. 李德桃. 柴油机涡流室内空气运动的研究. 江苏工学院学报,1983,4(1):1-14.

7. 李德桃. 压燃式发动机涡流燃烧室研究. 吉林工业大学学报,1983(4):107-110.

8. 李德桃. 关于涡流燃烧室内混合气形成和燃烧过程的研究. 拖拉机,1984(5):9-15.

9. 李德桃,等. 涡流燃烧室的结构设计对柴油机性能的影响. 小型内燃机,1985,15(6):20-27.

10. 李德桃,等. 涡流室式柴油机放热率的计算方法和计算程序的精确化研究. 内燃机学报,1986,4(1):313-324.

11. 孙平,李德桃. S195 型柴油机涡流室镶块稳态温度场的测量和计算分析. 拖拉机,1986(6):17-22.

12. 李德桃,朱晓光. 高速纹影摄影在分隔式柴油机工作过程中的应用. 第五届全国高速摄影与光学学会会议论文集, 西安,1987:84-85.

13. 李德桃,杨本洛,等. 涡流室式柴油机放热模型的应用研究. 江苏工学院学报,1988,9(2):1-8.

14. 朱广圣,李德桃,等. 运用韦伯函数分析分隔式发动机的燃烧过程. 中国内燃机学会燃烧专业委员会第四届年会论文,洛阳,1989.

15. 李德桃,朱晓光. 涡流室式柴油机起动孔作用机理和主燃烧室火焰扩展规律. 内燃机学报,1990,8(1):27-32.

16. 李树德,李德桃,等. 两极离心式 C 型调速器设计与计算. 中国内燃机学会中小功率专业委员会学术年会论文集, 长春,903040,1990.

17. 李德桃,等. 分隔式柴油机测试系统的建立和分析. 中国内燃机学会燃烧专业委员会第五届年会论文集,河南,1990.

18. 龙跃渊,李德桃. 涡流室式柴油机气缸压力、放热率和噪声的关系. 柴油机,1991(2):12-21.

19. 李德桃,等. 间喷式柴油机测试系统的建立和分析. 小型内燃机,1991(5):1-5.

20. 朱广圣,李德桃. 陶瓷发动机无环活塞的缸套间漏气量的理论和试验研究. 江苏工学院学报,1991,12(3):15-22.

21. 朱广圣,李德桃. 165F 柴油机装有无环活塞和陶瓷缸套时的侧向敲击分析计算. 特种发动机通讯,1992(1):167-177.

22. 朱广圣,李德桃,等. 涡流室柴油机连接通道流量系数的计算研究. 江苏工学院学报,1992,13(3):25-30.

23. 朱章宏,李德桃,等. S195 型柴油机冷起动条件下摩擦阻力的试验研究. 小型内燃机,1992,21(2):10-12.

24. 贾大锄,李德桃,等. 内燃机燃烧过程测量技术的进展. 中国内燃机学会测试技术分会第六届学术会议论文集,湖南,1992.

25. 严新娟,李德桃,等. 涡流室式柴油机工作过程计算中几个问题的探讨. 全国高等学校工程热物理第四届学术会议论文集,浙江,1992:1-4.

26. 计维斌,朱广圣,李德桃. 涡流室式柴油机连接通道流量系数的研究. 上海机械学院学报,1993,15(2):95-102.

27. 熊锐,李德桃,等. 柴油机喷油器陶瓷针阀体结构强度研究. 排灌机械,1993,11(6):32-34.

28. 李德桃,吴志新,等. 柴油机供油系统仿真及其在燃烧系统研究中的应用. 内燃机电脑应用,1994,1(1):25-33.

29. 李德桃,等. 涡流室式柴油机冷起动时非稳态燃烧分析. 内燃机学报,1994,12(1):29-35.

30. Zhu Yana, Li Detao. A thermodynamics model on IDI engine starting proceeding of COMODIA 94, Japan, 1994.

31. 熊锐,李德桃,等. 韦伯燃烧参数研究. 拖拉机与农用运输车,1994(4):31-35.

32. 熊锐,李德桃,等. 小缸径柴油机喷雾特性研究. 陆军船艇,1994(2):10-13.

33. 吴志新,李德桃,等. 小型高速直喷式柴油机燃烧系统的研究和发展. 陆军船艇,1995(2):31-38.

34. 王谦,李德桃,等. 柴油机涡流室内流场的实验和计算研究. 内燃机学报,1994,12(2):102-108.

35. 王谦,李德桃,等. 柴油机涡流室内涡流与紊流的测度与分析. 安徽工学院学报,1994,13(增刊).1994 年内燃机联合学术会议论文集:223-227.

36. 熊锐,李德桃,等. 柴油机涡流室温度场测试研究. 陆军船艇,1994(4):27-31.

37. 黄跃新,李德桃,等. 火花点火式发动机燃烧及其模型研究进展. 湖南大学学报,1995,22(5):57-60.

38. 吴志新,李德桃,等. 柴油机涡流室内非稳态温度场的测试和分析. 燃烧科学与技术,1995,1(4):312-318.

39 朱广圣,李德桃. 运用韦伯函数分析涡流室式柴油机的放热特性和性能. 小型内燃机. 1995,24(5):18-25.

40. 夏兴兰,李德桃,等. 一个简化的柴油机油滴燃烧模型. 小型内燃机,1996,25(3):7-11.

41. 夏兴兰,李德桃,等. JXX 型内燃机数据采集分析系统研制. 小型内燃机,1996,26(4):51-55.

42. 倪再恒,李德桃,等. 涡流室式柴油机瞬时传热系统研究. 江苏理工大学学报,1997,18(1):28-36.

43. 吴志新,李德桃,等. 柴油机燃烧喷注表面特性的研究. 江苏理工大学学报,1997,18(1):12-18.

44. 夏兴兰,李德桃,等. 内燃机燃烧过程数学模型的研究进展和动向. 江苏理工大学学报,1997,18(3):14-21.

45. 王谦,李德桃,等. 柴油机涡流室内空气运动的实验研究进展. 江苏理工大学学报,1997,18(4):33-41.

46. Gong Jinke, Li Detao, et al. Influence of ignition energy on gasoline engine performance. Proceedings of the International Conference on Internal Conbustion Engines, Wu-Han,China,1997:411-415.

47. 夏兴兰,李德桃,等. 涡流室柴油机燃烧压力预测的复合模型. 南昌大学学报,1997,19(1):5-10.

48. 夏兴兰,李德桃,等. 柴油机可用能分析. 南昌大学学报,1997,19(2):1-7.

49. 夏兴兰,李德桃,等. 涡流室式柴油机火用分析. 南昌大学学报,1997,19(3):54-59.

50. 夏兴兰,李德桃,等. 涡流室式柴油机准维相关火焰微元燃烧模型的研究. 燃烧科学与技术,1997,3(4):417-423.

51. 董刚,李德桃,等. 改进IDI柴油机燃烧室结构设计以降低排放的研究. 中国汽车工程学会发动机分会及中国内燃机学会汽油机煤气机分会九八年度学术论文集,江苏,1998:61-67.

52. Wang Qian, Li Detao and Yang Wenming. Study on establishment and improvement of emission regulations for motor vehicles in China. Proceedings of the fifth Tri-University International Joint seminar and Symposium, Thailand,1998:17-20.

53. 夏兴兰,李德桃,等. 涡流室式柴油机燃烧模型的研究进展. 江苏理工大学学报,1998,19(2).

54. 黎苏,李德桃,等. 船用中小缸径直喷式柴油机燃烧室中油线网分布的计算. 船舶工程,1998(5):14-17.

55. 朱亚娜,李德桃,等. 柴油机加速工况下燃烧特性的研究. 湖南大学学报,1999,26(2):24-28.

56. 熊锐,李德桃. 发动机温度测试方法介绍. 小型内燃机,1999,28(3):17-21.

57. 夏兴兰,李德桃. 涡流室式柴油机燃烧模型的研究进展和发展方向. 小型内燃机,1999,28(5):17-21.

58. 曹茉莉,李德桃,等. 6110型柴油机缸套变形的有限元计算与分析. 中国汽车工程学会发动机分会、中国内燃机学会汽油机煤气机分会学术年会论文集,江苏,1999:63-73.

59. 董刚,李德桃. 十六烷值改进剂对涡流室式柴油机排放特性影响的试验研究. 中国汽车工程学会发动机会会、中国内燃机学会汽油机煤气机分会学术年会论文集,江苏,1999.

60. 杨文明,李德桃. 影响涡流室式柴油机冷起动性能的因素分析. 燃烧科学与技术,1999,5(4):345-348.

61. 夏兴兰,李德桃,等. 涡流室式柴油机油滴破碎和喷雾碰壁三维数值模拟. 内燃机学报,1999,17(4):317-322.

62. 李德桃,朱亚娜,等. 柴油机非稳态燃烧的循环模拟与实测示功图分析. 江苏理工大学学报,2000,21(2):5-9.

63. 曹茉莉,李德桃,等. 柴油机缸套变形的有限元计算和实验分析. 江苏理工大学学报,2000,21(2):5-9.

64. 熊锐,李德桃,等. 柴油机燃烧过程涡流室内瞬态温度分布的研究. 中国机械工程,2000,11(4):469-472.

65. 李德桃,杨文明,等. 柴油机非稳态燃烧循环模拟. 江苏理工大学学报,2000,21(6):38-41.

66. 严新娟,李德桃. 涡流室式柴油机工作过程计算中若干问题研究. 江苏理工大学学报,2000,21(5):1-4.

67. 黎苏,李德桃,等. 汽油机非稳加速工况燃烧放热率计算模型修正. 内燃机学报,2000,18(3):275-278.

68. 杜爱民,李德桃,等. 涡流室式柴油机工作过程三维模拟程序. 燃烧科学与技术,2000,6(3):270-274.

69. 王谦,李德桃,等. 自由活塞式发动机 NO_x 排放的数值计算研究. 内燃机学报. 2001,19(6):562-566.

70. 胡林峰,李德桃. 柴油机共轨式电控喷油系统的发展动向. 小型内燃机,2001,30(3):29-32.

71. 梁凤标,李德桃,潘剑锋,等. 涡流室式柴油机性能的改进和排放的降低. 小型内燃机与摩托车,2001,30(4):32-35.

72. 李德桃,等. 柴油机非稳态燃烧的循环模拟及其实测示功图验证. 燃烧科学与技术,2002,8(3):238-240.

73. 胡林峰,李德桃,等. 柴油机电控共轨喷油系统的发展趋势及潜力. 中国汽车工程学会发动机分会与中国内燃机学会汽油机煤气机分会,2002 年度联合学术年会论文集,山东,2002:132-140.

74. 胡林峰,李德桃,等. 新型增压式柴油机共轨喷射系统的研究与开发. 内燃机工程,2002,23(1):1-4.

75. 何志霞,李德桃,等. Autocad 环境下离心泵性能预测软件的开发. 排灌机械,2002,20(4):14-17.

76. 李德桃. 热科学领域的高新技术及其产业化前景. 江苏大学学报:自然科学版,2002(1):14-16.

77. 胡林峰,李德桃,等. 柴油机电控共轨喷油系统的发展趋势及潜力. 中国汽车工程学会发动机分会与中国内燃机学会汽油机煤气机分会,2002 年度联合学术年会论文集,山东,2002:132-140.

78. 唐维新,李德桃,等. 计算机测控的风冷发动机风扇试验系统的研制. 内燃机工程,2003,24(3):9-10.

79. 潘剑锋 李德桃,等. 微热光电系统中的微燃烧研究. 高等学校工程热物理研究会第十届全国学术会议论文集,热科学与技术,2003,23(3):261-266.

80. 何志霞,李德桃,等. 基于神经网络技术的泵站机组性能预测. 江苏大学学报:自然科学版,2003,24(4):45-48.

81. 何志霞,李德桃,等. 柴油机燃油喷射系统模拟计算的发展与分析. 内燃机工程,2004,25(2):50-53.

82. 潘剑锋,李德桃,邓军,等. 微热光电系统中微燃烧的研究. 热科学与技术,2004,3(3):261-266.

83. 潘剑锋,李德桃,等. 发动机喷雾和燃烧过程的多功能动态可视化测试装置的研制和应用. 中国汽车工程学会发动机分会与中国内燃机学会汽油机煤气机分会,2004年度联合学术年会论文集:127-132. 内燃机工程,2004,25(6):7-10.

84. 梁凤标,李德桃,胡林峰. 高压共轨喷油系统中高速强力电磁阀的研究. 中国汽车工程学会发动机分会和中国内燃机汽油机煤气机分会学术会议集,浙江,2004:161-165.

85. 梁凤标,李德桃,等. 高压共轨式柴油机多次喷射的数值模拟. 农业机械学报,2006,37(4):40-43.

86. 唐维新,李德桃,袁文华,潘剑锋,邓军,黄俊,张孝友. 微型热光电系统中的微燃烧室研究. 邵阳学院学报:自然科学版,2004,1(3):68-70.

87. 何志霞,李德桃,等. 柴油机喷嘴内部空穴两相瞬态流动数值模拟分析. 中国内燃机学会中小功率柴油机分会学术年会论文集,湖南,2004.

88. 刘久斌,李德桃. 基于总和时间常数的送引风控制系统仿真. 计算机仿真,2004,21(7):149-151.

89. 梁凤标,李德桃,胡林峰. 柴油机高速强力电磁阀的数值模拟. 农业机械学报,2005,36(2):8-11.

90. 黄俊,薛宏,潘剑锋,李德桃. 微动力系统的若干研究动态和进展. 世界科技研究与发展,2005,27(1):5-9.

91. Pan Jianfeng, Li Detao , Yang Wenming, Xue Hong, Huang Jun. Micro-scale combustion research for micro-thermophotovoltaic systems. Heat Transfer Asian Research. 2005,34(6):369-379.

92. 袁文华,潘剑锋,李德桃,胡林峰,唐维新. 燃烧过程动态可视化装置的研制. 湖南大学学报,2005,32(3):74-78.

93. 何志霞,李德桃,袁文华,胡林峰,潘剑锋. 柴油机喷雾燃烧过程的动态可视化试验. 江苏大学学报:自然科学版,2005,26(5):397-400.

94. 何志霞,李德桃,梁凤标,等. 柴油机喷嘴结构对内部空穴流动的影响分析. 中国内燃机学会2005年学术年会暨APC2005年学术年会论文集,黑龙江,2005,10:214-218.

95. 何志霞,李德桃,袁建平,等. 柴油机喷嘴内部气液两相湍流场三维数值模拟. 内燃机工程,2005,26(6):18-21.

96. 潘剑锋,黄俊,刘久斌,李德桃,徐丰. 微型发动机的若干发展动态及分析. 小型内燃机与摩托车. 2006,35(6):46-50.

97. 李德桃,潘剑锋,薛宏,杨文明. 微动力机电系统的发展动态与展望. 江苏大学学报:自然科学版,2006,27(6):489-492.

98. Pan J F, Ding J N, Yang W M, Li D T, Xue H. Design conceits and testing of a prototype micro- thermophotovoltaic system. The First Annual IEEE International Conference on Nano/Micro-Engineered and Molecular Systems, ZhuHai, China,2006:144-148.

IEEE Review on Advances in Micro-，Nano，and Molecular Systems，2006，1：69-70

99. 梁凤标,潘剑锋,李德桃.高压共轨喷油系统中电磁阀的研究.江苏大学学报：自然科学版，2006,27(2):133-135.

100. 梁凤标,李德桃.高压共轨式燃油喷射系统中的电磁阀的数值模拟计算.小型内燃机与摩托车,2006,35(2):22-24.

101. 梁凤标,李德桃,胡林峰.新型增压式柴油机共轨喷射系统的研究.小型内燃机与摩托车,2006,34(5):31-34.

102. 黄俊,潘剑锋,李德桃,杨文明,徐丰.微热光电系统带环形翅片燃烧室的数值模拟.中国高等教育学会工程热物理专业委员会第十二届全国学术会议论文,重庆 2006.5.

103. 梁凤标,潘剑锋,李德桃,胡林峰.新型增压式共轨喷射系统的模拟计算分析和研究.小型内燃机与摩托车,2007,36(1):32-34.

104. 潘剑锋,黄俊,唐维新,李德桃,杨文明.微热光电系统燃烧室内截面突变对燃烧的影响.农业机械学报,2007,38(3):44-46.

105. 徐丰,潘剑锋,薛宏,李德桃,黄俊.基于 MEMS 的微动力系统的最新发展动态.机械与电子,2007(6):3-6.

106. 潘剑锋,李德桃.微热光电系统带双环形翅片燃烧室的数值模拟.中国高等教育学会工程热物理专业委员会第十三届全国学术会议,西安,2007.5.热科学与技术,2007,6(4):351-355.

107. 赵金刚,潘剑锋,李德桃,薛宏.六缸柴油机气缸盖冷却水道优化研究.中国高等教育学会工程热物理专业委员会第十三届全国学术会议论文,西安,2007.5.

108. 潘剑锋,李德桃,等.微 TPV 系统中多孔介质燃烧室的数值模拟.中国工程热物理学会燃烧分会 2007 年学术会议论文集,天津,2007:383-390.

109. 李德桃,等.能源对内燃机和其它动力装置发展的巨大影响.小型内燃机与摩托车,2008,37(4):88-90.

附录 B

著译作目录

李德桃. 柴油机. 南京：江苏人民出版社, 1974.

李德桃. 柴油机涡流燃烧室的研究与设计. 北京：机械工业出版社, 1986.

李德桃. 涡流室式柴油机燃烧过程和燃烧系统. 北京：科学出版社, 2000.

李德桃. 柴油机冷起动的基础研究和改善措施. 北京：科学出版社, 1998.

王谦, 李德桃, 潘剑锋. 燃烧学. 北京：中国科学文化出版社, 2002.

李德桃, 贺道德. 相似理论及其在热工上的应用（译自俄文）. 北京：科学出版社, 1962.

百 炼 成 钢

　　1952年,那是激情燃烧的岁月。解放后的新中国经济快速恢复,急需大量工业建设人才,国家除扩招外,还从部队和地方调大批有文化基础的年轻干部到大学学习。当时,内燃机权威戴桂蕊教授在湖南大学机械系任教,因此该校成为全国最早设立内燃机专业的三所大学之一,我们有幸成为首届内燃机班(60余人)的学生(王荣初曾任班长)。同学间年龄相差较大:不少人来自穷乡僻壤,经过了艰苦磨练,坚持赤脚晨跑,过冬仅穿两条单裤;思想和文化也有较大差别,但他们都有一个共同的理想——为祖国工业化贡献毕生精力。这些赤诚之心汇聚在岳麓山下,他们团结一致,互相帮助,很快成为一个和谐的大家庭,他们之间是那么坦诚,那么关爱,向着"三好"(身体好、学习好、工作好)的目标迈进。前后经过两次全国院系调整,先是成为华中工学院汽车内燃机系的内燃机专业班,后来集中到长春,成立汽车拖拉机学院(后更名为吉林工业大学,现并入吉林大学)。新成立的内燃机教研室有黄叔培、戴桂蕊、余克缙、徐硒祚、杨克刚等著名教授,无论师资还是设备在当时均属国内一流,在名师们的培养下,我们这个团结上进、生活俭朴、学习刻苦的集体中涌现出了一大批建设祖国的优秀人才。

　　李德桃就是其中的佼佼者。如今他不仅是一个有着非凡学术造诣的教授、博导,更是一位道德高尚、平易近人的学者。这个农民的儿子,当年是挟着一床薄被走进大学的,日常生活一贯俭朴。他认识到贫困的不只他一个,而是整个国家民族的现实,要改变落后的现状,必须人人奋起。他就是抱着为祖国工业建设贡献一生的理想而学习的,因此学习非常刻苦,对学习中发现的问题从不放过,直到与老师和同学讨论透彻为止。他基础扎实,成绩突出,解决实际问题的能力很强。更可贵的是,他在学生时代参加各种公益劳动时都是全身心投入,抢重担,走前头,使旁观者误认为他是掺入在学生中的码头工人或农民。1954年武汉抗特大洪灾,1960年困难时期长春支农,他都是标兵,受全校表扬。

　　李教授不仅在教书育人中谆谆善诱,治学上要求严格,而且在日常生活中像朋友一样对待学生。他待人诚恳忠实,谦虚谨慎,对任何人都是那样坦诚,那样和蔼可亲,对公私之间是公而忘私。从罗马尼亚回国时,生活上节约下一些外汇,他不是用来给自己买家电,而是买回大量图书。经济条件改善后,他生活仍保清贫,而把钱捐给家乡茶陵老区办教育,帮助一些困难的老同学渡难关。

　　不经一番寒彻骨,怎得梅花扑鼻香。李教授的成功是在艰苦磨炼中得来的,凭着他对祖国无限忠诚,靠着他的毅力与智慧,他才能在超越障碍、克服人为阻力中成为胜利者、成功者!

原吉林工业大学纪委书记　王荣初

原吉林工业大学教授　江国宪

大连理工大学教授　梁纶慧

2007年12月7日

桃李不言，下自成蹊

光阴荏苒，转眼间我博士毕业离校10年了。当年和李德桃先生一起的情景历历在目，多年来虽身处海外，却时时回想起先生的谆谆教诲。一路走来，深感受益匪浅。

1992年春，湖南大学在读硕士生的我，利用带本科生到上海实习的空隙，慕名拜访了李德桃教授并表达了希望攻读他的博士生的愿望。当时我为自己的冒昧有点忐忑，但他很热情地接待了我并给了我诸多指点。先生的鼓励给了我很大的勇气并坚定了我进一步求学的决心。那次的见面促成了我们以后的师生之缘，即使我后来到了北美，还数次与先生相见，并聆听他的教诲。

李教授自20世纪90年代初兼职湖南大学教授及博士生导师后，十多年如一日，甘于奉献，无怨无悔，为湖南大学内燃机专业的建设作出了巨大的贡献。当年先生经常给我们介绍内燃机学科国内外的最新信息，亲自为我们博士生授课。他从美国德州大学、大连理工、南京理工、江苏理工、吉林工大等多所知名大学请来教授讲学，使得学术气氛空前活跃；不仅如此，有着数十年发动机生产实践经验的李教授还组织和带领我们到生产实践中去。短短的几年时间，在他的指导下我参与了国家内燃机燃烧学重点实验室、湖南动力机厂、湘潭柴油机厂多个科研单位和厂家的科研课题，开阔了眼界，增强了动手能力，培养了团队协作精神，让我走出象牙塔，亲身经历了科研开发从理论到实践的整个过程。在我博士论文准备期间，李先生多次专程来湖南大学指点，我也前往江苏与那边的科研团队一起合作攻关。从讨论课题方向，制订试验计划，直至论文反复斟酌、不断修改。已经记不得有多少个日夜，先生利用休息时间指导我的研究工作，也不记得多少次李先生亲自过问并关心我的生活和待遇，只记得他为了学生们的成长呕心沥血，头发一根根地变白了。说到这里不能不提的是我每次到江苏大学，老师和师母都会热心地招待我，嘘寒问暖，给予了我无微不至的关心和帮助，让我内心无比的温暖。

在湖南大学当年提供的非常有限的科研和生活条件下，李先生依然为提高湖大内燃机学科的整体科研、教学水平，尤其是培养年轻科研力量不懈地努力。作为一名全国著名发动机专家、非直喷室式柴油机"泰斗"，李先生对学术发展要求从来很高，而对生活待遇几乎无所要求。在湖大建设村简陋的房间内，无论是严冬还是酷暑，李先生都抓紧时间为学科发展献计献策，并身体力行地开展工作。记得我曾经同先生一起出差到江西，当地发动机行业隆重接待，但先生总是力求简化，为了方便地开展工作宁可住在招待所，而婉谢入住豪华宾馆。他说："我们是来帮助人家解决实际技术难题的，不是来享受的。"后来我到过其他一些工厂，技术人员都这样评价李先生："他和别人不一样，要求的少而付出的多，是个难得的好老师啊。"很多工厂的技术人员也许以前没有听过李教授的课，但读过他的涡流室式柴油机专著和论文的却很多，大家对能称他一声"老师"而倍感荣幸。

后来我到了加拿大，恰逢先生访问底特律，我从多伦多赶来和他见面。先生仍然像当年那样关心着弟子的一切，想方设法把我介绍给其他在美国的学生，目的就是帮助我争取更多的发展机会。我后来能够到美国先进发动机技术公司工作，和李先生早前的推荐是分不开的。记得那次同李先生、彭立新博士、王谦博士一起到了芝加哥并登上了当时世界最高的西尔斯大厦的顶层,先生说:"希望我们的科研团队能像今天登上西尔斯大厦一样攀上科学的高峰,你们虽然人在海外,但仍然是我们科研团队的一员,大家一起来努力。"李先生就是这样,他时刻想着的就是我们这个集体。那次我们在芝加哥还得到了朱元宪博士的热情接待,朱博士说他和李先生很早以前有过一面之缘,那是他在博士学位答辩会时,李先生不辞辛苦、不顾严寒坐火车连夜赶到大连,这让他一直都很感动。李先生是那种只求付出不求回报的人,他的弟子和那些曾经得到他帮助的人怎能不铭记于心呢!

2006 年 7 月,我有幸在洛杉矶再次见到了李先生。出乎我意料的是,已经 70 高龄的他仍然是充满着科学探索的热忱,谈到他新的微型发动机研究和团队未来发展大计时,脸上充满了兴奋的神情,他那股对新事物的探究劲头,让我敬佩不已。

今天李先生已是桃李满天下,值此《选集》出版之际,广大弟子齐心协力共襄盛举,让我心潮澎湃、感慨万千。深信我们这个朝气蓬勃、充满活力的学术团队的各位弟子们,在未来的日子里会有更加杰出的成就,以我们的努力来回报恩师的辛勤培育。

美国先进发动机技术公司高级工程师、项目经理

黄跃欣 博士

2007 年 12 月 9 日于加州

在李老师科研团队中的一些感受

今年暑假回国拜会我的研究生导师李德桃教授时,得知由他的同事和学生提议要为他出一本论文选集,觉得真是一件众望所归的事情。李教授是国内涡流室式柴油机燃烧过程研究的首屈一指的专家,他和科研团队在动力机械领域的若干方面都取得了杰出成就。对涡流室柴油机的燃烧过程进行了一系列的基础研究和应用研究,成果多次获得国家级、省部级奖励。该论文选集的出版可以说是动力机械研究领域的一个重要贡献。作为李老师的学生,虽然自己未能在李老师的一系列的研究工作中做多少工作,但却为之感到无比的光荣。

我是 1986 年考入江苏工学院(现江苏大学)内燃机专业的,知道了李教授是内燃机燃烧领域的大家,是改革开放后出国最先获得内燃机博士学位者之一,第六、七届全国人大代表。同时也听说过李教授为专心研究而婉谢出任江苏省镇江市副市长等职,当时就很仰慕。后来本科毕业设计时,有幸接受他的指导,并且在工作两年后考取他的硕士研究生,有幸再次接受指导,从而近距离感受到李教授知识的渊博、治学的严谨和思维的敏锐,同时也感受到了他的正直不阿、平易近人、实事求是的高尚品格。

首先,感受最深的是李教授对真理的不懈追求。在我参加涡流室式柴油机的冷起动过程研究的一部分工作时,国内关于涡流室式柴油机的起动孔机理存在着两种观点。一种观点是在发动机的压缩过程中,空气通过起动孔进入涡流室,改善了燃油的喷雾状况,加快了低温时副室内的着火,从而改善了起动。另一种观点是副室内的部分喷雾可以通过起动孔直接进入冷起动时温度相对较高的主室,从而加快了着火过程。李教授觉得两种都有可能,但应通过实验验证。当时采用涡流燃烧室方式的柴油机面广量大,搞清其机理有着重要意义。于是,李教授对涡流室式柴油机的冷起动过程展开了一系列的基础研究。用高速摄影仪拍摄了冷起动时主副室的着火情况,同时测量了整个起动过程的压力变化,进行了放热计算。以后又通过多种不同结构设计的起动孔试验,证明了由于冷起动过程的非稳态特性,起动孔不仅可以使一部分燃油直接进入主室加快主室的燃烧,也可通过起动孔进入副室的压缩气流,改善副室内的着火条件从而改善起动过程,这一研究结果终结了国内学术界对该问题的争论。李教授的研究并没有停留在辨明起动孔的作用原理上,他将冷起动过程划分成若干典型的着火模式进行理论分析,为进一步改善该过程奠定了基础。1992 年去日本讲学时,这一研究成果得到了日本同行教授的高度赞赏。他还将此研究所获得的结果拓展到柴油机过渡工况时的燃烧过程的基础研究中。

其次,是李教授作为一名科学研究者的远见卓识和对科学的献身精神。1994 年他去美国进行合作研究,深深地感到发动机的排放问题已成为人类和地球健康发展中的一个重要议题,而国内内燃机领域还没有普遍的认识。于是将这一趋势和中科院院士史绍熙教授(现已故)交流,在国内内燃机的权威杂志《内燃机学报》上发表了《关于建立和完善我国车辆排放法规若干问题的探讨和建议》,呼吁重视该领域。十多年过去了,美国和欧洲的排放标准日趋严格,发动机的设计概念已和十多年前有了很大的区别。现在看来,当时的这一篇文章对国内排放问题来讲是具有划时代意义的。

从 20 世纪 80 年代末开始,由于对燃油消耗的重视,国内很多涡流室式柴油机厂家转为生产直喷式柴油机。可李教授并没有放弃对这一领域的研究。他认为涡流室式柴油机具有排放低的重要特点,它的存在对于发动机的排放来讲有着重要意义。如果大家都放弃这一领域的研究,那么以后开发此类发动机就会缺乏必要的理论基础。如今 20 年过去了,涡流室式燃烧方式依然存在。而且随着排放法规的严格化,有些国外的公司将以涡流室式柴油机为主来满足美国的 Tier3 排放法规,而 2009 年后在日本国内销售的单缸卧式柴油机也将以涡流室燃烧方式为主。

自己离开李教授 10 多年,看到这些发展过程,对李教授的这种远见卓识和献身精神而感到无比钦佩。

再次,就是李教授的挑战精神。这一点几乎感染了他的所有学生。在李教授的研究中,有着很多"首次"。如首次用激光摩尔偏折法测量涡流室内的温度分布;首次使用 LDA 实机测量了涡流燃烧室内的空气流动,等等。早在上世纪 70 年代,李教授和常州柴油机厂一起技术攻关,首次将涡流式室发动机的转速提高到 3 000 r/min,在我国自主开发内燃机的历史上起到了里程碑的作用。李教授经常用自己的亲身经历鼓励我们树立高目标,成为一个有作为的人。我想李教授的许多学生之所以后来在各自的领域里取得重要成绩,与从他那里接受到的这种训练有很大关系。

最后,留下深刻印象的是李教授的以身作则,对学生的百般爱护,关注学生的成长。尽管他是国内外知名专家,但从来没有以此自居。他经常给我们学生讲的是"消耗人民的尽可能地少,献给人民的尽可能地多"。李教授主持科研项目,提出研究思路,创造实验条件,亲自参与试验,并撰写学术论文,而通常总是让他人作为第一作者。他不仅注意我们学生在学问上的成长,同时也要求我们学生要有敢于追求真理,为社会作贡献的精神。这无疑给了我们最好的教育。

是李教授将我带到了科学研究的殿堂。正是从他那里学到的科学的思考方法,使我在科研和实际的开发工作中能够向前更进一步。每想至此,对李教授的百般敬意和感激之情总是油然而生。我想李教授的学生都会有类似的想法。衷心祝愿李教授的论文选集成功出版!

日本洋马发动机事业开发部高级工程师

田东波　博士

2008 年 2 月 18 日

忆 师 随 笔

1994 年,我被学校推荐接受研究生教育,并有幸投入李德桃老师门下。当时真是既高兴又紧张。一方面我知道李老师是我校很有名气的六位国务院学位委员会审批的博士生导师之一,能成为他的学生,我感到非常自豪,但另一方面,听说李老师要求非常严格,心里又有点忐忑不安,怕自己适应不了。在这种心境下,贝石颖老师带领我来到了李老师家。他家给我的印象是很简朴,几乎没有什么值钱的家具。唯一给我留下深刻印象的是几幅字画,当时我以为这可能是名人字画。后来才知道,这是他的老同学与老朋友写给他留作纪念的励志字画。

李老师留给我的第一印象是严肃、坦诚、不苟言笑。他先询问我的个人情况,并问我有没有什么困难。随后他介绍内燃机的发展现状与趋势,以及我可能要做的研究方向。这期间,他都是非常认真地谈论有关内燃机的专业知识,基本上没有说笑的时候。以后随着接触的增多,我才慢慢了解到李老师其实也是一个蛮亲切的人,而且还非常节俭。

1994 年 11 月初,李老师要去美国进行合作研究,师兄弟们让我送他去上海。那是我第一次与李老师单独相处,也因为这次机会,使我更加的了解到李老师的为人。一路上他给我讲了很多过去亲身经历的事情。谈到了我校一位已故的传奇人物——戴桂蕊教授,30—40 年代就已经是一位非常有名的教授了,并于 50 年代研制出我国第一台内燃水泵,当时在国际上也是处于领先水平。作为戴教授的助手,李老师从他身上学到了很多,特别是对科研的专注,并由此进入科学殿堂。70 年代,李老师在极其困难的生活和工作条件下,离开妻女,义无反顾地与常州柴油机厂的技术人员和工人长年累月的住在车棚,一起设计,一起做试验,最终开发出“低油耗、低排放、低爆压”的柴油机涡流燃烧室,为我国涡流室式柴油机的快速发展作出了巨大的贡献。有关论文在当时唯一的内燃机刊物上发表,引起广泛注意,成果无偿为数十厂家所用。前全国人大副委员长彭冲去常州考察时,还特意去了解这项成果,看到一身油污的李老师时,还以为是一个普通工人呢。后来当李老师参加全国人民代表大会遇到彭冲副委员长,谈起这事时,彭冲副委员长还记忆犹新。他总是希望我们能很好的继承老一辈吃苦耐劳的精神,并一代一代传承下去。老师常引用居里夫人的话“知识是人类之公产,应让人类共享才对”。

到上海后,我本以为李老师会找一家靠近机场的旅店住宿,可方便乘飞机赴美,但他却选择了上海机械专科学校招待所。他说靠近机场的旅馆通常价格很贵,这里不仅住宿费低,吃饭等开销也要便宜得多。虽然出国的费用可以报销,但他还是尽力节省每一分钱。我在生活中常看到李老师的这种平民心态和草根情怀。

随着与师兄弟们相处日久,我还得知了更多关于李老师的一些事。1979 年,李老师因为过去所取得的成绩,被选派去罗马尼亚进一步攻读博士学位,师从著名教授贝林单。这是我国文革后最早的一批公派出国留学生。在罗马尼亚留学期间,李老师虽然必须克服语言上的困难,但仍然非常出色的完成了博士论文。贝林单教授的评价是最确切的:“你是我所有学生中做的最好的三个学生之一。”1982 年在回国途中,还差点被当时的苏联截去。李老师坐火车回国途经前苏联,克格勃还上车劝说李老师留在前苏联进行研究,并许诺比中国好得多的研究条件,但这些都没有打动李老师报效祖国的心。回国后,李老师放弃了出任江苏省镇江市副市长、江苏省镇江市市政协副主席等职位的机会,他表示自己的专长是搞科研,不希望因为忙于公务而荒废自己心爱的研究工作。李老师追求事业、淡薄名利的精神使我深受感动。

李老师不仅有强烈的事业心,追求科研与教学工作卓越,而且有强烈的社会责任心,从以下所举数例可见一斑。

他在任第六、七届全国人大代表时,曾提出建议"农业机械化法"的议案(第51号,含农业机械化法建议稿),这是第七届江苏代表团仅立4项议案之一。为此,他曾深入农村、农机管理站,走访农机专家,调查研究,并多次召开座谈会,表现出作为一名人大代表的高度社会责任感。

江苏大学地处镇江市郊,很长一段时间使用自建的简易自来水系统。李老师回国后,深感学校饮水质量问题的严重性,除及时向校领导反映情况外,每次赴京参加人大会议,都抽空去机械工业部找各级领导反映此问题。后由机械工业部拨400万元专款,从市区直接接自来水管到学校,从而避免了不达标水质对全校师生健康的影响。

李老师作为一位我国较早研究内燃机排放的专家,早在1984年,就取得了降低涡流室式柴油机 NO_x 排放20%的成果。他深刻地认识到车辆排放的危害性。90年代初,他在美国进行合作研究期间,出于对我国环保问题的深切关注,与几位中美专家教授多次讨论了这一问题,后由他执笔写成《关于建立和完善我国车辆排放法规若干问题的探讨和建议》一文,此文由中科院院士史绍熙(现已故)和李老师等人联名发表。李老师又将此文附专函寄给国家高层领导,得到有关领导的支持,这些工作均促进了我国车辆排放问题的加速解决。

今天回想起来,为什么李老师有如此强烈的事业心和社会责任心?我认为主要是由于他有一种爱国爱民、吃苦耐劳、踏实做事、百折不挠的精神。这本论文集展示了这些精神。

新加坡国立大学研究员

杨文明　博士

2008 年 1 月 16 日

感 念 师 恩

我怀念着李德桃老师带我所走过的那段岁月,那点点滴滴的欢笑与温馨,无微不至的关怀与感动,如涓涓溪流,绵绵不绝。当时只道是寻常。如今,我与老师各自工作、生活在不同的城市,远隔万水千山,难以再有做学生时在老师身边随时聆听教诲的机会,一切往事都变成了对老师深切的感念。

我是 1992 年进入江苏工学院攻读博士学位时开始跟随李老师的。三年时光,对老师从陌生到熟悉再到敬重。感受的不仅是学者的严谨和渊博,不仅是大师的睿智和深刻,更是老师的高尚人格与无私品行。我所接触的李老师,不长于圆滑之道,也不擅于炒作自己。老师的为人,正直、善良、热心,忧国忧民,爱憎分明;老师的生活,简单、朴实、淡泊,却充盈着随遇而安的满足;老师的言行有时似乎是不合"时宜",却显示着他的真知灼见、正直无私和深切的人文关怀。

老师是内燃机学界的名家,他的学术之路,是由勤奋与坚毅打造的。他潜心研究非直喷式柴油机燃烧过程和燃烧系统达 40 余年,为该领域奠定了科学基础。老师以"消耗人民的尽可能地少,献给人民的尽可能地多"作为座右铭,用行动教给了我什么是对国家的忠诚,什么是对事业的执着。让我懂得了一个人不仅要把所从事的工作当作一种职业,更应该作为一项事业来对待。干工作不仅仅是为了养家糊口,更重要的是要在岗位上做出成绩来,得到大家的承认与肯定,这就是老师教导我们的要有事业心和责任感。

老师为人真诚,无论对亲人、朋友还是晚辈均热心相待、倾力支持,对我们这些弟子更像对自己的孩子一样爱护。当我的研究结论与专家的观点相左时,他鼓励我不要迷信权威,要带着疑问来思考。我的每一篇文章他都仔细阅读严格把关,一篇发表在《兵工学报》上的论文被要求改了 6 遍,面对着我的不解,老师解释说:"如果连自己导师这一关都通不过,又怎能让其他评委通过呢?"老师出国前还不忘托付多位教授指导我的课题,其后又赶在我论文付印前回国,一遍又一遍地修改我的毕业论文。为了让我在毕业答辩时取得好成绩,他专门为我举行预答辩,动员大家为我出谋划策。老师要求我们每周定时汇报学业进展、探讨存在的问题。为解决我们在校生的生活困难,忙着给我们找课题、找兼职、找实习机会。老师更是操心毕业生找工作的情况,为我们写推荐信联系接收单位。最让我难以忘怀的是 1994 年我在南京理工大学做实验的日日夜夜,老师陪伴着我,帮我联系实验场地、仪器和人员,随时指导我解决实验中的问题。为了将非常有限的经费用到实处,老师连 2 元钱的专线车都舍不得坐,同我一起去挤 2 角钱的公共汽车。和我一样住在由学生宿舍改成的 10 元钱一天的招待所里,我睡上铺,他睡下铺,在南京那像火炉般的夏季里,没有风扇,蚊帐也是千疮百孔,我们没有胶布去粘补蚊帐上的破洞,就把蚊帐上的棉线抽出来扎住洞口,不让蚊子咬人。这就是一位担任过多届全国人大代表的知名教授,这就是一位年过花甲的老人,为了我这个学生,他操足了心……行笔到此,我的眼睛不禁湿润了……

老师凡事都只为别人着想,不愿意给他人添麻烦。就是自己老家的政府请他回去作报告,为了不让接车的司机半夜到车站接他,他宁可放弃直达列车而选择分段转车,为此近七旬的老人累得大病一场,回到家就住进了医院。老师对生活要求不高,我们曾私下里笑谈过老师的"小气",别人寄给他的信封还要反过来折好粘好继续使用。我们在毕业后去看望他时,偶尔带些小礼物都要被老师责怪,反而带回来他更多的馈赠。

我们所能做到的,就是效仿老师的热心助人,同门一心,是逢年过节对老师的问候,是打电话的报喜不报忧。还记得毕业答辩我获得优秀成绩时老师眼中的欣慰与自豪,还记得我评上教授职称后给老师汇报时老师的爽朗笑声,还记得我到广州工作前向老师告别时老师对我的谆谆教诲和殷切希望。老师要求我们的,也是他最为期盼的,是我们的成绩斐然,是同门的亲密无间,是所有他关心和爱护的人都工作生活得更加美好。我的老师,不是一无所求,可是他所执著追求的,除却他的弟子们都能成为国家的栋梁之材,就仅仅是人海茫茫中的脉脉温情,是世态炎凉中的永恒真诚。

一位好老师,可以改变学生的命运。李老师曾经影响我并将继续影响和督导着我的,是他那"高调做事,低调做人"的作风,是那宽厚善良的心灵与高尚的人格魅力:没有居高临下、循章摘句式的说教。我们跟着老师,学着做人,学着做学问。很多师兄弟正是在老师的影响下,确立了一生奋斗的具体目标。我也是从老师身上深切感受到高尚人格的巨大感召力量。

时光匆匆,毕业转眼已十年有余。我也成为了大学教授,也在带自己的研究生。漫漫长路,老师仍然是我一生的榜样,伴随着我对真理的追求,对道德的探索。我期待着能早日带着成绩到老师面前,恭恭敬敬地向他行礼,把这段岁月向老师娓娓道来,共同感念隐于风中的欢笑和泪水。

李老师,对于您来说,我可能只是您无数个学生中普通的一个,但对于我而言,您对我心灵的震撼却足够我感念终生! 祝您平安,幸福!

广东工业大学汽车工程系主任、教授、学科带头人

熊 锐 博士

2007 年 8 月 30 日

恩师倡导的"四跨"合作研究和人才培养

每一条路都有起点,每一条江河都有源泉,每一座大厦都有根基,每一个人都有恩师相伴!一提到"恩师"这两个字,牵动我内心的,总是感动!是他陪我走过成长岁月,是他陪我度过人生灿烂年华,是他教会我责任与荣誉,是他鞭策我永远乐观前行……李德桃教授,我的恩师,就是这样一位德高望重、无私奉献于教育事业的老师。

他于1992年正式受聘为湖南大学兼职教授和博士生导师,由此开始了我们的师生缘。他倡导的"跨单位、跨地区、跨国界、跨学科"的学术团队十分和谐,也为湖南大学动力机械及工程学科的建设和发展准备了开阔的平台。在艰苦的条件下无私奉献,先后为湖南大学指导博士研究生近10名,争取科研课题6项,发表高水平学术论文20多篇。他在南京理工大学、河南科技大学、长沙内燃机研究所等单位兼职,也同样做了许多实实在在的工作。

印象中的李老师生活总是那么艰苦朴素,言谈之中总透出长者与父母般的关怀,让我们在学习、生活中处处都能感受到来自老师的温暖。他以身垂范,乐于奉献,既教导学生如何做人,又善于为学生排忧解难。当学生在学习、生活、工作或思想情感上遇到问题或困难时,都愿意向他倾诉和求助,在李老师那里总能得到中肯的指点和安慰,使我们能卸下负担,轻松应对生活和学习。他善于将自己的学术成就和国内外研究成果相结合并运用于课堂教学中,这样让学生在学习基本知识的时候就对国际国内的研究前沿有了比较系统的认识,开阔了研究视野。老师将这种先进的教学理念经过自己长期的改进、积累,形成了独特的教学风格,使无数学生领略到科研的魅力,并带领大家最终走上科研道路,在各个领域取得很好的成绩。"李教授不管培养什么学生,都会出现奇迹。"李老师周围的同事都是这样评价他的。

李老师一直在教学科研第一线,以严谨求实的态度追求科学真理,探索科学理论,解决科学难题,结出了丰硕的果实。他注重培养学生的独立性和创新性以及团队精神,遇事总是身体力行,亲自主持和参加各项研究与实验工作,带领研究生跑遍湖南和江西所有的内燃机厂,进行调研,推广研究成果。作为导师,他不是授人以鱼,而是授人以渔。给学生指明正确的方向和方法,任其跌打摔爬,在浮沉起伏中锻炼独立性和创造性。

李老师严谨的治学态度、渊博的专业知识、活跃的学术思想、平易近人的工作作风和忘我的工作热情深得同事们的一致好评,也赢得了学生们的尊敬和爱戴。我一直想用一句最简单的话语来概括我心中的恩师,但总是不满意,最后发现恩师的名字其实早已注解了他的人生——德艺双馨、桃李满园!

湖南大学教授、博导、学科带头人

龚金科　博士

2007年9月4日

短期的接触 长期的记忆

我真正接触李老师还是 2002 年，他给我的第一印象是认真与坦诚。他思维活跃、不拘一格。跟他交谈，理性的铺陈和感性的抒发都恰如其分，让人如沐春风，从中获得享受，得到提高。

2003 年至 2004 年我有幸在李老师的指导下做访问学者，加入了他的科研团队，参与了微型热光电系统的研究。耳闻目睹，感慨良多，深刻体会到李老师除了渊博的专业知识之外，还是一个特别富有人格魅力的人；他既是我们学业上的导师，也是我们情感上的挚友。他的严谨、博学以及敏锐的判断力，深深地影响着我，也通过我影响着我现在的学生。

老师"当过农民，下过工厂，留过洋"。20 世纪 60—70 年代，在开展科研极其困难的条件下，开发出"低油耗、低排放、低爆压"的柴油机涡流燃烧室，为我国涡流室式柴油机的快速发展作出了巨大的贡献，并获得国家发明奖。据说前人大副委员长彭冲去常州考察时，还特意去看了这项成果，看到一身油污的他，还以为是一个普通工人呢。

老师生活简朴，淡泊名利。他多次婉谢了组织要他当"官"的安排，是一位没有沾染官本位习气的纯粹的教师。在学术环境浮躁之风甚嚣尘上之际，他依然如故，工作异常勤奋、投入，踏踏实实地做学术。年近七旬的他，连星期天都很少休息。记得我有一篇论文拿给李老师审阅时，他百忙中挤时间指导我前前后后修改了好几遍，直到他基本满意为止，结果该文顺利地被一家核心期刊录用，EI 收录。

在学习和研究的过程中，李老师总是努力给我们创造了良好的学术氛围，通常每半月都要在他的研究室举行学术交流会。硕士、博士、访问学者都要参加，就自己课题的新进展或者所遇到的困难进行交流，极大地活跃了研究室的学术氛围。在李老师的带领和指导下，我们课题组建成了全国第一个微型热光电系统试验台，研究成果国内领先，还取得了国家专利。

老师对我们的生活也十分关心。我做访问学者时，寝室里的床比较窄，老师怕我睡不好，就把自己家的宽床搬来给我用。每逢节日，老师和师母都会把我们叫到他们家做客，真有一种回家的感觉。如今想来，那种温暖的感觉真是难以形容，叫我永难忘怀。

如今老师已是七旬老人了，依然念念不忘自己的师长，多次深情地回忆他的老师们。其中一位是戴桂蕊教授。作为科研助手，李老师从戴教授身上学到了很多，特别是对科研的专注。后来，李老师为戴教授子女生活问题的解决作了多方面的努力；也为自己其他老师的著作出版、经济困难提供了帮助。我读过李老师为记念贝林单教授逝世 10 周年而写的怀念文章，字里行间，浓浓的师生之情跃然纸上。李老师是一位真正做到"学师爱生"的人。

能够在访问学者期间得到李老师这样一位好导师的指导，是我一生的幸运！在江苏大学做完访问学者，回到邵阳学院又有三个年头了，我还一直得到李老师的指导。他勤奋刻苦的学习精神和认真负责的工作精神，使我受益匪浅，终生受用。

邵阳学院教授

唐维新

2007 年 8 月 8 日

回忆在工程热物理研究室
学习和工作的日子

　　我 1988 年师从李德桃老师攻读硕士学位,1995 年以讲师的身份离开,先后在研究室学习和工作了 8 年多。在这人生最美好的岁月,有幸能遇到李老师这样的好导师,给予了我学业、事业和生活上极大的指导和帮助,使我终生受益。

　　李老师曾两度入围工程院院士候选名单,他迄今所取得的学术成就有目共睹。作为李老师的一名学生,我想通过自己在学习和工作中亲历的几件小事,从几个侧面反映科研团队的一些情况和感受导师感人的一面。

　　至今我仍然能清晰地记得与李老师第一次见面的情形。那是 1988 年炎夏的晌午时分,我敲开了老师的家门,站在面前的是位年过半百、头发花白、精神矍铄的老人,他笑吟吟地把我让进门,让我一下子就消除了紧张的情绪。接着他仔细地询问了我的工作和学习情况,介绍了研究室的一些工作情况。由于缺人员,他让我尽早来研究室报到,最后还给了我一些资料以便尽快适应以后的学习。这第一次的见面就让我体会到老师慈祥又严格的作风。

　　那时的研究室,研究经费微薄,实验场地缺乏,实验设备,主要靠"四跨"来支撑。试验一般是在外单位进行,李老师几乎都亲临现场指导,要求我们反复测试,直至达到满意结果。我们有时都觉得结果很好了,老师却坚持再做直到最好。我在做模拟计算时,经常因为程序中一点不易察觉的小问题发生死循环,而当时计算机速度又奇慢,运行一次计算循环就要花很长时间,常常在电脑前耗时无数却收获全无,我经常恼火得不想继续尝试了,而李老师总是启发我寻找原因,鼓励我继续做下去。对于论文的写作,老师认真到苛刻的程度。那时我们工作条件很差,论文是用手写,原件上往往被改得面目全非,要一遍又一遍地誊写、修改。当时觉得很受挫折,现在回想起来才能深深感受到老师的良苦用心!

　　毕业了,终于毕业了! 我有幸能留在研究室做助教和研究工作。本以为工作会轻松,没想到做起课题来更是辛苦。对于纵向课题,面对众多的资料和爆炸似的信息,要大量阅读,大量地检索国内外的资料;在前人研究成果的基础上,创新试验方法和计算模型;和李老师一起再三检查和修改文稿后才投稿。对于横向课题,经常要到企业做试验,采集数据进行分析,真正做到解决实际问题。李老师经常为企业排忧解难,他到工厂时,都能受到工厂领导、技术人员和工人对他的热烈欢迎和尊重。

　　李老师一方面对学生的学习和工作严格要求,另一方面对学生的生活也非常关心,每个学生在生活上有困难他都会尽力帮助。有一次与老师坐江轮出差,我们不坐在同一个船仓内,在一个大的停靠站,他竟拎着一袋各色水果送到我的船仓内! 看着江风吹动着他的白发,我差一点感动得掉下眼泪! 还有一次,下雪后刚刚放晴,是拍雪景留影的好时候,校园内好多人争相踏雪拍照。当时我和一岁大的女儿正呆在家中,出乎意料的是,李老师竟拿着相机邀我们在门外的雪松树前拍了 2 张照片,这 2 张照片我一直珍藏着,经常拿出来与女儿一同观看,感受着老师对一个普通学生的关切之情!

　　8 年的学习和工作留下的回忆太多太多,不是一页纸可以承载的!我虽然离开了研究室,却依然与导师保持着联系。导师对我的帮助和教导将永远是我前进的动力,我会永远记住导师严谨的学风、慈祥的笑容,永远记住我们团队艰苦奋斗、勤俭办事的精神。

特灵空调公司高级工程师

严新娟

2007 年 12 月 28 日

编 后 记

李老师退休时,我们连个茶话会也没开,心中一直有愧;他70寿辰之际,有人提议开个庆祝会,又被婉谢。转眼4年过去了,我们的愧疚之情却有增无减。薛宏、杨文明两位教授建议编辑出版一本李老师的学术论文集,大家觉得是个好主意。对一位百折不挠、毕生从事学术研究的学者,将他几十年的学术论文汇聚成集,回顾一下李老师带领我们科研团队走过的这一段艰辛而欣慰的探索路程,既可为后来者提供有益的借鉴,也可作为我们对李老师最好的礼物。

李老师的学术人生是排除万难、奋力拼搏的一生,李老师的论文是精益求精的。这些不同时期的文章都是李老师和他的团队兢兢业业、在非常艰苦的工作条件下尽最大努力完成的。我们从中发挥的仅是裁缝做百衲衣的拼接作用。当然也容纳了我们在选择论文时的一些考虑,譬如论文的学术价值和创新性,为国家带来的经济效益、社会效益和环境效益,以及是否反映动力机械学科的发展前沿技术等。读者可以从《选集》中看到我们这一思路。我们希望这样的阐释没有偏离读者的愿望太远。其实,那些未选录的文章也是整个研究过程的有机组成部分,但为了节省篇幅,最后决定放在未选论文题录中,以供索引之用。

我们还特地选录了几篇李老师的好友和研究生专门为《选集》出版而撰写的短文。这些文章既可对所选论文进行某些解读,也从一个侧面反映了我们团队的美好情谊和人文情怀,谨作为科技与人文结合的一种尝试,奉献诸君。

本着"出精品"的理念,我们群策群力,整个出版过程体现了坚强的科研团队精神。入选论文分五部分,均按时间排列。由于文章的时间跨度很大,尽管我们对有关标准和名词术语进行了校订和加注,但仍未做到完全统一,不完善之处难免,欢迎读者批评指正。

最后,我们真挚地感谢江苏大学和能动学院领导的支持;对袁寿其校长带领学校深化内涵建设的眼光和魄力表示敬意。岑可法院士、90高龄的高良润教授、薛宏教授特地为本书作序,使之增色不少。江苏大学李汉中教授、林松教授,天津大学史连佑教授以及李老师的老同学、原吉林工业大学纪委王荣初书记、江国宪教授和大连理工大学梁纶慧教授在编辑过程中给与了莫大的关心和帮助,后者卧病在床,仍关心《选集》的事宜,令我们感动不已。也感谢其他所有关心论文选集出版的单位和朋友,对于江苏大学出版社为选集的出版所作的努力,也一并致谢。

顾子良	胡林峰	杨文明	王 谦	单春贤	潘剑锋
何志霞	唐爱坤	段 炼	李晓春	张梦云	赵金刚

编辑组: